Computer-Aided Drug Discovery and Design

Volume 3

Series Editor

Alan Talevi, Facultad de Ciencias Exactas, Universidad Nacional de La Plata, La Plata, Argentina

The series covers all the stages of drug discovery and development that use computational approximations, including bioinformatics, cheminformatics, structure-based approximations, ligand-based approximations and network-approaches. Drug discovery and development stages include target identification and validation, hit identification and hit-to-lead and lead optimization programs.

- The series covers the early drug discovery and development process in a comprehensive manner.
- It explains both the historical background and underlying principles of each methodology, as well as state-of-the-art innovations within each field. For instance, when dealing with structure-based approximations, both classical universal scoring functions and the novel trend in the field (tailored scoring functions) will be considered.
- Accordingly, both experts and students taking their first steps in computer-guided drug discovery will find the series of interest. Students can resort to the series volumes to get familiarized with the basics of the methodologies, and experts can turn to the series for trends in the field.
- No other series covers all the fields and approximations in a deep,updated and comprehensive manner.
- For an integrative perspective, the potential readers might acquire all the books that compose the series; on the contrary, people with a narrower, more specific scope can resort to a particular volume of choice

Vinícius Gonçalves Maltarollo
Editor

Computer-Aided and Machine Learning-Driven Drug Design

From Theory to Applications

Springer

Editor
Vinícius Gonçalves Maltarollo
Departamento de Produtos Farmacêuticos,
Faculdade de Farmácia
Universidade Federal de Minas Gerais
Belo Horizonte, Minas Gerais, Brazil

ISSN 2730-5457 ISSN 2730-5465 (electronic)
Computer-Aided Drug Discovery and Design
ISBN 978-3-031-76717-3 ISBN 978-3-031-76718-0 (eBook)
https://doi.org/10.1007/978-3-031-76718-0

This Springer imprint is published by the registered company Springer Nature Switzerland AG
The registered company address is: Gewerbestrasse 11, 6330 Cham, Switzerland

This book is dedicated to the most important women in my life. To Tereza, my mother, whose wisdom and nurturing spirit gave me the foundation to question and grow. To Renata, my wife and dearest companion, whose love and support have helped me become the best version of myself, guiding me to places I never imagined I could reach. And to Nina, my daughter, whose presence fills me with the strength to face every challenge with courage and resilience.

Foreword

For several decades, Computer-Aided Drug Design (CADD) and Machine Learning have been applied in drug discovery projects, and now there are several drugs in the market developed with the aid of such techniques. Computational techniques, methods, and concepts such as quantitative structure-activity relationships (QSAR), molecular docking, molecular dynamics, free-energy perturbation and quantum mechanics, and virtual screening of compounds databases, among several others, have been cornerstones in several drug discovery projects. These methods have now been boosted by data-driven artificial intelligence (AI) methods. Although it is anticipated that drugs cannot be designed and developed solely by computers, novel computational approaches are emerging, and existing techniques continue to evolve rapidly. Therefore, the scientific community must keep up to date with the most recent developments and practical applications.

The CADD research field comprises several different knowledge areas, and often, researchers are only familiar or experienced with a small fraction of them. Indeed, pharmaceutical industries and large academic groups rely on a broad range of professionals, including chemists, biologists, pharmacists, computer scientists, and other related ones. In this sense, being an expert in every CADD approach is challenging. Furthermore, well-established methods are constantly revisited, and novel approaches and modifications are introduced, such as machine-learning-based scoring functions for molecular docking.

The book "Computer-Aided and Machine Learning-Driven Drug Design" discusses theoretical and successful practical applications of computational techniques. The book, written by authors from diverse countries and geographical regions, is expected to serve the scientific community by providing an overview and solid concepts related to chemoinformatic, bioinformatics, and molecular modeling in the context of AI; key considerations at the interface between computational and the required experimental validation; and practical and recent applications of such technique to real-world drug discovery campaigns. The book can be used as a textbook for courses that include or are focused on CADD and as reference material for researchers and practitioners of current CADD and machine learning.

The editor and all contributor authors are very grateful to the Springer editorial staff for their support in developing and publishing the book.

DIFACQUIM Research Group,
Department of Pharmacy, School of
Chemistry, National Autonomous
University of Mexico, Mexico City,
Mexico

José L. Medina-Franco

Acknowledgments

All the chapters were subject of a blind peer-reviewing process and were approved for publication. Therefore, I would like to thank all the colleagues that put their hard efforts to review the technical content of this book prior to publication. Your contribution definitively enhanced the quality of the presented content and ensured the state of the art of each subject. Thank you very much.

Aaron Sweeney (Centre for Structural Systems Biology, Germany)
Adolfo Henrique de Moraes (Federal University of Minas Gerais, Brazil)
Albérico Borges Ferreira da Silva (University of São Paulo, Brazil)
Alessandro Silva Nascimento (University of São Paulo, Brazil)
Antti Poso (University of East Finland, Finland)
Azam Rashidian (University of Tübingen, Germany)
Diego Magno Martins (Federal University of Minas Gerais, Brazil)
Emmanuela Ferreira de Lima (Federal Institute of Paraíba, Brazil)
Fernanda Rodrigues Soares (Federal University of Triângulo Mineiro, Brazil)
Frederico Gualberto Ferreira Coelho (Federal University of Minas Gerais, Brazil)
Gabriel Corrêa Veríssimo (Federal University of Minas Gerais, Brazil)
Hongtao Zhao (AstraZeneca, Sweden)
Jadson Castro Gertrudes (Federal University of Ouro Preto, Brazil)
José Teófilo Moreira Filho (National Institute of Environmental Health Sciences, USA)
Karen Cacilda Weber (Federal University of Paraíba, Brazil)
Katarina Nikolic (University of Belgrade, Republic of Serbia)
Kathia Maria Honório (University of São Paulo, Brazil)
Lauren Hubert Jaeger (Federal University of Juiz de Fora, Brazil)
Lílian Sibelle Campos Bernardes (Federal University of Santa Catarina, Brazil)
Luan Carvalho (Europharma, Brazil)
Lucas Nicolás Alberca (National University of La Plata, Argentine)
Mateus de Sá Magalhães Serafim (Federal University of Minas Gerais, Brazil)
Maya Topf (Centre for Structural Systems Biology, Germany)

Michell de Oliveira Almeida (University of São Paulo, Brazil)
Paavo Honkakoski (University of East Finland, Finland)
Paula Homem-de-Mello (Federal University of ABC, Brazil)
Petr Pavek (Charles University, Czech Republic)
Rafael Lopes Almeida (Federal University of Minas Gerais, Brazil)
Rafaela Salgado Ferreira (Federal University of Minas Gerais, Brazil)
Renata Barbosa de Oliveira (Federal University of Minas Gerais, Brazil)
Renato Dias da Cunha (Federal University of ABC, Brazil)
Rodrigo Bentes Kato (Federal University of Minas Gerais, Brazil)
Saulo Fehelberg Pinto Braga (Federal University of Ouro Preto, Brazil)
Thales Kronenberger (University of Tübingen, Germany)

Contents

Part I
General Topics and Methods

Chapter 1
Echoes from the Past, Visions from the Future: A Journey into Medicinal Chemistry and Computational Drug Discovery

Vinicius Gonçalves Maltarollo, Ekaterina Shevchenko, Thales Kronenberger, and Ricardo José Alves

Abstract Drug discovery has evolved significantly since the early days of isolating natural compounds such as morphine from opium poppies in the early nineteenth century. Medicinal chemistry drastically changed with the development of aspirin, derived from willow tree extracts, marking the beginning of the synthetic chemistry era. As scientific knowledge expanded, so did the methodologies, transitioning from serendipitous findings to targeted chemical synthesis and high-throughput screening. In recent decades, the advent of computational techniques has revolutionized drug design, enabling researchers to model molecular interactions (mainly comparing structure-based drug design strategies) and predict biological activity (in special considering ligand-based drug design strategies) with unprecedented precision. This shift towards computational drug design has accelerated the discovery process,

V. G. Maltarollo · R. J. Alves (✉)
Departamento de Produtos Farmacêuticos, Faculdade de Farmácia, Universidade Federal de Minas Gerais, Belo Horizonte, Minas Gerais, Brazil
e-mail: dylancover@gmail.com

E. Shevchenko
Department of Pharmaceutical and Medicinal Chemistry, Institute of Pharmaceutical Sciences, Eberhard-Karls-Universität Tübingen, Tübingen, Germany

T. Kronenberger
Department of Pharmaceutical and Medicinal Chemistry, Institute of Pharmaceutical Sciences, Eberhard-Karls-Universität Tübingen, Tübingen, Germany

Tübingen Center for Academic Drug Discovery & Development (TüCAD2), Tübingen, Germany

Excellence Cluster "Controlling Microbes to Fight Infections" (CMFI), Tübingen, Germany

Partner-site Tübingen, German Center for Infection Research (DZIF), Tübingen, Germany

School of Pharmacy, Faculty of Health Sciences, University of Eastern Finland, Kuopio, Finland

V. G. Maltarollo (ed.), *Computer-Aided and Machine Learning-Driven Drug Design*, Computer-Aided Drug Discovery and Design 3,
https://doi.org/10.1007/978-3-031-76718-0_1

making it more efficient and allowing for the development of novel therapies with greater specificity and reduced side effects.

Keywords Computer-Aided drug discovery · Medicinal Chemistry · Drug Design · LBDD · SBDD

1 Historical Perspective on Drug Discovery and Computer-Aided Drug Discovery (CADD)

The process of drug discovery has drastically changed over the centuries adapting to the available knowledge and new technologies. Many early drugs were discovered through serendipity or empirical observation and a lot of ingenuity from scientists. Since ancient times, humankind searched for something to relieve pain and cure diseases. Among the several commonly used remedies were plant extracts, generally used as beverages. Several of them did work, others were inactive or even caused toxicity. With the scientific expansion in the nineteenth century, the development of analytical methods for isolation, purification, and characterization of pure compounds from plant extracts led to the discovery of several plant-based medicinal compounds. To mention some, morphine (Fig. 1.1) is one of the several alkaloid components of opium, the exudate obtained from *Papaver somniferum* was isolated by German apothecary Friedrich Sertürner in 1817. This fantastic painkiller is used even today to relieve severe pain [1].

From the extracts of willow tree (*Salix* spp.), salicin, a plant-derived glycoside used to relieve several types of pain, was isolated in 1826 by a French chemist named Henri Leroux. Chemical hydrolysis and oxidation furnished salicylic acid (Fig. 1.2), which was also a component of the plant. Salicylic acid was eventually used as a

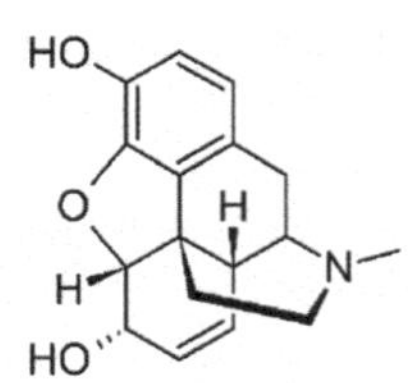

Fig. 1.1 Molecular structure of morphine

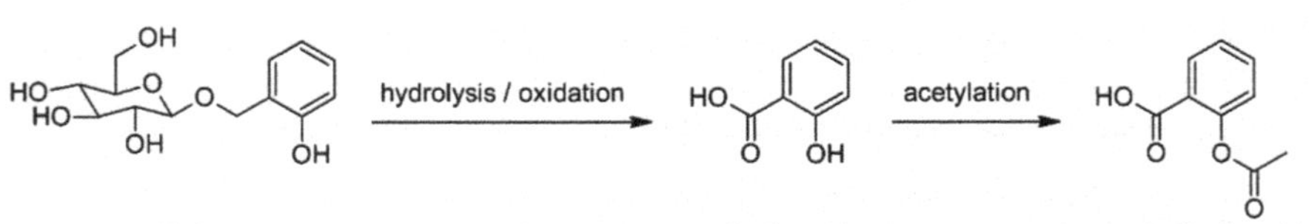

Fig. 1.2 Hydrolysis and oxidation of salicin into salicylic acid followed by acetylation into acetylsalicylic acid

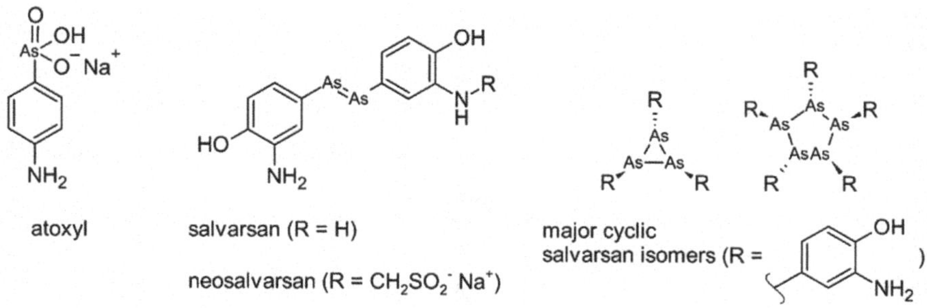

Fig. 1.3 Molecular structures of atoxyl, salvarsan, and neosalvarsan

Fig. 1.4 Molecular structures of prontosil rubrum and sulphanilamide

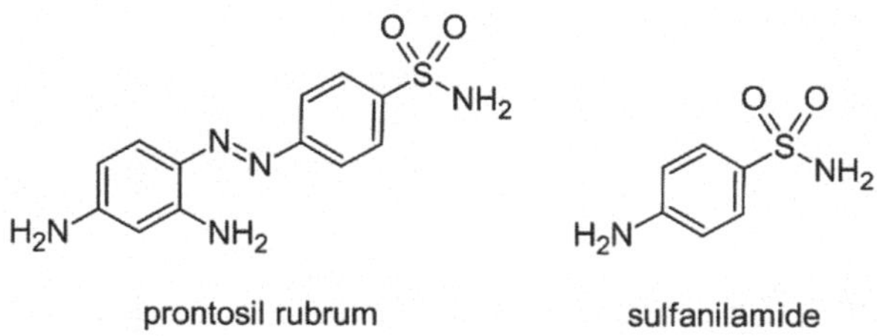

painkiller but was replaced by acetylsalicylic acid, a less irritant drug. Acetylsalicylic acid was first synthesized by French chemist Charles Frédéric Gerhardt in 1853. Later, Felix Hoffmann, a German chemist working for Bayer company, acetylated salicylic acid to create Aspirin, the trade name of acetylsalicylic acid which is also used nowadays [2]. Thus, acetylsalicylic acid is a synthetic drug inspired by a natural product, salicin.

Several drugs of synthetic origin were developed in the twentieth century due to advances in synthetic routes and analytical techniques. In 1905, the French pharmacist Pierre Jacques Antoine Béchamp synthesized atoxyl (Fig. 1.3, left), the first aromatic arsenical compound which displayed trypanocidal activity, although it was too toxic. In 1907, the German chemist Alfred Bertheim synthesized salvarsan (Fig. 1.3, centre). The structure of salvarsan was much later revised, and it was shown that the compound does not have a double bond between arsenic atoms and is composed of a mixture of cyclic species, majorly comprising 3- and 5-membered arsenical ring derivatives (Fig. 1.3, right structures) [3, 4].

Paul Ehrlich, another famous German scientist, is well known as the father of chemotherapy and who coined the term "magic bullet", meaning a compound that would cure disease selectively, without causing host toxicity. Bertheim and Paul Ehrlich, inspired by the work of Béchamp, synthesized more than six hundred arsenical compounds, among which salvarsan stood out as an effective drug to treat syphilis, being introduced into the market in 1910. Due to solubility issues, neosalvarsan (Fig. 1.3), a water-soluble derivative, was also prepared.

The dye industry was the origin of the discovery of sulphonamides, an important class of synthetic antibacterial drugs. Gerhard Johannes Paul Domagk, a German physician at that time working for Bayer, from the IG Farben cartel, described, in

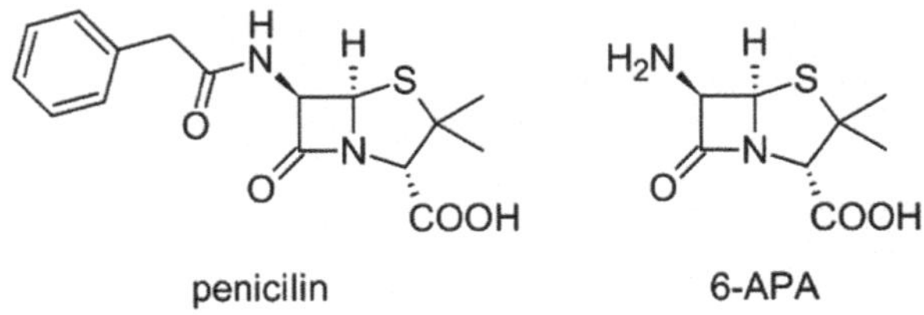

Fig. 1.5 Molecular structures of penicillin and 6-APA

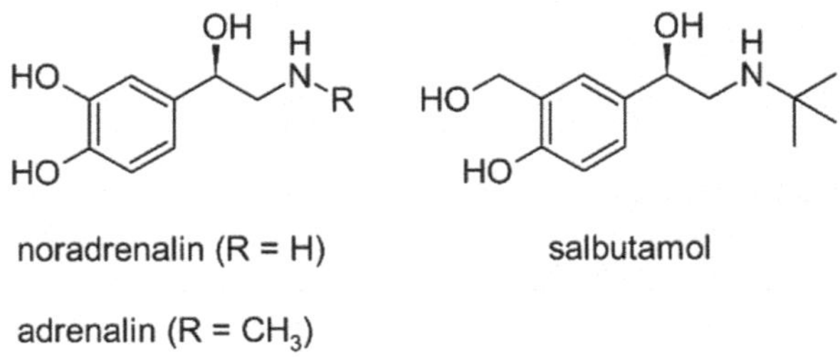

Fig. 1.6 Molecular structures of noradrenaline, adrenaline, and salbutamol

1935, that prontosil rubrum (Fig. 1.4), an azo-dye, although inactive against *Streptococci* in vitro, protected mice from infection. Later, it was discovered by French chemist Ernest Fourneau and colleagues that prontosil is converted in vivo to sulphanilamide (4-aminobenzenesulphonamide, Fig. 1.4), the first of a series of sulpha drugs [5].

The serendipitous discovery of penicillin (benzylpenicillin, Fig. 1.5) by Scottish physician and microbiologist Alexander Fleming in 1929 was a breakthrough in drug discovery. Penicillin saved the lives of thousands of soldiers in World War II. Besides, the production of 6-aminopenicillanic acid (6-APA, Fig. 1.5) from benzylpenicillin allowed for the preparation of several analogues that were resistant to acid hydrolysis and to beta-lactamases produced by penicillin-resistant bacteria. Finally, the discovery of penicillin set a new era in drug discovery, namely, the search for antibiotic compounds from fungi and other microorganisms [5].

The development of salbutamol (albuterol, Fig. 1.6), a beta-2 adrenergic agonist (bronchodilator) drug by Scottish pharmacologist David Jack in 1966, was an ingenuous process in a time where the biological function of adrenaline alog with the receptor occupation theory for the drug action was the only guide for researchers to infer the prototypical binding mode. Adrenaline acts on alfa, beta-1, and beta-2 adrenergic receptors inducing vasoconstriction, increase of heart beating, and bronchodilatation, respectively. Noradrenaline, which lacks the *N*-methyl group, has comparatively lower activity on beta receptors. Thus, it was rationalized that the presence of *N*-alkyl groups could enhance selectivity for the beta receptors. Replacing the *N*-methyl group successively with increasingly bulkier groups showed that the tert-butyl group was the best to confer selectivity to the beta-2 receptor. The modification of the catechol moiety, by replacement of the 3-hydroxy group for a hydroxymethyl group, makes the compound resistant to the action of the enzyme catechol *O*-methyl transferase (COMT), thus prolonging its action as compared to the catecholamines. Since its launch in 1969, salbutamol has ever since been used in the management of asthma [6, 7].

The history of computer methods applied to drug discovery dates from the early 1960s, marked by the discovery of quantitative structure–activity relationships studies reported by Prof. Corwin Hansch's research group. Those ligand-based studies were later supported (1970–1980s) by the rise in the number of 3D structures determined by X-ray crystallography. This paved the way for the design of novel structure-based drugs, focusing on human haemoglobin [8]. In that decade, there was an increase from 69 crystallographic structures, in 1980, to 365 in 1989, leveraging an average of 30 new structures deposited per year [9]. In other words, the amount of information regarding structures of proteins was, at the time, modest. The Protein Databank (PDB) [10] was created in 1971 and first established in the Brookhaven National Laboratory and the Cambridge Crystallographic Data Center [11]. This initiative has remained a beacon for structural information for more than 50 years.

As the data generated increased and became more available, so did the algorithms to process it. The computer methods also gained interest outside of the academic readership and, in October 1981, Fortune Magazine published the cover of an article entitled "Designing drugs with computers at Merck" calling it "The Next Industrial Revolution" [12]. The increasing number of projects and acceptance of computational methods as essential tools in drug design even led some authors to suggest a change from the term "computer-aided-" to "computer-driven drug design" (CDDD) [13].

Classical concepts, such as Emil Fischer's "lock-and-key" theory introduced in 1894 [14], influenced the development of most computational methods to predict protein-ligand binding. The lock-and-key derived computer-aided drug design (CADD) models not only encompass a good geometrical fit but also protein-ligand complementarity due to hydrophobic and polar interactions. Apart from interactions, modern molecular modelling considers both ligands' and binding sites' flexibility, binding, and distortion energies, solvation, and entropic effects [15].

Forty years later, the subject of magazines [16], newspapers [17], and even news pieces from prestigious journals, such as Nature Medicine [18], featured the rise of Artificial Intelligence (AI) applied to drug discovery. This wave was not met without its share of questions, scientists from the 1970s, investigated the actual impact of rational drug design by evaluating whether the incremental potency experimentally observed was related to interactions with the molecular target [19]. Similarly, in 2021, the role of AI-based drug design was also challenged [20] looking specifically at how AI needs to evolve and produce novel data to measure and ensure its actual impact on drug design.

Regardless of the method's nature itself, current CADD approaches are inserted in almost every step of drug discovery. They became integral to industry and academic research's backbone, given the brought layer of structural rationality to the hit identification and its further development into a lead [21] such as understanding of protein-ligand interactions as well as physicochemical and structural features of bioactive compounds, which are the foundation of medicinal chemistry. CADD

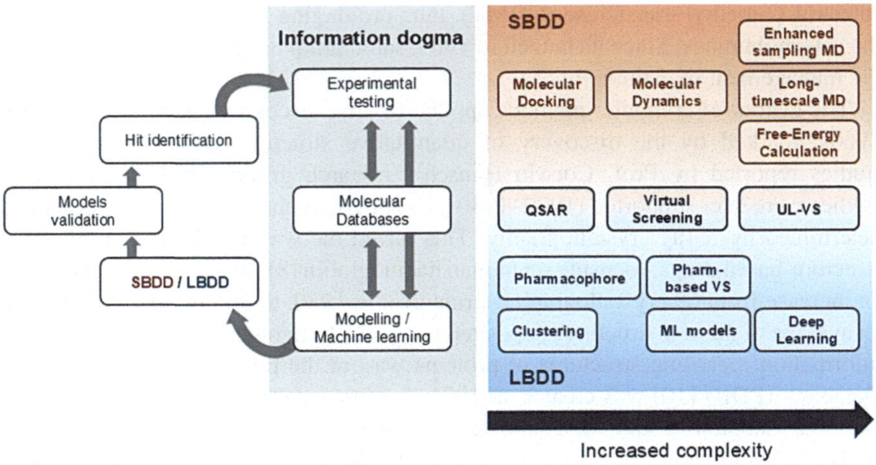

Fig. 1.7 Methods related to SBDD and LBDD strategies and the interchangeability of information between computational procedures, databases, and experimental testing

offers a wide range of methods and techniques to gain insights into protein-ligand binding and phenotypic activities for providing valuable data to support medicinal chemistry efforts [22–24].

CADD techniques provide insights into atomistic details beyond what can be obtained with the current experimental methods. Additionally, molecular modelling works with large pools of chemical data, allowing great chemical diversity to be reasonably analysed and generating new hypotheses capitalizing on large data sets. The research and development process is steadily modernized through the rapid adoption of technologies and techniques that have the potential to drastically improve the drug development pipeline. Despite the speed at which these approaches are emerging, a comprehensive understanding of their applicability and limitations is to be achieved [25].

For instance, CADD can be used as part of major drug design strategies in combination with experimental methods. Of course, the main purpose of computational methods is not to replace experimentation but to support multidisciplinary teams with rational and faster strategies. Considering this, they are often used in parallel or followed by chemical series synthesis and evaluation as well as structural determination using X-ray crystallography. In this sense, the drug design strategies could be classified as structure-based drug design (SBDD) or ligand-based drug design (LBDD) depending on the available information of the studied system [26, 27]. The molecular design process could comprise a single or several methods in the pipeline depending on the amount of available information (Fig. 1.7).

Despite the many computational methods to be mentioned in the next session and this book, one of the main key points is the development of reliable force fields. A force field is a computational equation model used to describe the forces between a collection of atoms. The force field works as a special case of energy functions or interatomic potentials and encompasses not only the calculation form but also the parameter sets used to calculate the atom's potential energy. These models can be used to derive potential energy and forces, and by consequence acceleration and speed, within molecules or between molecules. The main reason why computational methods can be used in drug design is due to the development of fast and reasonably reliable force fields, which are computationally inexpensive when compared to QM calculations.

The work from McCammon et al. [28] highlights the approximate nature of their potential energy function as the main limitation for what was the first MD simulation ever run. Later generalizations of these models included parameters derived from QM calculations resulting in the first force fields (this topic is extensively discussed in Chap. 8).

1.1 Structure-Based Drug Discovery (SBDD)

SBDD is a drug design method that utilizes 3D structures of molecular targets and focuses on the design and optimization of a ligand that fits accurately inside the binding pocket and results in beneficial target-ligand interactions [29]. Structure-based drug discovery is a rapidly growing area due to steadily increasing structural information available, arising not only from genomics and proteomics data but also from AI-augmented systems and novel machine learning approaches [30]. The SBDD process is iterative and frequently requires multiple cycles before an optimal initial hit compound can proceed to experimental validation. A key limitation in the early stages is the creation of misleading and biologically irrelevant models that can ruin the entire pipeline. Relevant biological phenomena to be considered are, for instance, different temperatures, pH, and the possibility that the target explores multimerization, as well as interaction with other proteins and macromolecules such as nucleic acids and/or membranes.

Those functional assemblies frequently lead to conformational changes in the target's substructures, which shift the size and available interactions in the binding site [31, 32]. In addition, understanding the relevant functional changes that a target undergoes upon modulation/activation is crucial to aligning the initial model representations with real-world experimental observations.

After, the target 3D structure can be obtained from databases (see Chap. 2) or, in case its crystal structure or NMR is yet to be resolved, the 3D structure can be predicted by leveraging other structures with the highest sequence similarity to the target or using machine learning (ML)-based techniques (see Chap. 4). The selection for an appropriate method ranges from homology modelling towards AI-powered approaches, such as AlphaFold, which are chosen according to the uniqueness of the target and the system's complexity.

Subsequently, the selection of the binding site determination needs to be identified or predicted to facilitate further optimization of hit compounds down the line.

Those sites can be active/catalytic sites or allosterically relevant binding pockets. In some cases, binding sites are well established within a protein family, providing structural features or specific residues that orient the binding pocket generation. As an example, in protein kinases, various substructural elements can be used as landmarks towards the classical ATP binding site such as the gatekeeper, DFG-motif, or G-rich loop. In addition, many structures have co-crystallized ligands available [33] .

Whether the binding site is unknown for the target family, computational methods can predict probable locations. One can choose between the methods relying on geometrical properties such as POCKET [34], PASS [35], LIGSITE [36], or combined with the physics approach such as PocketFinder [37] or SiteMap [38, 39]. It is interesting to consider that for less-defined (or transient) binding pockets, there is a chicken-and-egg situation. The presence of a high-affinity binding ligand would allow the experimental determination of that particular target conformation with a well-defined pocket, however, a well-defined pocket can be a requirement for the identification of even initial hits. Computational modelling, utilizing enhanced sampling, and well-designed long timescale MD simulation studies can untwist this loop on a case-by-case basis (see Chaps. 8 and 9).

The identification/selection of the binding site opens many potential avenues for the SBDD. For instance, one can establish a pharmacophore model and pharmacophore-based virtual screenings using relevant amino acids, determined by interaction points with ligands, sequence conservation, or prior knowledge. Alternatively, one can carry out structure-based docking (i.e. virtual screening (VS)) of a chemical library, which is often computationally more intensive than the pharmacophore counterpart but requires no prior knowledge. Once the method and the screening database are selected, the VS campaign can be performed.

Next, the initial hit list should be scored and can also be filtered according to the properties required for project purposes. From the ligand perspective, essential pharmacokinetic properties can include physicochemical parameters (MW, number of heavy atoms, hydrogen bond donors and acceptors, and rotatable bonds), lipophilicity (Log $P_{o/w}$), water solubility (log S), pharmacokinetics, drug-likeness (Lipinski violations), as well as an evaluation of the synthesis accessibility for medicinal chemistry [40]. From the protein-ligand perspective, docking poses are often visually inspected for relevant pharmacophoric points and interactions with key amino acids, and those poses can even be rescored/reranked (please, see Chap. 7). Data from smaller VS campaigns can in addition be utilized as input for ML-boosted docking models that will quickly screen for libraries a hundred times larger than the original data set (see Chap. 11). Once the initial hits satisfy the criteria, optional short-scale MD simulations (200–500 ns) can be performed to validate the ligand stability within the target binding pocket, resulting in selecting top hit compounds for experimental validation. Finally, the cycle between model validation and the selection of top hits is frequently repetitive and requires refinement, the determination of which steps can be refined/repeated greatly varies and the diversity of shapes in those loops is often just limited by the computational resources.

1.2 Ligand-Based Drug Design (LBDD)

The premise of ligand-based drug design originates from the foundation concept of medicinal chemistry, which asserts that molecules with a high degree of structural similarity possess comparable bioactivity [41]. LBDD approaches utilize primary data on active compounds (approved drugs, published reports, etc.) to predict or generate novel drug-like compounds with similar biological effects. The inactive compounds are also beneficial for the predictions, while they can be used to identify undesirable ligand features and validate the accuracy of the computational model. In this sense, the typical workflow starts with an initial set of compounds with known potency that proceeds to a chosen computational method for the similarity search. A similarity search may be conducted using a wide range of molecular descriptors or filters. Molecular descriptors are categorized based on the search's dimensionality. Molecular weight and log P are typical 1D descriptors, whereas topological indications and fingerprints are 2D descriptors. The characteristics covered by 3D descriptors encompass a broad range of properties from electrostatic potential to the 3D geometry of ligand moieties. Furthermore, the 3D descriptors are integrated into pharmacophore modelling. According to IUPAC recommendations, a pharmacophore is "an ensemble of steric and electronic features that is necessary to ensure the optimal supramolecular interactions with a specific biological target and to trigger (or block) its biological response" [42]. The ideal scenario for pharmacophore modelling is when a protein-ligand complex is co-crystallized, and the ligand shows sufficient potency and bioactivity. An additional step using Quantitative Structure-Activity Relationships (QSAR) (see Chap. 6) or machine learning (see Chap. 4) can be added to the pipeline if a comprehensive data set of active compounds is available. The most popular metric to quantify the similarity of new compounds to the initial set is the Tanimoto coefficient [43, 44].

1.3 Echoes from the Past, Visions from the Future

Considering all those contexts, opinions, and a brief historical background, we can conclude that CADD/CDDD strategies benefit from a multitude of methods and approaches. In this book, computational methods will be mainly described in the first section ("General Topics and Methods") with their particularities, physicochemical and mathematical fundamentals, validation procedures, and available tools. This section aims to introduce the methods as well as update them in terms of the state-of-the-art for both beginners and experienced researchers in the drug design field.

In addition, the second section of this book (named "The pitfalls between experimentation and simulation") provides insights on maybe one of the most important questions of using CADD for targeting to put a compound into clinics: what are the limitations and problems at the interface between experimentation and simulation? In this sense, the readership is stimulated to think about "Can we trust this simulation output?", "Which biological questions can those methods help to answer?", or "What is the confidence level to predict some experimental properties?" considering the errors and confidence of predictions as well as experimental errors and variations that we very often forget about.

Lastly, a third section entitled "From computer towards the clinical" introduces several case studies describing examples of CADD employment for bioactive compound design or cases where a posteriori interpretation of the results was made possible by this technology.

As we are moving beyond databases with millions and billions of data points, scientists studying those connections are being daily flooded with new information. We believe it is important to remember that knowledge is not just data accumulation. Building knowledge takes time, effort, and willingness to work in the interfaces of interdisciplinary fields. We hope this book can help you to develop a common language to speak with scientists from your complementary fields, ask the right questions, and take inspiration to build the necessary bridges.

Acknowledgments VGM would like to thank the Fundação de Amparo à Pesquisa do Estado de Minas Gerais - FAPEMIG (grants APQ-01818-21 and RED-00110-23). TK and ES are funded by the Fortune Initiative and from TüCAD2 and CMIF. TüCAD2 and CMIF are supported by the Federal Ministry of Education and Research (BMBF) and the Baden-Württemberg Ministry of Science as part of the Excellence Strategy of the German Federal and State Governments. As well as the German Center for Infection Research (DZIF, TTU06.716).

References

1. Serturner, F. (1817). Ueber das Morphium, eine neue salzfähige Grundlage, und die Mekonsäure, als Hauptbestandtheile des Opiums. *Annalen der Physik, 55*, 56–89.
2. Mahdi, J. G., Mahdi, A. J., Mahdi, A. J., & Bowen, I. D. (2006). The historical analysis of aspirin discovery, its relation to the willow tree and antiproliferative and anticancer potential. *Cell Proliferation, 39*, 147–155.
3. Riethmiller, S. (2005). From atoxyl to salvarsan: searching for the magic bullet. *Chemotherapy, 51*, 234–242.
4. Lloyd, N. C., Morgan, H. W., Nicholson, B. K., & Ronimus, R. S. (2005). The composition of Ehrlich's salvarsan: resolution of a century-old debate. *Angewandte Chemie, International Edition, 44*, 941–944.
5. Bentley, R. (2009). Different roads to discovery: Prontosil (hence sulfa drugs) and Penicillin (hence β-lactams). *Journal of Industrial Microbiology & Biotechnology, 36*, 775–786.
6. Jack, D. (1991). The 1990 Lilly Prize Lecture. A way of looking at agonism and antagonism: Lessons from salbutamol, salmeterol and other β-adrenoceptor agonists. *British Journal of Clinical Pharmacology, 31*, 501–514.

7. Lemke, T. L., Williams, D. A., Roche, V. F., & Zito, S. W. (2013). *Foye's principles of medicinal chemistry* (p. 1319). Lippincott Williams & Wilkins.
8. Beddell, C. R., Goodford, P. J., Norrington, F. E., Wilkinson, S., & Wootton, R. (1976). Compounds designed to fit a site of known structure in human haemoglobin. *British Journal of Pharmacology, 57*(2), 201.
9. Protein Data Bank. PDB statistics: Overall growth of released structures per year. Available at: https://rcsb.org/stats/growth/growth-released-structures
10. Burley, S. K., Berman, H. M., Kleywegt, G. J., Markley, J. L., Nakamura, H., & Velankar, S. (2017). Protein Data Bank (PDB): the single global macromolecular structure archive. In *Protein crystallography: methods and protocols* (pp. 627–641).
11. Protein Data Bank. (1971). Crystallography: Protein Data Bank. *Nature: New Biology, 233*, 223.
12. Bartusiak, M. (1981, October 5) Designing drugs with computers. *Fortune*, 47–50.
13. Frye, L., Bhat, S., Akinsanya, K., & Abel, R. (2021). From computer-aided drug discovery to computer-driven drug discovery. *Drug Discovery Today: Technologies, 39*, 111–117.
14. Fischer, E. (1894). The influence of configuration on enzyme activity. *Berichte der Deutschen Chemischen Gesellschaft, 27*, 2984–2993. (Translated from German).
15. Schneider, H.-J. (2003). Introduction to molecular recognition models. In *Protein-Ligand interactions* (pp. 21–50).
16. Heaven, W. D. (2023, March/April). AI is dreaming up drugs that no one has ever seen. Now we've got to see if they work *MIT Technology Review*. Available at: https://www.technologyreview.com/2023/02/15/1067904/ai-automation-drug-development/
17. Metz, C. (2019, February 5). Making new drugs with a dose of artificial intelligence. *The New York Times*. Available at: https://www.nytimes.com/2019/02/05/technology/artificial-intelligence-drug-research-deepmind.html
18. Arnold, C. (2023). Inside the nascent industry of AI-designed drugs. *Nature Medicine, 29*(6), 1292–1295.
19. Belleau, B. (1970). Rational drug design: mirage or miracle? *Canadian Medical Association Journal, 103*(8), 850.
20. Bender, A., & Cortés-Ciriano, I. (2021). Artificial intelligence in drug discovery: what is realistic, what are illusions? Part 1: Ways to make an impact, and why we are not there yet. *Drug Discovery Today, 26*(2), 511–524.
21. Prieto-Martinez, F. D., Lopez-Lopez, E., Euridice Juarez-Mercado, K., & Medina-Franco, J. L. (2019). Chapter 2 - Computational drug design methods—Current and future perspectives. In K. Roy (Ed.), *In Silico drug design* (pp. 19–44). Academic Press.
22. Zhao, L., Ciallella, H. L., Aleksunes, L. M., & Zhu, H. (2020). Advancing computer-aided drug discovery (CADD) by big data and data-driven machine learning modeling. *Drug Discovery Today, 25*(9), 1624–1638.
23. Vemula, D., Jayasurya, P., Sushmitha, V., Kumar, Y. N., & Bhandari, V. (2023). CADD, AI and ML in drug discovery: A comprehensive review. *European Journal of Pharmaceutical Sciences, 181*, 106324.
24. Chandershekar, A., Bhaskar, A., Mekkanti, M. R., & Rinku, M. (2020). A review on computer aided drug design (CAAD) and it's implications in drug discovery and development process. *International Journal of Health Care and Biological Sciences*, 27–33.
25. Osakwe, O. (2016). Chapter 5 - The significance of discovery screening and structure optimization studies. In O. Osakwe & S. A. A. Rizvi (Eds.), *Social aspects of drug discovery, development and commercialization* (pp. 109–128). Academic Press.
26. dos Santos Nascimento, I. J., & de Moura, R. O. (2023). Ligand and structure-based drug design (LBDD and SBDD): Promising approaches to discover new drugs. *Applied Computer-Aided Drug Design: Models and Methods*, 1.
27. Hassan Baig, M., Ahmad, K., Roy, S., Mohammad Ashraf, J., Adil, M., Haris Siddiqui, M., et al. (2016). Computer aided drug design: Success and limitations. *Current Pharmaceutical Design, 22*(5), 572–581.

28. McCammon, J. A., Gelin, B. R., & Karplus, M. (1977). Dynamics of folded proteins. *Nature, 267*(5612), 585–590.
29. Klebe, G. (2013). Protein modeling and structure-based drug design. In G. Klebe (Ed.), *Drug design: Methodology, concepts, and mode-of-action* (pp. 429–448). Springer.
30. Jumper, J., et al. (2021). Highly accurate protein structure prediction with AlphaFold. *Nature, 596*, 583–589.
31. Maltarollo, V. G., Shevchenko, E., Lima, I. D. D. M., Cino, E. A., Ferreira, G. M., Poso, A., & Kronenberger, T. (2022). Do go chasing waterfalls: Enoyl reductase (fabi) in complex with inhibitors stabilizes the tetrameric structure and opens water channels. *Journal of Chemical Information and Modeling, 62*(22), 5746–5761
32. Díaz-Holguín, A., Rashidian, A., Pijnenburg, D., Monteiro Ferreira, G., Stefela, A., Kaspar, M., . . . & Kronenberger, T. (2023). When two become one: Conformational changes in FXR/RXR heterodimers bound to steroidal antagonists. *ChemMedChem, 18*(4), e202200556
33. Martins, D. M., Fernandes, P. O., Vieira, L. A., Maltarollo, V. G., & Moraes, A. H. (2024). Structure-guided drug design targeting abl kinase: How structure and regulation can assist in designing new drugs. *ChemBioChem*, e202400296.
34. Levitt, D. G., & Banaszak, L. J. (1992). POCKET: A computer graphics method for identifying and displaying protein cavities and their surrounding amino acids. *Journal of Molecular Graphics, 10*, 229–234.
35. Brady, G. P., & Stouten, P. F. (2000). Fast prediction and visualization of protein binding pockets with PASS. *Journal of Computer-Aided Molecular Design, 14*, 383–401.
36. Hendlich, M., Rippmann, F., & Barnickel, G. (1997). LIGSITE: automatic and efficient detection of potential small molecule-binding sites in proteins. *Journal of Molecular Graphics & Modelling, 15*, 359–363, 389.
37. Huang, B., & Schroeder, M. (2006). LIGSITEcsc: Predicting ligand binding sites using the Connolly surface and degree of conservation. *BMC Structural Biology, 6*, 19.
38. Halgren, T. (2007). New method for fast and accurate binding-site identification and analysis. *Chemical Biology & Drug Design, 69*, 146–148.
39. Halgren, T. A. (2009). Identifying and characterizing binding sites and assessing druggability. *Journal of Chemical Information and Modeling, 49*, 377–389.
40. Opo, F. A. D. M., et al. (2021). Structure based pharmacophore modeling, virtual screening, molecular docking and ADMET approaches for identification of natural anti-cancer agents targeting XIAP protein. *Scientific Reports, 11*, 4049.
41. Martin, Y. C., Kofron, J. L., & Traphagen, L. M. (2002). Do structurally similar molecules have similar biological activity? *Journal of Medicinal Chemistry, 45*, 4350–4358.
42. Wermuth, C. G., Ganellin, C. R., Lindberg, P., & Mitscher, L. A. (1998). Glossary of terms used in medicinal chemistry (IUPAC Recommendations 1998). *Pure and Applied Chemistry, 70*, 1129–1143.
43. Bajusz, D., Racz, A., & Heberger, K. (2015). Why is Tanimoto index an appropriate choice for fingerprint-based similarity calculations? *Journal of Cheminformatics, 7*, 20.
44. Racz, A., Bajusz, D., & Heberger, K. (2018). Life beyond the Tanimoto coefficient: Similarity measures for interaction fingerprints. *Journal of Cheminformatics, 10*, 48.

Chapter 2
Molecular Databases

Daniela Quadros de Azevedo, Rachel Oliveira Castilho, Alejandro Gómez-García, and José L. Medina-Franco

Abstract Compound databases (DBs) aim to organize the information needed for the initial stages of drug discovery. The collection and organization of curated information from bibliographic searches in DBs helps the scientific community to develop multidisciplinary research areas. For example, the chemical and biological properties contained in compound libraries, including PubChem, ZINC, BindingDB, and ChEMBL, are widely used in drug discovery projects. The importance of DBs in these projects is continuously increasing, beyond their role as compound repositories. In fact, compound DBs and chemical datasets can be a centerpiece in pharmaceutical companies, as well as in academic and government research centers. Several research groups have recently used computational methodologies to screen large DBs of compounds before experimental screening and designing their experiments. The number of DB compounds in the public domain, including those for compounds of natural origin, is increasing. This is in line with the growing and synergistic combination of natural product research and chemoinformatics, for example, NuBBE_{DB} (Nuclei of Bioassays, Ecophysiology and Biosynthesis of Natural Products) and BIOFACQUIM (A Mexican Compound Database of Natural Products). Developing a database involves four steps. Step 1—Search for chemical or biological information in indexed databases that report the strategies used to find the information that will make up the DBs. Step 2—Curation: describes the database (DB) curation processes, automated or manual. Step 3—DB management and network visualization: describes different systems to process DBs. Step 4—Update and maintenance: The last and most important stage is that, in addition to the development of a DB, its maintenance requires various resources, both financial and human. The quality of the compound DBs is crucial to fulfill properly their role

D. Q. de Azevedo · R. O. Castilho
Departamento de Produtos Farmacêuticos, Faculdade de Farmácia, UFMG, Belo Horizonte, Minas Gerais, Brazil

A. Gómez-García · J. L. Medina-Franco (✉)
DIFACQUIM Research Group, Department of Pharmacy, National Autonomous University of Mexico, Mexico City, Mexico
e-mail: medinajl@unam.mx

V. G. Maltarollo (ed.), *Computer-Aided and Machine Learning-Driven Drug Design*, Computer-Aided Drug Discovery and Design 3,
https://doi.org/10.1007/978-3-031-76718-0_2

as drug design tools. For automated curation, some tools can be used, such as KNIME and Open Babel. Currently, and in the near future, these virtual libraries could also enrich the medically relevant chemical space for inhibitors and potential drug candidates. In addition, the incorporation of chemoinformatics tools, the automation of curation processes, and the allocation of financial and human resources for the maintenance of DBs will contribute significantly to the development of drug design discovery and development projects.

Keywords Molecular database · Drug discovery · Chemoinformatic · Virtual screening

1 Introduction

Until the middle of the nineteenth century, the medicines used by humans were mainly of natural origin (animal, vegetable, and/or mineral) [1, 2] and their active principles (drugs) were compounds isolated from natural plant products and secondary metabolites of microorganisms [2, 3]. Examples include morphine (extracted from the poppy, *Papaver somniferum*), quinine (extracted from *Cinchona* species), atropine (extracted from *Atropa belladonna*), and salicin (extracted from *Salix alba*). In 1860, Kolbe and Lauteman synthesized salicylic acid, and in 1888, Hofman synthesized acetylsalicylic acid, supporting the arising of the pharmaceutical industry began. Consequently, in the twentieth century, the era of synthesis began with the development and production of drugs such as barbital, phenobarbital, epinephrine, amylocaine, and procaine [4]. Arguably, the discovery of penicillin by Alexander Fleming in 1928 sparked a renewed interest in natural products as sources of inspiration for drug development. In this regard, sulfanilide, which was also widely employed during World War, was synthesized in 1930. This led to subsequent breakthroughs, in the so-called "golden age" of antibiotics such as actinomycin (1942), streptomycin (1943), chloramphenicol (1947), and neomycin and erythromycin A in 1949 and 1950, respectively. Finally, the 1960s included rifamycin, vancomycin, lincomycin, and amphotericin B [5].

The accelerated pace of drug discovery during this period was partly due to advances in biology and chemistry, apart from the serendipitous discovery of drugs, such as penicillin. The resulting effects were significant in improving the quality of life and increased longevity [6]. Most human diseases have available therapeutic options, such as diabetes, Alzheimer's disease, immunological disorders, human immunodeficiency virus (HIV) and its associated acquired immune deficiency syndrome (HIV) [7], neglected tropical diseases (NTDs), and rare diseases [8]. However, new drugs are important when considering new diseases, the development of drug resistance, and our increasing understanding of health conditions allowing for the treatment of previously untreatable conditions [9].

In this sense, several strategic approaches have been proposed to increase efficiency in the drug discovery and development process, which can be adopted and exploited in a variety of ways. These approaches include genomics and proteomics (details in Chap. 3), complementary phenotyping and target-based screening platforms, drug repurposing and repositioning, collaborative research between industry and academia, outsourcing, molecular modeling, and artificial intelligence (AI) (details in Chap. 4) [6].

Modeling involves the use of in silico simulations to predict various properties of a compound, including pharmacokinetic and pharmacodynamic profiles (more details in Chap. 13) [10]. Advances in computer science have enabled the development of software that allows the simulation of drug-receptor binding processes, as well as a subset of processes known as virtual screening (VS), which increase the efficiency of hit identification from small and large existing chemical databases (DBs). VS is a process that generates results that can be validated in vitro and guides the optimization of lead compounds, improving the affinity of the drug to the receptor and its pharmacokinetic properties. VS also facilitates rational drug design by generating new compounds to optimize receptor binding [11]. For more details on virtual screening, refer to Chap. 11.

Moreover, AI is also being increasingly applied to drug design and development. This is possible by the availability of large chemical and biological DBs, which are today an important strategy for the development of accurate predictive models. AI has the potential to revolutionize drug discovery, particularly for unmet clinical needs, by enabling the screening of potential compounds with confident identification and validation of biological targets [11]. For more details on AI and machine learning, refer to Chap. 4.

In the period 2020 to 2022, some compounds have been developed exclusively by AI. As most are in clinical trials, they are described by code. DSP-1181 is a potent, long-acting 5-HT_{1A} serotonin receptor agonist with indications for the treatment of obsessive-compulsive disorder; DSP-0038 has the function of treating Alzheimer's disease and psychosis; EXS-21546 is an immunological compound with indications for the treatment of various types of tumors [12].

Currently, DBs may combine medicinal chemistry knowledge with important AI approaches and applications, providing not only quality content but also the user's ability to interact with this data and integrate it more easily. However, it is worth mentioning that even if these tools are not included in DBs, they help the drug discovery process. ZINC [13] and PubChem [14] are examples of DBs that have the potential to be used to capture data for the generation of predictive models using machine learning. Another database (DB) that includes AI tools is the latest version of ChEMBL (version 32) [15], which incorporates tools for calculating physicochemical properties of pharmaceutical interest.

Examples of software used in pharmaceutical modeling and AI-driven drug discovery are listed in Table 2.1.

Drug DBs aim to organize the information needed for the initial stages of drug discovery. The collection and organization of curated information from bibliographic searches in DBs help the scientific community to develop multidisciplinary research areas such as drug discovery, medicinal chemistry, chemosystematics, ethnopharmacology, and "omics" approaches. For example, many genomic studies, such as human genome mapping, have used GenBank and the DNA Databank of

Table 2.1 Examples of pharmaceutical modeling and artificial intelligence-guided drug discovery tools

Tool	Application in drug discovery	Link
AlphaFold[a]	Target modeling	https://alphafold.ebi.ac.uk/
AutoDock[a]	Protein ligand-binding modeling	http://autodock.scripps.edu/
DeepTox[b]	Predictive toxicology	https://deeptox.co
DiscoveryStudio[b]	Pharmacophore modeling, target identification, lead optimization	https://www.3dsbiovia.com/products/collaborative-science/bioviadiscovery-studio/
Glide[a]	Combinatorial chemistry and docking studies	https://www.schrodinger.com/glide
PPB2[b]	Predictive pharmacology	http://ppb2.gdb.tools/
PotentialNet[a]	Protein-ligand binding and molecular properties modeling	https://pubs.acs.org/doi/full/10.1021/acscentsci.8b00507
DataWarrior[b]	Prediction of physicochemical properties of pharmaceutical interest, cheminformatics calculations, multivariate data analysis, and interactive visualization with dynamic plots	https://openmolecules.org/datawarrior/

[a]Tools that support drug design
[b]Tools that support drug design usually included in DBs

Japan (DDBJ) [16]. Many pharmacological, computational, and proteomic studies have employed the Protein Data Bank (PDB) [17], the Human Proteome Map, and the Peptide Atlas [18]. In addition, several metabolomic studies have relied on the Human Metabolome DB (HMDB) [19], Golm Metabolome DB [20], Global Natural Product Social Molecular Networking (GNPS) [21], Biological Magnetic Resonance Bank (BMRB) [22], and Mass Bank [23]. In addition, the chemical and biological properties contained in compound libraries, including PubChem [14], ChemSpider [24], ZINC [13], PK/DB [25], BindingDB [26], ChemBank [27], and ChEMBL [15], are widely used in drug discovery projects.

The importance of DBs in new drug discovery projects is continuously increasing, beyond their role as compounds repositories. In fact, compound DBs and chemical datasets can be a centerpiece in pharmaceutical companies, as well as in academic and government research centers [28]. Compounds present in DBs have already led to the development of drugs in clinical use to treat different diseases. Several research groups have recently used computational methodologies to screen large DBs of compounds before experimental screening and designing their experiments [29].

The number of DB compounds in the public domain, including those for compounds of natural origin, is increasing. This is in line with the growing and synergistic combination of natural product research and chemoinformatics [30], for example, $NuBBE_{DB}$ (Nuclei of Bioassays, Ecophysiology and Biosynthesis of Natural Products) [31], BIOFACQUIM (A Mexican Compound Database of Natural Products) [32], and NPACT (Naturally occurring Plant-based Anticancerous Compound-Activity-Target Database) [33]. In addition to the natural products

DBs, other DBs have emerged with the aim of drug design, such as the Data Repository of Antiviral Peptides and Proteins (DRAVP). This database provides information on the antiviral activity, structure, physicochemical properties, and literature data of the peptides and proteins that make up the repository [34].

The COVID-19 pandemic has also led to the emergence of DBs that have gathered information to optimize research into this new viral disease, including the search for new drugs, such as the Small Molecule Antiviral Compound Collection (SMACC), which includes bioactivity data available in ChEMBL for compounds that have assays for emerging viruses that pose the greatest potential threat to global human health [35].

Furthermore, other databases recently developed to SARS- CoV-2, such as CoV-RDB [36], CORDITE [37], DockCoV2 [38], H2V [39], SARS-CoV-2 3D [40], and ZINCPharmer [41], bring other functionalities: (i) CoV-RDB contains data on the neutralizing susceptibility of SARS-CoV-2 variants to monoclonal antibodies, convalescent plasma, and vaccinated plasma; (ii) CORDITE provides drug interactions for SARS-CoV-2 current drug options; (iii) DockCoV2 shows computational representation of molecular docking; (iv) H2V contemplates information how the human body responds to viral infections; (v) SARS-CoV-2 3D provides possible drug targets from the coronavirus proteome; and (vi) the ZINCPharme uses the ZINC database, employs the Pharmer pharmacophore search technology, and also provides tools to construct and refine pharmacophore hypotheses directly from molecular structure.

Arguably, the great challenge of these repositories is the standardization of procedures for the curation of information and compounds, as well as the provision of financial resources and researchers for their maintenance and sustainability [42]. In this sense, we discussed these matters in the following topics.

2 Databases and Curation

Compound DBs usually are built by universities and research institutes. The development of a compound DB usually involves major steps (Fig. 2.1): First, searching the literature for the chemical and/or biological information that will constitute the DBs. At this stage, other databases may also be used as a source of chemical and biological information, such as ZINC, PubChem, and HMDB. Examples of chemical information are structure and molecular weight, 2D and 3D structures, SMILES, ClogP, and information on in vitro, in vivo, and ex vivo biological activities. Second, curating the information and compounds, and finally, selecting the management system and creating the DB website.

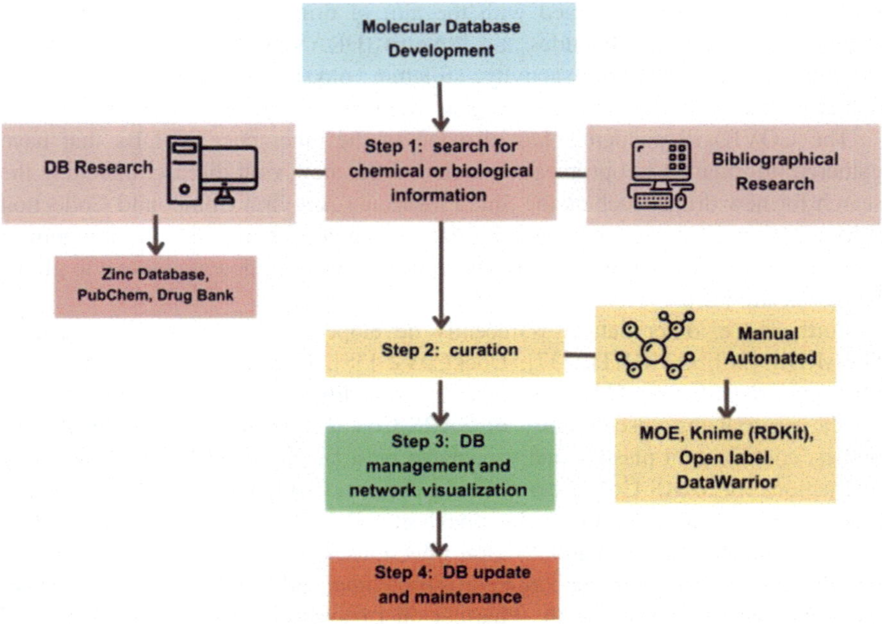

Fig. 2.1 The steps of the development of a molecular database. Step 1—search for chemical or biological information in indexed databases: reports the strategies used to find the information that will make up the DBs. Step 2—curation: describes the DB curation processes, automated or manual. Step 3—DB management and network visualization: describes different systems to process DBs. Step 4—update and maintenance: the last and most important stage is that, in addition to the development of a DB, its maintenance requires various resources, both financial and human

2.1 *First Step: Search for Chemical or Biological Information in Indexed Databases*

The strategy for feeding the compounds' information to the DB mainly uses data from indexed DBs. This approach was used in ChEMBL, NuBBE$_{DB}$, BIOFACQUIM, NPACT, and TCM Database@Taiwan. The information contained in a DB is extensive, as it may comprise 2D and 3D structures of compounds, with or without their biological activity. Antiviral Medicinal Plants and Natural Products DB (avMpNp DB) is a DB developed in Brazil and contains bioactive compounds from biodiversity with antiviral activity. The strategy for building the avMpNp DB consisted first of an extensive bibliographic search in academic DBs and compound DBs, to systematize information about the bioactive compounds included in the DB [43]. ChEMBL is a DB that was introduced in 2009 as an open-access resource and plays an important role in drug discovery and validation of computational tools. A large proportion of this bioactivity data in ChEMBL is currently manually extracted from scientific literature [15]. NPASS (Natural Product Activity and Species Source Database) [44], COCUNUT (COlleCtion of Open Natural ProdUcTs) [45], and

PHCS (Persian Herbal Constituents Database) [46] are databases of natural products that also employ manual information extraction of the literature.

The availability of public chemistry and bioactivity DBs, along with large-scale data-driven applications, has increased the community's attention to data curation and integrity issues, such as structure quality, name-to-structure fidelity, structure–activity mapping, activity data accuracy, assay description sufficiency, target assignment, author errors, and redundancy. Together, such factors can provide higher confidence assertions and therefore more robust applications and models from the available compounds' information in DBs.

In particular, the compounds contained in DBs have already led to the development of drugs in clinical use to treat various diseases and are contributing to the development of compounds in clinical development and basic research. Virtual libraries thus help to increase the success rate of the lead selection process by ensuring the quality, diversity, and consistency of the curated data. The number of public domain compound databases is increasing and the process of building and curating these libraries is critical as the data must be diverse and reliable to enable safe trials. Therefore, the assembly and analysis of compound DBs in terms of their structure and the legitimacy of the structures is crucial [47].

2.2 Second Step: Data Curation

Data curation includes, for example, elimination of salts, adjustment of protonation states, optimization of geometry by energy minimization, and elimination of duplicated molecules. This curation process can involve several manual and automated steps and aims to maximize data accessibility and comparability and improve data integrity and flag outliers, ambiguities, and potential errors. Standard protocols are used in manual DB curation processes. Although this step is not easy, it is feasible. It would be advisable for a research group to be responsible for this endeavor, using publicly available tools and scripts or workflows available on public repositories such as GitHub, where examples of database curation are freely available [48].

The automated curation process uses platforms with different functionalities. One example is Open Babel, a tool that provides a solution to the proliferation of different file formats in chemistry. It also contains conformer searching, 2D visualization, filtering, batch conversion, substructure, and similarity searching. For developers, it can be a programming library for chemical data handling in areas such as organic chemistry, drug design, materials science, and computational chemistry. It is freely available under an open-source license from http://openbabel.org [49].

ChEMBL, for example, is a database that uses both manual and automated curation strategies and implements their validation and standardization process using pipelining tools such as Pipeline Pilot [50] or the Konstanz Information Miner (KNIME) analysis platform [51]. These tools also allow for more flexibility such as new components that can be added or adapted as needs change. The ChEMBL DB providers routinely include a salt stripping process in their

Table 2.2 Some examples of tools of automatized curation: platforms, functionalities, and uses

Platform	Advantages	Functionalities for curation (examples)	Databases use
Molecular Operating Environment (MOE)	Integrated computer-aided molecular design platform-small molecules, peptides, biologics Help draw chemical structures and facilitate the storage and interconversion between standard file formats Free and open-source academic version	Disconnects salts and metals Removes simple components Recalculates states of protonation, determines wedge bonds for bonds from chiral centers Calculates missing chiral parities from existing wedge bond	$NuBBE_{DB}$ BIOFACQUIM
Konstanz information miner (KNIME)[a] RDKit[b]	[a]Offers over 300+ connectors to data sources, and integrations to all popular machine learning libraries [b]Manipulate molecular structures in Python [a]Freely available	Removal of characters encoding stereoisomerism in SMILES format (@; \; /) Removal of salts Neutralization of charges Maintenance of compounds containing only the following elements: (H, C, N, O, F, Br, I, Cl, P, S) Creation of compounds in InChI, InChIKey, and canonical SMILES formats from standardized compounds	PeruNPDB Super Natural II ChEMBL
Open Babel	Open, collaborative project Convert, analyze, or store data from molecular modeling, chemistry, biochemistry, or related areas	Removal of salts Adjustment of the protonation state of the structures Convert between molecule file formats	SWMD ChemDB

[a]Advantage of KNIME; [b]Advantage of RDKit

standardization based on a library of pharmaceutically relevant salts [15]. Table 2.2 shows examples of automatized curation, platform, functionalities, and their use.

$NuBBE_{DB}$ [31] and BIOFACQUIM [32] use Molecular Operating Environment (MOE) [52] for automatized curation. This node disconnects salts and metals, removes simple components, recalculates states of protonation, determines wedge bonds for bonds from chiral centers, and calculates missing chiral parities from existing wedge bonds. Using this same software, inorganic compounds can be eliminated, as well as duplicated compounds [53, 54].

Another example of a DB using automated curation strategies is the Seaweed Metabolite Database (SWMD), which comprises compounds, derived from seaweeds, and it has been curated using the Open Babel software. This strategy allowed the removal of salts and the adjustment of the protonation state of the structures. Duplicate structures were manually removed after detection in SMILES strings using Microsoft Excel 2016, followed by manual inspection [55]. Open Babel is a full-featured open chemical toolbox, designed to translate the many different representations of chemical data [49]. It allows anyone to search, convert, analyze, or store data from molecular modeling, chemistry, solid-state materials, biochemistry, or related areas. It provides both ready-to-use programs as well as a complete, extensible programmer's toolkit for developing cheminformatics software. In addition, the ChemDB is a small molecule database that also uses Open Babel to convert between molecule file formats [56].

KNIME analysis platform, which includes RDKit, is also used to curate chemical structures in DBs, such as PeruNPDB, the Peruvian Natural Products Database, and Super Natural II [57, 58]. RDKit is an open-source cheminformatics toolkit written in C++ that is also usable from Java or Python. It includes a collection of standard cheminformatics functionality for molecules, substructure searching, chemical reactions, coordinate generation (2D or 3D), fingerprinting, curation, as well as a high-performance database cartridge for working with molecules using the PostgreSQL DB [57–59].

DataWarrior is a multi-functional and interactive chemical data analysis and visualization tool. It provides interactive options for visualizing and curating data, assessing correlations, and extracting knowledge from large datasets [60]. The DataWarrior tool was used to eliminate duplicate structures in the different studies, such as in the identification of anti-schistosomal, anthelmintic, and antileishmanial compounds [61, 62].

DBs also use manual curation, for example, ChEBI (Chemical Entities of Biological Interest), which is a manually curated DB and ontology that organizes small molecule knowledge [63]. Last, PSC-db [64] and avMpNp DB [43] also employ manual curation using internal scripts.

2.3 *Third Step: Database Management and Network Visualization*

Molecular DBs contain a wide variety of data that may be processed by different systems, which, if unrelated, can lead to redundancy and inconsistency for the user. To solve this problem, the data needs to be stored only once on a platform that can be accessed and shared by all the systems involved. This solution results in a more complex software structure that requires the help of a database management system (DBMS) to maintain.

The choice of the DBMS model is fundamental to its development, as it is the basis for structuring a database (data types, relationships, and relevant constraints). The relational model is the most widely used model because of its greater flexibility and suitability for design and implementation, and many DB systems today are based on it. These characteristics are due to its structuring of data into relationships. Examples of DBs that are managed using a relational model include Super Natural II [65], ChEBI [63], Vietherb [66], ZINC [13], and NPACT [33].

Another option for implementing a molecular DB is to use a workflow-based management system (WBMS). WBMSs are data management systems that flexibly control the execution of a set of tasks, allowing this set of tasks to be modified without changing the system code. A major advantage of using a WBMS is its flexibility, changes to the stored information are reflected in changes into workflows, which makes it much easier to adapt the system and does not require changes in the system code. This is important because different bioactive compounds may have different types of data associated with them, and in a workflow-based WBMS, all this heterogeneous data can be easily managed [67].

2.4 Fourth Step: Updating and Maintenance

Usually, the DB function refers to compound repositories. In fact, compound DBs and their chemical datasets are a central part of pharmaceutical companies and private or government research centers. These DBs have been upgraded through the cooperation of chemoinformatics tools and the introduction of the new compound. A study that evaluated 52 DBs developed over the last 30 years found that most of them originated in the academic sector, such as in universities and research institutes, where maintenance and upgrading also take place. Private DBs are maintained by industry and their data is usually confidential [43] (Fig. 2.2).

3 Advantages and Disadvantages of Using Molecular Databases

Molecular DBs are useful resources in computer-aided drug design and play a central role in many chemoinformatics applications. It is possible to identify potential bioactive compounds with therapeutic activity through several chemoinformatic methodologies. Indeed, molecular DBs can provide access to hundreds, thousands, or even hundreds of thousands of compounds that can be virtually screened to predict which of them have the desired biological activity. Recently, the amount of freely accessible databases has increased, allowing access to an increasing number of compounds [68]. Molecular DBs not only provide access to chemical structures but also contain other useful information for the different drug design stages (Table 2.3).

As illustrated in Table 2.3, molecular databases are useful not only in drug discovery. For instance, SciFinder [78] allows users to check the reported synthesis pathways of a molecule and ChemSpider [24] contains spectroscopic data of the compounds. Other examples are ZINC [13], ChEMBL [15], and ChemSpider [24]. DBs provide information regarding the commercial availability of chemical compounds. Despite the above-mentioned advantages of using molecular DBs during the drug design process, there are still some associated deficiencies. The

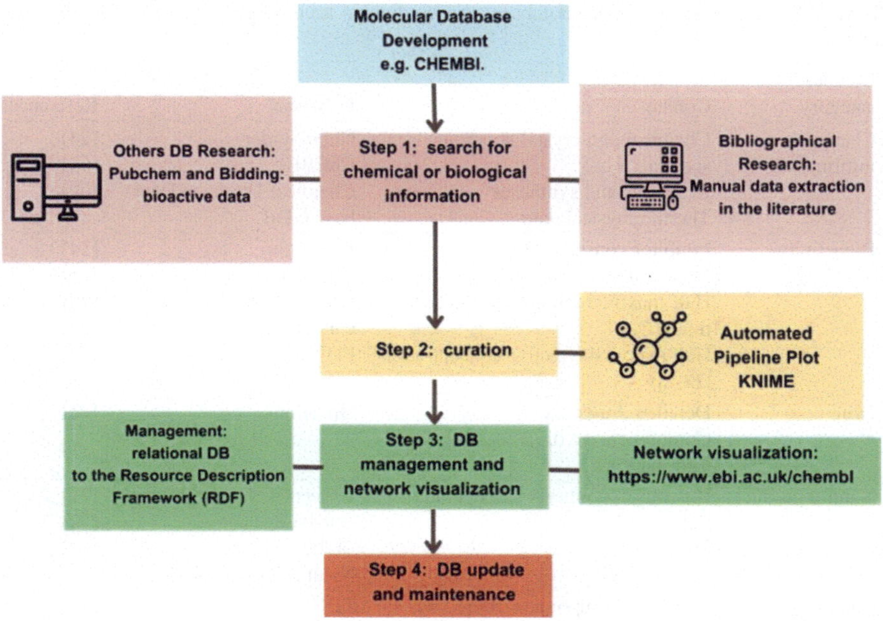

Fig. 2.2 Example of how to build a DB, CHEMBL. Step 1—Search for chemical or biological information in indexed databases: CHEMBL uses manual extraction of data from scientific literature. Step 2—Curation: CHEMBL DB uses tools such as KNIME for this purpose. Stage 3—DB management and network visualization: Access to the data in ChEMBL, a relational DB, is provided through a user interface, a set of web services, and a range of download formats including XML, JSON, and YAML. Stage 4—Updating and maintenance

structural integrity of the molecules is not always assured, and the annotations are not necessarily error-free. For instance, errors in the stereochemistry can be found in DBs, as valence issues and charge imbalances [79] and small structural errors can lead to significant losses of predictive abilities of quantitative structure–activity relationship (QSAR) models [80]. For more details on QSAR and machine learning predictors, please refer to Chap. 6. This is particularly common for publicly available databases, where it can be challenging to have a dedicated team to curate the chemical content and annotations in the compound DBs. Incomplete or wrong information provided in chemical DBs is not only limited to the chemical structures. Erroneous information regarding the biological activity of a molecule is something else found in chemical databases. For instance, it is not always reported the assay employed to determine the half maximal inhibitory concentration (IC_{50}) or the minimum inhibitory concentration (MIC) [81]. Besides, the assay conditions are not always reported, or the original references are missing [65]. Such incomplete or inaccurate data can lead to less reliable QSAR predictions. Thus, to improve the quality of chemical databases, periodic revisions and error reporting are recommended practices.

Table 2.3 Categories into which databases can be divided according to the type of information stored

Database category	Content	Database[a]	Reference
Chemical information	Chemical and crystal structures spectra Reactions and syntheses Thermophysical data	ChemSpider ChEBI Chemical Universe Database GDB	[24] [63] [69]
Bioactivity	Inhibitor constant (K_i) Dissociation constant (K_d) Half maximal inhibitory concentration (IC_{50}) Half maximal effective concentration (EC_{50})	PubChem ChEMBL BindingDB ChemBank PDBbind	[14] [15] [26] [27] [70]
Drug	Detailed drug data Comprehensive drug target information	DrugBank	[28]
Natural product	Pathways (synthesis and degradation) Structures	Universal Natural Product Database MeFSAT Natural Product Atlas	[71] [72] [73]
Chemical availability	Available compounds offered by chemical vendors	ZINC NCI	[13] [74]
Fragment	Physicochemical information Binding site preferences	FDB-17 Fragment Store PADFrag	[75] [76] [77]

[a]Representative examples. Some of the databases have more than one category that is not shown in the table, for example, PubChem (chemical information and chemical availability), ChEMBL (chemical information and drug), ChemSpider (bioactivity and chemical availability), and ZINC (chemical information and bioactivity)

4 Natural Product Databases for the Search and Development of New Drugs

Nature is a rich source of bioactive molecules that serve as therapeutic agents. These bioactive molecules are natural products, which can be defined as compounds produced by living beings and can be employed or proposed as therapeutic agents [82]. For instance, of the approved small molecules in the research area of cancer, from 1946 to 1980, 53% of the compounds that became new medicines corresponded to unaltered natural products or natural product derivatives. From 1981 to 2019, 64.9% of the approved small molecules were unaltered natural products or natural products inspired [83]. Moreover, natural products are an abundant source of privileged scaffolds: structures capable of providing useful ligands for more than one receptor [84]. Some examples of privileged scaffolds that come from natural products that are currently used in the design and development of new drug candidates are the terpenoid, polyketide, phenylpropanoid, and alkaloid structures [85]. Regarding natural product DBs, one application is the

Table 2.4 Representative natural product databases

Database	Number of compounds[a]	Accessibility	Outstanding features	References
Collection of Open Natural Products (COCONUT)	411,621	Open access	Predicted bioactivities	[45]
Universal Natural Product Database	~229,000	Open access	3D structures	[71]
SuperNatural 3.0	449,058	Open access	Toxicity Vendor information	[65]
ZINC	~80,000	Open access	Bioactivities Vendor information	[13]
Dictionary of Natural Products	~230,000	Commercial	Spectroscopic data	[82]
SciFinder	~300,000	Commercial	Reported synthesis routes	[78]
Reaxys	~200,000	Commercial	Bioactivity Toxicity Physicochemical data	[91]
TCM@Taiwan	~58,000	Open access	The largest database of natural products from Traditional Chinese Medicine	[92]
IMPPAT	~10,000	Open access	The largest database of natural products from traditional medicine in India	[93]
AfroDB	~1000	Open access	The largest database of natural products from traditional medicine in Africa	[94]
Phyto4Health	3128	Open access	Medicinal plants included in the Russian Pharmacopoeia Bioactivities Predicted bioactivities	[95]
$NuBBE_{DB}$	2223	Open access	Bioactivity Predicted spectroscopic data	[31, 96]
BIOFACQUIM	553	Open access	Taxonomic information of the producing organism	[32]

[a]Date of search: April 2024

design of pseudo-natural products, that is, molecules that retain the biological relevance of natural products yet exhibit structures and bioactivities not available in nature or in existing design strategies. Pseudo-natural products may display unexpected bioactivities that differ from the activities of the natural products from which their fragments are derived [86–88]. Besides, natural products usually have more structural diversity compared with synthesized small molecules [89].

Natural product DBs can be important tools for computer-aided drug design (CADD), providing access to a large number of diverse chemical structures. These DBs can be divided into commercial and open access (Table 2.4). Between 2000 and 2019, 123 commercial and open-access natural product collections have been published, of which 98 are somewhat accessible, 92 are open access, and only

50 contain molecular structures that can be retrieved for a chemoinformatic analysis [90]. For example, the Collection of Open Natural Products (COCONUT) [45] contains more than 411,000 natural product entries collected from 50 open-access natural product DBs. Similarly, the Universal Natural Product Database [71] is another compilation DB with more than 229,000 natural products. It provides 3D structures with stereochemical information and calculated molecular descriptors but is not yet accessible through the link in the original publication. Instead, it is available on another website [81]. Currently, SuperNatural 3 is the biggest open-access DB, which contains over 449,058 unique natural products and includes information about 2D structures, physicochemical properties, predicted toxicity class, and potential sellers, but it does not yet provide the option to download in bulk.

There are natural product databases comprised of molecules isolated and characterized in specific geographical regions. China is one of the regions with more natural product databases published [102 –110], because Traditional Chinese Medicine (TCM) is part of the Chinese public health system. There are two natural product DBs that comprise compounds that are part of the traditional medicine in India (Indian Ayurveda), such as IMPPAT [93] and MedPServer [106]. Regarding African Traditional Medicine, there are different natural product DBs published such as ConMedNP, p-ANAPL library, and others [107–111]. There is a database that contains natural products from medicinal Russian plants: Phyto4Health [95]. Latin America is a region that encompasses at least a third of global biodiversity [112]. All the published natural product databases of Latin America and their practical applications in the drug discovery area have been reviewed and discussed elsewhere [113]. Two representative examples of natural product databases from Latin America are $NuBBE_{DB}$ [31] and BIOFACQUIM [32].

5 Functionalization of Databases and Transformation into a System to Drug Design

The usage of molecular DBs as a drug design tool depends on the capacity of the chemoinformatic software to recognize the molecules. For this purpose, the simplified molecular input line entry system (SMILES) notation is the predominant input notation for the different chemoinformatic software packages [114]. Other notations that can be recognized by the chemoinformatic software overcome some disadvantages of the SMILES, such as the International Chemical Identifier (InChI) [115] and InChIKey [116]. SMILES arbitrary target specification (SMARTS) notation was developed to specify substructural patterns that allow matching molecules that contain a specified substructural pattern [117]. The different notations used by the chemoinformatic software, as well as their uses, advantages, and disadvantages, have been explained in detail elsewhere [118].

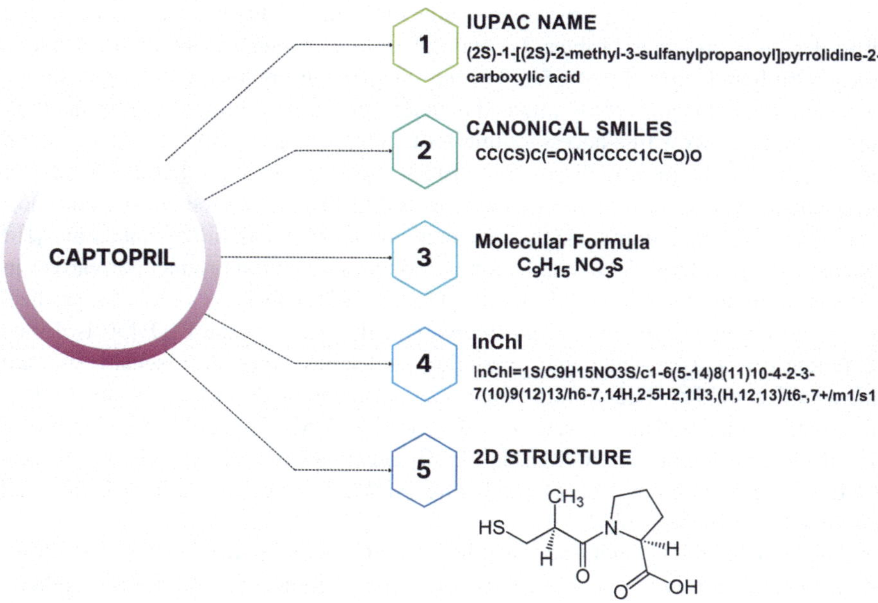

Fig. 2.3 Nomenclature and identification of compounds in a database, for example, captopril

Free-access online tools that allow the user to draw a chemical structure and convert it into either SMILES, InChI, InChIKey, or SMARTS can be found (Fig. 2.3), such as PubChem Sketcher [119]. Additionally, there are also tools that automatically recognize the chemical structures inside a .pdf file or image and transform into a SMILES string. Notwithstanding a manual inspection of the output molecules is required to detect possible mistakes in the output structures [120]. Molecules can be represented according to their 2D or 3D coordinates, and the two most commonly formats used are .mol2 and .sdf. These files can store the atomic coordinates of each atom of a molecule and the information of the connectivity among the atoms. Moreover, both formats allow the storage of physicochemical information of the molecule, while the .mol2 format allows the storage of the atomical charges of every atom [121, 122].

SMILES (Simplified Molecular-Input Line-Entry System) uses graph theory to represent structures. Each atom is represented by its symbol in the periodic table, brackets are used to indicate branching points, and numerical labels denote ring junctions. The basic grammar of SMILES also includes isotopic information, configuration on double bonds, and chirality, also known as isomeric SMILES. Later, IUPAC (International Union of Pure and Applied Chemistry) developed its own linear notation for representing chemical structures, InChIKey (International Union of Pure and Applied Chemistry Key), based on the InChI identifier. A compound 2D structure image can be generated using MarvinSketch [123].

Moreover, there are available open-source molecular DBs that can be used during drug design approaches (Tables 2.3 and 2.4 in this chapter). Once the database is already built and curated (see Sect. 2 of this chapter), or obtained either open source or commercial, it is possible to proceed to determine helpful information for the drug design process, with free-access online tools. One can cite NERDD [124], which offers access tools for drug discovery. NERDD hosts tools for predicting the sites of metabolism (FAME) [125] and metabolites (GLORY) [126] of small organic molecules, for flagging compounds that are likely to interfere with biological assays (Hit Dexter) [127], and for identifying natural products and natural product derivatives in large compound collections (NP-Scout) [128]. NERDD also has a tool that predicts if a small organic compound is an inhibitor of different human CYP450 isoforms (CYPlebrity) and another that classifies molecules into skin sensitizers and nonsensitizers [129]. These online tools also allow the prediction of the absorption, distribution, metabolism, excretion, and toxicity (ADMET) profile of a single or a list of molecules, such as ADMETlab [130] and PhaKinPro [131], among at least 18 free web servers capable of predicting ADMET properties, which have been discussed elsewhere [132].

Additionally, the synthetic accessibility score is another helpful metric that allows to determine the synthetic feasibility of given a molecule. With the synthetic accessibility score, it is possible for a user to select from a DB those compounds with the highest synthetic feasibility [133]. SYBA [134] and GASA [135] are two different synthetic accessibility scores that can be determined with free-access online tools. In addition, retrosynthetic planning can also be useful, as in AiZynthFinder, which recursively breaks down a molecule to its purchasable precursors [136].

Furthermore, one can also capture structural information from molecules. Two common molecular fingerprints employed to capture structural information are the Molecular ACCess System (MACCS) keys-166 bits [137] and the Extended Connectivity Fingerprint (ECFP6) [138]. MACCS keys and ECFP6 fingerprints encode the structure of the molecules in strings made of bits, that is, every molecule is represented with a string made by one and zeros. From the fingerprints, the structural similarity among the molecules can be measured using the Tanimoto coefficient (Tc) [139]. Herein, if two molecules share similar structures, then they will likely have similar bioactivities [140]. Thus, a molecule with a certain known biological activity can be identified using Tc calculations in a molecular database with similar compound's structure possibly having a similar bioactivity to the matched molecule. These fingerprints can be calculated, for example, by both CDK [141] or RDKit [142] nodes of the free-available software KNIME [51]. RDKit also can be used in the Python programming language. Last, another useful and free computational tool to calculate the fingerprints is PaDEL [143].

6 Chemoinformatics and Computational Tools for Supporting and Filtering Potential Drug Bioactive Compounds from Databases

The SMILES strings (see Sect. 5 in this chapter) available from the molecular DBs can also be used to determine physicochemical properties of interest, such as SlogP [144], molecular weight (MW), topological polar surface area (TPSA) [145], rotatable bonds (Rb), hydrogen bond acceptors (HBA), and hydrogen bond donors (HBD). The calculation of the physicochemical properties can be performed using the freely available software DataWarrior [146], KNIME [51] or using Python, employing the CDK [141] and RDKit [142] nodes (KNIME) or RDKit package (Python). One could argue that out of these three options, DataWarrior is the most user-friendly, as the list of SMILES strings can be imported into the program just with a copy/paste keyboard shortcut. Once the physicochemical properties have been calculated, the compounds can be filtered to a chosen numerical range for each property. There are some drug-likeness guidance parameters, also called rules of thumb by some authors, which are suggestions that help to choose the potential drug-like compounds such as Lipinski's rule of 5 (Ro5) [147, 148], Veber's rules [149], GlaxoSmithKline's (GSK) 4/400 rule [150], and Pfizer 3/75 rule [151] (Table 2.5). Compliance with either Lipinski's, Veber's, or GSK rules is commonly associated with good oral bioavailability. These suggestions or rules of thumb are based on the premise that certain physicochemical properties are directly associated with one or more parameters of the ADMET profile. Thus, if a group of molecules has a known ADMET profile, it is expected that other molecules would have a similar ADMET profile whenever they share similar physicochemical properties. Usually, these rules are helpful in filtering compounds that are expected to have a desirable ADMET profile [152]. Nonetheless, one should consider that the ADMET profile of a compound is influenced by many other aspects and not only by some physicochemical properties. Therefore, the fulfillment of one or more of these rules of thumb does not necessarily assure obtaining the expected desirable value of a parameter of the ADMET profile [153].

7 Databases in Drug Design and Discovery: Applicability, Challenges, and Successful Cases

The field of drug design and discovery still faces low efficacy, off-target delivery, as well as the time spent, and high cost associated [154]. In this scenario, the increase in biological and chemical data, including in vitro, in vivo, clinical studies, genomics studies, proteomics studies, metabolomics studies, gene ontology studies, and molecular pathway data, benefits from different data repositories that have been developed. For instance, ChemSpider, ChEMBL, ZINC, BindingDB, and PubChem are the essential DBs for compound synthesis and screening in the drug design and

Table 2.5 Rules of thumb or guides associated with drug-likeness

	Molecular weight	LogP	Number of hydrogen bond acceptors	Number of hydrogen bond donors	Sum hydrogen bond acceptors and donors	Topological polar surface area	Number of rotatable bonds	References
Lipinski's rule of 5	≤ 500	≤ 5	≤ 10	≤ 5				[147, 148]
Veber's rules					≤ 12	≤ 140	≤ 10	[149]
GlaxoSmithKline 4/400 rule	<400	<4						[150]
Pfizer 3/75 rule		>3				< 75		[151]

discovery process. ZINC database was the most preferred DB in virtual screening studies with an average use of 31.2% from 2015 to 2022 [133], selecting compounds according to their pharmacological and physicochemical properties of pharmaceutical interest [155].

Among the new drug design strategies, the search for information and computational tools is the most widely used due to its great ease of access and extremely low cost. A DB displays various techniques and information from, for example, in silico studies, such as molecular docking, molecular dynamics simulation, and VS. Therefore, a given DB is designed to include the information necessary for the first stage of drug discovery, including molecular structures, physicochemical properties, molecular properties of ligands, and their drug-likeness properties [156].

VS is an approach that benefits from virtual DBs containing a large number of compounds. VS allows the identification of novel hits as well as the prioritization of compounds for in vitro testing, resulting in a significant reduction in costs and attrition rates [154]. This pre-selection is done by virtually predicting the biological activity of interest using different structure-based drug design and/or ligand-based drug design (SBDD and/or LBDD) approaches. SBDD was the most prominently used type of VS and it accounted for an average of 57.6%, from 2015 to 2022 [133]. In the early stages of drug design, in silico studies represent a relevant strategy for reducing costs in this process [28]. DrugBank is one free-access website containing information on potential drugs and their targets that is usually accessed for such approaches. Drugbank is a bioinformatics and chemoinformatics resource that contains detailed information of drugs and their biological targets (29.785 targets). It contains a detailed description of these targets, such as their type, organism in which they occur, pharmacological function, and specific and general functions [156].

Despite successful drug discovery approaches accessing different DBs, it has been reported that between 0.1% and 3.4% of the chemical structures in chemical DBs are incorrect [157], including errors in molecular DBs regarding compounds' bioactivity data [158]. An error in the structure representation of the molecules can lead to inaccurate QSAR predictions, erroneous hazard and risk assessments, and wrong decisions in the early drug discovery process. In molecular DBs, it has been identified fundamental errors in stereochemistry, valency issues, and charge imbalances [157]. Unfortunately, still in 2023, an analysis identified a substantial number of errors in the identifiers and chemical structures in DBs, such as PubChem, CompTox Chemicals Dashboard, and European Chemicals Agency (ECHA). Further, the absence of these errors is not guaranteed even in government-funded DBs [159]. Therefore, it is important to identify and correct those wrong entries and structures in the molecular DBs.

There have been efforts with the aim of reverting the problem, such as an open-source chemical structure curation pipeline that was published and utilized successfully to standardize the nearly two million compounds in the ChEMBL DB [160]. In addition, compounds with critical issues were also identified so that they could be prioritized for manual curation [161]. The solution to these challenges requires coordinated strategies and a strong financial commitment by the government, funders, and institutions. It is estimated that the current global costs of maintaining public biomedical data repositories are under $300 million annually [162]. To maintain this, it is important to have a collaboration from universities, pharmaceutical industry, and institutions with the government to raise awareness about the importance of the situation with the capacity to maintain curate and create chemical DBs.

8 Perspectives

Based on research about DBs commonly used in drug discovery, we highlight the relevance of these libraries in this process. Natural product DBs, for example, are becoming promising accelerators in the development of drugs from bioactive compounds for various therapeutic areas such as cancer, viral, and neglected diseases. There is a growing trend toward such libraries, i.e., repositories of information on compounds with specific pharmacological activities. Currently, and in the near future, these virtual libraries could also enrich the medically relevant chemical space for inhibitors and potential drug candidates. In addition, the incorporation of chemoinformatics tools, the automation of curation processes, and the allocation of financial and human resources for the maintenance of DBs will contribute significantly to the development of drug design discovery and development projects.

9 Conclusion

The process of developing new bioactive compounds into drugs is long and complex. Scientific and technological advances at the interface of chemistry and biology have created remarkable opportunities and challenges for research and development. Computational simulations have been playing an important role in reducing costs and accelerating the drug design and discovery processes. These advantages, together with the large number of chemical entities available in DBs, represent a vast area of research for new potential drugs. The importance of DBs in new drug discovery projects is continuously increasing and is a centerpiece in pharmaceutical companies as well as in academic and government research centers. Notwithstanding, the quality of the compound DBs is crucial to properly fulfill their role as drug design tools. Therefore, the curation and maintenance of high-quality data is essential. Compounds discovered in existing DBs have already led to the development of drugs in clinical use to treat different diseases. This shows the feasibility of accessing different DBs that can continue to contribute to the field of drug design, discovery, and development.

Questions to answer when planning an experiment using databases

? What computational approaches are being used to accelerate drug discovery
What is a compound database?
What is the relevance of a compound database in drug discovery?
What are the greatest practical challenges to developing compound repositories?
What are the major steps to construct a compound database?
What are the largest compound databases annotated with biological activity?
What is data curation?
Mention examples of software used for data curation.
What are the major deficiencies or shortcomings in databases, particularly in the public domain?
Provide examples of at least five natural product databases publicly available.
What are the main linear notations to store the information of chemical structures in compound databases?
What are the main rules of thumb or empirical rules used in drug discovery projects?

References

1. Kubinyi, H. (2004). *Industrial Pharmacy*, 7.
2. Faller, B., Ottaviani, G., Ertl, P., Berellini, G., & Collis, A. (2011). *Drug Discovery Today, 16*, 976.
3. Viegas, C., Bolzani, V. S., & Barreiro, E. J. (2006). *Quimica Nova, 29*, 326.
4. Yunes, R. A., & Filho, V. C. (Org.). (2007). *Química de produtos naturais, novos fármacos e a moderna farmacognosia. 1 ed* (3003 p). Ed. Universidade do Vale do Itajaí.
5. Drews, J. (2000). Drug discovery: A historical perspective. *Science, 287*, 1960–1964.
6. Rotella, D. P. (2016). The critical role of organic chemistry in drug discovery. *ACS Chemical Neuroscience, 7*, 1315–1316.
7. Cummings, J. L., Morstorf, T., & Zhong, K. (2014). Alzheimer's disease drug development pipeline: Few candidates, frequent failures. *Alzheimer's Research & Therapy, 6*, 37.
8. Ridley, R. G. (2002). Medical need, scientific opportunity and the drive for antimalarial drugs. *Nature, 415*, 686–693.
9. Klebe, G. (2006). Virtual ligand screening: Strategies, perspectives and limitations. *Drug Discovery Today, 11*, 580–594.
10. Song, C. M., Lim, S. J., & Tong, J. C. (2009). Recent advances in computer-aided drug design. *Briefings in Bioinformatics, 10*, 579–591.
11. Chan, H. C. S., Shan, H., Dahoun, T., et al. (2019). Advancing drug discovery via artificial intelligence. *Trends in Pharmacological Sciences, 40*, 592–604.
12. Roda, C. I. N. (2022). *A inteligência artificial na descoberta de novos medicamentos* (Doctoral dissertation).
13. Sterling, T., & Irwin, J. J. (2015). ZINC 15–ligand discovery for everyone. *Journal of Chemical Information and Modeling, 55*(11), 2324–2337.
14. Kim, S., Thiessen, P. A., Bolton, E. E., Chen, J., Fu, G., Gindulyte, A., Han, L., He, J., He, S., Shoemaker, B. A., Wang, J., Yu, B., Zhang, J., & Bryant, S. H. (2016). PubChem substance and compound databases. *Nucleic Acids Research, 44*, D1202–D1213.
15. Papadatos, G., Gaulton, A., Hersey, A., & Overington, J. P. (2015). Activity, assay and target data curation and quality in the ChEMBL database. *Journal of Computer-Aided Molecular Design, 29*, 885–896.

16. Benson, D. A., Cavanaugh, M., Clark, K., Karsch-Mizrachi, I., Ostell, J., Pruitt, K. D., & Sayers, E. W. (2018). GenBank. *Nucleic Acids Research, 46*(D1), D41–D47.
17. Burley, S. K., Berman, H. M., Kleywegt, G. J., Markley, J. L., Nakamura, H., & Velankar, S. (2017). Protein data Bank (PDB): The single global macromolecular structure archive. *Protein Crystallography: Methods and Protocols*, 627–641.
18. Thul, P. J., & Lindskog, C. (2018). The human protein atlas: A spatial map of the human proteome. *Protein Science, 27*(1), 233–244.
19. Wishart, D. S., Guo, A., Oler, E., Wang, F., Anjum, A., Peters, H., et al. (2022). HMDB 5.0: The human metabolome database for 2022. *Nucleic Acids Research, 50*(D1), D622–D631.
20. Hummel, J., Selbig, J., Walther, D., & Kopka, J. (2007). The Golm metabolome database: A database for GC-MS based metabolite profiling. In *Metabolomics: A powerful tool in systems biology* (pp. 75–95). Springer.
21. Wang, M., Carver, J. J., Phelan, V. V., Sanchez, L. M., Garg, N., Peng, Y., et al. (2016). Sharing and community curation of mass spectrometry data with global natural products social molecular networking. *Nature Biotechnology, 34*(8), 828–837.
22. Ulrich, E. L., Akutsu, H., Doreleijers, J. F., Harano, Y., Ioannidis, Y. E., Lin, J., et al. (2007). BioMagResBank. *Nucleic Acids Research, 36*(suppl_1), D402–D408.
23. Horai, H., Arita, M., Kanaya, S., Nihei, Y., Ikeda, T., Suwa, K., et al. (2010). MassBank: A public repository for sharing mass spectral data for life sciences. *Journal of Mass Spectrometry, 45*(7), 703–714.
24. Pence, H. E., & Williams, A. (2010). ChemSpider: An online chemical information resource. *Journal of Chemical Education, 87*, 1123–1124.
25. Moda, T. L., Torres, L. G., Carrara, A. E., & Andricopulo, A. D. (2008). PK/DB: Database for pharmacokinetic properties and predictive in silico ADME models. *Bioinformatics, 24*(19), 2270–2271.
26. Liu, T., Lin, Y., Wen, X., Jorissen, R. N., & Gilson, M. K. (2007). BindingDB: A web-accessible database of experimentally determined protein-ligand binding affinities. *Nucleic Acids Research, 35*, D198–D201.
27. Seiler, K. P., George, G. A., Happ, M. P., Bodycombe, N. E., Carrinski, H. A., Norton, S., Brudz, S., Sullivan, J. P., Muhlich, J., Serrano, M., Ferraiolo, P., Tolliday, N. J., Schreiber, S. L., & Clemons, P. A. (2008). ChemBank: A small-molecule screening and cheminformatics resource database. *Nucleic Acids Research, 36*, D351–D359.
28. Miller, M. A. (2002). Chemical database techniques in drug discovery. *Nature Reviews Drug Discovery, 1*(3), 220–227.
29. Medina-Franco, J. L. (2015). Discovery and development of lead compounds from natural sources using computational approaches. In *Evidence-based validation of herbal medicine* (pp. 455–475). Elsevier.
30. Medina-Franco, J. L. (2020). Towards a unified Latin American natural products database: LANaPD. *Future Science OA, 6*(8), FSO468.
31. Pilon, A. C., Valli, M., Dametto, A. C., Pinto, M. E. F., Freire, R. T., Castro-Gamboa, I., et al. (2017). NuBBEDB: An updated database to uncover chemical and biological information from Brazilian biodiversity. *Scientific Reports, 7*(1), 7215.
32. Pilón-Jiménez, B. A., Saldívar-González, F. I., Díaz-Eufracio, B. I., & Medina-Franco, J. L. (2019). BIOFACQUIM: A Mexican compound database of natural products. *Biomolecules, 9.*
33. Mangal, M., et al. (2013). NPACT: Naturally occurring plant-based anti-cancer compound-activity-target database. *Nucleic Acids Research, 41*(D1), D1124–D1129.
34. Liu, Y., Zhu, Y., Sun, X., Ma, T., Lao, X., & Zheng, H. (2023). DRAVP: A comprehensive database of antiviral peptides and proteins. *Viruses, 15*, 820.
35. Martin, H. J., Melo-Filho, C. C., Korn, D., Eastman, R. T., Rai, G., Simeonov, A., et al. (2022). Small Molecule Antiviral Compound Collection (SMACC): a database to support the discovery of broad-spectrum antiviral drug molecules. *bioRxiv*.

36. Tzou, P. L., Tao, K., Pond, S. L. K., & Shafer, R. W. (2022). Coronavirus resistance database (CoV-RDB): SARS-CoV-2 susceptibility to monoclonal antibodies, convalescent plasma, and plasma from vaccinated persons. *PLoS One, 17*, e0261045.
37. Martin, R., Loechel, H. F., Welzel, M., Hattab, G., Hauschild, A. C., & Heider, D. (2020). CORDITE: The curated CORona drug InTERactions database for SARS-CoV-2. *Iscience, 23*(7).
38. Chen, T. F., Chang, Y. C., et al. (2021). DockCoV2: A drug database against SARS-CoV-2. *Nucleic Acids Research, 49*(D1), D1152–D1159.
39. Zhou, N., Bao, J., & Ning, Y. (2021). H2V: A database of human genes and proteins that respond to SARS-CoV-2, SARS-CoV, and MERS-CoV infection. *BMC Bioinformatics, 22*, 18.
40. Alsulami, A. F., Thomas, S. E., Jamasb, A. R., Beaudoin, C. A., Moghul, I., Bannerman, B., Copoiu, L., Vedithi, S. C., Torres, P., & Blundell, T. L. (2021). SARS-CoV-2 3D database: Understanding the coronavirus proteome and evaluating possible drug targets. *Briefings in Bioinformatics, 22*, 769–780.
41. Koes, D. R., & Camacho, C. J. (2012). ZINCPharmer: Pharmacophore search of the ZINC database. *Nucleic Acids Research, 40*, W409–W414.
42. Miranda-Salas, J., Peña-Varas, C., Martínez, I. V., Olmedo, D. A., Zamora, W. J., Chávez-Fumagalli, M. A., et al. (2023). Trends and challenges in chemoinformatics research in Latin America. *Artificial Intelligence in the Life Sciences, 100077.*
43. de Azevedo, D. Q., Campioni, B. M., Pedroz Lima, F. A. L., Medina-Franco, J., Castilho, R. O., & Maltarollo, V. G. (2024). A critical assessment of bioactive compounds databases. *Future Medicinal Chemistry*, 1–23.
44. Zeng, X., Zhang, P., He, W., Qin, C., Chen, S., Tao, L., et al. (2018). NPASS: Natural product activity and species source database for natural product research, discovery and tool development. *Nucleic Acids Research, 46*(D1), D1217–D1222.
45. Sorokina, M., Merseburger, P., Rajan, K., Yirik, M. A., & Steinbeck, C. (2021). COCONUT online: Collection of open natural products database. *Journal of Cheminformatics, 13*(1), 1–13.
46. Karimi-Jafari, M. H., Firouzi, R., Ashouri, M., & Poursoleiman, A. A. (2022). *A database of chemical compositions of Persian medicinal herbs.* https//:chemrxiv.org/engage/chemrxiv/articledetails/621e71035f1d9a5bb3ad2173. Accessed 20 Mar 2022.
47. Piccirillo, E., & Amaral, A. T. D. (2018). Busca virtual de compostos bioativos: conceitos e aplicações. *Química Nova, 41*, 662–677.
48. Chávez-Hernández, A. L., Sánchez-Cruz, N., & Medina-Franco, J. L. (2020). Fragment library of natural products and compound databases for drug discovery. *Biomolecules, 10*(11), 1518.
49. O'Boyle, N. M., Banck, M., James, C. A., Morley, C., Vandermeersch, T., & Hutchison, G. R. (2011). Open babel: An open chemical toolbox. *Journal of Cheminformatics, 3*(1), 1–14.
50. Warr, W. A. (2012). Scientific workflow systems: Pipeline pilot and KNIME. *Journal of Computer-Aided Molecular Design, 26*(7), 801–804.
51. Berthold, M. R., Cebron, N., Dill, F., Gabriel, T. R., Kötter, T., Meinl, T., Ohl, P., Thiel, K., & Wiswedel, B. (2009). KNIME—The Konstanz information miner. *SIGKDD Exploration Newsletter, 11*, 26.
52. Vilar, S., Cozza, G., & Moro, S. (2008). Medicinal chemistry and the molecular operating environment (MOE): Application of QSAR and molecular docking to drug discovery. *Current Topics in Medicinal Chemistry, 8*(18), 1555–1572.
53. Saldívar-González, F. I., Valli, M., Andricopulo, A. D., da Silva Bolzani, V., & Medina-Franco, J. L. (2018). Chemical space and diversity of the NuBBE database: A chemoinformatic characterization. *Journal of Chemical Information and Modeling, 59*(1), 74–85.
54. Durán-Iturbide, N. A., Díaz-Eufracio, B. I., & Medina-Franco, J. L. (2020). In silico ADME/Tox profiling of natural products: A focus on BIOFACQUIM. *ACS Omega, 5*(26), 16076–16084.

55. Al Sharie, A. H., El-Elimat, T., Al Zu'bi, Y. O., Aleshawi, A. J., & Medina-Franco, J. L. (2020). Chemical space and diversity of seaweed metabolite database (SWMD): A cheminformatics study. *Journal of Molecular Graphics and Modelling, 100*, 107702.
56. O'Boyle, N. M., Banck, M., James, C. A., Morley, C., Vandermeersch, T., & Hutchison, G. R. (2011). Open Babel: An open chemical toolbox. *Journal of Cheminformatics, 3*, 1–14.
57. Tuerkova, A., & Zdrazil, B. (2020). A ligand-based computational drug repurposing pipeline using KNIME and programmatic data access: Case studies for rare diseases and COVID-19. *Journal of Cheminformatics, 12*(1), 1–20.
58. Banerjee, P., et al. (2015). Super natural II—A database of natural products. *Nucleic Acids Research, 43*(D1), D935–D939.
59. Landrum, G. (2013). Rdkit documentation. *Release, 1*(1–79), 4.
60. Tandi, M., Tripathi, N., Gaur, A., Gopal, B., & Sundriyal, S. (2022). Curation and cheminformatics analysis of a Ugi-reaction derived library (URDL) of synthetically tractable small molecules for virtual screening application. *Molecular Diversity*, 1–14.
61. Coghlan, A., Padalino, G., O'Boyle, N. M., Hoffmann, K. F., & Berriman, M. (2022). Identification of anti-schistosomal, anthelmintic and anti-parasitic compounds curated and text-mined from the scientific literature. *Wellcome Open Research, 7.*
62. da Paixão, V. G., & da Rocha Pita, S. S. (2020). Novel scaffolds for Leishmania infantum trypanothione reductase inhibitors derived from Brazilian natural products biodiversity. *Anti-Infective Agents, 18*(4), 398–418.
63. Degtyarenko, K., de Matos, P., Ennis, M., Hastings, J., Zbinden, M., McNaught, A., Alcántara, R., Darsow, M., Guedj, M., & Ashburner, M. (2008). ChEBI: A database and ontology for chemical entities of biological interest. *Nucleic Acids Research, 36*, D344–D350.
64. Valdés-Jiménez, A., Peña-Varas, C., Borrego-Muñoz, P., Arrue, L., Alegría-Arcos, M., Nour-Eldin, H., et al. (2021). PSC-db: A structured and searchable 3d-database for plant secondary compounds. *Molecules, 26*(4), 1124.
65. Gallo, K., Kemmler, E., Goede, A., Becker, F., Dunkel, M., Preissner, R., & Banerjee, P. (2023). SuperNatural 3.0—A database of natural products and natural product-based derivatives. *Nucleic Acids Research, 51*(D1), D654–D659.
66. Nguyen-Vo, T., et al. (2018). VIETHERB: A database for Vietnamese herbal species. *Journal of Chemical Information and Modeling, 59*(1), 1–9.
67. Silva, T. S. (2018). *Desenvolvimento de banco de dados de pacientes submetidos ao transplante de células-tronco hematopoiéticas*. UFRS. Dissertação de Mestrado.
68. Yang, J., Wang, D., Jia, C., Wang, M., Hao, G., & Yang, G. (2019). Freely accessible chemical database resources of compounds for in silico drug discovery. *Current Medicinal Chemistry, 26*, 7581–7597.
69. Ruddigkeit, L., van Deursen, R., Blum, L. C., & Reymond, J.-L. (2012). Enumeration of 166 billion organic small molecules in the chemical universe database GDB-17. *Journal of Chemical Information and Modeling, 52*, 2864–2875.
70. Wang, R., Fang, X., Lu, Y., & Wang, S. (2004). The PDBbind database: Collection of binding affinities for protein-ligand complexes with known three-dimensional structures. *Journal of Medicinal Chemistry, 47*, 2977–2980.
71. Reymond, J. L., & Awale, M. (2012). Exploring chemical space for drug discovery using the chemical universe database. *ACS Chemical Neuroscience, 3*(9), 649–657.
72. Vivek-Ananth, R. P., Sahoo, A. K., Kumaravel, K., Mohanraj, K., & Samal, A. (2021). MeFSAT: A curated natural product database specific to secondary metabolites of medicinal fungi. *RSC Advances, 11*, 2596–2607.
73. van Santen, J. A., Poynton, E. F., Iskakova, D., McMann, E., Alsup, T. A., Clark, T. N., Fergusson, C. H., Fewer, D. P., Hughes, A. H., McCadden, C. A., Parra, J., Soldatou, S., Rudolf, J. D., Janssen, E. M.-L., Duncan, K. R., & Linington, R. G. (2022). The natural products atlas 2.0: A database of microbially-derived natural products. *Nucleic Acids Research, 50*, D1317–D1323.

74. Voigt, J. H., Bienfait, B., Wang, S., & Nicklaus, M. C. (2001). Comparison of the NCI open database with seven large chemical structural databases. *Journal of Chemical Information and Modeling, 41*, 702–712.
75. Visini, R., Awale, M., & Reymond, J.-L. (2017). Fragment database FDB-17. *Journal of Chemical Information and Modeling, 57*, 700–709.
76. Ahmed, J., Worth, C. L., Thaben, P., Matzig, C., Blasse, C., Dunkel, M., & Preissner, R. (2011). FragmentStore—A comprehensive database of fragments linking metabolites, toxic molecules and drugs. *Nucleic Acids Research, 39*, D1049–D1054.
77. Yang, J.-F., Wang, F., Jiang, W., Zhou, G.-Y., Li, C.-Z., Zhu, X.-L., Hao, G.-F., & Yang, G.-F. (2018). Padfrag: A database built for the exploration of bioactive fragment space for drug discovery. *Journal of Chemical Information and Modeling, 58*, 1725–1730.
78. Gabrielson, S. W. (2018). SciFinder. *Journal of the Medical Library Association*, 106.
79. Williams, A. J., & Ekins, S. (2011). A quality alert and call for improved curation of public chemistry databases. *Drug Discovery Today, 16*, 747–750.
80. Sorokina, M., Merseburger, P., Rajan, K., Yirik, M. A., & Steinbeck, C. (2021). COCONUT online: Collection of open natural products database. *Journal of Cheminformatics, 13*, 2.
81. ISDB. *A database of In-Silico predicted MS/MS spectrum of natural products*. Available online: http://oolonek.github.io/ISDB/. Accessed 12 June 2023.
82. Dictionary of Natural Products 31.1. Available online: https://dnp.chemnetbase.com/faces/chemical/ChemicalSearch.xhtml. Accessed 30 June 2022.
83. Newman, D. J., & Cragg, G. M. (2020). Natural products as sources of new drugs over the nearly four decades from 01/1981 to 09/2019. *Journal of Natural Products, 83*, 770–803.
84. Evans, B. E., Rittle, K. E., Bock, M. G., DiPardo, R. M., Freidinger, R. M., Whitter, W. L., Lundell, G. F., Veber, D. F., Anderson, P. S., & Chang, R. S. (1988). Methods for drug discovery: Development of potent, selective, orally effective cholecystokinin antagonists. *Journal of Medicinal Chemistry, 31*, 2235–2246.
85. Davison, E. K., & Brimble, M. A. (2019). Natural product derived privileged scaffolds in drug discovery. *Current Opinion in Chemical Biology, 52*, 1–8.
86. Karageorgis, G., Foley, D. J., Laraia, L., & Waldmann, H. (2020). Principle and design of pseudo-natural products. *Nature Chemistry, 12*, 227–235.
87. Karageorgis, G., Foley, D. J., Laraia, L., Brakmann, S., & Waldmann, H. (2021). Pseudo natural products-chemical evolution of natural product structure. *Angewandte Chemie, International Edition, 60*, 15705–15723.
88. Cremosnik, G. S., Liu, J., & Waldmann, H. (2020). Guided by evolution: From biology oriented synthesis to pseudo natural products. *Natural Product Reports, 37*, 1497–1510.
89. Saldívar-González, F. I., & Medina-Franco, J. L. (2020). Chemoinformatics approaches to assess chemical diversity and complexity of small molecules. In *Small molecule drug discovery* (pp. 83–102). Elsevier. ISBN 9780128183496.
90. Sorokina, M., & Steinbeck, C. (2020). Review on natural products databases: Where to find data in 2020. *Journal of Cheminformatics, 12*, 20.
91. Reaxys. Available online: https://www.reaxys.com. Accessed 30 June 2022.
92. Chen, C. Y. C. (2011). TCM database@ Taiwan: The world's largest traditional Chinese medicine database for drug screening in silico. *PLoS One, 6*(1), e15939.
93. Mohanraj, K., Karthikeyan, B. S., Vivek-Ananth, R. P., Chand, R. P. B., Aparna, S. R., Mangalapandi, P., & Samal, A. (2018). IMPPAT: A curated database of Indian medicinal plants, phytochemistry and therapeutics. *Scientific Reports, 8*, 4329.
94. Ntie-Kang, F., Zofou, D., Babiaka, S. B., Meudom, R., Scharfe, M., Lifongo, L. L., Mbah, J. A., Mbaze, L. M., Sippl, W., & Efange, S. M. N. (2013). AfroDb: A select highly potent and diverse natural product library from African medicinal plants. *PLoS One, 8*, e78085.
95. Ionov, N., Druzhilovskiy, D., Filimonov, D., & Poroikov, V. (2023). Phyto4Health: Database of phytocomponents from Russian pharmacopoeia plants. *Journal of Chemical Information and Modeling, 63*, 1847–1851.

96. Valli, M., dos Santos, R. N., Figueira, L. D., Nakajima, C. H., Castro-Gamboa, I., Andricopulo, A. D., & Bolzani, V. S. (2013). Development of a natural products database from the biodiversity of Brazil. *Journal of Natural Products, 76*, 439–444.
97. Li, B., Ma, C., Zhao, X., Hu, Z., Du, T., Xu, X., Wang, Z., & Lin, J. (2018). Ya TCM: Yet another traditional Chinese medicine database for drug discovery. *Computational and Structural Biotechnology Journal, 16*, 600–610.
98. Ru, J., Li, P., Wang, J., Zhou, W., Li, B., Huang, C., Li, P., Guo, Z., Tao, W., Yang, Y., Xu, X., Li, Y., Wang, Y., & Yang, L. (2014). TCMSP: A database of systems pharmacology for drug discovery from herbal medicines. *Journal of Cheminformatics, 6*, 13.
99. Kim, S.-K., Nam, S., Jang, H., Kim, A., & Lee, J.-J. (2015). TM-MC: A database of medicinal materials and chemical compounds in northeast Asian traditional medicine. *BMC Complementary and Alternative Medicine, 15*, 218.
100. Xu, H.-Y., Zhang, Y.-Q., Liu, Z.-M., Chen, T., Lv, C.-Y., Tang, S.-H., Zhang, X.-B., Zhang, W., Li, Z.-Y., Zhou, R.-R., Yang, H.-J., Wang, X.-J., & Huang, L.-Q. (2019). ETCM: An encyclopaedia of traditional Chinese medicine. *Nucleic Acids Research, 47*, D976–D982.
101. Fang, X., Shao, L., Zhang, H., & Wang, S. (2005). CHMIS-C: A comprehensive herbal medicine information system for cancer. *Journal of Medicinal Chemistry, 48*, 1481–1488.
102. Qiao, X., Hou, T., Zhang, W., Guo, S., & Xu, X. (2002). A 3D structure database of components from Chinese traditional medicinal herbs. *Journal of Chemical Information and Computer Sciences, 42*, 481–489.
103. Huang, J., Zheng, Y., Wu, W., Xie, T., Yao, H., Pang, X., Sun, F., Ouyang, L., & Wang, J. C. E. M. T. D. D. (2015). The database for elucidating the relationships among herbs, compounds, targets and related diseases for Chinese ethnic minority traditional drugs. *Oncotarget, 6*, 17675–17684.
104. Xu, J., & Yang, Y. (2009). Traditional Chinese medicine in the Chinese health care system. *Health Policy, 90*, 133–139.
105. Bultum, L. E., Woyessa, A. M., & Lee, D. (2019). ETM-DB: Integrated Ethiopian traditional herbal medicine and phytochemicals database. *BMC Complementary and Alternative Medicine, 19*, 212.
106. Potshangbam, A. M., Polavarapu, R., Rathore, R. S., Naresh, D., Prabhu, N. P., Potshangbam, N., et al. (2019). MedPServer: A database for identification of therapeutic targets and novel leads pertaining to natural products. *Chemical Biology & Drug Design, 93*(4), 438–446.
107. Ntie-Kang, F., Onguéné, P. A., Scharfe, M., Owono, L. C., Megnassan, E., Mbaze, L. M., Sippl, W., & Efange, S. M. N. (2014). ConMedNP: A natural product library from central African medicinal plants for drug discovery. *RSC Advances, 4*, 409–419.
108. Ibezim, A., Debnath, B., Ntie-Kang, F., Mbah, C. J., & Nwodo, N. J. (2017). Binding of anti-Trypanosoma natural products from African flora against selected drug targets: A docking study. *Medicinal Chemistry Research, 26*, 562–579.
109. Onguéné, P. A., Ntie-Kang, F., Mbah, J. A., Lifongo, L. L., Ndom, J. C., Sippl, W., & Mbaze, L. M. (2014). The potential of anti-malarial compounds derived from African medicinal plants, part III: An in silico evaluation of drug metabolism and pharmacokinetics profiling. *Organic and Medicinal Chemistry Letters, 4*, 6.
110. Ntie-Kang, F., Nwodo, J. N., Ibezim, A., Simoben, C. V., Karaman, B., Ngwa, V. F., Sippl, W., Adikwu, M. U., & Mbaze, L. M. (2014). Molecular modeling of potential anticancer agents from African medicinal plants. *Journal of Chemical Information and Modeling, 54*, 2433–2450.
111. Ntie-Kang, F., Amoa Onguéné, P., Fotso, G. W., Andrae-Marobela, K., Bezabih, M., Ndom, J. C., Ngadjui, B. T., Ogundaini, A. O., Abegaz, B. M., & Meva'a, L.M. (2014). Virtualizing the p-ANAPL library: A step towards drug discovery from African medicinal plants. *PLoS One, 9*, e90655.
112. Raven, P. H., Gereau, R. E., Phillipson, P. B., Chatelain, C., Jenkins, C. N., & Ulloa, C. (2020). The distribution of biodiversity richness in the tropics. *Science Advances, 6*.

113. Gómez-García, A., & Medina-Franco, J. L. (2022). Progress and impact of Latin American natural product databases. *Biomolecules, 12*.
114. Weininger, D. (1988). SMILES, a chemical language and information system. 1. Introduction to methodology and encoding rules. *Journal of Chemical Information and Modeling, 28*, 31–36.
115. Heller, S. R., McNaught, A., Pletnev, I., Stein, S., & Tchekhovskoi, D. (2015). Inchi, the IUPAC international chemical identifier. *Journal of Cheminformatics, 7*, 23.
116. Pletnev, I., Erin, A., McNaught, A., Blinov, K., Tchekhovskoi, D., & Heller, S. (2012). InChIKey collision resistance: An experimental testing. *Journal of Cheminformatics, 4*, 39.
117. Daylight Chemical Information System, Inc. *SMARTS—A language for describing molecular patterns*. Available online: https://www.daylight.com/dayhtml/doc/theory/theory.smarts.html. Accessed 3 June 2022.
118. Saldívar-González, F. I., Huerta-García, C. S., & Medina-Franco, J. L. (2020). Chemoinformatics-based enumeration of chemical libraries: A tutorial. *Journal of Cheminformatics, 12*, 64.
119. PubChem Sketcher. Available online: https://pubchem.ncbi.nlm.nih.gov/edit3/index.html. Accessed 15 Apr 2023.
120. Rajan, K., Brinkhaus, H. O., Sorokina, M., Zielesny, A., & Steinbeck, C. (2021). DECIMER-segmentation: Automated extraction of chemical structure depictions from scientific literature. *Journal of Cheminformatics, 13*, 20.
121. ChemicBook. Available online: https://chemicbook.com/2021/02/20/mol2-file-format-explained-for-beginners-part-2.html. Accessed 15 Apr 2023.
122. Structural Data Files. Available online: https://chem.libretexts.org/Courses/Intercollegiate_Courses/Cheminformatics/02%3A_Representing_Small_Molecules_on_Computers/2.05%3A_Structural_Data_Files. Accessed 15 Apr 2023.
123. Csizmadia, P. (1999). MarvinSketch and MarvinView: Molecule applets for the World Wide Web.
124. Stork, C., Embruch, G., Šícho, M., de Bruyn Kops, C., Chen, Y., Svozil, D., & Kirchmair, J. (2020). NERDD: A web portal providing access to in silico tools for drug discovery. *Bioinformatics, 36*, 1291–1292.
125. Šícho, M., Stork, C., Mazzolari, A., de Bruyn Kops, C., Pedretti, A., Testa, B., Vistoli, G., Svozil, D., & Kirchmair, J. (2019). FAME 3: Predicting the sites of metabolism in synthetic compounds and natural products for phase 1 and phase 2 metabolic enzymes. *Journal of Chemical Information and Modeling, 59*, 3400–3412.
126. de Bruyn Kops, C., Stork, C., Šícho, M., Kochev, N., Svozil, D., Jeliazkova, N., & Kirchmair, J. (2019). GLORY: Generator of the structures of likely cytochrome P450 metabolites based on predicted sites of metabolism. *Frontiers in Chemistry, 7*, 402.
127. Stork, C., Mathai, N., & Kirchmair, J. (2021). Computational prediction of frequent hitters in target-based and cell-based assays. *Artificial Intelligence in the Life Sciences, 1*, 100007.
128. Chen, Y., Stork, C., Hirte, S., & Kirchmair, J. (2019). NP-scout: Machine learning approach for the quantification and visualization of the natural product-likeness of small molecules. *Biomolecules, 9*.
129. Wilm, A., Norinder, U., Agea, M. I., de Bruyn Kops, C., Stork, C., Kühnl, J., & Kirchmair, J. (2021). Skin doctor CP: Conformal prediction of the skin sensitization potential of small organic molecules. *Chemical Research in Toxicology, 34*, 330–344.
130. Xiong, G., Wu, Z., Yi, J., Fu, L., Yang, Z., Hsieh, C., Yin, M., Zeng, X., Wu, C., Lu, A., Chen, X., Hou, T., & Cao, D. (2021). ADMETlab 2.0: An integrated online platform for accurate and comprehensive predictions of ADMET properties. *Nucleic Acids Research, 49*, W5–W14.
131. Rath, M., Wellnitz, J., Martin, H.-J., Melo-Filho, C., Hochuli, J. E., Silva, G. M., Beasley, J.-M., Travis, M., Sessions, Z. L., Popov, K. I., Zakharov, A. V., Cherkasov, A., Alves, V., Muratov, E. N., & Tropsha, A. (2024). Pharmacokinetics profiler (PhaKinPro): Model development, validation, and implementation as a web tool for triaging compounds with undesired pharmacokinetics profiles. *Journal of Medicinal Chemistry*.

132. Dulsat, J., López-Nieto, B., Estrada-Tejedor, R., & Borrell, J. I. (2023). Evaluation of free online ADMET tools for academic or small biotech environments. *Molecules, 28*.
133. Sabe, V. T., Ntombela, T., Jhamba, L. A., Maguire, G. E., Govender, T., Naicker, T., & Kruger, H. G. (2021). Current trends in computer aided drug design and a highlight of drugs discovered via computational techniques: A review. *European Journal of Medicinal Chemistry, 224*, 113705.
134. Voršilák, M., Kolář, M., Čmelo, I., & Svozil, D. (2020). SYBA: Bayesian estimation of synthetic accessibility of organic compounds. *Journal of Cheminformatics, 12*, 35.
135. Yu, J., Wang, J., Zhao, H., Gao, J., Kang, Y., Cao, D., Wang, Z., & Hou, T. (2022). Organic compound synthetic accessibility prediction based on the graph attention mechanism. *Journal of Chemical Information and Modeling, 62*, 2973–2986.
136. Genheden, S., Thakkar, A., Chadimová, V., Reymond, J.-L., Engkvist, O., & Bjerrum, E. (2020). AiZynthFinder: A fast, robust and flexible open-source software for retrosynthetic planning. *Journal of Cheminformatics, 12*, 70.
137. Kawakami, Y., Inoue, A., Kawai, T., Wakita, M., Sugimoto, H., & Hopfinger, A. J. (1996). The rationale for E2020 as a potent acetylcholinesterase inhibitor. *Bioorganic & Medicinal Chemistry, 4*(9), 1429–1446.
138. Bajad, N. G., Rayala, S., Gutti, G., Sharma, A., Singh, M., Kumar, A., & Singh, S. K. (2021). Systematic review on role of structure based drug design (SBDD) in the identification of anti-viral leads against SARS-Cov-2. *Current Research in Pharmacology and Drug Discovery, 2*, 100026.
139. Willett, P., Barnard, J. M., & Downs, G. M. (1998). Chemical similarity searching. *Journal of Chemical Information and Computer Sciences, 38*, 983–996.
140. Maggiora, G., Vogt, M., Stumpfe, D., & Bajorath, J. (2014). Molecular similarity in medicinal chemistry. *Journal of Medicinal Chemistry, 57*, 3186–3204.
141. Willighagen, E. L., Mayfield, J. W., Alvarsson, J., Berg, A., Carlsson, L., Jeliazkova, N., Kuhn, S., Pluskal, T., Rojas-Chertó, M., Spjuth, O., Torrance, G., Evelo, C. T., Guha, R., & Steinbeck, C. (2017). The chemistry development kit (CDK) v2.0: Atom typing, depiction, molecular formulas, and substructure searching. *Journal of Cheminformatics, 9*, 33.
142. Open-source chemoinformatics and machine learning. RDKit: Open-Source Cheminformatics Software. Available online: https://www.rdkit.org. Accessed 8 Feb 2023.
143. Yap, C. W. (2011). PaDEL-descriptor: An open source software to calculate molecular descriptors and fingerprints. *Journal of Computational Chemistry, 32*, 1466–1474.
144. Wildman, S. A., & Crippen, G. M. (1999). Prediction of physicochemical parameters by atomic contributions. *Journal of Chemical Information and Computer Sciences, 39*, 868–873.
145. Ertl, P., Rohde, B., & Selzer, P. (2000). Fast calculation of molecular polar surface area as a sum of fragment-based contributions and its application to the prediction of drug transport properties. *Journal of Medicinal Chemistry, 43*, 3714–3717.
146. Sander, T., Freyss, J., von Korff, M., & Rufener, C. (2015). DataWarrior: An open-source program for chemistry aware data visualization and analysis. *Journal of Chemical Information and Modeling, 55*, 460–473.
147. Lipinski, C. A., Lombardo, F., Dominy, B. W., & Feeney, P. J. (2001). Experimental and computational approaches to estimate solubility and permeability in drug discovery and development settings. *Advanced Drug Delivery Reviews, 46*, 3–26.
148. Lipinski, C. A. (2004). Lead- and drug-like compounds: The rule-of-five revolution. *Drug Discovery Today: Technologies, 1*, 337–341.
149. Veber, D. F., Johnson, S. R., Cheng, H.-Y., Smith, B. R., Ward, K. W., & Kopple, K. D. (2002). Molecular properties that influence the oral bioavailability of drug candidates. *Journal of Medicinal Chemistry, 45*, 2615–2623.
150. Gleeson, M. P. (2008). Generation of a set of simple, interpretable ADMET rules of thumb. *Journal of Medicinal Chemistry, 51*, 817–834.
151. Hughes, J. D., Blagg, J., Price, D. A., Bailey, S., Decrescenzo, G. A., Devraj, R. V., Ellsworth, E., Fobian, Y. M., Gibbs, M. E., Gilles, R. W., Greene, N., Huang, E., Krieger-Burke, T.,

Loesel, J., Wager, T., Whiteley, L., & Zhang, Y. (2008). Physiochemical drug properties associated with in vivo toxicological outcomes. *Bioorganic & Medicinal Chemistry Letters, 18*, 4872–4875.

152. Niu, Y., & Lin, P. (2023). Advances of computer-aided drug design (CADD) in the development of anti-Azheimer's-disease drugs. *Drug Discovery Today, 103665*.
153. Bagabir, S. A., Ibrahim, N. K., Bagabir, H. A., & Ateeq, R. H. (2022). Covid-19 and artificial intelligence: Genome sequencing, drug development and vaccine discovery. *Journal of Infection and Public Health, 15*(2), 289–296.
154. Kiriiri, G. K., Njogu, P. M., & Mwangi, A. N. (2020). Exploring different approaches to improve the success of drug discovery and development projects: A review. *Future Journal of Pharmaceutical Sciences, 6*(1), 1–12.
155. Gupta, R., Srivastava, D., Sahu, M., Tiwari, S., Ambasta, R. K., & Kumar, P. (2021). Artificial intelligence to deep learning: Machine intelligence approach for drug discovery. *Molecular Diversity, 25*, 1315–1360.
156. Knox, C., Wilson, M., Klinger, C. M., et al. (2024). DrugBank 6.0: The DrugBank knowledge base for 2024. *Nucleic Acids Research, 52*(D1), D1265–D1275.
157. Williams, A. J., Ekins, S., & Tkachenko, V. (2012). Towards a gold standard: Regarding quality in public domain chemistry databases and approaches to improving the situation. *Drug Discovery Today, 17*, 685–701.
158. Young, D., Martin, T., Venkatapathy, R., & Harten, P. (2008). Are the chemical structures in your QSAR correct? *QSAR and Combinatorial Science, 27*, 1337–1345.
159. Glüge, J., McNeill, K., & Scheringer, M. (2023). Getting the SMILES right: Identifying inconsistent chemical identities in the ECHA database, PubChem and the CompTox chemicals dashboard. *Environmental Science: Advances, 2*, 612–621.
160. Bento, A. P., Hersey, A., Félix, E., Landrum, G., Gaulton, A., Atkinson, F., Bellis, L. J., De Veij, M., & Leach, A. R. (2020). An open source chemical structure curation pipeline using RDKit. *Journal of Cheminformatics, 12*, 51.
161. Database under maintenance. (2016). *Nature Methods, 13*, 699–699.
162. Mullard, A. (2018). Re-assessing the rule of 5, two decades on. *Nature Reviews. Drug Discovery, 17*, 777.

Chapter 3
A Brief Introduction to Pharmacogenomics and Personalized Medicine in the Drug Design Context

Glaucio Monteiro Ferreira, Mario Hiroyuki Hirata, Thamires Pandolfi Cappello, Carolina Dagli-Hernandez, and André Rinaldi Fukushima

Abstract This chapter explores the field of in silico protein analysis and its growing importance at the interface of genomics and personalized medicine. Novel computational techniques improved the understanding of different genetic alterations in the protein's activity, interaction with other proteins, and structure.

This perspective discusses the novel applications of in silico methods in vaccine research, personalized medicine, and drug design. In addition, the chapter presents possible tools for manipulating protein model data, as well as discussing the moral and legal ramifications of these advances, especially with regard to pharmacogenomics and personalized medicine.

Keywords Protein analysis · Genomics · Computational biology · Protein structure prediction · Personalized medicine

G. M. Ferreira (✉) · M. H. Hirata
Department of Clinical and Toxicological Analyses, School of Pharmaceutical Sciences, University of Sao Paulo, São Paulo, SP, Brazil
e-mail: glauciom.ferreira@gmail.com

T. P. Cappello
Center for Health Law of University of São Paulo (USP), São Paulo, SP, Brazil

Faculdade de Ciência da Saúde IGESP (FASIG), São Paulo, SP, Brazil

C. Dagli-Hernandez
Faculty of Pharmaceutical Sciences, State University of Campinas, Campinas, SP, Brazil

A. R. Fukushima
Faculdade de Ciências da Saúde IGESP, São Paulo, SP, Brazil

V. G. Maltarollo (ed.), *Computer-Aided and Machine Learning-Driven Drug Design*, Computer-Aided Drug Discovery and Design 3,
https://doi.org/10.1007/978-3-031-76718-0_3

1 Introduction

In 2003, the Human Genome Project provided us with an extensive map of human DNA. This achievement significantly advanced biomedical research, fostering global collaborations among scientists and culminating in the so-called personalized medicine field and targeted treatments. Since the completion of the human genome, further advances in sequencing technologies enabled the determination of an individual's entire genome in a matter of days, making genomics a powerful tool in healthcare. The information provided by genomics can be used to diagnose genetic disorders, guide treatment decisions, and develop new drugs. From the diagnosis of genetic disorders to the development of tailored therapies for complex ailments like cancer, this convergence holds the promise of delivering more individualized, efficacious, and safer medical interventions.

Nevertheless, as with all transformative technologies, it also raises profound ethical, privacy, and accessibility considerations. As the exploration unfolds, a thorough analysis will be conducted on the potential, challenges, and assurances that genomics and personalized medicine offer to the future of healthcare.

Personalized medicine is particularly relevant in the treatment of complex diseases, such as cancer, where a one-size-fits-all approach may not be effective. By analyzing an individual's genetic makeup, doctors can identify specific mutations that drive the growth of a tumor and target those mutations with precision therapies. Personalized medicine can also help identify patients at risk of developing certain diseases, such as heart disease, and develop preventive measures tailored to their genetic profiles [1]. Its implementation deals with large amounts of genomic data, which on its own requires a dedicated processing computational infrastructure associated with sophisticated interpretative analysis, which relies on highly specialized personnel. In addition, there are ethical and privacy concerns related to the use of an individual's genetic information [2].

One of the most significant areas of research in genomics and personalized medicine is the development of precision therapies [3]. Precision therapies are treatments that target specific genetic mutations or pathways, making them more effective and less likely to cause adverse reactions. This approach is particularly relevant in cancer treatment, where targeted therapies have shown promising results in clinical trials.

Another area of research in genomics and personalized medicine is pharmacogenomics, which studies how an individual's genetic makeup affects their response to medications. By analyzing an individual's genetic profile, doctors can determine which medications are most likely to be effective and avoid medications that may cause adverse reactions. This approach has the potential to improve patient outcomes and reduce healthcare costs by minimizing the need for trial-and-error prescribing.

In addition to their potential applications in healthcare, genomics and personalized medicine also raise important ethical and social issues. There are concerns about the accessibility of genomic testing and personalized medicine, particularly for marginalized communities who may not have access to the latest technology or healthcare services. There are also questions about the privacy and ownership of genomic data and the potential for discrimination based on genetic information. By providing a deeper understanding of an individual's genetic makeup, personalized medicine has the potential to improve patient outcomes, reduce healthcare costs, and enhance our understanding of human biology. Several databases, such as the Catalogue Of Somatic Mutations In Cancer, COSMIC (https://cancer.sanger.ac.uk/cosmic), and cBioPortal for cancer genomics (https://www.cbioportal.org/), already display disease-genetic variation associations, especially in the context of highly mutating diseases such as cancer.

As the exploration of in silico protein analysis concludes, it becomes evident that the digital realm harbors immense potential in deciphering the complex language of proteins. A thorough investigation into computational methodologies has significantly broadened our understanding. However, standing on the threshold of this expansive domain, extensive real-world applications stemming from these insights are also emerging. The fusion of in silico techniques with tangible medical and technological solutions is not just a distant dream but an imminent reality. Transitioning into these applications in the following chapters, the transformative power of combining computational prowess with real-world biological challenges will be observed, ushering in an era of innovation and discovery.

2 In Silico Protein Analysis and Its Real-World Applications

The details of genetics and molecular biology involve the very essence of life, with proteins playing prominent roles. In today's technologically driven era, the digital realm provides a new stage for these performers. Nowadays, thanks to in silico simulations, it is possible to analyze, study, and even guess how proteins behave with different genetic tweaks. These computer-generated models, with precision, promise more than just academic revelations. They serve as a compass, guiding toward new advances in medical science, from personalized medicine to innovative drug design. This book contributes to the analysis of proteins in silico, revealing their transformative potential to shape the future of medicine and research. In the field of genetics and molecular biology, the impact of genetic variations on proteins is an extremely important topic. With the rapid evolution of technology, the demand for computational methods, specifically in silico simulations, is increasing to dissect and understand the nuances of these effects. In the realm of scientific research, the utilization of computer-based methods stands out for its ability to construct intricate 3D models of proteins, shedding light on their structures and functions. These insights offer tangible implications for real-world applications. This study investigates the advanced techniques used to analyze proteins in silico, exploring their broad relevance in modern medicine and research. By employing computational tools, researchers can gain deep insights into how genetic variations impact protein behavior, with potential implications for drug design and personalized medicine. One significant area where these methods find application is in vaccine development. By understanding the structural aspects of viral proteins, researchers can identify key antigenic regions crucial for triggering an immune response. Computational simulations aid in designing vaccines with improved efficacy and specificity, while also enabling the customization of immunotherapies based on individual genetic profiles. Moreover, in silico protein analysis contributes to optimizing vaccine delivery systems, allowing for the design of formulations that enhance stability, immunogenicity, and targeted delivery. These advancements hold promise for overcoming logistical challenges and improving vaccine accessibility, especially in underserved communities [4, 5].

In summary, the integration of computer-based approaches in protein analysis represents a transformative advancement in biomedical research. This study underscores the broad applicability of these techniques in vaccine development, heralding a future of more effective and tailored immunotherapies that address global health needs.

Equipped with precise tools and a large amount of data from estimated repositories such as the RCSB Protein Data Bank, it is possible to build detailed 3D models of proteins using Modeller (https://salilab.org/modeller/) and unravel their complex interactions using platforms such as ClusPro and FireDock. The AlphaFold methods have been also evaluated to study the impact of mutations on protein stability and pathogenicity, indicating that novel tools are being studied to solve genomics-related problems [6]. Please see Chap. 14 for more details on structural characterization and modelling of protein structures. However, research will not be limited to static structures. Using GROMACS 2019.1, it is possible to simulate the subtle movements of proteins. From this point on, the change and delta of the protein's evolution are captured and visualized, revealing the changes and twists that define its very essence.

2.1 Making and Matching Protein Models

Using existing data from the RCSB Protein Data Bank to build 3D models with Modeller. The best model will be chosen based on its energy efficiency, and its accuracy will be assessed using Ramachandran Plots. To understand how proteins interact, tools like ClusPro, FireDock, Haddock, and PatchDock will be used (Table 3.1).

Table 3.1 Main protein–protein docking software available

Software	Website	Availability
ClusPro	https://cluspro.bu.edu	Free
FireDock	https://www.cs.tau.ac.il//~ppdock/FireDock/	Free
Haddock	https://wenmr.science.uu.nl/haddock2.4/	Free
PatchDock	https://bioinfo3d.cs.tau.ac.il/PatchDock/	Free

Table 3.2 Main molecular dynamics software available

Software	Website	Availability
GROMACS	https://www.gromacs.org/	Free
AMBER	https://ambermd.org/	Free
DESMOND	https://www.deshawresearch.com/index.html	Academic license
OpenMM	https://openmm.org/	Free

Ab initio modeling, a technique in computational biology, predicts protein structures from fundamental principles, eliminating the reliance on pre-existing templates. It utilizes quantum mechanics or molecular mechanics to simulate atomic interactions and iteratively refines protein conformations. This method showed to be indispensable for understanding newly discovered or poorly characterized proteins, providing valuable insights into folding pathways and dynamics. Despite computational hurdles, ab initio modeling continues to serve as a crucial tool in structural biology, facilitating advancements in drug discovery and protein engineering.

2.2 *Simulating Protein Movements*

It is possible to use specialized molecular dynamics software, listed in Table 3.2, to simulate how these proteins move in a specific environment. These simulations will mimic real-life conditions, including temperature and pressure. In these simulations, proteins are commonly observed at the nanosecond scale, with the potential to extend the timeframe to a few microseconds depending on the structural effects that need to be simulated. This timeframe allows us to capture various conformational changes and transient interactions that occur within the protein structure. Additionally, it provides insights into the stability, flexibility, and functional dynamics of the proteins under investigation. For more molecular dynamics' details, please see Chap. 8 of this book.

2.3 *Analyzing Changes in Protein Shape*

To compare the original and altered proteins, various methods can be used to identify any noticeable changes in their structures and functions. Differences between the two protein structures will be quantified and analyzed, measuring changes in parameters such as secondary structure elements, solvent accessibility, and interatomic distances. It is possible to examine the patterns in the protein structures to identify any systematic changes resulting from the modifications. This may involve scrutinizing changes in folding motifs, hydrogen bonding networks, or spatial arrangements of key residues. This includes identifying key residues involved in binding sites, catalytic sites, or allosteric regulation, as well as evaluating changes in their interactions upon protein alteration. The PyMOL software (https://pymol.org/2/) makes it possible to visualize, analyze, and generate images of protein structures. This tool enables the creation of high-quality molecular graphics, facilitating the visual comparison and analysis of structural differences between the original and altered proteins. By integrating visual representation with quantitative analysis, a comprehensive understanding of the effects of protein modification can be achieved. To further highlight the importance of these strategies for understanding protein behavior and function, time must be devoted to analyzing changes in protein shape [7]. With the transition to these new topics, consider delving deeper into mechanics and molecular biology, illuminating the intricacies that lie at the heart of life itself [8].

These articles authored by Zhou et al. [9] Schreeck et al. [10] and Schwab and Schaeffeler [11] represent pivotal contributions from a prominent research group within the field of pharmacogenomics. Their collective research endeavors have illuminated crucial facets that shape the landscape of personalized medicine and drug development. In their comprehensive exploration, Zhou and colleagues [6] delve into the intricate realm of rare-variant pharmacogenomics, elucidating the manifold challenges and promising opportunities that accompany the study of genetic variations in drug response. Their insights provide valuable guidance for navigating the complexities inherent in tailoring pharmacotherapy to individual genetic profiles. Similarly, Schreeck et al. [10] offer invaluable insights into the realm of pediatric medicine and drug development within the context of pharmacogenomics. Their work underscores the importance of considering genetic factors in pediatric populations to optimize drug efficacy and safety, thereby advancing the field toward more tailored and effective therapeutic interventions for children. Furthermore, the seminal contribution by Schwab and Schaeffeler [11] highlights the foundational significance of pharmacogenomics in the broader context of personalized therapy. By emphasizing the pivotal role of genetic variation in drug response, they lay the groundwork for a paradigm shift toward precision medicine, wherein treatment strategies are tailored to individual genetic profiles to optimize therapeutic outcomes. Collectively, these articles not only contribute to the expanding body of knowledge in pharmacogenomics but also serve as guiding beacons for future research directions. They underscore the imperative of integrating genetic insights into clinical practice, fostering a deeper understanding of individualized drug response and paving the way for more effective and personalized therapeutic approaches.

3 Pharmacogenomics in Drug Development

The drug development process is onerous and expensive. A cross-sectional study estimated that the pharmaceutical industry spends around US$ 19 million per pivotal benefit trial and US$ 41 thousand per patient [12]. However, around 90% of clinical trials fail to find a candidate drug with good efficacy and safety profiles. In 40–50%

of cases, the failure is due to lack of efficacy, while 30% of trials fail due to unmanageable toxicity [13].

Additionally, the real-world efficacy and safety profiles differ from the pre-marketing phase. Many patients experience adverse drug reactions (ADRs), which are harmful and unintended reactions that result from medication use [14]. Approximately 8.3% of primary care patients have been affected with at least one ADR, of which 77% were nonpreventable [15]. Other patients experience treatment failure, even after adhering properly to treatment [16].

Recently, the pharmaceutical industry shifted the research and development (R&D) strategy from exploring drugs to treat chronic diseases to precision medicine models, where the investigational drug is aimed at specific targets and populations based on genetic information [17]. In this context, pharmacogenetics can be a useful tool for drug development.

Pharmacogenetics is the science that studies the impact of genetic variation on drug response, i.e., drug efficacy and safety. The term was first mentioned in 1959 by Friedrich Vogel in his book *Modern Problems of Human Genetics* [18] in reference to the hypothesis that genetic variation in drug-metabolizing enzymes plays a role in drug response. It was only after the development of technologies such as polymerase chain reaction (PCR) that pharmacogenetics started to be studied more profoundly, on a molecular level. After the completion of the Human Genome Project, in 2003, which sequenced around 90% of the human genome, [19] there was an increase in the number of studies that explored how single-nucleotide variations (SNVs) in DNA-affected drug response.

Pharmacogenomics and pharmacogenetics are terms that are often used interchangeably, although they are different concepts. Pharmacogenetics studies the impact of single variants on drug response, while pharmacogenomics focuses on the integral effect of multiple variants in genes involved in different parts of PK and PD [1]. As genomics research proliferates, pharmacogenomics emerges as a rapidly evolving field with significant potential in drug development. The majority of genetic variations influencing drug response are located within genes encoding proteins involved in drug pharmacokinetics (PK) and pharmacodynamics (PD). PK-related proteins encompass enzymes and drug transporters crucial for drug absorption, distribution, metabolism, and excretion (ADME), while PD involves receptors and other proteins mediating both desired therapeutic effects and undesirable adverse events. Consequently, numerous genes may impact drug response, rendering pharmacogenomics applicable across various stages of drug development, from preclinical to clinical studies. There are several main databases and sources for obtaining information related to pharmacogenomics, particularly regarding genetic variations of specific targets. Some of the key databases and sources include the following.

- *PharmGKB (Pharmacogenomics Knowledge Base)*: PharmGKB is a comprehensive resource that curates information on how genetic variation affects drug response. It provides data on pharmacogenomic associations, drug pathways, and clinical guidelines.

- *DrugBank*: DrugBank is a widely used database that provides comprehensive information on drug targets, pharmacology, and drug–drug interactions. It includes data on genetic variations that influence drug metabolism and response.
- *ClinVar*: ClinVar is a public archive of reports on the relationships between human variations and phenotypes, with a focus on clinically relevant information. It includes data on genetic variants associated with drug response and adverse reactions.
- *PubMed*: PubMed is a vast database of biomedical literature, including pharmacogenomic studies and reviews. It can be searched to find research articles on genetic variations of specific drug targets and their implications for drug response.
- *GWAS Catalog (Genome-Wide Association Studies Catalog)*: The GWAS Catalog provides a curated collection of published genome-wide association studies, including those related to pharmacogenomics. It can be used to identify genetic variants associated with drug response traits.

To search for information about genetic variations of certain drug targets, one can follow these step-by-step guidelines (Fig. 3.1).

1. *Identify the drug target of interest*: Determine the specific protein or genetic locus targeted by the drug for pharmacological effect.
2. *Access relevant databases*: Utilize databases such as PharmGKB, DrugBank, ClinVar, PubMed, and GWAS Catalog to search for information related to the genetic variations of the target.
3. *Perform a search*: Use keywords related to the drug target and pharmacogenomics, along with terms specifying genetic variations such as single nucleotide polymorphisms (SNPs) or gene mutations.
4. *Review search results*: Examine the search results to identify relevant studies, reports, and databases containing information on genetic variations associated with the drug target.
5. *Evaluate the evidence*: Assess the quality and relevance of the evidence presented in the retrieved information, including the study design, sample size, statistical significance, and clinical implications.
6. *Interpret findings*: Consider the implications of identified genetic variations on drug response, including potential effects on drug efficacy, toxicity, and recommended dosage.

In the pre-clinical phase, for instance, pharmacogenomics can help identify potential drug targets based on genetic variants associated with the pathophysiology of diseases. Using pharmacogenomics, targeted drugs can be designed considering a specific genetic profile. This targeted approach increases the likelihood of treatment success and reduces the risk of adverse reactions.

Genetic variations in the PCSK9 can impact its protein expression or enzymatic activity. Commonly, these alterations can lead to increased PCSK9 activity, resulting in higher degradation of the low-density lipoprotein receptor (LDLR). Consequently, this hampers the cells' ability to remove LDL cholesterol from the

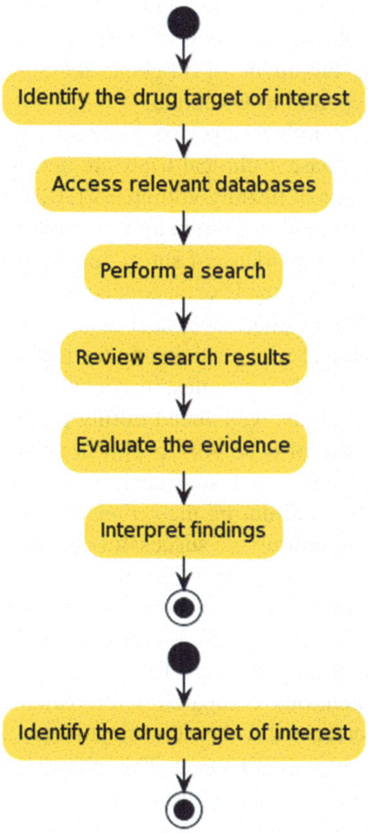

Fig. 3.1 Step-by-step to search for information about genetic variations of certain drug targets

bloodstream, leading to elevated plasma LDL cholesterol levels characteristic of familial hypercholesterolemia (FH). Chemically, variations in the PCSK9 gene may alter the amino acid sequence of the PCSK9 protein. These changes can affect the protein's structure, stability, and function, influencing its enzymatic activity and interaction with the LDL receptor. To gather information about these variations, genetic analyses like sequencing of the PCSK9 gene can be conducted. This helps in pinpointing specific mutations or polymorphisms within the gene associated with modified PCSK9 function. Additionally, functional studies can be carried out to evaluate the impact of these variations on PCSK9 activity and LDL receptor degradation in cell-based or animal models. It is essential to recognize that these variations manifest genetically as mutations, resulting in changes in the chemical composition of the PCSK9 protein. Therefore, modeling of these mutated proteins is crucial to grasp the structural and functional implications of these genetic changes. Simply seeking the wild-type PCSK9 structure in the Protein Data Bank (PDB) would not suffice in addressing the complexities introduced by genetic variations, as it does not account for the specific amino acid sequence alterations associated with mutations. Hence, computational modeling approaches are indispensable to comprehend the effects of these variations on protein structure and function, aiding in the development of targeted therapies like PCSK9 inhibitors for treating familial hypercholesterolemia.

Another step where pharmacogenomics can be useful is in clinical trial phase 1. Phase 1 clinical trials are designed to elucidate the PK profile of investigational drugs in healthy volunteers [20]. By sequencing the participants and using genomic information from healthy volunteers, pharmacogenomics can help to identify possible biomarkers of safety for the investigational drug. An unpublished survey from the Industry Pharmacogenomics Working Group (I-PWG) from 2017 revealed that 79% of pharmaceutical companies that are members of this group sequenced their clinical trial participants for performing pharmacogenomic studies of their investigational molecules, especially in the oncology area [2]. Exploring pharmacogenomics well ahead during phase 1 can help to better characterize the PK profile. It can also be very useful in subsequent phases 2 and 3, by improving participant selection and stratification for understanding the safety and efficacy outcomes observed during these phases.

Gefitinib is a successful example of the application of genomics in phase 3 clinical trials. Gefitinib is an epidermal growth factor receptor tyrosine kinase inhibitor (EFGR-TKI) used to treat nonsmall cell lung cancer (NSCLC). *EFGR* specific gene mutations are reported in about 10–20% of NSCLC patients. Some studies reported that the efficacy of gefitinib was enhanced in patients carrying *EGFR* mutations compared with patients with no mutations [21]. Clinical trials were performed to test this hypothesis, such as the Iressa Survival Evaluation in Lung Cancer (ISEL) trial. This phase 3 study pre-selected patients with *EGFR* mutations and verified that these mutations were predictive of gefitinib-related treatment effect over placebo on overall survival [22]. After several clinical trials, it was a consensus that gefitinib should be used in selected patients harboring these mutations, which is why NSCLC patients should now be genetically screened to verify if gefitinib therapy is indicated or not.

For more information on how pharmacogenomics can be applied to drug development, the reader is referred to excellent reviews on this topic [17, 23].

4 Case Studies of Genomics-Based Drug Design

Genomics-based drug design is a rapidly developing field that aims to use genomic data to design drugs that are tailored to the unique genetic profile of everyone. Case studies have shown the potential of this approach in predicting cancer cell sensitivity to drugs based on genomic and chemical properties [24]. Additionally, genome sequence variability has been shown to predict drug precautions and withdrawals from the market, highlighting the importance of genomics in drug development and safety [25].

Despite earlier efforts in genomics-based drug discovery, the overall clinical efficacy of developed drugs has remained unimpressive, owing in large part to the heterogeneous causes of disease [26]. However, recent technological and analytical advances in genomics have made it possible to rapidly identify and interpret the genetic variation underlying disease, leading to the development of personalized medicine [26].

Pharmacogenomics, the study of an individual's response to drugs as a result of their genetic makeup, has been merged with pharmacology and genomics to produce safe and effective drugs that are customized to the unique genetic profile of each individual [3]. The implementation of personalized medicine in clinical practice requires the integration of genomic data with clinical data, such as electronic medical records [3].

Advances in genomics have also facilitated drug development by providing genetic and genomic knowledge for target identification, understanding the biological relevance of a drug target, and prioritizing drug targets [27]. The CRISPR-Cas9 library screening approach has been used to detect survival-essential and drug-resistance genes via gain or loss of function, enabling genomic screening for gene activation or inhibition [28].

Predicting the response of a specific cancer to therapy is a primary goal in modern oncology, aiming for personalized treatment. Large-scale research has revealed relationships between genomic alterations and drug responses. Computational approaches have been proposed to predict sensitivity based on genomic features and the chemical properties of drugs. Machine learning models were developed to predict the response of cancer cell lines to treatment, based on both the genomic features of the cells and the chemical properties of the drugs [24].

The work reported by Spahn and colleagues [29] is one important example of a potential personalized medicine pipeline. In this work, the authors simulated different FGFR2 mutants identified on patient-derived cholangiocarcinoma samples. They demonstrated the ability of lenvatinib (an unspecific first-generation inhibitor) to more efficiently interfere in mutant-harboring lines, in comparison to the wild-type, in agreement with patient treatment data. The supporting in silico modeling, using long MD simulations, suggests that lenvatinib can adapt better to FGFR2 kinase mutations' binding sites than specific inhibitors, such as infigratinib or pemigatinib. This study highlights how different computational approaches can inform bed-side decisions, but more importantly, generate a framework that informs recidivists of common mutations/cases.

The genomic revolution has been transforming the field of medicine, enabling a more precise approach to drug discovery and development. With a deeper understanding of human genes and their functions, pharmacogenomics emerged, which harnesses advanced techniques to identify markers that determine a patient's response to specific treatments. This approach, centered on an individual's genetic profile, promises to optimize therapies, enhancing outcomes across various medical specialties and leading the way for more personalized treatments. However, the adoption of this new perspective demands ethical reflection and clear guidelines for its implementation in clinical practice [30].

In conclusion, genomics-based drug design has shown potential in predicting cancer cell sensitivity to drugs and predicting drug precautions and withdrawals from the market. Recent technological and analytical advances in genomics have made it possible to rapidly identify and interpret the genetic variation underlying disease, leading to the development of personalized medicine. The implementation of personalized medicine in clinical practice requires the integration of genomic data with clinical data. Advances in genomics have also facilitated drug development by providing genetic and genomic knowledge for target identification and prioritization [26].

4.1 Challenges and Opportunities in Genomics and Personalized Medicine for Drug Design

Genomics and personalized medicine have opened new opportunities and challenges in drug design. The use of genomic data in drug design has an immediate impact on structural proteomic/genomic projects, as well as on rational drug design [31]. The widespread application of personal genome sequencing in clinical settings for predictive and preventive medicine has been limited due to the lack of comprehensive computational analysis pipelines [32]. However, recent technological and analytical advances in genomics have made it possible to rapidly identify and interpret the genetic variation underlying disease, leading to the development of personalized medicine [33].

Pharmacogenomics, the study of an individual's response to drugs as a result of their genetic makeup, has been merged with pharmacology and genomics to produce safe and effective drugs that are customized to the unique genetic profile of each individual [30]. The amount of data generated in genomics fits the definition of big data and needs specific bioinformatics processing following standard steps: data collection, processing, analysis, and interpretation [34].

Genome-wide association studies (GWAS) have proved to be a beneficial method to identify novel common genetic variations not only for disease susceptibility but also for drug efficacy and drug-induced toxicity, creating a field of pharmacogenomics studies [35]. The identification of disease-causing biomolecules using genome sequencing has enabled the development of chemical probes of function, preclinical lead modalities, and ultimately FDA-approved drugs [36].

In conclusion, genomics and personalized medicine have opened up new opportunities and challenges in drug design. The use of genomic data in drug design has an immediate impact on structural proteomic/genomic projects, as well as on rational drug design. Recent technological and analytical advances in genomics have made it possible to rapidly identify and interpret the genetic variation underlying disease, leading to the development of personalized medicine. Pharmacogenomics addresses patient-to-patient variation in drug response and has the potential to produce safe and effective drugs that are customized to the unique genetic profile of everyone [37, 38].

4.2 *Ethical and Legal Issues in Genomics and Personalized Medicine for Drug Design*

Ethical and legal issues are important considerations in the development and implementation of genomics and personalized medicine for drug design. The emergence of genetic testing (GT) has raised ethical, legal, and social implications (ELSI) [39]. The adoption of personalized medicine will require changes in healthcare infrastructure, diagnostics and therapeutics business models, reimbursement policy from government and private payers, and a different approach to regulatory oversight [40]. Governance issues arise in the context of an ongoing dispersion of national regulatory power, and it has become impossible to govern society from a single center [41].

Pharmacogenomics, the study of an individual's response to drugs as a result of their genetic makeup, has the potential to produce safe and effective drugs that are customized to the unique genetic profile of each individual [38]. However, pharmacogenomics and pharmacogenetics raise ethical problems with specific nuances and subtleties and high complexity levels [34]. The use of genomic and molecular data contained in the patients' genotypes has allowed for the development of new personalized pharmaceutical products, but it is important to consider the informed consent of patients and the protection of their privacy [34].

In conclusion, ethical and legal issues are important considerations in the development and implementation of genomics and personalized medicine for drug design. The emergence of genetic testing has raised ethical, legal, and social implications. The adoption of personalized medicine will require changes in healthcare infrastructure, diagnostics and therapeutics business models, reimbursement policy from government and private payers, and a different approach to regulatory oversight. Governance issues arise in the context of an ongoing dispersion of national regulatory power. Pharmacogenomics and pharmacogenetics raise ethical problems with specific nuances and subtleties and high complexity levels, and it is important to consider the informed consent of patients and the protection of their privacy [42].

To guarantee safe society from the ethical and legal standpoint, law and bioethics should constantly evolve to follow up technological development and make sure that ethical and legal matters are in line with scientific advancements. The delay in establishing well-delimited ethical and legal criteria for application of advancements in healthcare sciences may give rise to a bioethical regress, jeopardizing rights earned throughout the years.

As broadly discussed in this chapter, pharmacogenomics represents an important advancement in the strategy to treat pathologies, enabling more effective pharmacological targeting and, thus, increasing patients' quality of life [43].

This marks a noteworthy advancement in personalized medicine, in which comprehension of an individual's genetic sequencing not only offers insights regarding their health status but also enables the prescription of highly personalized treatments, which may be safer and more effective. The fundamental purpose is to identify the suitable drug in the correct dose and at the right time, all based on deep knowledge of the patient's genomics.

To reach that goal, the collection of comprehensive patient information, including general personal data, genetic data regarding DNA sequencing, as well as information in connection with lifestyle and life habits, heredity, medical history, etc., is required. All such information is characterized by the confidential, intimate, and private nature thereof, which is essential and intrinsic to human rights. And these are the data and information that will allow establishment of pharmacokinetic profile of the patients, enabling individualized treatment [44].

Therefore, the ethical and legally robust application of pharmacogenomics faces multiple fundamental challenges within the scope of law and bioethics. This includes considerations regarding social, economic, legal, and ethical factors. Hence, a detailed analysis of existing ethical and legal precepts is imperatively required, aiming, if need be, at creating a general regulatory framework that establishes guidelines to protect genetic data and promote the use thereof in compliance with universal bioethical principles and specific laws of each nation.

Within the legal context, crucial issues arise mainly regarding the safeguard of highly sensitive personal data that have already been internationally regulated by means of the Data Protection Laws. Furthermore, inspection and restriction of use of such data for purposes that have not been duly consented to by the holders thereof is of the essence.

In this setting, two essential legal principles come in: (i) Data Protection and (ii) Free, Prior, and Informed Consent. This takes place because, by collecting information of such nature, a space is created for filing of highly sensitive data that may be employed for purposes beyond pharmacogenomics. Such undue use may, by itself, breach fundamental principles, such as equality and dignity of individuals.

The discussion about the use of genetic data is controversial and polemical because it is known that such information has the potential to lead to discrimination and improper use in selecting characteristics of humans, individuals, and embryos.

Additionally, bioethics concern comes under an even stronger spotlight, considering that countless abuses were committed in the name of science, jeopardizing human life and dignity, before global bioethical principles went on to guide scientific conduct. The main underlying mission of all such bioethical principles is, undoubtedly, full safeguard of human beings, their existence, and, above all, their dignity.

In contemporary society, the importance of ethics cannot be overstated, particularly within the realm of healthcare. Ethical considerations serve as the cornerstone for establishing guidelines and protocols that govern medical practice, ensuring the well-being and rights of patients are upheld. [45] eloquently define ethics as the science of conduct, presenting two fundamental concepts within this discipline. First, ethics is perceived as guiding human behavior toward predetermined ends, derived from an understanding of human nature. Second, it is recognized as the study of the variability of human conduct, aiming to regulate and guide behavior accordingly. This nuanced understanding underscores the complexity of ethical decision-making in healthcare, where practitioners must navigate diverse moral frameworks to provide optimal care while upholding ethical standards.

Ethics thus becomes an instrument to guide human conducts for the purpose of safeguarding the individual's dignity, autonomy, life, physical integrity, equality, and sameness in healthcare services, especially when no other guidelines to follow exist.

Bioethics, the concept of which is provided in the *Encyclopedia of Bioethics* as this conduct is examined in the light of moral values and principles, [46] steers the decisions affecting life, health, and dignity of human beings.

Although the work *Principles of Biomedical Ethics* [47] is not the pioneer in analyzing and establishing bioethical principles, it is deemed to be the "most influent" and "more widely broadcast" in study [48].

In the dynamic landscape of biomedical ethics, the contributions of [47] stand as a cornerstone, reshaping the discourse and practice of ethical decision-making in healthcare. Their seminal work, *Principles of Biomedical Ethics*, not only established a set of guiding principles but also offered a comprehensive framework for their practical application. Beauchamp and Childress recognized the need for more than theoretical constructs; they sought to provide clinicians and ethicists with a robust toolkit for navigating the complexities of ethical dilemmas in medical practice. By elucidating the operational intricacies of bioethical principles, they empowered stakeholders to engage in informed decision-making processes that prioritize patient welfare and ethical integrity. Thus, their work remains a foundational resource in shaping ethical standards and practices within the biomedical sciences.

Given the intricate nature of the subject, bioethics may not offer absolute solutions to the conflicts at hand; however, it may provide guidance that establishes guidelines that are reasonable and prudent [49]. The objective is to guarantee a fair, equitable, and universal distribution of the benefits from healthcare services, without discriminating [50].

Within the context of the healthcare area, however, the complexity of the conflicts at stake often requires making decisions that involve assessment of the associated costs and benefits. In the work *Principles of Biomedical Ethics*, the inherent complexity of moral decision-making, particularly within the healthcare domain, is emphasized. This complexity arises from the multitude of factors that must be considered when making healthcare-related decisions, which extend beyond purely medical considerations. Health-related decisions inherently involve the assessment of values that encompass both medical and nonmedical aspects, such as weighing the costs and benefits of medical interventions. Physicians routinely base their treatment judgments on the balance between potential benefits and harm to the patient.

Four fundamental principles of bioethics emerge from this context: autonomy, nonmaleficence, beneficence, and justice. These principles provide the foundation for robust bioethical guidelines that can guide the application of pharmacogenomics with legal rigor and adherence to universal ethics, irrespective of specific regulatory frameworks.

Autonomy, as one of these principles, underscores the importance of securing free and informed consent from all individuals involved. However, it is crucial to ensure that autonomy is genuinely free, without defects in consent or omission of information, which includes addressing issues related to personal vulnerability.

The principles of nonmaleficence and beneficence emphasize a commitment to improving the quality of life and treatment efficacy while minimizing side effects for both individual patients and the broader population. This commitment rejects medical approaches based on "trial and error" methods, prioritizing patient well-being and safety above all else, as highlighted by Brito et al. [50]. The principle of nonmaleficence, which originates from the Hippocratic oath whose motto is "primum non nocere," meaning "first, do not cause damage," keeps a direct connection with the essential objectives of pharmacogenomics, which strives to reach such goals, considering that its fundamental purpose is to improve humanity's quality of life [50].

In the realm of bioethics, Marcelo Quentin sheds light on the foundational principle of beneficence, emphasizing its pivotal role in promoting the well-being of individuals, particularly those who are ill. According to Quentin, beneficence entails a profound acknowledgment of the moral worth of others, driving healthcare professionals and society at large to prioritize the prevention of harm and the enhancement of wellness. This principle compels individuals to meticulously evaluate the potential risks and benefits associated with medical interventions, striving to maximize the benefits for patients while minimizing any potential harm. Furthermore, beneficence encompasses the complementary principle of "nonmaleficence," which underscores the imperative of avoiding actions that may cause harm to the patient. In essence, Quentin's perspective underscores the dual nature of ethical considerations, emphasizing the imperative of balancing benevolence with the obligation to do no harm [51].

Exploring the principle of justice unveils a complex sphere entangled with social, political, and economic dynamics, particularly poignant in the context of developing countries. Here, tests and treatments linked to genetic data remain prohibitively expensive, rendering them accessible only to a privileged minority. Remarkably, the principle of justice emerges as the most intricate in this landscape. The prevailing economic interests of pharmaceutical giants often prioritize profit maximization over addressing illnesses affecting marginalized populations, particularly those deemed unprofitable or lacking potential returns to the system. As noted, treatments for rare diseases, categorized as "orphan medications," are sidelined due to their low profitability and lack of investment appeal. Consequently, pharmacogenomics might inadvertently perpetuate a discriminatory environment, disproportionately excluding certain social groups, particularly in nations where genetic testing remains financially out of reach for a significant portion of the populace. This predicament poses a formidable challenge and threatens to undermine the principle of justice in its entirety.

On the flip side, as underscored by Jorge Alberto Iriart, the emergence of precision medicine unfolds within the intricate framework of globalized capitalism, characterized by what Rose terms "economies of vitality." This concept delineates a novel economic domain, known as bioeconomics, wherein biocorporations wield control over lives, thereby generating value. In this context, the manipulation of biological data and healthcare interventions becomes instrumental in the accumulation of a distinct form of capital known as biocapital. This perspective sheds light on the intersection of political and economic forces shaping the trajectory of precision medicine, highlighting the profound influence of capitalist dynamics on the healthcare landscape [52].

In simpler terms, pharmacogenomics emerges as a promising solution to fortify basic and preventive healthcare, potentially slashing costs linked with ineffective treatments. This perspective holds merit, provided concerns about the affordability and accessibility of genetic tests for marginalized social groups are adequately tackled. Nevertheless, within this context, it is crucial to acknowledge two branches of the principle of justice: one of utilitarian nature, striving to maximize benefits for both patients and society and another of egalitarian ethos, dedicated to ensuring equal value for individuals and equitable opportunities [53].

Based on this premise, compliance with the bioethical principle of justice is intrinsically connected to the guarantee of egalitarian access and fair opportunities to all citizens as regarding treatment.

In addition to the bioethical aspect of the matter, the legal aspect is associated to two fundamental pillars: (i) the patient's consent regarding the purpose, use, and disposal of data, and (ii) guaranteed protection of such data against leaks, with specific and clear guidelines regarding disposal thereof. Each country, considering their legal and regulatory structure in connection with medical secrecy, data protection, and patient's informed consent, should consider the particularities of such collections in implementing public policies that regulate the matter.

Genetic data are already collected for miscellaneous purposes worldwide, especially to map genetic diseases at birth, such as the newborn blood spot test. However, the manner in which such data are collected, stored, and protected is crucial to avoid breaches of various natures. Thus, the public power is accountable for guaranteeing safety in the collection, storage, and use of such genetic data.

Hence, pharmacogenomics represents a significant advancement in the healthcare area. Nevertheless, its ethical and legal application faces complex challenges that go beyond medicine, involving social, economic, ethical, and legal matters. To guarantee a safe society from the bioethical and legal standpoint, it is essential that law and bioethics continuously evolve to follow up technological development and establish solid criteria. This includes the protection of sensitive genetic data, compliance with fundamental bioethical principles, such as autonomy, nonmaleficence, beneficence, and justice, and the promotion of egalitarian access to treatments. Moreover, considering the safety and privacy of the genetic data is of the essence to guarantee that use thereof is consented to and does not breach the fundamental principles of human dignity. Ultimately, ethics and bioethics play a core role in providing guidance to such complex matters, promoting a fairer and safer society.

Questions to answer when planning genomics and personalized medicine in drug design studies?

? What is the genomic basis of the disease?
What are the target genes and pathways?
How do genetic variations affect drug response?
What are the biomarkers for disease and drug response?
How can genomics inform drug discovery and development?
What are the ethical, legal, and social implications?
What technologies and methodologies are required?
How can personalized medicine be integrated into clinical practice?
What are the challenges and limitations?
How to ensure collaboration among stakeholders?

References

1. Dere, W. H., & Suto, T. S. (2009). The role of pharmacogenetics and pharmacogenomics in improving translational medicine. *Clinical Cases in Mineral and Bone Metabolism, 6*, 13–16.
2. Bienfait, K., et al. (2022). Current challenges and opportunities for pharmacogenomics: Perspective of the industry pharmacogenomics working group (I-PWG). *Human Genetics, 141*, 1165–1173.
3. Ramayanam, N. R., Amarnath, R. N., & Vijayakumar, T. M. (2022). Pharmacogenetic biomarkers and personalized medicine: Upcoming concept in pharmacotherapy. *Research Journal of Pharmacy and Technology, 15*, 4289–4292.
4. Castro, K. M., Scheck, A., Xiao, S., & Correia, B. E. (2022). Computational design of vaccine immunogens. *Current Opinion in Biotechnology, 78*, 102821.
5. Gourlay, L., Peri, C., Bolognesi, M., & Colombo, G. (2017). Structure and computation in immunoreagent design: From diagnostics to vaccines. *Trends in Biotechnology, 35*(12), 1208–1220.
6. Pak, M. A., Markhieva, K. A., Novikova, M. S., Petrov, D. S., Vorobyev, I. S., Maksimova, E. S., et al. (2023). Using AlphaFold to predict the impact of single mutations on protein stability and function. *PLoS One, 18*(3), e0282689.
7. Modeling mutations in protein structures – Feyfant – 2007 – Protein Science – Wiley Online Library. https://onlinelibrary.wiley.com/doi/10.1110/ps.072855507.
8. Pandurangan, A. P., & Blundell, T. L. (2020). Prediction of impacts of mutations on protein structure and interactions: SDM, a statistical approach, and mCSM, using machine learning. *Protein Science, 29*, 247–257.
9. Zhou, Y., Tremmel, R., Schaeffeler, E., Schwab, M., & Lauschke, V. M. (2022). Challenges and opportunities associated with rare-variant pharmacogenomics. *Trends in Pharmacological Sciences, 43*, 852–865.
10. Schreeck, F., Ahne, G., Tremmel, R., Schaeffeler, E., & Schwab, M. (2022). Pharmacogenomics in pediatric medicine and drug development. *Pharmacogenomics, 23*, 709–712.
11. Schwab, M., & Schaeffeler, E. (2012). Pharmacogenomics: A key component of personalized therapy. *Genome Medicine, 4*, 1–3.
12. Moore, T. J., Heyward, J., Anderson, G., & Alexander, G. C. (2020). Variation in the estimated costs of pivotal clinical benefit trials supporting the US approval of new therapeutic agents, 2015–2017: a cross-sectional study. *BMJ Open, 10*, e038863.
13. Sun, D., Gao, W., Hu, H., & Zhou, S. (2022). Why 90% of clinical drug development fails and how to improve it? *Acta Pharmaceutica Sinica B, 12*, 3049–3062.

14. Coleman, J. J., & Pontefract, S. K. (2016). Adverse drug reactions. *Clinical Medicine, 16*, 481–485.
15. Insani, W. N., et al. (2021). Prevalence of adverse drug reactions in the primary care setting: A systematic review and meta-analysis. *PLoS One, 16*, e0252161.
16. Ray, S. (2014). Clopidogrel resistance: The way forward. *Indian Heart Journal, 66*, 530–534.
17. Yin, O., & Vandell, A. (2019). Incorporating pharmacogenomics in drug development. In *Pharmacogenomics* (pp. 81–101). Elsevier.
18. Vogel, F. (1959). Moderne Problem Der. *Humangenetik, 12*.
19. Lander, E. S., et al. (2001). Initial sequencing and analysis of the human genome. *Nature, 409*, 860–921.
20. Kandi, V., & Vadakedath, S. (2023). Clinical trials and clinical research: A comprehensive review. *Cureus, 15*, e35077.
21. Cappuzzo, F., et al. (2005). Epidermal growth factor receptor gene and Protein and Gefitinib sensitivity in non–small-cell lung cancer. *JNCI Journal of the National Cancer Institute, 97*, 643–655.
22. Hirsch, F. R., et al. (2006). Molecular predictors of outcome with Gefitinib in a phase III placebo-controlled study in advanced non–small-cell lung cancer. *Journal of Clinical Oncology, 24*, 5034–5042.
23. Singh, B., Jain, P., Devaraja, K., & Aggarwal, S. (2023). Chapter 3: Pharmacogenomics in drug discovery and development. In S. A. Ganie, A. Ali, M. U. Rehman, & A. Arafah (Eds.), *Pharmacogenomics* (pp. 57–96). Academic.
24. Menden, M. P., et al. (2013). Machine learning prediction of cancer cell sensitivity to drugs based on genomic and chemical properties. *PLoS One, 8*, e61318.
25. Lee, K. H., et al. (2016). Genome sequence variability predicts drug precautions and withdrawals from the market. *PLoS One, 11*, e0162135.
26. Dugger, S. A., Platt, A., & Goldstein, D. B. (2018). Drug development in the era of precision medicine. *Nature Reviews. Drug Discovery, 17*, 183–196.
27. Spreafico, R., Soriaga, L. B., Grosse, J., Virgin, H. W., & Telenti, A. (2020). Advances in genomics for drug development. *Genes, 11*, 942.
28. Chan, Y.-T., et al. (2022). CRISPR-Cas9 library screening approach for anti-cancer drug discovery: Overview and perspectives. *Theranostics, 12*, 3329–3344.
29. Spahn, S., Kleinhenz, F., Shevchenko, E., et al. (2024). The molecular interaction pattern of lenvatinib enables inhibition of wild-type or kinase-mutated FGFR2-driven cholangiocarcinoma. *Nature Communications, 15*, 1287.
30. Emilien, G., Ponchon, M., Caldas, C., & Isacson, O. Impact of genomics on drug discovery and clinical medicine. *QJM*.
31. Conejero-Muriel, M., Contreras-Montoya, R., Díaz-Mochón, J. J., de Cienfuegos, L. Á., & Gavira, J. A. (2015). Protein crystallization in short-peptide supramolecular hydrogels: A versatile strategy towards biotechnological composite materials. *CrystEngComm, 17*, 8072–8078.
32. Patowary, A., et al. (2012). Systematic analysis and functional annotation of variations in the genome of an Indian individual. *Human Mutation, 33*, 1133–1140.
33. Tyukavin, A. I., et al. (2021). Biodizine as a civilizational challenge of modern pharmaceuticals. *Pharmacy Formulations, 3*, 108–117.
34. Barrot, C.-C., Woillard, J.-B., & Picard, N. (2019). Big data in pharmacogenomics: Current applications, perspectives and pitfalls. *Pharmacogenomics, 20*, 609–620.
35. Low, S.-K., Takahashi, A., Mushiroda, T., & Kubo, M. (2014). Genome-wide association study: A useful tool to identify common genetic variants associated with drug toxicity and efficacy in cancer pharmacogenomics. *Clinical Cancer Research, 20*, 2541–2552.
36. Angelbello, A. J., et al. (2018). Using genome sequence to enable the design of medicines and chemical probes. *Chemical Reviews, 118*, 1599–1663.
37. Tsourounis, M., Stuart, J., Pignato, W., Toscani, M., & Barone, J. (2015). Current trends in personalized medicine and companion diagnostics: A summary from the DIA meeting on

personalized medicine and companion diagnostics. *Therapeutic Innovation & Regulatory Science, 49*, 530–543.
38. Madian, A. G., Wheeler, H. E., Jones, R. B., & Dolan, M. E. (2012). Relating human genetic variation to variation in drug responses. *Trend in Genetics, 28*, 487–495.
39. Cheung, N. Y. C., et al. (2021). Perception of personalized medicine, pharmacogenomics, and genetic testing among undergraduates in Hong Kong. *Human Genomics, 15*, 54.
40. Abrahams, E., Ginsburg, G. S., & Silver, M. (2005). The personalized medicine coalition: Goals and strategies. *American Journal of Pharmacogenomics : Genomics-Related Research in Drug Development and Clinical Practice, 5*, 345–355.
41. Prainsack, B., Naue, U. Relocating health governance: personalized medicine in times of 'global genes'. *Personalized Medicine, 3,* 349–355.
42. Gershon, E. S., Alliey-Rodriguez, N., & Grennan, K. (2014). Ethical and public policy challenges for pharmacogenomics. *Dialogues in Clinical Neuroscience, 16*, 567–574.
43. Mattevi, V. S., & Tagliari, C. F. (2017). Pharmacogenetic considerations in the treatment of HIV. *Pharmacogenomics, 18*, 85–98.
44. Rodríguez Duque, R., & Miguel Soca, P. E. (2020). Farmacogenómica: principios y aplicaciones en la práctica médica. *Revista Habanera de Ciencias Médicas*, e3128.
45. Abbagnano, N. (2007). Dicionário de Filosofia (5th ed). Ed. Martins Fontes.
46. Sgreccia, E. (1996). Manual de bioética: fundamentos e ética biomédica. In *Manual de bioética: fundamentos e ética biomédica* (pp. 686–686).
47. Beauchamp, T. L.; Childress, J. F. (2013) Principles of Biomedical Ethics. New York: *Oxford University Press.*
48. Ferrer, J. J., & Álvarez, J. C. (2005). *Para fundamentar a bioética: teorias e paradigmas teóricos na bioética contemporânea*. Edicoes Loyola.
49. Suarez-Kurtz, G. (2018). Pharmacogenetic testing in oncology: A Brazilian perspective. *Clinics (Sao Paulo), 73*, e565s.
50. Ruiz-Hornillos, J., et al. (2021). Bioethical concerns during the COVID-19 pandemic: What did healthcare ethics committees and institutions state in Spain? *Frontiers in Public Health, 9*, 737755.
51. Brito, M. (2015). A farmacogenética e a medicina personalizada. *Saúde e Tecnologia, 14,* 5–10.
52. Ortolan, A., et al. (2021). The genetic contribution to drug response in Spondyloarthritis: A systematic literature review. *Frontiers in Genetics, 12*, 703911.
53. Iriart, J. A. B. (2019). Medicina de precisão/medicina personalizada: análise crítica dos movimentos de transformação da biomedicina no início do século XXI. *Cadernos De Saude Publica, 35*, e00153118.

Chapter 4
Machine Learning and Neural Network Methods Applied to Drug Discovery

Daniel S. de Sousa, Aldineia P. da Silva, Rafaela M. de Angelo, Laise P. A. Chiari, Kathia M. Honorio, and Albérico B. F. da Silva

Abstract Throughout this chapter, we will explore how machine learning and neural networks are shaping the evolution of drug design, its fundamental applications, limitations, and the challenges that still need to be overcome. The revolutionary potential of this approach promises to continue contributing to the discovery of new therapies and advancing pharmaceutical science.

Keywords Machine learning · Drug design · Applications · Neural networks

1 Historical Background

The use of machine learning (ML) in the field of drug discovery represents a revolutionary approach that has profoundly transformed how scientists and researchers approach the development of new therapeutic compounds. Throughout history, the field of drug design has faced numerous challenges in identifying and optimizing molecules capable of treating diseases effectively, safely, and efficiently. However, the advent of ML has brought new perspectives and remarkable advances. Historically, the drug design process has predominantly relied on trial-and-error approaches, with scientists conducting exhaustive experiments to identify compounds with desirable properties [1]. Nevertheless, the increasing availability of data and the growing computational processing power have opened new possibilities for the application of ML algorithms [2]. Notable examples of this evolution include the use of Quantitative Structure-Activity Relationship (QSAR), which has enabled

D. S. de Sousa · L. P. A. Chiari · A. B. F. da Silva (✉)
São Carlos Institute of Chemistry, University of São Paulo, São Carlos, SP, Brazil
e-mail: alberico@iqsc.usp.br

A. P. da Silva · K. M. Honorio
School of Arts, Sciences and Humanities, University of São Paulo, São Paulo, SP, Brazil

R. M. de Angelo
Center for Natural Sciences and Humanities, Federal University of ABC, Santo André, SP, Brazil

V. G. Maltarollo (ed.), *Computer-Aided and Machine Learning-Driven Drug Design*, Computer-Aided Drug Discovery and Design 3,
https://doi.org/10.1007/978-3-031-76718-0_4

the prediction of biological activities based on the molecular structure, and the application of deep learning algorithms, such as neural networks, which have enhanced the ability to identify complex patterns in extensive data sets [3]. For more details on QSAR, see Chap. 6.

This transformation has significantly impacted applications in drug design. ML is now widely employed to expedite the identification of drug candidates, optimize the structure of existing molecules, and predict potential side effects, resulting in a more efficient and economically advantageous process. Furthermore, the capability to construct models of interactions between compounds and biological targets has improved the understanding of drug mechanisms, unveiling new perspectives for therapeutic development [1–3].

However, despite its notable achievements, the application of ML in drug design faces critical challenges and limitations. The need for high-quality and reliable data, which is not always readily available, is one of these barriers. Additionally, ML models' interpretability and domain expertise's incorporation remain ongoing challenges. Ensuring the safety and efficacy of drug candidates identified by ML algorithms is also essential [4, 5]. For more details on the application of ML models, see Chap. 6.

The history of ML and neural network development has been marked by numerous peaks and valleys, woven with both triumphs and setbacks over its expansive chronology. Currently, our primary focus does not involve an in-depth exploration of the historical context. Instead, our goal is centered on encapsulating this narrative concerning medicinal chemistry, highlighting key events involving ML that have been impactful in drug discovery and its evolution.

1.1 Timeline

Before we embark on the application of ML into our field of expertise, it is essential to trace its historical evolution. The initial concepts of ML emerged in 1943, albeit not in the concrete ML form we know today. Instead, these nascent ideas originated from an endeavor to understand the workings of neurons. It was the neurophysiologist Warren McCulloch and the mathematician Walter Pitts who ventured into creating a simplified neural network using electrical circuits [6]. Intriguingly, this development predated the advent of the first electronic digital computer, ENIAC, in 1946 [7]. This highlights that the quest for creating artificial intelligence (AI) has deep historical roots, evolving alongside the nascent field of computer science.

The initial connection between artificial intelligence (AI) and computers was established in 1950 when the mathematician and computer scientist Alan Turing introduced a test to evaluate a machine's capacity to exhibit human-like intelligence, known as "Turing test" [8]. However, the term "machine learning"only entered the lexicon in 1952, coined by computer scientist Arthur Samuel, a pioneer in the field [9]. Samuel developed a program capable of playing checkers at a championship level, utilizing an algorithm called "alpha-beta pruning." Nonetheless, it is worth noting that while situated within the realm of artificial intelligence, this program did not involve the process of learning from data [10].

The first authentic ML algorithm emerged in 1957, courtesy of the psychologist Frank Rosenblatt. This algorithm marked the prototype of Artificial Neural Network (ANN) and was named the "perceptron," a single-layer neural network [11]. Subsequently, a plethora of other methods were conceived, including the Nearest Neighbor Algorithm (1967), paving the way for the diverse methods that are encountered today [12].

With the advent of ML techniques, numerous fields of knowledge have been embraced and enriched. For instance, the vast domain of chemistry witnessed significant advancements with the introduction of the Dendral system, developed by Edward Feigenbaum and Joshua Lederberg in 1965. This innovative system's primary objective was to elucidate the chemical structures of compounds by intricately interpreting spectrographic data [13]. In the field of drug discovery, ML made its initial foray in the 1990s, and its applications have proliferated in the 2000s up to the present day [3–5, 14–18].

A significant milestone in the history of ML in drug design was the introduction of the "drug-likeness" concept in 1998 by Ajay et al. Their model was designed to predict with high accuracy whether a molecule could be categorized as a drug or not, employing Bayesian neural network algorithms [19]. In the ensuing years, during the 2000s, the creation of new methods, such as Random Forest, and the popularization of other ML techniques like SVM, decision trees, and Naive Bayes, among others, along with the increasing availability of data, led to the development of various strategies to enhance the drug discovery process [14–16, 20].

In the same decade, the first substantial databases were established, including ZINC and PubChem (2004) [21, 22]. This was followed by the creation of DrugBank in 2006 [23] and ChEMBL in 2008 [20, 24]. However, significant advances in the field of drug discovery only materialized in more recent times. In 2015, Atomwise introduced AtomNet, the pioneering deep learning neural network for structure-based drug design, utilizing three-dimensional representations of chemical interactions. The system identified chemical features like aromaticity, sp^3 carbons, and hydrogen bonding, akin to how image recognition networks comprehend spatially proximate features. AtomNet subsequently played notable importance in predicting novel candidate biomolecules for various disease targets, notably contributing to treatments for the Ebola virus and multiple sclerosis [25]. The culmination of progress in the field occurred in 2020, marked by two notable events: the discovery of halicin, an antibiotic [26] through deep learning models, and the introduction of the first planned machine learning drug candidate for the treatment of cancer and cardiovascular disease into preclinical testing [27]. Figure 4.1 illustrates the chronological progression of ML in the field of drug discovery, depicting key milestones.

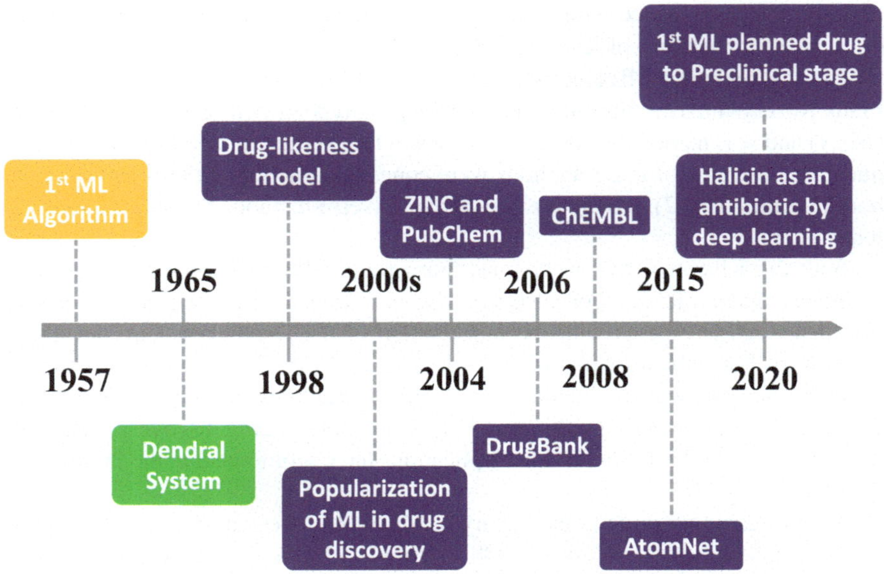

Fig. 4.1 Timeline with some events within the field of machine learning since its inception (yellow), spanning from chemistry (green) to drug discovery (purple)

2 Methodology Overview

ML is a fundamental branch of knowledge within the field of AI that stands out for its ability to enable computational systems to learn and improve from data, rather than being explicitly programmed. While AI encompasses a wide spectrum of techniques and approaches aimed at endowing machines with human-like intelligence and behavior, ML focuses on a system's capacity to acquire knowledge and enhance its performance through data analysis, pattern recognition, and continuous adaptation. This sets ML apart from traditional programming approaches, allowing autonomous systems and algorithms to make decisions and perform tasks independently based on past experiences [28–32].

Within the realm of ML, Neural Networks (NNs) assume a fundamental role. These computational models utilize interconnected layers of artificial neurons to analyze and recognize intricate patterns within data. NNs can be both shallow, with few layers, or deep, incorporating multiple layers, giving rise to what is known as Deep Learning (DL). As a specialized branch of ML, DL, places particular emphasis on employing deep neural networks to address complex tasks, such as image and speech recognition, natural language processing, and more [30–38]. This hierarchy of concepts illustrates the progressive depth and specialization that occurs within the broader domain of AI, as illustrated in Fig. 4.2.

In medicinal chemistry, numerous ML algorithms are essential for addressing diverse needs related to drug discovery. They are integral in tasks ranging from

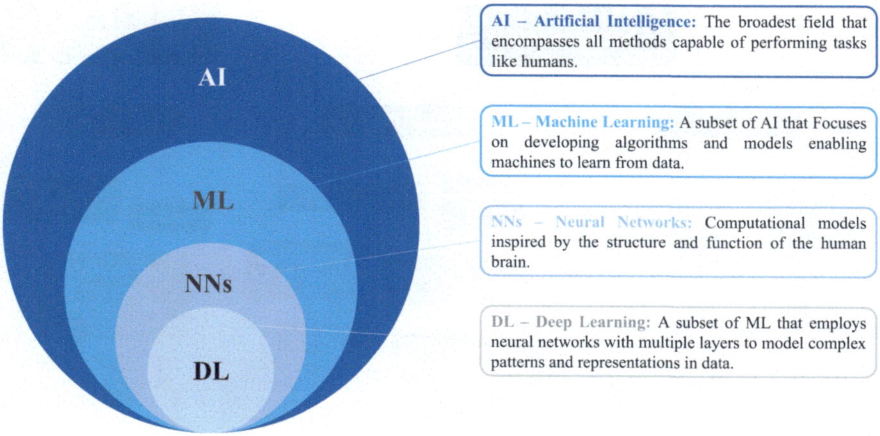

Fig. 4.2 Schematic representation of the hierarchy of concepts in the scope of artificial intelligence to deep learning

target identification to the optimization of clinical trials. Broadly, ML methods can be categorized into two major groups: supervised learning and unsupervised learning [14].

In supervised learning methods, algorithms are trained on labeled data sets where the relationship between inputs and outputs is known. This enables the prediction or classification of new data based on prior learning. On the other hand, unsupervised learning methods explore underlying structures in the data, either by grouping them into clusters or by reducing dimensionality. These methods are important for uncovering insights and patterns in unlabeled data, facilitating data segmentation into similar groups, and simplifying the representation of complex data [32, 35, 36, 39].

Furthermore, these methods can be combined to give rise to other approaches, such as semi-supervised learning and reinforcement learning. These approaches offer versatile solutions for a wide range of data sets, leveraging the strengths of both supervised and unsupervised learning paradigms. Semi-supervised learning, for instance, utilizes a combination of labeled and unlabeled data, capitalizing on the benefits of limited labeled information and the vastness of unlabeled data. On the other hand, reinforcement learning introduces a dynamic element by allowing models to learn through interaction with an environment, receiving feedback in the form of rewards or penalties to refine their decision-making processes. This amalgamation of methods contributes to the versatility and efficacy of ML applications in various domains, including drug discovery [14, 30, 32, 39].

The purpose of this chapter is not to delve into the detailed principles of operation for all methods, but rather to provide an overview of the broader context within the field of drug discovery, focusing on the commonly used ML methods for this purpose. Figure 4.3 presents the main ML algorithms employed in various drug design applications.

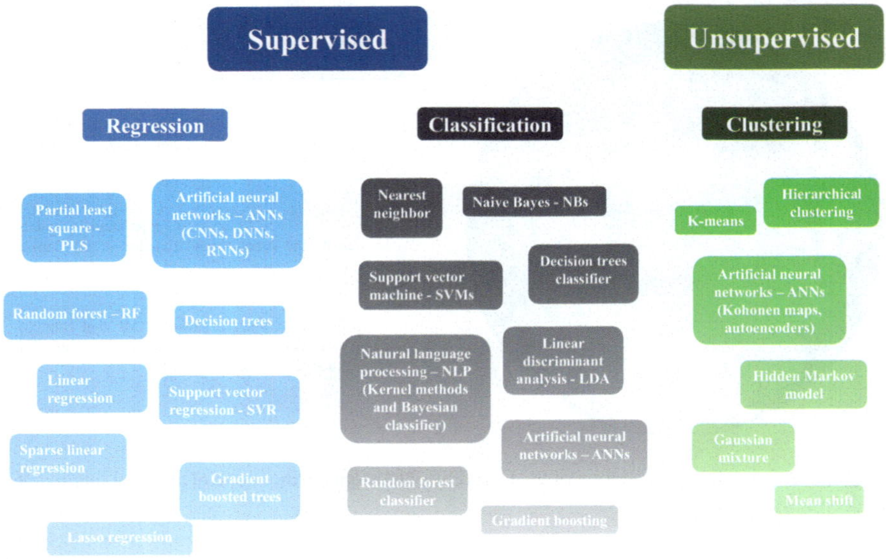

Fig. 4.3 Examples of machine learning methods used in drug discovery followed by their classification based on their use and nature

Among these methods, NNs, which will be discussed in more detail later, along with Random Forest (RF) and Support Vector Machine (SVM), are the most widely employed approaches in drug design. [15]

RF is an ensemble ML method that combines multiple decision trees to make predictions. This algorithm extends the capabilities of decision trees while addressing their limitations. Its distinctive feature resides in creating an ensemble of decision trees during the training process, with each tree derived from a random sample of the training data, a procedure commonly referred to as bootstrapping. Furthermore, when splitting nodes in the tree, only a random subset of features is considered, which reduces correlation and enhances the model's robustness. To make predictions, Random Forest employs a majority vote mechanism in classification tasks, selecting the most frequently predicted class by individual trees, and it averages tree predictions in regression tasks. A key advantage of Random Forest is its ability to mitigate overfitting. This is achieved by allowing individual trees to be highly customized to the training data, while the combination of predictions from multiple trees results in a more general and accurate model. Additionally, Random Forest provides a measure of feature importance for each attribute used in training, which facilitates feature selection and aids in understanding which variables have the greatest impact on predictions [15, 40].

SVM is a class of ML algorithms widely employed for classification and regression tasks. These models are rooted in convex optimization theory, aiming to find the decision hyperplane that best separates classes within a feature space. The primary goal of SVM is to maximize the margin between data points from different classes, which contributes to the robust generalization of the model to unseen data. In cases where data are not linearly separable, SVM can resort to kernel functions, which map data to a higher-dimensional space, enabling efficient separation of nonlinear classes. This ability to handle complex, nonlinearly separable data makes SVM a powerful tool in various ML and pattern recognition applications [15, 41, 42].

Both RF and SVM are relevant methods in drug design and related applications. They offer different approaches to modeling and solving complex problems, and their selection often depends on the specific characteristics of the data and the nature of the problem at hand. RF's ensemble nature enhances robustness, while SVM's emphasis on finding optimal decision boundaries makes it suitable for classification tasks, even with complex data sets [15, 42].

2.1 Neural Networks

The human brain is a marvel of complexity and power, exhibiting an astonishing capacity to process vast amounts of information in a short period. At its core, the brain operates through intricate networks of neurons, which are responsible for the transmission and processing of information. This remarkable mechanism has inspired numerous researchers to emulate its functionality, particularly the experiential learning process. This endeavor has given rise to what we know as neural networks, also called artificial neural networks (ANNs) [38].

ANNs have been meticulously designed to address a wide range of tasks spanning from the simplest to the most complex, all within a remarkably short period. This ability to solve a broad spectrum of issues promptly stems from the foundational design of ANNs, which closely mirror the biological functioning of the human brain. This structure comprises a multitude of interconnected artificial neurons, which operate together to process information and make decisions. Within the architecture of an ANN, each component performs similarly to the organization of the human brain [38]. To illustrate these components, we will begin by presenting different forms of ANNs, which vary in complexity and operating principles.

2.1.1 Perceptron

As previously pointed out, a perceptron is the simplest model of a neural network, with its main characteristic being that it has only a single layer, hence the name Single-Layer Artificial Neural Network. To understand how it operates, we can break down its core components and processes. Inputs (x) represent the numerical values provided to the perceptron (independent variables), while weights (w) are adjustable parameters that determine the relative importance of these inputs, which are fine-tuned during training. A bias (b) is a constant value added to the weighted sum of inputs, allowing control of the output. The operation of the perceptron can be summarized by the equation for the weighted sum (Z), which is calculated as the sum of the products of inputs and their corresponding weights, plus the bias, as represented in Eq. 4.1:

$$Z = \sum_{i=0}^{n} x_i w_i = \overbrace{x_0 w_0}^{b} + x_1 w_1 + \ldots + x_n w_n. \tag{4.1}$$

A node (neuron), which in the case of a perceptron is singular, serves as a processing unit that receives Z, as input and applies an activation function (f) which in this model is essential a step function. The primary purpose of this activation function is to generate the output, the dependent variable (Y), as established in Eq. 4.2 [43, 44].

$$Y = f(Z) = \begin{cases} 1 \text{ if } Z > 0, \\ 0 \text{ if } Z \leq 0. \end{cases} \tag{4.2}$$

The perceptron's outputs represent the final results, commonly used for tasks like binary classification. However, its main limitation is that, since it only consists of a single neuron, it is primarily suitable for addressing linearly separable problems. Nevertheless, understanding the operation of the perceptron is essential as it lays the foundation for comprehending more intricate neural network models with multiple layers and interconnected neurons. These advanced models have the capacity to express probabilities or continuous values, making them applicable to a wider range of complex tasks [43, 44]. Figure 4.4 illustrates a perceptron's structure, aiding in the visualization of its basic components.

2.1.2 Multilayer Neural Networks

A Multilayer Neural Network or multilayer perceptron (MLP) as the name suggests represents an advanced form of neural network that incorporates the presence of more than one layer of interconnected neurons, resulting in a more intricate architecture when compared to the traditional perceptron. The fundamental structure of an MLP closely resembles that of a perceptron, comprising an input layer that receives input data, one or more hidden layers that process and transform these data, and an output layer that furnishes the final network output following processing. These layers play distinct roles and perform specific functions within the network [45, 46].

The input layer is the initial layer of the network and is designed to accept input data, which may encompass features, characteristics, or information fed into the network for processing. This layer does not engage in any calculations but simply conveys the data to the intermediary layers [45, 46].

The hidden layers are the heart of the MLP and are tasked with processing input data through mathematical operations such as weight multiplication, bias addition, and applying activation functions. These layers are termed "hidden" because their

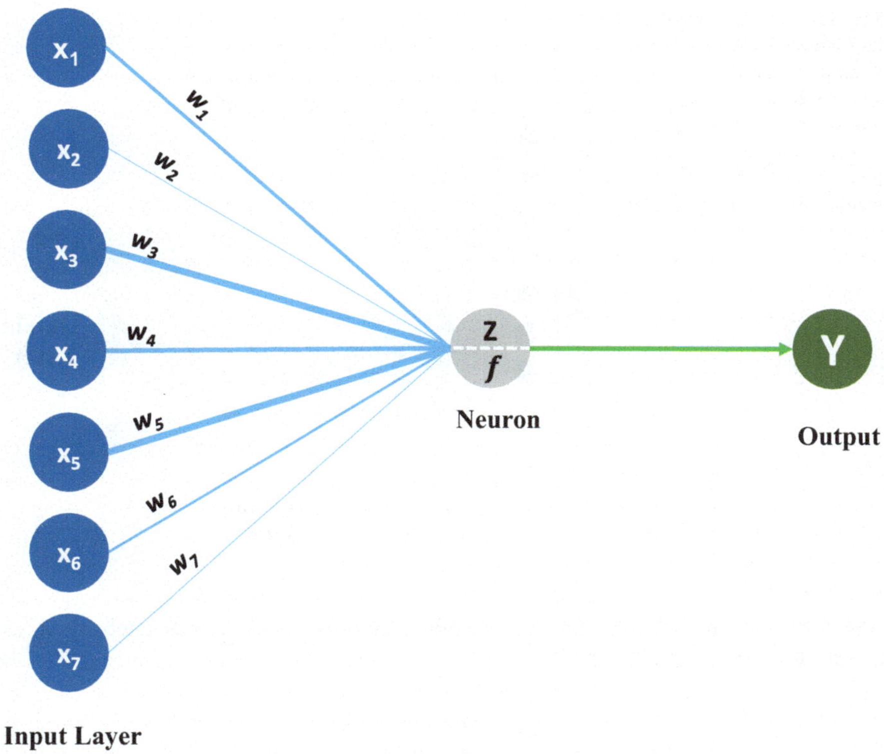

Fig. 4.4 Representative diagram of the architecture of a perceptron

operations are not directly observable in the network training process but are essential in feature extraction and data representation [45, 46].

The output layer constitutes the final layer of the network and generates the ultimate output, which may take the form of a prediction, classification, or any other desired outcome based on the specific problem at hand. The structure and the number of neurons in this layer depend on the nature of the task for which the network is being designed [45, 46]. Figure 4.5 illustrates an example of the generic structure of an MLP, providing a visual aid for understanding its fundamental components.

In the case of these models, the increase in the number of layers and neurons allows for the execution of more complex and nonlinear tasks, as the model's capacity for representation and learning becomes more sophisticated. This is due to the enhanced ability to express and adapt to intricate patterns in the data. As part of this process, new activation functions can be applied, with the most common ones being the ReLU and sigmoid functions [47, 48].

The ReLU (Rectified Linear Unit) activation function is widely used due to its simplicity and effectiveness in overcoming gradient vanishing problems. It

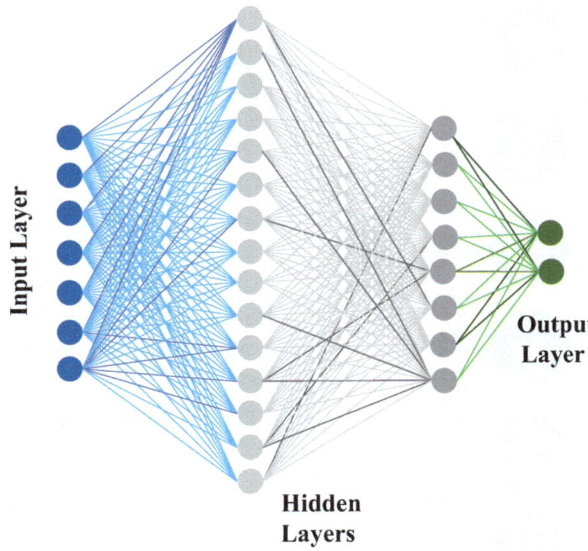

Fig. 4.5 Architecture of a multilayer neural network: 7 input neurons (blue), 15 neurons in the first hidden layer (gray), 8 neurons in the second hidden layer (gray), and 2 neurons in the output layer (green)

transforms negative values into zero while leaving positive values unchanged, as expressed in Eq. 4.3 [47, 48].

$$f(Z) = \begin{cases} Z \text{ if } Z \geq 0, \\ 0 \text{ if } Z < 0. \end{cases} \tag{4.3}$$

On the other hand, the sigmoid function is a classic activation function that is often used in output layers of neural networks, especially in binary classification problems. It maps input values to the range from 0 to 1, which is useful for estimating probabilities [47, 48]. The expression that represents this function is shown in Eq. 4.4.

$$f(Z) = \frac{1}{1 + e^{-z}} \tag{4.4}$$

Moreover, there are other activation functions, such as the hyperbolic tangent (tanH) and the parametric rectified linear unit (PReLU), which also are important in different contexts and can be chosen based on the characteristics of the problem being addressed [47, 48]. Table 4.1 summarizes the other activation functions.

2.1.3 Types of Neural Networks

Neural networks not only vary in the number of neurons and layers but can also take on various formats that directly influence their operating principles, opening a wider range of applications and distinct purposes. This gives rise to a variety of neural

Table 4.1 Mathematical formulations of other examples of activation functions

Name	Equation
Identity	$f(Z) = Z$
TanH	$f(Z) = \frac{2}{1+e^{-2Z}} - 1$
ArcTan	$f(Z) = \tan^{-1}(Z)$
PReLU	$f(Z) = \begin{cases} Z \text{ if } Z \geq 0, \\ \alpha Z \text{ if } Z < 0. \end{cases}$
ELU	$f(Z) = \begin{cases} Z \text{ if } Z \geq 0, \\ \alpha(e^Z - 1) \text{ if } Z < 0 \end{cases}$
SoftPlus	$f(Z) = \ln(1 + e^Z)$

network architectures [38, 45]. Figure 4.6 illustrates the most well-known types of neural networks.

It is possible to observe in Fig. 4.6 that neural network methods vary fundamentally in shape, neuron composition, operating direction, and activation functions. The purpose of this chapter is not to explain how all these methods work, so we will highlight some more widely used methods in the field of drug discovery. For this, we will divide them into two categories: Feedforward and Recurrent Neural Network.

Feedforward

The feedforward models are a fundamental category in the field of ANNs due to their simple and effective approach. Essentially, they follow the principle of a unidirectional flow of information, from input to output, without the formation of cycles or loops. This characteristic makes feedforward methods especially suitable for handling tasks that do not involve temporal or sequential dependencies, such as classification tasks. The perceptron itself, one of the pioneering concepts in the field of neural networks, is a classic example of this approach [38, 46, 49].

The operation of a feedforward method is relatively simple but effective. Initially, a set of input information is provided to the model. This information is then processed by a series of layers of neurons. Each neuron in a layer performs a weighted linear combination of the input values and applies a nonlinear activation function. This activation function is essential because it introduces nonlinearity into the network, allowing it to capture complex relationships in the data. This is important for the ability of neural networks to model real-world problems, which often involve nonlinear relationships [38, 46, 49].

The feedforward approach encompasses a variety of algorithms and architectures that are widely used in different application domains. Some notable examples include Radial Basis Function (RBF) networks and Convolutional Neural Networks (CNN) [38, 45, 50].

RBF

The structure of an RBF is typically the same as that of a Multilayer Perceptron, but it utilizes radial functions as activation functions [49], as described in Eq. 4.5.

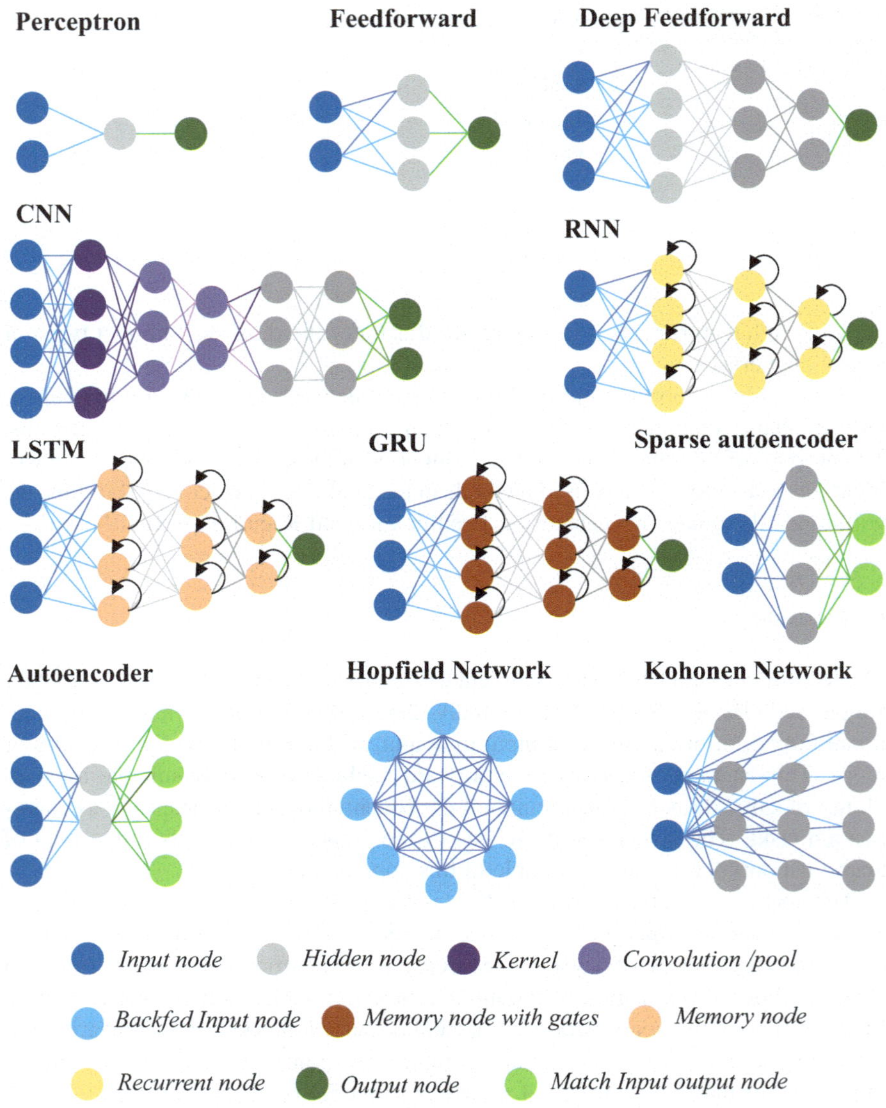

Fig. 4.6 Architecture representation of some types of neural networks

$$f(\boldsymbol{x}) = \sum_{i=1}^{n} w_i \varphi(\|\boldsymbol{x} - \boldsymbol{c}_i\|) \tag{4.5}$$

In this context, where the input is modeled as a vector $\boldsymbol{x}$ from the set of real numbers of N dimensions, n is the number of neurons in the hidden layer, c_i is the center vector for neuron i, and w_i is the weight of neuron i in the linear output neuron.

Functions that depend only on the distance from a center vector are radially symmetric about that vector, hence the name radial basis. In the basic form, all inputs are connected to each hidden neuron. The norm is typically taken to be the Euclidean distance, and the radial basis function is commonly taken to be Gaussian [50].

The main characteristic of RBF networks is their ability to approximate complex functions with an approach based on the nonlinear representation of data. This is particularly useful in problems where the relationship between input and output data cannot be easily modeled by linear functions. Additionally, RBF networks have the ability to adapt automatically to training data, adjusting the weights and centers of the neurons in the hidden layer [45, 50].

CNN

A Convolutional Neural Network stands as a widely employed neural network architecture in tasks linked to image and video processing. This technology proves indispensable in fields such as computer vision, pattern recognition, and ML applied to visual data [51–53].

Drawing inspiration from the workings of the human visual cortex, CNNs operate through convolutional layers, responsible for applying convolution operations to input images. This convolution entails the use of filters to extract features such as edges, textures, and relevant patterns in the image. Following the convolutional layers, pooling layers come into play, tasked with reducing the spatial dimension of feature maps, and preserving crucial information while decreasing the network's parameter count. Proceeding with the process, fully connected layers are employed to make decisions based on the previously extracted features. These layers resemble those found in traditional neural networks, where outputs are computed using adjustable weights [51–53].

Activation functions, such as ReLU, are applied at each layer to introduce non-linearities into the network, enabling the learning of more complex representations. Additionally, normalization layers, like Batch Normalization, can be incorporated to enhance the network's training [51–53].

Recurrent Neural Networks

Recurrent Neural Networks (RNNs) constitute a type of neural network architecture designed to handle sequential or temporal data. Unlike conventional feedforward networks, RNNs feature feedback connections, allowing the output of a layer to be fed back into the input of the same layer. This capability enables the network to maintain an internal memory [49, 54–56].

The key concept in RNNs is their ability to process sequences of data. Each unit in an RNN layer retains an internal state updated based on the current input and the previous state, enabling the network to capture contextual information over time.

However, traditional RNNs face certain limitations, such as the vanishing or exploding gradient problem, which hinders the training of long sequences. To overcome these issues, more advanced variants have emerged, such as Long Short-Term Memory (LSTM) and Gated Recurrent Unit (GRU). These architectures introduce memory control mechanisms that aid in preserving and forgetting information more effectively [54–57].

LSTM

Long Short-Term Memory (LSTM) Networks emerge as a solution to the challenges faced by conventional RNNs when dealing with sequential data. While RNNs are widely employed for such tasks, they often struggle to capture long-term relationships. LSTMs offer a more effective approach, enabling the network to memorize and access relevant information across extensive sequences, proving particularly valuable in contexts where understanding temporal relationships is central [49, 57, 58].

The basic structure of an LSTM includes memory units called cells, each equipped with three gates: the input gate, the forget gate, and the output gate. These gates function as control mechanisms that regulate the amount of information that enters, is forgotten, or is transmitted [57, 58].

The LSTM operates at each time step, making decisions based on the characteristics of sequential data and the previous memory. This allows the network to learn complex temporal dependencies over time [57, 58].

The fundamental advantage of LSTMs lies in their ability to selectively retain and discard information. This flexibility makes LSTMs effective in modeling patterns in sequential data, applicable in various areas such as pattern recognition, machine translation, and time series prediction [57, 58].

GRU

Similar to LSTMs, Gated Recurrent Units, or GRUs, were developed to overcome the limitations of traditional RNNs. GRUs simplify the structure of LSTMs by combining the concept of memory units and gates into a single unit. They retain the ability to capture long-term information but with a more streamlined architecture [49, 57, 59].

In a GRU, we also find two main gates: the update gate and the reset gate. The update gate determines how much of the previous memory should be retained, while the reset gate decides how much of the previous information should be forgotten. The simplicity of GRUs compared to LSTMs can result in more efficient computation and faster training in certain situations [57, 59].

2.2 Deep Learning

Deep learning, within the broader context of ML, is an approach that surpasses the limitations of conventional neural networks by incorporating three or more layers.

While traditional ML can be likened to a neural network with a single layer, deep learning expands upon this architecture, mimicking to some extent the complexity of the human brain [5, 18, 26, 33, 34].

By adding hidden layers, regardless of the method employed, the neural network gains the ability to learn and generalize from vast amounts of data. While a single neural layer can make basic predictions, additional layers enhance and refine the model's accuracy. This expansion not only increases the system's complexity but also intensifies its aptitude for learning [5, 18, 26, 33, 34].

As the neural network deepens, its capacity to extract abstract and hierarchical features from data significantly improves. This feature enables the model to identify complex patterns, contributing to a deeper and more refined understanding of the data. The richness of data used during training not only fortifies the model but also empowers it to handle a broader range of situations, making it more versatile and adaptable [5, 18, 26, 33, 34].

Comparatively, while traditional ML may be effective for specific tasks with smaller data sets, deep learning excels in scenarios where the complexity and diversity of the data demand a more sophisticated approach [5, 18, 26, 33, 34].

3 Using Machine Learning

ML and NNs, as discussed earlier, exhibit a variety of approaches in terms of operation, structure, and purposes, offering a wide range of application possibilities. To employ ML in a specific application, it is essential to thoroughly understand the structure of the involved stages. In summary, these stages involve defining the problem, data collection, data preprocessing, model selection, model training, validation, model tuning, and prediction (Fig. 4.7). Each of these steps is of fundamental importance in the development and successful application of solutions based on this approach.

3.1 Problem Definition

In the ML journey, the initial phase of defining the problem, despite its apparent simplicity, assumes elemental importance, serving as an essential foundation for the project's success. At this stage, it is imperative to cultivate a deep understanding not only of the available data but also of the inherent peculiarities within the domain where the model will be applied and the challenges to be faced.

Establishing a specific goal for the project is fundamental. Answering fundamental questions such as "What are we seeking to predict or classify?" is necessary, delineating the precise task that the model must perform. Additionally, it is necessary to identify the type of problem at hand, whether it is classification, regression, clustering, or anomaly detection.

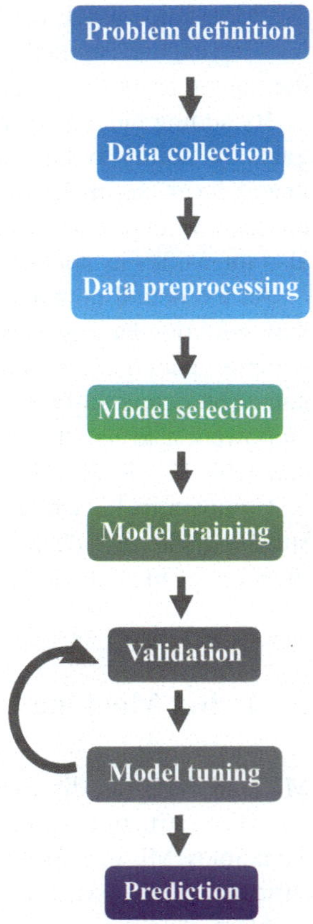

Fig. 4.7 Basic steps to build a machine learning model. Data management (blue), model management (green), validations and controls (gray), and final stage (purple)

Let us take an example. Initially, understanding the nature of the target disease, its molecular characteristics, and underlying biological mechanisms is elementary. From this understanding, the decision about what to do follows target discovery, ligand search, prediction of biological activity, or evaluation of ADMETox (Absorption, Distribution, Metabolism, Excretion, and Toxicology) properties. Careful analysis will determine whether it is appropriate to employ ML for any of these objectives.

Furthermore, it is necessary to define whether the appropriate approach would be regression, classification, or another method. For example, in ligand search, classification techniques may be suitable, while in property prediction, regression may be more indicated.

The decision about the type of problem and the specific task will guide the choice of the most appropriate ML algorithm. This selection is fundamental to solidifying the foundations of the successful development of the scientific project at hand.

3.2 Data Collection

Data collection is a fundamental element in the construction and performance of any model. The quality and quantity of the provided data are primary in the effectiveness of the resulting model. It is essential to consider various factors during this process to ensure the robustness and relevance of the information used [60–64].

Firstly, the quantity of data is a determining factor. The more data made available to the model, the greater its learning and generalization capabilities. However, the pursuit of quantity should not compromise quality. It is important to ensure that the collected data are relevant to the specific context of the problem at hand [60, 61].

Data representativeness is equally essential, especially in sampling cases. The data should faithfully reflect the diversity and complexity of the environment or phenomenon that the model aims to address. Otherwise, the model may develop biases and may not be able to handle situations outside the scope of the provided data [60, 61].

The choice of the database is a strategic step in data collection. Different types of data can be stored in various formats, and the selection of the appropriate database will directly influence the efficiency of the model. The structure and flexibility of the database should be considered to ensure effective data handling throughout the process. In the field of drug discovery, there are various databases designed for a variety of applications, which are highlighted in Table 4.6 under the "Resources and Tools" section.

3.3 Data Preprocessing

Data preprocessing is an integral step in ML, as the quality of data and the useful information derived from it directly affect the model's learning ability. In general, all raw data undergo several processing stages to enhance its quality. These stages can be simple steps that can be automated or even performed manually in some cases. These steps can be listed as follows:

Data Cleansing This step involves removing missing, inconsistent, or irrelevant data from data sets. Tasks within this phase may include eliminating records with missing values, addressing outliers, and rectifying data errors. For example, in a database of active compounds with various properties, some molecules may lack Log P values. It is essential to remove them because inferring that these values are null can significantly impair the model's performance. Moreover, outliers in molecular data can distort the overall analysis. For instance, an outlier could represent a compound with exceptionally high or low binding affinity to a target protein [65].

Normalization or Standardization It is often necessary to normalize or standardize data to ensure uniform scaling across all characteristics. This measure is important in preventing features with vastly different magnitudes from exerting undue

Table 4.2 Example of one-hot coding for types of molecules

Molecule type	Small molecule	Peptide	Antibody
Small molecule	1	0	0
Peptide	0	1	0
Antibody	0	0	1

influence during the training process. Consider a data set that includes molecular descriptors such as molecular weight, lipophilicity, and binding affinity to a target protein. These descriptors may naturally exhibit different scales, with molecular weight measured in Daltons, lipophilicity represented by log P values, and binding affinity often quantified in terms of IC_{50} or K_i values. During the normalization phase, each molecular descriptor is transformed to a common scale, scaling the values of each descriptor to a range between 0 and 1, making them comparable regardless of their original units [66]. This is very important because features with larger scales might otherwise dominate the learning process, potentially overshadowing the significance of other descriptors.

Encoding Categorical Variables When the data set includes categorical variables, such as different molecule types or protein categories, it becomes necessary to encode them into numerical values, often achieved through techniques like one-hot coding. Consider a data set containing information about various drug molecules, where a categorical variable represents the type of molecule, such as small molecules, peptides, or antibodies. To incorporate this categorical information into an ML model, one-hot coding is applied. In this technique, each category is assigned a binary value, and a new binary column is created for each category. The presence of a specific category is indicated by a "1" in the corresponding column, while the absence is denoted by a "0" [67]. An example of this process is shown in Table 4.2.

In this example, the categorical variable "Molecule Type" has been encoded into three binary columns using one-hot coding. Each row now represents a drug molecule and indicates its type through the binary columns. This encoding ensures that categorical variables do not introduce ordinal relationships that may mislead the ML model. By representing categorical information in a numerical format, the model can effectively incorporate these features into its learning process, contributing to more accurate predictions.

Dimensionality Reduction In high-dimensional data sets, such as those generated in genomics studies, it can be useful to apply dimensionality reduction techniques such as PCA (Principal Component Analysis) or t-SNE (t-distributed Stochastic Neighbor Embedding). For instance, consider a genomics data set where each sample is characterized by the expression levels of thousands of genes. The high dimensionality of these data can make it computationally intensive and may lead to the curse of dimensionality, where the model's performance decreases as the number of features increases. In the case of PCA, the algorithm identifies the principal components, which are linear combinations of the original features that capture the

Table 4.3 An example of the original data set

Sample	Gene 1	Gene 2	...	Gene 1000
1	0.5	1.0	...	0.8
2	0.8	0.7	...	0.2
...	...	...	...	...
N	0.9	0.6	...	1.0

Table 4.4 The same data set after applying PCA (reduced to 2 PCs)

Sample	PC1	PC2
1	0.3	0.5
2	0.1	0.8
...	...	...
N	0.6	0.2

maximum variance in the data. By retaining a subset of these principal components, one can effectively reduce the dimensionality of the data set while preserving the most critical information. This not only makes the data set more manageable but also helps in identifying patterns and relationships between samples. Similarly, t-SNE is a nonlinear dimensionality reduction technique that focuses on preserving the pairwise similarities between data points. It is particularly useful for visualizing high-dimensional data in lower-dimensional spaces, revealing the underlying structure and clusters within the data set [68, 69]. Tables 4.3 and 4.4 show a simplified example with an original data set and after applying PCA, respectively.

After applying PCA, the data set might be represented using a reduced set of principal components.

In this transformed representation, the dimensionality has been reduced to two principal components (PC1 and PC2).

Data Sampling This technique is used to enhance the performance of ML models when there is a significant imbalance between classes of interest. For instance, when predicting the biological activity of chemical compounds, it is not uncommon to encounter data sets where inactive compounds significantly outnumber their active counterparts. To mitigate this issue, two prevalent data sampling techniques are commonly used: undersampling and oversampling. Undersampling involves reducing the number of instances belonging to the majority class (in this case, inactive compounds) to match the number of instances of the minority class (active compounds). This approach helps balance the class proportions in the data set, ensuring that the model is not overwhelmed by the abundance of instances from the majority class. While undersampling can lead to a reduction in the amount of data available for training, it helps prevent the model from being biased toward the dominant class. Conversely, oversampling entails generating additional copies of instances from the minority class (active compounds) to balance the class proportions. This measure is implemented to ensure that the model has sufficient exposure to instances of the minority class, preventing it from exhibiting bias toward the majority class. Techniques such as duplicating existing instances, generating synthetic samples (using methods like SMOTE—Synthetic Minority Over-sampling Technique), or other sophisticated oversampling methods are employed to augment the representation of the minority class in the data set [70, 71]. Consider a data set for predicting the biological activity of chemical compounds, where only 10% of the compounds are labeled as active, while the remaining 90% are inactive. To address this class imbalance, an undersampling approach would involve randomly selecting 10% of the instances from the inactive class, creating a balanced data set. On the other hand, an oversampling approach might generate additional synthetic instances for the active class to match the size of the inactive class, ensuring that the model is exposed to a more balanced representation of both classes during training.

Feature Selection It is the process of discerning the most pertinent features for the specific task at hand. The primary goals of feature selection are to reduce data dimensionality and enhance model performance by focusing on the most informative attributes. In various data sets, especially those with a large number of features, not all features contribute equally to the predictive power of a model. Some features may be redundant, irrelevant, or even introduce noise, leading to overfitting. Feature selection helps address these issues by retaining only the most significant features, thereby streamlining the data representation and improving the efficiency and effectiveness of the ML model [72, 73]. For example, consider a data set for predicting disease outcomes based on patient profiles. The data set may include numerous features such as age, gender, blood pressure, cholesterol levels, and genetic markers. Feature selection would involve identifying which subset of these features is most informative for accurately predicting the disease outcome. By focusing on the most relevant features, the model becomes more interpretable, computationally efficient, and less prone to overfitting. The process of feature selection can be approached in various ways, including filter methods, wrapper methods, and embedded methods. Filter methods assess the relevance of features independently of the chosen ML algorithm, wrapper methods use the model's performance as a criterion for feature selection, and embedded methods incorporate feature selection as an integral part of the model training process.

Separation of Training, Validation, and Test Sets This practice is important to ensure robust evaluation and optimization of model performance. In this process, the data set is divided into three distinct subsets. The Training Set is used to train the model, allowing it to learn patterns and relationships in the data. During this stage, the model's parameters are iteratively adjusted to improve the accuracy of predictions. The Validation Set plays a key role in hyperparameter tuning and model evaluation during training, which will be discussed later. The Test Set is reserved for the final evaluation of the model's performance. It provides an unbiased assessment of how well the model generalizes to new, unseen data. After training and hyperparameter tuning, the model is evaluated on the Test Set to measure its ability to make accurate predictions in real-world scenarios. In drug discovery, allocating approximately 70–80% of the data to the training set, 10–15% to the validation set, and another 10–15% to the test set is a common practice [74]. The specific percentages may vary based on data characteristics, but this general split provides a balance between training the model effectively and assessing its generalization capabilities in drug discovery contexts.

Treatment of Temporal Data When dealing with temporal data, such as time series records of biological activity over time, it is essential to appropriately handle the data, taking into account temporal dependencies. Time series data often exhibit patterns and dependencies over sequential observations. To address this, techniques like lag features, rolling statistics, and time-based cross-validation can be employed. Lag features capture historical values, rolling statistics summarize trends, and time-based cross-validation ensure evaluation reflects the temporal nature of the data. Additionally, considering seasonality, trend decomposition, and incorporating time-aware models, like RNNs or LSTMs, enhances the modeling of temporal dynamics in biological activity data sets [75].

3.4 Model Selection

In the application of ML in the context of drug discovery, the model selection stage is important for identifying promising drug candidates. Let us consider a data set describing the molecular properties of different compounds and their effectiveness in treating a specific disease. In this scenario, choosing the appropriate model can be crucial for accurately predicting the biological activity of new compounds.

Simpler models, such as linear regression, may be used for problems where the relationships between molecular features and drug efficacy are predominantly linear. However, in more complex cases where molecular interactions are nonlinear and involve a variety of factors, more sophisticated models, such as NNs or DL methods, may be more appropriate.

Model selection should consider the ability to handle molecular nuances such as specific protein interactions, the three-dimensional structure of the molecule, and other complexities inherent in biochemistry. Experimentation with different algorithms and parameter adjustments becomes even more relevant, as drug discovery often requires highly specialized models tailored to the specific characteristics of the problem at hand [2–5, 38, 73]. The topic of applications addresses which environments in drug discovery are suitable for the most appropriate ML algorithms.

3.5 Model Training

The training stage in ML models is fundamental for empowering the algorithm to perform specific tasks based on the provided data. During this phase, the model is exposed to labeled data sets, where inputs (features) are associated with known outputs. The goal is to adjust the model's parameters to minimize the difference between the predicted outputs and the actual outputs present in the training data.

The training process can be divided into several iterations known as epochs. In each epoch, the model goes through the entire training set, makes predictions for each input, and compares these predictions with the actual outputs. Based on this discrepancy, an optimization algorithm adjusts the model's weights and biases, aiming to reduce error and improve prediction accuracy [38, 46, 74].

The loss function is notable in this context, quantifying the discrepancy between the model's predictions and the actual labels. During training, the objective is to minimize this loss function, resulting in a more accurate and generalizable model [76].

The backpropagation technique is widely used in this process. It involves calculating the gradient of the loss function with respect to the model's parameters, allowing weighted adjustments during the optimization phase [77]. This is essential for updating the weights of connections between units in a neural network.

It is important to mention that the batch size is also a critical aspect. Training can be performed in batches of data, where the model is adjusted based on a subset of the training set [78]. This not only reduces computational requirements but also introduces a form of regularization that can benefit the model's generalization to new data.

Suppose we have a data set of molecular information about compounds and their biological activities concerning a specific target such as a protein associated with a disease. During training, the model would be exposed to this set, adjusting its parameters to learn complex patterns that relate molecular features to biological activities. The loss function would be applied to assess the discrepancy between the activities predicted by the model and the actual activities observed in the training data. A common example would be the use of Mean Squared Error (MSE) as the loss function.

3.6 Validation

During the model evaluation phase, various methods are applied to ensure a comprehensive and accurate analysis of performance on unseen data. Initially, test data sets, consisting of examples not used during training, are employed. The model makes predictions for these data, and the discrepancy between predictions and actual labels provides a direct measure of its generalization capability.

In the context of classification, metrics such as precision, recall, and F1-score are fundamental. Precision measures the proportion of correct predictions to the total predictions, while recall (or sensitivity) assesses the proportion of true positives to the total actual positives. The F1-score provides a harmonic average between precision and recall, proving useful in situations with class imbalance. The confusion matrix is a visual tool that exposes correct predictions and confusion between classes [79].

For regression problems, metrics such as Mean Absolute Error (MAE) and Mean Squared Error (MSE), along with Y-randomization, are employed. MAE represents the average of absolute differences between predictions and actual labels, while MSE measures the average of squared differences between predictions and actual labels. Y-randomization is a technique used to assess the robustness of a regression model, evaluating whether the model captures the relationship between independent and dependent variables or merely adjusts to the training data. The technique involves randomizing the values of the dependent variable while keeping the values of independent variables unchanged. The model is then trained on the randomized data, and performance is compared with the model trained on the original data. If the model trained on randomized data performs similarly to the model trained on the original data, it suggests that the model is not capturing the relationship between independent and dependent variables [73, 79–82].

Cross-validation, such as the K-Fold Cross-Validation method, is another indispensable approach. This method divides the data set into K parts, trains the model on K-1 parts, and evaluates one part. This process is repeated K times, and the average of performance metrics provides a more robust insight into how well the model generalizes [60, 61].

Learning curves are valuable for understanding how the model's performance varies with the size of the training set. This helps determine whether the model would benefit from more data or has reached a saturation point.

ROC (Receiver Operating Characteristic) curves and AUC (Area under the Curve) are often used in classification problems. The ROC curve represents the true positive rate versus the false-positive rate for different decision thresholds. The AUC is a metric quantifying the discriminative ability of the model, with a higher AUC being desirable [83].

Residual analysis is applied to understand the differences between the model's predictions and actual values. Observing patterns in residuals can indicate trends or structures not captured by the model. Additionally, Bootstrap is a resampling technique that generates multiple samples with replacements from the original data set. This approach can be useful for evaluating the stability of performance estimates [80, 81].

Statistical tests are employed to assess whether performance differences between models are statistically significant. This is especially relevant when comparing alternative models to identify the most effective one.

In drug discovery, these methods are essential to assess the model's effectiveness. The choice of specific methods depends on the characteristics of the problem at hand, but the joint application of these techniques is necessary for a robust and reliable model evaluation.

3.7 Tuning

The model tuning stage, also known as hyperparameter tuning, takes place after the validation phase in the development process of the ML model. While validation is crucial for assessing the model's performance on a separate data set not used during training, the tuning stage aims to further optimize the model's performance by adjusting its hyperparameters [84].

Hyperparameters are external elements to ML models, essential for influencing their performance but not adjusted during training. Proper selection of these parameters is crucial to ensure a well-performing model. Below are some common hyperparameters and considerations on how to adjust them. The learning rate controls the size of steps taken during optimization. It can be manually adjusted, starting with a small value. To find a learning rate that results in stable and efficient training, grid search or random search methods can be used. Grid search tests combinations of hyperparameters in a predefined grid, while random search tests combinations randomly. The number of epochs represents how many times the algorithm goes through the entire training set during training. If the model is overfitting, it may be necessary to reduce the number of epochs. Batch size refers to the number of training samples used in one iteration. Larger sizes can speed up training but demand more memory. Smaller sizes may result in smoother convergence [84–86].

In the neural network architecture, the number of layers and neurons in each layer is a significant hyperparameter. It is often advisable to initiate with a simple architecture and gradually escalate complexity based on the specific demands of the task. Employing cross-validation becomes essential in this iterative process to identify the architecture that exhibits optimal generalization performance across diverse data sets [87].

To mitigate the risk of overfitting in NNs, regularization-related hyperparameters, such as L1 and L2 regularization terms, are highly recommended. These terms introduce penalties for large weight magnitudes, promoting a more generalized model. Experimentation with different values for regularization terms is recommended, with larger values intensifying the penalty and aiding in the prevention of overfitting [87–89].

Another pertinent hyperparameter is the dropout rate, denoting the percentage of neurons randomly "turned off" during training to enhance robustness and prevent overfitting. Commonly ranging from 0.2 to 0.5, the dropout rate can be finetuned empirically based on the characteristics of the data. Careful adjustment of this parameter contributes to the network's ability to learn robust features while avoiding reliance on specific neurons, thereby promoting better generalization to unseen data [87].

In ensemble models like RF or gradient boosting, the number of trees is a significant hyperparameter. A larger number generally improves performance, but there is a point of diminishing returns. The maximum tree depth is relevant in ensembles. Deeper depths in a model can potentially result in overfitting. Therefore, it is advisable to experiment with various depth values and employ cross-validation as a means of assessing performance [90].

In the context of KNN (K-Nearest Neighbors) algorithms, the choice of the number of neighbors is an elementary hyperparameter. Opting for small values can make the model excessively sensitive to noise, potentially leading to overfitting. On the other hand, selecting large values may result in an overly smoothed model, potentially missing important patterns in the data [91].

In the case of SVM, the kernel type (e.g., linear, polynomial, and radial) and its associated parameters are determined in model performance. The kernel defines the transformation applied to the input data, and variations in kernel types and parameters can significantly affect the model's ability to capture complex relationships within the data. Therefore, a thoughtful exploration of different kernel types and their respective parameter settings, possibly through techniques like grid search, is essential to ensure the SVM model is appropriately configured for the specific characteristics of the data set at hand [92].

It is important to emphasize that the evaluation and parameter tuning process take place through iterative cycles. These cycles encompass ongoing analysis of the model's performance on validation data sets, facilitating the detection of potential enhancements. Following each cycle, the validation process is concluded by evaluating the final model using an independent test set. The inclusion of this test set is crucial to guarantee that the model not only adapts well to the training data but also demonstrates effective generalization to new data. This ensures the robustness and reliability of the model across diverse scenarios [73, 80, 85, 86].

3.8 Prediction

The prediction phase in the ML process is the step where the trained model is used to make predictions or inferences on unseen data. After completing the training, validation, and hyperparameter tuning steps, the model is ready to be applied to new data for making predictions or classifications.

During the prediction phase, input data are fed into the model, and the model utilizes the patterns learned during training to generate predictions or inferences. The specific nature of the prediction task can vary widely, depending on the type of problem being addressed. Additionally, the interpretability of the model can be explored to understand how the model makes decisions. In critical tasks, such as in healthcare, the interpretability of the model is often as important as predictive performance. The prediction phase is the culmination of the ML process, transforming the knowledge acquired during training into practical insights and actions for real-world applications.

4 Limitations

ML is a powerful approach to solving a variety of complex problems, but it is determined to recognize and understand its limitations. Like any tool, ML has constraints that can affect its effectiveness in different contexts. The following are some of the most significant limitations associated with ML.

4.1 Bias

A significant limitation of ML lies in the unavoidable presence of biases. These biases can be inadvertently incorporated into models due to the nature of the data used in training. When data sets reflect historical inequalities or mirror the biases of their creators, ML algorithms tend to reproduce and, in some cases, amplify these biases [93, 94].

The source of biases often stems from human decisions during the data collection, feature selection, and labeling process. Additionally, ML algorithms can unintentionally magnify existing biases in the data, resulting in discriminatory outcomes in areas such as drug efficacy predictions, patient stratification, or adverse effect profiling [93–95].

The complexity of these issues is exacerbated by the fact that ML algorithms often operate as black boxes, making it difficult to understand how decisions are made. This makes it challenging to identify and rectify biases, especially when results are presented without a clear explanation [93–95].

Mitigating biases in ML models requires a holistic approach that involves developers' awareness of ethical challenges, the implementation of more equitable data collection practices, and the exploration of advanced methods to identify and correct biases existing in models [95].

4.2 Overfitting and Underfitting

Overfitting and underfitting are two opposing phenomena that can occur when training ML models. Overfitting occurs when a model is trained too well on the training data but fails to generalize to new, unseen data. In other words, the model "learns" the training data so well that it ends up capturing specific patterns from that set, including noise, instead of learning general patterns that would apply to other data sets. This often results in lower performance when the model is confronted with data that were not used during training [89, 93–95].

On the other hand, underfitting occurs when a model is too simple to capture the complexity of the training data. This means that the model fails to learn the underlying patterns and cannot adjust adequately to the training data. As a result, the model performs poorly on both the training data and new data [96–98].

Finding the right balance between overfitting and underfitting is fundamental for developing an effective ML model. This often involves adjusting the model's complexity using techniques such as regularization, data augmentation, cross-validation, and hyperparameter tuning. Cross-validation, for example, can help evaluate the model's performance on unseen data and fine-tune its behavior to ensure proper generalization [93–95].

4.3 *Interpretability*

Interpretability is an essential characteristic in ML models, referring to the ability to understand and explain the decisions made by the model. As complex algorithms, such as deep neural networks, gain popularity, the ability to interpret how and why a model makes a particular prediction becomes increasingly important. The lack of interpretability in ML models can be a barrier, especially in critical applications such as target identification, drug design, and pharmacokinetic profile. The inability to explain model decisions can lead to distrust in the results, hindering the acceptance and adoption of these technologies in sensitive environments [93–95, 99].

Several challenges contribute to the lack of interpretability. Complex models, like deep neural networks, often operate as black boxes, making it difficult to understand how a specific input translates into an output. Additionally, interpretability is often sacrificed in pursuit of performance, especially in more advanced models. Approaches to improve interpretability include the use of simpler models, such as decision trees, which can be more easily understood. Additionally, post hoc interpretability methods, such as Lime and SHAP, have been developed to provide insights into model decisions, even when they are intrinsically complex [100, 101].

Interpretability is not just a technical concern but also an ethical issue. In sectors where decisions directly impact people's lives, it is decisive that users can understand and trust the model predictions. Therefore, advancing research and practices in interpretability is essential to ensure the responsible and ethical use of artificial intelligence [100, 101].

4.4 *Computational Cost*

Computational cost is a decisive consideration in many aspects of computer science and software development. It refers to the resource-intensive nature of resources such as the processing time of the Central Processing Unit (CPU), memory, and energy associated with the execution of operations or algorithms on a computer system. As problems and data sets addressed by ML algorithms increase in scale and complexity, computational cost becomes a central concern. Advanced models, such as deep neural networks, may require specialized hardware and substantial computing resources for efficient training.

In addition, the utilization of big data, particularly in virtual high-throughput screening and deep learning applications, especially CNNs, necessitates efficiency to ensure prompt and effective responses. Strategies for addressing computational costs encompass algorithm optimization, task parallelization, judicious selection of hardware architectures, and the implementation of techniques such as pruning (eliminating insignificant connections in neural networks) to streamline model complexity [102].

The evolution of hardware technology, such as Graphics Processing Units (GPUs) and Tensor Processing Units (TPUs), has played a significant role in mitigating computational costs in data-intensive tasks like ML model training. Computational cost is not just a technical consideration but also has economic implications, especially in cloud environments where resources are often paid for based on usage. Therefore, computational efficiency becomes necessary for optimizing costs and ensuring the economic viability of ML-based solutions [102, 103].

4.5 Data Dependency

Data dependency is a central concept in ML, emphasizing the critical influence that input data have on the performance and effectiveness of models. The quality and representativeness of the data used during model training directly impact the model's ability to generalize to new, unseen data. If the training data are not representative of the real-world domain of the problem or is biased in some way, the model may fail to make accurate predictions in real-world situations [93–95].

Furthermore, data dependency is also related to the need for sufficient data to train models effectively. Complex models, such as deep neural networks, often require large data sets to learn meaningful patterns and avoid overfitting. The quality and diversity of data are also determining for addressing potential biases. If the data reflect existing inequalities or prejudices, the model may perpetuate or amplify these biases, leading to unfair or discriminatory outcomes. Data dependency is not limited to the training phase; it is also relevant during the inference phase when the model makes predictions or decisions based on input data. If the input data are of low quality, incorrect, or incomplete, the model's predictions can be inaccurate or inappropriate [93–95].

4.6 Robustness

Robustness in ML refers to the ability of a model to make accurate and consistent predictions across a variety of conditions and situations. However, ML models face several limitations in terms of robustness. One of the primary limitations is sensitivity to perturbations in input data. Small changes or noise in the data can lead to significant variations in model predictions, especially in complex models like deep neural networks. This makes models more prone to errors when confronted with data that slightly differ from the training data [93–95].

Furthermore, the robustness of ML models is often compromised by adversarial attacks. Adversaries can intentionally manipulate input data subtly to deceive the model and induce errors. This vulnerability to adversarial attacks is a significant concern in critical applications such as security, health, and finance. Another limitation is related to the distribution of data. If the model is trained on a data set that does not fully represent the real-world application domain, it may struggle to generalize to new data, resulting in a drop in performance. Additionally, robustness is also affected by the presence of outliers in the data. Models sensitive to outliers may have impaired performance when faced with atypical or extreme examples [93–95].

5 Applications in Drug Discovery

The drug discovery process is marked by complex stages, precisely because it deals with biological systems that possess numerous pieces of information, many of which are often not completely mapped. This information is now being systematically measured and mined at unprecedented levels using a plethora of 'omics techniques (proteomics, genomics, epigenomics, metabolomics, and transcriptomics). Certainly, these areas have a high volume of data and variables, and therefore, the application of ML techniques is of great importance [14, 15, 38, 99, 104].

In the drug development stages, one can highlight the identification of targets, lead discovery, pre-clinical development, and clinical development. ML algorithms are applied in all these stages, each with a distinct function and interest, with their applicability defined by the specific characteristics of each algorithm. Figure 4.8 illustrates this relationship in a summarized way [14, 15, 38, 99, 104].

5.1 Target Identification

Target identification is an essential step in the drug discovery process, involving the identification and validation of specific biological molecules or pathways that play a key role in a disease. This intricate process requires a comprehensive understanding of the molecular mechanisms underlying the pathological condition. Researchers employ various techniques such as genomics, proteomics, and bioinformatics to pinpoint potential targets, seeking proteins or genes that are aberrantly expressed or dysregulated in the diseased state. Successful target identification lays the foundation for the development of targeted therapies, allowing for the design of drugs that precisely intervene in the disease-associated pathways, potentially leading to more effective and tailored treatments with minimal side effects [14, 15, 20, 99].

Undoubtedly, the process of target identification involves establishing a causal link between the identified target and the associated disease, necessitating the

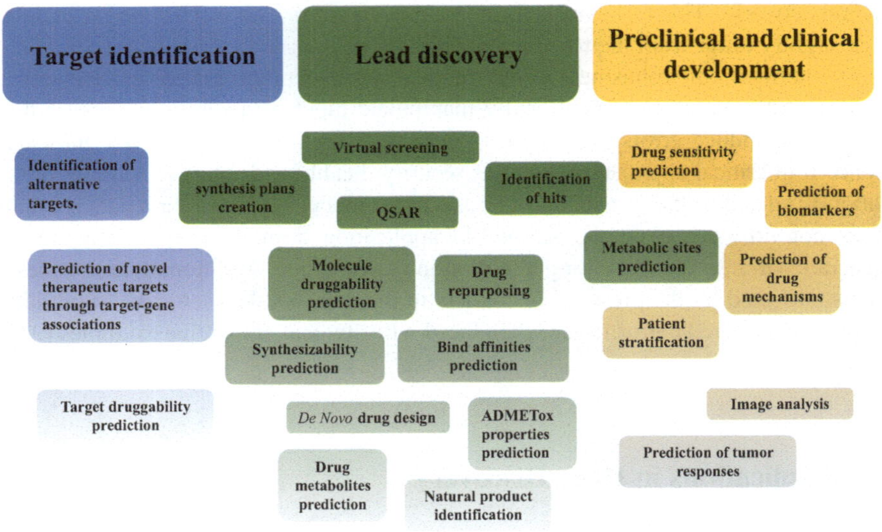

Fig. 4.8 Mainly applications of ML algorithms in the drug discovery process

demonstration that manipulating the target yields a discernible impact on the disease. This approach demands diverse data sets with varying dimensionalities and magnitudes. In addition to high-reliability literature data, it requires comprehensive omics data representing both disease and normal states. The application of ML algorithms further facilitates the treatment, understanding, and elucidation of the causal relationship between the disease and the target [14, 15, 20, 99].

In pursuit of this objective, commonly employed ML algorithms include classifiers such as SVM, Nearest Neighbors, gradient boosting, and Natural Language Processing (NLP) methods. These classifiers are primarily applied for establishing associations among target-disease-drug, identifying druggable and nondruggable targets for specific diseases, and predicting novel therapeutic targets through target–gene associations. Additionally, regression methods, including ANNs, and RF, among others, are applied for target identification. These methods are typically utilized to assess the druggability of targets based on pharmacokinetic properties, protein structure or sequence, and exploration of novel targets such as those associated with Huntington's disease and the identification of alternative targets [14, 15].

5.2 *Lead Discovery*

Lead discovery represents one of the most extensive stages in the drug discovery process, typically the primary focus of medicinal chemists. This stage encompasses the discovery of hits, progressing from hit-to-lead to lead optimization. Its focus is on the identification of promising chemical compounds with the potential to become the foundation for new therapeutic agents. This intricate stage involves screening vast libraries of compounds to pinpoint molecules exhibiting desirable biological activities against a specific target or pathway associated with a disease. Traditionally, the lead discovery process involves extensive laboratory assays that are time-consuming and expensive. The application of ML in this phase optimizes the process, accelerating the identification of promising drug candidates [14, 15].

In the first phase corresponding to the identification of hits, virtual screening and drug repurposing stand out. Drug repurposing aims to identify new purposes for existing drugs. In this process, the use of Deep Neural Networks (DNNs) is common, but methods such as naive Bayes, SVM, KNN, and RF can also be found. Additionally, these methods can be applied to predict binding affinities in the virtual screening process based on score functions from docking, without the need to apply physical concepts and functions (which are more complex). Algorithms based on deep learning, such as CNNs, shine in handling 3D structures [14, 15].

In the hit-to-lead process, two approaches heavily involve ML algorithms, namely QSAR and De Novo design. In QSAR, essentially any regression algorithm can be used for predicting biological activity, showcasing the technique's versatility across various methods. In De Novo drug design, RNNs play a significant role in creating new compounds with desirable biological activity [14, 15]. For more details on the application of QSAR and ML models, see Chap. 6.

In the final phase corresponding to lead optimization, various methods are used for predicting ADME properties, toxicity, and ultimately the druggability of biologically active compounds. The most employed methods fall within the realm of deep learning such as DNNs and CNNs. Conventional ML methods like SVM can also be found, but their performance is markedly inferior to DL methods. Another determining factor is assessing synthesizability and creating synthesis plans for the envisioned compounds, a process often performed by reinforcement learning algorithms. Furthermore, ML algorithms find applications in predicting drug metabolites and metabolic sites, as well as natural product identification [14, 15].

5.3 *Preclinical and Clinical Development*

The evolution of ML application in clinical and pre-clinical stages is constantly progressing, and already showing significant advancements. In the pre-clinical phase, ML algorithms stand out for their ability to identify and predict biomarkers, predominantly utilizing regression and classification approaches. These biomarkers are of utmost relevance, enabling a deeper understanding of drug mechanisms and efficiently differentiating them for specific patients. Commonly used traditional ML methods for this purpose include SVM and RF. These models are applied in the initial stages of clinical trials, aiming to reduce both the time and costs associated with the final stages [14, 15].

In the clinical context, the application of ML models for biomarker discovery and prediction of therapeutic responses continues to stand out. The emphasis on classification approaches is evident, where algorithms are employed to stratify patients, identify potential indications, and suggest drug mechanisms [14, 15].

Another critical application is the construction of drug sensitivity models, where supervised learning techniques, such CNNs, are employed. These models are particularly applicable in digital pathology, where image analysis allows the prediction of tumor responses to different treatments. Drug sensitivity models and their corresponding biomarkers are validated by independent test data sets, covering both pre-clinical trials and early clinical stages [14, 15].

6 Resources and Tools

The discovery of new drugs represents a complex challenge, but recent advances in ML implicated accelerating this process. Various software and online platforms have been developed to integrate ML algorithms in the analysis of vast sets of biological and chemical data, providing valuable insights for scientists and researchers. These tools offer a variety of applications, ranging from target identification to drug prioritization, prediction of properties, and assistance in generating synthesis pathways. In this dynamic and challenging field, the diversity of available tools is notable, as detailed in the accompanying Table 4.5, which provides a comprehensive overview of various platforms, their availability, software format, and specific applications.

Beyond tools, databases are fundamental in constructing suitable models for drug discovery, serving as a critical starting point for the ML process. These repositories of information are necessary for the collection and organization of relevant data, providing a robust foundation for the implementation of ML algorithms. Table 4.6 presents the main databases along with their respective descriptions used in the field of drug discovery for building ML models.

7 Challenges and Perspectives

The integration of ML into drug discovery processes holds immense promise, offering unprecedented advancements in identifying potential therapeutic compounds and accelerating the development of new drugs. However, this transformative approach is not without its challenges and ethical considerations. Among the challenges, stand out:

Experimental Validation While ML offers powerful predictive capabilities, it is not a substitute for traditional experimental methods. Experimental validation remains indispensable to confirm and interpret ML predictions accurately. The integration of ML with experimental approaches can synergize the drug discovery process. ML can efficiently analyze vast data sets and propose potential drug

Table 4.5 Some key ML-based tools used in the drug discovery process

Name	Availability	Software format	Application
Open targets [105]	Free	Web platform	Target identification and drug prioritization
admetSAR [106]	Free	Web platform	Properties Prediction
ACD/Percepta [107]	Commercial	Installable	Properties Prediction
pkCSM [108]	Free	Web platform	Properties Prediction
IBM RXN [109]	Free/ Commercial	Web platform	Synthesis pathway generation
DeepChem [110]	Free	Library	Multi-application in drug design
RDKit [111]	Free	Library	Multi-application in drug design
AutoQSAR [112]	Commercial	Installable	QSAR modeling
PaDEL-Descriptor [113]	Free	Installable	Molecular descriptors calculation/QSAR modeling
SYBYL-X [114]	Commercial	Installable	Molecular descriptors calculation/QSAR modeling
MatLab [115]	Commercial	Installable	QSAR modeling

candidates, but human researchers must validate and interpret these findings. The collaborative efforts of ML and human expertise can optimize the drug discovery process, leading to the accelerated development of new drugs [127]. For more details on experimental assays, see Chap. 12.

Regulation and Ethics The integration of ML in drug discovery raises ethical considerations and regulatory challenges. Decision-making processes impacting people's health, such as drug development choices and clinical trial selections, are increasingly influenced by AI algorithms. Ensuring fairness, transparency, and unbiased outcomes becomes paramount. Biases in AI algorithms could lead to unequal access to medical treatment, contradicting principles of equality and justice. Additionally, the potential automation of certain tasks in the pharmaceutical industry may raise concerns about job losses. Ethical AI implementation involves regulatory compliance, periodic audits for bias, and the development of protocols ensuring data privacy and security [128].

Generalization The challenge of generalization involves the ability of ML models to apply learned knowledge to new, unseen data. ML models trained on specific data sets may struggle to generalize their predictions to diverse patient populations or different experimental conditions. Overfitting to a particular data set can limit the model's adaptability. Techniques such as transfer learning, where models leverage knowledge gained from one domain to enhance performance in another, can help address generalization challenges in drug discovery ML [15].

Table 4.6 Main databases used in the drug discovery process

Name	Description	URL
ChEMBL [24]	Information on bioactive molecules with drug-like properties	https://www.ebi.ac.uk/chembl/g/
DrugBank [23]	Information on drugs and drug targets	https://go.drugbank.com/
PubChem [22]	Information on chemical structures, identifiers, properties, activities, patents, safety, toxicity data	https://pubchem.ncbi.nlm.nih.gov/
UniProt [116]	Information on protein sequences and their functions, many entries from genome sequencing projects	https://www.uniprot.org/
RCSB PDB [117]	Archive of 3D structure data for large biological molecules (proteins, DNA, RNA)	https://www.rcsb.org/
DrugCentral [118]	Comprehensive information on drugs and their targets	https://drugcentral.org/
ADReCS [119]	Adverse Drug Reaction Classification System	http://bioinf.xmu.edu.cn/ADReCS/
CTD [120]	Comparative Toxicogenomics Database	https://ctdbase.org/
DisGeNET [121]	Discovering gene–disease associations	https://www.disgenet.org/
UK BioBank [122]	United Kingdom Biobank: Large-scale biomedical database	https://www.ukbiobank.ac.uk/
STRING database [123]	Protein–protein interaction information	https://string-db.org/
ZINC [21]	Free database of commercially available compounds for virtual screening	https://zinc.docking.org/
Therapeutic Target Database [124]	Information on the known and explored therapeutic protein and nucleic acid targets	https://idrblab.net/ttd/
Promiscuous 2.0 [125]	Database of promiscuous proteins and compounds	https://bioinf-applied.charite.de/promiscuous2/index.php
ClinicalTrials.gov [126]	Database of privately and publicly funded clinical studies conducted worldwide	https://clinicaltrials.gov/

Data Integration In the dynamic landscape of drug discovery, the integration of data emerges as a critical hurdle, encompassing both heterogeneous inputs from "omics" domains and the complexities of homogeneous databases. Robust pipeline architectures, powered by Extract Transform Load (ETL) tools, are indispensable for orchestrating the seamless convergence of diverse data sets from public repositories. This integration process demands not only technical prowess but also the application of advanced technologies, including ML and big data analytics. Even within homogeneous data sets, challenges persist, spanning testing, logical considerations, cross-platform normalization, and statistical intricacies. ML and big data analytics serve as invaluable assets in navigating these challenges, providing a cohesive framework for integrating diverse strands of information. In the pharmaceutical realm, where research spans from molecular insights to individual patient data, the challenge of integrating heterogeneous data becomes particularly pronounced. The need for a sophisticated level of AI becomes evident, ensuring a comprehensive understanding of data gathered across varying contexts and scales. Modern data connectors are essential for centralizing dissimilar data, offering a strategic solution to allocate and manage this diverse information effectively [15].

7.1 Future Trends

At the crossroads of science and technology, drug discovery is undergoing a significant revolution, thanks to the increasing application of ML techniques. This revolution not only accelerates the process of identifying new drugs but also provides a deeper and more personalized understanding of therapies. In this context, several emerging trends act as promising catalysts for the future of drug discovery.

Transfer Learning This approach stands out as a dominant trend in the field, characterized by the widespread adoption of pre-trained machine learning models. This methodology involves utilizing models that have been initially trained on extensive data sets, which may not necessarily pertain to the domains of biology or medicine. However, these pre-trained models can be fine-tuned and adapted to specific tasks within drug discovery. By capitalizing on the knowledge acquired in a diverse range of domains, these models facilitate the accelerated analysis of intricate biological data. This, in turn, provides a substantial advantage in pinpointing potential therapeutic targets and identifying promising compounds [129].

Explainable AI (XAI) The reliance on ML-driven drug discoveries crucially depends on the ability to interpret and understand model decisions. The rise of Explainable Artificial Intelligence (XAI) is, therefore, a response to this need. As ML models become more complex, ensuring transparency and interpretability becomes paramount. XAI techniques provide insights into how models make decisions, allowing researchers to validate and better understand predictions, ensuring more reliable outcomes [130].

Use of Imaging Data and Molecular Structure The integration of imaging and molecular structure data marks another innovative frontier in drug discovery. Advances in imaging techniques, such as high-resolution microscopy and computed tomography, are now complemented by ML algorithms that can extract crucial information from these images. Similarly, the analysis of complex molecular structures, aided by ML models, allows for a more refined understanding of interactions between chemical compounds and biological targets [14].

Deep Learning Advancements Advancements in Deep Learning continue to shape the landscape of drug discovery. More complex models, such as deep neural networks, have the ability to learn more abstract hierarchical representations, capturing subtle nuances in data. This capability for extracting complex features is particularly valuable when dealing with the diversity and complexity of biological data, enabling a deeper understanding of causal relationships and underlying patterns [5].

Generative AI This field is revolutionizing drug discovery through advanced models like deep neural networks. These models efficiently explore complex molecular structures, predicting drug candidates with unprecedented precision. Their nuanced understanding of biological data uncovers hidden patterns, ushering in a new era of personalized medicine and targeted drug development. The future of drug discovery is shaped by the transformative capabilities of generative AI [131].

As we move toward the future, these converging trends at the intersection of ML and drug discovery promise not only to transform the efficiency of the process but also to open new horizons in therapy personalization and the development of innovative treatments. The marriage between biology and artificial intelligence is shaping an exciting era where therapeutic innovation is driven by the synergy between the human mind and the analytical capabilities of machines.

8 Guiding Questions for Implementing Machine Learning in Drug Design

Before we conclude, the following set of questions is designed to guide the effective use of machine learning in drug design. These questions cover various aspects, including the primary objectives, data availability and quality, model selection, performance evaluation, and integration into the drug development process. Additionally, they address challenges related to data preprocessing, model interpretation, and resource requirements. By thoroughly addressing these questions, researchers and practitioners can ensure that their machine learning projects in drug design are well founded, robust, and capable of overcoming specific challenges in this field. Guiding questions for implementing machine learning in drug design

? **What is the primary objective of using machine learning in drug design**?
Identification of new molecules?
Optimization of existing molecules?
Prediction of efficacy and toxicity?
Target identification/optimization?
What data are available for training machine learning models?
Chemical structures?
Biological assay data?
Genomic data?
What are the quality and quantity of this data?
Is there enough data for effective training?
Is the data well annotated and standardized?

(continued)

What types of machine learning models will be used?
Supervised learning models (e.g., *regression and classification)?*
Unsupervised learning models (e.g., *and dimensionality reduction)?*
Reinforcement learning models?
How will the performance of the models be evaluated?
What performance metrics will be used (e.g., *accuracy, recall, and AUC-ROC)?*
What validation techniques will be applied (e.g., *cross-validation and data split)?*
How will imbalanced data be handled, if present?
Resampling techniques (e.g., *oversampling and undersampling)?*
How will the results of the models be interpreted?
Feature importance analysis?
Model explanations and visualizations?
How will machine learning models be integrated into the drug design process?
Integration with virtual screening platforms?
Integration with in vitro *and* in vivo *experimentation pipelines?*
What are the main challenges and limitations expected in the application of machine learning to drug design?
Interpretability issues?
Generalization to new data?
How will the lifecycle of machine learning models be managed?
Model updates and retraining?
Post-implementation performance monitoring?
What computational resources will be needed?
Required hardware and software?
Use of GPUs or TPUs to accelerate training?
How will the impact of machine learning models on the drug development process be measured?
Cost and time reduction?
Increased success rate in clinical trials?

9 Conclusions

In conclusion, the exploration of ML within the realm of AI offers a transformative lens through which we understand, navigate, and revolutionize complex processes like drug discovery. As we traverse the intricate stages of problem definition, model selection, training, and validation, it becomes evident that the success of ML applications hinges on the careful consideration of various factors, including problem specificity and algorithm selection. However, with the power of ML, especially within the domain of neural networks and deep learning, we witness a paradigm shift in how we approach challenges in drug discovery.

Yet, this transformative journey is not without its challenges. The omnipresence of biases, the delicate balance between overfitting and underfitting, the imperative need for interpretability, computational costs, data dependencies, and the quest for robust models all contribute to the nuanced landscape of ML in drug discovery. Each limitation serves as a reminder of the ethical responsibilities, technical intricacies, and economic considerations that accompany the integration of ML into such critical domains.

The application of ML in drug discovery, exemplified through target identification, lead discovery, and preclinical and clinical development, underscores the potential for accelerated advancements in the pharmaceutical industry. The careful orchestration of various ML algorithms, from classifiers to regression models, empowers researchers to sift through vast data sets and extract meaningful insights that shape the trajectory of drug development.

As we navigate through the challenges and leverage the emerging trends of transfer learning, explainable AI, and the integration of imaging data and molecular structures, we glimpse into a future where the marriage between biology and artificial intelligence paves the way for unprecedented innovation. The continued advancements in deep learning and the collaborative synergy between human expertise and machine analytical capabilities promise not just efficiency but also a profound understanding of the intricate interplay between therapeutic compounds and biological targets.

In essence, the union of ML and drug discovery heralds a promising era where scientific exploration is propelled by the fusion of human intellect and the computational prowess of machines. As we stand at the crossroads of science and technology, the journey forward promises not just accelerated drug development but a deeper, more personalized understanding of therapies, fostering a landscape where innovation thrives at the intersection of AI and the intricacies of life sciences.

References

1. Gelijns, A. C. (2014). *Technological innovation: Comparing development of drugs, devices, and procedures in medicine*. National Academies Press.
2. Taye, M. M. (2023). Understanding of machine learning with deep learning: Architectures, workflow, applications and future directions. *Computers, 12*(5), 91.
3. Niazi, S. K., & Mariam, Z. (2023). Recent advances in machine-learning-based chemoinformatics: A comprehensive review. *International Journal of Molecular Sciences, 24*(14), 11488.
4. Vora, L. K., Gholap, A. D., Jetha, K., Thakur, R. R. S., Solanki, H. K., & Chavda, V. P. (2023). Artificial intelligence in pharmaceutical technology and drug delivery design. *Pharmaceutics, 15*(7), 1916.
5. Askr, H., Elgeldawi, E., Aboul Ella, H., Elshaier, Y. A. M. M., Gomaa, M. M., & Hassanien, A. E. (2023). Deep learning in drug discovery: An integrative review and future challenges. *Artificial Intelligence Review, 56*(7), 5975–6037.
6. McCulloch, W. S., & Pitts, W. (1943). A logical calculus of the ideas immanent in nervous activity. *The Bulletin of Mathematical Biophysics, 5*, 115–133.
7. Weik, M. H. (1961). The ENIAC story. *Ordnance, 45*(244), 571–575.
8. Turing, A. M. (1950). I.—Computing machinery and intelligence. *Mind, LIX*(236), 433–460.

9. McCarthy, J., & Feigenbaum, E. A. (1990). In memoriam: Arthur samuel: Pioneer in machine learning. *AI Magazine, 11*(3), 10.
10. Samuel, A. L. (1959). Some studies in machine learning using the game of checkers. *IBM Journal of Research and Development, 3*(3), 210–229.
11. Rosenblatt, F. (1957). *The perceptron, a perceiving and recognizing automaton Project Para*. Cornell Aeronautical Laboratory.
12. Cover, T., & Hart, P. (1967). Nearest neighbor pattern classification. *IEEE Transactions on Information Theory, 13*(1), 21–27.
13. E. A. Feigenbaum, B. G. Buchanan, and J. Lederberg, 1970. "On generality and problem solving: A case study using the DENDRAL program,"
14. Vamathevan, J., et al. (2019). Applications of machine learning in drug discovery and development. *Nature Reviews. Drug Discovery, 18*(6), 463–477.
15. Dara, S., Dhamercherla, S., Jadav, S. S., Babu, C. H. M., & Ahsan, M. J. (2022). Machine learning in drug discovery: A review. *Artificial Intelligence Review, 55*(3), 1947–1999.
16. Zhang, L., Tan, J., Han, D., & Zhu, H. (2017). From machine learning to deep learning: Progress in machine intelligence for rational drug discovery. *Drug Discovery Today, 22*(11), 1680–1685.
17. Stephenson, N., et al. (2019). Survey of machine learning techniques in drug discovery. *Current Drug Metabolism, 20*(3), 185–193.
18. Popova, M., Isayev, O., & Tropsha, A. (2018). Deep reinforcement learning for de novo drug design. *Science Advances, 4*(7), eaap7885.
19. Ajay, Walters, W. P., & Murcko, M. A. (1998). Can we learn to distinguish between 'drug-like' and 'nondrug-like' molecules? *Journal of Medicinal Chemistry, 41*(18), 3314–3324.
20. Carracedo-Reboredo, P., et al. (2021). A review on machine learning approaches and trends in drug discovery. *Computational and Structural Biotechnology Journal, 19*, 4538–4558.
21. Irwin, J. J., & Shoichet, B. K. (2005). ZINC− a free database of commercially available compounds for virtual screening. *Journal of Chemical Information and Modeling, 45*(1), 177–182.
22. Wang, Y., Xiao, J., Suzek, T. O., Zhang, J., Wang, J., & Bryant, S. H. (2009). PubChem: A public information system for analyzing bioactivities of small molecules. *Nucleic Acids Research, 37*(suppl_2), W623–W633.
23. Wishart, D. S., et al. (2008). DrugBank: A knowledgebase for drugs, drug actions and drug targets. *Nucleic Acids Research, 36*(suppl_1), D901–D906.
24. Gaulton, A., et al. (2012). ChEMBL: A large-scale bioactivity database for drug discovery. *Nucleic Acids Research, 40*(D1), D1100–D1107.
25. Wallach, I., Dzamba, M., & Heifets, A. (2015). AtomNet: a deep convolutional neural network for bioactivity prediction in structure-based drug discovery. *arXiv preprint arXiv:1510.02855*.
26. Stokes, J. M., et al. (2020). A deep learning approach to antibiotic discovery. *Cell, 180*(4), 688–702.
27. Burki, T. (2020). A new paradigm for drug development. *Lancet Digit Health, 2*(5), e226–e227.
28. Whitby, B. (2009). *Artificial intelligence*. The Rosen Publishing Group, Inc.
29. Winston, P. H. (1984). *Artificial intelligence*. Addison-Wesley Longman Publishing, Inc.
30. Ramesh, A. N., Kambhampati, C., Monson, J. R. T., & Drew, P. J. (2004). Artificial intelligence in medicine. *Annals of the Royal College of Surgeons of England, 86*(5), 334.
31. Hunt, E. B. (2014). *Artificial intelligence*. Academic.
32. El Naqa, I., & Murphy, M. J. (2015). *What is machine learning?* Springer.
33. Rusk, N. (2016). Deep learning. *Nature Methods, 13*(1), 35.
34. LeCun, Y., Bengio, Y., & Hinton, G. (2015). Deep learning. *Nature, 521*(7553), 436–444.
35. Bonaccorso, G. (2017). *Machine learning algorithms*. Packt Publishing Ltd.
36. Mahesh, B. (2020). Machine learning algorithms-a review. *International Journal of Science and Research (IJSR), 9*(1), 381–386.
37. Zhou, Z.-H. (2021). *Machine learning*. Springer Nature.

38. Agatonovic-Kustrin, S., & Beresford, R. (2000). Basic concepts of artificial neural network (ANN) modeling and its application in pharmaceutical research. *Journal of Pharmaceutical and Biomedical Analysis, 22*(5), 717–727.
39. Morales, E. F., & Escalante, H. J. (2022). A brief introduction to supervised, unsupervised, and reinforcement learning. In *Biosignal processing and classification using computational learning and intelligence* (pp. 111–129). Elsevier.
40. Rigatti, S. J. (2017). Random forest. *Journal of Insurance Medicine, 47*(1), 31–39.
41. Lorena, A. C., & De Carvalho, A. C. (2007). Uma introdução às support vector machines. *Revista de Informática Teórica e Aplicada, 14*(2), 43–67.
42. Siemers, F. M., & Bajorath, J. (2023). Differences in learning characteristics between support vector machine and random forest models for compound classification revealed by Shapley value analysis. *Scientific Reports, 13*(1), 5983.
43. Rosenblatt, F., & Papert, S. (2021). *Perceptron, 9*.
44. Gallant, S. I. (1990). Perceptron-based learning algorithms. *IEEE Transactions on Neural Networks, 1*(2), 179–191.
45. Ruck, D. W., Rogers, S. K., & Kabrisky, M. (1990). Feature selection using a multilayer perceptron. *Journal of Neural Network Computing, 2*(2), 40–48.
46. Popescu, M.-C., Balas, V. E., Perescu-Popescu, L., & Mastorakis, N. (2009). Multilayer perceptron and neural networks. *WSEAS Transactions on Circuits and Systems, 8*(7), 579–588.
47. Sharma, S., Sharma, S., & Athaiya, A. (2017). Activation functions in neural networks. *Towards Data Science, 6*(12), 310–316.
48. A. D. Rasamoelina, F. Adjailia, and P. Sinčák, "A review of activation function for artificial neural network," in 2020 IEEE 18th World Symposium on Applied Machine Intelligence and Informatics (SAMI), IEEE, 2020, pp. 281–286.
49. Leijnen, S., & van Veen, F. (2020). The neural network zoo. *Proceedings of the West Mark Educational Association conference, 47*(1).
50. Hassoun, M. H. (1995). *Fundamentals of artificial neural networks*. MIT Press.
51. O'Shea, K., & Nash, R. (2015). An introduction to convolutional neural networks. *arXiv preprint arXiv:1511.08458.*
52. Chauhan, R., Ghanshala, K. K., & Joshi, R. C. (2018). Convolutional neural network (CNN) for image detection and recognition. In *2018 first international conference on secure cyber computing and communication (ICSCCC)* (pp. 278–282). IEEE.
53. Albawi, S., Mohammed, T. A., & Al-Zawi, S. (2017). Understanding of a convolutional neural network. In *2017 international conference on engineering and technology (ICET)* (pp. 1–6). IEEE.
54. Grossberg, S. (2013). Recurrent neural networks. *Scholarpedia, 8*(2), 1888.
55. Medsker, L. R., & Jain, L. C. (2001). Recurrent neural networks. *Design and Applications, 5*(64–67), 2.
56. Medsker, L., & Jain, L. C. (1999). *Recurrent neural networks: Design and applications*. CRC Press.
57. Nosouhian, S., Nosouhian, F., & Khoshouei, A. K. (2021). *A review of recurrent neural network architecture for sequence learning: Comparison between LSTM and GRU.*
58. Sherstinsky, A. (2020). Fundamentals of recurrent neural network (RNN) and long short-term memory (LSTM) network. *Physica D, 404*, 132306.
59. Dey, R., & Salem, F. M. (2017). Gate-variants of gated recurrent unit (GRU) neural networks. In *2017 IEEE 60th international Midwest symposium on circuits and systems (MWSCAS)* (pp. 1597–1600). IEEE.
60. Martín-Guerrero, J. D., & Lamata, L. (2022). Quantum machine learning: A tutorial. *Neurocomputing, 470*, 457–461.
61. Chao, W.-L. (2011). *Machine learning tutorial*. Digital Image and Signal Processing.
62. Ball, N. M., & Brunner, R. J. (2010). Data mining and machine learning in astronomy. *International Journal of Modern Physics D, 19*(07), 1049–1106.

63. Emmanuel, T., Maupong, T., Mpoeleng, D., Semong, T., Mphago, B., & Tabona, O. (2021). A survey on missing data in machine learning. *Journal of Big Data, 8*(1), 1–37.
64. Kumar, S. A., et al. (2022). Machine learning and deep learning in data-driven decision making of drug discovery and challenges in high-quality data acquisition in the pharmaceutical industry. *Future Medicinal Chemistry, 14*(4), 245–270.
65. Ridzuan, F., & Zainon, W. M. N. W. (2019). A review on data cleansing methods for big data. *Procedia Computer Science, 161*, 731–738.
66. Ali, P. J. M., Faraj, R. H., Koya, E., Ali, P. J. M., & Faraj, R. H. (2014). Data normalization and standardization: A technical report. *Machine Learning Technical Report, 1*(1), 1–6.
67. Potdar, K., Pardawala, T. S., & Pai, C. D. (2017). A comparative study of categorical variable encoding techniques for neural network classifiers. *International Journal of Computers and Applications, 175*(4), 7–9.
68. Kurita, T. (2019). Principal component analysis (PCA). In *Computer vision: A reference guide* (pp. 1–4).
69. Wattenberg, M., Viégas, F., & Johnson, I. (2016). How to use t-SNE effectively. *Distill, 1*(10), e2.
70. Chawla, N. V., Bowyer, K. W., Hall, L. O., & Kegelmeyer, W. P. (2002). SMOTE: Synthetic minority over-sampling technique. *Journal of Artificial Intelligence Research, 16*, 321–357.
71. Barandela, R., Valdovinos, R. M., Sánchez, J. S., & Ferri, F. J. The imbalanced training sample problem: Under or over sampling? In *Structural, Syntactic, and Statistical Pattern Recognition: Joint IAPR International Workshops, SSPR 2004 and SPR 2004, Lisbon, Portugal, August 18-20, 2004. Proceedings* (pp. 806–814). Springer, 2004.
72. Cai, J., Luo, J., Wang, S., & Yang, S. (2018). Feature selection in machine learning: A new perspective. *Neurocomputing, 300*, 70–79.
73. Raschka, S. (2018). Model evaluation, model selection, and algorithm selection in machine learning. *arXiv preprint arXiv:1811.12808.*
74. da Silva, A. P., Chiari, L. P. A., Guimaraes, A. R., Honorio, K. M., & da Silva, A. B. F. (2021). Drug design of new 5-HT6R antagonists aided by artificial neural networks. *Journal of Molecular Graphics & Modelling, 104*, 107844.
75. Kim, Y.-J., & Chi, M. (2018). Temporal belief memory: Imputing missing data during RNN training. In *Proceedings of the 27th international joint conference on artificial intelligence (IJCAI-2018).*
76. Christoffersen, P., & Jacobs, K. (2004). The importance of the loss function in option valuation. *Journal of Financial Economics, 72*(2), 291–318.
77. Wythoff, B. J. (1993). Backpropagation neural networks: A tutorial. *Chemometrics and Intelligent Laboratory Systems, 18*(2), 115–155.
78. Devarakonda, A., Naumov, M., & Garland, M. (2017). Adabatch: Adaptive batch sizes for training deep neural networks. *arXiv preprint arXiv:1712.02029.*
79. Humphrey, A., et al. (2022). Machine-learning classification of astronomical sources: Estimating F1-score in the absence of ground truth. *Monthy Notices of the Royal Astronomical Society Letters, 517*(1), L116–L120.
80. Kuhn, M. (2014). Futility analysis in the cross-validation of machine learning models. *arXiv preprint arXiv:1405.6974.*
81. Roy, K., Das, R. N., Ambure, P., & Aher, R. B. (2016). Be aware of error measures. Further studies on validation of predictive QSAR models. *Chemometrics and Intelligent Laboratory Systems, 152*, 18–33.
82. Rücker, C., Rücker, G., & Meringer, M. (2007). Y-randomization and its variants in QSPR/QSAR. *Journal of Chemical Information and Modeling, 47*(6), 2345–2357.
83. Narkhede, S. (2018). Understanding auc-roc curve. *Towards Data Science, 26*(1), 220–227.
84. Feurer, M., & Hutter, F. (2019). Hyperparameter optimization. In *Automated machine learning: Methods, systems, challenges* (pp. 3–33).
85. Bergstra, J., Bardenet, R., Bengio, Y., & Kégl, B. (2011). Algorithms for hyper-parameter optimization. In *Advances in neural information processing systems* (Vol. 24).

86. Kim, Y., & Chung, M. (2019). An approach to hyperparameter optimization for the objective function in machine learning. *Electronics (Basel), 8*(11), 1267.
87. Kadhim, Z. S., Abdullah, H. S., & Ghathwan, K. I. (2022). Artificial neural network hyperparameters optimization: A survey. *International Journal of Online & Biomedical Engineering, 18*(15), 59.
88. Mazilu, S., & Iria, J. (2011). L1 vs. l2 regularization in text classification when learning from labeled features. In *2011 10th international conference on machine learning and applications and workshops* (pp. 166–171). IEEE.
89. Kuhn, M., Johnson, K., Kuhn, M., & Johnson, K. (2013). Over-fitting and model tuning. In *Applied predictive modeling* (pp. 61–92).
90. Probst, P., Wright, M. N., & Boulesteix, A. (2019). Hyperparameters and tuning strategies for random forest. *Wiley Interdisciplinary Reviews: Data Mining and Knowledge Discovery, 9*(3), e1301.
91. Ghawi, R., & Pfeffer, J. (2019). Efficient hyperparameter tuning with grid search for text categorization using kNN approach with BM25 similarity. *Open Computer Science, 9*(1), 160–180.
92. Friedrichs, F., & Igel, C. (2005). Evolutionary tuning of multiple SVM parameters. *Neurocomputing, 64*, 107–117.
93. Mastorakis, G. (2018). Human-like machine learning: limitations and suggestions. *arXiv preprint arXiv:1811.06052*.
94. Malik, M. M. (2020). A hierarchy of limitations in machine learning. *arXiv preprint arXiv:2002.05193*.
95. Kenge, R. (2020). Machine learning, its limitations, and solutions over IT. *International Journal of Information Technology Modeling and Computing, 11*, 73.
96. Jabbar, H., & Khan, R. Z. (2015). Methods to avoid over-fitting and under-fitting in supervised machine learning (comparative study). *Computer Science, Communication and Instrumentation Devices, 70*(10.3850), 978–981.
97. Bilmes, J. (2020). *Underfitting and overfitting in machine learning* (UW ECE course notes) (Vol. 5).
98. Ying, X. (2019). An overview of overfitting and its solutions. In *Journal of physics: Conference series* (p. 022022). IOP Publishing.
99. Lavecchia, A. (2015). Machine-learning approaches in drug discovery: Methods and applications. *Drug Discovery Today, 20*(3), 318–331.
100. Carvalho, D. V., Pereira, E. M., & Cardoso, J. S. (2019). Machine learning interpretability: A survey on methods and metrics. *Electronics (Basel), 8*(8), 832.
101. Gramegna, A., & Giudici, P. (2021). SHAP and LIME: An evaluation of discriminative power in credit risk. *Frontiers in Artificial Intelligence, 4*, 752558.
102. Justus, D., Brennan, J., Bonner, S., & McGough, A. S. (2018). Predicting the computational cost of deep learning models. In *2018 IEEE international conference on big data (Big Data)* (pp. 3873–3882). IEEE.
103. Miller, T., et al. (2023). Advancements in artificial intelligence circuits and systems (AICAS). *Electronics (Basel), 13*(1), 102.
104. Elbadawi, M., Gaisford, S., & Basit, A. W. (2021). Advanced machine-learning techniques in drug discovery. *Drug Discovery Today, 26*(3), 769–777.
105. Koscielny, G., et al. (2017). Open targets: A platform for therapeutic target identification and validation. *Nucleic Acids Research, 45*(D1), D985–D994.
106. Yang, H., et al. (2019). admetSAR 2.0: Web-service for prediction and optimization of chemical ADMET properties. *Bioinformatics, 35*(6), 1067–1069.
107. Advanced Chemistry Development Inc. (2016). *(ACD/Labs) ACD/PERCEPTA Version 2015*. Frankfurt am Main. www.acdlabs.com/pka
108. Pires, D. E. V., Blundell, T. L., & Ascher, D. B. (2015). pkCSM: Predicting small-molecule pharmacokinetic and toxicity properties using graph-based signatures. *Journal of Medicinal Chemistry, 58*(9), 4066–4072.

109. Schwaller, P., Strobelt, H., Laino, T., & Hoover, B. *IBM RXN for Chemistry: Unveiling the grammar of the organic chemistry language*.
110. Ramsundar, B. et al., DeepChem: Democratizing deep-learning for drug discovery, quantum chemistry, materials science and biology.
111. Landrum, G. (2013). RDKit: A software suite for cheminformatics, computational chemistry, and predictive modeling. *Greg Landrum, 8*, 31.
112. Dixon, S. L., Duan, J., Smith, E., Von Bargen, C. D., Sherman, W., & Repasky, M. P. (2016). AutoQSAR: An automated machine learning tool for best-practice quantitative structure–activity relationship modeling. *Future Medicinal Chemistry, 8*(15), 1825–1839.
113. Yap, C. W. (2011). PaDEL-descriptor: An open source software to calculate molecular descriptors and fingerprints. *Journal of Computational Chemistry, 32*(7), 1466–1474.
114. Sybyl, X. (2011). *Molecular modelling software* (Vol. 1). Tripos Certara.
115. Matlab, S. (2012). *Matlab*. The MathWorks.
116. The UniProt Consortium. (2017). UniProt: The universal protein knowledgebase. *Nucleic Acids Research, 45*(D1), D158–D169.
117. Burley, S. K., Berman, H. M., Kleywegt, G. J., Markley, J. L., Nakamura, H., & Velankar, S. (2017). Protein Data Bank (PDB): The single global macromolecular structure archive. In *Protein crystallography: Methods and protocols* (pp. 627–641).
118. Avram, S., et al. (2023). DrugCentral 2023 extends human clinical data and integrates veterinary drugs. *Nucleic Acids Research, 51*(D1), D1276–D1287.
119. Cai, M.-C., et al. (2015). ADReCS: An ontology database for aiding standardization and hierarchical classification of adverse drug reaction terms. *Nucleic Acids Research, 43*(D1), D907–D913.
120. Davis, A. P., Wiegers, T. C., Johnson, R. J., Sciaky, D., Wiegers, J., & Mattingly, C. J. (2023). Comparative Toxicogenomics database (CTD): Update 2023. *Nucleic Acids Research, 51*(D1), D1257–D1262.
121. Piñero, J., et al. (2015). DisGeNET: A discovery platform for the dynamical exploration of human diseases and their genes. *Database, 2015*, bav028.
122. UK Biobank. (2014). About UK biobank.
123. Szklarczyk, D., et al. (2023). The STRING database in 2023: Protein–protein association networks and functional enrichment analyses for any sequenced genome of interest. *Nucleic Acids Research, 51*(D1), D638–D646.
124. Zhou, Y., et al. (2024). TTD: Therapeutic target database describing target druggability information. *Nucleic Acids Research, 52*(D1), D1465–D1477.
125. Gallo, K., Goede, A., Eckert, A., Moahamed, B., Preissner, R., & Gohlke, B.-O. (2021). PROMISCUOUS 2.0: A resource for drug-repositioning. *Nucleic Acids Research, 49*(D1), D1373–D1380.
126. Zarin, D. A., Tse, T., Williams, R. J., Califf, R. M., & Ide, N. C. (2011). The ClinicalTrials. Gov results database—Update and key issues. *New England Journal of Medicine, 364*(9), 852–860.
127. Gonoskov, A., Wallin, E., Polovinkin, A., & Meyerov, I. (2019). Employing machine learning for theory validation and identification of experimental conditions in laser-plasma physics. *Scientific Reports, 9*(1), 7043.
128. Vayena, E., Blasimme, A., & Cohen, I. G. (2018). Machine learning in medicine: Addressing ethical challenges. *PLoS Medicine, 15*(11), e1002689.
129. Cai, C., et al. (2020). Transfer learning for drug discovery. *Journal of Medicinal Chemistry, 63*(16), 8683–8694.
130. Ponzoni, I., Páez Prosper, J. A., & Campillo, N. E. (2023). Explainable artificial intelligence: A taxonomy and guidelines for its application to drug discovery. *Wiley Interdisciplinary Reviews: Computational Molecular Science, 13*(6), e1681.
131. Bian, Y., & Xie, X.-Q. (2021). Generative chemistry: Drug discovery with deep learning generative models. *Journal of Molecular Modeling, 27*, 1–18.

Chapter 5
Clustering of Small Molecules

Alan Talevi, Lucas Alberca, and Carolina Bellera

Abstract Clustering of small molecules finds a diversity of applications in chemistry and, in particular, in the fields of cheminformatics and drug discovery. It may be used directly as an unsupervised machine-learning strategy to identify existing patterns in a chemical data set or libraries or integrated into supervised machine-learning studies to partition a sample of compounds into representative subsamples (e.g., training and validation data). It may also be applied to select which in silico hits from a virtual screening campaign will be submitted to experimental confirmation, or to define which hits emerging from a "wet" screening campaign will be prioritized for further development or characterization. Here, we review general strategies to validate the output of a clustering algorithm and discuss current challenges and possible future directions in the field of small molecule clustering.

Keywords Cluster replication · Clustering, External validity · Hierarchical clustering · Internal validity measures · Multi-view clustering · Non-hierarchical clustering · Relative validity · Small molecules · Stability measures · Subspace clustering · Supervised learning · Unsupervised learning · Validation · Validity

A. Talevi (✉)
Laboratory of Bioactive Compound Research and Development (LIDeB), Faculty of Exact Sciences, University of La Plata (UNLP), Buenos Aires, Argentina

Consejo Nacional de Investigaciones Científicas y Técnicas (CONICET), CCT La Plata, La Plata, Buenos Aires, Argentina

Boolzi SA, Buenos Aires, Argentina
e-mail: alantalevi@gmail.com

L. Alberca · C. Bellera
Laboratory of Bioactive Compound Research and Development (LIDeB), Faculty of Exact Sciences, University of La Plata (UNLP), Buenos Aires, Argentina

Consejo Nacional de Investigaciones Científicas y Técnicas (CONICET), CCT La Plata, La Plata, Buenos Aires, Argentina

V. G. Maltarollo (ed.), *Computer-Aided and Machine Learning-Driven Drug Design*, Computer-Aided Drug Discovery and Design 3,
https://doi.org/10.1007/978-3-031-76718-0_5

1 Historical Background

Human beings have developed the ability to detect higher-level organizations and group objects and experiences into functional categories, which is a precondition of sophisticated thought [45]. Categorization provides the basis to recognize and respond instantly and appropriately to objects and situations even if they are not identical to the ones that we have previously met. In other words, *categorization is a way of anticipating behavior based on prior experience with similar objects or scenarios*. This is a central proposition to understand the application of the approaches described in this chapter.

Perceptual grouping is performed at a remarkable speed based on principles of *proximity* (objects that are close to each other in space are perceived as part of a group), *similarity* (grouping elements by shape, size, and color), *continuity* (a relation is naturally perceived between objects that are arranged in continuous lines or curves), and *common fate* (objects that move at the same speed and/or direction tend to be grouped), among others [9]. The tendency of human beings to group things together is reflected in their language: common nouns name people, animals, or things of the same class or species, whereas proper nouns name particular objects individualized from any other of the same species (a particular person, a particular animal, a particular city, etc.). In fact, categories of knowledge that have a common noun in a language are more efficiently captured by the eye than those that do not (which is known as *conceptual grouping*) [31]. Moreover, collective nouns express functional and/or proximity relationships between similar objects (named by the same common noun), for example, a pack of wolves, a crowd, a school of fish, and a choir.

A fundamental challenge in the field of artificial intelligence is how to teach a machine to perform grouping or categorization as good as (or better than) evolution has taught human beings to do [41]. Let us take, for instance, the molecular structures in Fig. 5.1. Even if the observer has no specific training in chemistry, it is likely that they will perceive the similarities between the two compounds that share the benzodiazepine nucleus. However, when dealing with large-scale chemical data sets or repositories, visually analyzing and comparing every possible pair of molecules to produce a hypothesis on possible clusters that may be derived from it

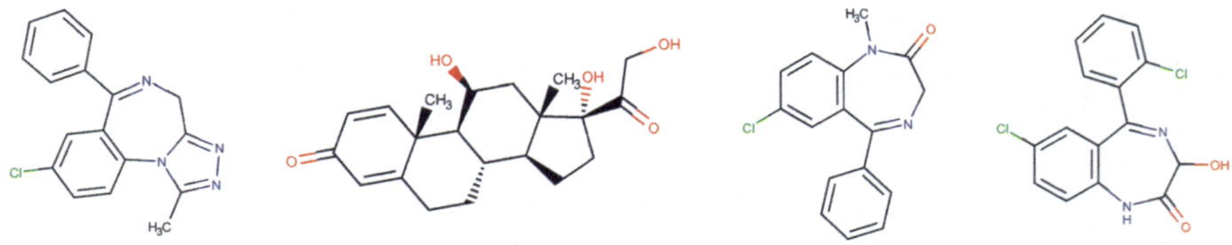

Fig. 5.1 The human brain performs perceptual grouping at a remarkable speed based on principles such as similarity. It is likely that even people with no training in chemistry will cluster the three compounds that share the benzodiazepine nucleus (alprazolam, diazepam, and lorazepam) apart from the corticosteroid (prednisolone)

may be highly impractical. Moreover, some similarities between objects may not be obvious from a fast visual inspection.

Computer-aided clustering of chemical entities therefore has three potential advantages: (a) automation, (b) high-throughput, and (c) the possibility of detecting clustering patterns that exceed prior knowledge and prior judgments regarding which molecular features are relevant for meaningful and informative clustering.

1.1 Applications in Drug Discovery

The clustering of small molecules is of substantial importance in the field of computer-guided drug discovery and in conducting virtual and wet screening experiments. Depending on the pursued goals, the structure of chemical data in suitable subspaces of the chemical space can be relevant to sample chemical data sets in a representative or biased manner.

First, the stratified split of data sets into representative training and validation sets is regularly performed when building quantitative structure–activity relationships (QSAR) and, in general, for any kind of machine-learning application. Such a practice can respond to two different (and, in a certain sense, opposite) purposes: (a) ensure that the training data present maximum coverage of the chemical space covered by the data set, in order to expand the generalizability and the applicability domain of the inferred models [21, 30], and (b) avoid compound biases and overoptimistic validation output by allocating very similar compounds in both the training and validation sets ([32, 58]). For more details, see Chap. 6.

Second, structuring data in the chemical space can be useful in investigating unexplored regions for novel bioactive compounds [16, 52]. On the contrary, focusing on specific already-explored regions that congregate active compounds may be of interest to generate preferred screening collections or focused compound libraries ([6, 59]), or simply to increase the probability of finding a bioactive compound (at the expense of chemical novelty).

Third, the clustering of small molecules can be useful to guide the selection of subsets of in silico hits that will be submitted for experimental confirmation [38]. Because of funding limitations, the number of virtual hits that arise from in silico screening experiments most often exceeds the number that can be physically accessed (synthesized, isolated, or acquired from vendors) and evaluated experimentally. Clustering and subsequent representative sampling can guarantee the coverage of as much chemical diversity as possible with those relatively scarce in silico hits submitted to wet assays. Later, those experimentally confirmed hits can be enriched by testing other compounds present in the library with the same or similar scaffolds. Likewise, the same principle can be used to expedite structure-based screening without exhaustive docking of every molecule in ultra-large chemical libraries [54, 55].

1.2 Classification

Broadly speaking, clustering approaches in general (and also in the particular case of small molecule clustering) can be classified as *hierarchical* or *nonhierarchical* methods.

In hierarchical clustering, the data are partitioned successively, in a nested way [18]. *Bottom-up* or *agglomerative* approaches proceed by a series of consecutive fusions of the N objects into groups (eventually merging them all together in the last step of the procedure). *Top-down* or *divisive* approaches, in contrast, serially divide the N objects into smaller groups (eventually resulting in isolated elements). Hierarchical clustering is most often graphically represented as a dendrogram or tree (a representation of clusters at multiple levels). A disadvantage of these approaches is that once the individual elements have been merged (into supra-clusters, *agglomerative approaches*) or segregated (into subclusters, *divisive approaches*), this cannot be reverted or revised in subsequent steps, that is, they cannot be optimized. Typical examples of hierarchical methods are single, complete, and Ward linkage, with the latter being frequently used in the context of chemical library clustering [37]. Another example is the *Maximal Common Substructure* (MCS) approach, which groups compounds based on their common subgraph of the greatest cardinality. It has been implemented at Chemaxon's Jchem (https://chemaxon.com/).

Nonhierarchical methods, for their part, do not resort to a tree-like structure, and usually require, on the part of the user, to predefine the number of clusters that they wish to obtain or the initial data points that will be assigned to each cluster, which generally requires prior knowledge regarding the structure of the data. Moreover, as hierarchical methods decide locally, step by step, which objects should be merged into the same clusters or split into different clusters, they avoid hard combinatorial optimization problems. In contrast, optimization approaches, such as those used in nonhierarchical methods, often involve iterative processes to minimize an objective function, which can be computationally intensive. The clusters are iteratively refined by either minimizing or maximizing some numerical criterion, as in the popular K-means [60] (where each cluster is represented by its centroid or center of gravity, which in general does not match actual data point) or K-medoids algorithms [26] (where each cluster is represented by actual data points that minimize the sum of distances to other members of the same cluster). Both K-means and K-medoids algorithms can be regarded as *prototype-based* clustering, as the data are arranged around representative points for each partition. Sphere-exclusion algorithms, a type of *contiguity-based* clustering, as in the popular method by Butina [10], represent another example of a frequently used nonhierarchical approach.

The preceding classification is, however, extremely basic, and only one of many other possible classification schemes [1, 14, 46, 47, 56]. For instance, we may consider *density-based algorithms* (which construct density functions that reflect the spatial distribution of the data points: each cluster corresponds to a high-density region separated from other high-density regions by sparsely populated regions of the space), *distribution-* or *model-based clustering* (in which the algorithms assume

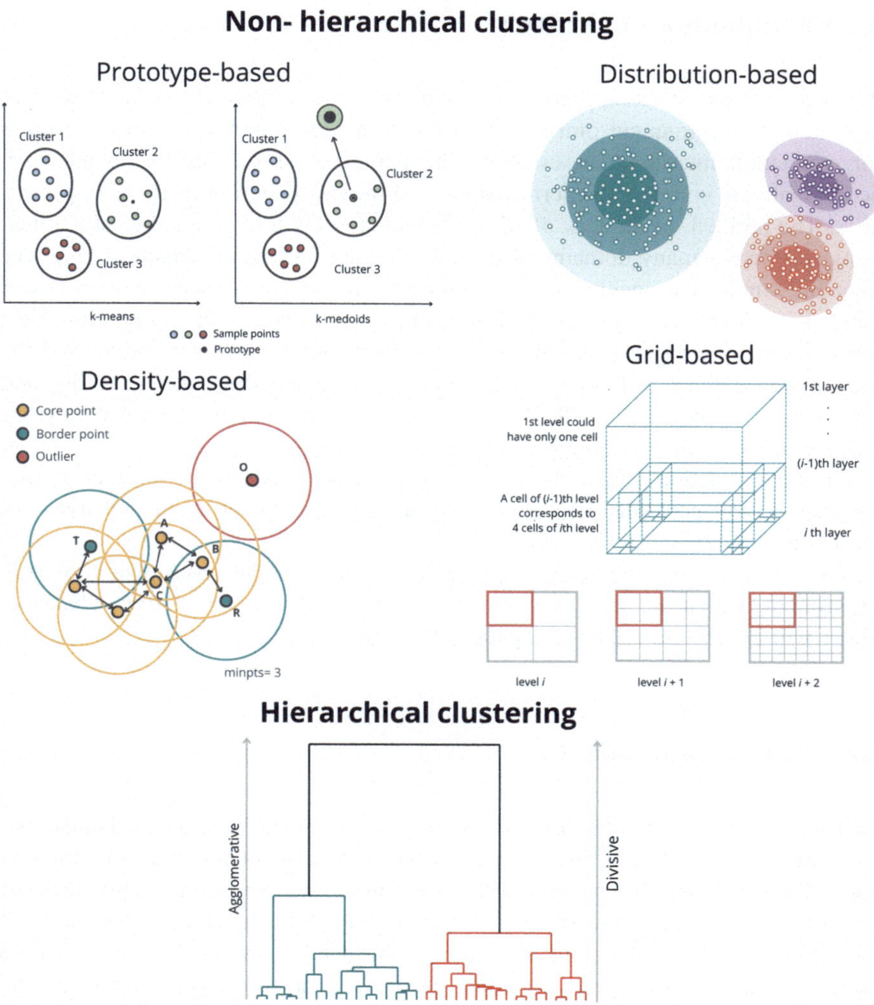

Fig. 5.2 Schematic representation of some of the types of clustering mentioned in the main text

that the data are generated from a mixture of underlying probability distributions, e.g., Gaussian); *grid-based algorithms* (which quantize the space into a finite number of cells for the sake of efficiency), *fuzzy* or *soft clustering* (which, in line with fuzzy logic, considers that a given object belongs to multiple clusters, in contrast to traditional *hard* or *exclusive clustering*, in which each object exclusively corresponds to a single cluster), or *partial clustering* (which considers that some object may be outliers that may not belong to any cluster, in contrast with *complete clustering*), among other possible classifications. Some of these approximations are graphically schematized in Fig. 5.2.

2 Validations and Controls

A common issue in the applications of clustering techniques in the field of small molecules is the frequent omission of validation procedures and validity indexes (or evaluation measures) to establish the goodness of the results obtained or, occasionally, to compare the performance of different clustering methods or parameters. The problem of the lack of adequate validation in a large number of applications cuts across many domains of cheminformatics and bioinformatics (e.g., very frequently molecular docking applications do not resort to any sort of scoring power validation). An exception is the field of quantitative structure–activity relationships, where recognized experts and international organizations have long advocated the use of well-defined techniques and metrics to evaluate the generalizability and robustness of models [20, 48–50], which have even been used with regulatory scope in the field of toxicology.

Cluster validation should be part of any cluster analysis because clustering approaches will generally find clusters in any given data set, even if it does not present a "natural" cluster structure.

We can mention three general approaches to validate the result of a clustering experiment, which are [8, 47] *internal*, *external*, and *relative cluster validation*. These will be next described in separate subsections.

2.1 Internal Validation

Internal clustering validation measures how good and stable the proposed clustering structure is, without resorting to any criterion external to the data set. Internal validation is possibly the most objective (or "pure") way to judge the goodness of a given clustering structure since it does not obey any external criteria unrelated (or diffusely connected) to the structure of the data. It should be noted that, in principle, any external criterion could respond to a subjective interest of the researcher and, therefore, bias what is considered a good result toward one that confirms a previous hypothesis about the data. In fact, cluster techniques are usually described as unsupervised machine-learning techniques, precisely to highlight that they are not based on any type of prior judgment, information, or assumptions about the data. Therefore, it sounds quite reasonable to resort to unsupervised methods to evaluate the validity of an unsupervised method. However, as discussed extensively in Sect. 3, the product of a cluster analysis is usually rather evaluated in practical terms, based on its usefulness or functionality for specific objectives.

An ideal clustering structure is usually recognized as one that provides highly cohesive clusters (i.e., with small within-cluster or intra-cluster distances) well separated in space from each other (i.e., with large intercluster or between-cluster distances). Connectedness, that is, to what extent an object is assigned to the same cluster as its nearest neighbor in the data space, is also sometimes assessed. On the other hand, a stable clustering structure would be one that is not substantially modified when the data set is subjected to perturbation (which would suggest that the structure of the data that has been found obeys a natural underlying structure, if such a structure exists).

To begin with, we could point out that cohesion can be calculated based on the sum of distances or the sum of proximities between elements assigned to the same cluster *Ci*, usually weighting it by the size of the corresponding cluster, while separation could be calculated in an analogous way but considering the distances or proximities of elements assigned to different clusters *Ci* and *Cj* [47], where x and y are elements of clusters:

$$\text{Cohesion}(C_i) = \sum_{\substack{x \epsilon C_i \\ y \epsilon C_i}} \text{proximity}(x, y) \tag{5.1}$$

$$\text{Separation}\left(C_i, C_j\right) = \sum_{\substack{x \epsilon C_i \\ y \epsilon C_j}} \text{proximity}(x, y) \tag{5.2}$$

Eventually, cohesion can also be calculated as the sum of distances between an object and the centroid or medoid of the corresponding cluster, whereas separation can be estimated as the distance between centroids or medoids of different clusters, or the minimal distance between elements from different clusters.

The metrics most used to evaluate the goodness of the clustering structure produced by a certain method simultaneously reflect, in a single variable, the cohesion and separation of the clustering, which makes sense since cohesion and separation show opposite trends versus the number of clusters (cohesion increases with the number of clusters considered, while separation decreases). The Dunn index [17] and the silhouette width [43] are two popular examples of internal validity measures used to validate nonhierarchical clustering approaches.

The Dunn index *D* is computed as the ratio between the minimal between-cluster distance to the maximal within-cluster distance (which is also regarded as the maximal diameter):

$$D = \frac{\text{separation}}{\text{diameter}} \tag{5.3}$$

The Dunn index can assume values between zero and infinity and must be maximized.

The silhouette width or silhouette index can be calculated as the average of the silhouette value $S(i)$ across all the elements of the clustered data set.

If we call $a(i)$ the average distance of the ith element of the data set with respect to all the other elements of its own cluster, and if we call $b(i)$ the minimum average distance between the *i*th element and the elements of any other cluster than its own (in other words, the average distance between element *i* and those elements assigned to the closest foreign cluster), then $S(i)$ is calculated as

$$S(i) = \frac{a(i) - b(i)}{\max\{a(i), b(i)\}} \tag{5.4}$$

A negative $S(i)$ indicates that, probably, the *i*th element has been assigned to an incorrect cluster since it is on average closer to the objects of a foreign cluster than to those of its own. An $S(i)$ close to zero suggests that element *i* is positioned equidistantly between two clusters. The closer $S(i)$ is to 1, the more confidence there will be that the cluster assignment has been correct. The silhouette width ranges from -1 to 1, and a value equal to 1 reflects ideal clustering: dense and well-spaced clusters, with a correct assignment of all elements to their respective clusters.

Other internal validation measures that combine within- and between-cluster distances are the Calinski and Harabasz index, also known as the variance ratio criterion [11], McClain and Rao index [33], the Gamma index [5], the Davies–Bouldin index [15], and the SD and SDbw indexes [14], among others. The Frey index [61], for its part, also combines measures of cohesion and separation, but can only be used in the context of hierarchical clustering techniques. This index calculates the ratio of difference scores from two successive levels in the hierarchic tree (level *j* and level $j + 1$). The difference between the average between-cluster distance for each of the two hierarchical levels is placed in the numerator, while the difference between the average within-cluster distance for the two successive levels is placed in the denominator:

$$K = \frac{\overline{d}_{bj+1} - \overline{d}_{bj}}{\overline{d}_{wj+1} - \overline{d}_{wj}} \tag{5.5}$$

where $\underline{d}_b$ represents the mean between-cluster distance and $\underline{d}_w$ represents the mean within-cluster distance. Its authors proposed using a ratio score of 1.00 to identify the correct cluster level. Other validation metrics commonly used in the context of hierarchical clustering, such as the cophenetic correlation coefficient, measure the degree of agreement between the cophenetic matrix and the proximity matrix. The cophenetic distance between two objects, in the framework of an agglomerative clustering approach, is the distance at which the algorithm merges those objects in the same cluster at the same time. In the cophenetic matrix, the entries are the cophenetic distances between the different pairs of objects. Further information on these and other internal measures can be found in [14].

Stability measures, on the other hand, compare the results of clustering when it is performed using the entire data set or after excluding one or more data points (i.e., they are based on a re-sample and re-cluster strategy to assess the significance of the cluster derived from the original data set) [8, 22]. In some ways, the procedure used to assess stability is like cross-validation procedures used in the context of supervised machine learning techniques (such as leave-one-out or leave-some-out cross-validation), only that in this case other numerical measures to quantitatively evaluate the robustness of the originally generated partitions [23]. For example, the Average Proportion of Nonoverlap measures the average proportion of observations not placed in the same cluster by clustering based on the complete data set and clustering based on the fraction of the data set that remains after the removal of one or more data points. Other options include calculating and comparing the mean within- and between-cluster distances before and after data removal. Naturally, the greater the perturbation that a data structure resists, without substantial modifications, the more stable such a data structure will be considered. In this case, the size of the perturbation would be related, fundamentally, to the proportion of data that are removed.

Cluster replicability is another similar approach that can be used to assess the robustness of a given data structure. This may be done, for instance, by separately clustering two subsamples of a dataset and comparing the resulting clusters (e.g., in the case of prototype-based clustering, by measuring the distances between the corresponding centroids or medoids). Alternately, after clustering the first subsample, the objects of the second subsample can be assigned to the cluster represented by the nearest centroid and, later, used to conduct a second independent clustering analysis on the second subsample and compare the degree of agreement between the "predicted" and "actual" cluster structures through a quantitative measure, such as the Rand index or a variant of it [29]. Replication of the results not only across entities but also across techniques and molecular features [40] would be indicative of high-order stability. Importantly, Krieger and Green [29] showed that under certain circumstances (large sample sizes and highly correlated clustering variables), internal validation may fail as criteria of goodness of clustering. Moreover, using synthetic data, it was shown that replication experiments tend to provide high Rand indexes at low numbers of clusters (below 5) even if the "true" number of clusters in the data is higher [7].

2.2 External Validation

External validation uses an external criterion from outside the clustering process (a label or dependent variable that is not used to cluster the data) to verify the ability of the cluster to produce meaningful results [8]. But first, it is necessary to understand what is considered significant. When working with synthetic data, it is relatively easy to establish an underlying, "true" structure of the data. On the other hand, when working with real data, defining a truth criterion can be more complicated and, as we discuss in the next section, subject to the researcher's personal interests or ideas. As Milligan [36] accurately stated several decades ago: "... the validity of the external criterion variable in an applied situation is usually open to question. Failure to find significant group differences may be due to either a lack of cluster structure or an invalid criterion variable."

For example, in the field of drug discovery, the external data label may be related to a biological activity or other pharmaceutically relevant property. Such properties are undeniably related to the molecular structure of the data set instances; however, the inability of a clustering algorithm to replicate the structure imposed by the external criterion does not necessarily mean that the clustering algorithm is failing. It is possible that the algorithm has detected a "true" cluster structure unrelated with the chosen external criterion, or that the molecular characteristics used to partition the data do not correlate well with the chosen class label.

Classification and regression metrics for model validation in the field of supervised machine learning have been reviewed in Chap. 6.

2.3 Relative Cluster Validation

As many clustering algorithms (e.g., the K-means and K-medoids algorithms) require input from the user on the number of clusters to consider, a frequent incognita when using these approaches is how to decide on such a parameter (and, eventually, on other parameters used as input). This is often done in a rather systematic way, by computing the value of a validity measure (e.g., an internal validity index) across a range of values of the parameter to optimize (e.g., the number of partitions to consider). Therefore, relative validation does not involve a particular type of measure but uses supervised or unsupervised measures commented on in the previous subsections for a specific purpose [47]: to optimize the value of a parameter or hyperparameter.

Occasionally, more than one complementary measure of validity can be considered, with the subsequent visualization of solutions in two- or three-dimensional space. This can be particularly useful for the visualization of the results obtained using conflicting measures [22].

The elbow method is an example of a commonly used graphical approach to decide on the number of clusters. It is usually performed in the context of the K-means algorithm, by plotting the within-cluster-sum-of-squares (or other measure of goodness) versus different possible values of K. The optimal K value is the point at which an "elbow" or "knee" (i.e., a relatively sharp drop in variance) is formed in the graph. The elbow method (like other visual approaches) has often been criticized because of its subjective nature [62] as, at times, a clear elbow may not appear in the plot (suggesting no "natural" group structure in the data?) or more than one elbow may be observed (suggesting more than one "natural" cluster structure fits the data); even if the optimal value of the parameter under optimization is found by more numerical (rather than just visual means), such as slope analysis, the result can be sensitive to parameter range and pre-processing (e.g., scaling) [44]. Depending on the criterion of goodness under use, the relationship between the validity criterion and the number of clusters may not increase or decrease monotonically, and a "peak," instead of a sharp variance drop, could be found at the optimal value. Various studies comparing the capacity of various clustering validity indexes to identify the ground-truth cluster structure of synthetic and real-world data sets have been published. In general, validity indices that combine some measure of within-cluster and between-cluster distances are among the best-performing indices. The seminal work of Milligan and Cooper (1985) compared the ability of 30 clustering validity measures available at the time to identify the true structure of 108 synthetic data sets. The Calinski and Harabasz index was the one that demonstrated the best performance (it is worth noting that at that time the silhouette index had not yet been reported). A much more recent similar study [3] also compared 30 clustering validity indexes, in this case against a much broader spectrum of data that included 720 synthetic data sets and 20 real data sets. The silhouette index obtained the best results, closely followed by the Davies–Bouldin and Calinski–Harabasz indexes. Despite the temporal distance between them, both studies agreed on the relatively few studies in which the results of a clustering experiment are evaluated using statistical tests (which is also commonplace in the field of cheminformatics), a situation that contrasts with what occurs in the field of supervised machine learning. It should be noted that Arbelaiz et al. observed that the performance of validity metrics to identify data structure dropped substantially when noisy data sets or data sets with overlapping clusters were analyzed. Surprisingly, introducing just 10 percent of noisy data dropped the average value of the indices by two-thirds.

Two recently reported small molecular clustering approaches, iRaPCA and deep clustering, use visual inspection of the silhouette width versus K to select the adequate number of clusters [21, 38] (Fig. 5.3).

Another way in which the optimal K can be selected as the initial input for algorithms that require pre-definition of the number of clusters involves a two-step hybrid hierarchical plus nonhierarchical approach in which the hierarchical algorithm is used to define the number of clusters and provides an initial guess of cluster centroids (a stochastically or systematically chosen seed for each cluster, taken from

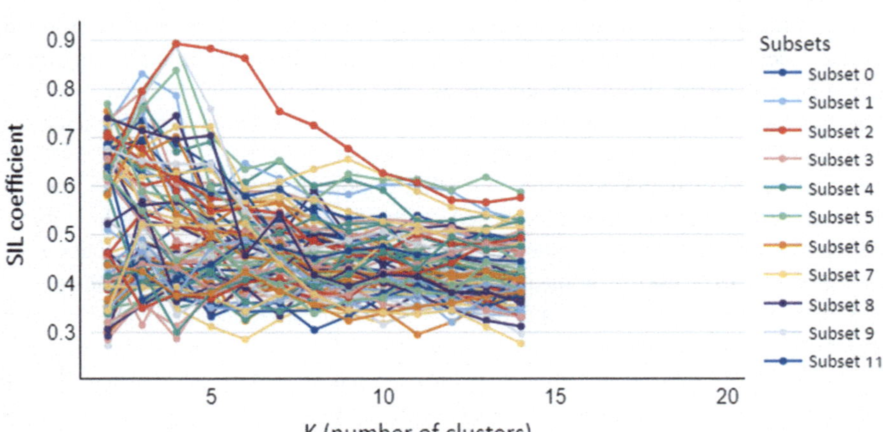

Fig. 5.3 Silhouette width versus K plots from iRaPCA. One plot is generated per random subspace. In the example, subset 2 (red plot) provides the best cluster structure according to the silhouette index (about 0.9) for $K = 4$

the partitions of the dendrogram); these data points then serve as the starting points for subsequent nonhierarchical clustering [18, 28]. It has been known for decades that randomly selected seeds typically lead to poorer recoveries of known partitions than seed selection obtained via initial hierarchical clustering [35].

A selection of freely available resources to perform small molecule clustering is included in Table 5.1, with the authors' appraisal of their advantages and disadvantages. Some questions to answer when planning a clustering study are listed in Table 5.2.

3 Challenges and Perspectives

Suppose that we were assigned the "simple" task of clustering the following three objects: a goldfish, a dolphin, and a human. Some people may argue that the goldfish and the dolphin should be grouped together, based on criteria of shape, living environment, external organs, etc. Other people may suggest that dolphins and humans are more similar based, for instance, on the similarities between their internal organs (lungs, brains, etc.). Both arguments are sound. The previous example illustrates that *the cluster structure of a dataset strongly depends on the features that have been used to partition the data*: if different sets of features are considered, we will possibly obtain different results.

It is commonly argued, on the other hand, that the cluster structures that are most interesting are those that are "meaningful," a concept that is usually used as a synonym for "useful." For instance, it could be said that a classification of books based on subject matter, literary genre, or geographical origin of the author(s) would be of much wider use than a classification based on the color of the book's binding [18]. However, if we Google "noir books," we may be surprised to see that the colors red and black predominate in the cover images that the search returns. Another curious fact that contributes to this line of argument: in Italy, the books belonging to the genre of crime fiction are known as *libri gialli* (literally, "yellow books"), because back in the 1920s the prestigious publishing house Mondadori characterized a popular series of crime fiction with yellow covers.

This second example illustrates two other important points. The first is that the features that a researcher chooses based on the "natural" structure that he subjectively anticipates in the data can omit valuable (but unexpected) features that could, in turn, lead to a useful or meaningful cluster structure (e.g., cover colors as a possible feature to group books) and that, at first glance, would surely be discarded in favor of more "reasonable" features. Second, if the structure of the data depends on the variables used to partition it, if unsuspected variables can lead to valuable clusters, and if the value criterion varies according to the objective of the clustering experiment, it is possible that there is no natural underlying structure (ground truth) to structure the data.

Although the existence of such ground truth is assured when working with synthetic data, it is not so evident when working with real data. The personal view

Table 5.1 Resources to perform small molecules

Name	Clustering method	Pros and cons	Reference
ChemBioServer 2.0 (available as web app)	Hierarchical and affinity propagation clustering	Pros: Compound fingerprints can be provided by the user or generated on-site using 166-bit MACCS Open Babel fingerprint from .sdf or .mol files. In the case of hierarchical clustering, the user can select among different distances, linkage approaches, and thresholds. A tutorial is available on the site. Results are stored for a week Cons: Despite the tutorial and the availability of a supporting paper, the information on the fundamentals of the affinity propagation clustering may be a bit scarce	[25]
ChemMine tools (available as web app)	Hierarchical clustering, multidimensional scaling (MDS), and binning clustering	Pros: Compounds can be inputted using SMILES notation, as . sdf or using PubChem CIDs. They can also be drawn online. The user can tune different parameters. A tutorial is available online. Depending on the clustering methods, the output can be exported graphically or as tables/.csv file. Past jobs are accessible. Free support available Cons: Even for small sized data sets, jobs run rather slowly. Despite the tutorial and the availability of a supporting paper, the details of the procedures are a bit scarce, though most of them use very well-documented functions implemented in R	[4]
Deep-Clustering (available as Python code)	VAE (Variational Auto Encoder) & K-means AE (Auto encoder) & K-means	Pros: The clustering procedure exploits both local and global molecular features. It can be applied to large-scale chemical libraries. The supporting paper describes the clustering procedure in detail Cons: Unavailability as web application	[21]

(continued)

Table 5.1 (continued)

Name	Clustering method	Pros and cons	Reference
ChemmineR (R package)	Hierarchical clustering: Maximum Common Substructure (MCS) Search. Nonhierarchical clustering: K-means clustering	Pros: ChemmineR is a popular cheminformatics package, with plenty of documentation and an active community forum and blog available through their developers. The developers welcome community-contributed resources	[12]
hclust (R function)	Different hierarchical clustering methods (Ward, single, complete, and average linkage, among others)	Pros: This is a well-documented R function to perform hierarchical clustering based on different combinations of distances and linkage methods. In fact, it has been used to build some other clustering resources listed here	[53]
RDKIT (collection of cheminformatics and machine-learning software written in C++ and Python)	Sphere exclusion and fuzzy clustering	Pros: Large community of users. Community forum is available. Well documented Cons: Time-consuming, these methods can be slow to run and are best used on small sets (no more than a few hundred molecules) of small molecules	[24]
iRaPCA (available as web app and Python code)	Hybrid clustering method using molecular descriptors and K-means	Pros: The developers have validated their approach across 29 data sets of different sizes, obtaining better metrics than several classic approximations. The user can tune different parameters and input their own molecular descriptor sets to perform the clustering. Allows iterative sub-clustering. Free support is available. The supporting paper describes the clustering procedure in detail Cons: Long computer times for large-scale libraries. Low interpretability	[38]
SOMoC (available as a web app and Python code)	Non-hierarchical clustering using EState1 fingerprints and UMAP	Pros: The developers have validated their approach across 29 data sets of different sizes and obtained better metrics than several classic approximations. The user can tune different parameters. Free support is available. The supporting paper describes the clustering	[38]

(continued)

Table 5.1 (continued)

Name	Clustering method	Pros and cons	Reference
		procedure in detail. Compatible with large-scale libraries. Cons: Relatively low interpretability	
clValid (R package)	Variety of methods for validating the results from a cluster analysis.	–	[8]
NbClust (R package)	Clustering validity indexes can be applied to outputs of two clustering algorithms: K-means and hierarchical agglomerative clustering (HAC), by varying all combinations of a number of clusters, distance measures, and clustering methods	Pros: Provides 30 indexes that determine the number of groups in a data set. The user can simultaneously evaluate several clustering schemes while varying the number of clusters. Nbclust offers functionalities to visualize the results of various indices	[14]
cclust (R package)	Convex Clustering methods include the K-means algorithm, the On-line Update algorithm (Hard Competitive Learning), and the Neural Gas algorithm (Soft Competitive Learning)	Pros: cclust offers a variety of convex clustering methods and provides several clustering indexes. It can handle complex data sets with large numbers of data points and dimensions. Cons: The interpretation of clustering results can be difficult, especially for high-dimensional data sets or those with complex structures	[63]
MASSA (Python code)	Explores the biological, physicochemical, and structural spaces of molecules using PCA, hierarchical clustering analysis and K-modes	Initial evaluation of the algorithm in the context of QSAR/QSPR modeling to split data sets into training and test sets produced models with low variability and better values for validation metrics, even when the descriptors used in the QSAR/QSPR were different from those used in the separation of training and test sets, Generates graphical representations that can provide further insights into the data	[51]

Table 5.2 Questions to answer when planning a clustering study

Have you checked relevant literature on the subject?
Is there some prior clue about the structure of the data that can serve as an external validation criterion?
If so, do you trust the data? Can you do anything to minimize data noise and data mislabeling?
Would it be convenient for you to include a feature-selection step?
What internal cluster validation index(es) will you include?
Would you address cluster replicability?
Has any bias been reported for the clustering method(s) of choice?

of the authors is that in the real world a ground truth, a ground cluster structure of real data, does not exist (especially if we use clustering as a truly unsupervised approximation, with no preconception or prior labeling of the data points). This view seems in line with the notion of *clusterability*. As pointed out by Ackerman and Ben-David [2], "if a data set is hard to cluster then it does not have a meaningful clustering structure." However, we accept that some representations of the data are more useful for specific purposes, and their usefulness depends on the value criteria of the user.

This leads us to the problem of variable selection, an old debate that has not yet been fully settled. Ketchen et al. [27] describe three fundamental approaches to the selection of relevant features for a clustering experiment: *inductive*, *deductive*, and *cognitive*.

The inductive approach considers as many variables as possible because it is not known a priori which features provide the best differentiation among observations; thus, using as many clustering variables is likely to maximize the chances of discovering significant differences [28, 34]. In contrast, the *deductive* and *cognitive* approaches (pre)select the feature space based on either theoretic knowledge or expert judgment. The (often vast) universe of possible features is thus pruned to a more limited and, in principle, meaningful number based on previous knowledge.

The preceding concepts are captured, in one way or another, by clustering approaches that incorporate dimensionality-reduction and/or feature-selection steps, although the latter have rarely been applied in the field of small molecule clustering [42]. Small molecules are often represented by high-dimensional vectors (i.e., a large pool of global and/or local molecular features), which, for different reasons (to facilitate visual representation, reduce computational cost, and/or capture orthogonal and significant feature projections), are often pre-processed using dimensionality reduction techniques, such as Principal Component Analysis (PCA) [38] or the Uniform Manifold Approximation and Projection (UMAP) approaches [24, 38]. Hadipour et al. [21] recently reported an open-source deep clustering approach, in which dimensionality-reduction techniques were implemented serially. First, the authors resorted to PCA on sets of global and local molecular features, leading to a representation comprising 243 features, which was then subjected to further dimensionality reduction using autoencoders.

In particular, McKelvey's view is well embodied by subspace clustering approximations, which attempt to find relevant cluster structures in different subspaces of the feature space. In this way, large pools of variables could be exploited through smaller subsets, either through systematic exploration (computationally very demanding for high numbers of features) or stochastic, until combinations of variables useful for the problem in question are found, without a priori selection of variables biased by the researcher's prior knowledge (and lack of knowledge!). For example, starting from descriptor sets provided by the user, the iRaPCA approach [38] explores stochastic subsets of molecular descriptors where cohesive and well-separated clusters can be found, as judged by the silhouette coefficient or other validity measures. Each of the so-obtained clusters can be further divided, iteratively, into sub-clusters. iRaPCA has shown, without iterations, consistent and almost optimal behavior in benchmarking experiments across 29 data sets of variable sizes [38, 39].

On the other hand, multi-view clustering has also attracted much attention lately [13, 19, 57] and is better aligned with the idea of an underlying ground-truth structure of the data (in which case any view of the data would originate from an underlying latent space). However, it has largely been overlooked in the field of cheminformatics. Different views of the data may be integrated or combined because of their complementarity (this can be of particular interest when the clustering procedure is supervised via external labeling of the data) or based on their consensus (returning to the idea that high-level stability should be verified across different algorithms, features, and entities).

4 Conclusions

The principle of similarity ("guilty by association") has been and is of great importance in the field of cheminformatics and drug discovery, in which molecular similarity (apparent or not) is used to identify new bioactive scaffolds, develop a series of analogs around such scaffolds, and make predictive interpolations related to physicochemical and biological properties of interest. However, in several ways, the field of supervised machine learning has matured more rapidly than its unsupervised counterpart. Not only is supervised learning much more widely used, but rigorous internal and external validation of supervised models is common practice, while validation of small molecule clustering experiments is much more sporadic. Furthermore, while supervised machine-learning procedures in the cheminformatics field have rapidly assimilated the newest algorithms, small molecule clustering still does not fully exploit recent progress such as subspace clustering and multi-view clustering (with exceptions). A good part of the bibliography used in this chapter comes from disciplines as distant and dissimilar as psychology, organizational management or marketing, image recognition, bioinformatics, etc. In addition, of course, to pure mathematics. The progress that clustering theory has experienced in other disciplines has spilled over, rather very slowly, to cheminformatics.

In a chemical universe that is expanding exponentially, as demonstrated by the currently available ultra-large chemical libraries, the development and application of specific clustering methods will allow the detection of unforeseen patterns in the data and will accelerate the exploitation of this virtually infinite chemodiversity. It is no longer possible to trust the human ability to detect molecular patterns, not only because the vast number of accessible molecules cannot be encompassed humanly but also because in such richness and chemical diversity, it is likely that relationships that are not apparent and unnoticed to the naked eye will appear, the detection of which requires higher levels of abstraction and the integration of chemical knowledge and advanced mathematics. Subspace clustering and multi-view clustering, still largely unexplored in the field of chemistry, will provide solutions to the problem of automatic or semi-automatic analysis of ultra-large chemical libraries.

Acknowledgments The three authors are members of CONICET and UNLP. They thank these institutions and FONCYT (PICTs 2019-0984, 2019-1075, 2021-0720) for their financial support.

References

1. Adnan, M., Slavic, G., Martin Gomez, D., Marcenaro, L., & Regazzoni, C. (2023). Systematic and comprehensive review of clustering and multi-target tracking techniques for LiDAR point clouds in autonomous driving applications. *Sensors (Basel), 23*, 6119.
2. Ackerman, M., & Ben-David, S. (2009). Clusterability: A theoretical study. *Proceedings of the Twelfth International Conference on Artificial Intelligence and Statistics, PMLR, 5*, 1–8.
3. Arbelaitz, I., Gurrutxaga, I., Muguerza, J., Pérez, J. M., & Perona, I. (2013). An extensive comparative study of cluster validity indices. *Pattern Recognition, 46*, 243–256.
4. Backman, T. W., Cao, Y., & Girke, T. (2011). ChemMine tools: An online service for analyzing and clustering small molecules. *Nucleic Acids Research, 39*, W486–W491.
5. Baker, F. B., & Hubert, L. J. (1975). Measuring the power of hierarchical cluster analysis. *Journal of the American Statistical Association, 70*, 31–38.
6. Böcker, A., Derksen, S., Schmidt, E., Teckentrup, A., & Schneider, G. (2005). A hierarchical clustering approach for large compound libraries. *Journal of Chemical Information and Modeling, 45*, 807–815.
7. Breckenridge, J. N. (2000). Validating cluster analysis: Consistent replication and symmetry. *Multivariate Behavioral Research, 35*, 261–285.
8. Brock, G., Pihur, V., Datta, S., & Datta, S. (2008). clValid: An R package for cluster validation. *Journal of Statistical Software, 25*, 1–22.
9. Brooks, J. L. (2014). Traditional and new principles of perceptual grouping. In J. Wagemans (Ed.), *The Oxford handbook of perceptual organization* (pp. 57–87). Oxford University Press.
10. Butina, D. (1999). Unsupervised data base clustering based on daylight's fingerprint and tanimoto similarity: A fast and automated way to cluster small and large data sets. *Journal of Chemical Information and Computer Sciences, 39*, 747–750.
11. Calinski, R. B., & Harabasz, J. (1974). A dendrite method for cluster analysis. *Communications in Statistics, 3*, 1–27.
12. Cao, Y., Charisi, A., Cheng, L. C., Jiang, T., & Girke, T. (2008). ChemmineR: A compound mining framework for R. *Bioinformatics, 24*(15), 1733–1734.
13. Cao, Z., & Xie, X. (2024). Structure learning with consensus label information for multi-view unsupervised feature selection. *Expert Systems with Applications, 238*, 121893.

14. Charrad, M., Ghazzali, N., Boiteau, V., & Niknafs, A. (2014). NbClust: An R package for determining the relevant number of clusters in a data set. *Journal of Statistical Software, 61*, 1–36.
15. Davies, D., & Bouldin, D. W. (1979). A cluster separation measure. *IEEE Transactions on Pattern Analysis and Machine Intelligence, 1*, 224–227.
16. Domingo-Fernández, D., Gadiya, Y., Mubeen, S., Healey, D., Norman, B. H., & Colluru, V. (2023). Exploring the known chemical space of the plant kingdom: Insights into taxonomic patterns, knowledge gaps, and bioactive regions. *Journal of Cheminformatics, 15*, 107.
17. Dunn, J. C. (1974). Well-separated clusters and optimal fuzzy partitions. *Journal of Cybernetics, 4*, 95–104.
18. Everitt, B. S., Landau, S., Leese, M., & Stahl, D. (2011). *Cluster analysis* (5th ed., p. 71). Wiley.
19. Guo, J., Sun, Y., Gao, J., Hu, Y., & Yin, N. (2022). Rank consistency induced multiview subspace clustering via low-rank matrix factorization. *IEEE Transactions on Neural Networks and Learning Systems, 33*, 3157–3170.
20. Gramatica, P. (2013). On the development and validation of QSAR models. *Methods in Molecular Biology, 930*, 499–526.
21. Hadipour, H., Liu, C., Davis, R., Cardona, S. T., & Hu, P. (2002). Deep clustering of small molecules at large-scale via variational autoencoder embedding and K-means. *BMC Bioinformatics, 23*(Suppl. 4), 132.
22. Handl, J., Knowles, J., & Kell, D. B. (2005). Computational cluster validation in post-genomic data analysis. *Bioinformatics, 21*, 3201–3212.
23. Hawkins, D. M., Basak, S. C., & Mills, D. (2003). Assessing model fit by cross-validation. *Journal of Chemical Information and Computer Sciences, 43*, 579–586.
24. Hernández-Hernández, S., & Ballester, P. J. (2023). On the best way to cluster NCI-60 molecules. *Biomolecules, 13*, 498.
25. Karatzas, E., Zamora, J. E., Athanasiadis, E., Dellis, D., Cournia, Z., Spyrou, G. M., Thomas, J. B., & Snow, C. C. (2020). ChemBioServer 2.0: An advanced web server for filtering, clustering and networking of chemical compounds facilitating both drug discovery and repurposing. *Bioinformatics, 36*(8), 2602–2604.
26. Kaufman, L., & Rousseeuw, P. J. (1990). Partitioning around medoids (program PAM). In *Finding groups in data: An introduction to cluster analysis*. Wiley.
27. Ketchen, D. J., Thomas, J. B., & Snow, C. C. (1993). Organizational configurations and performance: A comparison of theoretical approaches. *Academy of Management Journal, 36*, 1278–1313.
28. Ketchen, D. J., & Shook, C. L. (1996). The application of cluster analysis in strategic management research: An analysis and critique. *Strategic Management Journal, 17*, 441–458. https://doi.org/10.1002/(SICI)1097-0266(199606)17:6<441::AID-SMJ819>3.0.CO;2-G
29. Krieger, A. M., & Green, P. E. (1999). A cautionary note on using internal cross validation to select the number of clusters. *Psychometrika, 64*, 341–353.
30. Leonard, J. T., & Roy, K. (2006). On selection of training and test sets for the development of predictive QSAR models. *QSAR and Combinatorial Science, 25*, 235–251.
31. Lupyan, G. (2008). The conceptual grouping effect: Categories matter (and named categories matter more). *Cognition, 108*, 566–577.
32. Mayr, A., Klambauer, G., Unterthiner, T., Steijaert, M., Wegner, J. K., Ceulemans, H., Clevert, D. A., & Hochreiter, S. (2018). Large-scale comparison of machine learning methods for drug target prediction on ChEMBL. *Chemical Science, 9*, 5441–5451.
33. McClain, J. O., & Rao, V. R. (1975). CLUSTISZ: A program to test for the quality of clustering of a set of objects. *Journal of Marketing Research, 12*, 456–460.
34. McKelvey, B. (1975). Guidelines for empirical classification of organizations. *Administrative Science Quarterly, 20*, 509525.
35. Milligan, G. W. (1980). An examination of the effect of six types of error perturbation on fifteen clustering algorithms. *Psychometrika, 45*, 325–342.
36. Milligan, G. W. (1981). A Monte Carlo study of thirty internal criterion measures for cluster analysis. *Psychometrika, 46*, 187–199.

37. Murtagh, F., & Contreras, P. (2017). Algorithms for hierarchical clustering: An overview. *Wiley Interdisciplinary Reviews: Data Mining and Knowledge Discovery, 2*, e1219.
38. Prada Gori, D. N., Llanos, M. A., Bellera, C. L., Talevi, A., & Alberca, L. N. (2022a). iRaPCA and SOMoC: Development and validation of web applications for new approaches for the clustering of small molecules. *Journal of Chemical Information and Modeling, 62*, 2987–2998.
39. Prada Gori, D. N., Alberca, L. N., Rodriguez, S., Llanos, M. A., Bellera, C. L., & Talev, A. (2022b). LIDeB tools: A Latin American resource of freely available, open-source cheminformatics apps. *Artificial Intelligence in the Life Sciences, 2*, 100049.
40. Punj, G. N., & Stewart, D. W. (1983). Cluster analysis in marketing research: Review and suggestions. *Journal of Marketing Research, 20*, 134–148.
41. Risser-Maroix, O., Marzouki, A., Djeghim, H., Kurtz, C., & Lomenie, N. (2021, September). *Learning an adaptation function to assess image visual similarities*. Paper presented at ORASIS 2021, Centre National de la Recherche Scientifique. Available from https://hal.science/hal-03339731v2/document. Accessed 17 Nov 2023.
42. Rivera-Borroto, O. M., Marrero-Ponce, Y., García-de la Vega, J. M., & Grau-Ábalo, R. C. (2011). Comparison of combinatorial clustering methods on pharmacological data sets represented by machine learning-selected real molecular descriptors. *Journal of Chemical Information and Modeling, 51*, 3036–3049.
43. Rousseeuw, P. (1987). Silhouettes: A graphical aid to the interpretation and validation of cluster analysis. *Journal of Computational and Applied Mathematics, 20*, 53–65.
44. Schubert, E. (2023). Stop using the elbow criterion for k-means and how to choose the number of clusters instead. *ACM SIGKDD Explorations Newsletter, 25*, 36–42.
45. Seger, C. A., & Miller, E. K. (2010). Category learning in the brain. *Annual Review of Neuroscience, 33*, 203–219.
46. Sheikholeslami, C., Chatterjee, S., & Zhang, A. (2000). WaveCluster: A multi-resolution clustering approach for very large spatial database. *VLDB Journal, 8*, 289–304.
47. Tan, P. N., Steinbach, M., & Kumar, V. (2005). Cluster analysis: Basic concepts and algorithms. In *Introduction to data mining*. Addison-Wesley Longman Publishing.
48. Tichý, M., & Rucki, M. (2009). Validation of QSAR models for legislative purposes. *Interdisciplinary Toxicology, 2*, 184–186.
49. Tropsha, A. (2010). Best practices for QSAR model development, validation, and exploitation. *Molecular Informatics, 29*, 476–488.
50. Tropsha, A., Gramatica, P., & Gombar, V. K. (2003). The importance of being earnest: Validation is the absolute essential for successful application and interpretation of QSPR models. *QSAR and Combinatorial Science, 22*, 69–77.
51. Veríssimo, G. C., Pantaleão, S. Q., Fernandes, P. O., Gertrudes, J. C., Kronenberger, T., Honorio, K. M., & Maltarollo, V. G. (2023). MASSA Algorithm: An automated rational sampling of training and test subsets for QSAR modeling. *Journal of Computer-Aided Molecular Design, 37*, 735–754.
52. Virshup, A. M., Contreras-García, J., Wipf, P., Yang, W., & Beratan, D. N. (2013). Stochastic voyages into uncharted chemical space produce a representative library of all possible drug-like compounds. *Journal of the American Chemical Society, 135*, 7296–7303.
53. Voicu, A., Duteanu, N., Voicu, M., Vlad, D., & Dumitrascu, V. (2020). The rcdk and cluster R packages applied to drug candidate selection. *Journal of Cheminformatics, 12*(1), 3.
54. Yang, Y., Yao, K., Repasky, M. P., Leswing, K., Abel, R., Shoichet, B. K., & Jerome, S. V. (2021). Efficient exploration of chemical space with docking and deep learning. *Journal of Chemical Theory and Computation, 17*, 7106–7119.
55. Yu, L., He, X., Fang, X., Liu, L., & Liu, J. (2023). Deep learning with geometry-enhanced molecular representation for augmentation of large-scale docking-based virtual screening. *Journal of Chemical Information and Modeling, 63*, 6501–6514.
56. Zhang, C., Huang, W., Niu, T., Liu, Z., Li, G., & Cao, D. (2023). Review of clustering technology and its application in coordinating vehicle subsystems. *Automotive Innovation, 6*, 89–115.

57. Zhang, C., Fu, H., Hu, Q., Cao, X., Xie, Y., Tao, D., & Xu, D. (2020). Generalized latent multi-view subspace clustering. *IEEE Transactions on Pattern Analysis and Machine Intelligence, 41*, 86–99.
58. Lopez-Del Rio, A., Nonell-Canals, A., Vidal, D., and Perera-Lluna, A. (2019). Evaluation of cross-validation strategies in sequence-based binding prediction using deep learning. J. Chem. Inf. Model 59, 1645–1657.
59. Harris, C. J., Hill, R. D., Sheppard, D. W., Slater, M. J., and Stouten, P. F. (2011). The design and application of target-focused compound libraries. Comb. Chem. High. Throughput Screen 14, 521–531.
60. MacQueen, J. (1967). "Some methods for classification and analysis of multivariate observations," in Proceedings of the 5th Berkeley Symposium on Mathematical Statistics and Probability Volume 1. Nerkeley. Editors L. M. Le Cam, and J. Neyman (University of California Press), 281–297.
61. Frey, T., & Van Groenewoud, H. (1972). A cluster analysis of the D-squared matrix of white spruce stands in Saskatchewan based on the maximum-minimum principle.Journal of Ecology, 60, 873–886.
62. Milligan, G.W., Cooper, M.C. An examination of procedures for determining the number of clusters in a data set. Psychometrika 50, 159–179 (1985).
63. Dimitriadou E (2023) cclust: convex clustering methods and clustering indexes http://CRAN.R-project.org/package=cclust. R package version 0.6-26

Chapter 6
QSAR and Machine Learning Predictors

Philipe Oliveira Fernandes and Vinicius Gonçalves Maltarollo

Abstract This chapter delves into the fundamental principles and applications of quantitative structure–activity relationship (QSAR) and machine learning (ML)-based predictors in the realm of drug design and chemical biology. QSAR establishes a quantitative relationship between the chemical structure of molecules and their biological activities or physicochemical properties. The evolution of QSAR from its first reports to its modern applications was covered comprising the theoretical foundations, encompassing descriptors, mathematical models (followed by brief examples of ML applied to this field), and statistical validation techniques employed in QSAR analysis. Interestingly, many recognized and accepted good practices and validation protocols align with OECD guidelines for QSAR applications for regulatory purposes. In this sense, notably, the QSAR field became important outside of the academic boundaries. Additionally, this chapter discusses current challenges and emerging trends in QSAR research, including the incorporation of machine learning algorithms and big data analytics for enhanced predictive accuracy and applicability. Overall, this chapter serves as a comprehensive guide for researchers and practitioners in understanding and leveraging QSAR as a pivotal tool in rational drug design and chemical biology.

Keywords QSAR · Machine learning · Molecular descriptors · OECD principles

1 Historical Background

The Quantitative Structure–Activity Relationship (QSAR) field was introduced when researchers tried to correlate the molecular structure of similar compounds to a specific biological activity. This process can be traced back to 1863 when Cros first reported a correlation between the toxicity of primary aliphatic alcohols and their water solubility [1], a pioneer work in this field. Another notable work from the

P. O. Fernandes · V. G. Maltarollo (✉)
Departamento de Produtos Farmacêuticos, Faculdade de Farmácia, Universidade Federal de Minas Gerais, Belo Horizonte, Minas Gerais, Brazil

V. G. Maltarollo (ed.), *Computer-Aided and Machine Learning-Driven Drug Design*, Computer-Aided Drug Discovery and Design 3,
https://doi.org/10.1007/978-3-031-76718-0_6

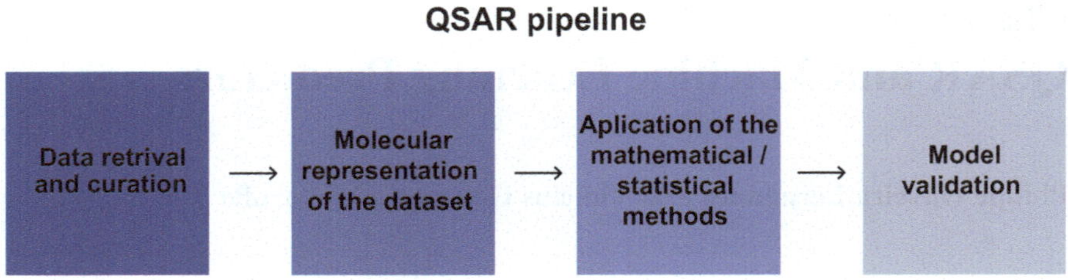

Fig. 6.1 Simplified process to build and validate a QSAR model

nineteenth century was the correlation between melting- and boiling points described by Edmund J. Mills in 1884 [2]. Since then, the field expanded, and in 1962, the work from Hansch et al. [3] was considered a milestone for the modern QSAR period. A relationship between phenoxyacetic acid derivatives and their biological activity as plant growth regulators was established. In this work, they reported the famous equation relating the biological activity (y) and the descriptors (X) multiplied by specific coefficients that could be interpreted as individual importance to modulate the modeled effect. In the following years, Corwin Hansch participated in two other works [4, 5] that establish the basis of QSAR modeling used today, especially the work from Fujita et al. [5] were described as a new substituent constant. Pioneering works of these using multi-parametric regression are worth mentioning to establish the structure–activity correlation. Later, the evolution of computers allowed the use of new ways of representing molecules through theoretical descriptors, and today QSAR is an inseparable part of the drug design/discovery process [6].

As a mature research field, QSAR modeling has well-defined procedures for exploring the relationship of chemical compounds to biological activity or other properties. In this sense, the model that predicts other properties than the biological activity is also called the quantitative structure–property relationship (QSPR) [7–9], and other names are used for different purposes such as quantitative structure–reactivity relationships (QSRR) [10–12] and quantitative structure–toxicity relationship (QSTR) [13–15].

The construction of a model to predict a biological activity as a function of the molecular structure is the primary goal, and it can be explained simply in four general steps (Fig. 6.1). The first one is obtaining the data set, a group of molecules with the endpoint (the biological activity or other property that will be modeled) experimentally measured. These data can be obtained by in-house assays, nevertheless, it is more common to be retrieve datasets from the literature. Later, these data should be carefully inspected to ensure their quality. Fourches et al. [16] in the paper "Trust, but verify: on the importance of chemical structure curation in cheminformatics and QSAR modeling research" discuss the errors in structural representations observed in medicinal chemistry publications and how erroneous structures and/or duplicate entries negatively influence the QSAR models, reducing their statistical power and even leading to a complete failure. This and the follow-up

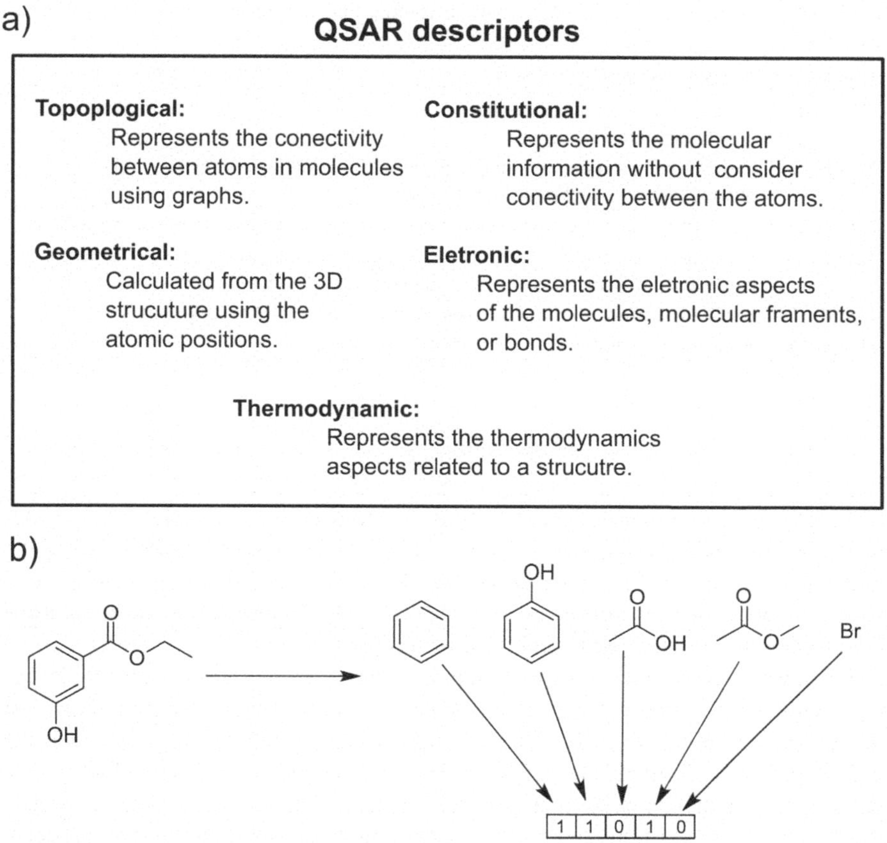

Fig. 6.2 Descriptors and fingerprints. (**a**) Representation of the molecular descriptors classification by the information described by the class. (**b**) Schematic representation of a molecular fingerprint, where the molecule is compared to specific fragments, and this information is combined in a binary vector, called a fingerprint

work [17] described a pipeline to process the information to properly perform a data curation.

Later, a mathematical representation of the molecular structure should be done as a descriptor, a fingerprint, physicochemical properties, a graph, etc. A descriptor is a numerical representation of a molecule using predefined rules and it can be obtained from the molecular formula, bidimensional, or three-dimensional structure [18, 19]. A way to classify them is related to the information obtained by using the descriptor such as geometrical descriptors that reflect the 3D structure of the molecule [19] (Fig. 6.2a). Similar to a descriptor, a fingerprint is also a representation of the molecule based on specific rules, but in a vector, where each position (*bin*) carries a different information [20] (Fig. 6.2b).

The descriptors can be obtained from the molecular formula, bidimensional, or three-dimensional structure. Usually, the descriptors classify the QSAR model

Table 6.1 Classification of the QSAR models according to their dimensionality of the molecular descriptor

Class	Examples of used descriptors
1D-QSAR	Descriptors based on global molecular properties such as pk_a, LogP, molecular weight, and others
2D-QSAR	Descriptors based on structural patterns such as connectivity indices, 2D pharmacophores, and molecular fragments
3D-QSAR	Descriptors based on noncovalent interaction fields around 3D molecular models
4D-QSAR	Descriptors that include multiple ligand conformations to the 3D QSAR model, where the conformations can be generated by molecular dynamics
5D-QSAR	Explicitly includes several induced-fit models to the 4D QSAR model
6D-QSAR	Adds the solvation model to the 5D QSAR model

according to their dimensionality [21, 22] (Table 6.1). For example, a 2D-QSAR model is built using descriptors calculated from the bidimensional structure of the molecules. A 3D-QSAR model uses the three-dimensional representation of a molecule, while higher-dimensional QSAR (4D, 5D, 6D, and 7D) includes multiple conformations and other structural information [23].

Once represented, the set of the endpoint vector (y) and descriptors matrix (X) representing the compounds (samples) are used to build a model, but first, they must be split into training test sets. The training set compounds are used to teach the algorithm what descriptors are important to predict a given endpoint, in other words, to tune the relationship between X and y variables; and the test set is used to validate the model built [24] (more about this process will be discussed in further sections).

QSAR modeling already has well-defined protocols and procedures to perform the application of the techniques, especially with the growing collection of biological data available in databases such as ChEMBL [25] and PubChem [26]. The work titled "Best practices for QSAR model development, validation and exploration" developed by Alexander Tropsha [24] is an important starting point to adequately build a QSAR model. Furthermore, the Organisation for Economic Co-operation and Development (OECD) [27] principles also provide valuable checkpoints that must be followed by the QSAR model for regulatory purposes.

From the Hansch [3, 4] and Fujita [5] works, the use of linear models was well adopted to build the relationships. One application example from that time is the work of Di Paolo [28] who had built QSAR models to correlate the anesthetic activity of ether derivatives. Another contemporaneous example is the one published by Hansch and Klein [29] where they explore enzyme ligand interactions applying QSAR techniques.

Over time, several QSAR methodologies have been developed and some of them are worth mentioning. The first one is the one based on the Molecular Interaction Fields (MIF) Comparative Molecular Field Analysis (CoMFA) [30, 31], Comparative Molecular Similarity Indices Analysis (CoMSIA) [32], and hologram quantitative structure-activity relationship (HQSAR) [33, 34]. These three QSAR

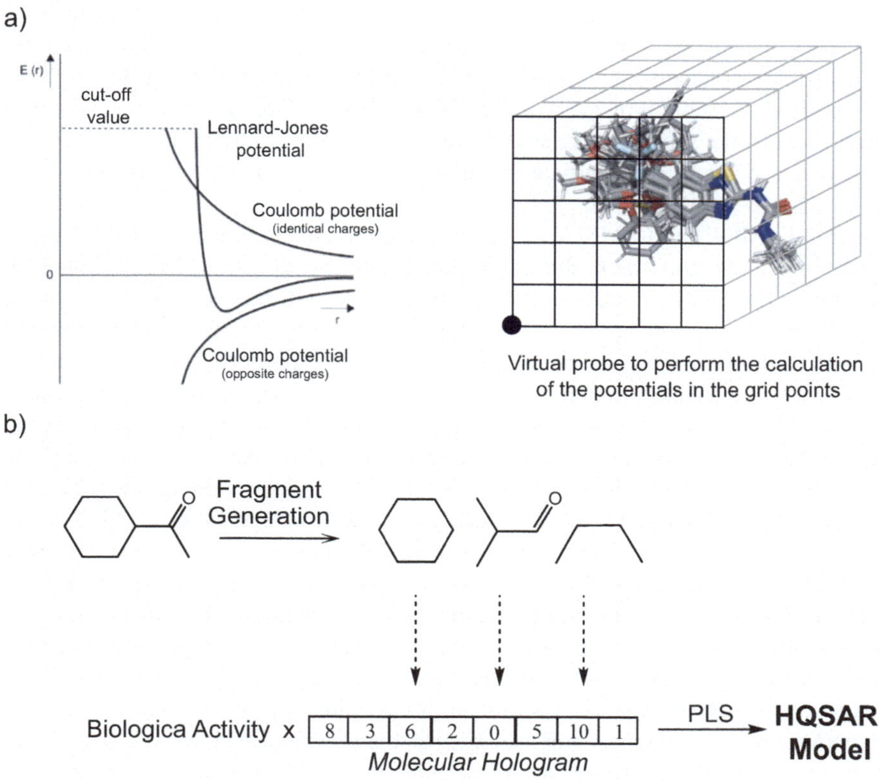

Fig. 6.3 Different QSAR methods: (**a**) schematic example of the grid-box used to calculate the MIFs, a crucial step for performing CoMFA and CoMSIA QSAR. (**b**) Pipeline of the HQSAR process. A molecular structure is fragmented, and these portions are combined into an array. Then, it is correlated to the biological activity using a PLS model

methodologies were successfully used in the field and they became a product implemented in the SYBYL software, commercialized by Tripos Inc.

The CoMFA QSAR is based on the correlation of the steric and electrostatic fields around the molecule and its biological activity. These two fields are calculated using a virtual probe atom (usually a sp^3 C^+ atom) to model the stereochemical interactions by the Lenard-Jones potential and to model the electrostatic interactions by the Coulombic potential in specific x, y, z coordinates points into a lattice intersection (also known as the grid box) (Fig. 6.3a). These two sterical and electrostatic MIFs in the absence of a molecular target knowledge could be interpreted as “the negative” of ligand-receptor representation, giving insights for molecular optimization aiming to increase the potency of novel compounds. Due to the three-dimensional requirements and the need to analyze all compounds in the same position, molecular alignment is a key step to building a CoMFA model. A PLS regression is done using the energy calculated from these fields in each point of the grid (*X*) and the endpoint to be modeled (usually, a biological activity, *y*) [30]. Similarly, the CoMSIA QSAR is based on the same principles of the CoMFA but computing similarity indices at each point of the same grid, related to steric, electrostatic, and hydrophobic potentials obtained by a Gaussian-type function. Also, the PLS regression is used in the CoMSIA methodology [32].

Both CoMFA and CoMSIA are dependent on the three-dimensional structure of the compounds and their alignment and, therefore are considered 3D-QSAR methods. The HQSAR was an alternative to work with bidimensional structures, using structural fragments of a set of molecules (Fig. 6.3b). This is a proprietary QSAR method implemented by Tripos in the SYBYL platform [33]. In this methodology the molecular structure is described as a hologram iteratively generated with a fingerprint composed of molecular fragments built from each molecule structure. Then, the PLS regression is applied to generate the QSAR model. Despite the evolution of the QSAR field and the popularity of machine learning-based QSAR, CoMFA, CoMSIA, and HQSAR are still used in drug design campaigns [35–40].

The QSAR field is continuously growing through the development of new methods and applications. Since 2007 more than 1000 relevant articles have been published annually [41]. Machine learning methods (see Chap. 4) have significantly advanced QSAR research by enabling the understanding of complex relationships not observed by linear models. This evolution did not just happen in the academia, but today QSAR are valuable instrument in the industry [42]. For that reason, it is important to follow the best practices such as the OECD principles, and to promote high quality in QSAR modeling, those guidelines will be discussed in the next section.

2 OECD Principles

Usually, the QSAR modeling prioritizes optimizing the model's performance to accurately replicate experimental outcomes. This approach leads to the development of models with limited transparency and interpretability, commonly referred to as "black boxes" such as Artificial Neural Networks (ANN), algorithms inspired by the natural neural networks [43]. An extensive review of ANN-based algorithms is available in Chap. 4. Despite their utility in predictive applications, these models have not sufficiently supported the generation of reliable decisions within regulatory applications [44], due to this behavior. This behavior and other misconceptions of the QSAR modeling resulted in the publication of the "Guidance document on the validation of (quantitative)structure-activity relationships [(Q)SAR] models" by the OECD [45]. In this document, five principles to be addressed before the application of a QSAR model (Fig. 6.4) were established. This publication marked a notable advancement in the development of in silico models, emphasizing the necessity to explicitly explain the model's intended purpose, construction methodology, and

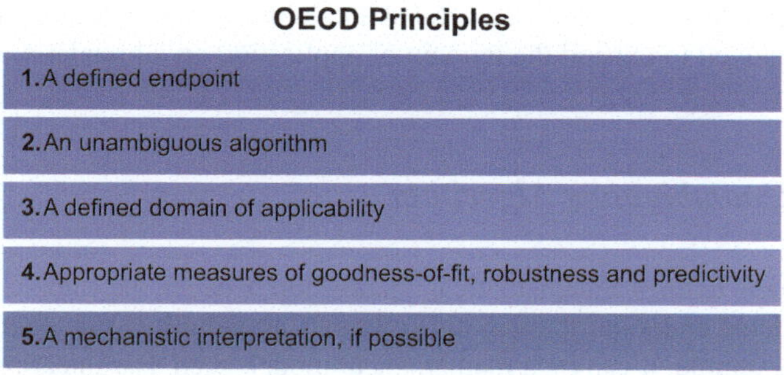

Fig. 6.4 Five principles defined by the OECD to develop and validate a QSAR model for reglementary purposes

performance metrics [44]. The following section will present in a more detailed, but not intended to be exhaustive, way.

2.1 A Defined Endpoint

The first principle defines that a QSAR model should have a defined endpoint and ensure the clarity of the prediction. The OECD suggests how specific an endpoint should be to ensure reliability based on the data used to build the model. It also describes examples associated with the OECD Test Guidelines such as physico-chemical properties (e.g., melting point, boiling point, and water solubility) environmental fate (e.g., biodegradation, hydrolysis, and bioaccumulation), ecological effects (e.g., acute fish toxicity, and alga toxicity), and human health effects (e.g., acute oral toxicity, acute dermal toxicity, genotoxicity, and organ toxicity).

This transparency in the data is an essential requirement for regulatory agencies to validate the application of the QSAR model for a particular problem. This principle also highlighted the importance of the quality of the data obtained and recommends that ideally all QSAR should be developed using experimental data generated by a single experimental protocol. This recomendation tries to avoid the introduction of errors related to the interexperimental variability. For more details on experimental assays, please refer to Chap. 12. Cronin and Schultz [46] also suggested that variations in the analyst and the equipment could introduce noise to the biological data. When it is not possible, they recommend a training set representative of the different protocols [27].

For the endpoints that may result from diverse chemical mechanisms, QSAR models should either be developed separately for each mechanism and applied to narrowly defined classes of chemicals, or a broader QSAR relationship can be established based on shared observations across noncongeneric chemical classes with distinct mechanisms. Either approaches can be integrated, or a statistical approach capable of simultaneous global modeling across multiple mechanisms must be employed to expand the domain of application on a broader scale.

2.2 An Unambiguous Algorithm

A QSAR model defines a mathematical relationship between the chemical structures and the modeled endpoint, the way this relationship is built is the algorithm of the model. For example, it can be a mathematical model, such as a linear regression or PLS, or a set of knowledge-based rules, such as a decision tree. In this sense, an unambiguous algorithm is capable of describing how the value was estimated and can be reproduced if desired.

The OECD document presents some algorithms that can perform regression, classification, or clustering (discussed in Chap.). The ability to understand how the estimation is performed contributes to the transparency of the model. Another important aspect is the distinction between the transparency of the algorithm and the ability to interpret it. For example, a linear regression can be transparent in terms of the variables' coefficients and how they are used to perform a prediction, however the variables themselves may not have an understandable physicochemical meaning based on the endpoint modeled. Two-dimensional autocorrelations and 3D Morse descriptors [47] are one example of this problem. These descriptors are widely used to build QSAR models and, despite the authors trying to give them a physical meaning, it is very shallow and hard to understand how they affect the biological activity.

Machine learning methods are more complex in comparison with traditional regression methods (such as linear regression) due to the number of mathematical operations and complex functions applied in the original descriptor values making it hard to perform a straightforward interpretation. Due to this machine learning researchers are working to develop strategies to easily interpret the variables of these so-called black-box models. Recently, Rodríguez-Pérez and Bajorath [48] described in a perspective article methodologies for better understanding the machine learning models and their individual predictions, as well the current challenges for the integration of these techniques in medicinal chemistry in this field.

2.3 A Defined Domain of Applicabilit

This principle relies on the necessity to establish the scope and limitations of a model, and it is based on the structure and physicochemical properties of the training set. It is expected that a model can give reliable predictions for compounds similar to those used to build the model and predictions outside this boundary are less likely to be reliable because they are extrapolations of the model.

Due to the nature of the applicability domain (AD), each model has a specific one based on the data used to train it. Also, the outcome of an AD can be categorical (e.g., yes or no) or quantitative, determining the degree of similarity between the predicted compound and the training data. There are different methods of similarity to determine if a compound falls within the AD of a model. The OECD document suggests some strategies to assess the AD, but the simplest one is observing the range of the descriptors used in the model. If the predicted compound is between the minimum and maximum values observed for each descriptor in the training set, it is inside the domain. For the model displaying few descriptors, it is easy to observe; nevertheless, it can be more complicated in complex models.

2.4 Appropriate Measures of Goodness-of-Fit, Robustness, and Predictivity

This OECD principle advocates for the necessity of statistical validation to ensure the predictability of the model. The OECD document divides the assessment of model performance (or statistical validation) into two stages: the internal performance (goodness-of-fit and robustness) and the external performance (predictivity). The combination of these two methods is used to analyze the model, avoiding being overfitted and underfitted. The target is a model not so simple that lacks information and is not so complex to model the noise in the data [49, 50].

The statistical validation enables comparison between different models in order to select the most predictive model. Another function of this process is to avoid "spurious" models based on correlations by chance, which are not meaningful and not predictive [51–53]. The internal validation is a measure of the model's performance using only the training set samples, which is the data used to build the model. In this sense, models that could not predict the data used in the training process could be discarded in the early stages of QSAR modeling. External validation is a way to measure the performance using compounds unseen by the model (the section validation and controls will further discuss how statistical metrics can be applied to perform these validations) and, therefore, represents an application of QSAR in actual drug discovery pipeline mimicking the prediction of compounds in commercial libraries. For that reason, it is important to assess the applicability domain to ensure a test set representative of the training set.

2.5 A Mechanistic Interpretation, if Possible

Historically, the first QSAR models were developed using congeneric compounds (molecules from the same chemical class with minor variations in substituents), and the activity was related to biological activities produced by the compounds' molecular structure. The statistical methods were applied to describe the relationships and support chemical knowledge, beginning the idea of mechanistic interpretation. This OECD principle is not mandatory but is desirable, once a QSAR model is consistent if the knowledge of the chemical/toxicological process provides more credibility and acceptance of the predictions [27]. Furthermore, the interpretation of a model allows an "extra layer" of validation since the proposed mechanism could be checked according to its chemical meaning and if it is expected or not. For example, it is expected that hydrogen bond descriptors positively influence the aqueous solubility of a QSPR model [54]; otherwise, there is a high possibility that the model is not correct.

The "if possible" clause added in the principle is due to the interactive nature of the modeling process. Usually, data exploration and modeling lead to the generation and testing of a hypothesis, and a useful QSAR model may lack mechanistic interpretation due to different reasons. Nevertheless, this principle motivates the modeler to seek a mechanistic interpretation that can contribute to the understanding of the statistic validation [27].

3 Software and Tools

Unless you are performing your own assays, the first step to performing QSAR modeling is retrieving information from the literature. This process can be made manually by compiling published works relevant to the project or using available databases. The second approach is very common today since there are several initiatives available. One popular example is the ChEMBL database [55], a large-scale database comprising several bioactivity information. The compounds' chemical structure and detailed information about the assays are provided, comprising binding measurements, functional assays, pharmacokinetic properties, and toxicity. Besides ChEMBL there are other large databases such as PubChem [56] and BindingDB [57]. A detailed explanation of available chemical information sources is described in Chap. 2.

The next important step in the development of a QSAR model is the representation of the molecules. To obtain these representations, the molecular descriptors and/or fingerprints are calculated using specific software and web servers. A nonexhaustive list of available tools to perform descriptors/fingerprint calculations is described in Table 6.2.

Another emerging way to represent molecules is using learned embeddings such as convolutional and graph encoding to represent the molecular structures. This approach is focused on models capable of learning generalizable representations from a molecular data set. Convolutional encodings perform mathematical operations compacting the space containing the variables. One example is the work from Coley et al. [74] where atom and bond attributes were used in a convolutional neural network to represent the molecules. The authors applied the methodology to predict aqueous solubility, octanol solubility, melting point, and toxicity. Another

Table 6.2 Software and web servers available for descriptors and fingerprint calculation

Name	Desc.	FP	Source
Dragon [58]	5270	1	https://www.talete.mi.it/
CDK [59]	275	9	https://cdk.github.io/
E-dragon [60]	1600	–	https://vcclab.org/lab/edragon/
Mold2 [61]	779	–	https://www.fda.gov/science-research/bioinformatics-tools/mold2
Pybel [62]	24	4	https://github.com/pybel/pybel
PaDEL [63]	1875	12	https://www.yapcwsoft.com/dd/padeldescriptor/
RDKit	198	8	https://www.rdkit.org/
PyDPI [64]	615	7	https://pypi.org/project/pydpi/
Chemopy [65]	1135	7	https://github.com/ifyoungnet/Chemopy
ChemDes [66]	3679	59	http://www.scbdd.com/chemdes/
Rcpi [67]	308	10	https://bioconductor.org/packages/release/bioc/html/Rcpi.html
BioTriangle [68]	540	7	http://biotriangle.scbdd.com/
ChemSAR [69]	783	10	http://chemsar.scbdd.com/
Mordred [70]	1825	–	https://github.com/mordred-descriptor/mordred
PyBioMed [71]	775	19	https://github.com/gadsbyfly/PyBioMed
alvaDesc [72]	566	3	https://www.alvascience.com/alvadesc/
BioMedR [73]	293	13	https://github.com/wind22zhu/BioMedR

Desc. number of calculated descriptors; *FP* number of calculated fingerprint sets

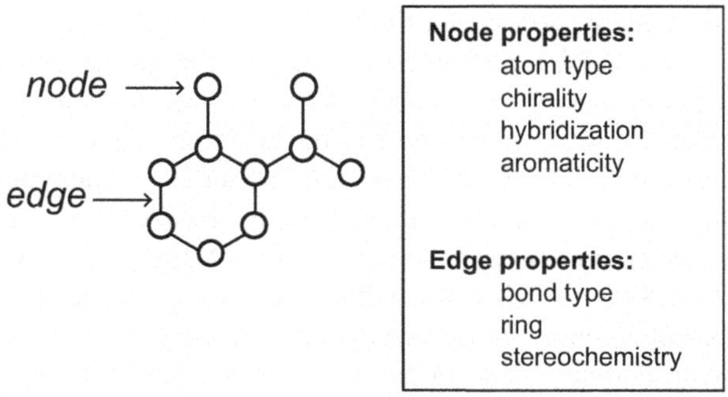

Fig. 6.5 Simplified example of a graph representation of a molecule

application of a convolutional encoding is the use of 3D information to predict protein-ligand binding affinity [75]. In the graph approach, atoms are represented as nodes and the bonds are represented by edges; both nodes and edges have associated features such as the atomic number, formal charge, and bond type (Fig. 6.5). It produces a 2D data structure; nevertheless, 3D information such as chirality or stereochemistry can be included in nodes or edges [76]. One example of graph usage is the Chemprop [77], a machine-learned package that has been used to build models to predict biological activities [78, 79], molecular properties [80, 81], and other chemical endpoints [82]. The work from McGibbon et al. [18] titled "From intuition to AI: evolution of small molecule representations in drug discovery" provides a comprehensive approach on how the molecular representations evolved over time.

After the descriptor calculation, the division of the data set into training and test sets is a critical step to build a robust and generalizable model. This process can be performed randomly; however, a rational choice can perform better. [83–87] Generally, the aim of rational sampling is to maximize the descriptors space diversity of the training data and also maintaining the test set representative. There are many algorithms used in the QSAR context to perform this selection such as sphere exclusion, Kennard-Stone, k-means clustering, hierarchical clustering analysis (HCA), minimal test set dissimilarity, Self-Organizing Maps (SOM), and others [40, 83, 84, 88–91]. One tool available for performing it is the MASSA Algorithm, a Python software to provide an automated rational sampling using HCA, a clustering algorithm. [92] For more details on clustering algorithms, please refer to Chap. 5.

To apply the mathematical/statistical methods, a variety of options can be used. The one with more freedom is through programming, especially using Python, a popular language in cheminformatics. Several Python libraries are already available to deal with chemical structure such as RDKit [93]. Also, the Scikit-learn [94] and SciPy [95] libraries are capable of building and validating QSAR models using different statistical and machine-learning methods. For those who are not familiar with programming in Python, the TeachOpenCADD [96] is an amazing tool to learn about QSAR and other computer-aided drug design techniques. Also, platforms with a graphical user interface (GUI), such as KNIME [97], Orange [98], and Weka [99], may be used to build and validate the QSAR model. Both KNIME and Weka had available workflows for QSAR modeling in the literature [100, 101].

The SYBYL platform was a popular tool for developing different QSAR models. It is proprietary, nevertheless, there are open tools for performing QSAR modeling. There is software to perform the whole process of QSAR modeling such as the webserver 3D-QSAR.com [102]. The server can preprocess the ligand files, perform the molecular alignment, and apply the CoMFA strategy combined with PLS. Another similar example is the 3D-QSARpy [103], a Python software for building 3D-QSAR models using MIFs. The fields available are based on the Lennard-Jones potential, the Coulomb electrostatic interactions, hydrogen bond forming atoms, and hydrophobic atoms. In this software machine learning algorithms are available to build the QSAR models.

Another example is the QSARINS [104], a software to perform de build, validate, and analyze QSAR models. This program applies an MLR approach to build the model and has a series of visualization tools to perform a graphical inspection. Nevertheless, the descriptors should be calculated using external software such as the ones described in Table 6.2. Furthermore, the LQTA-QSAR software [105] allows the construction of 4D-QSAR models from molecular dynamics trajectories, building MIFs using different types of atoms, ions, and functional groups. After, the variables are selected using the Ordered Predictor Selection (OPS) algorithm [106] and then the PLS is applied to build the QSAR models. Other software for performing QSAR modeling are described in Table 6.3.

Table 6.3 Example of available software for QSAR modeling

Name	Description	Source
VCCLAB [60]	A web server containing several methods to calculate molecular descriptors. It also can apply PLS and neural networks to modeling the data	https://vcclab.org/lab/
QSAR modeling [107]	The software uses PLS-based regression and has methods for variable selection and model validation	https://lqta.iqm.unicamp.br/
CORAL-QSAR/QSPR [108]	The software builds QSAR models from structures in the SMILE format using the Monte Carlo technique for function optimization	http://www.insilico.eu/coral/
BILIN	The software is suitable for calculating linear regression and performing some nonlinear analyses	https://www.kubinyi.de/bilin-program.html
BuildQSAR [109]	The software can perform the mathematical modeling, hypothesis testing, and graphics/correlation analysis	–

4 Validations and Controls

Statistical validation is not the same as experimental validation. For example, QSAR models are extensively validated by statistical strategies and metrics but its proof-of-concept for drug design purposes should be the designing of a new chemical entity followed by the experimental evaluation of the predicted activity/property.

4.1 Internal and External Validation

The early QSAR modeling relies on the division of data sets into two smaller subsets: training and test sets, usually with 80 and 20% of the total data, respectively. Of course, this proportion could be different ranging from 90/10 to 50/50.

The training set compounds are employed to teach the algorithm or to extract information to build the equation that correlates biological activity with the descriptors. However, this very same subset is often used to validate internally the ability of the QSAR model to make predictions (in this case, often called the robustness of the model). It is expected that the predictions are highly correlated with experimental data used to train the model (measured as r^2). However, the training set should be divided into smaller subsets to validate the predictions. This technique is called cross-validation (or k-fold cross-validation) and consists of a split training set k times into k sub-subsets, and using one subgroup for validation purposes and the remaining groups for training the model. In this sense, 5-fold and 10-fold crossvalidation as well as leave-one-out (LOO) crossvalidation are the most employed strategies. As the name suggests, LOO leaves one compound out of the training set, generates a model, predicts the activity for this compound, stores the predicted value, and then, returns this sample for the original training set, removes

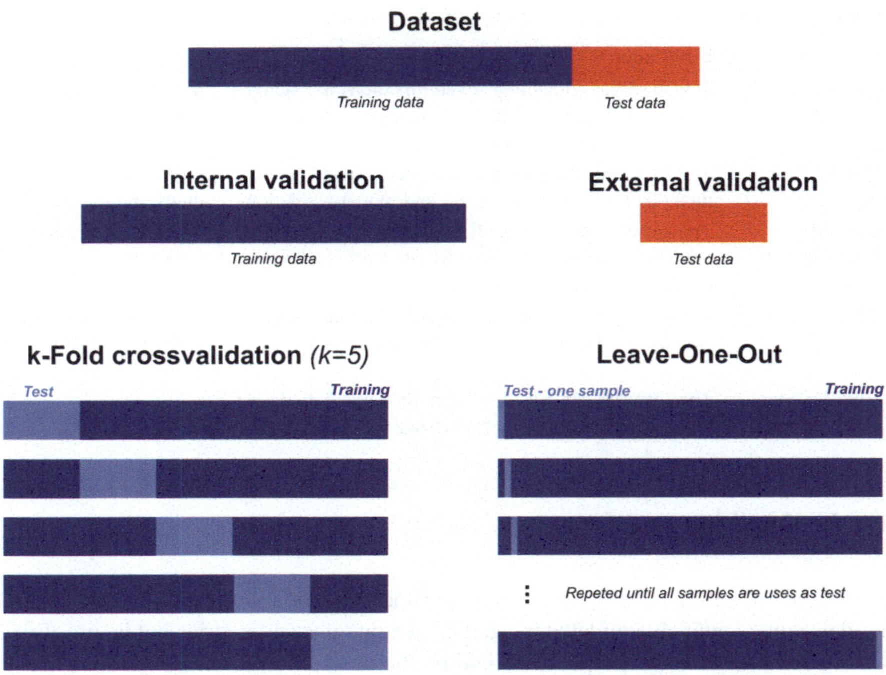

Fig. 6.6 Validations and controls for training and test/repeated k-Fold/Leave-One-Out (LOO)

another compound, and does this set of tasks until all compounds were used for prediction. Finally, the stored predicted values are used to calculate errors of prediction and correlation coefficients (Fig. 6.6).

In contrast, 5-fold crossvalidation does this process only five times instead of leaving one single compound, this strategy leaves 1/5 (or 20%) of samples out at the same time, 5 times (Fig. 6.5). Likewise, 10-fold crossvalidation does this sub-subset division 10 times, leaving 1/10 (10%) of samples out for internal validation 10 times.

Some works validate the final reported models using the strategy leave-many-out, which consists of varying the sample ratio in sub-subset division in several replicates (generally, 3 or 5 replicates), systematically to cover small and large sub-subsets of test compounds. In that sense, this strategy could evaluate the consistency of the model to keep its robustness throughout variation of sub-training and sub-test sizes and suggest that the generated model was not obtained as an artifact of a single replicate of a specific data set composition.

4.1.1 Regression Metrics

For regression models, there are several metrics (Table 6.4) that should be calculated to assess both robustness and predictability. Robustness could be defined as the ability to maintain its predictive performance across different data sets and is

Table 6.4 Regression metrics commonly used to validate QSAR models

Criteria	Metrics
Internal Q^2	$Q^2 = 1 - \left(\frac{\sum_{i=1}^{n_{TR}} \left(\widehat{y_i} - y_i\right)^2}{\sum_{i=1}^{n_{TR}} (y_i - \bar{y})^2}\right)^2 \approx 1$
Golbraikh and Tropsha criteria	$R^2 = \left(\frac{\sum_{i=1}^{n_{EXT}} (y_i - \bar{y})\left(\widehat{y_i} - \bar{\widehat{y}}\right)}{\sqrt{\sum_{i=1}^{n_{EXT}} (y_i - \bar{y})^2 \sum_{i=1}^{n_{EXT}} \left(\widehat{y_i} - \bar{\widehat{y}}\right)^2}}\right)^2 \approx 1$ $R_0^2 = 1 - \frac{\sum_{i=1}^{n_{EXT}} \left(\widehat{y_i} - y_i^{r0}\right)^2}{\sum_{i=1}^{n_{EXT}} \left(\widehat{y_i} - \bar{\widehat{y}}\right)^2} \approx R^2$; $R_0'^2 = 1 - \frac{\sum_{i=1}^{n_{EXT}} \left(y_i - y_i^{r0}\right)^2}{\sum_{i=1}^{n_{EXT}} (y_i - \bar{y})^2} \approx R^2$ $k = \frac{\sum_{i=1}^{n_{EXT}} y_i \widehat{y_i}}{\sum_{i=1}^{n_{EXT}} \widehat{y_i}^2} \approx 1$; $k' = \frac{\sum_{i=1}^{n_{EXT}} y_i \widehat{y_i}}{\sum_{i=1}^{n_{EXT}} y_i^2} \approx 1$; $y_i^{r0} = k\widehat{y}_i$; $\widehat{y}_i^{r0} = k' y_i$ R_0^2 and $R_0'^2$ are calculated forcing the regression line to pass through the origin, k and k' are the slope of the regression lines
External Q^2 functions	$Q_{F1}^2 = 1 - \frac{\sum_{i=1}^{n_{EXT}} \left(y_i - \widehat{y_i}\right)^2}{\sum_{i=1}^{n_{EXT}} (y_i - \bar{y}_{TR})^2} = 1 - \frac{PRESS}{TSS_{EXT(\bar{y}_{TR})}}$ $Q_{F2}^2 = 1 - \frac{\sum_{i=1}^{n_{EXT}} \left(y_i - \widehat{y_i}\right)^2}{\sum_{i=1}^{n_{EXT}} (y_i - \bar{y}_{EXT})^2} = 1 - \frac{PRESS}{TSS_{EXT(\bar{y}_{EXT})}}$ $Q_{F3}^2 = 1 - \frac{\left[\sum_{i=1}^{n_{EXT}} \left(y_i - \widehat{y_i}\right)^2\right]/n_{EXT}}{\left[\sum_{i=1}^{n_{EXT}} (y_i - \bar{y}_{EXT})^2\right]/n_{TR}} = 1 - \frac{PRESS/n_{EXT}}{TSS_{EXT(\bar{y}_{EXT})}/n_{TR}}$
Concordance correlation coefficient (CCC)	$CCC = \frac{2\sum_{i=1}^{n_{EXT}} (y_i - \bar{y})\left(\widehat{y_i} - \bar{\widehat{y}}\right)}{\sum_{i=1}^{n_{EXT}} (y_i - \bar{y})^2 + \sum_{i=1}^{n_{EXT}} \left(\widehat{y_i} - \bar{\widehat{y}}\right)^2 + n_{EXT}\left(\widehat{y_i} - \bar{\widehat{y}}\right)^2}$
Roy et al. criteria	$r_m^2 = r^2\left(1 - \sqrt{r^2 - r_0^2}\right) > 0.5$ $\overline{r_m^2} = \frac{\left(r_m^2 + r'^2_m\right)}{2}$ $\Delta r_m^2 = \left\lvert r_m^2 - r'^2_m \right\rvert < 0.2$ r^2 and r_0^2 are respectively the determination coefficients of the regression function, calculated using the experimental and the predicted data of the prediction set, forcing respectively the origin of the axis (r^2) or not (r_0^2). r_m^2 is calculated using experimental values on the ordinate axis, using them on the abscissa
Root mean square error	$RMSE = \sqrt{\frac{\sum_{i=1}^{n_{EXT}} \left(y_i - \widehat{y_i}\right)^2}{n_{EXT}}}$
Mean absolute error	$MAE = \frac{\sum_{i=1}^{n_{EXT}} \left\lvert y_i - \widehat{y_i}\right\rvert}{n_{EXT}}$

TR Training set; *EXT* External set, y_i Experimental data values; $\widehat{y}_i$ Predicted data values; $\bar{y}$ Average of the experimental data values; $\bar{\widehat{y}}$ Average of the predicted values

calculated using different internal validation strategies. The predictability is the predictive performance of a QSAR model for unseen new molecules (test set). For example, R^2-derived metrics, such as Q^2 calculated with LOO or other internal validation strategy, a metric to estimate the robustness of a model. Usually, some authors report an R^2 (calibration coefficient) metric which is calculated with the same formula as Q^2 but with no internal validation. Together, R^2 and Q^2 could be used as a guide to prevent overfitting: ideally, the difference between those two metrics should be lower than 0.3 ($R^2 - Q^2 > 0.3$) [93, 94].

In addition, errors such as RMSE and/or MAE metrics could be calculated in both internal and external validation procedures. The Root Mean Squared Error (RMSE), derived from MSE, offers an interpretable measure by maintaining the same units as the target variable. The Mean Absolute Error (MAE), which measures the average magnitude of errors in predictions without considering their direction, provides a straightforward interpretation of prediction accuracy.

Nowadays, it is recommended that a consensus of all metrics should be used in the evaluation of a given QSAR model's quality [110]. Those metrics are Q^2 coefficients discussed by Consonni and colleagues [112], r^2_m parameters were introduced by Roy and collaborators [111], and CCC from Gramatica work [113] as well as the errors of predictions.

4.1.2 Classification Metrics

Usually, classification models use the amount of true positive and negative (TP and TN) as well as false-positive and -negative (FP and FN) predictions to calculate access robustness and predictivity. Those classes of predictions are commonly organized in the confusion matrix (Fig. 6.7).

		Predicted condition	
		Predicted positive	*Predicted negative*
Actual condition	*Positive*	True positive (**TP**)	False negative (**FN**)
	Negative	False positive (**FP**)	True negative (**TN**)

Fig. 6.7 Elements of a confusion matrix

Table 6.5 Classification metrics applied to validate classification models using the elements of the confusion matrix

Metric	Formula
True Positive Rate (TPR), recall, sensitivity	$\mathrm{TPR} = \frac{2\mathrm{TP}}{2\mathrm{TP}+\mathrm{FP}+\mathrm{FN}}$
True Negative Rate (TNR), specificity, selectivity	$\mathrm{TNR} = \frac{\mathrm{TN}}{\mathrm{TN}+\mathrm{FP}}$
Accuracy (ACC)	$\mathrm{ACC} = \frac{\mathrm{TP}+\mathrm{TN}}{P+N}$
Balanced Accuracy (bACC)	$\mathrm{bACC} = \frac{\mathrm{TPR}+\mathrm{TNR}}{2}$
Matthews' correlation coefficient (MCC)	$\mathrm{MCC} = \frac{(\mathrm{TP}\times\mathrm{TN})-(\mathrm{FP}\times\mathrm{FN})}{\sqrt{(\mathrm{TP}+\mathrm{FP})+(\mathrm{TP}+\mathrm{FN})+(\mathrm{TN}+\mathrm{FP})+(\mathrm{TN}+\mathrm{FN})}}$
F1-Score	$F1\ \mathrm{Score} = \frac{2\mathrm{TP}}{2\mathrm{TP}+\mathrm{FP}+\mathrm{FN}}$
CK	$\mathrm{CK} = \frac{2\times(\mathrm{TP}\times\mathrm{TN}+\mathrm{FP}\times\mathrm{FN})}{(\mathrm{TP}+\mathrm{FP})(\mathrm{TP}+\mathrm{FN})(\mathrm{TN}+\mathrm{FP})(\mathrm{TN}+\mathrm{FN})}$
Enrichment factor at X% of the screened dataset	$\mathrm{EF}_{X\%} = \frac{\frac{\mathrm{actives}_{X\%}}{\mathrm{dataset}_{X\%}}}{\frac{\mathrm{actives}_{\mathrm{total}}}{\mathrm{dataset}_{\mathrm{total}}}}$

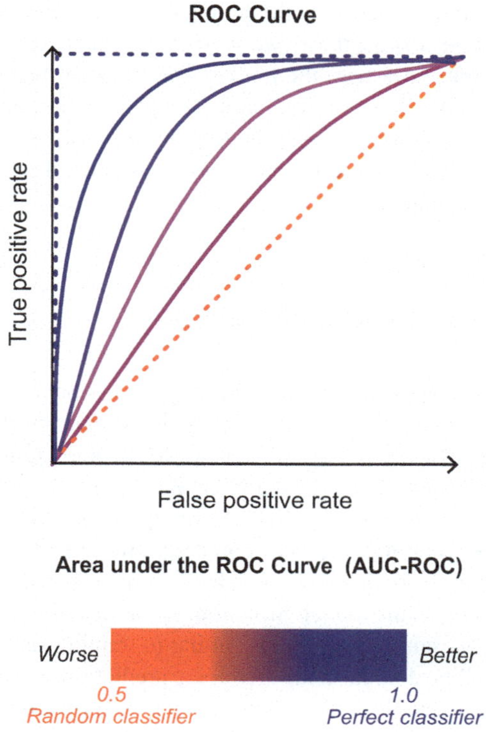

Fig. 6.8 Schematic example of a Receiver Operating Characteristic (ROC) Curve and the interpretation of the Area Under the ROC Curve (AUC-ROC)

Among the metrics to validate classification models (Table 6.5) and quantify their performance, accuracy (ACC) is a fundamental metric that represents the proportion of correct predictions out of the total predictions made and may be the most employed one. However, accuracy alone can be misleading, especially with imbalanced data sets (a set of compounds with more samples present in one class than others). To address this, recall or true positive rate (TPR) and sensitivity or true negative rate (TNR) are also used to avoid models with rates of false positives and negative predictions. In this sense, those metrics could cover the ability of a given model to predict in each class. In the same sense, the Matthews' Correlation Coefficient (MCC) is a metric used to evaluate the performance of classification models, particularly in binary classification tasks taking into consideration all four categories of the confusion matrix: true positives (TP), true negatives (TN), false positives (FP), and false negatives (FN).

Additionally, the Receiver Operating Characteristic (ROC) Curve and the Area Under the Curve (AUC) (Fig. 6.8) offer insights into the model's ability to distinguish between classes across different threshold settings, with the AUC summarizing the overall performance. While the area under the ROC curve (AUC-ROC) provides an overall assessment of the ability to distinguish between classes (active and inactive), the enrichment factor metrics and AUC-BEDROC offer information about this distinguishing ability in the initial fractions (X% of the data set) of the curve. Usually, authors define X% as equal to 0.5, 1.0, 2.0, 5.0, and/or 10.0%. In other words, these two metrics evaluate the protocol's ability to distinguish between molecules with the highest likelihood of being biologically active according to the probabilities of prediction of a given classification model [114–116].

As for regression models, an ensemble of all metrics from the confusion matrix as well as from ROC curves are considered the state-of-the-art analyses for the model's evaluation.

4.2 Applicability Domain

The applicability domain (AD) assessment is another analysis that is not properly a validation but is extremely important to ensure that external validations were performed adequately as well as predictions on unseen databases (e.g., virtual screening) are trustable or not. This importance is also highlighted and highly recommended by OECD principles for QSAR prediction rule number three. AD is a modeling step that compares the test set and/or external sets (e.g., libraries for future applications) with the training set considering their similarity in the space used to train the model. Structural similarity using fingerprints is also used for this task. In this sense, compounds similar to training set ones are considered inside the applicability domain and, therefore, predictions done to this sample could be considered trustworthy rather than just obtained by chance. In contrast, dissimilar compounds are considered outside the applicability domain and, therefore, the predictions made for this sample could be considered just a guess or random prediction from the model. There are a few methods to estimate the applicability domain of a given model such as based on the descriptor ranges, the knowledge density in the descriptors space around a query compound, or using a convergence of elements to support the prediction [117].

An interesting example was reported by Serafim and colleagues [118], discussing the implications of analyzing the applicability domain and illustrating it with an unsuccessful virtual screening campaign. In this work, hits were submitted to experimental validation and failed in this step due to the very dissimilar structures and properties in comparison to the training set of the model employed in the predictions.

4.3 Randomization Tests

After the internal and external validations, it is important to ensure that the observations in the models are genuine, and not an artifact of random correlation. This validation is made by shuffling the data, and it is expected that the model using these scrambled data performs worse than the original model.

Y-scrambling is a method to evaluate if the generated model was obtained by chance. In this method, a model is trained using the same training set of compounds, and the same set of descriptors but the activity (the Y property) is randomized. As expected, this new "random" model should perform worse than the original unscrambled model. In other words, correlation coefficients calculated values should be lower than acceptable thresholds of quality, and errors of prediction should be higher than acceptable. Of course, it is recommended that this task should be carried out several times, usually, reported works use from 20 to 200 runs of Y-scrambling tests.

Other metrics that should be considered in this validation are the correlation between q2 calculated with internal validation and the correlation coefficient between the original Y values and scrambled Y values. In this sense, it is expected that the degree of Y-randomization is correlated with the degree of "quality loss" (or decrease in q2, for example) of scrambled models.

There are some variations such as progressive scrambling implemented with the original CoMFA in the Sybyl package. This method progressively randomizes the Y values in distinct ranges of biological activity. For example, if a data set compound's pIC50 values range from 4 to 10, this method generates several bin windows for randomization: first, it splits the data set into six bins of 1 log unity (randomizes the pIC50 values of compounds inside that bin); then, in three bins of 2 log unities; and so on until randomizing the biological activity values in one single bin (the entire data set such as the traditional Y-scrambling).

Lastly, X-scrambling validation is usually made for classification methods since the Y value is binary (generally, 0 for inactive and 1 for actives). In this sense, X-scrambling aims to evaluate the same feature of the model (possibility to be obtained by chance) but randomizing the X variables.

5 Interpretation

Seeking the improvement of performance and generality, the QSAR models increase their complexity, however, this gain in complexity is only reasonable if the simpler models cannot handle the problem [119]. The prediction understanding promotes transparency and integrates with expert knowledge promoting useful insights learned by the observed patterns, and the interpretation can be made in two different ways, global and individual prediction [48] (Fig. 6.9).

The global model interpretations are related to what were the prioritized variables and how they are used to perform the predictions. A classic example of global model interpretations is linear regressions. Analyzing the coefficients in the equation, it is possible to observe how each variable contributes to the model (positively or negatively) and their magnitude. The same analogy can be applied in linear Support Vector Machine models, observing the feature weights. Tree-based algorithms are

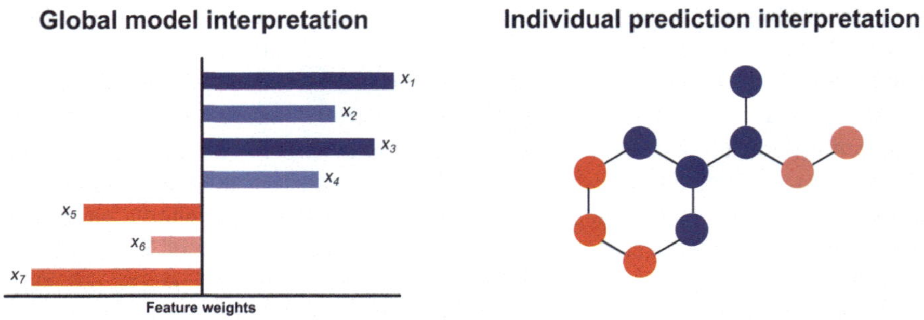

Fig. 6.9 Strategies to interpret a QSAR model. The first one is observing globally how the features impact the model, and the second one is using an individual prediction to see how the feature in the molecule impacts the prediction

also globally interpreted in terms of information theory, estimating the magnitude of a contribution [48]. Besides the algorithm interpretation, the feature permutation is a viable and method-independent strategy for interpreting the model globally [120].

In contrast to the global interpretation are the individual predictions. This approach tries to understand what patterns and features are important in the prediction of a specific compound. Not necessarily a globally important feature will be important in the prediction of a single compound and both can be complementary. Several agnostic methods have been used to interpret individual predictions such as Locally Interpretable Model-Agnostic Explanations (LIME) [121] and Shapley Additive Explanations (SHAP) [122]. LIME generates local explanations by perturbing input features and observing how these perturbations affect the model's predictions, resulting in a contribution that explains how each feature influenced the prediction. SHAP also quantifies the contribution of each feature to the model's output but is based on the principle of Shapley values derived from cooperative game theory. The importance of each feature is accessed based on its contribution to the prediction relative to all possible feature combinations. SHAP can provide both local interpretations and global insights. The local interpretation is obtained when Shapley values are calculated for each feature to explain the contribution of that specific prediction and globally when this process is made across all predictions in the data set.

6 Practical Advice During QSAR Modeling

The process of modeling can be quite challenging in the first look. In summary, based on the work experience, some questions are described in Table and the answers may help guide when planning a QSAR study.

Useful considerations during the QSAR model development

? Questions to answer when planning a QSAR modeling
Data collection and preprocessing
How were the experimental data generated? Is there a unified protocol? How to deal with these differences?
Is the training/test data representative in terms of protocol and terms of the descriptors used?
How will missing values, outliers, and redundant descriptors be handled?
Will the data need to be standardized or transformed?
Model selection
Which modeling techniques (e.g., linear regression, neural networks, and support vector machines) will be employed?
How will the model's complexity be determined and optimized?
Model validation
What validation methods (e.g., cross-validation, external validation) will be used to assess model performance?
How will overfitting be addressed?
Interpretation and visualization
How will the QSAR model results be interpreted in the context of chemical and biological knowledge?
What visualization techniques will be utilized to aid in understanding the relationships between chemical structure and activity?
Application ***and deployment***
How will the QSAR model be applied to predict the activity of new compounds?
How will the models be accessible to nonexperts?
What are the limitations and assumptions of the model, and how will they be communicated?
Iterative improvement
How will the pipeline be iteratively improved based on feedback and new data?
Are there opportunities to incorporate additional data sources or enhance modeling techniques?

7 Application

As examples of QSAR applications in the drug discovery field, five works published between 2023 and 2024 were described in detail. As selection criteria, the selected articles used computational techniques to identify active compounds that were experimentally validated.

Moreira-Filho et al. [123] used classification and regression machine learning models to predict the schistosomicidal activity of compounds that had not been experimentally tested. Subsequently, the prioritized compounds underwent testing on both schistosomula and adult stages of *Schistosoma mansoni*. Among them, four compounds demonstrated substantial activity against schistosomula, with 50% effective concentration values ranging from 9.8 to 32.5 μM, while showing no toxicity in animal and human cell lines.

A virtual screening comprising multiple computational methods was reported by Mushtaq and colleagues [124] to identify compounds with anti-interleukin-2 activity. The authors validated a molecular docking protocol and applied it in combination with pharmacophore filtering starting from a library containing 11.9 million compounds from ZINC and resulting in 24 compounds that had their activities predicted with a properly validated CoMFA model. Then, only nine compounds were submitted to experimental validation and three of them were considered promising hits due to the IL-2 inhibitory effect.

An interesting application of machine-learning-based virtual screening was reported by Barbosa et al. [125] In this work, they generated and validated kNN and Random Forest models to predict activity against *Trypanosoma cruzi* and used the models to screen a natural products database. After the selection of a virtual hit, they identified a plant that produced this hit as a metabolite (*Cymbopogon schoenanthus*), performed the isolation and identification of the selected compound, a diterpenoid called andrographolide, and tested it. As a result of the experimental validation, this compound showed IC_{50} values of 29.4 and 2.9 μM against trypomastigote and amastigote forms of *T. cruzi* and a selectivity index equal to 32, comparable with the positive control.

In 2024, Fernandes and colleagues reported a machine learning-based virtual screening [126] or discovering antibacterial compounds against methicillin-susceptible and resistant strains of *Staphylococcus aureus*. In this work, they generated descriptor-based QSAR classification models using several diverse machine learning methods for three data sets: one comprised of compounds with activity against susceptible strains of *S. aureus*; another one comprised of compounds with activity against resistant strains; and a third one comprised of compounds with activity against both strains. This last data set was very important to weigh the consensus selection of hits for experimental testing since it comprised both modeled activities. In this sense, the hit rate of models generated from this data set was higher than the other two models.

Wong et al. [127] experimentally screened 39,312 compounds against a methicillin-susceptible *S. aureus* strain (RN4220) as a model for antibacterial activity and against human liver carcinoma cells (HepG2), human primary skeletal muscle cells (HSkMCs), and human lung fibroblast cells (IMR-90) as models for cytotoxicity. After, they used Chemprop to train a graph neural network to predict a binary classification task for all properties (activity and toxicities). After training and validation of models, the authors screened two libraries for obtaining compounds with predicted antibacterial activity and no cytotoxic profile. Following this, potential false-positive compounds were removed using PAINS rules and undesired compounds due to reactivity, metabolic instability, and generalized toxicity were also removed by using Brenk structural alerts. Lastly, the authors selected compounds with similarity scores equal or lower than 0.5 in comparison with data set compounds. In this sense, they started with approximately 12 million compounds, and, after filtering, they selected 1261 compounds. As a strategy to interpreting the models, the authors used Monte Carlo tree searches to explore the chemical space and understand the smallest portion of a molecule responsible for their classification as active. With this approach, it was possible to highlight important structural features responsible for predictions. Then, using the rational structural predictions they filtered the hits according to analogs that match known antibacterial classes such as quinolone, cephalosporins, and β-lactams, then selecting 9 compounds for experimental validations. Four of the nine tested compounds were actives reaching a 44% success rate.

This last example is not an actual application of QSAR for drug design discovery, but a free platform available for early stages ADME profile prediction [128]. The Pharmacokinetics Profiler (PhaKinPro, available at: https://phakinpro.mml.unc.edu/) is a web server with QSAR models to predict hepatic stability, microsomal half-life in sub-cellular and tissue, renal clearance, blood–brain barrier (BBB) permeability, central nervous system (CNS) activity, Caco-2 permeability, plasma protein binding, plasma half-life, microsomal intrinsic clearance, and oral bioavailability. Despite the validation metrics of each model, the predictions at the web server provide the confidence of prediction, if the compound is inside of AD or not, and the importance of molecular fragments for the prediction as the interpretation of the model.

8 Challenges and Perspectives

As mentioned, statistical methods and algorithms have a poor ability to handle data from different sources and, consecutively, biological data from different protocols. In that sense, ML methods emerge as promising methods to model data with noise due to the superior ability of generalization. Indeed, this task has naturally evolved since biology and chemistry joined the big data concept with large databases. However, it desired methods to ensure how the data noise affects the quality of the predictions.

Another related issue is better attention and report of imbalanced data sets for classification models and potential gaps in modeled activities for regression models. Very often, authors underreport this aspect of data sets and, of course, biased data sets produce biased models which make biased predictions. The balance between classes introduces a bias in the modeled activity. However, other sources of bias must be avoided and well reported in QSAR modeling protocols, such as structural and physicochemical biases which could be solved (or, at least, used to warn potential users of reported models) with the combination of data set characterization and proper AD definition.

Multi-task models, in other words, models with the ability to predict more than one property (y) are well established in the literature; however, authors very often report parallel individual models. Maybe, the lack of ready-to-use software with this ability and the need for coding limit the spreading of this modality of modeling. In the same way, transfer learning is a set of methods that transfer knowledge from one trained and validated model to another. Of course, both models must share mechanisms or similarities in the modeled property, for example, models to predict the binding affinity of ligands to close homolog and structurally similar enzymes. This concept of transfer learning is extremely useful since it saves enormous amounts of time in training steps for similar task models. But, as multi-task models, transfer learning in QSAR is underreported in comparison to regular standard QSAR models.

Finally, not only the QSAR field, but all subjects across chemistry should feel the impact of large language models (LLM) because now they have been closing the gap between machines and humans across many knowledge domains [129] and its potential already has been demonstrated to understand complex molecular distribution [130]. The recent interest of the worldwide community on language models such as Generative Pre-trained Transformer 3 (GPT-3) from OpenAI popularized with the ChatGPT API has shed light on the subject. In this sense, LLM has been explored to predict chemical tasks as well as is in the spotlight of novel methods to be explored.

Acknowledgements The authors would like to thank the Fundação de Amparo à Pesquisa do Estado de Minas Gerais - FAPEMIG (grants APQ-01818-21 and RED-00110-23).

References

1. Cros, A. (1863). *Action de l'alcohol Amylique Sur l'organisme* (PhD Thesis, Thesis). Strasbourg, University of Strasbourg.
2. Mills, E. J. (1884). XXIII. On melting-point and boiling-point as related to chemical composition. *The London, Edinburgh, and Dublin Philosophical Magazine and Journal of Science, 17*(105), 173–187.
3. Hansch, C., Maloney, P. P., Fujita, T., & Muir, R. M. (1962). Correlation of biological activity of phenoxyacetic acids with Hammett substituent constants and partition coefficients. *Nature, 194*(4824), 178–180.
4. Hansch, C., Muir, R. M., Fujita, T., Maloney, P. P., Geiger, F., & Streich, M. (1963). The correlation of biological Activity of plant growth regulators and Chloromycetin derivatives with Hammett constants and partition coefficients. *Journal of the American Chemical Society, 85*(18), 2817–2824.
5. Fujita, T., Iwasa, J., & Hansch, C. (1964). A new substituent constant, π, derived from partition coefficients. *Journal of the American Chemical Society, 86*(23), 5175–5180.
6. Soares, T. A., Nunes-Alves, A., Mazzolari, A., Ruggiu, F., Wei, G.-W., & Merz, K. (2022). The (Re)-Evolution of Quantitative Structure–Activity Relationship (QSAR) studies propelled by the surge of machine learning methods. *Journal of Chemical Information and Modeling, 62*(22), 5317–5320.
7. Shi, Y., Yu, M., Liu, J., Yan, F., Luo, Z.-H., & Zhou, Y.-N. (2022). Quantitative structure–property relationship model for predicting the propagation rate coefficient in free-radical polymerization. *Macromolecules, 55*(21), 9397–9410.
8. Yu, M., Shi, Y., Liu, X., Jia, Q., Wang, Q., Luo, Z.-H., Yan, F., & Zhou, Y.-N. (2023). Quantitative Structure-Property Relationship (QSPR) framework assists in rapid mining of highly thermostable polyimides. *Chemical Engineering Journal, 465*, 142768.
9. Czub, N., Szlęk, J., Pacławski, A., Klimończyk, K., Puccetti, M., & Mendyk, A. (2023). Artificial intelligence-based quantitative structure–property relationship model for predicting human intestinal absorption of compounds with serotonergic activity. *Molecular Pharmaceutics, 20*(5), 2545–2555.

10. Sterling, A. J., Zavitsanou, S., Ford, J., & Duarte, F. (2021). Selectivity in organocatalysis—From qualitative to quantitative predictive models. *WIREs Computational Molecular Science, 11*(5), e1518.
11. Eckhoff, M., Diedrich, J. V., Mücke, M., & Proppe, J. (2024). Quantitative structure–reactivity relationships for synthesis planning: The benzhydrylium case. *The Journal of Physical Chemistry. A, 128*(1), 343–354.
12. Paradies, J. (2023). Structure-reactivity relationships in borane-based FLP-catalyzed hydrogenations, dehydrogenations, and cycloisomerizations. *Accounts of Chemical Research, 56*(7), 821–834.
13. Wang, L.-L., Ding, J.-J., Pan, L., Fu, L., Tian, J.-H., Cao, D.-S., Jiang, H., & Ding, X.-Q. (2021). Quantitative structure-toxicity relationship model for acute toxicity of organophosphates via multiple administration routes in rats and mice. *Journal of Hazardous Materials, 401*, 123724.
14. Mukherjee, R. K., Kumar, V., & Roy, K. (2022). Ecotoxicological QSTR and QSTTR modeling for the prediction of acute oral toxicity of pesticides against multiple avian species. *Environmental Science & Technology, 56*(1), 335–348.
15. Rai, M., Paudel, N., Sakhrie, M., Gemmati, D., Khan, I. A., Tisato, V., Kanase, A., Schulz, A., & Singh, A. V. (2023). Perspective on quantitative structure–toxicity relationship (QSTR) models to predict hepatic biotransformation of xenobiotics. *Liver, 3*(3), 448–462.
16. Fourches, D., Muratov, E., & Tropsha, A. (2010). Trust, but Verify: On the importance of chemical Structure curation in cheminformatics and QSAR modeling research. *Journal of Chemical Information and Modeling, 50*(7), 1189–1204.
17. Fourches, D., Muratov, E., & Tropsha, A. (2016). Trust, but Verify II: A practical guide to chemogenomics data curation. *Journal of Chemical Information and Modeling, 56*(7), 1243–1252.
18. McGibbon, M., Shave, S., Dong, J., Gao, Y., Houston, D. R., Xie, J., Yang, Y., Schwaller, P., & Blay, V. (2024). From intuition to AI: Evolution of small molecule representations in drug discovery. *Briefings in Bioinformatics, 25*(1), bbad422.
19. Khan, A. U. (2016). Descriptors and their selection methods in QSAR analysis: Paradigm for drug design. *Drug Discovery Today, 21*(8), 1291–1302.
20. Cereto-Massagué, A., Ojeda, M. J., Valls, C., Mulero, M., Garcia-Vallvé, S., & Pujadas, G. (2015). Molecular fingerprint similarity search in virtual screening. *Methods, 71*, 58–63.
21. Verma, J., Khedkar, V. M., & Coutinho, E. C. (2010). 3D-QSAR in drug design - A review. *Current Topics in Medicinal Chemistry, 10*(1), 95–115.
22. G. Damale, M., N. Harke, S., A. Kalam Khan, F., B. Shinde, D., & N. Sangshetti, J. (2014) Recent advances in multidimensional QSAR (4D-6D): A critical review. *Mini Reviews in Medicinal Chemistry, 14*(1), 35–55.
23. Polanski, J. (2009). Receptor dependent multidimensional QSAR for modeling drug-receptor interactions. *Current Medicinal Chemistry, 16*(25), 3243–3257.
24. Tropsha, A. (2010). Best practices for QSAR model development, validation, and exploitation. *Molecular Informatics, 29*(6–7), 476–488.
25. Gaulton, A., Hersey, A., Nowotka, M., Bento, A. P., Chambers, J., Mendez, D., Mutowo, P., Atkinson, F., Bellis, L. J., Cibrián-Uhalte, E., Davies, M., Dedman, N., Karlsson, A., Magariños, M. P., Overington, J. P., Papadatos, G., Smit, I., & Leach, A. R. (2017). The ChEMBL database in 2017. *Nucleic Acids Research, 45*(D1), D945–D954.
26. Wang, Y., Xiao, J., Suzek, T. O., Zhang, J., Wang, J., Zhou, Z., Han, L., Karapetyan, K., Dracheva, S., Shoemaker, B. A., Bolton, E., Gindulyte, A., & Bryant, S. H. (2012). PubChem's BioAssay database. *Nucleic Acids Research, 40*(D1), D400–D412.
27. Organisation for Economic Co-operation and Development. *OECD principles for the validation, for regulatory purposes, of (Quantitative) structure-activity relationship models*. https://www.oecd.org/chemicalsafety/risk-assessment/37849783.pdf. Accessed 2021-12-07.
28. Di Paolo, T. (1978). Structure-activity relationships of anesthetic ethers using molecular connectivity. *Journal of Pharmaceutical Sciences, 67*(4), 564–566.

29. Hansch, C., & Klein, T. E. (1986). Molecular graphics and QSAR in the study of enzyme-ligand interactions. On the definition of bioreceptors. *Accounts of Chemical Research, 19*(12), 392–400.
30. Cramer, R. D., Patterson, D. E., & Bunce, J. D. (1988). Comparative Molecular Field Analysis (CoMFA). 1. Effect of shape on binding of steroids to carrier proteins. *Journal of the American Chemical Society, 110*(18), 5959–5967.
31. Clark, M., Cramer, R. D., Jones, D. M., Patterson, D. E., & Simeroth, P. E. (1990). Comparative Molecular Field Analysis (CoMFA). 2. Toward its use with 3D-structural databases. *Tetrahedron Computer Methodology, 3*(1), 47–59.
32. Klebe, G., Abraham, U., & Mietzner, T. (1994). Molecular similarity indices in a comparative analysis (CoMSIA) of drug molecules to correlate and predict their biological Activity. *Journal of Medicinal Chemistry, 37*(24), 4130–4146.
33. Lowis, D. R. (1997). HQSAR: A new, highly predictive QSAR technique. *Tripos Technical Notes, 1*(5), 17.
34. Seel, M., Turner, D. B., & Willett, P. (1999). Effect of parameter variations on the effectiveness of HQSAR analyses. *Quantitative Structure-Activity Relationships, 18*(3), 245–252.
35. Abdizadeh, R., Hadizadeh, F., & Abdizadeh, T. (2020). QSAR analysis of Coumarin-based Benzamides as histone deacetylase inhibitors using CoMFA, CoMSIA and HQSAR methods. *Journal of Molecular Structure, 1199*, 126961.
36. Ding, H., Xing, F., Zou, L., & Zhao, L. (2024). QSAR analysis of VEGFR-2 inhibitors based on machine learning, Topomer CoMFA and molecule docking. *BMC Chemistry, 18*(1), 59.
37. Edache, E. I., Uzairu, A., Mamza, P. A., Shallangwa, G. A., & Ibrahim, M. T. (2024). Design of some potent non-toxic autoimmune disorder inhibitors Based on 2D-QSAR, CoMFA, molecular docking, and molecular dynamics investigations. *Intelligent Pharmacy*.
38. Abdizadeh, R., Hadizadeh, F., & Abdizadeh, T. (2020). Molecular modeling studies of anti-Alzheimer agents by QSAR, molecular docking and molecular dynamics simulations techniques. *Medicinal Chemistry, 16*(7), 903–927.
39. Lino, C. I., Gonçalves de Souza, I., Borelli, B. M., Silvério Matos, T. T., Santos Teixeira, I. N., Ramos, J. P., Maria de Souza Fagundes, E., de Oliveira Fernandes, P., Maltarollo, V. G., Johann, S., & de Oliveira, R. B. (2018). Synthesis, molecular modeling studies and evaluation of antifungal activity of a novel series of Thiazole derivatives. *European Journal of Medicinal Chemistry, 151*, 248–260.
40. Veríssimo, G. C., Menezes Dutra, E. F., Teotonio Dias, A. L., de Oliveira Fernandes, P., Kronenberger, T., Gomes, M. A., & Maltarollo, V. G. (2019). HQSAR and random forest-based QSAR models for Anti-T. Vaginalis activities of nitroimidazoles derivatives. *Journal of Molecular Graphics and Modelling, 90*, 180–191.
41. Wu, Z., Zhu, M., Kang, Y., Leung, E. L.-H., Lei, T., Shen, C., Jiang, D., Wang, Z., Cao, D., & Hou, T. (2021). Do we need different machine learning algorithms for QSAR modeling? A comprehensive assessment of 16 machine learning algorithms on 14 QSAR data sets. *Briefings in Bioinformatics, 22*(4), bbaa321.
42. Brown, F. K., Sherer, E. C., Johnson, S. A., Holloway, M. K., & Sherborne, B. S. (2017). The evolution of drug design at Merck Research Laboratories. *Journal of Computer-Aided Molecular Design, 31*(3), 255–266.
43. Loyola-Gonzalez, O. (2019). Black-box vs White-box: Understanding their advantages and weaknesses from a practical point of view. *IEEE Access, 7*, 154096–154113.
44. Barber, C., Heghes, C., & Johnston, L. (2024). A framework to support the application of the OECD Guidance documents on (Q)SAR model validation and prediction assessment for regulatory decisions. *Computational Toxicology, 30*, 100305.
45. Organisation for Economic Co-operation and Development. (2007). *Guidance document on the validation of (quantitative) structure-activity relationship [(Q) SAR] models*. Organisation for Economic Co-operation and Development.
46. Cronin, M. T. D., & Schultz, T. W. (2003). Pitfalls in QSAR. *Journal of Molecular Structure: THEOCHEM, 622*(1), 39–51.

47. Borota, A., Mracec, M., Gruia, A., Rad-Curpăn, R., Ostopovici-Halip, L., & Mracec, M. (2011). A QSAR study using MTD method and dragon descriptors for a series of selective ligands of α2C adrenoceptor. *European Journal of Medicinal Chemistry, 46*(3), 877–884.
48. Rodríguez-Pérez, R., & Bajorath, J. (2021). Explainable machine learning for property predictions in compound optimization. *Journal of Medicinal Chemistry, 64*(24), 17744–17752.
49. Jouan-Rimbaud, D., Massart, D. L., & De Noord, O. E. (1996). Random correlation in variable selection for multivariate calibration with a genetic algorithm. *Chemometrics and Intelligent Laboratory Systems, 35*(2), 213–220.
50. Hawkins, D. M. (2004). The problem of overfitting. *Journal of Chemical Information and Computer Sciences, 44*(1), 1–12.
51. Topliss, J. G., & Edwards, R. P. (1979). Chance factors in studies of Quantitative Structure-Activity Relationships. *Journal of Medicinal Chemistry, 22*(10), 1238–1244.
52. Wold, S., & Dunn, W. J. (1983). Multivariate Quantitative Structure-Activity Relationships (QSAR): Conditions for their applicability. *Journal of Chemical Information and Computer Sciences, 23*(1), 6–13.
53. Clark, M., & Cramer, R. D., III. (1993). The probability of chance correlation using partial least squares (PLS). *Quantitative Structure-Activity Relationships, 12*(2), 137–145.
54. Schaper, K.-J., Kunz, B., & Raevsky, O. A. (2003). Analysis of water solubility data on the basis of HYBOT descriptors. *QSAR & Combinatorial Science, 22*(9–10), 943–958.
55. Mendez, D., Gaulton, A., Bento, A. P., Chambers, J., De Veij, M., Félix, E., Magariños, M. P., Mosquera, J. F., Mutowo, P., Nowotka, M., Gordillo-Marañón, M., Hunter, F., Junco, L., Mugumbate, G., Rodriguez-Lopez, M., Atkinson, F., Bosc, N., Radoux, C. J., Segura-Cabrera, A., Hersey, A., & Leach, A. R. (2019). ChEMBL: Towards direct deposition of bioassay data. *Nucleic Acids Research, 47*(D1), D930–D940.
56. Kim, S., Chen, J., Cheng, T., Gindulyte, A., He, J., He, S., Li, Q., Shoemaker, B. A., Thiessen, P. A., Yu, B., Zaslavsky, L., Zhang, J., & Bolton, E. E. (2023). PubChem 2023 update. *Nucleic Acids Research, 51*(D1), D1373–D1380.
57. Gilson, M. K., Liu, T., Baitaluk, M., Nicola, G., Hwang, L., & Chong, J. (2016). BindingDB in 2015: A public database for medicinal chemistry, computational chemistry and systems pharmacology. *Nucleic Acids Research, 44*(D1), D1045–D1053.
58. Mauri, A., Consonni, V., Pavan, M., Todeschini, R., et al. (2006). Dragon software: An easy approach to molecular descriptor calculations. *Match, 56*(2), 237–248.
59. Steinbeck, C., Han, Y., Kuhn, S., Horlacher, O., Luttmann, E., & Willighagen, E. (2003). The Chemistry Development Kit (CDK): An open-source Java library for chemo- and bioinformatics. *Journal of Chemical Information and Computer Sciences, 43*(2), 493–500.
60. Tetko, I. V., Gasteiger, J., Todeschini, R., Mauri, A., Livingstone, D., Ertl, P., Palyulin, V. A., Radchenko, E. V., Zefirov, N. S., Makarenko, A. S., Tanchuk, V. Y., & Prokopenko, V. V. (2005). Virtual computational chemistry laboratory—Design and description. *Journal of Computer-Aided Molecular Design, 19*(6), 453–463.
61. Hong, H., Xie, Q., Ge, W., Qian, F., Fang, H., Shi, L., Su, Z., Perkins, R., & Tong, W. (2008). Mold2, molecular descriptors from 2D structures for chemoinformatics and toxicoinformatics. *Journal of Chemical Information and Modeling, 48*(7), 1337–1344.
62. O'Boyle, N. M., Morley, C., & Hutchison, G. R. (2008). Pybel: A Ppython wrapper for the OpenBabel cheminformatics toolkit. *Chemistry Central Journal, 2*(1), 5.
63. Yap, C. W. (2011). PaDEL-descriptor: An open source software to calculate molecular descriptors and fingerprints. *Journal of Computational Chemistry, 32*(7), 1466–1474.
64. Cao, D.-S., Liang, Y.-Z., Yan, J., Tan, G.-S., Xu, Q.-S., & Liu, S. (2013). PyDPI: Freely available python package for chemoinformatics, bioinformatics, and chemogenomics studies. *Journal of Chemical Information and Modeling, 53*(11), 3086–3096.
65. Cao, D.-S., Xu, Q.-S., Hu, Q.-N., & Liang, Y.-Z. (2013). ChemoPy: Freely available Python package for computational biology and chemoinformatics. *Bioinformatics, 29*(8), 1092–1094.

66. Dong, J., Cao, D.-S., Miao, H.-Y., Liu, S., Deng, B.-C., Yun, Y.-H., Wang, N.-N., Lu, A.-P., Zeng, W.-B., & Chen, A. F. (2015). ChemDes: An integrated web-based platform for molecular descriptor and fingerprint computation. *Journal of Cheminformatics, 7*(1), 60.
67. Cao, D.-S., Xiao, N., Xu, Q.-S., & Chen, A. F. (2015). Rcpi: R/Bioconductor package to generate various descriptors of proteins, compounds and their interactionsc. *Bioinformatics, 31*(2), 279–281.
68. Dong, J., Yao, Z.-J., Wen, M., Zhu, M.-F., Wang, N.-N., Miao, H.-Y., Lu, A.-P., Zeng, W.-B., & Cao, D.-S. (2016). BioTriangle: A web-accessible platform for generating various molecular representations for chemicals, proteins, DNAs/RNAs and their interactions. *Journal of Cheminformatics, 8*(1), 34.
69. Dong, J., Yao, Z.-J., Zhu, M.-F., Wang, N.-N., Lu, B., Chen, A. F., Lu, A.-P., Miao, H., Zeng, W.-B., & Cao, D.-S. (2017). ChemSAR: An online pipelining platform for molecular SAR modeling. *Journal of Cheminformatics, 9*(1), 27.
70. Moriwaki, H., Tian, Y.-S., Kawashita, N., & Takagi, T. (2018). Mordred: A molecular descriptor calculator. *Journal of Cheminformatics, 10*(1), 4.
71. Dong, J., Yao, Z.-J., Zhang, L., Luo, F., Lin, Q., Lu, A.-P., Chen, A. F., & Cao, D.-S. (2018). PyBioMed: A python library for various molecular representations of chemicals, proteins and DNAs and their interactions. *Journal of Cheminformatics, 10*(1), 16.
72. Mauri, A. (2020). alvaDesc: A tool to calculate and analyze molecular descriptors and fingerprints. In K. Roy (Ed.), *Ecotoxicological QSARs* (pp. 801–820). Springer US.
73. Dong, J., Zhu, M.-F., Yun, Y.-H., Lu, A.-P., Hou, T.-J., & Cao, D.-S. (2021). BioMedR: An R/CRAN package for integrated data analysis pipeline in biomedical study. *Briefings in Bioinformatics, 22*(1), 474–484.
74. Coley, C. W., Barzilay, R., Green, W. H., Jaakkola, T. S., & Jensen, K. F. (2017). Convolutional embedding of attributed molecular graphs for physical property prediction. *Journal of Chemical Information and Modeling, 57*(8), 1757–1772.
75. Wang, Y., Wu, S., Duan, Y., & Huang, Y. (2022). A point cloud-based deep learning strategy for protein–ligand binding affinity prediction. *Briefings in Bioinformatics, 23*(1), bbab474.
76. David, L., Thakkar, A., Mercado, R., & Engkvist, O. (2020). Molecular representations in AI-driven drug discovery: A review and practical guide. *Journal of Cheminformatics, 12*(1), 56.
77. Heid, E., Greenman, K. P., Chung, Y., Li, S.-C., Graff, D. E., Vermeire, F. H., Wu, H., Green, W. H., & McGill, C. J. (2024). Chemprop: A machine learning package for chemical property prediction. *Journal of Chemical Information and Modeling, 64*(1), 9–17.
78. Stokes, J. M., Yang, K., Swanson, K., Jin, W., Cubillos-Ruiz, A., Donghia, N. M., MacNair, C. R., French, S., Carfrae, L. A., Bloom-Ackermann, Z., et al. (2020). A deep learning approach to antibiotic discovery. *Cell, 180*(4), 688–702.
79. Jin, W., Stokes, J. M., Eastman, R. T., Itkin, Z., Zakharov, A. V., Collins, J. J., Jaakkola, T. S., & Barzilay, R. (2021). Deep learning identifies synergistic drug combinations for treating COVID-19. *Proceedings of the National Academy of Sciences, 118*(39), e2105070118.
80. Lim, M. A., Yang, S., Mai, H., & Cheng, A. C. (2022). Exploring deep learning of quantum chemical properties for absorption, distribution, metabolism, and excretion predictions. *Journal of Chemical Information and Modeling, 62*(24), 6336–6341.
81. Lenselink, E. B., & Stouten, P. F. W. (2021). Multitask machine learning models for predicting Lipophilicity (logP) in the SAMPL7 challenge. *Journal of Computer-Aided Molecular Design, 35*(8), 901–909.
82. McGill, C., Forsuelo, M., Guan, Y., & Green, W. H. (2021). Predicting infrared spectra with message passing neural networks. *Journal of Chemical Information and Modeling, 61*(6), 2594–2609.
83. Masand, V. H., Mahajan, D. T., Nazeruddin, G. M., Hadda, T. B., Rastija, V., & Alfeefy, A. M. (2015). Effect of information leakage and method of splitting (rational and random) on external predictive ability and behavior of different statistical parameters of QSAR model. *Medicinal Chemistry Research, 24*(3), 1241–1264.

84. Martin, T. M., Harten, P., Young, D. M., Muratov, E. N., Golbraikh, A., Zhu, H., & Tropsha, A. (2012). Does rational selection of training and test sets improve the outcome of QSAR modeling? *Journal of Chemical Information and Modeling, 52*(10), 2570–2578.
85. Puzyn, T., Mostrag-Szlichtyng, A., Gajewicz, A., Skrzyński, M., & Worth, A. P. (2011). Investigating the influence of data splitting on the predictive ability of QSAR/QSPR models. *Structural Chemistry, 22*(4), 795–804.
86. Esbensen, K. H., & Geladi, P. (2010). Principles of proper validation: Use and abuse of re-sampling for validation. *Journal of Chemometrics, 24*(3–4), 168–187.
87. Hawkins, D. M., Basak, S. C., & Mills, D. (2003). Assessing model fit by cross-validation. *Journal of Chemical Information and Computer Sciences, 43*(2), 579–586.
88. Andrada, M. F., Vega-Hissi, E. G., Estrada, M. R., & Garro Martinez, J. C. (2017). Impact assessment of the rational selection of training and test sets on the predictive ability of QSAR models. *SAR and QSAR in Environmental Research, 28*(12), 1011–1023.
89. Wu, W., Walczak, B., Massart, D. L., Heuerding, S., Erni, F., Last, I. R., & Prebble, K. A. (1996). Artificial neural networks in classification of NIR spectral data: Design of the training set. *Chemometrics and Intelligent Laboratory Systems, 33*(1), 35–46.
90. Kronenberger, T., Windshügel, B., Wrenger, C., Honorio, K. M., & Maltarollo, V. G. (2018). On the relationship of Anthranilic derivatives Structure and the FXR (Farnesoid X receptor) agonist Activity. *Journal of Biomolecular Structure and Dynamics, 36*(16), 4378–4391.
91. Gomes, R. A., Genesi, G. L., Maltarollo, V. G., & Trossini, G. H. G. (2017). Quantitative structure–activity relationships (HQSAR, CoMFA, and CoMSIA) studies for COX-2 selective inhibitors. *Journal of Biomolecular Structure and Dynamics, 35*(7), 1436–1445.
92. Veríssimo, G. C., Pantaleão, S. Q., de Fernandes, P. O., Gertrudes, J. C., Kronenberger, T., Honorio, K. M., & Maltarollo, V. G. (2023). MASSA algorithm: An automated rational sampling of training and test subsets for QSAR modeling. *Journal of Computer-Aided Molecular Design, 37*(12), 735–754.
93. *RDKit*. https://www.rdkit.org/. Accessed 2022-03-08.
94. Pedregosa, F., Varoquaux, G., Gramfort, A., Michel, V., Thirion, B., Grisel, O., Blondel, M., Prettenhofer, P., Weiss, R., Dubourg, V., Vanderplas, J., Passos, A., Cournapeau, D., Brucher, M., Perrot, M., & Duchesnay, É. (2011). Scikit-learn: Machine learning in Python. *Journal of Machine Learning Research, 12*(85), 2825–2830.
95. Virtanen, P., Gommers, R., Oliphant, T. E., Haberland, M., Reddy, T., Cournapeau, D., Burovski, E., Peterson, P., Weckesser, W., Bright, J., van der Walt, S. J., Brett, M., Wilson, J., Millman, K. J., Mayorov, N., Nelson, A. R. J., Jones, E., Kern, R., Larson, E., Carey, C. J., Polat, İ., Feng, Y., Moore, E. W., VanderPlas, J., Laxalde, D., Perktold, J., Cimrman, R., Henriksen, I., Quintero, E. A., Harris, C. R., Archibald, A. M., Ribeiro, A. H., Pedregosa, F., & van Mulbregt, P. (2020). SciPy 1.0: Fundamental algorithms for scientific computing in python. *Nature Methods, 17*(3), 261–272.
96. Sydow, D., Morger, A., Driller, M., & Volkamer, A. (2019). TeachOpenCADD: A teaching platform for computer-aided drug design using open source packages and data. *Journal of Cheminformatics, 11*(1), 29.
97. Berthold, M. R., Cebron, N., Dill, F., Gabriel, T. R., Kötter, T., Meinl, T., Ohl, P., Sieb, C., Thiel, K., & Wiswedel, B. (2008). KNIME: The Konstanz information miner. In C. Preisach, H. Burkhardt, L. Schmidt-Thieme, & R. Decker (Eds.), *Data analysis, machine learning and applications* (Studies in classification, data analysis, and knowledge organization) (pp. 319–326). Springer Berlin Heidelberg.
98. Demšar, J., Curk, T., Erjavec, A., Gorup, Č., Hočevar, T., Milutinovič, M., Možina, M., Polajnar, M., Toplak, M., Starič, A., Štajdohar, M., Umek, L., Žagar, L., Žbontar, J., Žitnik, M., & Zupan, B. (2013). Orange: Data mining toolbox in Python. *Journal of Machine Learning Research, 14*, 2349–2353.
99. Frank, E., Hall, M. A., & Witten, I. H. (2016). *The WEKA workbench*. Morgan Kaufmann.

100. Joshi, R., Zheng, Z., Agarwal, P., Hatmal, M. M., Chang, X., Seidler, P., & Haworth, I. S. (2024). KNIME workflows for applications in medicinal and computational chemistry. *Artificial Intelligence Chemistry, 2*(1), 100063.
101. Nantasenamat, C., Worachartcheewan, A., Jamsak, S., Preeyanon, L., Shoombuatong, W., Simeon, S., Mandi, P., Isarankura-Na-Ayudhya, C., & Prachayasittikul, V. (2015). AutoWeka: Toward an automated data mining software for QSAR and QSPR studies. In H. Cartwright (Ed.), *Artificial neural networks* (pp. 119–147). Springer.
102. Ragno, R. (2019). www.3d-Qsar.Com: A web portal that brings 3-D QSAR to all electronic devices—The Py-CoMFA web application as tool to build models from pre-aligned datasets. *Journal of Computer-Aided Molecular Design, 33*(9), 855–864.
103. de Silverio, P. S. S. N., de Viana, J. O., & Barbosa, E. G. (2023). 3D-QSARpy: Combining variable selection strategies and machine learning techniques to build QSAR models. *Brazilian Journal of Pharmaceutical Sciences, 59*, e22373.
104. Gramatica, P., Chirico, N., Papa, E., Cassani, S., & Kovarich, S. (2013). QSARINS: A new software for the development, analysis, and validation of QSAR MLR models. *Journal of Computational Chemistry, 34*(24), 2121–2132.
105. Martins, J. P. A., Barbosa, E. G., Pasqualoto, K. F. M., & Ferreira, M. M. C. (2009). LQTA-QSAR: A new 4D-QSAR methodology. *Journal of Chemical Information and Modeling, 49*(6), 1428–1436.
106. Teófilo, R. F., Martins, J. P. A., & Ferreira, M. M. C. (2009). Sorting variables by using informative vectors as a strategy for feature selection in multivariate regression. *Journal of Chemometrics, 23*(1), 32–48.
107. Martins, J. P. A., & Ferreira, M. M. C. (2013). QSAR modeling: um novo pacote computacional open source para gerar e validar modelos QSAR. *Química Nova, 36*, 554–560.
108. Toropova, A. P., & Toropov, A. A. (2014). CORAL software: Prediction of carcinogenicity of drugs by means of the Monte Carlo method. *European Journal of Pharmaceutical Sciences, 52*, 21–25.
109. de Oliveira, D. B., & Gaudio, A. C. (2000). BuildQSAR: A new computer program for QSAR analysis. *Quantitative Structure-Activity Relationships, 19*(6), 599–601.
110. Veerasamy, R., Rajak, H., Jain, A., Sivadasan, S., Varghese, C. P., & Agrawal, R. K. (2011). Validation of QSAR models-strategies and importance. *International Journal of Drug Design & Discovery, 3*, 511–519.
111. Pratim Roy, P., Paul, S., Mitra, I., & Roy, K. (2009). On two novel parameters for validation of predictive QSAR models. *Molecules, 14*(5), 1660–1701.
112. Consonni, V., Ballabio, D., & Todeschini, R. (2009). Comments on the definition of the Q2 parameter for QSAR Validation. *Journal of Chemical Information and Modeling, 49*(7), 1669–1678.
113. Gramatica, P., & Sangion, A. (2016). A historical excursus on the statistical validation parameters for QSAR Models: A clarification concerning metrics and terminology. *Journal of Chemical Information and Modeling, 56*(6), 1127–1131.
114. Venkatraman, V., Chakravarthy, P. R., & Kihara, D. (2009). Application of 3D Zernike descriptors to shape-based ligand similarity searching. *Journal of Cheminformatics, 1*(1), 19.
115. Avram, S. I., Crisan, L., Bora, A., Pacureanu, L. M., Avram, S., & Kurunczi, L. (2013). Retrospective group fusion similarity search based on eROCE evaluation metric. *Bioorganic & Medicinal Chemistry, 21*(5), 1268–1278.
116. Castillo-González, D., Mergny, J.-L., De Rache, A., Pérez-Machado, G., Cabrera-Pérez, M. A., Nicolotti, O., Introcaso, A., Mangiatordi, G. F., Guédin, A., Bourdoncle, A., Garrigues, T., Pallardó, F., Cordeiro, M. N. D. S., Paz-y-Miño, C., Tejera, E., Borges, F., & Cruz-Monteagudo, M. (2015). Harmonization of QSAR best practices and molecular docking provides an efficient virtual screening tool for discovering new G-Quadruplex ligands. *Journal of Chemical Information and Modeling, 55*(10), 2094–2110.

117. Hanser, T., Barber, C., Marchaland, J. F., & Werner, S. (2016). Applicability domain: Towards a more formal definition$. *SAR and QSAR in Environmental Research, 27*(11), 865–881.
118. Serafim, M. S. M., Pantaleão, S. Q., da Silva, E. B., McKerrow, J. H., O'Donoghue, A. J., Mota, B. E. F., Honorio, K. M., & Maltarollo, V. G. (2023). The importance of good practices and false hits for QSAR-driven virtual screening real application: A SARS-CoV-2 Main protease (Mpro) case study. *Frontiers in Drug Discovery, 3.*
119. Rudin, C. (2019). Stop explaining black box machine learning models for high stakes decisions and use interpretable models instead. *Nature Machine Intelligence, 1*(5), 206–215.
120. Breiman, L. (2001). Random forests. *Machine Learning, 45*, 5–32.
121. Ribeiro, M. T., Singh, S., & Guestrin, C. (2016). "Why should I trust you?": Explaining the predictions of any classifier. In *Proceedings of the 22nd ACM SIGKDD International Conference on Knowledge Discovery and Data Mining* (pp. 1135–1144). ACM.
122. Lundberg, S. M., & Lee, S.-I. (2017). A unified approach to interpreting model predictions. In *Advances in neural information processing systems* (Vol. 30). Curran Associates, Inc.
123. Moreira-Filho, J. T., Neves, B. J., Cajas, R. A., de Moraes, J., & Andrade, C. H. (2023). Artificial intelligence-guided approach for efficient virtual screening of hits against Schistosoma Mansoni. *Future Medicinal Chemistry, 15*(22), 2033–2050.
124. Mushtaq, M., Usmani, S., Jabeen, A., Nur-e-Alam, M., Ahmed, S., Ahmad, A., & Ul-Haq, Z. (2023). Identification of potent anti-immunogenic agents through virtual screening, 3D-QSAR studies, and in vitro experiments. *Molecular Diversity.*
125. Barbosa, H., Espinoza, G. Z., Amaral, M., de Castro Levatti, E. V., Abiuzi, M. B., Veríssimo, G. C., de Fernandes, P. O., Maltarollo, V. G., Tempone, A. G., Honorio, K. M., & Lago, J. H. G. (2024). Andrographolide: A diterpenoid from Cymbopogon schoenanthus identified as a new hit compound against Trypanosoma cruzi using machine learning and experimental approaches. *Journal of Chemical Information and Modeling, 64*(7), 2565–2576.
126. Fernandes, P. O., Dias, A. L. T., dos Santos Júnior, V. S., Sá Magalhães Serafim, M., Sousa, Y. V., Monteiro, G. C., Coutinho, I. D., Valli, M., Verzola, M. M. S. A., Ottoni, F. M., de Pádua, R. M., Oda, F. B., dos Santos, A. G., Andricopulo, A. D., da Silva Bolzani, V., Mota, B. E. F., Alves, R. J., de Oliveira, R. B., Kronenberger, T., & Maltarollo, V. G. (2024). Machine learning-based virtual screening of antibacterial agents against methicillin-susceptible and resistant staphylococcus aureus. *Journal of Chemical Information and Modeling, 64*(6), 1932–1944.
127. Wong, F., Zheng, E. J., Valeri, J. A., Donghia, N. M., Anahtar, M. N., Omori, S., Li, A., Cubillos-Ruiz, A., Krishnan, A., Jin, W., Manson, A. L., Friedrichs, J., Helbig, R., Hajian, B., Fiejtek, D. K., Wagner, F. F., Soutter, H. H., Earl, A. M., Stokes, J. M., Renner, L. D., & Collins, J. J. (2023). Discovery of a structural class of antibiotics with explainable deep learning. *Nature, 626*(7997), 177–185.
128. Rath, M., Wellnitz, J., Martin, H.-J., Melo-Filho, C., Hochuli, J. E., Silva, G. M., Beasley, J.-M., Travis, M., Sessions, Z. L., Popov, K. I., Zakharov, A. V., Cherkasov, A., Alves, V., Muratov, E. N., & Tropsha, A. (2024). Pharmacokinetics Profiler (PhaKinPro): Model development, Validation, and implementation as a web tool for triaging compounds with undesired pharmacokinetics profiles. *Journal of Medicinal Chemistry, 67*(8), 6508–6518.
129. White, A. D. (2023). The future of chemistry is language. *Nature Reviews Chemistry, 7*(7), 457–458.
130. Flam-Shepherd, D., Zhu, K., & Aspuru-Guzik, A. (2022). Language models can learn complex molecular distributions. *Nature Communications, 13*(1), 3293.

Chapter 7
Molecular Docking: State-of-the-Art Scoring Functions and Search Algorithms

Rafaela M. de Angelo, Daniel S. de Sousa, Aldineia P. da Silva, Laise P. A. Chiari, Albérico B. F. da Silva, and Kathia M. Honorio

Abstract This study addresses the technique of molecular docking in drug design, emphasizing the importance of advanced scoring functions and search algorithms. It explores the generation of detailed models involving protein targets and ligand molecules and the evaluation of multiple ligand conformations to identify favorable molecular interactions. Additionally, it underscores the prediction of the ligand's most stable and energetically favorable conformation to interact with the target protein. Molecular docking plays a fundamental role in the early stages of drug discovery and development, assisting in selecting promising drug candidates and estimating the affinity between candidate molecules and target proteins. The key contributions of this chapter include the exploration of the importance of molecular docking in drug design and the analysis of advanced scoring functions and search algorithms. These topics are essential components for successful molecular docking. In addition, this chapter will discuss the relevance of result validation and the need for accuracy in the predictions based on docking simulations. Furthermore, it evaluates the challenges faced in molecular docking and reflects on using machine learning techniques in this context. It provides insights into the interaction between molecules and target proteins, which is essential for understanding the mechanisms of action of bioactive substances. It also emphasizes the acceleration in drug discovery and design, resource management, and the enhancement of understanding of molecular interactions as significant outcomes. These contributions underscore

R. M. de Angelo
University Federal of ABC (UFABC), Center for Natural and Human Sciences (CCNH), Santo André, SP, Brazil

D. S. de Sousa · L. P. A. Chiari · A. B. F. da Silva
University of São Paulo (USP), São Carlos Institute of Chemistry (IQSC), São Carlos, SP, Brazil

A. P. da Silva · K. M. Honorio (✉)
University of Sao Paulo (USP), School of Arts, Sciences and Humanities (EACH), Sao Paulo, Sao Paulo, Brazil
e-mail: KMHONORIO@USP.BR

V. G. Maltarollo (ed.), *Computer-Aided and Machine Learning-Driven Drug Design*, Computer-Aided Drug Discovery and Design 3,
https://doi.org/10.1007/978-3-031-76718-0_7

the importance of molecular docking as a fundamental tool in pharmaceutical research and drug development.

Keywords Molecular docking · Scoring functions · Search algorithms · Drug design

1 Molecular Docking

Molecular docking is a sophisticated computational technique extensively utilized across various scientific domains, including chemistry, pharmacy, and molecular biology. Its primary objective is to simulate the interaction between a molecule and the three-dimensional structure of a specific target. The docking procedure goal is to generate detailed models of the target macromolecule (e.g., protein) together with the small (ligand) molecule. It often starts from multiple ligand conformations are assessed to pinpoint the position and orientation that yield favorable molecular interactions, which are quantified using a scoring function (SF). These interactions encompass hydrogen bonds, van der Waals forces at the protein's active site, and other relevant interactions crucial for the binding. Regardless of the scoring function employed, these interactions play a significant role in determining the docking outcomes [1].

In a nutshell, the primary objective of the docking is to predict a favorable target-ligand conformation, maximizing the number of interactions and minimizing the predicted binding energy (derived from the SF). This information is crucial in the search for drug candidates, as it helps select promising bioactive substances, optimize structures of existing molecules, and estimate the affinity between the candidate molecule and the target protein. In summary, the docking technique is crucial in the early stages of drug discovery and development [2].

Kuntz and colleagues pioneeringly introduced the docking methodology in 1982, with the article titled "A Geometric Approach to Macromolecule-Ligand Interactions." [3] In this work, the authors proposed that molecular recognition between molecules, based on chemical and geometric aspects, could be explored through three-dimensional models of both the ligand and the target. Since then, molecular docking has been widely applied as a rapid approach to estimate how a particular compound binds to a biological target and to predict/estimate the affinity of this interaction.

In this chapter, we will discuss the relevance of docking in drug design and discovery, advances in scoring functions and search algorithms (including machine learning, ML), calculations involved in docking simulations, essential components of effective docking programs, limitations of the technique, and validation of docking results (including the consensus technique). We end the chapter discussing perspectives from different experts, the ML integration in molecular docking, and giving a perspective on the most recent challenges encountered in the field.

1.1 Importance/Relevance of Docking in the Drug Design and Discovery

Traditionally, the search for new drugs starting from the initial hits involves the synthesis and testing of many compounds; this process is error prone and resource intensive. Docking can help to deprioritize compounds with a low likelihood of success, saving valuable time and resources. Docking results can also be used to optimize the structure of existing hit molecules, during the hit-to-lead phase, improving their effectiveness and selectivity to the target protein [2].

Molecular docking provides detailed insights into the protein–ligand interactions at the atomic level. The early identification of hits as drug candidates with a high probability of success can save significant resources in the long term by avoiding investments in compounds that will not work or have severe side effects. It is worth mentioning that the docking technique is also relevant in the era of personalized medicine, where drugs are tailored to individual patient characteristics, helping design medications that precisely fit variant or mutant proteins [4].

When one employs molecular docking and faces some of its complexities, it becomes crucial to recognize the pivotal role played by the choice of scoring functions and search algorithms. These elements constitute the essential foundations of docking tools, guiding the investigation of molecular interactions and determining optimal binding conformations. Before delving into recent advancements in scoring functions and search algorithms, it is relevant to briefly contextualize these components' importance within the molecular docking realm.

2 Advances in Scoring Functions and Search Algorithms

Molecular docking is an ever-evolving field and state-of-the-art scoring function and search algorithms continue to advance rapidly. It is crucial to highlight significant trends and recent advancements within these domains, delineated through the presentation of three principal categories of scoring functions: force-field-based, empirical, and knowledge-based [5]. Figure 7.1 summarizes the main classes of scoring functions used to estimate protein–ligand interactions, categorizing them into three principal topics: types of functions, datasets (concerning already obtained data), and applications for different functions.

In recent years, several studies related to docking simulations have focused on the development and refinement of scoring functions [6], such as:

- *Machine learning-based scoring functions:* Machine learning techniques to develop more accurate scoring functions have gained prominence in recent years. Deep learning algorithms are being applied to model complex protein–ligand interactions [7]. Figure 7.2 illustrates some machine learning-based algorithms that can be applied to scoring functions and optimizing protein–ligand

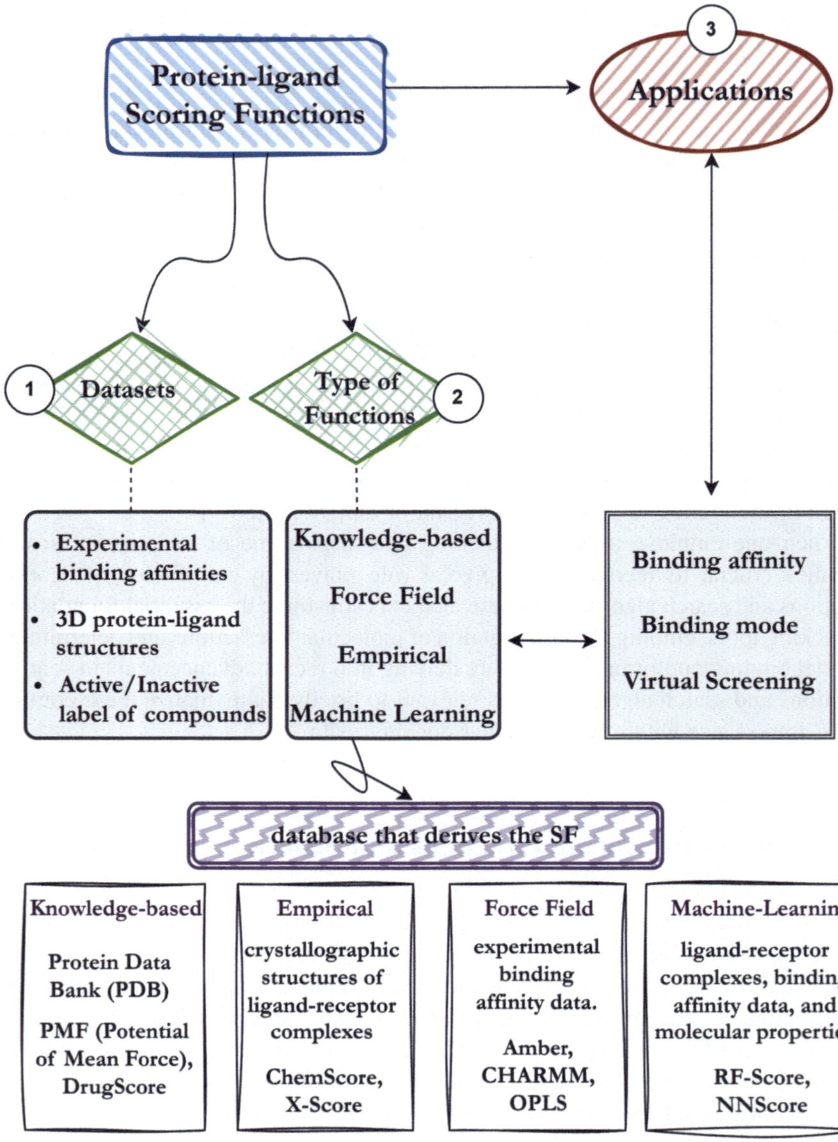

Fig. 7.1 Diagram of categories, datasets, and applications of scoring functions employed during the protein–ligand docking. (1) Datasets can be used to obtain scoring function (SF), which can estimate binding affinities, for example. (2) Regarding function types, SF can be classified into three categories: knowledge-based, force field-based, and empirical; below we find the databases that derive the SF. (3) SF can be applied to estimate interactions between a large number of compounds and proteins, optimizing the search for molecules of interest

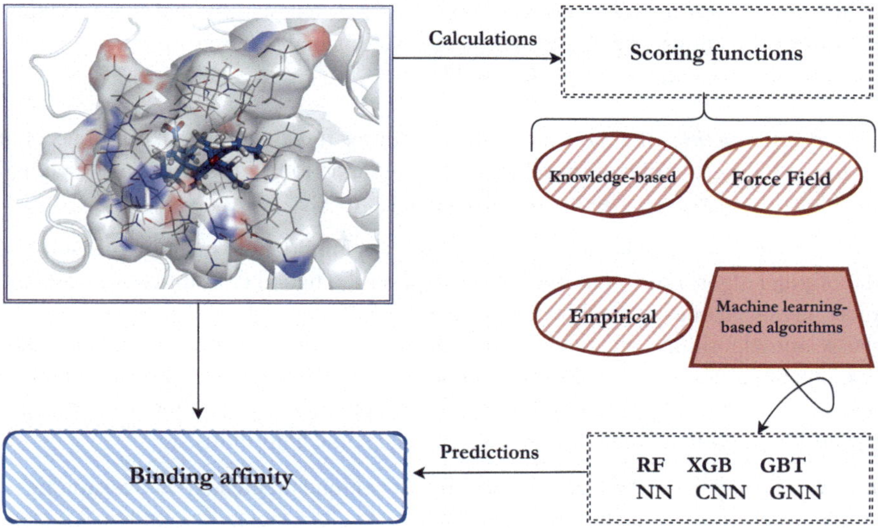

Fig. 7.2 Machine learning-based algorithms used in scoring functions, which are present in docking programs. Each acronym represents a unique approach in the field of machine learning, tailored to different types of data and specific tasks. RF (Random Forest); XGB (XGBoost); GBT (Gradient Boosting Trees); NN (Neural Network); CNN (Convolutional Neural Network); GNN (Graph Neural Network)

interaction calculations. The limitations of generic ML scoring functions include the possibility of not capturing specific nuances of protein–ligand interaction, resulting in less precise predictions. Additionally, the reliance on broad datasets can make models less sensitive to the unique features of specific interactions. On the other hand, the advantages of generic ML scoring functions lie in their ability to generalize and computational efficiency. These functions can be trained on large datasets to capture general protein–ligand interaction patterns, making them applicable to various biological systems. Furthermore, using ML algorithms may enable the discovery of complex relationships between molecule features and their interactions, which traditional approaches may not readily identify.

- *Multiscale scoring functions:* Scoring functions that consider multiple scales of interactions, such as local and global interactions, are being developed to improve the accuracy of binding predictions. An example of global interactions cited in this work is related to the long-range electrostatic interactions between charged residues on the protein surface and polar or charged groups at the ligand molecule. These can play a significant role in determining the overall stability of the interactions and the binding affinity of the protein–ligand complex [8].

- *Integration of structural and energetic data:* The integration of structural information, such as structural biology data, with energetic data in scoring functions is becoming more common, enhancing the accuracy of docking simulations. While this integration improves docking accuracy, it is noteworthy that many original scoring functions like ALP/ChemScore were actually based on energy results from pharmaceutical companies [9–11].

- *Scoring functions and solvation effect:* Scoring functions that take into account the influence of solvent on protein–ligand interactions are being developed and refined to improve accuracy in solvated systems [12, 13].

Taking into account the search algorithms, it is important to mention that some of the original algorithms focused on just sampling the ligand conformation, while the more advanced ones expand on the protein–ligand sampling as a single process. So, the search algorithms integrated into molecular docking platforms serve as robust computational tools for exploring the conformational space of ligands for protein active sites. Each algorithm embodies distinct methodologies tailored to optimize ligand-binding interactions. Below, we can cite some algorithms:

Random Forest (RF): An ensemble learning technique predicated on decision trees, RF excels in classification and regression tasks. Its adaptability renders it conducive to identifying promising ligand poses within docking studies [14].

XG-Boost (XGB): An optimized implementation of gradient boosting, XGB is proficient in handling classification and regression challenges. While powerful, vigilant parameter regularization is essential to mitigate overfitting risks [15].

Gradient Boosted Trees (GBT): Analogous to XGB but with a more generalized boosting methodology, GBT offers efficacy in classification and regression endeavors. Prudent parameter tuning is requisite to circumvent overfitting concerns [16].

Neural Networks (NN): NNs, comprising interconnected layers of neurons, exhibit prowess in capturing intricate data representations. In docking, NNs are instrumental in discerning the intricate relationship between ligand structure and protein binding affinity [17].

Convolutional Neural Networks (CNN): Specialized in grid-structured data processing, such as images, CNNs are adept at extracting local patterns pertinent to ligand–protein interactions. Their application in molecular docking facilitates comprehensive structural analysis [18].

Graph Neural Networks (GNN): Tailored for data represented as graphs, GNNs excel in modeling complex atomistic interactions within molecular systems. Leveraging their capabilities, GNNs offer versatile representations of molecular structures, facilitating spatial relationship elucidation [19].

These algorithms, either employed individually or synergistically, augment the precision and efficiency of molecular docking endeavors. The selection of an appropriate algorithm hinges upon the specific attributes of the docking problem and the nature of the data under investigation [20].

About the search algorithms implemented in docking programs, the main characteristics of these algorithms include:

- *Enhancement of global search algorithms:* Global search algorithms, such as genetic algorithms and particle swarm optimization (heuristic), are being improved to explore ligand conformation spaces more effectively [21]. It is worth noting that the diversity of available search functions has increased, underscoring the importance of selecting the appropriate algorithm for molecular docking, as it is not always a one-size-fits-all problem [22].

- *Hardware acceleration:* The use of high-performance hardware, such as GPUs (Graphics Processing Unit) and TPUs (Tensor Processing Unit), accelerates the search for binding conformations, making the docking process faster and more efficient [23].

- *Intelligent local search methods:* Local search algorithms are being enhanced with intelligent strategies to increase the probability of finding high-affinity binding conformations [24–26].
- *Integration of molecular dynamics:* Integrating molecular dynamics with docking is becoming more common, allowing for a more realistic consideration of molecular flexibility during the search [26, 27].

- *Flexible conformation simulations:* Algorithms that enable simulations of flexible conformation are being developed, allowing ligands and proteins to change their conformations during the docking process [28, 29].

- *Rescoring approaches:* Processes in which ligands are re-ranked after the initial search can help improve the accuracy of predictions [4, 5].

- *Integration of experimental information:* Algorithms that integrate docking and experimental information, such as mass spectrometry or electron microscopy data, are being developed to enhance validation and prediction accuracy [6].

It is important to note that docking simulations continue to benefit from multidisciplinary collaboration among computer scientists, structural biologists, chemists, pharmacologists, and others. Furthermore, applying machine learning techniques and using high-performance computing resources significantly drives the state-of-the-art in these areas. As new technologies and approaches are developed, research in molecular docking continues to advance, providing new insights into drug candidate discovery/planning and understanding protein–ligand interactions.

2.1 Search Algorithms, Scoring Functions, and Machine Learning

Machine learning algorithms can be used to optimize parameters in molecular docking techniques, enhancing their ability to predict more accurate molecular interactions. In most of the cases, ML-based scoring functions represent a methodology based on supervised learning, employing structural data with labels reflecting experimentally measured binding affinities. Initially, this ML-integration relies on traditional approaches such as Support Vector Machines (SVM), Random Forest (RF), and Gradient Boosting Trees (GBT), aiming to improve scoring performance on reference test sets. The inputs for these models included manually designed descriptors such as molecular fingerprints, ligand features, atom pair terms, and force field terms [30, 31]. Currently, there are other approaches to incorporate machine learning techniques into scoring functions, which can improve simulation performance and/or optimize the types of result analyses. Below, we will describe some machine learning-based scoring functions presented in Table 7.1.

2.2 Critical Characteristics of Search Algorithms

Search algorithms are classified into three main groups, according to the methodology applied to explore ligand flexibility: systematic, deterministic, and stochastic [47]. Table 7.2 presents the critical characteristics of each algorithm.

As expected, an essential factor in the molecular docking process is the choice of the search algorithm that will position the compound in the active site/binding site of the target. Among the primary algorithms, Genetic Algorithms (GA) and Lamarckian Genetic Algorithms (LGA) stand out as stochastic methods and explore the conformations of the receptor–ligand complex by allowing for random conformational modifications of the ligand and, occasionally, specific protein residues [53]. It is noteworthy to mention the primary reference for the GOLD program, which employs GA considering the ligand flexibility and the flexibility of some residue side chains.

GA is based on Darwin's theory of evolution, considering the ligand's degrees of freedom as "genes," composing "chromosomes" referring to ligand poses. Changes leading to the generation of new conformations are determined by mutations, which randomly alter the "genes," and by "crossover," allowing the exchange between two "genes" that will be evaluated by the scoring function, considering the new generations those with the highest estimated affinity [54].

Still, in line with evolutionary ideas from biology, the LGA algorithm incorporates the same biological concepts but is applied to computational processes. This approach starts with a population of ligand conformations, evaluates their fitness based on interaction energy, and selects the fittest ligands as parents for the next generation. The crossover and mutation processes introduce diversity while the

Table 7.1 Description of some machine learning-based scoring functions

Algorithm	Name	Input feature	Dataset	Reference
	RF-score	Counting pairs of protein–ligand atom types within a predefined distance cut-off	PDBbind v2007	[31]
RF ǀ Random Forest	$SFCscore^{RF}$	Descriptors related to ligand-specific interactions and surface area	PDBbind v2007	[32]
	ΔVinaRF20	Vina empirical and surface area terms	PDBbind v2014 CSAR dataset	[33]
	ΔVinaXGB	Vina empirical factors, surface area components, ligand stability factors, and bridge water considerations	PDBbind v2016 CSAR dataset	[34]
XGB ǀ eXtreme Gradient Boosting	ΔLinF9XGB	A sequence of Gaussian terms defining protein–ligand interactions, surface area components, ligand characteristics, bridge water considerations, and pocket attributes	PDBbind CSAR dataset BindingDB	[35]
	AGL-Score	Features of protein–ligand complexes based on algebraic graph theory	PDBbind	[36]
GBT ǀ Gradient Boosting Tree	ECIF-GBT	Counting pairs of protein–ligand atom types while taking into account the connectivity of each atom	PDBbind v2016	[37]
	NNScore 1.0	Descriptors of specific interactions and ligand-dependent	MOAD PDBbind	[38]
NN ǀ Nearest Neighbors	NNScore 2.0	Vina empirical factors, counts of protein–ligand atom-type pairs within a predefined distance threshold	MOAD PDBbind	[39]
	AtomNet	Local structure-based 3D grid from protein–ligand structures	DUD-E	[40]
	Pafnucy	Atom property-based 3D grid from protein–ligand structures	PDBbind v2016	[41]
CNN ǀ Convolutional Neural Network	Kdeep	Atom type-based 3D grid from protein–ligand structures	PDBbind v2016	[42]
				[43]
	OnionNet	Element-specific contacts between protein and ligand atoms without the need for rotation, categorized by different distance ranges	PDBbind v2016	
	PotentialNet	Atom node feature and distance matrix	PDBbind v2007	[44]
	graphDelta			[45]

(continued)

Table 7.1 (continued)

Algorithm	Name	Input feature	Dataset	Reference
GNN I Graph Neural Network		Atom node features considering local environment and distance matrix	PDBbind v2018	
	SIGN	Distance matrix of atom nodes and angle matrix of bond edges	PDBbind v2016	[46]

Table 7.2 Key characteristics of the main categories of search algorithms

Systematic	The systematic approach is the only one mentioned earlier that guarantees convergence to the global minimum. It systematically and exhaustively explores all ligand degrees of freedom, traversing the energy surface. The main problem related to systematic methods is that computational cost grows explosively with the addition of degrees of freedom in the system. This approach becomes prohibitive for larger ligands (i.e., with many degrees of freedom). This approach is not always suitable for applications that rely on more agile methods [48]
Deterministic	The deterministic approach uses energy minimization and molecular dynamics methods. These methods employ the gradient (i.e., the direction of the steepest ascent on a surface) and system movement over time, respectively. Therefore, they always ensure convergence to an energy minimum. However, such a minimum cannot be guaranteed to correspond to the global minimum of the energy surface. These methods are strictly dependent on the initial conditions of the system and often get trapped in local minima with high energy barriers [49]
Stochastic	This approach involves heuristic methods in exploring the energy surface. Stochastic methods randomly vary all ligand degrees of freedom (translational, rotational, and conformational) at each step, generating a wide diversity of solutions. These solutions are evaluated based on a probabilistic criterion to decide whether they will be rejected. Examples of this approach include Monte Carlo and Evolutionary Algorithm [50, 51]. The main disadvantage of this methodology is that convergence to the global minimum is not guaranteed, requiring multiple independent runs of the algorithm to maximize the probability of finding an optimal result [52]

fitness of the descendants is re-evaluated. The process is repeated for several generations until stop criteria are met. The method is called "Lamarckian" because it allows information learned in previous generations to be passed on to future generations, leading to convergence in favorable ligand conformations. In this case, an individual (ligand) with ideal interaction characteristics would pass this information to the new individual or population (group of ligands) [54]. It is essential to highlight that the evolution and the widespread use of molecular docking techniques have allowed this computational tool to be used to screen new potential ligands against a large variety of biological targets [55]. Figure 7.3 summarizes the search methodologies that can be explored to understand and improve ligand flexibility in a docking simulation.

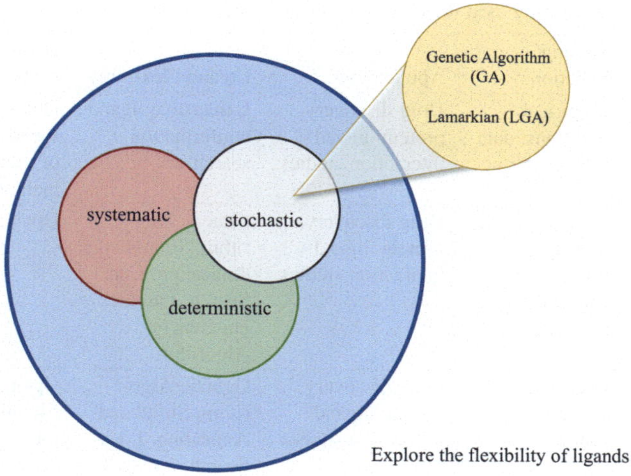

Fig. 7.3 Diagram representing the search algorithms that can be used to explore ligand flexibility in a docking simulation

2.3 *Docking Programs and Scoring Functions*

Scoring functions are mathematical algorithms used in molecular docking simulations to assess and rank different ligand conformations in the target protein. The main goal of scoring functions is to estimate the affinity or binding energy between the ligand and the protein in a given conformation. Table 7.3 lists some of the most commonly used molecular docking software, the scoring functions implemented in each, and their potential applications [56].

It is important to note that most docking programs offer options to customize scoring functions or allow users to define their energy terms, depending on the specific needs of their research. Additionally, the choice of software and scoring function should be based on the characteristics of the target protein, the ligand of interest, and the objectives of the docking simulation [71]. Below is an overview of scoring functions and the calculations involved. Scoring functions are based on an energy function that models interactions between the ligand and the protein. This energy function consists of terms describing various molecular effects, such as:

- *van der Waals energy:* related to London dispersion forces representing attraction between atoms due to temporary fluctuations in their electron clouds, as well as steric repulsion forces preventing ligand atoms from overlapping with protein atoms, ensuring physically feasible binding.
- *Hydrogen bonding energy:* considers hydrogen bonds formed between the ligand and critical residues in the protein's active site.
- *Electrostatic energy:* related to interactions between electric charges, such as ionic interactions.

Table 7.3 Most commonly used docking programs and their scoring functions

Program	Scoring Functions	Applications	Ligand Flexibility	Distinctive Features
Schrödinger Suite (Glide) [57]	Empirical functions, data-driven scoring	Drug discovery, protein–ligand interaction studies	Exhaustive ligand conformation search	High accuracy and speed, integration of multiple scoring methods
AutoDock [58]	Energy-based scoring functions, empirical, grid-based docking	Drug discovery, protein–ligand interaction studies	Genetic Algorithm, Simulated Annealing, Local Search, Lamarckian Genetic Algorithm	Open source
Autodock Vina* [59]	Energy-based scoring functions	Drug discovery, protein–ligand interaction studies	Genetic Algorithm, Simulated Annealing, Local Search, Lamarckian Genetic Algorithm	Open source, user-friendly graphical interface, GPU compatibility
GOLD (Genetic Optimization for Ligand Docking) [60]	Energy-based scoring functions, solvent potential	Drug discovery, protein–ligand interaction studies	Genetic algorithm	Flexibility in defining force fields, ligand encapsulation method, advanced parameterization options
FlexX [61]	Energy-based scoring functions, stereochemical penalization	Drug discovery, protein–ligand interaction studies	Incremental construction algorithm	Use of atomistic potentials learned by Ma chine Learning
Surflex-Dock [62]	Energy-based scoring functions, ligand selection methods	Drug discovery, protein–ligand interaction studies	Incremental construction algorithm	High-precision ligand selection methods, combination of diverse scoring functions
Dock [63]	Force field-based and empirical scoring functions	Drug discovery, protein–ligand interaction studies	Incremental construction algorithm	Scores based on force field and empirical methods, flexibility in developing docking protocols
Dock4 [64]	Physical-, empirical-, and knowledge-based scores	Drug discovery, protein–ligand interaction studies	Incremental construction algorithm	Combine grid-based docking approaches with molecular optimization and custom docking methods.
Dock6 [65]	Interaction energy, solvation, and	Drug discovery, protein–ligand interaction studies	Incremental construction algorithm	Efficient search algorithms, improved scoring

(continued)

Table 7.3 (continued)

Program	Scoring Functions	Applications	Ligand Flexibility	Distinctive Features
	penalties for sterically unfavorable	(particularly useful for large compound libraries)		methods, and improvements in the handling of flexible ligands. It is also highly modular and extensible.
LibDock [66]	Data-driven scoring functions, atom pair clustering	Prediction of high-affinity ligands, protein– ligand interaction studies	Incremental construction algorithm	Use of simplified force fields, focus on predicting pharmacological ligands
Molecular Operating Environment (MOE) [67]	Energy-based scoring functions	Drug discovery, protein–ligand interaction studies	Search algorithms, Genetic algorithms, Verlet's algorithm, Quantum chemistry algorithms	Intuitive graphical interface, integrated database, algorithm integration
PLANTS [68]	Interaction energy, solvation, and penalties for sterically unfavorable	Drug discovery, protein–ligand interaction studies	Incremental construction algorithm	Efficient search algorithms, improved scoring methods, and improvements in the handling of flexible ligands

Smina: This is an enhanced AutoDock Vina version designed for improved performance and flexibility. Smina retains Vina's core functionality while introducing optimizations for multi-threading, energy minimization, and scoring function customization [69]
Gnina: Gnina is an extension of Smina that integrates deep learning (in particular, convolutional neural networks—CNNs) in molecular docking. By leveraging CNNs, Gnina aims to enhance ligand pose prediction and scoring accuracy and efficiency. It offers advanced features such as atom typing, symmetry detection, and ensemble docking [70]
*These implementations represent advancements in the Vina docking framework, incorporating state-of-the-art methodologies to address the challenges of molecular docking more effectively

- *Desolvation terms:* considers how ligand binding affects water molecules around it.
- *Global scoring function:* considers an overall score for each ligand conformation in the protein. This score is an estimate of the total binding energy in that conformation.
- *Entropic terms:* scoring functions may also consider entropic terms reflecting changes in entropy during the binding process. This is important because ligand binding to the protein can affect the flexibility of the involved molecules.

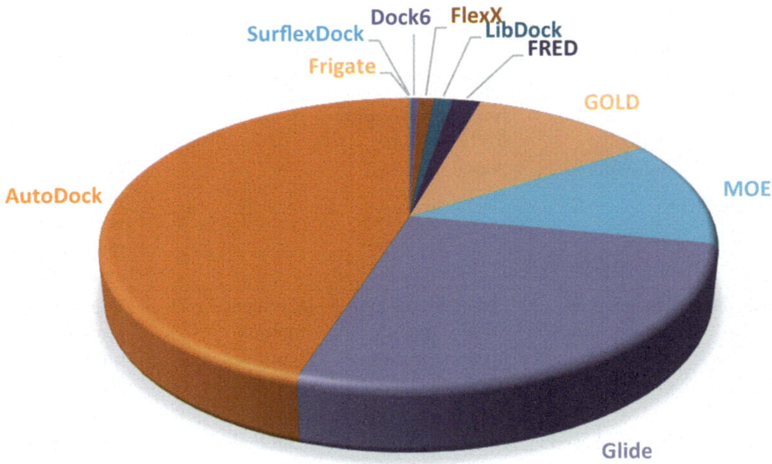

Fig. 7.4 The most used molecular docking programs in the last years (2012–2022). The search was carried out on Web of Science, with the keywords: "software name" and docking ($N = 8496$; the search was based on the following topics in the Web of Science = title and abstract)

- *Optimization:* to find the best binding conformation, the docking program scans for different positions and orientations of the ligand in the protein. At each scan step, the scoring function is calculated to assess the conformation's quality in terms of binding energy.
- *Ranking:* conformations can be ranked based on the calculated scores. The conformation with the lowest score (most favorable energy) is usually considered the most likely. It is expected to represent the most realistic interaction of the ligand complexed with the protein.

- *Results:* at the end of the docking simulation, the program lists the most promising ligand conformations in the target protein, ranked by their scores obtained via the scoring function. This helps researchers identify the most likely and favorable conformations for interaction.

Each docking program may offer various options for different scoring functions, with varied weights in terms of energy and entropy [72]. The choice of the scoring function can significantly affect docking results, and developers often adjust and validate these functions to optimize program performance on known and experimental datasets. Figure 7.4 presents the main docking programs employed in the last years (2012–2022).

The four most widely used docking programs in the last years primarily focus on drug discovery and studying interactions between proteins and ligands. The AutoDock program is renowned for its versatility and effectiveness in predicting interactions between ligands (molecules) and target proteins. Moreover, it is one of the fastest and most widely used open-source molecular docking programs [73, 74]. The Glide program (by Schrödinger) is known for its computer efficiency and accuracy in docking simulations, making it a popular choice in the pharmaceutical industry [75]. The GOLD program stands out for its genetic optimization algorithm, allowing efficient exploration of the conformational space of molecules during docking. It is widely used in drug candidate discovery and identifying ligand poses (conformations) that fit well into target proteins [76]. Finally, the MOE program is a software suite that includes docking and molecular modeling features and tools for visualization and analysis of chemical structures. In addition to docking simulations, MOE is used in molecular design studies, molecular dynamics simulations, and structure analysis [77].

It is worth noting that each docking program has its characteristics and advantages, and the choice between them often depends on the project's specific needs, resource availability, and the user's familiarity with the tool. Accuracy, calculation speed, accessibility, and compatibility with the system and target proteins should be considered when selecting a docking program for a specific task.

3 Calculations Performed During Docking Simulations

Molecular docking simulations involve complex calculations to predict how a molecule (ligand) fits into a target protein. The essential calculations performed during the docking process include estimating the several energy terms related to the binding between the ligand and receptor [78]. This energy is the sum of contributions from various interactions, as presented in Table 7.4.

Molecular docking simulations entail a series of intricate calculations to predict how a molecule (ligand) interfaces with a target protein. Critical calculations performed during docking involve estimating several energy terms pertinent to the binding between the ligand and receptor [78]. This energy is the cumulative result of various interactions, as outlined in Table 7.4. In addition to considerations regarding binding energy, another critical aspect to address is the flexibility of structures involved in the simulation. Many proteins and ligands exhibit high flexibility, implying significant structural changes during interaction. Consequently, in flexible docking, computations must account for determining the most stable conformations of the protein and ligand throughout the binding process while accommodating the flexibility of specific residues and the ligand. This can render the docking process impractical due to hardware limitations in scanning the conformational space of the entire system.

It is worth noting that the calculations mentioned above are carried out iteratively to ascertain the most probable binding conformation. The efficacy of docking simulations hinges on the accuracy of molecular models utilized, the quality of energy parameters, and efficient search strategies. These simulations are indispensable tools in drug candidate design and discovery, facilitating an understanding of crucial interactions between ligands and biological targets.

Table 7.4 Energy terms related to the binding between the ligand and receptor

van der Waals	*Dispersion Energy:* represents the attraction between atoms due to fluctuations in electron clouds. *van der Waals Radius:* defines the contact distance between atoms. *Lennard-Jones Potential:* describes the attractive and repulsive interaction between atoms. *Steric Repulsion Energy:* represents repulsion when atoms come too close. These properties are crucial for assessing the complementarity between ligands and proteins in docking simulations.
Hydrogen Bonding	*Hydrogen Bonding Energy:* measures the strength of the interaction between hydrogen atoms and hydrogen bond donors or acceptors. *Hydrogen Bonding Distance:* defines the ideal distance between donor and acceptor hydrogen atoms. *Hydrogen Bonding Angle:* evaluates the orientation of hydrogen atoms in a hydrogen bond. *Free Energy of Hydrogen Bonding:* represents the total contribution of hydrogen bonds to the stability of the ligand-protein interaction.
Electrostatic	*Electrostatic Potential:* describes the resulting electric field from charges on atoms and influences interaction direction in docking. *Coulombic Interaction Energy:* measures the contribution of Coulomb forces in ligand–protein interactions. *Potential Surface Maps:* visually represent the distribution of electrostatic properties around atoms. These properties are essential for understanding electrostatic interactions in ligand–protein docking.
Solvation	*Free Solvation Energy:* measures the contribution of solvation to ligand–protein affinity. *Free Desolvation Energy:* evaluates the energetic cost of removing solvation around the ligand and protein during docking. *Ligand Solvation Free Energy:* calculates the specific contribution of the ligand to solvation energy. *Protein Solvation Free Energy:* represents the contribution of the protein to the total solvation energy. These properties aid in understanding interactions with the aqueous environment during docking.
Desolvation	In this step, consideration is given to how the protein and ligand affect the structure of water molecules around them. This is important because ligand binding to the protein often involves the displacement of water molecules from the interface.
Entropic Terms	*Conformational Entropy:* measures the freedom of movement of the ligand and protein, affecting system entropy. *Translational Entropy:* refers to the variation in entropy due to changes in the position of molecules in space. *Rotational Entropy:* represents the entropy resulting from molecular rotation. *Vibrational Entropy:* evaluates the variation in entropy due to molecular vibrations. These entropic terms influence the stability and affinity of ligand–protein interactions during docking simulations.

It is important to note that the list of interaction/energy terms provided is not comprehensive, and docking scores frequently miss capturing all interactions. An illustrative instance of extending a scoring function to encompass significant missing interactions and generate more precise docking results can be found here: [79].

4 Essential Components for a Good Docking Program

"A docking protocol requires extensive sampling of the ligand's conformational space to position it at the binding site of a protein, generating a large number of potential orientations and conformations of a ligand, known as poses. A good positioning algorithm samples 'all' possible binding modes, while the scoring system ranks all solutions and identifies the most probable 'binding mode' of the ligand." [80].

A well-designed molecular docking program should incorporate essential components to provide accurate and valuable results. The following are the critical components that contribute to an effective docking program [81, 82]:

- *Accurate molecular modeling:* The program must be able to accurately represent the three-dimensional structures of the target protein and the ligand. This includes considering the flexibility of the protein and the ligand.
- *Precise energy parameters:* The accuracy of binding energy calculations is crucial. Energy parameters should be well-calibrated and reflect the key molecular interactions of real/experimental systems, such as hydrogen bonds, van der Waals interactions, and electrostatics.
- *Efficient search algorithms:* The program should use efficient search algorithms to explore the conformational space for the most stable conformations of the protein–ligand complex. Algorithms like Monte Carlo, genetic algorithms, or optimization methods are most common in docking programs.
- *Accurate scoring function:* An appropriate scoring function is crucial for evaluating and ranking ligand conformations in the protein. The scoring function should accurately reflect the binding energy and consider entropic factors.
- *Desolvation:* Considering desolvation is important to assess how ligand binding affects water molecules around. This step is critical for accurate models.
- *Molecular flexibility:* In many cases, target protein and ligand structures exhibit high degrees of freedom. Thus, the program should handle the flexibility of the systems, allowing optimization of molecular conformations during the docking process.
- *User-friendly interface:* An intuitive user interface is desirable for researchers to set parameters, visualize results, and interpret data easily.
- *Adequate computational resources:* Docking calculations can be computer intensive, especially in high-precision and complex simulations. The program should be able to use suitable hardware or cluster computing systems.

- *Validation and benchmarking:* A reliable docking program should be validated and benchmarked on known/experimental datasets to ensure consistent and accurate results.

- *Customization and extensibility:* The program should allow customization and application extension to meet the specific needs of each research project.

- *Integration with other software:* The ability to export and import data from other related software, such as molecular viewers and analysis tools, is desirable.

A well-designed molecular docking program should consider all the mentioned components to provide reliable and valuable results related to drug candidate discovery and design processes. It is important to note that the choice of the program should be based on the specific needs of the project and the research context.

5 Limitations of the Docking Technique

The molecular docking technique has significant limitations that need to be acknowledged by the user. They start with the fact that docking may not always accurately predict a compound's binding conformation in the active site of a protein. This happens because the search and scoring functions used may not capture all the complexities related to the molecular docking [83].

Another point to consider is related to the fact that proteins and ligands have flexible structures, but most docking programs treat molecules as rigid during calculations. This can lead to inaccurate results, in particular for the systems with a high number of degrees of freedom. For this kind of systems, an option would be to use flexible and/or ensemble docking. Additionally, most scoring functions do not handle entropic terms well, meaning they do not adequately consider changes in entropy during the binding process. Ignoring these changes can lead to inaccurate predictions regarding binding affinity.

Regarding the proper treatment of solvation (how water molecules are around the protein and ligand), it can be said that this is still a challenge and can significantly affect docking results. Besides, for particular systems, the addition of structural water is very important to properly describe the action mechanisms of the protein. In other cases, for example, a system with a low structural information/quality, the addition of solvation is not the most important feature to be considered in the initial analysis. The choice of initial conformations can also be a problem, as the quality of docking results largely depends on the choice of the initial conformations of the ligand and protein. If these conformations are not representative, the results may be distorted [84]. Furthermore, the technique relies on high-quality three-dimensional structures of the protein and ligand. When these structures are unavailable or of lower quality, docking simulations may be less reliable.

It is important to note that the results of docking simulations should be experimentally validated, which can be a costly and time-consuming step. Experimental validation is crucial to determine the accuracy of predictions. In summary, molecular docking is a valuable tool, but input data must be reliable, and results should be interpreted cautiously, considering its intrinsic limitations. It is vital to use docking in conjunction with other experimental and computational techniques for a more comprehensive assessment of protein–ligand interactions and to increase the reliability of predictions.

6 Validation of Docking Results

The validation of docking results is a critical step in determining the reliability of predictions generated by these simulations. Various approaches and validation methods can be employed to assess the performance of docking techniques, for example [85, 86]:

- *Cross-docking validation* refers to the simulation (docking) of the ligand of a particular biological target to the other tridimensional structures of the same receptor. This practice provides the accuracy of the docking algorithm to predict the best ligand pose at the binding site of the biological target.

- *Benchmarking with experimental data:* comparing docking results with experimental data is a crucial form of validation. This may involve comparing predicted binding affinity with experimentally measured affinity, overlaying predicted binding conformations with experimental structures, or analyzing predicted interactions against experimentally observed interactions [87, 88].
  - The validation and benchmarking of docking programs are crucial to ensure their reliability, consistency, and accuracy in predicting ligand–protein interactions. Several methods can be employed to achieve this point, including:
  - Experimental data comparison: docking results can be compared with experimental data, such as crystallographic structures of ligand–protein complexes. If the predicted binding poses and binding affinities align well with experimental observations, it indicates the reliability of the docking algorithm. For example, the Directory of Useful Decoys (DUD) dataset provides a benchmark for assessing docking performance against experimentally validated ligand–protein complexes [89].
  - Cross-docking studies: in cross-docking studies, ligands are docked into multiple protein structures to evaluate the program's ability to discriminate between binding sites and protein conformations. This assesses the docking program's robustness and versatility. The Astex Diverse Set (ADS) is a commonly used dataset for cross-docking studies due to its diverse protein–ligand complexes [90].

   - Scoring function evaluation: The performance of scoring functions within docking programs can be assessed by comparing predicted binding affinities with experimental binding constants or free energies. Statistical metrics such as root mean square error (RMSE) or Pearson correlation coefficient can quantify the agreement between predicted and experimental values [91].
   - Virtual screening campaigns: virtual screening campaigns involve the docking of large compound libraries against a target protein to identify potential lead compounds. The enrichment factor, receiver operating characteristic (ROC) curve, and area under the curve (AUC) are metrics used to evaluate virtual screening performance. The DUD-E dataset [40] is often used for virtual screening benchmarking [92].
   - Community challenges and collaborative efforts: participation in community challenges, such as the D3R Grand Challenge or the Critical Assessment of PRediction of Interactions (CAPRI), allows docking programs to be evaluated in blind tests against undisclosed protein–ligand complexes. These challenges provide an independent assessment of program performance in a competitive setting [93].
   - By employing these approaches of validation and benchmarking, docking programs can demonstrate their reliability and accuracy in predicting ligand–protein interactions across diverse datasets and scenarios. Additionally, publications in peer-reviewed journals often report detailed validation studies and benchmarking results for docking programs, providing valuable insights into their performance characteristics [92].

- *Retrospective validation:* using historical datasets where binding activity is known and comparing docking predictions with these known results is the focus of this approach. This is useful for evaluating the accuracy and performance of the method under controlled conditions.

- *Negative control:* this step involves performing docking simulations where the ligand has no affinity for the target protein. This helps identify false positives, i.e., ligands that are erroneously predicted to have affinity for the target protein.

- *Statistical analysis:* conducting statistical analyses, for example, via Receiver Operating Characteristic (ROC) and Precision-Recall (PR) curves, to evaluate the performance of docking simulations in classifying active and inactive ligands.

- *Reproducibility tests:* it is important to repeat docking simulations with different configurations or parameters to verify the consistency of results. This helps determine if the results are robust.

- *Validation with analog ligands:* in this step, we perform docking simulations with a congener series of analog ligands and compare predictions based on their structural similarity. This can be used to assess the method's ability to recognize structural similarities.

- *Prospective validation:* in this step, the goal is to apply the docking method to systems not yet experimentally studied and compare predictions with subsequent experimental results.

- *Blind docking:* in this type of validation, the aim is to perform docking simulations without prior knowledge of the location of the binding site on the target protein. This tests the method's ability to predict the location of known binding sites and estimate unknown sites.

- *Comparative validation:* here, the focus is on comparing the performance of the docking method with other protein–ligand affinity prediction methods, such as force-field-based docking, molecular dynamics, or approaches employing machine learning techniques.

It is important to note that there is no single validation method universally applicable in docking studies, as the choice of validation method depends on the context, research objectives, and available data. In many cases, a combination of validation methods is used to obtain a comprehensive assessment of docking technique performance. Additionally, collaboration among researchers and the availability of well-defined experimental datasets play a fundamental role in the successful validation of docking predictions [94].

6.1 Success in Docking Predictions and the Consensus Technique

Using a consensus employing multiple docking algorithms is a common and valuable strategy in research involving drug discovery and studies of protein–ligand interactions. This approach involves conducting multiple independent docking simulations for a protein–ligand complex and then combining the results to obtain a more reliable and accurate prediction of the binding conformation and affinity of the compound at the receptor site.

Docking simulations can be sensitive to various variables and parameters, and this diversity may influence individual results. By achieving a consensus of results obtained from different docking algorithms, it is possible to reduce the impact of outliers or inaccuracies, obtaining a more accurate estimate. The consensus takes into account the inherent uncertainty in docking simulations. Different simulations may produce slightly different results, but the consensus provides a robust measure reflecting the intrinsic variability of docking algorithms [86, 95].

The consensus process can help identify binding conformations consistent across multiple simulations, increasing confidence that these conformations are more likely to occur in reality. By verifying if various simulations produce consistent results, the consensus helps assess the reliability of docking predictions. Conformations or interactions that are consistent in multiple simulations are more likely to be reliable.

The consensus allows researchers to analyze protein–ligand interactions better by considering a variety of conformations, and possible Consensus helps reduce false positives (ligands that appear to bind but are not relevant) and false negatives (ligands that are mistakenly discarded) in docking predictions [86].

In drug discovery research, decisions regarding selecting potential bioactive compounds and designing new molecules can be guided by consensus-based docking results, providing a solid foundation for decision-making with lower error chances. It is important to note that consensus from results via different docking algorithms is not a definitive solution to all challenges associated with molecular docking, and its application should be carefully planned and validated. Additionally, how to obtain the consensus and the number of simulations to be performed may vary depending on the system of interest and research objectives. However, in many cases, a consensus from various docking programs is a practical approach to improve the reliability of predictions and the quality of results.

7 Inappropriate Use of Validation Methods in Docking

The inappropriate use of validation methods is a significant concern when applying molecular docking techniques. Validation is essential to determine the reliability of predictions generated in docking simulations. Improper use can lead to erroneous conclusions and excessive confidence in results that are not genuinely predictive [95]. Several factors may contribute to less accurate conclusions about docking results, such as:

- *Lack of experimental data:* Initial validation of docking algorithms relies on the availability of relevant experimental data. Validation becomes compromised if high-quality experimental data are unavailable for comparison with docking results.
- *Validation on biased datasets:* Researchers sometimes validate their results on datasets previously used to develop or tune scoring functions. This can overestimate the method's accuracy, as training data may be "biased" to fit the algorithms used.
- *Lack of diversity in molecular targets:* Some docking studies may focus only on a specific type of target protein, limiting the generalization of results. Proper validation should include a variety of molecular targets to determine applicability in different contexts.
- *Failure to consider system flexibility:* If the docking algorithm does not account for the flexibility of the protein and/or ligand, validation may be inaccurate, especially in highly flexible systems.

- *Neglecting entropic terms:* The exclusion of entropic terms in estimating binding energy (terms describing changes in entropy during the binding process) can lead to inaccurate validation, as these changes are critical for estimating protein–ligand affinity [2].

- *Validation limited to single metrics:* Validation often focuses on a single metric, such as binding affinity. However, other metrics, such as ligand specificity and correct binding conformation, are also essential and should be considered.

- *Ignoring solvent influence:* Validation may be inadequate if the solvent is not adequately treated in docking simulations, as solvation is a critical factor in protein–ligand interactions.

- *Failure to conduct repeatability studies:* Docking results may vary depending on execution conditions and parameters. Repeating studies is vital to assess result robustness.

A comprehensive approach should be adopted to overcome the previously cited issues and appropriately use validation methods in the molecular docking process. This includes selecting independent validation datasets, considering different performance metrics, evaluating flexibility and solvation, and repeating experiments to ensure result consistency. Collaboration within the scientific community is crucial for sharing data and knowledge and promoting rigorous validation practices [85, 86, 95].

8 Evaluation from an Expert's Perspective

Expert evaluation is essential when analyzing results obtained through molecular docking simulations, and this is due to a series of fundamental reasons. The results of these simulations can be incredibly complex, involving various metrics, binding energy data, conformations, and structural information. In this context, the expertise of a specialist is necessary, enabling the precise interpretation of these complex data. Experts can assess these results in light of the biological and chemical context, considering the target protein's function and the specific application. They can discern whether the identified conformations are relevant in the biological and chemical context [81].

Experimental validation is critical to confirming predictions made through docking. In this regard, a specialist can conceive and conduct experiments that confirm or refute predictions, becoming a crucial component in the validation process. Assessing the accuracy of docking simulations is a task that requires technical and scientific knowledge. A specialist can identify potential sources of error, such as the quality of protein and ligand structures, the effectiveness of scoring functions, and the appropriateness of simulated conditions.

In areas such as drug design and discovery, the analysis of docking results often directly affects strategic decisions, such as the selection of drug candidates and the design of potential bioactive compounds. In this context, the guidance of a specialist is of utmost importance. As mentioned earlier, docking simulations have their limitations and uncertainties. An expert can identify and communicate these limitations clearly, providing a balanced and realistic view of the conclusions. As proteins and ligands can exhibit flexibility, specialized analysis is necessary to understand how this flexibility affects docking predictions and consider how dynamic conformations may influence binding. Experts can assess whether docking results align with current knowledge about the target protein's biology, its function in the organism, and its relationship with the overall biological context [82].

In this way, based on their knowledge, an expert can draw solid conclusions and provide recommendations on the next steps of research, such as additional validation experiments, ligand optimization, or structural modifications. Expert evaluation is an essential step in translating raw data into scientific knowledge, experimental validation, and guiding strategic decisions in research related to drug discovery, molecular biology, and chemistry.

9 Use of Machine Learning in Molecular Docking

Machine learning and molecular docking have an increasingly close relationship in the field of drug discovery and studies of protein–ligand interactions. The application of machine learning techniques throughout molecular docking simulations has the potential to enhance the accuracy and efficiency of predictions. Below are listed some ways in which these two areas relate [96–98]:

- *Improvement of scoring functions:* Scoring functions used in docking simulations are crucial for estimating the binding affinity between protein and ligand. Machine learning techniques can be used to develop more accurate and specific scoring functions, incorporating a broader range of molecular features and interactions [30].

- *Drug candidate selection:* Machine learning algorithms can analyze large libraries of chemical compounds and predict which ones are more likely to be promising ligands for a specific target protein. This can save time and resources in virtual screening. For example, a study [74] on COVID-19 utilized about 2.178 million unique protein/ligand binding site combinations available in databases to optimize the search for potent inhibitors, applying ML techniques to facilitate and guide *in silico* studies.

- *Classification of active and inactive ligands:* Machine learning techniques can be applied to classify ligands as active or inactive based on their chemical and structural properties. This is useful for the virtual screening of compound libraries. In a study by Salimi et al. [99], a screening pipeline was built using machine learning algorithms integrated with checks for similarity to approved drugs to find new inhibitors for the vascular endothelial growth factor receptor-2 signaling pathway.

- *Prediction of protein–ligand interactions:* Machine learning algorithms can predict specific interactions between proteins and ligands, identifying critical amino acid residues for binding or preferred binding sites [97].

- *Modeling molecular flexibility:* Molecular flexibility is a significant challenge in docking simulations. Machine learning methods can be applied to model the flexibility of proteins and ligands, allowing more realistic predictions. Harmalkar and Gray [100] studied the difficulties in predicting protein interactions and ways to apply machine learning to optimize these simulations. The study showed that docking simulations in the CAPRI [101] challenge included a wide variety of target types, with 11 out of 28 "easy" targets achieving high-quality structure. However, for the 17 "difficult" targets, the intrinsic flexibility of biomolecules remains a challenge, with only 2 achieving high quality.
- *Enhancement of validation:* Machine learning techniques can be employed to enhance the validation of docking results, identifying more reliable and effective metrics to assess the method's performance. In this case, a study by Zhang et al. [102] presented an internal and external dataset used to cross-validate eight machine learning methods. The results showed that the extremely random tree model performed better and was adopted as the first step in virtual screening.

- *Increase in computational efficiency:* Machine learning methods can be used to accelerate the processing of large volumes of data in docking simulations, making the process more efficient [98].

- *Discovery of structure–activity relationships:* Machine learning can help discover complex relationships between the chemical structure of ligands and their biological activity, aiding in compound optimization. For example, in Hermansyah's study [103], selective DPP-4 inhibitors against DPP-8 and DPP-9 were identified using an AI-based quantitative structure–activity relationship (QSAR) workflow, enabling faster screening of millions of molecules for the DPP-4 target compared to other screening methods.

- *Integration of diverse data:* Machine learning techniques enable the integration of data from various sources, such as high-throughput screening data, structural biology information, and gene expression data, for a more comprehensive system analysis and optimization of simulations [104].

To assess the evolution in the use of machine learning in docking simulations, publications indexed in the "Web of Science" database between 2012 and 2022 were compiled, relating to the use of molecular docking with machine learning (Fig. 7.5).

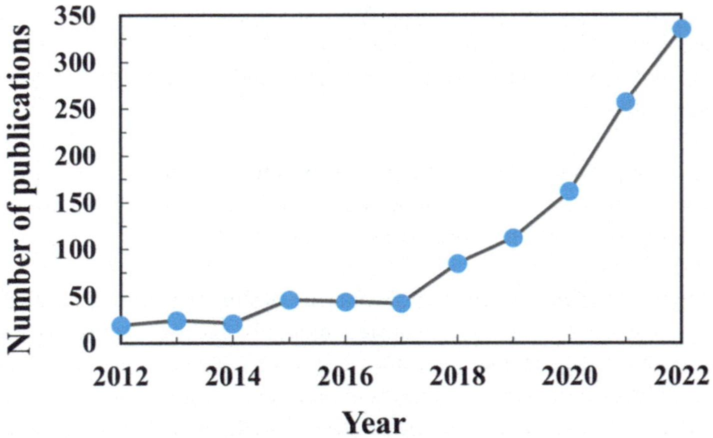

Fig. 7.5 Number of studies (2012–2022) considering the use of machine learning techniques and docking simulations for the design and discovery of drug candidates. Research conducted on the "Web of Science" platform on October 15, 2023, with the combination of the following keywords: "machine learning" AND "molecular docking"

A significant increase, approximately 14 times, in the number of publications using ML techniques to enhance docking studies can be observed.

In summary, integrating machine learning techniques in molecular docking studies can significantly enhance the accuracy and efficiency of predictions, accelerating the discovery of drug candidates and enabling a deeper understanding of protein–ligand interactions. Collaboration among computer scientists, structural biologists, and chemists is crucial for the success of this approach.

10 Advancements and Improvements in Computational Resources

Recent advancements in computational resources have led to significant progress in molecular docking, playing a crucial role in drug discovery. A rising technique is structure-based virtual screening, which benefits from the increasing availability of high-resolution target structures and large-scale virtual compound libraries, such as REAL combinatorial libraries. A REAL combinatorial library collects virtually generated chemical compounds representing a wide structural diversity. The term "REAL" stems from "Readily Available for Synthesis," indicating their readiness for laboratory synthesis. These libraries are designed to offer a vast array of molecules that can be synthesized efficiently and economically in the laboratory [105].

However, new approaches are necessary to keep pace with these libraries' exponential growth. In this regard, based on a modular approach, the V-SYNTHES method emerges as a promising solution. This method conducts hierarchical screening of a REAL library containing over 11 billion compounds. Initially, V-SYNTHES identifies the most promising scaffold–synthon combinations as seeds for iterative growth, refining them to select complete molecules with the best docking scores. This approach enables efficient detection of high-scoring compounds in an extensive chemical space while docking only a minimal fraction of the library. This is highly advantageous in terms of computational and economic efficiency. Furthermore, the method has been experimentally validated in synthesizing and testing cannabinoid antagonists, demonstrating significant improvement over conventional virtual screenings [105].

Another notable advancement lies in incorporating molecular strain as an additional parameter in evaluating ligand scores during the docking process. A recent approach utilizes information on relative torsional populations from the Cambridge Structural Database to precalculate these strain energies, resulting in a more accurate and efficient evaluation. Retrospective studies have shown that including these strain energies significantly improves success rates by preferentially excluding false high-scoring molecules. This approach, independent of the scoring function used, stands out for its speed and practicality, and it is applicable even in large-scale compound libraries [106].

Moreover, careful selection of small molecule libraries is crucial to this process. Libraries capable of synthesizing billions of compounds, such as the Examine library, are available. Finally, despite the advancements represented by AlphaFold2 in predicting protein structures with high precision, studies have shown that the quality of predicted structures does not always translate into satisfactory performance during the docking process. Removing low-confidence regions from the predicted structure and flexibilizing side chains are promising strategies to optimize docking results, underscoring the importance of refined adjustments for successfully applying these models in drug discovery. It is also important to mention that recent approaches have been developed to consider the refolding around the ligand. New models like AlphaFold3 and Neuraplex have showed excellent performance in this area [107].

11 Challenges

Improving molecular docking techniques faces various technical and scientific challenges as the pursuit of greater accuracy and applicability is ongoing. Some of the main challenges include enhancing the accuracy of scoring functions and adequately considering the flexibility of the systems under consideration. Cur rent functions may not accurately capture all ligand–receptor interactions, especially in highly flexible systems or those with weak interactions. Integrating the flexibility of both the protein and the ligand in simulations is a complex task. Developing effective methods to handle flexibility, including molecular dynamics coupled with docking, is a significant challenge [2]. Considering entropic terms in scoring functions is another complicating factor but essential for accurately predicting protein–ligand affinity. Correctly modeling changes in entropy during the binding process is equally challenging. The influence of solvent on protein–ligand interactions must be accurately addressed as well. Developing more precise solvation methods is a considerable challenge. Additionally, experimental validation is crucial for refining docking techniques. However, it is not always easy to conduct due to the complexity of protein–ligand interactions and the availability of high-quality experimental data [4].

Extending the applicability of docking to more diverse targets, such as membrane proteins and protein–protein complexes, is another challenge, as these systems can be highly complex and require specific approaches. Integrating the docking process with other computational techniques, such as molecular dynamics, machine learning, and quantum simulations, is an evolving approach to improving the accuracy and comprehensiveness of predictions. It is worth noting that docking simulations on multiple targets are complex and challenging, as they require considering the competition between different ligands at a single or multiple binding sites [71].

Other factors that make docking simulations somewhat challenging involve the following factors: validation and prediction of weak molecular interactions, such as those involved in allosteric inhibitors or protein modulators; docking simulations can be computationally intensive, requiring significant resources; the need for the development of efficient algorithms to reduce computation time; better understanding of the complexities of biology involved in protein–ligand interactions may be essential to enhance docking techniques and make them more realistic; availability of high-quality protein and ligand structures is essential for the quality of docking results; use of high-quality reference experimental datasets to validate and improve docking techniques.

Overcoming the challenges mentioned above and others not described here regarding docking simulations requires multidisciplinary collaboration among computer scientists, structural biologists, chemists, and pharmacologists. Continuous research and development of more advanced methods will improve molecular docking techniques and their application in various areas, including drug candidate discovery.

12 Conclusions

Molecular docking techniques play a crucial role in drug discovery and candidate drug design worldwide and have undergone significant technological advances. Table 7.5 presents the key reasons why molecular docking algorithms are essential in the drug discovery and design process and how various improvements have contributed to the advancement and quality of docking analyses.

Molecular docking techniques are vital in global drug discovery and design. They help expedite the entire process, save resources, and enhance accuracy in identifying drug candidates with lower error rates. Ongoing technological advancements in molecular docking further broaden its impact and potential in pharmaceutical and biological researches.

Table 7.5 Importance and challenges/advancements in docking algorithms

Fact	Importance
Accelerates drug candidate discovery	The design of drug candidates is lengthy, costly, and risky. Molecular docking enables the virtual screening of compounds, expediting the identification of drug candidates with higher success rates. This saves time and resources
Resource savings	The use of docking techniques helps reduce the number of compounds to be synthesized and experimentally tested, saving financial and laboratory resources
Expands research scope	Molecular docking algorithms make it possible to investigate a wide variety of compounds, including those not easily accessible through experimental approaches
Assists in selectivity studies	Docking techniques aid in designing compounds that can selectively bind to specific targets, minimizing unwanted side effects
Understanding molecular mechanisms	Analyzing protein–ligand interactions through docking contributes to understanding the molecular mechanisms underlying diseases, leading to the development of more effective treatments
Personalization of therapies	Docking can also be used to develop drugs for personalized therapies tailored to individual patient needs
	Challenge/Advancement
Increased accuracy of scoring functions	Scoring functions used in docking simulations have been enhanced, becoming more accurate and representative of ligand–receptor interactions
Inclusion of flexibility	Improved docking software allows for considering certain flexibility in the protein and the ligand, enhancing prediction accuracy
Integration with Machine Learning	Integrating machine learning techniques and docking algorithms improves the prediction of protein–ligand interactions, making simulations more effective
Use of supercomputers	Supercomputers and high-performance clusters allow the execution of more complex docking simulations
Quality data related to structural biology	The availability of high-quality structural biology data, such as protein structures determined by crystallography or nuclear magnetic resonance, enhances the accuracy of docking simulations
More accurate experimental validation	Technological advances in experimental validation, such as high-resolution mass spectrometry and cryo-electron microscopy, assist in confirming predictions from docking simulations

Questions to Answer When Planning a Docking Experiment

? What is the main objective of the docking experiment?
What are the features and limitations of the biological receptor's structure?
What are the main molecular properties of the ligands under study?
Have the three-dimensional structures been verified and optimized?
Are there structural or conformational errors in the selected molecules?
Which docking software will be used? Is it suitable for the system under study?
What is the best scoring function to use for the study in question?
Are there any licensing limitations for the tools being used?
What positive and negative controls will be used to validate the docking protocol?
How will the results be analyzed? What criteria will be used to evaluate the estimated binding energy and molecular interactions?
What recent advancements in scoring functions and docking flexibility can be leveraged?
Is there a possibility to integrate machine learning techniques to improve the prediction of protein–ligand interactions?
Is high-quality structural biology data available to enhance the accuracy of docking simulations?
What experimental methods will be used to validate the results of the docking simulations?
What are the limitations of molecular docking simulations?

References

1. Fan, J., Fan, J., Fu, A., & Zhang, L. (2019). Progress in molecular docking. *Quantitative Biology, 7*, 83–89.
2. Winkler, D. A. (2020). Ligand entropy is hard but should not be ignored. *Journal of Chemical Information and Modeling, 60*(10), 4421–4423.
3. Kuntz, I. D., Blaney, J. M., Oatley, S. J., Langridge, R., & Ferrin, T. E. (1982). A geometric approach to macromolecule-ligand interactions. *Journal of Molecular Biology, 161*(2), 269–288.
4. Stanzione, F., Giangreco, I., & Cole, J. C. (2021). Use of molecular docking computational tools in drug discovery. In *Progress in medicinal Chemistry (Capítulo Quatro)* (pp. 273–343). Elsevier.
5. Li, J., Fu, A., & Zhang, L. (2019). An overview of scoring functions used for protein– Ligand interactions in molecular docking. *Interdisciplinary Sciences, Computational Life Sciences, 11*, 320–328.
6. Zheng, L., Meng, J., Jiang, K., Lan, H., Wang, Z., Lin, M., Li, W., Guo, H., Wei, Y., & Um, Y. (2022). Improving protein–ligand docking and screening accuracies by incorporating a scoring function correction term. *Briefings in Bioinformatics, 23*(3).
7. Shen, C., Hu, Y., Wang, Z., Zhang, X., Pang, J., Wang, G., Zhong, H., Xu, L., Cao, D., & Hou, T. (2021). Beware of the generic machine learning-based scoring functions in structure-based virtual screening. *Briefings in Bioinformatics, 22*(3), bbaa070.
8. Nguyen, D. D., & Wei, G.-W. (2019). AG-score: Algebraic graph learning score for protein–ligand binding scoring, ranking, docking, and screening. *Journal of Chemical Information and Modeling, 59*(7), 3291–3304.
9. Britt, H. M., Cragnolini, T., & Thalassinos, K. (2022). Integration of mass spectrometry data for structural biology. *Chemical Reviews, 122*(8), 7952–7986.

10. Sun, H., Zhao, L., Peng, S., & Huang, N. (2014). Incorporating replacement free energy of binding-site waters in molecular docking. *Proteins: Structure, Function, and Bioinformatics, 82*(9), 1765–1776.
11. Libouban, P. Y., Aci-Sèche, S., Gómez-Tamayo, J. C., Tresadern, G., & Bonnet, P. (2023). The impact of data on structure-based binding affinity predictions using deep neural networks. *International Journal of Molecular Sciences, 24*(22), 16120.
12. Miranda-Quintana, R. A., & Smiatek, J. (2022). Application of fundamental chemical principles for solvation effects: A unified perspective for interaction patterns in solution. *The Journal of Physical Chemistry B, 126*(43), 8864–8872.
13. Englebienne, P., & Moitessier, N. (2009). Docking ligands into flexible and solvated macromolecules. 4. Are popular scoring functions accurate for this class of proteins? *Journal of Chemical Information and Modeling, 49*(6), 1568–1580.
14. Cano, G., Garcia-Rodriguez, J., Garcia-Garcia, A., Perez-Sanchez, H., Benediktsson, J. A., Thapa, A., & Barr, A. (2017). Automatic selection of molecular descriptors using random forest: Application to drug discovery. *Expert Systems with Applications, 72*, 151–159.
15. Sikander, R., Ghulam, A., & Ali, F. (2022). XGB-DrugPred: Computational prediction of druggable proteins using eXtreme gradient boosting and optimized features set. *Scientific Reports, 12*, 5505.
16. Zhang, Y., Wang, Y., Zhou, W., Fan, Y., Zhao, J., Zhu, L., Lu, S., Lu, T., Chen, Y., & Liu, H. (2019). A combined drug discovery strategy based on machine learning and molecular docking, *93*(5), 685–699.
17. Bongini, P., Bianchini, M., & Scarselli, F. (2021). Molecular generative graph neural networks for drug discovery. *Neurocomputing, 450*, 242–252.
18. Vaz, J. M., & Balaji, S. (2021). Convolutional neural networks (CNNs): Concepts and applications in pharmacogenomics. *Molecular Diversity, 25*, 1569–1584.
19. Wieder, O., Kohlbacher, S., Kuenemann, M., Garon, A., Ducrot, P., Seidel, T., & Langer, T. (2020). A compact review of molecular property prediction with graph neural networks. *Drug Discovery Today: Technologies, 37*, 1–12.
20. Wang, L., Zhou, Z., Yang, X., Shi, S., Zeng, X., & Cao, D. (2024). The present state and challenges of active learning in drug discovery. *Drug Discovery Today, 29*(6), 103985.
21. Ewing, T. J. A., & Kuntz, I. D. (1997). Critical evaluation of search algorithms for automated molecular docking and database screening. *Journal of Computational Chemistry, 18*(9), 1175–1189.
22. Rosenfeld, R., Vajda, S., & DeLisi, C. (1995). Flexible docking and design. *Annual Review of Biophysics and Biomolecular Structure, 24*, 677–700.
23. Solis-Vasquez, L., Tillack, A. F., Santos-Martins, D., Koch, A., LeGrand, S., & Forli, S. (2022). Benchmarking the performance of irregular computations in AutoDock-GPU molecular docking. *Parallel Computing, 109*, 102861.
24. Fu, Y., Wu, X., Chen, Z., Sun, J., Zhao, J., & Xu, W. (2015). A new approach for flexible molecular docking based on swarm intelligence. *Mathematical Problems in Engineering, 2015*, 540186., 10 pages.
25. Tavares, J., Tantar, A.-A., Melab, N., & Talbi, E.-G. (2008). The impact of local search on protein-ligand docking optimization. In *2008 eighth international conference on hybrid intelligent systems* (pp. 53–58).
26. Leonhart, P. F., & Dorn, M. (2019). A biased random key genetic algorithm with local search chains for molecular docking. In P. Kaufmann & P. Castillo (Eds.), *Applications of evolutionary computation. EvoApplications 2019* (Vol. 11454, pp. 1–14). Springer.
27. Vidal-Limon, A., Aguilar-Toalá, J. E., & Liceaga, A. M. (2022). Integration of molecular docking analysis and molecular dynamics simulations for studying food proteins and bioactive peptides. *Journal of Agricultural and Food Chemistry, 70*(4), 934–943.
28. Vant, J. W., Lahey, S.-L. J., Jana, K., Shekhar, M., Sarkar, D., Munk, B. H., Kleinekathöfer, U., Mittal, S., Rowley, C., & Singharoy, A. (2020). Flexible fitting of small molecules into

electron microscopy maps using molecular dynamics simulations with neural network potentials. *Journal of Chemical Information and Modeling, 60*(5), 2591–2604.
29. Davis, I. W., & Baker, D. (2009). RosettaLigand docking with full ligand and receptor flexibility. *Journal of Molecular Biology*.
30. Yang, C., Chen, E. A., & Zhang, Y. (2022). Protein–ligand docking in the machine-learning era. *Molecules, 27*, 4568.
31. Ballester, P. J., & Mitchell, J. B. (2010). A machine learning approach to predicting protein–ligand binding affinity with applications to molecular docking. *Bioinformatics, 26*, 1169–1175.
32. Zilian, D., & Sotriffer, C. A. (2013). Sfcscore rf: A random forest-based scoring function for improved affinity prediction of protein–ligand complexes. *Journal of Chemical Information and Modeling, 53*, 1923–1933.
33. Wang, C., & Zhang, Y. (2017). Improving scoring-docking-screening powers of protein–Ligand scoring functions using random forest. *Journal of Computational Chemistry, 38*, 169–177.
34. Lu, J., Hou, X., Wang, C., & Zhang, Y. (2019). Incorporating explicit water molecules and ligand conformation stability in machine-learning scoring functions. *Journal of Chemical Information and Modeling, 59*, 4540–4549.
35. Yang, C., & Zhang, Y. (2022). Delta machine learning to improve scoring-ranking-screening performances of protein–ligand scoring functions. *Journal of Chemical Information and Modeling, 62*, 2696–2712.
36. Nguyen, D. D., & Wei, G.-W. (2019). Agl-score: Algebraic graph learning score for protein–ligand binding scoring, ranking, docking, and screening. *Journal of Chemical Information and Modeling, 59*, 3291–3304.
37. Sánchez-Cruz, N., Medina-Franco, J. L., Mestres, J., & Barril, X. (2021). Extended connectivity interaction features: Improving binding affinity prediction through chemical description. *Bioinformatics, 37*, 1376–1382.
38. Durrant, J. D., & McCammon, J. A. (2010). NNScore: A neural-network-based scoring function for the characterization of protein–ligand complexes. *Journal of Chemical Information and Modeling, 50*, 1865–1871.
39. Durrant, J. D., & McCammon, J. A. (2011). NNScore 2.0: A neural-network receptor– Ligand scoring function. *Journal of Chemical Information and Modeling, 51*, 2897–2903.
40. Warren, G. L., Andrews, C. W., Capelli, A.-M., Clarke, B., LaLonde, J., Lambert, M. H., Lindvall, M., Nevins, N., Semus, S. F., Senger, S., Tedesco, G., Wall, I. D., Woolven, J. M., Peishoff, C. E., & Head, M. S. (2006). A critical assessment of docking programs and scoring functions. *Journal of Medicinal Chemistry, 49*(20), 5912–5931.
41. Stepniewska-Dziubinska, M. M., Zielenkiewicz, P., & Siedlecki, P. (2018). Development and evaluation of a deep learning model for protein–ligand binding affinity prediction. *Bioinformatics, 34*, 3666–3674.
42. Jiménez, J., Skalic, M., Martinez-Rosell, G., & De Fabritiis, G. (2018). K deep: Protein–Ligand absolute binding affinity prediction via 3D-convolutional neural networks. *Journal of Chemical Information and Modeling, 58*, 287–296.
43. Zheng, L., Fan, J., & Mu, Y. (2019). OnionNet: A multiple-layer intermolecular-contact-based convolutional neural network for protein–ligand binding affinity prediction. *ACS Omega, 4*, 15956–15965.
44. Feinberg, E. N., Sur, D., Wu, Z., Husic, B. E., Mai, H., Li, Y., Sun, S., Yang, J., Ramsundar, B., & Pande, V. S. (2018). PotentialNet for molecular property prediction. *ACS Central Science, 4*, 1520–1530.
45. Karlov, D. S., Sosnin, S., Fedorov, M. V., & Popov, P. (2020). graphDelta: MPNN scoring function for the affinity prediction of protein–ligand complexes. *ACS Omega, 5*, 5150–5159.

46. Li, S., Zhou, J., Xu, T., Huang, L., Wang, F., Xiong, H., Huang, W., Dou, D., & Xiong, H. (2021). Structure-aware interactive graph neural networks for the prediction of protein-ligand binding affinity. In *Proceedings of the 27th ACM SIGKDD Conference on Knowledge Discovery & Data Mining*, Singapore, 14–18 August 2021 (pp. 975–985).
47. Brooijmans, N., & Kuntz, I. D. (2003). Molecular recognition and docking algorithms. *Annual Review of Biophysics and Biomolecular Structure, 32*, 335–373.
48. Halperin, I., Ma, B., Wolfson, H., & Nussinov, R. (2002). Principles of docking: An overview of search algorithms and a guide to scoring functions. *Proteins, 47*(4), 409–443.
49. Fuhrmann, J., Rurainski, A., Lenhof, H.-P., & Neumann, D. (2010). A new Lamarckian genetic algorithm for flexible ligand-receptor docking. *Journal of Computational Chemistry, 31*, 1911–1918.
50. Sang, Y., Zhu, Y., Zhao, H., & Tang, M. (2016). Study on an interactive truck crane simulation platform based on virtual reality technology. *International Journal of Distance Education Technologies, 14*(2), 64–78.
51. Merlitz, H., Burghardt, B., & Wenzel, W. (2003). Application of the stochastic tunneling method to high throughput database screening. *Chemical Physics Letters, 370*(1–2), 68–73.
52. Ding, F., Yin, S., & Dokholyan, N. V. (2010). Rapid flexible docking using a stochastic Rotamer library of ligands. *Journal of Chemical Information and Modeling, 50*(9), 1623–1632.
53. Ferreira, L. G., Dos Santos, R. N., Oliva, G., & Andricopulo, A. D. (2015). Molecular docking and structure-based drug design strategies. *Molecules, 20*, 13384–13421.
54. Wellock, C., & Ross, B. J. (2001). An examination of Lamarckian g enetic algorithms. Technical Report #CS-01-01, Department of Computer Science, Brock University, St. Catharines, Ontario, Canada.
55. Schneider, G., & Böhm, H. J. (2002). Virtual screening and fast automated docking methods. *Drug Discovery Today, 7*(Supplement 1), 64–70.
56. Chang, Y., Hawkins, B. A., Du, J. J., Groundwater, P. W., Hibbs, D. E., & Lai, F. (2023). A guide to in silico drug design. *Pharmaceutics, 15*, 49.
57. Schrödinger, LLC. (2023). *Glide: Software de docking molecular*. https://www.schrodinger.com/
58. AutoDock. (2010). *AutoDock (Versão 1.1.2). [Software de modelagem molecular]*. The Scripps Research Institute. http://vina.scripps.edu/
59. Eberhardt, J., Santos-Martins, D., Tillack, A. F., & Forli, S. (2021). AutoDock Vina 1.2.0: New docking methods, expanded force field, and Python bindings. *Journal of Chemical Information and Modeling, 61*(8), 3891–3898.
60. GOLD. (1993). *GOLD (Versão 5.8). [Software de modelagem molecular]*. The Cambridge Crystallographic Data Centre. https://www.ccdc.cam.ac.uk/solutions/csd-discovery/components/gold/
61. FlexX. (1997). *FlexX (Versão 4.1). [Software de modelagem molecular]*. BioSolveIT GmbH. https://www.biosolveit.de/FlexX/
62. Surflex-Dock. (2010). *Surflex-Dock (Versão 2.0). [Software de docking molecular]*. Molecular Insights Pharmaceuticals, Inc. https://www.molecularinsights.com/surflex-dock/
63. Dock. (2005). *Dock (Versão 4.0). [Software de docking molecular]*. University of California, San Francisco. https://ccsb.scripps.edu/dock/
64. Ewing, T. J., Makino, S., Skillman, A. G., & Kuntz, I. D. (2001). DOCK 4.0: Search strategies for automated molecular docking of flexible molecule databases. *Journal of Computational Aided Molecular Design, 15*(6), 411–428.
65. Allen, W. J., Balius, T. E., Mukherjee, S., Brozell, S. R., Moustakas, D. T., Lang, P. T., Case, D. A., Kuntz, I. D., & Rizzo, R. C. (2015). DOCK 6: Impact of new features and current docking performance. *Journal of Computational Chemistry, 36*(15), 1132–1156.

66. LibDock. (2010). *LibDock (Versão 2.1). [Software de docking molecular]*. ACME Pharmaceuticals. https://www.acmepharmaceuticals.com/libdock
67. *MOE (Molecular Operating Environment) (Versão 2021.09)*. Chemical Computing Group. 2021. https://www.chemcomp.com/Products.htm
68. Korb, O., Stützle, T., & Exner, T. E. (2009). Empirical scoring functions for advanced protein–ligand docking with PLANTS. *Journal of Chemical Information and Modeling, 49*(1), 84–96.
69. Mirzaei, H. H., Jassbi, A. R., Pirhadi, S., & Firuzi, O. (2020). Study of the mechanism of action, molecular docking, and dynamics of anticancer terpenoids from salvia lachnocalyx. *Natural Product Research, 34*(1), 24–33.
70. McNutt, A. T., Francoeur, P., Aggarwal, R., et al. (2021). GNINA 1.0: Molecular docking with deep learning. *Journal of Cheminformatics, 13*, 43.
71. Jakhar, R., Dangi, M., Khichi, A., & Chhillar, A. K. (2020). Relevance of molecular docking studies in drug designing. *Nome da Revista, 15*(4), 270–278.
72. Halperin, I., Ma, B., Wolfson, H., & Nussinov, R. (2002). Principles of docking: An overview of search algorithms and a guide to scoring functions. *Proteins: Structure, Function, and Bioinformatics, 47*(4), 409–443.
73. (2021). *Journal of Chemical Information and Modeling, 61*(8), 3891–3898. Publication Date: July 19, 2021
74. Bitencourt-Ferreira, G., Pintro, V. O., & de Azevedo, W. F. (2019). Docking with AutoDock4. In W. de Azevedo Jr. (Ed.), *Docking Screens for Drug Discovery. Methods in Molecular Biology* (Vol. 2053). Humana.
75. Hosseini, M., Chen, W., Xiao, D., & Wang, C. (2021). Computational molecular docking and virtual screening revealed promising SARS-CoV-2 drugs. *Precision Clinical Medicine, 4*(1), 1–16.
76. de Paula, A., Angelo, R. M., & Honorio, K. M. (2021). Topomer-CoMFA proposed as a tool to construct dual EGFR/HER-2 models. *Journal of Molecular Modeling, 27*, 239.
77. Abumelha, H. M., Al-Fahemi, J. H., Althagafi, I., et al. (2020). Deliberate-characterization for Ni(II)-Schiff Base complexes: Promising in-vitro anticancer feature that matched MOE docking-approach. *Journal of Inorganic and Organometallic Polymers and Materials, 30*(12), 3277–3293.
78. Agarwal, S., & Mehrotra, R. (2016). Mini review: An overview of molecular docking. *JSM Chemistry, 2*, 1024.
79. Xiao, Y., & Woods, R. J. (2023). Protein–ligand CH−π interactions: Structural informatics, energy function development, and docking implementation. *Journal of Chemical Theory and Computation, 19*(16), 5503–5515.
80. Santos, L. H. (2021). Docagem molecular: em busca do encaixe perfeito e acessível. *BIOINFO—Revista Brasileira de Bioinformática*, Edição #01, Julho. .
81. Fischer, A., Smieško, M., Sellner, M., & Lill, M. A. (2021). Decision making in structure-based drug discovery: Visual inspection of docking results. *Journal of Medicinal Chemistry, 64*(5), 2489–2500.
82. Tripathi, A., & Bankaitis, V. A. (2017). Molecular docking: From lock and key to combination lock. *Journal of Molecular Medicine: Clinical Applications, 2*(1).
83. Chen, Y. C. (2015). Beware of docking! *Trends in Pharmacological Sciences, 36*(2), 78–95.
84. Santos, L. H. S., Ferreira, R. S., & Caffarena, E. R. (2019). Integrating molecular docking and molecular dynamics simulations. *Methods in Molecular Biology, 2053*, 13–34.
85. Bajorath, J. (Ed.). (2012). *Rational drug design: Methods and protocols (Vol. 928).*
86. Schlick, T. (2010). *Molecular modeling and simulation: An interdisciplinary guide (Vol. 21).*
87. Neves, M. A. C., Totrov, M., & Abagyan, R. (2012). Docking and scoring with ICM: The benchmarking results and strategies for improvement. *Journal of Computational Aided Molecular Design, 26*(8), 675–686.

88. Damm-Ganamet, K. L., Smith, R. D., Dunbar, J. B., Jr., Stuckey, J. A., & Carlson, H. A. (2013). CSAR benchmark exercise 2011–2012: Evaluation of results from docking and relative ranking of blinded congeneric series. *Journal of Chemical Information and Modeling, 53*(8), 1853–1870.
89. Yang, J., Shen, C., & Huang, N. (2020). Predicting or pretending: Artificial intelligence for protein-ligand interactions lack of sufficiently large and unbiased datasets. *Frontiers in Pharmacology, 11.*
90. Ravindranath, P. A., Forli, S., Goodsell, D. S., Olson, A. J., & Sanner, M. F. (2015). AutoDockFR: Advances in protein-ligand docking with explicitly specified binding site flexibility. *PLoS Computational Biology, 11*(12), e1004586.
91. Ashtawy, H. M., & Mahapatra, N. R. (2015). A comparative assessment of predictive accuracies of conventional and machine learning scoring functions for protein-ligand binding affinity prediction. *IEEE/ACM Transactions on Computational Biology and Bioinformatics, 12*(2), 335–347.
92. Xia, J., Tilahun, E. L., Reid, T.-E., Zhang, L., & Wang, X. S. (2015). Benchmarking methods and data sets for ligand enrichment assessment in virtual screening. *Methods, 71*, 146–157.
93. Orengo, C., Velankar, S., Wodak, S., Zoete, V., Bonvin, A. M. J. J., Elofsson, A., Feenstra, K. A., Gerloff, D. L., Hamelryck, T., Hancock, J. M., Helmer-Citterich, M., Hospital, A., Orozco, M., Perrakis, A., Rarey, M., Soares, C., Sussman, J. L., Thornton, J. M., Tuffery, P., Tusnady, G., Wierenga, R., Salminen, T., & Schneider, B. (2020). A community proposal to integrate structural bioinformatics activities in ELIXIR (3D-Bioinfo community). *F1000Research, 9*, ELIXIR-278.
94. Collins, A., Koehler, M., & Lynch, C. (n.d.). Methods that support the validation of agent-based models: An overview and discussion. *Journal of Artificial Societies and Social Simulation, 27*(1), 11. https://www.jasss.org/27/1/11.html
95. Sethi, A., Joshi, K., Sasikala, K., & Alvala, M. (2019). Molecular docking in modern drug discovery: Principles and recent applications. *Drug Discovery and Development- New Advances, 2*, 1–21.
96. Shen, C., Ding, J., Wang, Z., Cao, D., Ding, X., & Hou, T. (2020). From machine learning to deep learning: Advances in scoring functions for protein–ligand docking. *WIREs Computational Molecular Science, 10*, e1429.
97. Crampon, K., Giorkallos, A., Deldossi, M., Baud, S., & Steffenel, L. A. (2022). Machine-learning methods for ligand–protein molecular docking. *Drug Discovery Today, 27*(1), 151–164.
98. Kashyap, J., & Datta, D. (2022). Drug repurposing for SARS-CoV-2: A high-throughput molecular docking, molecular dynamics, machine learning, and DFT study. *Journal of Materials Science, 57*, 10780–10802.
99. Salimi, A., Lim, J. H., Jang, J. H., & Lee, J. Y. (2022). The use of machine learning modeling, virtual screening, molecular docking, and molecular dynamics simulations to identify potential VEGFR2 kinase inhibitors. *Scientific Reports, 12*(1), 18825.
100. Harmalkar, A., & Gray, J. J. (2021). Advances to tackle backbone flexibility in protein docking. *Current Opinion in Structural Biology, 67*, 178–186.
101. Janin, J., Henrick, K., Moult, J., Eyck, L. T., Sternberg, M. J., Vajda, S., Vakser, I., & Wodak, S. J. (2003). CAPRI: A critical assessment of PRedicted interactions. *Proteins: Structure, Function, and Genetics, 52*(1), 2–9.
102. Zhang, Y., Wang, Y., Zhou, W., Fan, Y., Zhao, J., Zhu, L., Lu, S., Lu, T., Chen, Y., & Liu, H. (2019). A combined drug discovery strategy based on machine learning and molecular docking. *Chemical Biology & Drug Design.*
103. Hermansyah, O., Bustamam, A., & Yanuar, A. (2021). Virtual screening of dipeptidyl peptidase-4 inhibitors using quantitative structure–activity relationship-based artificial intelligence and molecular docking of hit compounds. *Computational Biology and Chemistry, 95*, 107597.

104. Khamis, M. A., Gomaa, W., & Ahmed, W. F. (2015). Machine learning in computational docking. *Artificial Intelligence in Medicine, 63*(3), 135–152.
105. Sadybekov, A. A., Sadybekov, A. V., Liu, Y., Iliopoulos-Tsoutsouvas, C., Huang, X. P., Pickett, J., Houser, B., Patel, N., Tran, N. K., Tong, F., Zvonok, N., Jain, M. K., Savych, O., Radchenko, D. S., Nikas, S. P., Petasis, N. A., Moroz, Y. S., Roth, B. L., Makriyannis, A., & Katritch, V. (2022). Synthon-based ligand discovery in virtual libraries of over 11 billion compounds. *Nature, 601*(7893), 452–459.
106. Gu, S., Smith, M. S., Yang, Y., Irwin, J. J., & Shoichet, B. K. (2021). Ligand strain energy in large library docking. *Journal of Chemical Information and Modeling, 61*(9), 4331–4341.
107. Holcomb, M., Chang, Y. T., Goodsell, D. S., & Forli, S. (2023). Evaluation of AlphaFold2 structures as docking targets. *Protein Science, 32*(1), e4530.

Chapter 8
Drug Design in Motion: Concepts and Applications of Classical Molecular Dynamics Simulations

Ekaterina Shevchenko, Stefan Laufer, Antti Poso, and Thales Kronenberger

Abstract Molecular dynamics (MD) simulations have transformed the landscape of drug design, being one of the first computational techniques to provide insights into the behaviour of biomolecules at the atomic level. Nowadays, MD play a crucial role in predicting ligand binding affinities, exploring the protein–ligand complexes, facilitating drug binding studies, understanding protein–protein interactions, and discovering structural states and binding sites. The integration of MD simulations

---

E. Shevchenko (✉)
Department of Pharmaceutical and Medicinal Chemistry, Institute of Pharmaceutical Sciences, Eberhard-Karls-Universität Tübingen, Tübingen, Germany

Tuebingen Center for Academic Drug Discovery & Development (TüCAD2), Tübingen, Germany
e-mail: ekaterina.shevchenko@uni-tuebingen.de

S. Laufer
Department of Pharmaceutical and Medicinal Chemistry, Institute of Pharmaceutical Sciences, Eberhard-Karls-Universität Tübingen, Tübingen, Germany

Tuebingen Center for Academic Drug Discovery & Development (TüCAD2), Tübingen, Germany

Cluster of Excellence iFIT (EXC 2180) "Image-Guided and Functionally Instructed Tumor Therapies", University of Tübingen, Tübingen, Germany

A. Poso
Department of Pharmaceutical and Medicinal Chemistry, Institute of Pharmaceutical Sciences, Eberhard-Karls-Universität Tübingen, Tübingen, Germany

Tuebingen Center for Academic Drug Discovery & Development (TüCAD2), Tübingen, Germany

Cluster of Excellence iFIT (EXC 2180) "Image-Guided and Functionally Instructed Tumor Therapies", University of Tübingen, Tübingen, Germany

Excellence Cluster "Controlling Microbes to Fight Infections" (CMFI), Tübingen, Germany

Partner-site Tübingen, German Center for Infection Research (DZIF), Tübingen, Germany

School of Pharmacy, Faculty of Health Sciences, University of Eastern Finland, Kuopio, Finland

V. G. Maltarollo (ed.), *Computer-Aided and Machine Learning-Driven Drug Design*, Computer-Aided Drug Discovery and Design 3,
https://doi.org/10.1007/978-3-031-76718-0_8

with experimental techniques accelerates the identification and optimisation of potential drug candidates. In this chapter, we discuss applications of MD simulations that cover a wide range of possibilities within in silico drug design. Among others are the docking postprocessing of protein–ligand complexes, drug binding studies, protein–protein interactions, and the discovery of structural states and binding sites.

Keywords Long timescale · Principal component analyses · Conformational changes · Molecular dynamics simulations

Abbreviations

AUC	area under curve
CADD	computer-aided drug design
CD	circular dichroism
CPU	central processing unit
FRET	fluorescence resonance energy transfer
GPU	graphics processing unit
HAC	heavy atom count
Log Po/w	lipophilicity, n-octanol-water partition coefficient
Log S	water solubility
MD	molecular dynamics
MM/GBSA	molecular mechanics energies combined with the generalised Born and surface area continuum solvation
MMFF94	Merck Molecular Force Field
MSM	Markov State Modelling
NpT	isothermal-isobaric ensemble
NVE	microcanonical ensemble
NVT	canonical ensemble
OPLS	optimised potentials for liquid simulations
PCA	principal component analysis
PC	periodic conditions
PBC	periodic boundary conditions
PDB	Protein Data Bank
PSA	polar surface area

T. Kronenberger (✉)
Department of Pharmaceutical and Medicinal Chemistry, Institute of Pharmaceutical Sciences, Eberhard-Karls-Universität Tübingen, Tübingen, Germany

Excellence Cluster "Controlling Microbes to Fight Infections" (CMFI), Tübingen, Germany

Partner-site Tübingen, German Center for Infection Research (DZIF), Tübingen, Germany

School of Pharmacy, Faculty of Health Sciences, University of Eastern Finland, Kuopio, Finland
e-mail: thales.kronenberger@uni-tuebingen.de

QM/MM	quantum mechanics/molecular mechanics
RAS	rat sarcoma virus protein family
ROC	receiver operating characteristics
RMSD	root-mean-square deviation
RMSF	root-mean-square fluctuation
SASA	solvent accessible area
SBDD	structure-based drug discovery
TIP3P	three-site rigid water molecule
TR-FRET	time-resolved fluorescence energy transfer
UFF	universal force field
VdW	van der Waals interaction energy
VS	virtual screening

1 A Brief Introduction to In Silico Structure-Based Drug Discovery

Since its debut in the early 1980s, in silico or computer-aided drug design (CADD) has gained recognition as a methodology that influences almost every step of drug discovery. CADD has become integral to the industry and academic research since the constant development of its techniques brings a layer of structural rationality to the hit identification and development [1]. The relevance of molecular modelling and its fundamentals can be found in the understanding of protein–ligand interactions, which is also the foundation of medicinal chemistry. From the original lock-and-key concept [2], CAAD has been expanded to not only consider geometrical fit but also complementarity within the hydrophobic and polar protein–ligand interactions. Apart from interactions, molecular modelling often considers ligand and binding site flexibility, binding and distortion energies, solvation effects, entropy, and complementarity of the molecular force field [3].

Additionally, CADD works with large pools of chemical data, allowing great chemical diversity to be reasonably analysed and generating new hypotheses capitalising on large datasets. The adoption of technologies and techniques have the potential to drastically improve the drug development pipeline. Despite the speed at which these approaches are emerging, a comprehensive understanding of their applicability and limitations remains an ongoing conversation [4]. However, this brief description tells only a part of the story, providing a glance into molecular modelling, emphasising structure-based and molecular dynamics techniques.

Structure-based drug discovery (SBDD) is a drug design set of approaches which utilises the target's 3D structures and focuses on the design and optimisation of a ligand that accurately fits inside the binding pocket and results in beneficial protein–ligand interactions [5]. Structure-based drug discovery is a rapidly growing area due to steadily increasing structural information available, arising not only from

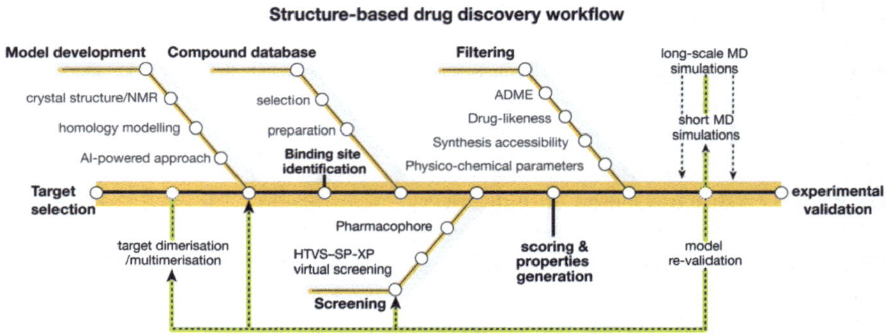

Fig. 8.1 An example of a structure-based drug discovery workflow. The thick yellow line illustrates the path from the target selection to experimental validation, while the brunches highlight the variety of methods that might be used along the journey. The dashed green arrows indicate that SBDD is an iterative process, emphasising the necessity of extra computations and revalidation along the way

genomics and proteomics data but also powered with AI systems and novel machine learning approaches [6]. The SBDD process is iterative and requires optimisation of multiple cycles to yield a final lead compound. Figure 8.1 describes the essential steps involved in the structure-based pipeline.

One of the key factors in the initial target selection step is to avoid the creation of misleading and/or biologically irrelevant models that would result in early phase failure. Of note, one should consider the target's multimerisation state, as well as interaction with other proteins and macromolecules, such as nucleic acids and/or membranes. The formation of functional assemblies frequently leads to geometrical/ conformational changes, which can cause the shift of size or interactions near the binding site. Understanding the functional changes that a target undergoes upon modulation/activation is crucial for creating an initial representative model that aligns with real-world conditions.

In the next step, the target protein 3D structure can be obtained, upon availability, from the Protein Data Bank (PDB). In case the crystal structure or NMR is not resolved to date, the 3D structure can be predicted by leveraging other high-quality structures with the highest sequence similarity to the target. The tools of choice at this step are homology modelling or, recently, AI-powered approaches, such as AlphaFold, chosen according to the template availability. For more details on how to obtain protein structures, see Chap. 14.

Following that, the ligand's binding site needs to be identified, which is often well established within a protein family. It provides the structural features or specific residues that can be a starting point for the target binding pocket generation. For instance, in protein kinases, various subcultural elements can be used as landmarks within the binding site, such as the gatekeeper, DFG-motif, or G-rich loop. If applicable, the binding site can be obtained with the localisation of natural substrate, for example, ATP binding site, in the event of kinases. Whether the binding site is unknown for the target protein and other members within the protein family of interest, computational methods can be applied to suggest the probable locations. One can choose between the methods relying on geometrical properties, such as POCKET [7], PASS [8], LIGSITE [9], or combined with the physics approach, such as PocketFinder [10] or SiteMap [11, 12].

There are several options to pursue after the binding site is identified. For instance, one can establish a pharmacophore model, which offers an additional opportunity to validate the model before proceeding with pharmacophore-based virtual screening. Another option is to carry out the docking (see Chap. 9) of a substantial chemical library. Once the method is chosen, and the screening database (see Chap. 2 for more detailed information about molecular databases) is selected and prepared according to the requirements of the method, the virtual screening (VS) can be utilised. Next, when the initial hits are obtained, the compound set should be scored and filtered according to the properties required for project purposes. The essential pharmacokinetic properties for the hits list validation include physicochemical parameters (molecular weight, number of heavy atoms, hydrogen bond donors and acceptors, and rotatable bonds), lipophilicity (Log Po/w), water solubility (log S), pharmacokinetics, drug-likeness, and chemical synthesis accessibility [13]. The cycle between model validation and the top hits choice is frequently repeated aiming for refinement. Once the initial hits satisfy the criteria, optional short-scale MD simulations (200–500 ns) can be performed to validate the ligand stability within the binding pocket, ensuing in selecting a list of top hits for experimental validation. The relevance of MD simulations in multiple stages of this pipeline is discussed in the next sections of this chapter.

2 The Early Beginnings of Molecular Dynamics Simulations

Molecular dynamics (MD) simulations have transformed the landscape of drug design and development, providing valuable insights into the behaviour of biomolecules at the atomic level. The roots of MD can be traced back to the mid-twentiethcentury when the exploration of methods to study the dynamics of molecules and their interactions *in silico* started to gain popularity. The early development of MD can be attributed to the pioneering work of Dr. Martin Karplus and Dr. Michael Levitt, who in the early 1970s, independently laid the foundation for the theoretical framework of MD simulations [14, 15]. Their seminal papers explored the concept of using classical mechanics to describe the motion of atoms and molecules over time. These fundamental contributions ignited interest in the scientific community and paved the way for further advancements.

In 1975, Levitt and Warshel introduced a simplified representation of protein structures that reduced the degrees of freedom and the number of interaction centres by averaging over groups of atoms [16]. They represented each residue using only two centres, the Cα atom and the centroid of the side chain and assumed interactions to occur only between side chains. The model reduced the dimensionality of the conformational space and led to smoother side chain representations. They used effective time-averaged potential functions to account for the fine details and rapidly changing variables. The simplified model was tested on a bovine pancreatic trypsin inhibitor (PTI) and successfully simulated the stable conformation of a folded protein. The authors also proposed a hierarchical approach, starting with simplified models and gradually incorporating more atoms, to understand and simulate complex biological assembly processes [17, 18]. Their work laid the foundation for MD simulations in protein folding and biomolecular studies.

The 1980s witnessed significant progress in MD simulations, as computational power and algorithms improved. Scientists began to apply MD to study the dynamics of small molecules and biomolecules such as proteins and nucleic acids [19]. Driven by advancements in both hardware and software, MD simulations became increasingly popular for exploring the structural and dynamic properties of biological systems.

In the 1990s, MD simulations evolved into a versatile tool for understanding complex biological phenomena and interactions [20, 21]. Researchers started using MD to investigate protein-ligand binding, protein folding, and protein–protein interactions, among other biologically relevant processes [22–25]. These studies provided valuable insights that complemented experimental findings and aided in rational drug design.

With the advent of more sophisticated force fields, parallel computing, and efficient algorithms, MD simulations entered a new era in the twenty-first century [26–30]. Simulations of larger biomolecular systems, such as membrane proteins and nucleic acid complexes, became feasible [31]. The combination of MD with other computational methods, like quantum mechanics/molecular mechanics (QM/MM) (see Chap. 10), further expanded the capabilities of drug design approaches [32, 33].

Nowadays, MD simulations have become an indispensable tool in modern drug discovery and development. They play a crucial role in predicting ligand binding affinities, exploring the protein–ligand complexes [34, 35], facilitating drug binding studies [36, 37], understanding protein–protein interactions [38], and discovering structural states and binding sites [39]. The integration of MD simulations with experimental techniques accelerates the identification and optimisation of potential drug candidates.

MD simulations can empower the studying of the structure within molecular systems at the atomic scale in a dynamic manner. The MD simulation generates a trajectory that allows observing an atom's individual movement as a function of time. The forces between interacting atoms are estimated using a force field, and the system's total energy is computed according to Newton's law of classical mechanics. The integration of Newton's laws of movements produces subsequent configurations of the developing system during MD simulations, producing trajectories that describe the locations and velocities of the particles throughout time. The applications of MD simulations cover a wide range of possibilities within *in silico* drug design. Among others are the docking postprocessing of protein–ligand complexes [34, 35], drug binding studies [36, 37], protein–protein interactions [38], and the discovery of structural states and binding sites [39].

3 System Preparation for MD Simulations

A basic pipeline for setting up a system for MD simulation requires the protein of interest to be prepared by the simulation software and chosen force field. System preparation follows the structure preparation with additional examination of crystallisation artefacts, correct bond orders, the assignment of correct protonation/ionisation states to protein residues and the presence of ions and associated molecules. On occasion, the modelling of missing residues in pdb structures also takes place here. In the next sessions we discuss, timescales and replicas, followed by ensemble models, parametrisation, and force fields, followed by solvation and periodic conditions.

3.1 Solvation and Microensemble

Next, the solvation model is to be chosen. Many types of water models have been developed regarding the system solvation with water. These models are categorised based on several important factors, including the number of interaction points the molecule can make (site), the inclusion of polarisation effects, and flexibility or rigidity. One of the most common water models is TIP3P, a three-site rigid water molecule with assigned Lennard-Jones parameters and potential [40].

A thermodynamics ensemble used to characterise the system must be chosen during the system preparation step. The thermodynamics ensemble is an idealisation of the model system composed of multiple replicas of this system, all taken into account simultaneously, and each reflects a potential state in which the actual system may be [41]. In other words, the thermodynamics ensemble is a statistical approximation that allows to extrapolate the fundamental features of the thermodynamic system through classical and quantum mechanics [42]. The ensemble classification is based on how the model system is separated from the outer environment. The first category is the microcanonical ensemble, which depicts the model system as entirely isolated from the environment (Fig. 8.2a). The microcanonical ensemble refers to the acronym NVE, which means that the total number of particles **N**, volume **V**, and energy **E** remains constant [43]. As a factor that assumes the thermodynamic interaction of the model system with its surroundings, the temperature cannot be defined for the NVE ensemble. NVE is used for describing states of a system with defined total energy. The next ensemble category is termed canonical and provides the option to determine the system temperature, therefore referring to the acronym

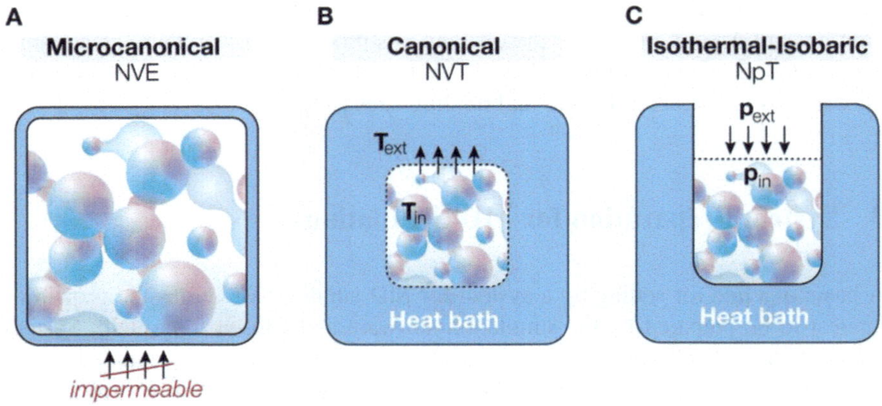

Fig. 8.2 Thermodynamic ensembles in molecular dynamics. (**a**) A microcanonical (NVE) ensemble is an isolated system with a fixed number of particles, total volume, and energy. (**b**) Canonical (NVT) ensemble as a system with fixed total volume but with temperature-dependent boundary energy transfer. (**c**) Isothermal-Isobaric (NpT) ensemble as a system allowing volume alteration for the pressure equilibration. Ext external, environmental, in internal, model system value

NVT, where N stands for the total number of particles, V for volume, and T for temperature. NVT allows the energy transfer between the boundaries of the model system and the environment while prohibiting substance exchange (Fig. 8.2b) [43].

In other words, this ensemble addresses the model system's potential states that are in thermal equilibrium with the environment. This ensemble model can be visualised as a heat bath with constant temperature, which is several magnitudes larger than the model system [44]. The difference in size ensures that no amount of heat produced by the model system will cause the heat bath's temperature to rise considerably. As a result of the thermal contact between the model system and the environment, the system will now transmit heat to and across the environment until they reach thermal equilibrium [45]. The last ensemble category is the isothermal-isobaric, which as with the canonical ensemble, allows the energy transfer between the model system and environment, but allows the volume change. The system's volume fluctuates to equilibrate the system's internal pressure with the pressure applied to the system by its environment (Fig. 8.2c) [45]. The acronym for the isothermal-isobaric ensemble is NpT, where **N** stands for the persistent total number of particles, **p** for pressure, and **T** for temperature.

Moreover, NpT is the most used ensemble. However, prior to the start of MD, the so-called model relaxation step is conducted. The model relaxation is an essential step which ensures the proper system equilibration and removal of the high strain degree from the newly constructed system. The model relaxation involves several phases, where a particular ensemble is combined with specific pressure, temperature, and timestep until the equilibration is achieved. The placement and selection of ions and the specification of temperature and pressure are all components of the system preparation process. It is frequently desirable to have an electrically neutral system for MD simulation. However, it is not strictly necessary, for example, with the Desmond engine [46], which applies a uniform background charge distribution to neutralise the system in the Ewald summation [47]. Moreover, systems can be set up in a salt solution rather than a pure solvent or even include other small molecules as buffers in a mixed-solvent model [48].

Finally, periodic boundary conditions (PBC) are then applied to the system by selecting the simulation box's form and cut-off. It is necessary to note that to account for potential protein movements, the distance from the simulation box's edge to the protein must be considerable. If the simulation box size is insufficient, MD-derived findings can frequently be artefactual or misleading.

Once the box shape and dimensions, the total number of atoms, number of water molecules, implicit or explicit membranes or solvents, salt concentration, lipid composition, cofactors, and metals have been determined, the force field should be chosen to continue with the MD simulation.

3.2 Force Fields: General Concept and Relevant Choices

While in classical physics, force fields evaluate a system's potential energy, the fundamental distinction between the two in molecular modelling is that the energy landscape is depicted as an energy gradient dispersed across the particle positions [76]. Reconstructing a realistic simulation of a molecular system on an atomistic level is strongly dependent on force field parameters, which are the core for deriving meaningful structural information and relative energies from MD simulation.

To date, additive or non-polarisable force fields are the most prevalent in small molecule drug design, frequently sharing the basic potential energy function and the parameters comprising the function itself [49]. The name "additive" originates from Coulomb's equation for electrostatic interactions, which states that a system's potential energy is the sum of all atom–atom individual interactions (Eq. 8.1) [49]. The potential functions, which are essentially an array of equations representing the potential energy, and its components, form the basis of the force field core. The parameters in this set of equations are another component distinguishing a force field.

In the vast majority of condensed-phase simulations, the total potential energy of a cohort of molecules is determined as a sum of inter- and intra-molecular interaction energies between all components of the system (Eq. 8.1) [50, 51].

$$E\left(\vec{X}\right) = \sum\nolimits_{a<b} Eab + \sum\nolimits_{a} E_a^{\text{int}} \tag{8.1}$$

The total potential energy of a set of constituents *ab* with coordinates $\vec{X}$. Composed by the bonded terms (first term) and nonbonded interactions (second term), *ab* refers to a set of molecules and/or ions. This notation applies to two exemplary particles *a* and *b*, which in a real calculation condition would be expanded to multiple pairs of particles.

The following terms commonly describe the intramolecular potential energy: harmonic bond stretching (Eq. 8.2), angle bending (Eq. 8.3), torsional angle definition as Fourier series (Eq. 8.4), Coulomb electrostatics and Lennard-Jones potential (Eq. 8.5) [51].

$$E_{\mathrm{bond}} = \sum\nolimits_i k_{b,i} \left(r_i - r_{0,i}\right)^2 \tag{8.2}$$

The first force field term defines bond stretching energies in harmonic (ideal) conditions, where k_b is the bond-stretching constant, which regulates the rigidity of the bond spring.

$$E_{\mathrm{bend}} = \sum\nolimits_i k_{\vartheta,i} \left(\vartheta_i - \vartheta_{0,i}\right)^2 \tag{8.3}$$

The second force field term defines the angle bending energy, where k_ϑ is the angle bending constant, determining the rigidity of the spring's angle.

$$E_{\mathrm{torsion}} = \sum\nolimits_i \left\{ \frac{V_{1,i}(1 + \cos\varphi_i)}{2} + \frac{V_{2,i}(1 - \cos 2\varphi_i)}{2} + \frac{V_{3,i}(1 + \cos 3\varphi_i)}{2} \right\} \tag{8.4}$$

The third force field term represents the torsional energies as a Fourier series.

$$E_{nb} = \sum\nolimits_{i<j} \left\{ \frac{q_i q_j e^2}{r_{ij}} + 4\varepsilon_{ij} \left[\frac{\sigma_{ij}{}^{12}}{r_{ij}} - \frac{\sigma_{ij}{}^{6}}{r_{ij}} \right] \right\} \tag{8.5}$$

The fourth force field term represents the nonbonded energy between all atom pairs, where σ_{ij} is the equilibrium distance and r_{ij} is the distance between interacting atoms.

Force fields can be roughly divided into three major classes. One of the defining characteristics of class one is the utilisation of harmonic movements to depict bond stretching and angle bending. According to the assumption, the amount of restoring force is proportional to the displacement from the equilibrium position for class one force fields [52]. The approximation in class one force fields is referred to as quadratic because the square of the displacement energy is linearly associated with the energy of the harmonic oscillator [53]. Moreover, the parametrisation of bond stretching and the angle bending often approach harmonic behaviour only close to the equilibrium. The most famous examples of the class one force fields are the Optimised Potentials for Liquid Simulations (OPLS) [54], AMBER [55, 56], CHARMM [57], and GROMOS [58]. One of the first techniques to be established with constant parameter optimisation for the propagation of thermodynamic characteristics in the liquid state applied to small molecules was OPLS, which stands for optimised potentials for liquid simulations [54]. On the OPLS core, the next generation of force fields was built: OPLS3 [28], OPLS3e[29], and OPLS4 [30] with the maintenance of the nonbonded parameters. In general, AMBER refers to a collection of force fields that can be split according to the simulated biomolecular system. For instance, for protein simulations, AMBER suggests ff19SB [59], and its evolution ff19SB-ILDN (with improved side-chain torsion potentials [60]); for lipids or complex membrane simulations—LIPID21 [61]; q and for nucleic acids—DNA OL15 [62], and RNA OL3 [63]. As AMBER, CHARMM also represents a set of force fields, for instance, the all-atom CHARMM22 [64] and extended atom force field CHARMM19 [65], which are annually updated (see https://www.academiccharmm.org/program/versions for a current list, accessed on February 2024). Last but not least, the very popular GROMOS force-fields (not to be confused with the GROMOS software), which are united atom force fields, i.e., without explicit aliphatic (non-polar) hydrogens and are considered an "all purposed" set, with accurate parameters for phosphorylation and others post-translational modifications (see [66] for the most recent version and https://www.gromos.net/#Reif2012, for current list and modifications, accessed on February 2024).

Class two provides additional anharmonic cubic and quartic components for the potential energy of bonds, resulting in more detailed geometrical modelling of the vibrations of bonds. In addition, these force fields include cross-terms that describe the interactions between nearby located angles and dihedrals. Class two force fields include MMFF94 (Merck Molecular Force Field [67]), which parameters are primarily derived from quantum calculations rather than experimental data [68].

Another example is UFF [69] (Universal Force Field), whose original application was somewhat restricted due to the parameters, particularly for metals and inorganic substances [41]. The UFF initiative seed the idea recently followed up, in 2018, by the Open Force Field consortium (OFF, https://docs.openforcefield.org/en/latest/, accessed on February 2024). On their own words, they aim to develop automated and systematic data-driven techniques to parameterise and assess new generations of more accurate force fields. Despite their efforts, the most recent benchmarking shows that public force fields (i.e., OpenFF Parsley and Sage, GAFF and CGenFF) had comparable accuracy, while OPLS3e was significantly more accurate [70].

Class three includes force fields that contain extended parameters applicable to organic chemistry, such as the Jahn-Teller effect or stereoelectronic effects. For example, AMOEBA [71] is a polarisable force field that employs atomic-induced dipole to model polarisation while assuming that averaging polarisation is insufficient [72]. Another example is DRUDE [73], which uses non-polarisable force fields to leverage atom-to-atom Coulomb electrostatic interactions as its core while integrating polarisation effects via NAMD and a dual-Langevin thermostat approach [74].

Important to highlight another class of force fields, known as coarse-grained force fields, which employ a distinct strategy in molecular dynamics simulations. The idea behind the coarse-grained approach is the reduction of the number of degrees of freedom within a system. This is achieved by parametrising the most significant interactions with the force field while representing a particular set of atoms as a single bead. The definition of the most significant interactions might be intricate depending on the parametrisation method, hence tabulated potentials are frequently employed. The purpose of coarse-grained models is to replicate specific characteristics of a given system, which can encompass an atomistic protein model or experimental data. The properties one intends to replicate in the model determine the classification of the coarse-grained force fields. For instance, free energy conservation is the focus of the MARTINI [40] force field and the simplex method [73]. Another example is inverse Monte Carlo with structure-based coarse-grained modelling, emphasising the radial distribution [75].

In the case no parameters are available for a molecule in that particular force field, one can resource to reparameterisation or development of new parameters, which will be force field compatible. In order to ensure reproducibility, though, the new data would need to be validated against experimental measurements or high-level QM data (for more information on quantum mechanics, see Chap. 10). It is also important to provide a description of bonded and nonbonded potential parameters, by atom and provide the topology(-ies) used for that determination. Software tools such as the Antechamber (from the Amber package [76]) or web server interfaces like LigParGen (for OPLS force-fields [77]) facilitate the topology analyses, writing the input for the corresponding simulation engine [78].

3.3 The Concept of Replicas and Timescale

One of the essential aspects that can be carefully considered when the MD simulation is conducted is the timescale, as different types of protein motions occur at distinct timescales. For instance, the side-chain rotamer movements can be observed in the range of ps to μs, followed by the loop motions in the field of ns to μs, with more significant domain movements starting to be evident in μs + timescale [79]. Thus, one should consider the reasonable timescale of the MD simulation in accordance with the movements to be observed to deliver valuable results (Fig. 8.3).

To obtain meaningful and comprehensive results from MD simulations, one should carefully consider one of the essential aspects of this methodology. The selection of an appropriate timescale directly influences the range of molecular motions and events that can be captured within the simulated system. Shorter timescales are suitable for studying fast processes such as bond vibrations and side-chain rotamer movements (fs to ns). In contrast, longer timescales are required to investigate slower and more complex events, such as loop motions (ns to μs), significant domain movements and large-scale conformational transitions (μs+). However, excessively long simulations conducted without a clear research purpose, may not only become computationally expensive but also lead to diminishing returns in terms of new insights. To strike a balance between accuracy, computational feasibility and deriving meaningful results, one should align the simulation timescale requirements with the specific research question to be answered through MD.

The current model for understanding protein folding and dynamics asks us to imagine the protein dynamics process as a large funnel, in which the surface would represent the diversity of protein conformation, and its deepness would be proportional to the energy systems. The deeper pockets of this funnel constitute global

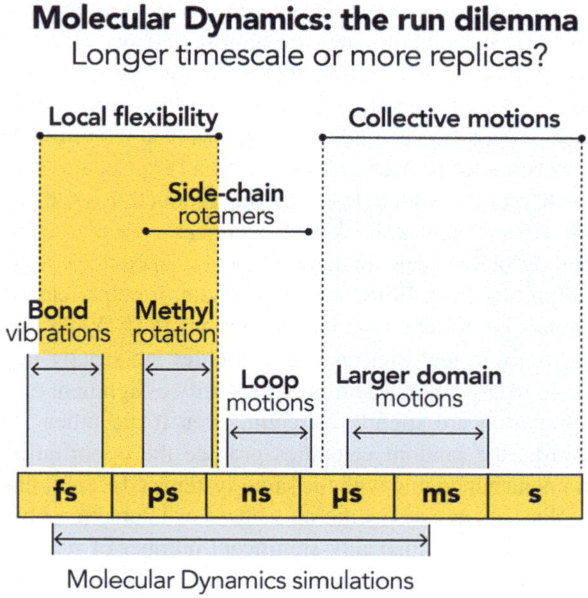

Fig. 8.3 A spectrum of dynamic processes in proteins across timescale occurring in molecular dynamics simulation. (Adapted from Henzler-Wildman et al. [79], dynamic personalities of protein [79])

low-energy protein conformations. This funnel is rather wrinkly, with smaller wells, which represent the energy minima of different protein conformations. What the MD simulation does in this model is to navigate these inner surfaces and unbiasedly sample around the protein conformational space.

Considering the funnel model and the problem of timescale for protein dynamics, we understand that classical unbiased MD simulations might not be always suitable to sample molecular events involving a transition between those energy barriers, such as large conformational changes. Alternatively, given enough justification and experimental evidence, one can apply enhanced sampling simulations, such as nonadaptive biasing potential methods (e.g., Gaussian-accelerated MD), adaptive bias simulations (e.g., metadynamics), replica exchange methods, etc. Examples of experimental evidence can rise from (TR-)FRET assays or site-direct mutagenesis in combination with photocrosslinking, showing that two portions of the protein are in close proximity, multiple diverse crystal structures, showing different conformations and even highlighting large "missing" portions due to high flexibility.

If enhanced sampling simulations are used, the choices for sampling methodology, parameters, and, most importantly, the convergence criteria (where to stop sampling and when to consider that the generated conformations are artificial) should be described. We refer to the recently published guidelines for reporting MD simulation data [80].

The other crucial point for comprehensive MD simulation is the number of replicas of the single system to be conducted to avoid false positive conclusions [81]. The term "replica" refers to the simulations of the identical system, sharing the same number of atoms, initial structure, and preparation protocol, repeated several times. Given the principle of energy conservation within the ensemble required for the MD calculations and the almost derisory effects arising from changes in the floating-point precision or hardware, multiple runs of these systems would lead to very similar thermodynamical and, consequently, conformational properties. In this sense, the difference between replicas is in the initial velocities generated randomly according to the Maxwell distribution [81]. As a result, the velocities are unique for each replica, which leads to the production of different simulation trajectories. However, even replicas with identical velocities can produce distinct trajectories for various reasons, including machine-specific settings and the specification of the compiling [81]. In theory, the ergodic principle claims that the velocities have no impact on infinite dynamic simulations [82]. While molecular dynamics are not run eternally in real life, random velocities are strictly important, especially for short-scale MDs. The random initial velocity assignment ensures that the results of every simulation are slightly different, even if the other settings are identical. In other words, the random velocities provide the opportunity to observe real-world phenomena happening with the same system at different time points. Moreover, multiple replicas ensure a specific movement or interaction observed not in a single simulation but in a statistically significant number of replicas is not biased or artificially generated but related to the real-world evidence. The choice between the multiple simulation replicas over a single but long-scale one is frequently a question that should be answered before the initiation of the MD according to the aims and resources of the project.

The work from Knapp et al. [81] performed 100 identically parametrised replicas of 3 μs each for a small 10 amino acid long peptide, and also 100x replicas of 100 ns each for the T-cell receptor/MHC system (827 amino acids). Using those simulations as show-case, they compared randomly chosen subgroups within these replica sets, being able to estimate the reproducibility and reliability that could be achieved by a given number of replicas at a given simulation time. They highlight two major points: i) conclusions drawn from single simulations are not reproducible and ii) observations drawn from multiple shorter replicas were more reliable than using a single longer trajectory. Based on their data, a minimum of 5–10 replicas were required [81]. Their comment that the actual number of replicas needed will depend on the level of reliability sought and, herein we add that is highly dependent on the size and intrinsic flexibility of the system.

Our next example applied multiple short simulations on the conformational space of aptamers [83]. They combine the enhanced sample approach of starting from high-temperature equilibration runs (six in total), leading to higher flexibility, with independent replicas from the output of each of these systems (10 × 100 ns per output). They analysed the recurrence rate of found conformations and properties among the different starting points, showing a clear dependence on initial conformation. This highlights the need of using different initial configurations as simulation starting points to avoid being trapped in energy minima. Especially, since analyses of convergence of those unbiased trajectories might not allow the detection of slow transitions between kinetically trapped metastable states.

4 Molecular Dynamics Analysis

The vast spectrum of MD analysis methods can be divided into three perspectives, which encompass the joint analysis of protein and ligand, then the ligand and the protein analysed separately (Fig. 8.4). Further, each subcategory is described with examples of frequently applied techniques.

4.1 *Protein Perspective*

4.1.1 Protein Root Mean Square Deviation (RMSD)

RMSD is a measurement of the deviation of group coordinates from a reference position. The RMSD variation between given groups of atoms, such as Cα atoms or protein backbone, is commonly used in analyses of MD [84]. RMSD provides insights into the deviations in the protein conformation along the MD trajectory, which correlates with the indication of protein stability. Namely, the values are calculated for atomic coordinates at each frame of the MD trajectory, which allows seeing the value variation along the timescale (Fig. 8.5a, blue). One crucial factor in RMSD is the choice of the reference point to which the subsequent RMSD values are compared. This is because the quality and reliability of the reference point determine how accurate the RMSD calculations will be. For instance, the simulation's starting frame is commonly designated as the reference point in MD. Further, if the starting frame contains poor contacts, such as clashes or ambiguities in the ligand pose, one will obtain high RMSD values. High values in this regard point back to the poor

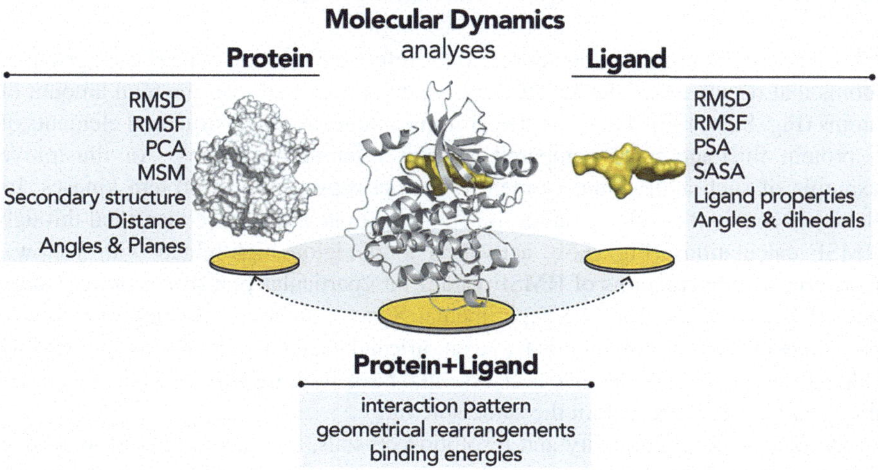

Fig. 8.4 Three approaches of Molecular Dynamics trajectory analysis and their sample methodologies. These different techniques are discussed in detail in the next sessions

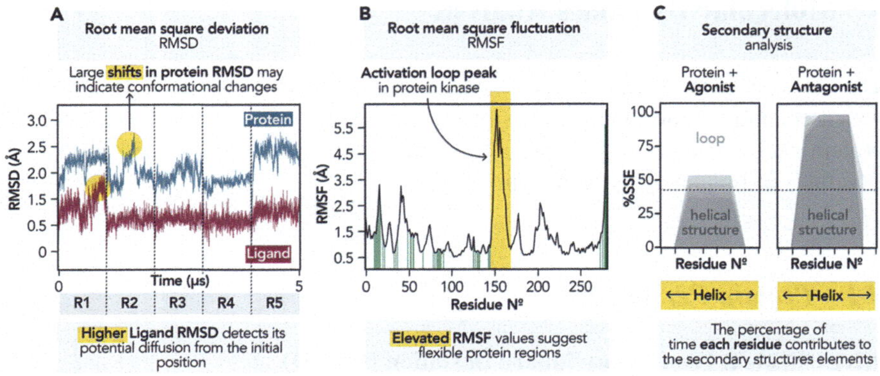

Fig. 8.5 Illustrative showcase examples of RMSD (**a**), RMSF (**b**), and secondary structure analysis (**c**) on a molecular dynamic *trajectory*. R1–5 indicates simulation replicas 1–5 accordingly. In (**c**) the replicas are shown with the shades of overlayed grey. In (**b**) green vertical lines indicate protein residues that are in contact with the ligand

choice of the reference point rather than necessarily indicating ligand instability or poor model quality. Moreover, RMSD indicates the approximate timeline of system equilibration, which can be observed from higher-than-average RMSD values at the beginning of MD. RMSD can be calculated in the majority of drug discovery software, such as GROMACS [46], Maestro Simulation interaction analysis tool, or for complex selections can be scripted with MDAnalysis.analysis.rms [85] module for Python.

4.1.2 Protein Root Mean Square Fluctuation (RMSF)

RMSF is a measurement of deviation from a reference position of a selected group of atoms that occurs along the MD trajectory and is averaged over the total amount of atoms (Fig. 8.5b) [86]. Thus, the RMSF values indicate which structural elements of a protein fluctuate most compared to their reference positions. An illustrative example of such a structural element is the activation loop in protein kinases. Its flexible behaviour is often evident in MD trajectories and can be identified through RMSF calculations (Fig. 8.5b, activation loop region highlighted with yellow). Conceptually, fluctuations of RMSF values in a particular part of the protein align with B-factor values from the experimental X-ray structures, allowing direct comparison between the computational and empirical data. Generally, RMSF and RMSD calculations are bound together as an essential basis for trajectory analysis; therefore, they can be calculated within the same software.

Despite using high-quality and resolution crystals, one has to bear in mind that diffraction data is not homogeneous across the system and, specifically in the ligand binding pocket, it is important to verify whether certain atoms/bonds are correctly assigned and supported by proper electron densities. When working with low-available data's projects, where structural information is sparse, this may serve as a warning that lowly supported and defined regions will lead to artificially highly flexible in the simulations.

4.1.3 Radius of Gyration (R_g)

Since its original postulation on crystal structures, the radius of gyration determines the protein structure compactness. The lowest radius of gyration and, accordingly, the tightest packing would correlate with a higher number of intra-molecular protein contacts and could be used as a folding measurement [87]. The R_g can be used as a numerical statistical indicator of large conformational change along the protein trajectory (specially in fibre-like and semi-flexible polymer proteins), but also to monitor the protein stability as a quality control, for globular proteins.

4.1.4 Protein Secondary Structure Analysis

The initial classification of the secondary structure was based on distinct hydrogen bonding modes. It encompassed three elements: helix, strand, and coil [88], further, this classification was extended to eight states: α-helix, 3_{10} helices, π-helix, β-strand, β-bridge, β-turn, bend, and loop or others [89]. As the physical process of protein folding is tightened up with the various biological events and processes [90], deriving secondary structure information from MD trajectories can provide insights into the structural behaviour of the protein (Fig. 8.5c). Although to date, the possibility of computational prediction is limited to helix, strand, or loop (coil), the results can highlight valuable changes in the investigated structure correlated with the impact of the ligand, mutation or normal protein function. For instance, a notable shift in the secondary structure composition of a specific helix could be observed when the protein interacts with an agonist (Fig. 8.5c, left chart; the helix maintains a helical structure for around 50% of the simulation time across 5 replicas) compared to antagonist (Fig. 8.5c, right chart; the helix keeps a helical structure for about 95% of the simulation time across 5 replicas). This observation can provide a hint for further in-depth system analysis. Computational results of the secondary structure prediction can be further supported with circular dichroism (CD), which provides an experimental evaluation of proteins' secondary structure. Up to this point, secondary structure calculations along MD trajectories are somewhat rare and frequently require custom coding for a system.

4.1.5 Principal component Analysis (PCA)

PCA is a prominent approach for evaluating large datasets with several dimensions or variables per observation. It enhances the data's readability, while maintaining the most information, besides facilitating visualisation. In other words, PCA retains the greatest amount of statistically significant data within a given dataset that results in dimensionality while reducing the "noisy" data that causes valuable statistical trends to be overseen. This is conducted by the linear translation of the dataset to a new coordinate system, which produces a two- or three-dimensional plot with data points that allow deriving the clusters of nearby data points visually. The number of principle components chosen for the visualisation is the main difference between 2D and 3D PCA. The construction of principle components in PCA aims to capture the most variety in the dataset: PC1 depicts the greatest variance, PC2 describes the next-greatest variety, and so forth. Most of the variation is usually captured by the first two to three PCAs, and the remaining ones can be ignored without losing valuable data. The linear translation (linear reduction) is a process of finding new uncorrelated variables that represent linear functions of the initial set and sequentially optimising the variance [91]. In addition, PCAs are widely used in a variety of disciplines where large datasets must be statistically analysed. Among these disciplines are genomics, astrophysics, machine learning, and data science.

Moreover, PCA is widely used to analyse MD trajectories, where it applies the same principle: a linear transformation that diagonalises the covariance matrix and eliminates the spontaneous linear correlations between the coordinates [92, 93]. PCA results can be utilised for the free energy landscape construction, which describes the metastable conformational states and the transition states [94]. These states contribute to understanding functionally relevant motions within the investigated system, which can be hard to derive by visual examination due to numerous other motions simultaneously occurring within the MD trajectory. In other words, PCA's application to MD trajectory is instrumental in identifying dominant protein motions, such as hinge-like movements or domain rotations. By projecting MD frames onto eigenvectors corresponding to principal components, one can visualise how protein conformations evolve over time and uncover correlations among atoms. The PCA analysis can be conducted, for instance, with GROMACS [46] covariance tool (gmx covar and gmx anaeig scripts).

While PCA provides quantitative analysis, direct visual observation remains valuable for comprehending protein dynamics. The input selection for PCA impacts the scope and specificity of the results and can range from all atomic coordinates to Cα atoms or particular residues. While using all atomic coordinates provides a comprehensive view of global motions and conformational changes, this will incur higher computational costs, due to the curse of dimensionality. While it captures global motions, interpreting such high-dimensional data might require advanced visualisation and analysis techniques. Therefore, to gain insights into the overall protein dynamics, one may consider selecting alpha carbon atoms as an input for PCA analysis. This strategy simplifies the analysis while capturing essential backbone motions, for instance, effectively highlighting hinge-like movements and other significant conformational changes, while disregarding a fine representation of side-chain dynamics.

Meanwhile, focusing on specific residues for PCA narrows down the analysis to regions of interest. This approach can well capture local dynamics, bearing in mind that global motions can still be influenced, to some extent, by these regions, even if they are not directly analysed. Thus, this choice offers insights into the behaviour of selected regions, the choice of which is usually connected with a portion of specific protein machinery. Examples of such regions can be α-C helix in protein kinases, switch-I and II in RAS proteins, or transmembrane helixes 5–6 in class A GPCRs.

PCA can be used to investigate variations in the dynamic behaviour across different systems. For instance, our work illustrates the use of PCA on different simulated kinase states (multimerisation and phosphorylation) to unravel the impact of different phosphorylation states on kinase behaviour [95]. As a result, in PCA score plots each system showcased a distinct trajectory, highlighting the impact of phosphorylation on the kinase conformational landscape. The phosphorylation states exhibited a direct correlation with the profiles observed in the principal components. Further, complementing the quantitative PCA results with a visual inspection of the simulation trajectories revealed system-specific shifts in the activation segment, distinctly deviating from the crystal conformation. Therefore, generating the basis for future protein analysis.

4.1.6 Markov State Modelling

Markov State Modelling (MSM) plays a significant role in the modelling and interpretation of MD simulation data, while it allows pinpointing of statistically significant events in the MD trajectories [96]. MSM is a robust framework for analysing long-scale MDs, where it becomes more challenging to visually observe changes in protein dynamics and substructural geometrical variations.

MSM stands for a master equation framework, indicating that the system's complete dynamics can be highlighted using just the MSM. The master equation formalism has also been applied in a wide range of scientific disciplines [97]. By definition, MSM is a $n \times n$ square matrix (often referred to as transition probability matrix), in which the n states represent the total number of configurations that the system may exist in [98]. The dynamical change of the system can be observed by defining the states at time points separated by the lag of time (τ). This separation of the states ensures the lag of time to be a Markov process [98]. The term "Markov process" refers to a random process in which, given the present, the future is independent of the past [96]. In terms of MD, the Markov process means that the probability of the system transition from one state to another does not depend on where the system was before this initial state.

The obstacles in building an MSM may be divided into two main categories: the definition of the states in a kinetically relevant order and the effective employment of the state decomposition for building an efficient transition matrix [99]. Additionally, a bias-variance problem related to the choice of the number of states is a dilemma with MSMs. When utilising a limited number of states, one knows analytically that the predicted value of the relaxation timescales is less than the real value and that this bias gets less as the number of states increases. On the other hand, with a given dataset, as the number of states rises, the statistical error in the MSM grows [99]. Although algorithms have been developed to deal with this problem, to date, there are no algorithms in the literature that automatically and effectively balance these conflicting sources of errors [100].

As in application of PCA on MD trajectory, MSM requires careful consideration of the input data. There are three common input choices for MSM calculation: atomic coordinates, alpha carbon atoms, and functional groups. Using atomic coordinates as input data offers a potentially detailed review of protein dynamics. However, this level of granularity and complexion can lead to computational challenges due to the sheer volume of data. Alternatively, selecting alpha carbon atoms simplifies the input data while retaining backbone motions. This approach is particularly useful for capturing large-scale conformational changes while reducing computational complexity. Another strategy involves focusing on specific functional groups or key residues. This allows one to focus the calculation on the role of these groups in driving conformational transitions. Regardless of the input choice, constructing an MSM involves estimating transition probabilities between different conformational states. These probabilities help to understand the kinetics of state-to-state transitions and have a glance into the protein's conformational landscape.

MSMs have been applied effectively to various molecular dynamics studies and beyond. For instance, the study by Meng et al. [101] generated Markov state models for the inhibited Abl Tyrosine Kinase [101]. Through extensive simulations totalling around 400 μs, achieved through multiple runs of 400–750 ns each, they revealed novel conformations previously not disclosed by X-ray crystallography.

In a similar vein, the study by Maltarollo et al. [102] demonstrated the application of MSM to explore dynamic interactions and collaborative events within protein binding sites for the FabI protein [102]. By utilising replicas of apo structures with varying timescales and starting conformations, they generated a total concatenated trajectory of 60 μs. Analysing this trajectory using MSMs revealed notable aspects such as the most probable states, relative free energies, and key structural features. This analysis underscored the dynamic interplay and cooperation between binding sites, elaborating their functional dynamics and sequence of geometrical reorganisations. One has to bear in mind that even this long timescale might not be sufficient to capture all the large conformational changes, and their work mostly focused on describing C-terminal and interface conformational changes, where larger full-domain rearrangements would require even longer timescales or enhanced sampling. Interestingly, the process of MSM generation can highlight under sampled zones of the phase space, suggesting where to focus further simulation, adapting the sampling.

In a separate study, Roux's group demonstrated the utility of MSMs in the study of tyrosine kinases [103, 104]. They contrasted models from apostructure simulations with those from a diverse ensemble of initial structures, namely DFG-in, DFG-out, and apostructure states. Notably, even apo-kinase simulations unveiled novel structures in uncharacterised kinases and accurately restored known 3D inhibited-like conformations. This approach, grounded in comprehensive sampling and equilibrium probabilities, expands the utility of MSMs beyond traditional methods. As an additional conclusion, MSM can complement virtual screening strategies. By identifying potentially druggable pockets through Markov states, MSMs enable the detection of sites that might be overlooked by conventional methods, elevating the potency and selectivity of screened target compounds.

4.1.7 Distance Calculations

Distance analysis is a powerful tool that enables the exploration of changes in relevant distances within a system across the course of MD simulation. This technique involves measuring the distances between specific pairs of atoms, residues, or structural elements within a biomolecular system. For instance, these selections could involve comparing the distances between two atoms (such as Cα atoms),

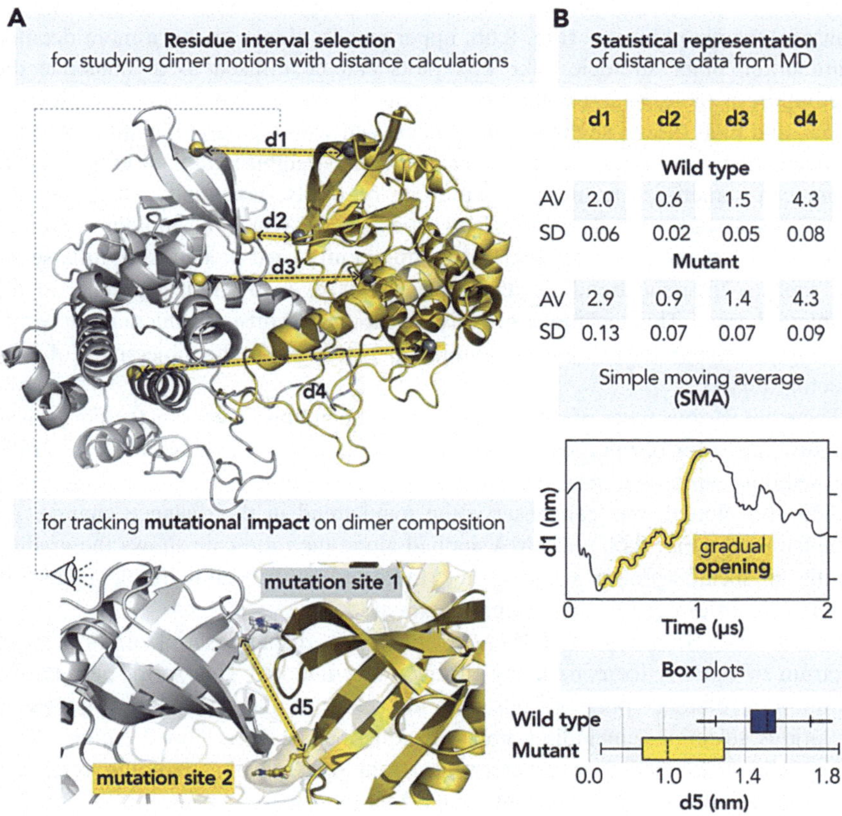

Fig. 8.6 Example of distance calculation applied on MD trajectory to track various events within protein structure (**a**) and ways of its statistical representation (**b**). AV average, SD standard deviation. Statistical boxplot representation of angle values, derived from MD trajectory. Boxes display the quartiles of the dataset (25–75%) and whiskers the rest of the data within 1.5 times the interquartile range (IQR). (Adapted from Shevchenko et al. [95])

two residues, or even larger structural features like helices, chains, or lobes within a protein (Fig. 8.6a) [95, 102, 105]. Such analysis provides insights into the dynamic behaviour of various components of the system.

Distance analysis can be employed to understand the dynamic behaviour of multimeric assemblies and the interdependence of their constituent monomeric units across different systems (Fig. 8.5a) [106]. Selecting several intervals for calculating distance can provide comprehensive data to track the co-dependent movements of monomeric substructures (Fig. 8.6a, d1–d4). Another useful application of distance analysis is in understanding the impact of mutations. For instance, d5 on Fig. 8.6a describes the change of N-lobe composition upon mutation.

It is important to note that distance analysis generates lots of data, which can rapidly accumulate with longer simulation times. To extract meaningful trends from this data and avoid delving into the noise, statistical methods are often employed to explore and visualise the information (Fig. 8.6b). Common approaches include calculating averages and standard deviations to provide a basic numerical assessment of dynamic processes (Fig. 8.6b, upper panel). However, for a more detailed examination, other strategies like box plots can be utilised to demonstrate the distribution of motions across various systems (Fig. 8.6b, box plots highlighted in yellow, and blue showcase the difference between wild type and mutant protein).

When tracking distance changes over time, a straightforward plotting of MD output can sometimes result in a multitude of peaks, potentially obscuring the primary statistical trends. In such cases, employing a moving average (MA) technique can prove beneficial. MA, frequently used in analysing time-series data such as financial trends, calculates averages over different segments of the complete dataset. This technique effectively captures short-term fluctuations while focusing on the broader trends—aligning perfectly with the objectives of MD trajectory analysis.

There are various types of moving averages, depending on various factors such as data structure—for our purpose we employ simple moving average (SMA), which uses a sliding window to take the average over a set number of time periods. Once applied and plotted, one can observe the main trend in the distance change. As demonstrated in Fig. 8.6b, the SMA applied along the timescale shows the gradual opening of specific protein substructures (indicated by the yellow line), suggesting studying the origin of this diverging movement.

In practical terms, with GROMACS [107] gmx distance script offers a broad spectrum of options for conducting distance calculations. This script permits the definition of reference positions based on the centre of mass or centre of geometry for various selections, including atoms, residues, or custom-defined entities. This variety allows extracting relevant results from dynamically changing reference selection, which is highly important for the MD trajectory analysis.

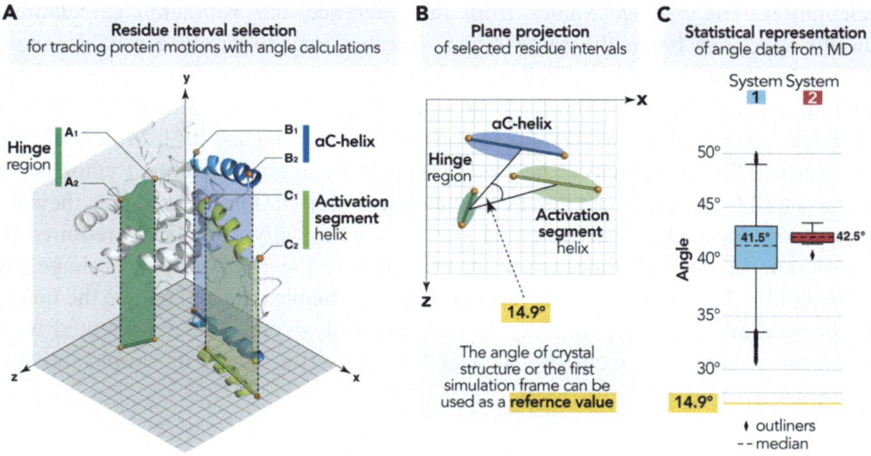

Fig. 8.7 Example of angle calculation applied on MD trajectory to track movement within a protein structure. (**a**) selection of residue intervals for angle construction. Selected residue intervals represent protein substructures, which movements should be tracked (B_1–B_2 and C_1–C_2) with a stable apex point (A_1–A_2). (**b**) Plane projection of selected residue intervals. Apex points defined by centre of geometry (cog). Cog is visualised on the plane projection with semi-transparent ovals of the same colours as displayed residue intervals. (**c**) Statistical boxplot representation of angle values, derived from MD trajectory. Boxes display the quartiles of the dataset (25–75%) and whiskers the rest of the data within 1.5 times of the interquartile range (IQR). The reference angle value is highlighted in yellow in (**b**) and (**c**) to underline the dynamic change one can observe, when angle calculation is applied on MD trajectory. (Adapted from Shevchenko et al. [95])

4.1.8 Angle and Plane Calculations

Angle or plane calculations are beneficial to support visually perceived motions within the MD trajectory with the numerical values. There are various examples of possible practical applications of this method. For instance, angle calculations can be utilised to define the protruding outward movement of αC-helix in kinases quantitatively, justify the opening or closure of Switch-II in RAS proteins or assist the hit-to-lead identification by defining the degree of pocket opening in MD simulations [108, 109].

The reference positions of the vertex and apex points can be defined using several methods, as seen with the distance calculations—namely, the centre of mass or geometry for an atom, residue, or substructure. For instance, Fig. 8.7a shows the selection of residue intervals (A_1–A_2, B_1–B_2, and C_1–C_2) for further angle calculation. Next, these intervals are projected on a plane, where the apex points are calculated as centres of geometry for each interval (Fig. 8.7b). The validation of the residues chosen for the computation of the angle value is an optional step that allows to verify if the calculated value is biased or whether a particular selection has no impact on the average angle value along the trajectories. This is accomplished by selecting the neighbouring to the reference selection residues and repeating the calculation. The average values from the reference and validation calculations should appear to be in the same range. Another way to produce reliable angle calculation between moving helices or domains is to choose the stable residue interval for the apex. On Fig. 8.7a the hinge region (A_1–A_2) defines the apex of the angle, as one of the most rigid regions of a protein kinase [95]. Moreover, the MD-derived RMSD values can be used to support the interval stability. Python is the primary tool for carrying out these computations for an MD trajectory since they are highly dependent on the investigated system. Additionally, the classes required to access data in the MD trajectories are provided by the MDAnalysis package for Python [110, 111]. Further, to represent the most valuable data and reduce the noise, the derived angle values, similarly to distance analysis, are best represented with statistical visualisation methods, such as box plots (Fig. 8.7c) or simple moving average (Fig. 8.6b).

4.2 *Protein–Ligand Perspective*

4.2.1 Protein–Ligand's Interaction Pattern Determination

The analysis of the interaction network that the ligand is forming with the protein is the fundamental part of the simulation analysis as the non-covalent interactions, namely hydrogen bonds and hydrophobic interactions, are the core of the ligand-protein binding [112]. Unlike docking, MD-derived interaction analysis gives not a single-point calculation but the statistical range of the simulation time when the ligand is in contact with a particular protein residue. Moreover, the interaction can be plotted as a function of time. The derived knowledge can be further applied for SAR analysis and hit-to-lead optimisation. Finally, distinct software may have different definitions of distance and angle thresholds that characterise specific interactions, particularly in the case of hydrophobicity.

4.2.2 Distances and Ligand-Induced Geometry Rearrangements

As in protein perspective, distance analysis comes in hand as a tool for understanding ligand-induced geometry rearrangements within proteins [113]. By quantifying distances between key residues for protein functioning and ligands, it is possible to unlock additional insights into the dynamic interactions that govern binding events [102]. These distance fluctuations not only reveal the stability and flexibility of protein–ligand complexes but also highlight allosteric effects and potential druggable sites that can be further utilised in the drug design process [114]. For instance, in the context of G protein-coupled receptors, distance analyses help to spot communication between ligands and the receptor's transmembrane helices, which engagement in turn can lead to agonistic or antagonistic effects, depending on a particular helix involved in ligand binding [115–119]. Moreover, distance analyses can be used in drug resistance studies, where the distance between ligands and the mutated residues can explain how mutations affect ligand binding [95, 120–122].

Moreover, distance analyses have led to the discovery of cryptic binding pockets, often overlooked in static structures, which can be targeted to increase the potency and selectivity of designed compounds [123]. Frequently, ligand binding at one site influences the conformational rearrangements and interactions at a distant site on the protein. By tracking distances between ligand-bound residues and residues at the allosteric site, it is possible to quantify the extent of this communication and predict how ligand binding triggers changes in distant regions. For instance, in nuclear receptor studies distance calculations can be used to spot the changes in cofactor binding or dimerisation interface alterations, induced by ligand binding [105, 124].

4.2.3 Predicting Protein-Ligand Binding Energies Through MD Trajectory

There are several methods employed to predict protein-ligand free binding energies, among those we highlight the free energy calculations, such as free energy perturbation (FEP [125]), relative binding energy prediction, thermal integration (TI [126]), and the Born and surface area continuum solvation (MM/GBSA). In general, the FEP and TI methods are among the most rigorous calculations of the binding free energy, with the trade-off on their high time-consumption and the limitations in the calculation of the relative free energy for highly flexible systems. As a comprehensive guideline for protein-ligand binding energy prediction, please consult the work from Ruiter and Oostenbrink [127]. We focus on the MM/GBSA calculations, more details on FEP can be found in Chap. 10.

MM/GBSA Energy Calculations Molecular mechanics energies combined with the generalised Born and surface area continuum solvation (MM/GBSA) allow estimating binding free energy ΔG, which determines the strength of protein-ligand binding [128] (Fig. 8.8a illustrates the protocol involved in the calculation). This technique has been effectively utilised to explain experimental results as well as to enhance virtual screening and docking outcomes [129]. This method is computationally more efficient than rigorous alchemical perturbation methods (e.g., free energy perturbation, beyond the scope of this chapter), but still more robust than simple molecular docking based on scoring functions. When derived from the MD trajectory, MM/GBSA offers data on the interactions the ligand employs to anchor itself inside the binding pocket and offers the advantage of better incorporating the explicit solvent dynamics' influence in the ligand binding, therefore accurately estimating their entropic contribution and allowing the analyses of protein-water-mediated interactions. These calculations may be used in the hit-to-lead approach to guide synthetic chemists' efforts toward improving the weakest feature of the ligand or to indicate which moiety should be modified to boost binding affinity or to

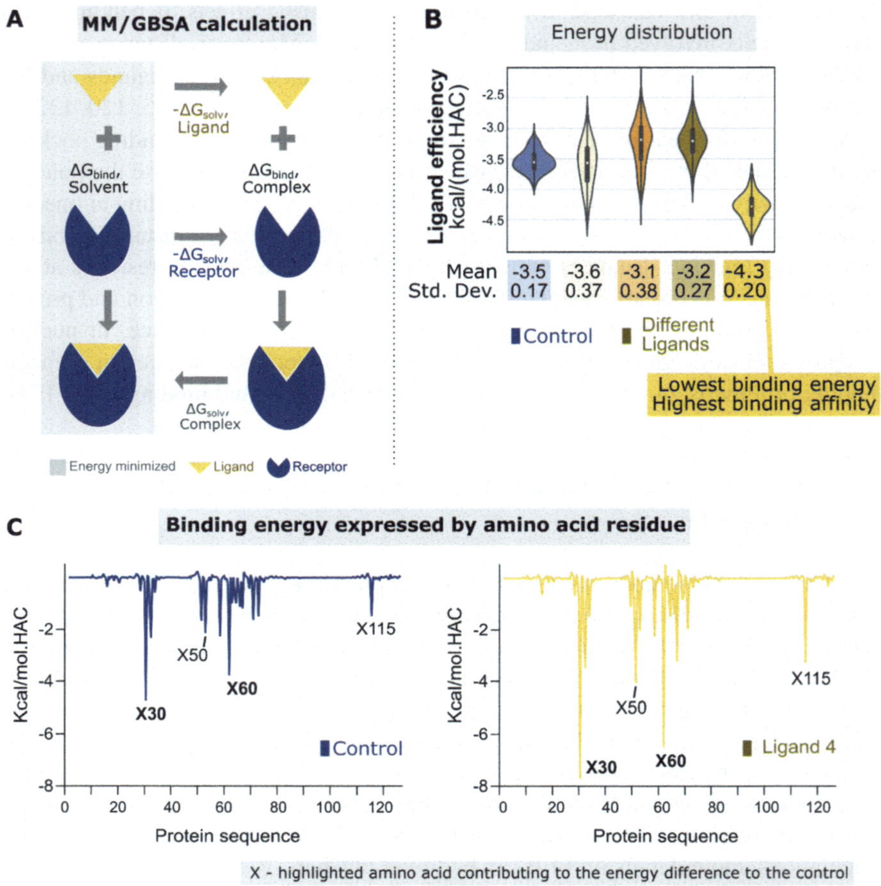

Fig. 8.8 MM/GBSA calculations can help to rank and rationalise protein–ligand complexes. (**a**) schematic representation of MM/GBSA calculations protocol and respective equation for binding energy calculation. Ligand efficiency binding energy (normalising binding energy by the number of heavy atoms). MM/GBSA calculations start with the ligand being extracted from the optimised complex and an energy calculation is run on it without minimisation (see cartoon), to get the energy of the ligand as optimised in the binding pocket. Next, energy minimisation is run on the ligand outside of the receptor. Both calculations are being done with the ligand alone in the solution. The energy difference is the ligand strain energy. The same procedure is performed for the receptor counterpart, generating the receptor energy. The differences between those two calculations generate ligand binding energy, which can be decomposed by residues or by properties. In all violin plot graphics (**b**), the median of the calculated energies is displayed below, together with its standard deviation, and free energy binding calculation (Kcal/mol normalised by the Heavy Atoms Count, HAC), can be decomposed as the average per residue of the protein sequence (**c**), where one can observe which residues contribute to the binding differences. (Adapted from Rashidian et al. [137])

evaluate the bioaccumulation behaviours of non-tested compounds and support their risk assessment. In computational chemistry, Binding free Energy (ΔG) is frequently calculated per the thermodynamic cycle (Eq. 8.6) [130].

$$\Delta G_{\text{bind,solv}} = \Delta G_{\text{bind,vacuum}} + \Delta G_{\text{solv,complex}} - \left(\Delta G_{\text{solv,ligand}} + \Delta G_{\text{solv,receptor}}\right) \quad (8.6)$$

Binding free energy in computational approaches (see Fig. 8.8a for a schematic representation). ΔG_{solv} indicate calculations on solvent (for ligand- and receptor-only, as well as their complex) while ΔG_{vacuum} refers to calculations without solvent.

In turn, the Solvation free energy (ΔG_{solv}) is described as a sum of the polar (ΔG_{pol}) and non-polar (ΔG_{nonpol}) components. When MM/GBSA is applied on the MD trajectory, the calculations are performed for every simulation frame following Eq. 8.7.

$$\Delta G_{\text{bind}} = < \Delta G^{\text{complex}}(i) - \Delta G^{\text{protein}}(i) - \Delta G^{\text{ligand}}(i) >_i \quad (8.7)$$

Binding free energy applied for MD trajectory, where $<\ldots>_i$ indicates that calculations are conducted over i simulation frames.

Next, each ΔG in Eq. 8.7 is decomposed to the terms described in Eq. 8.8.

$$\Delta G = E_{\text{int}} + E_{\text{ele}} + E_{\text{vdw}} + \Delta G_{\text{pol}} + \Delta G_{np} - T\Delta S \quad (8.8)$$

Decomposition of Binding free energy into physicochemical terms, where T represents temperature and the ΔS the variation in entropy (see [128] for further theory).

In Eq. 8.8, E_{int} stands for the molecular mechanic internal energy, E_{ele} for electrostatic energy, and E_{vdw} for van der Waals energies. The last three terms of this equation mean the following: The ΔG_{pol} and ΔG_{np} are the polar and nonpolar Solvation free energies, T is the absolute temperature, and ΔS represents the entropy estimate. While in MD, the protein is frequently solvated in a water box, before the MM/GBSA computations, the water molecules are removed and replaced with an implicit representation [131]. The generalised Born (GB, reflects term MM/GBSA) or Poisson-Boltzmann (PB, reflects MM/PBSA) can be used as implicit representation [132]. For the GB model, the polar solvation energy is described with Equation 9 [130].

$$\Delta G_{\text{pol}} = \sum_{ij} \Delta G_{tj}^{\text{pol}} \approx -\frac{1}{2}\left(\frac{1}{\epsilon_{\text{in}}} - \frac{1}{\epsilon_{\text{out}}}\right)\frac{1}{1+\beta\alpha}\sum_{ij} q_i q_j \left(\frac{1}{f_{ij}^{GB}} + \frac{\alpha\beta}{A}\right) \quad (8.9)$$

The polar solvation energy for the GB model.

In Eq. 8.9, $\varepsilon_{\text{in}} = 1$ (dielectric constant of the solute) and $\varepsilon_{\text{out}} = 80$ (dielectric constant of the solvent); $\beta = \varepsilon_{\text{in}}/\varepsilon_{\text{out}}$; $\alpha = 0.571412$. The A corresponds to the molecule's electrostatic dimensions. The f_{ij}^{GB} functional form (Eq. 8.10) describes the distance between atomic charges (r_{ij}) and effective Born radii (R), which in turn indicates the depth with how each atom is buried in the solvent (Eq. 8.11, references highlight the constant definitions [133, 134]).

$$f_{ij}^{GB} = \left[r_{ij}^2 + R_i R_j \exp\left(- \frac{r_{ij}^2}{4 R_i R_j} \right) \right]^{\frac{1}{2}} \quad (8.10)$$

Distance-Dependent Electrostatic Solvation Function is used to estimate the contribution of each atom pair's electrostatic interaction to the overall energy of the molecular system.

$$R_i^{-3} = \left(-\frac{1}{4\pi} \oint \frac{r - r_i}{|r - r_i|^6} \cdot dS \right) \quad (8.11)$$

Effective Born Radii Contour integral is limited to the dielectric boundary (∂V) of the molecule, r and r_i describe the molecule's position per the surface vector element *dS*.

The ΔG_{np} from Eq. 8.8 is estimated as proportional to the molecules' solvent accessible area (SASA) multiplied by the factor γ [135]. Once conducted, MM/GBSA calculations generate plenty of energy properties. These properties, which are broken down into contributions from different components in the energy expression, report energies for the ligand, receptor, and complex structures, as well as energy changes related to strain and binding.

Below we will briefly introduce the most used properties that one can derive from MM/GBSA.

Before diving into energy contributions, one important aspect of MM/GBSA should be mentioned: the impact of ligand size on overall energy and its contributions. When comparing ligands with notable differences in molecular weight, one may consider normalising the results by the Heavy Atom Count (HAC, i.e., number of non-hydrogen atoms in the ligand molecule) to account for differences in molecular sizes. In general, normalising energy values by HAC also enhances the comparability of ligands' binding affinities, irrespective of their sizes. This approach is based on the assumption that the ligand–protein interactions scale linearly with the number of heavy atoms in the ligand, which has been shown to hold reasonably well for smaller ligands. This normalised binding energy, known as Ligand Efficacy, is a frequently employed parameter in this context. However, HAC normalisation has limitations and should not be the only criterion for assessing the binding affinity of a ligand. Factors like shape and electrostatic complementarity should also be considered.

While the absolute energy values derived from MM/GBSA calculations should be interpreted with care, the focus should be on energy difference ratios when comparing ligands or ligand-WT and ligand-mutated protein systems, bearing in mind the relevance of the experimental design in including "control" simulations. Comparing ΔG Binding Free Energy values with ligand efficacy values (normalised by HAC) can reveal variations in absolute values across specific systems. These variations originate from using different scales (ΔG in kcal/mol and ligand efficacy in kcal/mol per heavy atom count) or disregarding molecular sizes (Fig. 8.8b).

A notable observation one can make is the ratios of the values between ligands remain consistent across different systems. This consistency in the distribution trend is noteworthy and provides an additional layer of validation when analysing MM/GBSA results. Validity is best conveyed by presenting energy calculations as a statistical distribution (i.e., box plots), rather than single data points (i.e., mean/average). The derived values from MM/GBSA fall into two major categories: ΔG with its energy components and Ligand strain and its constituents. In this context, ΔG normalised by HAC (ligand efficacy) belongs to the first category. The most frequently derived energy components from MM/GBSA can be categorised as follows (see Eqs. 8.7 and 8.8 for exemplary definitions):

- *Coulombic Interaction Energy*: Accounts for the electrostatic interactions between charged atoms. It includes both attractive (negative) and repulsive (positive) contributions. Attractive interactions occur between opposite charges, while repulsive interactions occur between like charges.
- *Hydrogen Bond Interaction Energy*: Quantifies the energy of hydrogen bonds formed between the ligand and protein. These bonds involve a hydrogen atom bonded to an electronegative atom and a nearby electronegative atom in the protein, contributing favourably to binding energy.
- *Packing Interaction Energy*. This term considers the steric complementarity between the ligand and the protein binding site. It accounts for favourable interactions when hydrophobic or nonpolar groups in the ligand and protein come into close contact, creating a well-packed binding interface.
- *Van der Waals (VdW) Interaction Energy*: The van der Waals forces are attractive forces that arise from fluctuations in electron distributions around atoms. This term represents the energy associated with favourable interactions between non-polar groups in the ligand and protein. It includes both attractive van der Waals forces and repulsive steric clashes.

Next, moving to second category, associated with Ligand Strain Energy. Ligand Strain Energy can provide valuable insights into the structural and energetic distortions that a ligand undergoes upon binding. It quantifies the energetic cost associated with adapting or rearranging the ligand's conformation to achieve optimal interactions within the binding pocket. In more detail, the Ligand Strain Energy accounts for conformational changes (rotations, bending, or stretching of the ligand's bonds), steric interactions or clashes, distortion of ligand electronic structure (adjustments in bond lengths, angles, and dihedrals), and hydrogen bonding and interaction matching. In turn, Ligand Strain Energy is constituted by the sum of their own different terms/components, namely Lipophilic, Coulomb, and Solvation energy. While Coulomb and Solvation may often appear quite similar, they still provide valuable insights into the ligand's strain energy. Here's how one can approach reading the values.

- *Ligand Strain Lipophilic Interaction Energy*: represents the strain energy associated with the ligand's conformation in the context of lipophilic interactions. It quantifies the distortion or deformation of the ligand molecule as it interacts with lipophilic regions in the binding site. A higher value indicates a greater distortion or deformation of the ligand molecule as it interacts with lipophilic regions in the binding site. It highlights the strain caused by the ligand's interaction with hydrophobic regions of the protein.
- *Ligand Strain Coulombic Interaction energy*: refers to the Coulombic energy term associated with the ligand in the binding site of a protein, quantifies the electrostatic interactions between the ligand and the surrounding environment, including the protein and solvent molecules. It accounts for the strain energy arising from the repulsion or attraction of charged particles. While the Coulomb values may appear similar to Ligand Strain Solvation Energy, they provide information about the strain resulting specifically from electrostatic interactions.
- *Ligand Strain Solvation Energy*: refers to the strain energy of the ligand due to solvation effects, quantifies the strain or distortion of the ligand's molecular structure caused by the presence of solvent molecules. This energy term captures the energetic cost associated with reorganising the ligand's conformation or adjusting its shape to accommodate solvent molecules.

As mentioned, the predicted binding energy (using MM/GBSA) can rank ligands more accurately than docking scores. The comparison between crystal structures/complexes and short MD simulations showed that [136] though the timescale impacted the predictions, the shorter evaluated simulations (~5 ns) have not led to better predictions. This is consistent with the observation that larger conformational changes required a higher number of representative frames to establish accurate predictions. We propose to observe the distribution of predicted energy values (Fig. 8.8b), as a control for this relationship between stable poses and predicted energy, where monomodal distributions would indicate reliable binding modes. Our recommendation is to compute the difference between the free energies of the targeted states (ligands of interest) in comparison to references, multiple crystal structures, and/or relevant models, as the quality of prediction can greatly vary among the different targets.

Additionally, the predictions calculated in the article were sensitive to changes in the solute dielectric constant, suggesting that this parameter should be carefully determined according to the characteristics of the protein/ligand binding interface. This concern could be dealt with by running simulations with known binders as controls for determining an optimal setup. Additionally, the decomposition of predicted energy among the individual contributing amino acid residues (Fig. 8.8c) provides data that can be used for further hit-to-lead optimisation or to interpret conformational changes in the protein.

4.3 *Ligand Perspective*

4.3.1 Ligand Properties

Plenty of ligand properties or quantitative descriptors are routinely applied for compound filtering before or after virtual screenings, and large-scale dockings can be used for calculation on an MD trajectory. However, the number of properties that are worth calculating as a function of time is quite limited and primarily used for purposes in industrial drug development. The typical examples of such descriptors are polar surface area (PSA) or solvent-accessible surface area (SASA). PSA is used to examine such parameters as oral absorption or blood–brain barrier permeation [138].

4.3.2 Ligand Root Mean Square Deviation

The theoretical background of RMSD calculations is the same for both ligand and protein, although the RMSD results are interpreted somewhat differently when a ligand is involved. For instance, if RMSD values of the ligand can be aligned on the ligand itself, the graph will reproduce the ligand's internal fluctuations as a function of time (Fig. 8.4a, red). When a shift of values is observed, the ligand's conformation changes during the trajectory (Fig. 8.4a, shift in RMSD indicated with a yellow circle). Alternately, the ligand can be aligned on the protein, in which case the plot will demonstrate how the ligand fluctuates concerning the protein. An expressive shift in those values may indicate that the ligand is unstable inside the binding site or moving outside of it. Additionally, RMSD calculations can be used to compare the difference between a docking output to a known binding pose from a crystal structure to validate the docking precision.

4.3.3 Ligand Root Mean Square Fluctuation

Ligand RMSF shares the theoretical background with the protein RMSF and indicates the atom fluctuations in the ligand. Similar to ligand RMSD, ligand RMSF can be aligned either on the ligand itself or on the protein. The data in the first case indicates ligand fluctuations, whereas, in the second, it depicts fluctuations in correspondence to the binding site. The flexibility or conformational alterations of a ligand inside a binding site are often evaluated using the ligand RMSF.

4.3.4 Angles and Dihedrals

The analysis of ligand torsional angles and dihedrals can supplement the understanding of ligand–receptor interaction and binding dynamics. Torsional angles refer to the rotations around a single bond, while dihedrals define the angle between two planes defined by four atoms following one after the other in the ligand structure. Practical applications of this analysis include identifying ligand key binding conformations, determining preferred binding modes, mapping the binding pocket, and characterising ligand-induced conformational changes in the receptor.

Moreover, it can supplement the identification of specific torsional angles that are responsible for favourable interactions with the target receptor, supporting the rationalisation of compounds' activity. Further, the discovered reliable torsional angles can be integrated into a pharmacophore or employed directly as filters in virtual screening campaigns to select compounds with a higher likelihood of binding effectively. Furthermore, torsional angles provide information on compound flexibility that can assist medicinal chemists to design and synthesise analogues with optimised torsional angles to enhance binding affinity and specificity of existing compound series. Ligand angles analysis can be conducted with tools such as CPPTRAJ [139] in Amber [57], MDAnalysis [110] in Python, GROMACS [107, 140], or Simulation Interaction Diagram in Maestro [47]. These software packages offer functionalities for calculating and visualising torsional angles and dihedrals in different formats, such as plots, radar charts, or heatmaps, providing valuable insights into ligand behaviour over time.

5 MD Simulations Together with Other Relevant SBDD Techniques

5.1 Protein Structure Prediction and Preparation

This step has become an emerging issue in computational chemistry, especially when refining low-resolution crystal structures, a task, which is further amplified by the continuous growth of structural data. Factors like missing hydrogens, tautomer uncertainties, and crystal packing complement the need for structure preparation. Hydrogen bond (H-bond) networks play a critical role in shaping the quality and stability of the future model, while also impacting binding specificity studies. This is especially relevant on X-ray structures, which often lack hydrogens [141, 142]. Other factors, such as crystal packing can lead to distortions in protein side-chain conformations, thereby influencing the interpretation of binding pockets [143, 144]. Accurate protein preparation involves such essential components as addressing missing hydrogens and optimising hydrogen bond networks. These steps are integral to various structure preparation protocols offered by software such as GROMACS [107], AutoDock [145], HADDOCK [146, 147], Chimera [148]/Chimera X, and Protein Preparation Wizard [149] (Schrödinger LLC). Finally, applying a chosen force field during the minimisation process ensures the accurate refinement of a protein structure. By attentively handling these aspects, reliable protein-ligand models can be created, preventing misleading outcomes in subsequent studies.

In terms of structure prediction, for instance, homology modelling can be a powerful technique for predicting the protein's secondary structures based on the sequence similarity to template structures. Combined with molecular dynamics, this method can enable in silico mutagenesis experiments, studying conformational changes, and allosteric effects.

The process involves model construction, which is based on frequently several experimentally established structures sharing common structural features with the target protein. If the sequence similarity between the target protein and existing templates is low, one can rely on well-established common features within the protein family of interest. For instance, class A GPCRs with unknown crystal structures can still be modelled using the classical seven transmembrane helices as a template [116].

Homology modelling can be divided into several steps, including template selection, sequence alignment, model construction, and refinement, which finalise generating a competent 3D structure of the protein query. The evaluation step is crucial for model quality, and various software and techniques are described in the respective chapter (see Chap. 14).

Since homology modelling can yield different results based on template choice, it is important to construct multiple models using various templates and compare them for accuracy. Several tools and online resources can facilitate homology modelling tasks, with the latter being the latest artificial intelligence-based method for predicting protein structures from amino acid sequences.

5.2 *Molecular Docking*

The docking methodology has been a pioneering computational tool in small molecule drug design since the early 1980s [150]. It serves to predict the orientation of ligands within a protein's binding site. In general, docking involves several stages, such as ligand and protein preparation, initial pose generation, and pose scoring [151]. Scoring functions can be empirical, knowledge-based, or force field-based [150]. Diverse approaches can be found in molecular docking software. Some require separate preparation of the protein structure and ligands in a relevant force field (e.g., Glide [152, 153]), while others combine all stages under a unified protocol (e.g., HADDOCK [146], AutoDock Vina [154, 155]). Regardless of the approach, ligand preparation protocols typically include generating 3D geometry and predicting tautomer and ionisation states. After ligand poses are generated, the crucial step of results validation takes place.

Validation may involve including known active compounds in the docking set, as they are expected to exhibit suitable binding modes and receive high docking scores when the procedure works accurately. Additionally, including known negatives or decoys in the ligand dataset helps assess the model's precision and reduces false positives, avoiding artificial enrichment [156]. While the enrichment factor is valuable, it has limitations, notably its dependence on the number of known actives in the initial dataset [156, 157]. An alternative approach is using ROC curves (receiver operating characteristics) combined with the area under the curve (AUC) to evaluate docking efficiency and determine the level of enrichment [156]. Nevertheless, ROC curves' restrictions [32] and trade-offs are known [158]. Each docking software follows a unique protocol. One can rapidly screen a substantial compound library, prioritising speed over quality with reduced conformations, just to have an additional thorough step with just the high-ranking candidates, aiming to balance resource.

Lastly, utilising derived docking poses as an initial frame for molecular dynamics (MD) simulations allows to explore dynamic interactions, gaining valuable insights into binding mechanisms and energy components. However, before conducting an MD simulation, it is crucial to assess docking accuracy. This is a result of a myriad of factors, relying on high-quality initial data and model, as well as carefully curating information on the protein family, ligand conformations, and protonation states. Additionally, prior data on active and decoy compounds should be considered. It is also important to acknowledge the reliance on force fields and scoring functions. Integrating docking and MD has proven successful in drug design, enabling the identification of lead compounds, and designing potent inhibitors in various scientific disciplines [159].

We cite the work from Chen [160], which critically review pitfalls of docking with further interesting insights in the MD simulation application. They propose that protein–ligand complexes with stable interactions in simulations would lead to a better representation of the binding modes. We also believe that the so-called unstable systems can be equally informative in showing protein–ligand interactions that can be changed. Relevant approaches directly integrating short explicit solvent MD simulations during the docking pipeline [161], generating an ensemble of protein conformations. This approach would, to some extent, account for the ligand-induced conformational changes.

Alternatively, the selection of relevant conformations from MD-trajectories can be performed the ability of each conformation to enrich known binders from known non-binders [162], which is relevant for targets with established decoy sets (see DUD-E database for readily available sets and the possibility to generate datasets [163]). In this approach the structures/conformations would be ranked by their discrimination power (using ROC and enrichment factors). It is relevant to notice that the increased enrichment seems to be dependent on a limited number of MD structures, when compared to crystal-exclusive docking [164, 165]. This alternative to exclusively use experimentally determined structures, as was the state-of-art for classical ensemble dockings, can be interesting in low-data scenarios. More details on molecular docking and sampling the protein conformational space are available in Chap. 9.

6 Concluding Remarks and Outlook

The addressed above in silico drug design approaches fall into two main categories: one focuses on comprehending the processes and subprocesses involved in disease development, and the other offers brand-new methods for disease intervention. In other words, the first is the use of computing to aid in elucidating the molecular mechanisms underlying abnormal biological processes. Specifically, locating previously undiscovered biomolecular events that might be connected to the pathophysiological mechanisms in potentially curable diseases, determining the ligand binding mode at a particular binding site with potential modulation of a protein's function for a specific circumstance, and investigating the mutational consequences related to disease progression.

It is important to emphasise additional issues that can be accomplished with the *in silico* approach while addressing the second point of new molecular strategy identification. These include the design of novel and highly selective drugs via virtual screening campaigns combined with experimental techniques; prediction and simultaneous optimisation of biopharmaceutical properties, such as solubility and toxicity predictions; discovery of novel and potentially druggable protein pockets. In addition, the stabilisation of a specific protein substructure or residue that encourages system-specific interactions can be used for drug development employing ligand feature identification. Providing the knowledge that fills the understanding gaps, *in silico* methods emphasise the development of new therapeutics as composed of minor details, each of which contributes to the broader picture.

Originally, one major drawback from MD application lied on the resources, as not every research group could afford a supercomputer. In recent years, the popularisation of multi-CPU clusters hosted by supercomputer centres give access to both infrastructure and expertise. The expansion of GPU-developed code sky-rockets the timescales from short nanoseconds to several milliseconds, with tera-scale performances on GPU clusters. However, one point that still lags is the experimental design and interpretation of the MD simulation. The relevance of interdisciplinary teams in deciding which systems are biologically/pharmacologically relevant to be simulated cannot be overstated. We hoped to illustrate some examples of analyses and comparative discussions along this chapter while providing a solid base for the MD interpretation.

Questions to answer when planning a MD

? Do you plan to observe small molecules or domain's movements? Be mindful of your timescale.
Does your recording time allow you to capture the required transitions, and do you have enough storage for those trajectories?
Given your available computational resources, what is the largest number of replicas per system you could generate?
Do you provide clear reference points (sequences and secondary structures) and definitions for your analyses to ensure reproducibility?
Do you provide clear parametrisation, force-field definitions, and meta data to allow reproduction of your simulations?

Acknowledgements The authors wish to acknowledge CSC—IT Center for Science, Finland, for the very generous computational resources provided in multiple of the examples cited within this chapter, as well as their striking enthusiasm in discussing scientific matters with our group. T.K., E. S., and A.P. are funded by the TüCAD2 and CMIF. TüCAD2 and CMIF are funded by the Federal Ministry of Education and Research (BMBF) and the Baden-Württemberg Ministry of Science as part of the Excellence Strategy of the German Federal and State Governments. S.L. and T.K. would like to acknowledge the cluster of Excellence iFIT (EXC 2180) "Image-Guided and Functionally Instructed Tumor Therapies".

Conflict of Interest The authors declare that they have no known competing financial interests or personal relationships that could have appeared to influence the work reported in this chapter.

Glossary

Free energy of binding within the context of ligand–protein complexes in drug design, the free energy of binding is defined as the free energy difference between the ligand-bound state (complex) and the free unbound states (free protein and free ligand).

Periodic boundary condition (PBC) is a method used in MD simulations to eliminate the issues concerning boundary effects, arising from finite size, by treating the system as infinite with the help of a unit cell.

Stochastic algorithms is a sampling method that incorporates random changes to the ligand in transitional, rotational, and conformational space to identify the most suitable ligand binding conformation.

Systematic search is a sampling method that utilises all degrees of freedom to sample the ligand-binding conformations.

Central processing unit (CPU) is the core processing unit of a computer that performs general-purpose computations required for various tasks, including molecular modelling. In molecular dynamics simulations, the CPU is traditionally used to handle complex calculations involving force fields, energy minimisation, trajectory analysis, and traditional docking protocols.

Graphics processing unit (GPU) is a specialised hardware unit designed for accelerating graphics rendering. In molecular modelling, GPUs have gained significant prominence due to their parallel processing capabilities. They are particularly advantageous for running computationally intensive tasks, such as molecular dynamics simulations.

References

1. Prieto-Martínez, F. D., López-López, E., Eurídice Juárez-Mercado, K., & Medina-Franco, J. L. (2019). Chapter 2: Computational drug design methods—Current and future perspectives. In K. Roy (Ed.), *In Silico drug design* (pp. 19–44). Academic Press.
2. Behr, J P. (1994) *Perspectives in Supramolecular Chemistry: The Lock-and-Key Principle. Volume I* (p. 30). Wiley.
3. Schneider, H.-J. (2003). Introduction to molecular recognition models. In *Protein-ligand interactions* (pp. 21–50). Wiley.
4. Osakwe, O. (2016). Chapter 5: The significance of discovery screening and structure optimization studies. In O. Osakwe & S. A. A. Rizvi (Eds.), *Social aspects of drug discovery, development and commercialization* (pp. 109–128). Academic.
5. Klebe, G. (2013). Protein modeling and structure-based drug design. In G. Klebe (Ed.), *Drug design: Methodology, concepts, and mode-of-action* (pp. 429–448). Springer.
6. Jumper, J., et al. (2021). Highly accurate protein structure prediction with AlphaFold. *Nature, 596*, 583–589.
7. Levitt, D. G., & Banaszak, L. J. (1992). POCKET: A computer graphics method for identifying and displaying protein cavities and their surrounding amino acids. *Journal of Molecular Graphics, 10*, 229–234.
8. Brady, G. P., & Stouten, P. F. (2000). Fast prediction and visualization of protein binding pockets with PASS. *Journal of Computer-Aided Molecular Design, 14*, 383–401.
9. Hendlich, M., Rippmann, F., & Barnickel, G. (1997). LIGSITE: Automatic and efficient detection of potential small molecule-binding sites in proteins. *Journal of Molecular Graphics & Modelling, 15*, 359–363, 389.
10. Huang, B., & Schroeder, M. (2006). LIGSITEcsc: Predicting ligand binding sites using the Connolly surface and degree of conservation. *BMC Structural Biology, 6*, 19.
11. Halgren, T. (2007). New method for fast and accurate binding-site identification and analysis. *Chemical Biology & Drug Design, 69*, 146–148.
12. Halgren, T. A. (2009). Identifying and characterizing binding sites and assessing Druggability. *Journal of Chemical Information and Modeling, 49*, 377–389.
13. Opo, F. A. D. M., et al. (2021). Structure based pharmacophore modeling, virtual screening, molecular docking and ADMET approaches for identification of natural anti-cancer agents targeting XIAP protein. *Scientific Reports, 11*, 4049.
14. Levitt, M., & Lifson, S. (1969). Refinement of protein conformations using a macromolecular energy minimization procedure. *Journal of Molecular Biology, 46*, 269–279.
15. The Nobel Prize in Chemistry 2013. NobelPrize.org. https://www.nobelprize.org/prizes/chemistry/2013/karplus/facts/
16. Levitt, M., & Warshel, A. (1975). Computer simulation of protein folding. *Nature, 253*, 694–698.
17. Warshel, A., & Levitt, M. (1976). Theoretical studies of enzymic reactions: Dielectric, electrostatic and steric stabilization of the carbonium ion in the reaction of lysozyme. *Journal of Molecular Biology, 103*, 227–249.
18. McCammon, J. A., Gelin, B. R., & Karplus, M. (1977). Dynamics of folded proteins. *Nature, 267*, 585–590.
19. Smith, B. (1989). Dynamics of proteins and nucleic acids: By JA McCammon and SC Harvey. p 234. Cambridge University Press, NY. 1987. $19.95 (pbk). *Biochemical Education, 17*, 220.
20. Karplus, M., & Petsko, G. A. (1990). Molecular dynamics simulations in biology. *Nature, 347*, 631–639.
21. Frauenfelder, H., Sligar, S. G., & Wolynes, P. G. (1991). The energy landscapes and motions of proteins. *Science, 254*, 1598–1603.
22. York, D. M., Darden, T. A., & Pedersen, L. G. (1993). The effect of long-range electrostatic interactions in simulations of macromolecular crystals: A comparison of the Ewald and truncated list methods. *The Journal of Chemical Physics, 99*, 8345–8348.

23. Pearlman, D. A., et al. (1995). AMBER, a package of computer programs for applying molecular mechanics, normal mode analysis, molecular dynamics and free energy calculations to simulate the structural and energetic properties of molecules. *Computer Physics Communications, 91*, 1–41.
24. Berendsen, H. J. C., van der Spoel, D., & van Drunen, R. (1995). GROMACS: A message-passing parallel molecular dynamics implementation. *Computer Physics Communications, 91*, 43–56.
25. Humphrey, W., Dalke, A., & Schulten, K. (1996). VMD: Visual molecular dynamics. *Journal of Molecular Graphics, 14*, 33–38, 27–33–38, 28.
26. Bowers, K. J., et al. (2006). Scalable algorithms for molecular dynamics simulations on commodity clusters. In *Proceedings of the 2006 ACM/IEEE conference on supercomputing 84-es*. Association for Computing Machinery.
27. Klauda, J. B., et al. (2010). Update of the CHARMM all-atom additive force Field for lipids: Validation on six lipid types. *The Journal of Physical Chemistry. B, 114*, 7830–7843.
28. Harder, E., et al. (2016). OPLS3: A force Field providing broad coverage of drug-like small molecules and proteins. *Journal of Chemical Theory and Computation, 12*, 281–296.
29. Roos, K., et al. (2019). OPLS3e: Extending force Field coverage for drug-like small molecules. *Journal of Chemical Theory and Computation, 15*, 1863–1874.
30. Lu, C., et al. (2021). OPLS4: Improving force Field accuracy on challenging regimes of chemical space. *Journal of Chemical Theory and Computation, 17*, 4291–4300.
31. Feig, M., & Brooks, C. L. (2004). Recent advances in the development and application of implicit solvent models in biomolecule simulations. *Current Opinion in Structural Biology, 14*, 217–224.
32. Senn, H. M., & Thiel, W. (2009). QM/MM methods for biomolecular systems. *Angewandte Chemie, International Edition, 48*, 1198–1229.
33. Lin, H., & Truhlar, D. G. (2007). QM/MM: What have we learned, where are we, and where do we go from here? *Theoretical Chemistry Accounts, 117*, 185–199.
34. Cavalli, A., Bottegoni, G., Raco, C., De Vivo, M., & Recanatini, M. (2004). A computational study of the binding of Propidium to the peripheral anionic site of human Acetylcholinesterase. *Journal of Medicinal Chemistry, 47*, 3991–3999.
35. Robustelli, P., et al. (2022). Molecular basis of small-molecule binding to α-Synuclein. *Journal of the American Chemical Society, 144*, 2501–2510.
36. Dror, R. O., Jensen, M. Ø., Borhani, D. W., & Shaw, D. E. (2010). Exploring atomic resolution physiology on a femtosecond to millisecond timescale using molecular dynamics simulations. *The Journal of General Physiology, 135*, 555–562.
37. Shan, Y., et al. (2022). How does a small molecule bind at a cryptic binding site? *PLoS Computational Biology, 18*, e1009817.
38. Mysore, V. P., et al. (2021). A structural model of a Ras–Raf signalosome. *Nature Structural & Molecular Biology, 28*, 847–857.
39. Wang, Y., Lupala, C. S., Liu, H., & Lin, X. (2018). Identification of drug binding sites and action mechanisms with molecular dynamics simulations. *Current Topics in Medicinal Chemistry, 18*, 2268–2277.
40. Marrink, S. J., Risselada, H. J., Yefimov, S., Tieleman, D. P., & de Vries, A. H. (2007). The MARTINI force Field: Coarse grained model for biomolecular simulations. *The Journal of Physical Chemistry. B, 111*, 7812–7824.
41. Heinz, H., Koerner, H., Anderson, K. L., Vaia, R. A., & Farmer, B. L. (2005). Force Field for mica-type silicates and dynamics of Octadecylammonium chains grafted to Montmorillonite. *Chemistry of Materials, 17*, 5658–5669.
42. Chaudhuri, G., & Gupta, S. D. (2007). Specific heat and bimodality in canonical and grand canonical versions of the thermodynamic model. *Phys. Rev. C, 76*, 014619.
43. Dunkel, J., & Hilbert, S. (2006). Phase transitions in small systems: Microcanonical vs. canonical ensembles. *Physica A: Statistical Mechanics and its Applications, 370*, 390–406.

44. Gao, X. (2022). The mathematics of the ensemble theory. *Results in Physics, 34*, 105230.
45. Appendix, H. (2002). Statistical mechanics of the Gibbs "ensemble". In D. Frenkel & B. Smit (Eds.), *Understanding molecular simulation* (2nd ed., pp. 563–571). Academic.
46. Kurzweil, R. (2006). Keynote speech: Raymond Kurzweil. In *Proceedings of the 2006 ACM/IEEE conference on supercomputing 0-es*. Association for Computing Machinery.
47. Desmond User Manual. 178.
48. Chan, W. K. B., Carlson, H. A., & Traynor, J. R. (2023). Application of mixed-solvent molecular dynamics simulations for prediction of allosteric sites on G-protein-coupled receptors. *Molecular Pharmacology*.
49. Lin, F.-Y., & MacKerell, A. D. (2019). Force fields for small molecules. *Methods in Molecular Biology (Clifton NJ), 2022*, 21–54.
50. Appendix, A. (2002). Equations of motion from the Lagrangian or Hamiltonian. In D. Frenkel & B. Smit (Eds.), *Understanding molecular simulation* (2nd ed., pp. 481–494). Academic.
51. Jorgensen, W. L., & Tirado-Rives, J. (2005). Potential energy functions for atomic-level simulations of water and organic and biomolecular systems. *Proceedings of the National Academy of Sciences, 102*, 6665–6670.
52. Guvench, O., & MacKerell, A. D. (2008). Comparison of protein force fields for molecular dynamics simulations. In A. Kukol (Ed.), *Molecular modeling of proteins* (pp. 63–88). Humana Press.
53. Ponder, J. W., & Case, D. A. (2003). Force fields for protein simulations. In *Advances in protein chemistry* (Vol. 66, pp. 27–85). Academic.
54. Jorgensen, W. L., Maxwell, D. S., & Tirado-Rives, J. (1996). Development and testing of the OPLS all-atom force Field on conformational energetics and properties of organic liquids. *Journal of the American Chemical Society, 118*, 11225–11236.
55. Salomon-Ferrer, R., Case, D. A., & Walker, R. C. (2013). An overview of the Amber biomolecular simulation package. *WIREs Computational Molecular Science, 3*, 198–210.
56. Case, D. A., et al. (2005). The Amber biomolecular simulation programs. *Journal of Computational Chemistry, 26*, 1668–1688.
57. Lee, J., et al. (2020). CHARMM-GUI supports the Amber force fields. *The Journal of Chemical Physics, 153*, 035103.
58. Christen, M., et al. (2005). The GROMOS software for biomolecular simulation: GROMOS05. *Journal of Computational Chemistry, 26*, 1719–1751.
59. Tian, C., et al. (2020). ff19SB: Amino-acid-specific protein backbone parameters trained against quantum mechanics energy surfaces in solution. *Journal of Chemical Theory and Computation, 16*, 528–552.
60. Lindorff-Larsen, K., et al. (2010). Improved side-chain torsion potentials for the Amber ff99SB protein force field. *Proteins: Structure, Function, and Bioinformatics, 78*, 1950–1958.
61. Dickson, C. J., Walker, R. C., & Gould, I. R. (2022). Lipid21: Complex lipid membrane simulations with AMBER. *Journal of Chemical Theory and Computation, 18*, 1726–1736.
62. Galindo-Murillo, R., et al. (2016). Assessing the current state of Amber force Field modifications for DNA. *Journal of Chemical Theory and Computation, 12*, 4114–4127.
63. Zgarbová, M., et al. (2011). Refinement of the Cornell et al. nucleic acids force Field based on reference quantum chemical calculations of Glycosidic torsion profiles. *Journal of Chemical Theory and Computation, 7*, 2886–2902.
64. MacKerell, A. D., et al. (1998). All-atom empirical potential for molecular modeling and dynamics studies of proteins. *The Journal of Physical Chemistry. B, 102*, 3586–3616.
65. Brooks, B. R., et al. (2009). CHARMM: The biomolecular simulation program. *Journal of Computational Chemistry, 30*, 1545–1614.
66. Reif, M. M., Hünenberger, P. H., & Oostenbrink, C. (2012). New interaction parameters for charged amino acid side chains in the GROMOS force Field. *Journal of Chemical Theory and Computation, 8*, 3705–3723.
67. Halgren, T. A., & Nachbar, R. B. (1996). Merck molecular force field. IV. Conformational energies and geometries for MMFF94. *Journal of Computational Chemistry, 17*, 587–615.

68. Halgren, T. A. (1999). MMFF VI MMFF94s option for energy minimization studies. *Journal of Computational Chemistry, 20*, 720–729.
69. Rappe, A. K., Casewit, C. J., Colwell, K. S., Goddard, W. A. I., & Skiff, W. M. (1992). UFF, a full periodic table force field for molecular mechanics and molecular dynamics simulations. *Journal of the American Chemical Society, 114*, 10024–10035.
70. Hahn, D. F., Gapsys, V., Groot, B. L. de, Mobley, D. L. & Tresadern, G. J. (2023). Current state of open source force fields in protein-ligand binding affinity predictions. Preprint at
71. Wu, J. C., Chattree, G., & Ren, P. (2012). Automation of AMOEBA polarizable force field parameterization for small molecules. *Theoretical Chemistry Accounts, 131*, 1138.
72. Jing, Z., et al. (2019). Polarizable force fields for biomolecular simulations: Recent advances and applications. *Annual Review of Biophysics, 48*, 371–394.
73. Li, H., et al. (2017). Drude polarizable force Field for molecular dynamics simulations of saturated and unsaturated Zwitterionic lipids. *Journal of Chemical Theory and Computation, 13*, 4535–4552.
74. Jiang, W., et al. (2011). High-performance scalable molecular dynamics simulations of a polarizable force Field based on classical Drude oscillators in NAMD. *Journal of Physical Chemistry Letters, 2*, 87–92.
75. Dunn, W. L. (1981). Inverse Monte Carlo analysis. *Journal of Computational Physics, 41*, 154–166.
76. Dodda, L. S., Cabeza de Vaca, I., Tirado-Rives, J., Jorgensen, W. L., & LigParGen web server: an automatic OPLS-AA parameter generator for organic ligands. (2017). *Nucleic Acids Research, 45*, W331–W336.
77. Decherchi, S., Bottegoni, G., Spitaleri, A., Rocchia, W., & Cavalli, A. (2018). BiKi life sciences: A new suite for molecular dynamics and related methods in drug discovery. *Journal of Chemical Information and Modeling, 58*, 219–224.
78. Huang, L., & Roux, B. (2013). Automated force field parameterization for non-polarizable and polarizable atomic models based on AB initio target DATA. *Journal of Chemical Theory and Computation, 9*.
79. Henzler-Wildman, K., & Kern, D. (2007). Dynamic personalities of proteins. *Nature, 450*, 964–972.
80. Soares, T. A., et al. (2023). Guidelines for reporting molecular dynamics simulations in JCIM publications. *Journal of Chemical Information and Modeling, 63*, 3227–3229.
81. Knapp, B., Ospina, L., & Deane, C. M. (2018). Avoiding false positive conclusions in molecular simulation: The importance of replicas. *Journal of Chemical Theory and Computation, 14*, 6127–6138.
82. Grossfield, A., & Zuckerman, D. M. (2009). Quantifying uncertainty and sampling quality in biomolecular simulations. *Annual Reports in Computational Chemistry, 5*, 23–48.
83. Yan, S., Peck, J. M., Ilgu, M., Nilsen-Hamilton, M., & Lamm, M. H. (2020). Sampling performance of multiple independent molecular dynamics simulations of an RNA Aptamer. *ACS Omega, 5*, 20187–20201.
84. Sargsyan, K., Grauffel, C., & Lim, C. (2017). How molecular size impacts RMSD applications in molecular dynamics simulations. *Journal of Chemical Theory and Computation, 13*, 1518–1524.
85. Theobald, D. L. (2005). Rapid calculation of RMSDs using a quaternion-based characteristic polynomial. *Acta Crystallographica. Section A, 61*, 478–480.
86. Martínez, L. (2015). Automatic identification of Mobile and rigid substructures in molecular dynamics simulations and fractional structural fluctuation analysis. *PLoS One, 10*, e0119264.
87. Lobanov, M. Y., Bogatyreva, N. S., & Galzitskaya, O. V. (2008). Radius of gyration as an indicator of protein structure compactness. *Molecular Biology, 42*, 623–628.
88. Pauling, L., Corey, R. B., & Branson, H. R. (1951). The structure of proteins: Two hydrogen-bonded helical configurations of the polypeptide chain. *Proceedings of the National Academy of Sciences, 37*, 205–211.

89. Kabsch, W., & Sander, C. (1983). Dictionary of protein secondary structure: Pattern recognition of hydrogen-bonded and geometrical features. *Biopolymers, 22*, 2577–2637.
90. Lyu, Z., Wang, Z., Luo, F., Shuai, J., & Huang, Y. (2021). Protein secondary structure prediction with a reductive deep learning method. *Frontiers in Bioengineering and Biotechnology, 9.*
91. Jolliffe, I. T., & Cadima, J. (2016). Principal component analysis: A review and recent developments. *Philosophical Transactions of the Royal Society A: Mathematical, Physical and Engineering Sciences, 374*, 20150202.
92. Sittel, F., Jain, A., & Stock, G. (2014). Principal component analysis of molecular dynamics: On the use of Cartesian vs. internal coordinates. *The Journal of Chemical Physics, 141*, 014111.
93. Jolliffe, I. T. (2002). Principal component analysis for time series and other non-independent Data. In I. T. Jolliffe (Ed.), *Principal component analysis* (pp. 299–337). Springer.
94. Onuchic, J. N., Luthey-Schulten, Z., & Wolynes, P. G. (1997). THEORY OF PROTEIN FOLDING: The energy landscape perspective. *Annual Review of Physical Chemistry, 48*, 545–600.
95. Shevchenko, E., Poso, A., & Pantsar, T. (2020). The autoinhibited state of MKK4: Phosphorylation, putative dimerization and R134W mutant studied by molecular dynamics simulations. *Computational and Structural Biotechnology Journal, 18*, 2687–2698.
96. Sarich, M., Prinz, J.-H., & Schütte, C. (2014). Markov model theory. In G. R. Bowman, V. S. Pande, & F. Noé (Eds.), *An introduction to Markov state models and their application to long timescale molecular simulation* (pp. 23–44). Springer.
97. Klippenstein, S. J., Pande, V. S., & Truhlar, D. G. (2014). Chemical kinetics and mechanisms of complex systems: A perspective on recent theoretical advances. *Journal of the American Chemical Society, 136*, 528–546.
98. Konovalov, K. A., Unarta, I. C., Cao, S., Goonetilleke, E. C., & Huang, X. (2021). Markov state models to study the functional dynamics of proteins in the wake of machine learning. *JACS Au, 1*, 1330–1341.
99. Pande, V. S., Beauchamp, K., & Bowman, G. R. (2010). Everything you wanted to know about Markov state models but were afraid to ask. *Methods, 52*, 99–105.
100. McGibbon, R. T., Schwantes, C. R., & Pande, V. S. (2014). Statistical model selection for Markov models of biomolecular dynamics. *The Journal of Physical Chemistry. B, 118*, 6475–6481.
101. Meng, Y., et al. (2018). Predicting the conformational variability of Abl tyrosine kinase using molecular dynamics simulations and Markov state models. *Journal of Chemical Theory and Computation, 14*, 2721–2732.
102. Maltarollo, V. G., et al. (2022). Do go chasing waterfalls: Enoyl Reductase (FabI) in complex with inhibitors stabilizes the Tetrameric structure and opens water channels. *Journal of Chemical Information and Modeling, 62*, 5746–5761.
103. Paul, F., Thomas, T., & Roux, B. (2020). Diversity of long-lived intermediates along the binding pathway of Imatinib to Abl kinase revealed by MD simulations. *Journal of Chemical Theory and Computation, 16*, 7852–7865.
104. Paul, F., Meng, Y., & Roux, B. (2020). Identification of Druggable kinase target conformations using Markov model metastable states analysis of apo-Abl. *Journal of Chemical Theory and Computation, 16*, 1896–1912.
105. Mejdrová, I., et al. (2023). Discovery of novel human constitutive Androstane receptor agonists with the Imidazo[1,2-a]pyridine structure. *Journal of Medicinal Chemistry, 66*, 2422–2456.
106. Calabrese, A. N., et al. (2020). Inter-domain dynamics in the chaperone SurA and multi-site binding to its outer membrane protein clients. *Nature Communications, 11*, 2155.
107. Bauer, P., Hess, B., & Lindahl, E. (2022). *GROMACS 2022.3 Manual.*
108. Pantsar, T. (2019). The current understanding of KRAS protein structure and dynamics. *Computational and Structural Biotechnology Journal, 18*, 189–198.

109. Johnson, C. W., et al. (2022). Regulation of GTPase function by autophosphorylation. *Molecular Cell, 82*, 950–968.e14.
110. Gowers, R. J., et al. (2016). MDAnalysis: A python package for the rapid analysis of molecular dynamics simulations. In *Proceedings of the 15th python in science conference* (pp. 98–105).
111. Michaud-Agrawal, N., Denning, E. J., Woolf, T. B., & Beckstein, O. (2011). MDAnalysis: A toolkit for the analysis of molecular dynamics simulations. *Journal of Computational Chemistry, 32*, 2319–2327.
112. Patil, R., et al. (2010). Optimized hydrophobic interactions and hydrogen bonding at the target-ligand Interface leads the pathways of drug-designing. *PLoS One, 5*, e12029.
113. Ayaz, P., et al. (2023). Structural mechanism of a drug-binding process involving a large conformational change of the protein target. *Nature Communications, 14*, 1885.
114. Ye, L., et al. (2018). Mechanistic insights into allosteric regulation of the A2A adenosine G protein-coupled receptor by physiological cations. *Nature Communications, 9*, 1372.
115. Hedderich, J. B., et al. (2022). The pocketome of G-protein-coupled receptors reveals previously untargeted allosteric sites. *Nature Communications, 13*, 2567.
116. Bayrak, A., et al. (2022). Discovery and development of first-in-class ACKR3/CXCR7 Superagonists for platelet degranulation modulation. *Journal of Medicinal Chemistry, 65*, 13365–13384.
117. Nygaard, R., et al. (2013). The dynamic process of β2-adrenergic receptor activation. *Cell, 152*, 532–542.
118. Ibrahim, P., Wifling, D., & Clark, T. (2019). Universal activation index for class A GPCRs. *Journal of Chemical Information and Modeling, 59*, 3938–3945.
119. Bruzzese, A., Dalton, J. A. R., & Giraldo, J. (2020). Statistics for the analysis of molecular dynamics simulations: Providing P values for agonist-dependent GPCR activation. *Scientific Reports, 10*, 19942.
120. Zhu, M., Wang, D. D., & Yan, H. (2021). Genotype-determined EGFR-RTK heterodimerization and its effects on drug resistance in lung Cancer treatment revealed by molecular dynamics simulations. *BMC Molecular and Cell Biology, 22*, 34.
121. Wang, R., & Zheng, Q. (2020). Multiple molecular dynamics simulations of the inhibitor GRL-02031 complex with wild type and mutant HIV-1 protease reveal the binding and drug-resistance mechanism. *Langmuir, 36*, 13817–13832.
122. Sun, H., Li, Y., Tian, S., Wang, J., & Hou, T. (2014). P-loop conformation governed Crizotinib resistance in G2032R-mutated ROS1 tyrosine kinase: Clues from free energy landscape. *PLoS Computational Biology, 10*, e1003729.
123. Sabanés Zariquiey, F., et al. (2022). Divide and conquer. Pocket-opening mixed-solvent simulations in the perspective of docking virtual screening applications for drug discovery. *Journal of Chemical Information and Modeling, 62*, 533–543.
124. Díaz-Holguín, A., et al. (2023). When two become one: Conformational changes in FXR/RXR heterodimers bound to steroidal antagonists. *ChemMedChem, 18*, e202200556.
125. Bash, P. A., Field, M. J., & Karplus, M. (1987). Free energy perturbation method for chemical reactions in the condensed phase: A dynamic approach based on a combined quantum and molecular mechanics potential. *Journal of the American Chemical Society, 109*, 8092–8094.
126. Beveridge, D. L., & DiCapua, F. M. (1989). Free energy via molecular simulation: Applications to chemical and biomolecular systems. *Annual Review of Biophysics and Biophysical Chemistry, 18*, 431–492.
127. de Ruiter, A., & Oostenbrink, C. (2024). Binding free energy calculations in drug discovery. In *Computational drug discovery* (pp. 1–20). Wiley.
128. Kollman, P. A., et al. (2000). Calculating structures and free energies of complex molecules: Combining molecular mechanics and continuum models. *Accounts of Chemical Research, 33*, 889–897.
129. Genheden, S., & Ryde, U. (2015). The MM/PBSA and MM/GBSA methods to estimate ligand-binding affinities. *Expert Opinion on Drug Discovery, 10*, 449–461.

130. Forouzesh, N., & Mishra, N. (2021). An effective MM/GBSA protocol for absolute binding free energy calculations: A Case study on SARS-CoV-2 spike protein and the human ACE2 receptor. *Molecules, 26*, 2383.
131. Godschalk, F., Genheden, S., Söderhjelm, P., & Ryde, U. (2013). Comparison of MM/GBSA calculations based on explicit and implicit solvent simulations. *Physical Chemistry Chemical Physics, 15*, 7731–7739.
132. Still, W. C., Tempczyk, A., Hawley, R. C., & Hendrickson, T. (1990). Semianalytical treatment of solvation for molecular mechanics and dynamics. *Journal of the American Chemical Society, 112*, 6127–6129.
133. Aguilar, B., & Onufriev, A. V. (2012). Efficient computation of the Total solvation energy of small molecules via the R6 generalized born model. *Journal of Chemical Theory and Computation, 8*, 2404–2411.
134. Aguilar, B., Shadrach, R., & Onufriev, A. V. (2010). Reducing the secondary structure bias in the generalized born model via R6 effective radii. *Journal of Chemical Theory and Computation, 6*, 3613–3630.
135. Gohlke, H., & Case, D. A. (2004). Converging free energy estimates: MM-PB(GB)SA studies on the protein–protein complex Ras–Raf. *Journal of Computational Chemistry, 25*, 238–250.
136. Hou, T., Wang, J., Li, Y., & Wang, W. (2011). Assessing the performance of the MM/PBSA and MM/GBSA methods. 1. The accuracy of binding free energy calculations based on molecular dynamics simulations. *Journal of Chemical Information and Modeling, 51*, 69–82.
137. Rashidian, A. et al. (2023). Filling the blank space: Branched 4-nonylphenols isomers are responsible for robust constitutive androstane receptor (CAR) activation by nonylphenol. Preprint at
138. Clark, D. E. (2011). What has polar surface area ever done for drug discovery? *Future Medicinal Chemistry, 3*, 469–484.
139. Roe, D. R., & Cheatham, T. E. I. (2013). PTRAJ and CPPTRAJ: Software for processing and analysis of molecular dynamics trajectory Data. *Journal of Chemical Theory and Computation, 9*, 3084–3095.
140. Abraham, M. J., et al. (2015). GROMACS: High performance molecular simulations through multi-level parallelism from laptops to supercomputers. *SoftwareX, 1–2*, 19–25.
141. Bitencourt-Ferreira, G., Veit-Acosta, M., & de Azevedo, W. F. (2019). Hydrogen bonds in protein-ligand complexes. In W. F. de Azevedo Jr. (Ed.), *Docking screens for drug discovery* (pp. 93–107). Springer.
142. Kuster, D. J., Liu, C., Fang, Z., Ponder, J. W., & Marshall, G. R. (2015). High-resolution crystal structures of protein helices reconciled with three-centered hydrogen bonds and multipole electrostatics. *PLoS One, 10*, e0123146.
143. Jacobson, M. P., Friesner, R. A., Xiang, Z., & Honig, B. (2002). On the role of the crystal environment in determining protein side-chain conformations. *Journal of Molecular Biology, 320*, 597–608.
144. Lovell, S. C., Word, J. M., Richardson, J. S., & Richardson, D. C. (1999). Asparagine and glutamine rotamers: *B* -factor cutoff and correction of amide flips yield distinct clustering. *Proceedings of the National Academy of Sciences, 96*, 400–405.
145. Morris, G. M., et al. (2009). AutoDock4 and AutoDockTools4: Automated docking with selective receptor flexibility. *Journal of Computational Chemistry, 30*, 2785–2791.
146. Dominguez, C., Boelens, R., & Bonvin, A. M. J. J. (2003). HADDOCK: A protein−protein docking approach based on biochemical or biophysical information. *Journal of the American Chemical Society, 125*, 1731–1737.
147. van Zundert, G. C. P., et al. (2016). The HADDOCK2.2 web server: User-friendly integrative modeling of biomolecular complexes. *Journal of Molecular Biology, 428*, 720–725.
148. Pettersen, E. F., et al. (2004). UCSF Chimera—A visualization system for exploratory research and analysis. *Journal of Computational Chemistry, 25*, 1605–1612.

149. Madhavi Sastry, G., Adzhigirey, M., Day, T., Annabhimoju, R., & Sherman, W. (2013). Protein and ligand preparation: Parameters, protocols, and influence on virtual screening enrichments. *Journal of Computer-Aided Molecular Design, 27*, 221–234.
150. Kitchen, D. B., Decornez, H., Furr, J. R., & Bajorath, J. (2004). Docking and scoring in virtual screening for drug discovery: Methods and applications. *Nature Reviews. Drug Discovery, 3*, 935–949.
151. Brooijmans, N., & Kuntz, I. D. (2003). Molecular recognition and docking algorithms. *Annual Review of Biophysics and Biomolecular Structure, 32*, 335–373.
152. Friesner, R. A., et al. (2006). Extra precision glide: Docking and scoring incorporating a model of hydrophobic enclosure for protein–ligand complexes. *Journal of Medicinal Chemistry, 49*, 6177–6196.
153. Friesner, R. A., et al. (2004). Glide: A new approach for rapid, accurate docking and scoring. 1. Method and assessment of docking accuracy. *Journal of Medicinal Chemistry, 47*, 1739–1749.
154. Eberhardt, J., Santos-Martins, D., Tillack, A. F., & Forli, S. (2021). AutoDock Vina 1.2.0: New docking methods, expanded force Field, and python bindings. *Journal of Chemical Information and Modeling, 61*, 3891–3898.
155. Trott, O., & Olson, A. J. (2010). AutoDock Vina: Improving the speed and accuracy of docking with a new scoring function, efficient optimization and multithreading. *Journal of Computational Chemistry, 31*, 455–461.
156. Verdonk, M. L., et al. (2004). Virtual screening using protein–ligand docking: Avoiding artificial enrichment. *Journal of Chemical Information and Computer Sciences, 44*, 793–806.
157. Huang, N., Shoichet, B. K., & Irwin, J. J. (2006). Benchmarking sets for molecular docking. *Journal of Medicinal Chemistry, 49*, 6789–6801.
158. Clark, R. D., & Webster-Clark, D. J. (2008). Managing bias in ROC curves. *Journal of Computer-Aided Molecular Design, 22*, 141–146.
159. Salmaso, V., & Moro, S. (2018). Bridging molecular docking to molecular dynamics in exploring ligand-protein recognition process: An overview. *Frontiers in Pharmacology, 9*.
160. Chen, Y.-C. (2015). Beware of docking! *Trends in Pharmacological Sciences, 36*, 78–95.
161. Miller, E. B., et al. (2021). Reliable and accurate solution to the induced fit docking problem for protein-ligand binding. *Journal of Chemical Theory and Computation, 17*, 2630–2639.
162. Rueda, M., Bottegoni, G., & Abagyan, R. (2010). Recipes for the selection of experimental protein conformations for virtual screening. *Journal of Chemical Information and Modeling, 50*, 186–193.
163. Mysinger, M. M., Carchia, M., Irwin, J. J., & Shoichet, B. K. (2012). Directory of useful decoys, enhanced (DUD-E): Better ligands and decoys for better benchmarking. *Journal of Medicinal Chemistry, 55*, 6582–6594.
164. Tian, S., et al. (2014). Assessing an ensemble docking-based virtual screening strategy for kinase targets by considering protein flexibility. *Journal of Chemical Information and Modeling, 54*, 2664–2679.
165. Nichols, S. E., Baron, R., Ivetac, A., & McCammon, J. A. (2011). Predictive power of molecular dynamics receptor structures in virtual screening. *Journal of Chemical Information and Modeling, 51*, 1439–1446.

Chapter 9
Conformational Sampling of Proteins: Methods for Simulate Protein Plasticity and Ensemble Docking

Ana Ligia Scott, Simone Queiroz Pantaleão, and Eric Allison Philot

Abstract The understanding of protein function by emphasizing the critical role of structural dynamics, or protein plasticity, in biological processes is a field of knowledge in constant evolution. It highlights that proteins are not static entities but exhibit intrinsic, nonrandom movements crucial for their interactions with other molecules. Advances in cryoelectron microscopy (cryo-EM) and crystallography have provided snapshots of proteins in various conformations, revealing the dynamic nature of macromolecules. Different mechanisms of molecular recognition, such as conformational selection and induced fit, could be simulated by using hybrid methods that integrate molecular dynamics and normal mode analysis to study protein conformational spaces. These approaches have significant implications for drug discovery, as they allow for more accurate modeling of protein–ligand interactions by considering the inherent flexibility of protein structures. Methods such as MDeNM, which integrates normal modes into molecular dynamics simulations, is highlighted for its ability to promote significant conformational changes, facilitating the flexible fitting of atomic structures into cryo-EM maps and revealing complex protein motions. Collective molecular dynamics (coMD) and ClustENM are also discussed, emphasizing their roles in conformational sampling and protein–ligand interactions. Ensemble docking strategies, including the recent essential dynamic ensemble docking (EDED) protocol, are reviewed for their potential in drug design, particularly in improving the accuracy of molecular docking by considering the flexibility of both ligands and target proteins. These methodologies, combined with advancements in force fields and postdocking techniques, contribute to a more comprehensive understanding and prediction of biomolecular interactions.

A. L. Scott · S. Q. Pantaleão (✉)
Center for Mathematics, Computing and Cognition, Federal University of ABC, Santo André, São Paulo, Brazil
e-mail: simone.qrzp@gmail.com

E. A. Philot
Educational and Research Institute, Molecular Oncology Research Center, Barretos Cancer Hospital, Barretos, São Paulo, Brazil

V. G. Maltarollo (ed.), *Computer-Aided and Machine Learning-Driven Drug Design*, Computer-Aided Drug Discovery and Design 3,
https://doi.org/10.1007/978-3-031-76718-0_9

Keywords Protein plasticity · Conformational sampling · Normal modes · Ensemble docking

1 Introduction

For a long time, the idea that the protein structure defines its function has been well accepted. In recent decades, this concept has been adjusted to include a new element: protein structural dynamics or plasticity. This chapter seeks to discuss the importance of protein local and global movements and their relevance for interaction with other macromolecules or small molecules.

Proteins are not static entities; they have nonrandom intrinsic movements that are extremely important for their function, often because of the interaction with different binding partners. Recent developments in cryoelectron microscopy (cryo-EM) and crystallography techniques have revealed multiple snapshots of increasingly large and flexible systems [51]. Cryo-EM has increasingly been used in structural biology to characterize not only a single structure but an ensemble of conformations, often relevant to biological function, particularly for supramolecular assembly [8, 48]. Thanks to radically new technological advances in both microscope hardware and image processing software, termed the resolution revolution, cryo-EM has dramatically improved in the last years and it can now capture information about biomolecular structure in multiple states to near-atomic resolution [44]. Macromolecules adopt many conformations in solution depending on their structure and shape, which determine their dynamics and function. These conformations exhibit unique structure-encoded dynamic properties that have a profound influence on their biological function [5, 58]. Spectacular progress has been achieved in cryogenic electron microscopy (cryo-EM) through technological advances in both microscope hardware and image processing software (the so-called resolution revolution). Cryo-EM can now capture information about biomolecular structures in multiple states to near-atomic resolution [44]. Not only has the number of structures solved by Cryo-EM increased steadily in recent years, but it has also allowed the characterization of ensembles of conformations.

Macromolecules can adopt multiple conformations in solution depending on their structure and shape, which are directly linked to their function. These conformations exhibit unique structure-encoded dynamic properties that have a profound influence on their biological function [5, 58]. Several studies have indicated that these dynamic properties are primarily determined by the topology of native contacts [36].

In the last decade, a series of studies established the role of structural dynamics, also called intrinsic dynamics, facilitating, if not driving, the interactions, and functioning of biomolecular systems in the cell [5, 51, 58]. Therefore, it is crucial to explore different conformations to model dynamical processes and determine stable states [71]. Such studies require both experimental and computational efforts. In this context, theoretical methods are becoming essential for understanding how different conformers interconvert to mediate biological function due to the accumulated structural information. A large number of studies based on coarse-grained

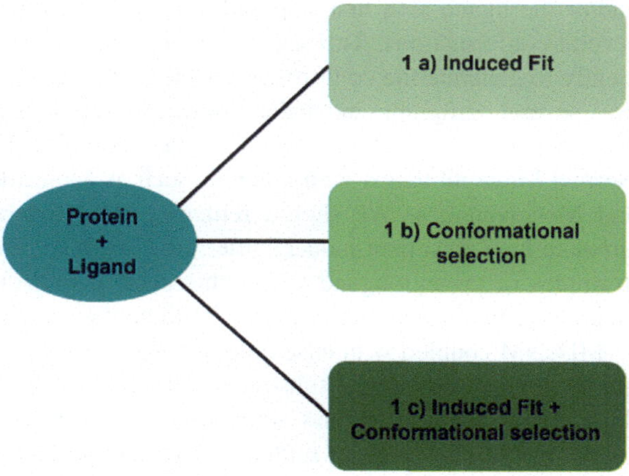

Fig. 9.1 Molecular recognition mechanisms

models (such as the Elastic Network Model, ENM) or full atomic normal mode analysis (NMA) have been applied to investigate macromolecular structures and the molecular mechanisms of their conformational changes [36]. Integrated methods that combine NMA (using ENM or complete atomic models), molecular dynamics, and experimental data from biophysical techniques, such as X-ray crystallography, are useful for studying the structural dynamics of macromolecules.

Related to molecular recognition mechanisms, it is necessary to better understand whether the interaction of proteins with binding partners occurs through possibilities: (i) conformational selection; (ii) induced adjustment, or even (iii) a combination of both. Figure 9.1 illustrates these three situations. In the first, the target visits several conformations energetically accessible by intrinsic motions and the ligand chooses the best (Fig. 9.1a). In the induced adjustment, the ligand binding to the target and provokes a conformational change to a lower energy level (Fig. 9.1b). The third consists of the combinations of both: there is a conformational change before the ligand binding to DNA after, as illustrated in Fig. 9.1c.

There are several examples illustrating each of these three cases and it is important to consider their role in the molecular interaction to improve the drug discovery methods/software. In several cases, it is very important to consider this target's plasticity when we perform *in silico* studies, such as molecular docking (please, see Chap. 7). In recent years, the importance of considering fitting effects in molecular docking calculations was widely recognized in the molecular modeling community. Although small-scale protein side chain movements are now accounted for in many high-end docking strategies, the explicit modeling of large-scale protein movements is still a challenging task. Ensemble-based methods have been introduced taking into account several protein conformations in Drug Design and/or Discovery. Some software couples a technique to sample the target with its docking algorithm.

We can use two distinct strategies to perform an ensemble docking (simulating the conformational selection): (a) use software that integrates sampling and docking algorithms and (b) use a software/method to sample the conformations for the Target and/or the ligand as a first step and after use software to simulate the docking of several conformations. This second strategy allows the use of methods more efficiently to generate the conformers for the target and/or ligand. In this case, we can use several different methods: Molecular Dynamics (please, see Chap. 8), Metadynamics, and Hybrid methods that combine Elastic Network Models or Normal Mode all atoms with methods such as Molecular Dynamics, Monte Carlo, and Metadynamics. We should remember that several times it is important to simulate the adjustment induced after the complexes are formed. There are several examples of protein–ligand where this kind of molecular motion is important to make the complexes more stable or facilitate the catalyses of substrates.

MDeNM coupled with ensemble docking helped in elucidating the mechanisms guiding the recognition of substrates and inhibitors by the sulfotransferase [21]. In the next section, we discuss some of these hybrid methods and applications to simulate ensemble docking protein–ligand and protein–protein.

2 Hybrid Methods to Sample the Protein Conformational Space

The protein plasticity, corresponding to the collective and noncollective motions, has become important in understanding the relationship between sequence, structure and function; and consequently, various aspects of biological processes such as [36]:

(i) The interaction of proteins with other macromolecules or with small molecules.
(ii) The molecular mechanisms involved in metabolic cycles.
(iii) The effect of mutations on the structure and function of proteins and others.

These protein movements can be classified according to different aspects. Yang and collaborators classify protein movements [68] into three classes: (1) local movements of protein fragments; (2) domain movements; and (3) movements involving more than one unit.

In another way, Bahar's group proposed the following classification [41]:

(i) Evolution of motions in the global (the three motions with lower frequency).
(ii) Low-frequency motions (from mode 4 until 20).
(iii) Low-to-intermediate frequency (from mode 21 to 60).
(iv) High-frequency (modes >60).

This classification is very interesting and useful for some kinds of analysis [41]. But in this text, we will use a generic classification using the nomenclature: collective and noncollective movements.

Collective Movements involve a considerable part of the molecule moving in a coordinated manner, leading to a global conformational change. For example, the ordered collective movements of one domain or more than one domain [6, 62]. In the case of involving more than one domain, we can imagine that this movement is composed of several collective (coordinated) movements, independent or not [37]. The collective one can be considered the first 20 motions.

Noncollective movements involve a small portion of the protein generally the result of a stimulus. As an example, we can mention an allosteric conformational change due to a small ligand binding. Another example that can be cited involves secondary structure transitions that result in local folding or unfolding (change in secondary structure) of the molecule, as can occur in the case of the Prion protein [22, 42].

Despite the great advances in scalable codes, graphics processing units (GPUs), and parallelization of simulation algorithms [51, 53, 56] allow us to simulate increasingly larger systems and therefore longer biological times such as entire bacterial cytoplasm in the submicrosecond range [69]. Still, for most proteins, these time scales cover a small part of the structural picture, and longer simulations are only accessible with special-purpose supercomputers like Anton (https://www.psc.edu/resources/anton-3/) [20, 63]. In addition to these technical aspects, there is a fundamental "sampling problem."

Several studies are showing that the way configuration spaces are sampled can be more critical than the simulation length. Normal Modes calculation using all-atom models or coarse-grained methods such as Elastic Network Models can predict with impressive accuracy, just from the general shape of a protein, not only the experimentally observed conformational changes but also entire sequences of intermediates structures [52]. The hybrid approaches allow for efficient conformational space sampling, considering both slow and fast movements [36]. In recent years, several hybrid methods have been developed to take into account local and global movements in conformational sampling such as MDeNM, CoMD, Dynomics, VMOD, and others [12, 36].

These methods have been used for many purposes [15, 23], such as: (i) efficiently sampling the conformational space of macromolecules; (ii) refining biomolecular complexes and assemblies; (iii) integrating experimental data for improving the determination of conformational ensembles; and (iv) conducting biomedical applications such as the study of allosteric movements, docking, and the impact of mutations.

Efficient conformational sampling is essential to identify stable structures as well as transition states of macromolecules and to calculate free energy differences [7]. Kaynak et al. [35] provided a comparative analysis of such hybrid methods developed for conformational space efficient sampling and the possible transitions between functional states [35]. They compared four methods: ClustENM [39], its recent extension ClustENMD [35], collective MD (CoMD) [27], and MD with excited NMs (MDeNM) [12]. The first three hybrid methods combine Elastic Network Models with minimizations and/or molecular dynamics to sample the conformational space of macromolecules or macromolecular complexes. The last method, MDeNM, combines normal mode calculation based on a force field (either with all atoms or just their alpha carbons) with MD to sample the conformers. In this chapter, the authors show that these methods can be more efficient than classical molecular dynamics, sampling some open, closed and intermediate states of

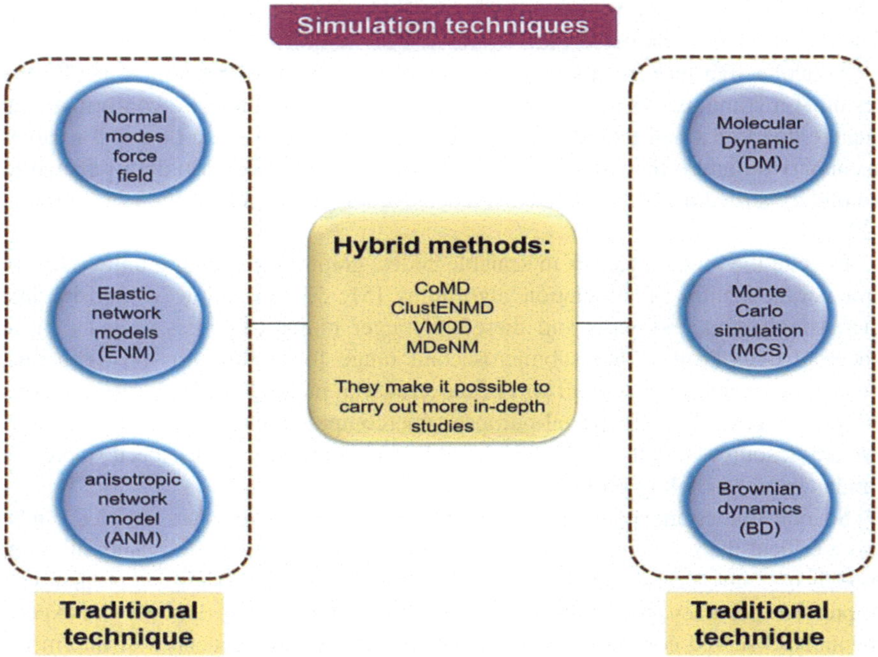

Fig. 9.2 Key combinations of methods for protein structural studies of proteins. Hybrid methods may combine elastic network model (ENM)-based or force-field (FF)-based or anisotropic network models NMA with various kinds of simulations such as Molecular Dynamics (MD) simulations, Monte Carlo and can improve the heterogeneity conformational of proteins

macromolecules. Another hybrid method that considers all-atom normal mode calculation is VMOD. The main methods present in hybrid approaches that use normal modes are summarized in Fig. 9.2.

2.1 *MDeNM*

The method MD with excited Normal Modes (MDeNM) was developed to promote large conformational changes during molecular dynamics simulations by incorporating normal modes. This technique is very useful for obtaining conformational changes that are rarely achieved by standard MD simulations unless excessively long simulations are carried out [12]. The MDeNM approach consists of a multireplica protocol designed to enhance conformational exploration in a subspace defined by a set of low-frequency normal modes, considering the dockings with the localized motions occurring within the Cartesian space [14]. A proof-of-concept study has shown that MDeNM can be a very efficient tool for performing flexible fitting of atomic structures into cryo-EM maps [13]. Each replica starts from an equilibrated initial structure whose normal modes are calculated, and the system is driven dynamically along a direction obtained by a linear combination of selected lowest-frequency modes. An efficient conformational sampling is essential for identifying stable, transient, and transition states of macromolecules and calculating free energy differences.

A recent MDeNM new version updates the direction used to move the system during simulations, taking into account the structural and energetic constraints imposed by the system itself and the environment, which allows the system to explore new paths, especially considering complex movements [14]. Without the need to recalculate the NMs like other methods (such as ClustENM, ACM, ACM-PCA, NMA-ITS), the advantage of this approach is to describe alternative paths in the conformational landscape using only a few initial modes as input.

The adaptive MDeNM (aMDeNM—[59]) has results compatible with several available methods when simulating simple large-scale movements such as a hinge or twisting movements while showing better performances considering complex movements involving reorganizations between protein domains [35]. Quiroz et al. combined the use of MDeNM with the Markov Model—PyEMMA [57] to study the effects of phosphorylation on the meta-state transition of the Human Dopamine Transporter (hDAT) and obtained relevant results [57].

Duda et al. used MDeNM to elucidate molecular mechanisms that guide the recognition of diverse substrates and inhibitors by SULT1A1. The hybrid method allowed them to explore the conformational space in a broad way for SULT1A1 in complex and to clarify a little more the mechanisms of binding of the protein to the substrate and the inhibitor [21].

2.2 *Collective Molecular Dynamics (coMD)*

The coMD method uses an Anisotropic Network Model (ANM) to guide the molecular dynamics for the system. ANM modes are selected with the help of a Monte Carlo (MC)/Metropolis scheme that allows the system to occasionally diverge from the shortest path and bypass energy barriers. This allows movements not to be limited to proceeding along a subset of low-frequency modes, which is quite interesting and may include a stochastic factor in the generation of conformations.

The use of Molecular Dynamics with a complete model of atoms at the step of the ANM allows us to generate coordinates of perfectly compacted side chain atoms and avoid nonphysical distortions in the geometry of the protein: As the ANM is defined exclusively by the contact topology between residues, changes in geometry affect the adaptively generated ANM modes in the next cycle of the simulations. Therefore, the basic approach is to deform the structure collectively along the modes predicted by the anisotropic network model, upon selecting them via a Monte Carlo/ Metropolis algorithm from among the complete pool of all accessible modes. The

CoMD method was proposed by [27] and implemented in the package Prody developed by Bahar's group [71, 72].

2.3 ClustENM and ClustENMD

ClustENM [39] is a fully automated conformational sampling method composed of multiple generations/cycles consisting of each of the following steps: (1) generating conformers by deformation along global NMs calculating using the lattice model anisotropic (ANM) [4], (2) clustering of generated conformers and (3) relaxation of cluster representatives. ANM modes are updated for each parent conformer in each generation. A series of deformations along the ± directions of some global modes (3–5) are performed, occurring a specific root mean square deviation (RMSD) for deformation (Step 1). Representative conformers selected to calculate the next generation of conformers are relaxed by energy minimization (EM) in implicit solvent (Step 3). Kurkcuoglu and Bovin discussed (pre and postdocking sampling of conformational changes using ClustENM and HADDOCK for protein–protein and protein–DNA). They applied ClustENM combined with HADDOCK software to perform pre and postdocking sampling to incorporate the effect of plasticity in protein–protein and protein–DNA complexes [28]. The authors concluded that conformational selection (presampling and docking) has the greatest impact on the quality of the final models especially in the case of protein–protein complexes. They also observed that for protein–DNA complexes, in contrast, the induced tuning stage of the pipeline (postsampling) significantly improved the quality of the final product models.

3 Ensemble Docking

Ensemble docking is a strategy that carries multiple independent molecular docking calculations (please, see Chap. 7), each of them using different target conformations, either from side-chains or larger backbone variation. This ensemble mimics the macromolecule movements, despite using individual rigid conformations. It is possible to discuss Ensemble Docking in two distinct contexts but similar in some aspects: treating problems involving complex of protein–ligand; and protein–protein interaction. These problems are important for the problem of Drug Design and/or Discovery.

Currently, protein–protein dockings also have an important guideline for success, which consists of understanding the modulation of protein–protein interactions, i.e., understanding which are primordial and which are advisory, even acting in different regions, with allosteric mechanisms. Facts like this can be observed through the stabilization of flexible loops by extensive sampling of the protein's conformational space.

Molecular dynamics have aspects that need to be constantly observed for biological systems, from adequate force fields to realistic sampling. In the study by [3], the author also mentions that an improvement in the force fields applied with the inclusion of electronic polarization together with quantum mechanics simulations also helps to provide good information about the biological system.

A computer program for coupling flexible ligands to flexible proteins using a genetic algorithm, FITTED 1.0, was developed by [11]. This tool includes the ProCESS and SMART modules which correspond to the protein/ligand configuration steps. Its validation was obtained by molecular docking of biological systems involved with protein inhibitors such as HIV-1 protease, among others.

There are two very distinct strategic approaches for Drug Design, the first consists of structure-based drug design (SBDD) and the other of ligand-based drug design (LBDD). Both can be applied to analyze the properties and characteristics of bioactive ligands. In the following session we discuss the main differences between their uses and applications in the docking of molecules.

Ensemble docking represents a molecular docking strategy in which the first step is to generate an ensemble of structural conformations of a biological target, usually obtained by molecular dynamics simulations, in which candidate molecules are subsequently docked. More than the number of couplings/target structures used, what determines the success of the approach is the selection of which structures are the most representative of the system of interest. When information such as the flexibility and thermal fluctuation of the atoms is taken into account, it is possible to explore conformational substates of these complexes that may correspond to important specific moments in the shape of the binding pockets. The need for more realistic simulations has led to the development of studies using a combination of techniques, which form the basis of ensemble docking. As reported by [3], the computational design of ligands based on the structure has been strategically improving since 1970. With computational advances, it has become possible to carry out more complete simulations, considering a greater number of structures with diverse states in terms of flexibility and the chemical space of the ligands. The author points out that some historical events have marked the field, such as in 1999 when work began to consider the flexibility of ligand pharmacophores. The main key points of the ensemble docking technique are illustrated in Fig. 9.3.

McKay et al. [43] proposed a new proposal for ensemble coupling, known as essential dynamic ensemble docking (EDED), the authors proposed this protocol for agonists of the PAC1 receptor, a class B GPCR, where three million compounds from the ZINC database were screened, with experimental validation of 23 compounds for PAC1 targeting. Its differential is that this technique can reduce the number of false negatives in screening and improve precision in the search for bioactive molecules. In this case, an important point is to obtain and select chemically significant receptor models for ligand docking. The technique consists of parameterizing the ligand with an OPLS3 force field. The trajectory frames with Principal Components (PCAs) values closest to the centers by RMSD measurement were adopted as receptors for the ensemble docking.

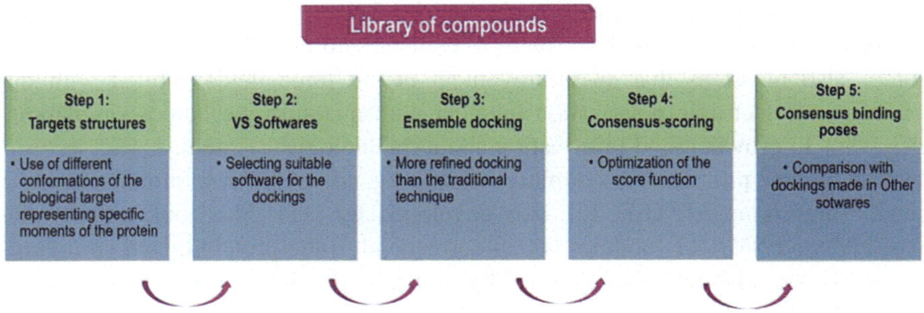

Fig. 9.3 Key points of the ensemble docking technique

The difference with other techniques is that this approach allowed us to arrive at a minimum of relevant receptors in the region of the binding pocket, as opposed to other techniques which generate many conformations of protein structures but not all of them relevant to the region in which the ligand will act, demonstrating that it is also possible to arrive at good docking results using a homology model.

Authors who generally follow the LBDD line also involve molecular activities, but related to other techniques, such as QSAR models with subsequent steps of molecular dynamics and study of postdocking normal modes. Two distinct lines can be interesting depending on the focus of each research. In the LBDD approach, the focus on knowledge generation seeks to elucidate the relationship between the structure and physicochemical attributes of molecules with their biological activity. The structure–activity relationship (SAR) considers pharmacophore models of the ligands that identify the main characteristics responsible for the activity, the quantitative structure–activity relationships (QSAR), providing quantitative predictions of activity, similarity, and combinations of chemical descriptors. In the work of Shim et al. [64], it is mentioned that 2D-QSAR is based on the calculation of physicochemical, electronic, topological and shape descriptors and is not closely related to the molecular docking technique. These are models that require preprocessing, the values are normalized and correlated, or redundant descriptors can be eliminated using the partial least squares (PLS) technique or principal component analysis (PCA), which require cross-validation, randomization or a set of external tests. The author, [64], also mentions other types of QSARs, now related to molecular docking, as they use 3D descriptors in bioactive conformations, such as in comparative molecular field analysis (CoMFA) and comparative molecular similarity indices (CoMSIA), which depend on the chemical and spatial characteristics of the molecules, such as molecular volume, surface area, solvation ΔG, dipole moments, HOMO and LUMO, based on the molecular interaction field in a 3D grid around the molecules with polar probes. The molecular docking step is used to align the ligands in their bioactive conformation. Statistical methods such as PLS or PCA are applied and related to biological activities. There is a type of model, known as 4D-QSAR, which takes flexibility into account and uses multiple conformations. 5D-QSAR, on the other hand, builds a pseudoreceptor based on information from the ligand to diversify the locations of the grid points in order to simulate a favorable induced fit of the system. Compared to 6D-QSAR, this model considers solvation energy terms. There is even another type, the 7D-QSAR, built with 3D structural data of the biological target, from X-ray crystallography or NMR, but they cannot be considered as an LBDD strategy because the starting point is not a specific descriptor of the ligand.

Pre and postdocking sampling related to conformational changes in protein–protein systems and protein–DNA systems have some very peculiar aspects and have contributed to various studies, as illustrated in the work of [38]. Simulating the dynamics of biological systems is challenging because it is necessary to consider the extent and direction of conformational changes that actually occur, it consists of both obtaining and selecting a series of conformations that represent the induced adjustment of the proteins studied, for this it is useful to apply combined methods capable of providing a conformational sampling in which the extent of the deformation is considered with its energy cost, as in the ClustENM technique—a modeling procedure based on elastic networks—using the HADDOCK software [18, 19, 67]. With regard to molecular coupling, it is possible to carry out pre and postcoupling steps as a strategy to produce models of adequate quality for the representability of biological systems.

In his work, the author [38] demonstrates the applicability of effective joint techniques, combining the generation and selection of conformational sampling with multidomain analysis of regions of interest, resulting in more realistic molecular couplings by considering the flexibility of the region, which can be applied to protein-protein or protein-DNA studies if the technique known as ClustENM-HADDOCK is used. So far, this technique is considered to have superior performance compared to traditional rigid coupling for protein-protein systems, but inferior in terms of the flexibility of the multidomain region. As far as the protein-DNA system is concerned, the simulations prove to be adequate in most cases.

In addition to raising discussion about the gaps in the molecular docking technique, the author [38], proposed an interesting pipeline to meet the challenges of the area, where he combined simulations before and after complexation, containing the following steps: (1) energy minimization of the structure subjected to elastic network modeling, with ClustENM and deformation test to categorize the RMSD deformation; (2) generation of conformers in which clusters with parent structures are excluded; (3) selection of 20 conformers based on the minimization energy, to perform the coupling, whereby, the initial structure is also present regardless of the energy value; (4) use of HADDOCK for the semi-flexible docking of structures; (5) selection of the best scoring coupled model for the second round of ClustENM (which is considered post-sampling); (6) verification and parameterization of ClustENM for each structure; (7) generation of final models, obtained after applying the technique for two more generations; (8) verification and classification of the quality of the final models, using the HADDOCK score.

The combined use of techniques that provide simulations of biological systems containing a longer profile, that is conformations before and after a molecular docking, has also been applied for large-scale docking (virtual screenings), as demonstrated in the work of [10]. A study was conducted with the dihydroorotate dehydrogenase (DHODH) protein, which is considered a key enzyme in the pyrimidine biosynthesis pathway and is involved in processes related to cancers, and some autoimmune and viral diseases.

The study by [10] followed important steps that provided good results: (1) selection and preparation of 38 hDHODH structures complexed with inhibitors of resolution <2 Å, superimposed with the ICM [1], with selection of the representative set of structures, based on clustering by the root mean square deviation (RMSD) with the aid of the bio3D R software [24]. The parameterization included the addition of polar hydrogen atoms corresponding to protonation at pH $= 7$, using OpenBabel 2.4.0 [50]. (2) Benchmarking of datasets and evaluation metrics, formation of a set/library using enhanced data (DUD-E) [47], with 67 active compounds and 1933 inactive compounds. The Boltzmann enhanced ROC curve (BEDROC, with $\alpha = 20$) was used and enrichment factors (EF) were calculated using the balancer [40].

The software and molecular docking parameters used were AutoDock Vina [66], LeDock [70], rDOCK [61], and ICM [1]. These tools were chosen because of their high overall performance, differences in the methods used to predict/score the poses generated, and availability. The consensus stage for generating various conformations of the ligand is just as important as the richness of generating the various conformations of the target protein used. According to the author's data, the success of the set fitting and consensus scoring lies precisely in the way the sets are selected and analyzed, with different conformations, not forgetting to normalize the fitting score. As for the consensus bond pose approach, all possible combinations of molecular docking were evaluated, and the compound that scored well in each docking program was considered the final candidate. The authors point out that, whenever possible, it is recommended to work with pre-validated data and the selection of suitable software and structures (representative of the system of interest) to increase the performance of simple molecular docking and virtual screening.

According to [22], a tool that can improve molecular docking results is HTP SurflexDock, which allows the simulation of the receptor's implicit flexibility. Considering a post-processing step that allows the reclassification of favorable compounds by exploring the conformational space and free energy of binding by MM/PBSA (Molecular mechanics Poisson-Boltzmann surface area). This is a suitable tool when information on the receptor's contact area is incomplete or specific optimization is required for a given ligand.

Considering a representative set of conformations of the biological target before molecular docking makes it possible to simulate the implicit flexibility of the receptor; however, to correct the accuracy of the scoring functions of the compounds, a post-processing step is necessary which explores the conformational space of the active site, generates new poses and estimates the binding free energy of the protein-ligand interaction, such as the MM/PBSA method, which takes into account the relative binding free energy obtained from the protein-ligand molecular dynamics in aqueous solvent, the change in potential energy in vacuum, the desolvation energy of the system, and the entropy of the complex in the gas phase. It is a well-known method for rescoring molecules.

With the HTP SurflexDock, it is possible to load a three-dimensional structure of the receptor and a set of small molecules (ligands) in PDBQT format. The molecules are classified according to the ΔG calculated for the ensemble, an ensemble with the original structure of the receptor is then built and three conformations are obtained from five-nanosecond molecular dynamics using Gromacs 5.1.5 [2] (a relaxation for other conformations), the trajectory is followed by clustering, and the conformations with a maximum RMSD of the binding site of 0.10–0.20 nm and grouped by the GROMOS algorithm [16]. A representative structure from the three most representative groups is included in the new set. AutoDock [49] and ADT scripts [46] are used, and ten poses are generated for the complexes [45]. HTP SurflexDock presents a score table with the ΔG of the best pose, and it is possible to apply two post-processing options: (I) select 10 compounds and expand the conformational space for a new molecular docking experiment cycle; (II) explore the conformational space of 30 new poses.

The latest version of DockThor-VS (available at: www.dockthor.lncc.br, last accessed on July, 2024) provides users with 3D structures for performing molecular docking of various ligands in selected structures of the wild-type and mutated proteins: Nsp3; Nsp5; Nsp12; Nsp15; and Spike. Encouraging the importance of using several conformations of the biological target in virtual screening, as according to the authors, using a single conformation of the protein would generate an impaired classification of the ligands in the complexes. A list of some interesting ensemble docking tools is presented in Table 9.1.

Another interesting study [65], which presents a protocol for generating various conformations of cyclin-dependent protein kinase 2 (CDK2), used normal mode analysis (DIMB [54] with harmonic constraints during minimization, followed by shifting by 25 lower frequency vectors until the mass-weighted mean square deviation (MRMSD) to 2 Å or –2 Å. The authors selected the structures for their conformational diversity and binding site topology, corresponding to five open conformations with more space in the protein's binding site to docking ligands. An innovative study [21] used hybrid molecular dynamics methods, and excited normal modes to obtain conformations of sulfotransferases (SULT) for docking molecular 132 ligands, helping to understand the mechanism of molecule binding at the site of this protein. Some studies involving proteins can be more structural, with a focus on understanding the structure itself and which regions are most important, such as the one conducted by Yang et al., [68], or more applied, such as the one by [41], where the focus is on demonstrating the impact of the composition of the protein structure. Both are relevant. Dudas et al. performed MD and MDeNM simulations for the SULT1A1/PAPS as well as MD and docking simulations with the substrate's estradiol and fulvestrant. The authors demonstrated that large conformational changes of the PAPS-bound SULT1A1 can occur and would be sufficient to accommodate large substrates, e.g. fulvestrant, independently of the co-factor movements. In this work, the structural displacements were successfully detected by the MDeNM simulations and suggest that a wider range of drugs could be recognized by PAPS-bound SULT1A1. The Hybrid method called MDeNM enables an extended sampling of the conformational space by running multiple

Table 9.1 Main tools of the ensemble docking technique

Tool ensemble docking	Key points	Docking	References
HEX	Combined strategy between molecular dynamics and molecular docking with the program HEX, which performs a systematic search of 6 degrees of freedom and classification of orientations by interaction energy	Protein-protein	Ritchie et al. [60]
HTP SurflexDock	It has two strategies: ensemble coupling with implicit receptor flexibility simulation and post-processing step with new scoring of promising compounds considering the conformational space or estimating the binding free energy using the MM/PBSA protocol	Protein-ligand/protein-protein	Filho et al. [22]
EDock-ML	It uses machine learning to decide whether compounds are good candidates for docking with different receptor conformations, considering the receptor's flexibility	Protein-ligand	Chandak et al. [9]
MDock	Automated docking that can dock ligands into multiple conformations of the receptor, automated, using an ensemble docking algorithm [47, 48]	Protein-ligand	Huang et al. [29] Huang et al. [30] Huang et al. [31] Huang et al. [32] Huang et al. [33] Huang et al. [34]
MDR SurFlexDock	Uses for molecular docking a discrete/representative set of receptors contact surfaces obtained from the clustering of molecular simulation trajectories, for simulation of the intrinsic flexibility of the ligand contact surface, the classification of ligands can be by the inhibition constant (Ki) and is also applicable to receptors originated by homology modeling	Protein-ligand	De Almeida Filho et al. [17]
DockThor-VS	It uses structures of different conformations complexed with ligands, taking into account the flexibility of the receptor and different states of protonation	Protein-ligand	Guedes et al. [25, 26]

short MD simulations during which motions described by a subset of low-frequency Normal Modes are kinetically excited. It was possible to detect "open"-like conformations of SULT1A1 efficiently.

Finally, there are several hybrid methods proposed in the literature; some of them integrate simulation methods such as molecular dynamics, Monte Carlo, optimization with elastic network models or normal modes analysis based on force field as MDeNM, CoMD, ClustENM, CLusteNMD and VMOD [14, 35, 42, 55]. These methods can be combined with the software listed in Table 9.1 which allows us to obtain conformations of protein structures that are representative of interesting states for the molecular coupling of bioactive ligands. Using structures from different points, considering the flexibility of a biological target, brings more realism to computer simulations, and is a good strategy for understanding the mechanism of action in the target and developing new molecules.

4 Good Practices for Simulations and Sample the Conformational Space

Molecular simulation techniques play a crucial role in our quest to understand and predict the properties, structure, and function of molecular systems. They are a fundamental tool as we aim to enable predictive molecular design. Simulation methods are useful for studying the structure and dynamics of complex systems that are too complicated for traditional theoretical approaches, helping to interpret experimental data in terms of molecular movements. We list some questions that can be important to plan your simulation on Table.

Questions to plan the simulations

? Guiding questions for implementing conformational sampling in drug design
What is the quality of your initial structures or models?
What are the protonation and phosphorylation conditions for your simulation?
How do you scale your simulation with multiple CPUs/GPUs?
Which force field should be used?
Which kind of motions is important for your problem: local motions or global motions?
Do I have computational resource and time to use explicit solvent and all-atoms models?
How can I optimize the parallelization of the simulations?
How representative are the structures obtained when you project them against principal components space?
How diverse is your sampling of the conformational space in terms of structural (RMSD) and energetic aspects?

References

1. Abagyan, R., Totrov, M., & Kuznetsov, D. (1994). ICM - A new method for protein modeling and design: Applications to docking and structure prediction from the distorted native conformation. *Journal of Computational Chemistry, 15*, 488–506.
2. Abraham, M. J., Murtola, T., Schulz, R., et al. (2015). GROMACS: High performance molecular simulations through multi-level parallelism from laptops to supercomputers. *SoftwareX, 1*, 19–25.

3. Amaro, R. E., Baudry, J., Chodera, J., Demir, Ö., McCammon, J. A., Miao, Y., & Smith, J. C. (2018). Ensemble docking in drug discovery. *Biophysical Journal, 114*(10), 2271–2278.
4. Atilgan, C. (2018). Computational methods for efficient sampling of protein landscapes and disclosing allosteric regions. *Advances in Protein Chemistry and Structural Biology, 113*, 33–63. . Epub 2018 Jul 25. PMID: 30149905.
5. Bahar, I., Lezon, T. R., Yang, L. W., & Eyal, E. (2010). Global dynamics of proteins: bridging between structure and function. *Annual Review of Biophysics, 39*, 23–42.
6. Bahar, I., Jernigan, R., & Dill, K. (2017). *Protein actions: Principles & modeling*. Garland Science, Taylor & Francis Group. ISBN: 9780815341772.
7. Black, K. A., He, S., Jin, R., Miller, D. M., Bolla, J. R., Clarke, O. B., Johnson, P., Windley, M., Burns, C. J., Hill, A. P., Laver, D., Robinson, C. V., Smith, B. J., & Gulbis, J. M. (2020). A constricted opening in Kir channels does not impede potassium conduction. *Nature Communications, 11*(1), 3024.
8. Callaway, E. (2020). Revolutionary cryo-EM is taking over structural biology. *Nature, 578* (7794), 201.
9. Chandak, T., & Wong, C. F. (2021). EDock-ML: A web server for using ensemble docking with machine learning to aid drug discovery. *Protein Science, 30*(5), 1087–1097. . Epub 2021 Mar 25.
10. Chilingaryan, G., Abelyan, N., Sargsyan, A., et al. (2021). Combination of consensus and ensemble docking strategies for the discovery of human dihydroorotate dehydrogenase inhibitors. *Scientific Reports, 11*, 11417.
11. Corbeil, C. R., Englebienne, P., & Moitessier, N. (2007). Docking ligands into flexible and solvated macromolecules. 1. Development and validation of FITTED 1.0. *Journal of Chemical Information and Modeling, 47*(2), 435–449.
12. Costa, M. G. S., Batista, P. R., Bisch, P. M., & Perahia, D. (2015). Exploring free energy landscapes of large conformational changes: Molecular dynamics with excited normal modes. *Journal of Chemical Theory and Computation, 60*(5), 2419–2423.
13. Costa, M. G. S., Fagnen, C., Vénien-Bryan, C., & Perahia, D. (2020). A new strategy for atomic flexible fitting in Cryo-EM maps by molecular dynamics with excited normal modes (MDeNM-EMFit). *Journal of Chemical Information and Modeling, 60*(5), 2419–2423.
14. Costa, M. G. S., Batista, P. R., Gomes, A., Bastos, L. S., Louet, M., Floquet, N., Bisch, P. M., & Perahia, D. (2023). MDexciteR: Enhanced sampling molecular dynamics by excited normal modes or principal components obtained from experiments. *Journal of Chemical Theory and Computation.*
15. Das, A., Gur, M., Cheng, M. H., Jo, S., Bahar, I., & Roux, B. (2014). Exploring the conformational transitions of biomolecular systems using a simple two-state anisotropic network model. *PLoS Computational Biology, 10*(4), e1003521.
16. Daura, X., Gademann, K., Jaun, B., et al. (1999). Peptide folding: When simulation meets experiment. *Angewandte Chemie International Edition, 38*(1–2), 236–240.
17. De Almeida Filho, J. L., & Fernandez, J. H. (2020). MDR SurFlexDock: A semi-automatic webserver for discrete receptor-ensemble docking. In L. Kowada & D. de Oliveira (Eds.), *Advances in bioinformatics and computational biology. BSB 2019* (Lecture Notes in Computer Science) (Vol. 11347). Springer.
18. De Vries, S. J., van Dijk, M., & Bonvin, A. M. (2010). The HADDOCK web server for data-driven biomolecular docking. *Nature Protocols, 5*(5), 883–897.
19. Dominguez, C., Boelens, R., & Bonvin, A. M. J. J. (2003). HADDOCK: A protein-protein docking approach based on biochemical or biophysical information. *Journal of the American Chemical Society, 125*(7), 1731–1737.
20. Dror, R. O., Dirks, R. M., Grossman, J. P., Xu, H., & Shaw, D. E. (2012). Biomolecular simulation: a computational microscope for molecular biology. *Annual Review of Biophysics, 41*, 429–452.

21. Dudas, B., Toth, D., Perahia, D., Nicot, A. B., Balog, E., & Miteva, M. A. (2021). Insights into the substrate binding mechanism of SULT1A1 through molecular dynamics with excited normal modes simulations. *Scientific Reports, 11*, 13129.
22. Filho, E. D. (org.), Scott, A. L., Philot, E. A., Lima, A. N. (2019). *Métodos computacionais no estudo de macromoléculas biológicas*, 2nd ed. (pp. 69–90). Editora Livraria da Física.
23. Fratev, F., & Sirimulla, S. (2019). An improved free energy perturbation FEP+ sampling protocol for flexible ligand-binding domains. *Scientific Reports, 9*(1).
24. Grant, B. J., Rodrigues, A. P. C., ElSawy, K. M., McCammon, J. A., & Caves, L. S. D. (2006). Bio3d: An R package for the comparative analysis of protein structures. *Bioinformatics, 22*, 2695–2696.
25. Guedes, I. A., et al. (2021). Drug design and repurposing with DockThor-VS web server focusing on SARS-CoV-2 therapeutic targets and their non-synonym variants. *Scientific Reports, 11*, 5543.
26. Guedes, I. A., Barreto, A. M. S., Marinho, D., Krempser, E., Kuenemann, M. A., Sperandio, O., Dardenne, L. E., & Miteva, M. A. (2021). New machine learning and physics-based scoring functions for drug discovery. *Scientific Reports, 11*(1), 3198.
27. Gur, M., Zomot, E., & Bahar, I. (2013). Global motions exhibited by proteins in micro- to milliseconds simulations concur with anisotropic network model predictions. *The Journal of Chemical Physics, 139*, 121912.
28. Haliloglu, T., & Bahar, I. (2015). Adaptability of protein structures to enable functional interactions and evolutionary implications. *Current Opinion in Structural Biology, 35*, 17–23.
29. Huang, S. Y., & Zou, X. Q. (2006). An iterative knowledge-based scoring function to predict protein-ligand interactions: I. Derivation of interaction potentials. *Journal of Computational Chemistry, 27*, 1866–1875.
30. Huang, S. Y., & Zou, X. Q. (2006). An iterative knowledge-based scoring function to predict protein-ligand interactions: II. Validation of the scoring function. *Journal of Computational Chemistry, 27*, 1876–1882.
31. Huang, S. Y., & Zou, X. Q. (2007). Ensemble docking of multiple protein structures: Considering protein structural variations in molecular docking. *Proteins, 66*, 399–421.
32. Huang, S. Y., & Zou, X. Q. (2007). Efficient molecular docking of NMR structures: Application to HIV-1 protease. *Protein Science: A Publication of the Protein Society, 16*, 43–51.
33. Huang, S. Y., & Zou, X. Q. (2008). An iterative knowledge-based scoring function for protein-protein recognition. *Proteins, 72*, 557–579.
34. Huang, S. Y., & Zou, X. Q. (2013). A non-redundant structure dataset for benchmarking protein-RNA computational docking. *Journal of Computational Chemistry*.
35. Kaynak, B. T., Krieger, J. M., Dudas, B., Dahmani, Z. L., Costa, M. G., Balog, E., Scott, A. L., Doruker, P., Perahia, D., & Bahar, I. (2022). Sampling of protein conformational space using hybrid simulations: A critical assessment of recent methods. *Frontiers in Molecular Biosciences, 9*, 832847.
36. Krieger, J. M., Doruker, P., Scott, A. L., Perahia, D., & Bahar, I. (2020). Towards gaining sight of multiscale events: Utilizing network models and normal modes in hybrid methods. *Current Opinion in Structural Biology, 64*, 34–41.
37. Krieger, J. M., Sorzano, C. O. S., Carazo, J. M., & Bahar, I. (2022 Apr 1). Protein dynamics developments for the large scale and cryoEM: case study of ProDy 2.0. *Acta Crystallographica Section D: Structural Biology, 78*(Pt 4), 399–409.
38. Kurkcuoglu, Z., & Bonvin, A. M. J. J. (2020). Pre- and post-docking sampling of conformational changes using ClustENM and HADDOCK for protein-protein and protein-DNA systems. *Proteins, 88*(2), 292–306.
39. Kurkcuoglu, Z., Bahar, I., & Doruker, P. (2016). ClustENM: ENM-based sampling of essential conformational space at full atomic resolution. *Journal of Chemical Theory and Computation, 12*, 4549–4562.

40. Lätti, S., Niinivehmas, S., & Pentikäinen, O. T. (2016). Rocker: Open source, easy-to-use tool for AUC and enrichment calculations and ROC visualization. *Journal of Cheminformatics, 8*, 45.
41. Li, H., Chang, Y. Y., Lee, J. Y., Bahar, I., & Yang, L. W. (2017). DynOmics: Dynamics of structural proteome and beyond. *Nucleic Acids Research, 45*(W1), W374–W380.
42. Lima, A. N., de Oliveira, R. J., Braz, A. S. K., de Souza Costa, M. G., Perahia, D., & Scott, L. P. B. (2018). Effects of pH and aggregation in the human prion conversion into scrapie form: A study using molecular dynamics with excited normal modes. *European Biophysics Journal, 47* (5), 583–590.
43. McKay, K., Hamilton, N. B., Remington, J. M., Schneebeli, S. T., & Li, J. (2022). Essential dynamics ensemble docking for structure-based GPCR drug discovery. *Frontiers in Molecular Biosciences, 9*, 879212.
44. Mitra, A. K. (2019). Visualization of biological macromolecules at near-atomic resolution: Cryo-electron microscopy comes of age. *Acta Crystallographica. Section F, Structural Biology Communications, 75*(Pt 1), 3–11.
45. Morris, G. M., Goodsell, D. S., Halliday, R. S., et al. (1998). Automated docking using a Lamarckian genetic algorithm and an empirical binding free energy function. *Journal of Computational Chemistry, 19*(14), 1639–1662.
46. Morris, G. M., Huey, R., Lindstrom, W., et al. (2009). AutoDock4 and AutoDockTools4: Automated docking with selective receptor flexibility. *Journal of Computational Chemistry, 30* (16), 2785–2791.
47. Mysinger, M. M., Carchia, M., Irwin, J., & Shoichet, B. K. (2012). Directory of useful decoys, enhanced (DUD-E): Better ligands and decoys for better benchmarking. *Journal of Medicinal Chemistry, 55*, 6582–6594.
48. Nogales, E. (2016). The development of cryo-EM into a mainstream structural biology technique. *Nature Methods, 13*, 24–27.
49. Norgan, A. P., Coffman, P. K., Kocher, J.-P. A., Katzmann, D. J., & Sosa, C. P. (2011). Multilevel parallelization of AutoDock 4.2. *Journal of Cheminformatics, 3*(1), 12.
50. O'Boyle, N. M., et al. (2011). Open babel: An open chemical toolbox. *Journal of Cheminformatics, 3*, 33.
51. Orellana, L. (2019). Large-scale conformational changes and protein function: Breaking the in silico barrier. *Frontiers in Molecular Biosciences, 6*.
52. Orellana, L., Yoluk, O., Carrillo, O., et al. (2016). Prediction and validation of protein intermediate states from structurally rich ensembles and coarse-grained simulations. *Nature Communications, 7*, 12575.
53. Páll, S., Abraham, M. J., Kutzner, C., Hess, B., & Lindahl, E. (2015). Tackling exascale software challenges in molecular dynamics simulations with GROMACS. In S. Markidis & E. Laure (Eds.), *Solving software challenges for exascale. EASC 2014. Lecture notes in computer science* (Vol. 8759). Springer.
54. Perahia, D., & Mouawad, L. (1995). Computation of low-frequency normal modes in macromolecules: Improvements to the method of diagonalization in a mixed basis and application to hemoglobin. *Computers & Chemistry, 19*(3), 241–246.
55. Philot, E. A., Perahia, D., Braz, A. S., Costa, M. G., & Scott, L. P. (2013). Binding sites and hydrophobic pockets in Human Thioredoxin 1 determined by normal mode analysis. *Journal of Structural Biology, 184*(2), 293–300.
56. Pouya, I., Pronk, S., Lundborg, M., & Lindahl, E. (2017). Copernicus, a hybrid dataflow and peer-to-peer scientific computing platform for efficient large-scale ensemble sampling. *Future Generation Computer Systems, 71(-)*, 18–31.
57. Quiroz, R. C. N., Philot, E. A., General, I. J., Perahia, D., & Scott, A. L. (2023). Effect of phosphorylation on the structural dynamics, thermal stability of human dopamine transporter: A simulation study using normal modes, molecular dynamics and Markov State Model. *Journal of Molecular Graphics & Modelling, 118*, 108359.

58. Resende-Lara, P. T., Perahia, D., Scott, A. L., & Braz, A. S. K. (2020). Unveiling functional motions based on point mutations in biased signaling systems: A normal mode study on nerve growth factor bound to TrkA. *PLoS One., 15*(6), e0231542.
59. Resende-Lara, P. T., Costa, M. G. S., Dudas, B., & Perahia, D. (2022). Adaptive collective motions: A hybrid method to improve conformational sampling with molecular dynamics and normal modes. biorxiv [Preprint]. Available at: https://www.biorxiv.org/.
60. Ritchie, D. W., & Kemp, G. J. (2000). Protein docking using spherical polar Fourier correlations. *Proteins, 39*(2), 178–194.
61. Ruiz-Carmona, S., et al. (2014). rDock: A fast, versatile and open source program for docking ligands to proteins and nucleic acids. *PLoS Computational Biology, 10*, e1003571.
62. Schay, G., Herényi, L., Fidy, J., & Osváth, S. (2013). Role of domain interactions in the collective motion of phosphoglycerate kinase. *Biophysical Journal, 104*(3), 677–682.
63. Shaw, D. E., Dror, R. O., Salmon, J. K., et al. (2009). Millisecond-scale molecular dynamics simulations on Anton. In *Proceedings of the Conference on High Performance Computing Networking, Storage and Analysis (SC09), Portland, OR, USA* (pp. 1–11).
64. Shim, J., & Mackerell, A. D., Jr. (2011). Computational ligand-based rational design: Role of conformational sampling and force fields in model development. *MedChemComm, 2*(5), 356–370.
65. Sperandio, O., Mouawad, L., Pinto, E., Villoutreix, B. O., Perahia, D., & Miteva, M. A. (2010). How to choose relevant multiple receptor conformations for virtual screening: A test case of Cdk2 and normal mode analysis. *European Biophysics Journal, 39*(9), 1365–1372.
66. Trott, O., & Olson, A. J. (2010). AutoDock Vina: Improving the speed and accuracy of docking with a new scoring function, efficient optimization, and multithreading. *Journal of Computational Chemistry, 31*, 455–461.
67. van Zundert, G. C. P., Rodrigues, J. P. G. L. M., Trellet, M., Schmitz, C., Kastritis, P. L., Karaca, E., Melquiond, A. S. J., van Dijk, M., de Vries, S. J., & Bonvin, A. M. J. J. (2016). The HADDOCK2.2 web server: User-friendly integrative modeling of biomolecular complexes. *Journal of Molecular Biology, 428*(4), 720–725.
68. Yang, L., Song, G., & Jernigan, R. L. (2007). How well can we understand large-scale protein motions using normal modes of elastic network models? *Biophysical Journal, 93*(3), 920–929. . Epub 2007 May 4.
69. Yu, I., Mori, T., Ando, T., Harada, R., Jung, J., Sugita, Y., & Feig, M. (2016). Biomolecular interactions modulate macromolecular structure and dynamics in atomistic model of a bacterial cytoplasm. *Elife, 5*, e19274.
70. Zhang, N., & Zhao, H. (2016). Enriching screening libraries with bioactive fragment space. *Bioorganic & Medicinal Chemistry Letters, 26*, 3594–3597.
71. Zhang, S., Krieger, J. M., Zhang, Y., Kaya, C., Kaynak, B., Mikulska-Ruminska, K., Doruker, P., Li, H., & Bahar, I. (2021). ProDy 2.0: Increased scale and scope after 10 years of protein dynamics modelling with Python. *Bioinformatics, 37*(20), 3657– 3659.
72. Zhang, Y., Krieger, J., Mikulska-Ruminska, K., Kaynak, B., Sorzano, C. O. S., Carazo, J.-M., Xing, J., & Bahar, I. (2021). State-dependent sequential allostery exhibited by chaperonin TRiC/CCT revealed by network analysis of Cryo-EM maps. *Progress in Biophysics and Molecular Biology, 160*, 104–120.

Chapter 10
Free Energy Perturbation and Free-Energy Calculations Applied to Drug Design

Deborah Antunes, Lucianna Helene Santos, Ana Carolina Ramos Guimarães, and Ernesto Raul Caffarena

Abstract Free energy perturbation (FEP) is a computational technique used to evaluate ligand-protein binding affinities for computer-aided drug optimization. FEP has been shown to be a valuable tool in both academic and pharmaceutical settings for optimizing drug candidates in a rational, efficient, and cost-effective manner. Recent advancements in algorithms, software tools, hardware capabilities, and machine learning integration have significantly improved the scope, applicability, and reliability of FEP calculations. In this chapter, we review recent developments in force field parameterization, software platforms, and automated workflows that have consolidated FEP as an essential methodology for structure-based drug discovery and have resulted in FEP calculations becoming more accessible to non-specialists, as well as applicable to a broad range of scenarios. We also describe the utility of the FEP technique in diverse contexts, including validating the binding modes and optimizing allosteric and covalent inhibitors. We illustrate its potential through vignettes and in-depth case studies, demonstrating its integration into machine-learning frameworks for predicting binding energies based on molecular structures. Furthermore, this chapter discusses the remaining challenges in sampling sufficiency and scalability to ultra-large compound libraries as well as emerging solutions through cloud computing and machine learning.

Keywords Free energy perturbation · Drug design · Bbinding affinity · Computational chemistry · Machine learning · Molecular structures

D. Antunes · A. C. R. Guimarães
Oswaldo Cruz Institute, Rio de Janeiro, Brazil

L. H. Santos (✉)
Institut Pasteur de Montevideo, Montevideo, Uruguay
e-mail: lsilva@pasteur.edu.uy

E. R. Caffarena (✉)
Scientific Computing Program, Rio de Janeiro, Brazil
e-mail: ernesto.caffarena@fiocruz.br

V. G. Maltarollo (ed.), *Computer-Aided and Machine Learning-Driven Drug Design*, Computer-Aided Drug Discovery and Design 3,
https://doi.org/10.1007/978-3-031-76718-0_10

1 Introduction

Drug design is the process of creating new pharmaceutical compounds or optimizing existing ones to target specific biological molecules or disease-related processes in the human body. Its main goal is to develop safe and effective medications that interact with specific molecular targets, such as proteins, enzymes, or receptors, and modulate their activity to treat or prevent diseases.

The drug discovery process consists of several crucial stages including target identification and validation, hit identification, hit expansion, hit-to-lead, and lead optimization. In the first stage, target identification and validation, the focus is on identifying the target molecule, usually a protein or another molecule involved in the disease process. The second stage, hit identification, involves testing thousands of compounds to identify potential candidates for therapeutic treatment. Compounds are evaluated through high-throughput screening and other methods to determine their efficacy against the target. In the hit expansion stage, promising compounds from the hit identification phase undergo further evaluation and testing to expand their potential as lead candidates. This stage involves refining the selection of compounds based on their activity, selectivity, and other desirable properties. The hit-to-lead phase involves optimizing the selected compounds to enhance their affinity, selectivity, efficacy, metabolic stability, and oral bioavailability. Lead optimization is the final stage, in which lead compounds undergo further optimization to improve their drug-like properties and enhance their potential as viable drug candidates. This phase involves iterative cycles of medicinal chemistry to refine the lead compounds and address issues related to their efficacy, safety, and pharmacokinetic properties.

The utilization of free-energy methods can yield essential thermodynamic and kinetic data through rigorous computational approaches. However, until a few years ago, high computing costs, sampling method restrictions, force field limitations, and a lack of automation prevented the widespread use of free-energy methods.

Over the last decade, considerable emphasis has been placed on enhancing the efficiency and feasibility of these approaches using workflows that utilize modern CPU- and GPU-based architectures. Therefore, from both academic and industrial standpoints, the utilization of free-energy calculation workflows has emerged as a compelling computational tool to facilitate the progress of drug discovery.

Today, scientists use a variety of methods to calculate drug design binding free energies. These include alchemical calculations, endpoint methodologies, empirical scoring functions, knowledge-based potentials, and quantum mechanics. Free energy perturbation (FEP) and thermodynamic integration (TI) methods are considered the "gold standard" for accurate in silico potency predictions, as they use alchemical transformations to calculate the relative binding free energies between two ligands. Second-group methods, such as molecular mechanics Poisson-Boltzmann surface area (MM-PBSA) and generalized Born surface area (MM-GBSA, see Chap. 8), estimate the binding free energy from the difference between bound and unbound free energies, making them computationally less expensive than alchemical methods [1].

Empirical scoring functions can also be used to evaluate the binding affinity using weighted terms representing hydrogen bonding, van der Waals interactions, and desolvation [2]. While they are fast, they are less accurate than other approaches. Additionally, knowledge-based statistical potentials can be derived from the analysis of known protein–ligand complex structures, which are computationally efficient but require structural data [3]. Finally, quantum mechanical methods such as the fragment molecular orbital (FMO) method use quantum mechanical calculations to evaluate the binding free energy, making them more precise but computationally intensive than classical force fields [4].

Other techniques utilize enhanced sampling of the free energy landscape, such as Metadynamics, ABF, and umbrella sampling, to explore the complex energy landscapes of biomolecular systems. These methods are more effective than standard molecular dynamics simulations for overcoming the limitations imposed by rare events [5].

Out of these above-mentioned methods, the methodology called Free Energy Perturbation (FEP) plays a significant role in the computational drug design step, helping researchers select and optimize potential drug candidates with improved binding properties and selectivity.

In drug design, FEP has emerged as a valuable tool for estimating the binding free energy of ligands to proteins, which is a critical determinant of their potential as successful drug candidates. This represents a robust and thermodynamically rigorous computational approach capable of predicting the binding affinity of small molecules, given that there is a direct relationship between the free energy and affinity of a ligand, as shown in Eq. (10.1):

$$\Delta G_{\text{bind}} = -kT \ln K_i \tag{10.1}$$

where ΔG represents the binding free energy, calculated as the difference in free energy between the bound and unbound states of the ligand–target complex, k is the Boltzmann constant, K_i is the inhibition constant (although the dissociation constant K_d could also be used), and T is the absolute temperature.

FEP's application in drug design extends the exploration and optimization of potential drug candidates across a vast chemical space, thereby facilitating the enhancement of multiple properties. Moreover, FEP's computationally driven approach can significantly reduce the cost and time associated with experimental testing by enabling the screening of less promising candidates in silico.

Many theoretical methodologies have significantly paved the way for the development of FEP, particularly prior to the seminal computational breakthroughs. Although most scientific sources attribute the FEP method to Zwanzig's [6], previous works by Kirkwood [7] and De Donder [8] contributed to setting the theoretical basis by introducing the generalized-extent parameter (λ), which reconciles statistical mechanics with the degree of evolution in a chemical reaction. Zwanzig derived an expression for the free energy difference between the two states of a system in terms of the probability distribution of the perturbation energy. The free energy difference for going from state A to state B is obtained from the Zwanzig equation (Eq. 10.2)

$$\Delta F(A \rightarrow B) = F_B - F_A = -k_B T \ln \left\langle \exp\left(\frac{U_B - U_A}{k_B T}\right) \right\rangle_A, \tag{10.2}$$

where F denotes the free energy, U represents the internal energy of the system, T is the temperature, k_B is Boltzmann's constant, and the angular brackets denote the average over a simulation run for state A. Briefly, in statistical mechanics, the phase space encompasses every conceivable arrangement of positions and momenta for atoms in a simulated system [9]. FEP utilizes a succession of intermediate overlapping states, referred to as λ-windows, which are determined by a coupling parameter λ that connects the potential energy functions of states A and B. The transition from A to B is accomplished by incrementally adjusting λ from 0 to 1 in a specified number of discrete steps, both forward and backward.

Subsequently, Valleu and Card [10] introduced a stratification strategy that connects the reference and target states, breaking the total free-energy difference into the sum of the finite free-energy differences between the intermediate states with increased overlap. In 1976 [10], Bennett independently developed a method called the Bennett Acceptance Ratio (BAR) and improved the efficiency and reliability of FEP calculations using a weighted average of forward and reverse transformations.

Soon after the beginning of the 1980s, almost 30 years after Zwanzig's equation was presented to the scientific community, the first successful attempt to use the FEP appeared. Postma et al., in 1982 [11], reported FEP calculations on forming a cavity in an explicit water simulation box. In 1985, Jorgensen et al. [12] used alchemical transformations of alkanes into alcohols to calculate their hydration-free energies with high precision. This method involves gradually transforming the ligand into the final compound, and then calculating the free energy change associated with each transformation step. This work is considered a cobblestone of FEP application in drug design because it hints at its potential given that the relative solvation-free energies play a major role in determining the relative binding free energy of two ligands at a common receptor site.

The introduction of the OPLS force field by Jorgensen and Tirado-Rives in 1988 [13] improved the accuracy and transferability of the molecular simulations. Similarly, other force fields developed simultaneously have incorporated the FEP methodology into the calculations [14, 15]. In the early 1990s, Kollman et al. [16] employed FEP methodology integrated into the AMBER program to investigate the binding energies of a range of thermolysin inhibitors.

Among the many contributions that consolidate Structure-Based Drug Discovery, such as the understanding of molecular interactions and the design of novel drug candidates based on structural information or molecular docking, which cleared the way for virtual screening, FEP calculations can be considered as one of the early and fundamental contributions to the field. Extensive reviews have thoroughly detailed the development timeline for a comprehensive historical account of FEP methodology [10, 11].

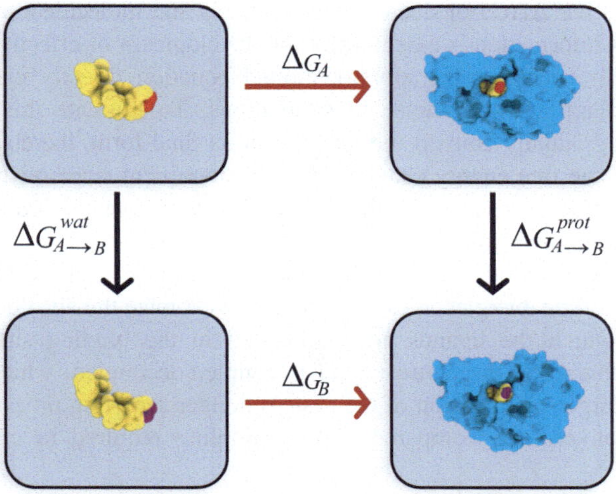

Fig. 10.1 The thermodynamic cycle in which molecule *A* alchemically transforms into molecule *B* at the protein binding site and the solvent (water)

From the beginning of 2000, the FEP technique has demonstrated a persistent trajectory of evolution and enhancement. Researchers have developed novel methodologies for the computation of ligand-protein binding free energy, thereby innovating approaches that harness FEP for novel drug design. This escalating computational potency has propelled FEP to become an indispensable instrument in the realm of drug discovery.

The predominant approach for implementing FEP in drug design entails the utilization of the alchemical transformation technique, which involves a thermodynamic cycle that establishes a correlation between the bound and free states. In thermodynamics, the free energy is a state function; consequently, the total variation throughout the cycle is zero. For example, if two ligands (A and B) are bound to an identical receptor, the free energy difference between them can be estimated by transforming one ligand into another through a nonphysical process, considering the medium in which they are immersed (Fig. 10.1). The thermodynamic cycle is an effective method for determining relative binding free energies.

$$\Delta G_B - \Delta G_A = \Delta G^{\text{prot}}_{A\rightarrow B} - \Delta G^{\text{wat}}_{A\rightarrow B} \tag{10.3}$$

The terms on the left-hand side of Eq. (10.3) reflect the variation in the ΔG values of interest ($\Delta\Delta G$), whereas the subsequent term indicates the alchemical transformation used to compute the difference. Calculating the $\Delta\Delta G$ is crucial in drug design because it offers precise information on how structural modifications or replacements in compounds affect the energy difference. By determining the sign of $\Delta\Delta G$, it is possible to ascertain whether modifications in compounds result in an increase (negative $\Delta\Delta G$) or decrease (positive) in the molecule's affinity for the receptor. This information is essential for the development of effective drugs.

Undoubtedly, utilizing the aforementioned equation directly can lead to the emergence of artifacts and noise in simulations. To mitigate this impact, it is imperative to gradually convert the ligand into its final form, thereby enabling the calculation of the free energy changes at each subsequent stage of the transformation. Consequently, the sum of these free-energy differences culminates in an estimate of the binding free energy.

Although the thermodynamic cycle is more commonly used for determining the relative binding free energies, it can be adapted to estimate the absolute binding free energies, in which the ligands are annihilated in the binding site and solvent [19, 20]. In this case, the entire ligand is coupled/decoupled, which results in a significantly larger perturbation of the system, longer sampling times are necessary to achieve convergence compared to the sampling required to converge $\Delta\Delta G$ estimates.

The FEP approach is highly reliable and produces accurate results compared with experimental data [21]. Nevertheless, successful implementation of this approach requires a systematic and meticulous methodology.

This chapter provides an overview of the latest advancements and challenges in the application of free-energy perturbation in drug design. It covers topics such as improved utilities for FEP calculations, including force fields and enhanced approaches for system preparation, strategies for addressing solvent, allosteric, and covalent ligand cases in FEP, and the utilization of Machine Learning (ML) in FEP calculations.

1.1 Advantages, Disadvantages, Innovations, and Challenges

FEP offers numerous advantages for drug design and serves as a reliable technique for computing the binding free energy between proteins and diverse compounds when ABFE is calculated, or a congeneric series of compounds for RBFE calculations. Owing to its thermodynamic rigor, the FEP method can yield more precise predictions than other computational methods [22], aiding in the creation of more powerful and selective drugs.

However, implementing FEP in realistic systems, such as protein–ligand complexes, poses notable challenges and limitations that require resolution. One prominent hurdle is the substantial demand for computational resources for FEP calculations, which renders it a costly and time-intensive methodology. Consequently, their widespread application in large-scale virtual screening remains limited (see Chap. 11 for fundamentals and other aspects of ultra-large virtual screening). Another challenge arises from the reliance on different force fields, which can lead to results that are difficult to comprehend and compare [23].

It is crucial to note that FEP typically involves nonphysical ("alchemical") perturbations [24], gradually transforming one ligand into another through a series of steps. However, although FEP shows satisfactory performance with ligands that have minimal structural changes, its accuracy decreases when dealing with substantial modifications [25].

For instance, when substantial structural rearrangements are essential, FEP calculations may not always produce outcomes independent of the structure. This is particularly apparent when intricate changes, such as loop and backbone movements, are crucial for the ligand-binding process [26].

Accurate sampling is a pivotal factor in FEP simulations. However, a notable gap exists in comprehensive studies aimed at defining an optimal sampling duration, representing a primary limitation of the FEP calculations [25]. Inadequate equilibration and coexistence of various stable binding conformations further underline the critical areas that require enhancement. In many cases, FEP calculations may be contingent upon the structure and may lack reliability, especially when simulations start from unknown or undefined binding poses. Sensitivity to the choice of force field parameters and various other considerations further accentuates the potential limitations of FEP calculations in certain scenarios [27]. For example, choosing an appropriate initial ligand–protein structure is crucial for achieving accuracy in FEP, especially in dynamic systems, when exploring an ensemble of structures to identify the optimal receptor conformation [22]. Furthermore, it is critical to determine the equilibration timescale for each window to ensure that the free energies converge. The coupling parameter λ influences the alchemical transformation between states, and it should be sampled with sufficient intermediate windows to capture the free energy changes accurately [27].

Recent years have witnessed notable progress and innovations in free-energy perturbation and its role in drug design. These advancements have been propelled by enhancements in both computational hardware and FEP methodologies, enabling the extensive evaluation of the accuracy and dependability of FEP calculations in drug discovery initiatives. Key developments encompass the establishment of FEP+ technology by Schrödinger Inc., detailed in https://www.schrodinger.com/science-articles/free-energy-methods-fep and [28]. FEP+ integrates cutting-edge force fields, advanced sampling techniques, and GPU acceleration to facilitate precise and reliable computations of protein–ligand binding free energies, thereby finding applications in drug discovery.

Moreover, FEP+ incorporates the latest OPLS force fields, such as OPLS4 and OPLS5, which offer comprehensive coverage of the chemical space for both drug discovery and materials science applications. These force fields build upon the extensive coverage and accuracy achieved in previous OPLS versions by improving the accuracy of functional groups, which have presented significant modeling challenges in the past. The new OPLS5, in particular, is a polarizable force field that improves the relative binding accuracy in the FEP+ and Desmond models by adding explicit polarization for polarizable atoms, molecular ions, and cation–pi interactions [29].

An additional advancement is the FEP/REST method [30], which introduced an efficient λ-hopping protocol meticulously designed to sample local structural rearrangements, facilitating the assessment of relative protein–ligand binding affinity within manageable simulation durations. REST is particularly helpful in cases where there are significant binding site rearrangements upon ligand binding, or when studying a series of diverse ligands.

Additionally, REST2 [31], an advancement in enhanced sampling, was designed to speed up the traversal of the phase space and accelerate convergence, which is especially beneficial for flexible binding sites and those exhibiting a relevant degree of induced fit. REST2 significantly reduces CPU demands compared to regular replica exchange, substantially enhancing sampling efficiency, particularly when addressing substantial solute conformational changes in aqueous protein solutions. It is important to note that the computational load may vary between CPU and GPU implementation. This innovation has been seamlessly integrated into FEP+, thereby broadening its scope and effectiveness.

A recent extension of the FEP+ method has enabled handling of challenging perturbations, including core-hopping transformations, macrocycle modifications, and optimization of reversible covalent inhibitors [28]. Specifically, in the domain of macrocyclic drugs, FEP has demonstrated its utility in the design and optimization of such drugs by capturing their conformational flexibility and diversity.

Furthermore, the use of FEP to study membrane proteins is promising. FEP aids in predicting ligand-binding affinity and specificity of membrane proteins by accounting for intricate environmental factors and interactions present in the membrane environment [32]. Significant progress has also been made in the application of FEP to reversible covalent inhibitors, a class of drugs that form reversible bonds with target proteins to modulate their activity and optimize drug function.

Finally, the Machine Learning (ML, see Chap. 4) methods have significantly impacted the realm of FEP applied to drug design. Notably, a recent study integrated cloud-based FEP calculations with synthetically aware enumerations and goal-directed generative machine learning to facilitate extensive chemical exploration and optimization on a large scale [33]. Cloud-based free-energy calculations are a computational methodology employed to estimate the binding affinity of molecules to a target protein. This approach harnesses the computational power of cloud computing, allowing parallel execution of large-scale simulations. Consequently, it reduces computational expenses and time. Moreover, these calculations can be synergistically combined with other methodologies such as synthetically aware enumerations and goal-directed generative machine learning. This combined approach enables exploration and optimization across vast chemical spaces [33].

1.2 Recent Advances in Accessible FEP Software Tools

FEP has become one of the most promising approaches for accurately predicting ligand-binding affinities and is a crucial application in computer-aided drug design and discovery [34]. Through simulation of the transformation between the initial and final end states along a coupling parameter λ, FEP allows for rigorous calculation of changes in the binding free energy based on statistical thermodynamics [34]. Although the theory behind FEP has been firmly established, practical applications on a large scale have faced hindrances owing to complexities in system preparation, simulation protocols, and convergence of computations [17, 18].

Nevertheless, noteworthy developments have been made in recent years to broaden the horizons and decrease the barriers to utilizing FEP in real-world drug discovery campaigns. Open-source tools, such as TIES 2.0 [35], aid in free-energy calculations through flexible alignment algorithms and automated workflows accessible via web portals. TIES 2.0 offers the capability to perform both FEP calculations and thermodynamic integration (TI), which is another method for computing free energy differences. TIES 2.0 specifically applies a dual-topology method to predict relative binding free energies using TI, as demonstrated using sets of congeneric ligands [36]. Although the practical aspects of the FEP and TI are similar, the underlying theory and implementation of the partial derivatives of the Hamiltonian with respect to the coupling parameter λ differ between the two methods. Platforms such as BRIDGE [4] enhance the reproducibility and sharing of FEP protocols by integrating codes from GROMACS and YANK into Docker containers, which operate seamlessly across various computing environments. The effectiveness of BRIDGE's capabilities was demonstrated by the discovery of drug targets, including cyclin-dependent kinase 2 (CDK2) and ST3Gal-I, where both absolute and relative FEP were combined [36].

Notably, efforts have been made to simplify the intricate setup process of FEP simulations, even with numerous variations in ligands or receptors [35]. CHARMM-GUI modules have been expanded to generate input files and analysis scripts for FEP via various molecular dynamics engines such as NAMD, GENESIS, and AMBER [37–39]. Automated workflow capabilities were verified by testing diverse ligand solvation and protein binding systems. Notably, the AMBER implementation facilitates a range of force-field combinations and advanced options, such as hydrogen mass repartitioning, to expedite FEP convergence [39]. Studies, including those benchmarked with BACE1, have demonstrated the essential role of multiple independent FEP runs in achieving statistically reliable binding free energies despite minor protocol variations [40]. These tools provide high-throughput FEP and make the methodology accessible to nonspecialist researchers [37–39].

Additional efforts have been dedicated to improving the automation of GROMACS simulation packages, one of the most widely used molecular dynamics software programs. The open-source Python tool PyAutoFEP enables adaptable configuration of alchemical free-energy calculations in GROMACS, enabling perturbation mapping between numerous ligand states, and integration of advanced sampling techniques such as replica exchange [41]. As demonstrated by a set of Farnesoid X receptor compounds, PyAutoFEP enables large-scale prediction of relative binding affinities that are comparable to those of the top-performing methods. In addition to PyAutoFEP, the SMArt Python package provides automation capabilities via a perturbation topology builder that employs graph theory algorithms [42]. By identifying the maximum number of common substructures,

SMArt defines an optimal transformation pathway and generates compatible topology files for GROMACS and GROMOS simulation packages. When applied to a set of lysine post-translational modifications, the perturbation topologies generated by SMArt yielded consistent free energy differences between simulations performed with the two simulation packages, except for perturbations involving net charge changes. The thermodynamic cycle closures obtained from these calculations were robust, with a value of 0.5 ± 0.3 kJ mol^{-1} for GROMOS and 0.2 ± 0.2 kJ mol^{-1} for GROMACS, indicating that SMArt produces precise and compatible alchemical transformations for both simulation packages.

These tools offer automated workflows tailored to GROMACS, making them more accessible. PyAutoFEP and SMArt effortlessly manage topology generation and analysis, making alchemical free energy methods easier to apply. The incorporation of improved sampling and multistate capabilities expands the range and dependability of free-energy predictions and, as open-source platforms, stimulates additional development and customization. PyAutoFEP and SMArt are prime examples of targeted endeavors to unleash the potential of FEP in practical applications.

Precise prediction of subtle energy differences between structurally similar ligand conformations or binding modes has become an essential tool for computer-aided drug optimization, and recent advances have strengthened its potential [43]. Open-source, user-friendly platforms permit the evaluation and exchange of optimal FEP protocols, whereas the automation of system preparation minimizes workflow barriers on a large scale. Combined with increased computational capabilities and improved force fields, FEP has immense potential for significantly enhancing and hastening molecular discovery and design.

1.3 Applications of FEP in Industry and Consortiums

The application of FEP and free-energy calculations has expanded beyond the academic realm, becoming an integral part of the drug discovery and development processes within the pharmaceutical industry. These computational techniques have emerged as powerful tools for predicting the binding affinities of drug candidates for their target proteins, thereby streamlining the overall drug discovery pipeline.

In the industrial context, FEP is extensively employed in the optimization of lead compounds. For example, the FEP+ tool developed by Schrödinger has been successfully integrated into the workflow of numerous pharmaceutical companies, enabling the accurate prediction of relative binding free energies and guiding medicinal chemists towards compounds with improved efficacy and selectivity [40]. Major pharmaceutical companies, including Merck, Novartis, and Pfizer, have integrated FEP calculations into their drug design pipelines, leveraging these methods to prioritize compounds for synthesis and biological testing, thereby enhancing the efficiency of drug development [44, 45].

Furthermore, industry–academia collaborations such as the Drug Design Data Resource (D3R) initiative have fostered the application and advancement of FEP methodologies in drug discovery [46]. These collaborative endeavors have led to significant advancements in computational techniques and their validation using experimental data, thereby underscoring the growing recognition of the utility of FEP in the industry.

The adoption of FEP in the pharmaceutical industry is driven by its unique advantages. Primarily, the cost-effectiveness of FEP is evident in its ability to reduce the number of compounds that require physical synthesis and experimental testing, leading to substantial savings. Second, the speed at which FEP can provide reliable predictions of binding affinities accelerates drug discovery. Finally, the enhanced precision of FEP in lead optimization aids in the development of more effective and selective drug candidates [47].

2 Expanding the Potential of FEP Calculations

The FEP methodology can be utilized to obtain a comprehensive understanding of the intricate biological circumstances that necessitate a thorough analysis of the underlying context. For instance, this method can be utilized to validate the accuracy of molecular binding positions by calculating the relative changes in the binding affinity between a group of molecules through nonphysical modifications. The accuracy of FEP calculations relies on several parameters, including the structural integrity of the protein, precise positioning of the ligand, and affinity range and appropriateness of the ligands used for FEP calculations. The presence of water molecules at the binding site can drastically affect the estimation of the Gibbs free energy (ΔG). Furthermore, the prediction of the affinity is significantly affected by whether the binding is covalent. The structural attributes of a ligand set can determine the level of easiness or difficulty of using this technology. The need for computational resources is another important factor to consider when using FEP, mainly because of the extensive sampling of the relevant configurations of the system. Hence, although FEP is a powerful instrument for clarifying intricate biological situations, its utilization requires meticulous deliberation of a particular context and accessible resources.

Currently, machine learning (ML) is incorporated into free energy perturbation (FEP) computations to enhance precision and effectiveness. In particular, one approach utilizes ML-derived correction terms to improve the accuracy of FEP predictions, as demonstrated in Scheen et al. [48], where ML was used to generate corrections for hydration free-energy calculations, leading to more precise results than standalone FEP methods. Another significant application involves the use of active learning (AL), a special case of ML, to prioritize molecules from large datasets for FEP calculations. Thompson et al. [49] showcased an AL framework that efficiently identified top-scoring molecules by iteratively sampling and updating the ML model, thereby optimizing the selection process and reducing the computational burden associated with exhaustive FEP simulations. This combination of AL and FEP allows for improved sampling efficiency by focusing computational resources on the most promising candidates. In the following sections, we provide an overview of various intriguing applications of FEP in a range of scenarios that pose challenges to the methodology.

2.1 Validating Binding Poses

The FEP approach is commonly employed for binding mode validation, delivering mechanistic insights into binding events and confirming the binding poses of ligands at the receptor site. However, to accurately predict the binding free energies of a congeneric ligand series, it is crucial to know at least one ligand's binding pose within the series. Failure to precisely adhere to the experimentally confirmed binding mode may result in diminished accuracy in binding free energy predictions through FEP calculations. Therefore, the correct prediction of the observed binding free energies validates the binding mode [50].

Additionally, this methodology can be used to elucidate different binding mode hypotheses for a given ligand at the receptor site. By comparing the free energies associated with different binding poses, FEP calculations can be used to identify the correct binding pose of the ligand [51].

2.2 Dealing with Solvent

Free energy methods play a crucial role in drug discovery by identifying affinity-binding molecules for specific proteins. To comprehend protein–ligand interactions in aqueous environments, it is imperative to analyze the contributions of proteins, ligands, and solvent particles, particularly emphasizing the influential role of binding site water molecules. One way to address this issue is to perform MDmix simulations, which have been applied to various biological systems, including protein–ligand interactions, protein–protein interactions, and protein folding. They have been shown to be particularly useful in identifying binding hotspots, allosteric sites, and cryptic pockets on protein surfaces, which can be exploited in drug design and development.

Structure-based drug discovery increasingly uses cosolvent or mixed-solvent MD (MDmix) simulations as a useful, effective, and widely used computational tool in SBDD for its accuracy in identifying ligand-binding sites and understanding molecular interactions [52]. Because polar interactions are sensitive to local surroundings and water molecules, features such as protein flexibility and solvation effects limit predictions, even though accurate representation of binding sites is essential for directing drug discovery efforts. To overcome these problems, MDmix is presented as a technique that provides more accurate interaction maps than conventional potentials, such as GRID [53], and among its facilities, it allows water displaceability predictions that are superior to techniques depend only on pure water solvation theories [54]. However, problems arise from presumptions regarding the transferability of free energy values and the precision of computed binding free energy maps (ΔG_{bind}), particularly for cosolvent molecules that are significantly affected by solvent environments and are only transferable to larger drug-like molecules. Decomposing the molecular binding free energy into precise atomic contributions and accounting for the impact of the remaining molecules improves the suitability of the free energy grids for directing rational ligand design by exposing both known and hidden binding hotspots with potential for drug development [55]. Remarkably, MDmix simulations calculate atomic contributions to the binding free energy, enabling precise and transportable maps for drug design.

Another topic that deserves attention is the perturbations in the structure and energetics of the binding site water network resulting from minor chemical modifications of small-molecule ligands, which are highly relevant, especially during the hit-to-lead and lead optimization phases of structure-based drug discovery. This entails sampling various configurations of the protein–ligand complex in an aqueous environment, following the Boltzmann distribution. However, certain drug design projects encounter challenges when the crystal structure of the binding pocket reveals a cavity that lacks a water-exchange channel. Conformational fluctuations that sporadically open the cavity pose significant timescale discrepancies with current simulation methods, posing difficulties in accurately calculating the binding free energy when pocket hydration undergoes changes between the initial and final states.

Conventional molecular dynamics (MD, see Chap. 8) simulations are unable to accurately capture the changes in hydration that occur within buried cavities. This shortcoming leads to a static number of water molecules that fail to adjust to ligand modifications, resulting in one of the challenges faced during calculations, which involves equilibrating water molecules between the bulk solvent and the buried cavities.

An insufficient sampling of water movements within the binding site during perturbations between these states results in pronounced hysteresis, introducing structural inaccuracies and potential errors in the computed free energies, particularly when running perturbations in both the forward and backward directions. For example, in alchemical calculations of absolute binding free energy (ABFE), decoupling of the ligand from the binding site results in an unoccupied binding pocket that typically becomes solvated. Cheng and Roux aimed to overcome this issue by calculating the absolute binding free energy on a system formed by a camphor molecule and cytochrome P450 and employing the GCMC/MD (grand canonical Monte Carlo/molecular dynamics) [56]. The Grand Canonical Monte Carlo (GCMC) is a computational method used to sample the solvent environment around a solute molecule and is mostly used to study the thermodynamics and kinetics of solvation processes [57]. In GCMC simulations, the number of solvent molecules is not fixed but rather fluctuates according to the chemical potential of the solvent, allowing solvent exchange between the bulk and solute environments.

Using GCMC solvent sampling and the double decoupling method, before the alchemical ligand transformation, the authors reduced the hysteresis and improved the accuracy of the FEP calculations for alchemical ligand transformations with water displacement. The Grand Canonical Monte Carlo (GCMC) method is implemented in FEP+ [28].

Conversely, Relative Binding Free Energy (RBFE) calculations provided reliable predictions of $\Delta\Delta G$ values when applied to ligands with chemical similarity. However, this approach encounters sampling limitations when computing the reversible path of transforming one ligand into another within the binding site, particularly in cases where the ligands vary in size or polarity, because the perturbation may allow additional water molecules to occupy the region of the binding site. In a study conducted in 2009, Michel et al. [58] highlighted that the growth of functional groups displacing buried water molecules leads to their entrapment in energetically unfavorable states. One way to address this issue is to decouple the water molecule before the alchemical transformation, incorporating the associated free energy of the displaced water molecule as a correction to the relative binding free energy. This approach requires prior knowledge of the confined water, which contrasts with the sampling done in the Grand-Canonical ensemble.

In another study, a computational RBFE experiment was conducted to examine the effect of adding a functional group to a molecule, which leads to the displacement of buried water molecules at the binding site [11]. FEP calculations applied to six sets of congeneric ligands demonstrated a strong dependence of the predicted $\Delta\Delta G$ value on the initial solvation state of the binding site, specifically concerning the displacement of the water molecules. They also showed that current FEP protocols with even long timescales do not solve the problem of trapped water molecules and lack re-solvation. So, a proposed alternative approach would be to incorporate a Grand Canonical Monte Carlo (GCMC) solvent sampling stage before the λ-hopping step to enhance the solvent's equilibration. In this line, Ben-Shalom et al. [59] incorporated Monte Carlo (MC) steps that allow water molecule exchange into and out of the binding pocket without following a physically realizable path to overcome the inadequate sampling of water moves. They developed a hybrid Monte Carlo/MD (MC/MD) method for the equilibration of water between the bulk solvent and buried cavities while sampling from the intended distribution of states. A hybrid Grand Canonical/Molecular Dynamics (GC/MD) approach during ligand perturbations in FEP calculations reduces hysteresis and improves accuracy compared to pure MD simulations and experimental data in several test cases. However, this also introduces complexities associated with the chemical potential of water and fluctuations in the number of water molecules in the simulated system. The AMBER simulation package offers this method, alternating between blocks of standard MD steps and blocks of translational MC water movement attempts within a rectangular region overlapping the protein interior and the bulk solvent. These translational moves enable water molecules to exchange between the bulk and buried cavities while maintaining a Boltzmann distribution of states. Specifically, water molecules can enter or exit the buried binding site as a ligand undergoes growth or contraction during the alchemical process of RBFE calculation. This approach is particularly valuable in drug discovery applications, especially in scenarios in which prior knowledge of the number of buried waters is unavailable, and it aims to address the problem of occluded binding sites.

Alternatively, the Grand Canonical Alchemical Perturbation (GCAP) method is a simulation technique that combines the grand canonical ensemble (μVT) with conventional alchemical free energy methods to calculate the relative binding affinities between ligands at a chosen level of hydration. This allows for the dynamic adaptation of water networks during ligand binding simulations, enabling the isolation of the contribution of water network displacement to relative affinity [60]. This method can be performed in one or two dimensions using a two-dimensional approach, allowing the transformation of the ligand along two coupling parameters. Although initially developed for ligand binding simulations, the GCAP method shows potential for application in other simulations where water molecules, such as protein–protein or protein–nucleic acid interactions, significantly affect the free energy of the system [60]. The effectiveness of its application in these contexts should be evaluated on a case-by-case basis, which requires further research.

2.3 FEP and Allostery

The FEP methodology has also proven its value in cases where the ligand binds to allosteric sites instead of to orthosteric sites. Allosteric inhibitors are molecules that bind to a receptor at a site different from the active site and alter the shape and activity of the receptor. They can slow or stop enzymatic reactions by preventing the substrate from fitting into the active site. Allosteric inhibitors offer several benefits over orthosteric inhibitors, including enhanced specificity and fewer side effects [61].

In 2021, Liang et al. [62] applied FEP to evaluate ten SHP2 allosteric inhibitors that differ in their chemical structures. SHP2 is a protein that regulates various cellular processes, and its dysfunction is associated with cancer, diabetes, and autoimmune diseases [63]. However, designing and optimizing SHP2 allosteric inhibitors is challenging because slight structural variations can lead to significant changes in their biological activities [64]. FEP is particularly suitable for calculating the binding free energy because it can capture the structural sensitivity of the SHP2 allosteric inhibitors. In this work, the authors tested three working schemes using a different number of aaaaaa-windows and different times per window. Particularly, they tested the methodology's performance using 16, 32, and 64 intermediate aaaaaa-states, considering for each scheme simulation times of 16 ps, 32 ps, and 64 ps per window. Ultimately, they achieved an error of 0.5 kcal mol^{-1} between the predicted and experimental values. They also compared FEP and MMPBSA calculations and confirmed that FEP provides more precise predictions of the absolute free energy of the SHP2 allosteric inhibitors. It is important to recognize that accurately sampling large-scale movements to induce allosterism is a difficult task, especially when working within a limited timeframe. Despite this, the calculated RMSE was remarkably low, which suggests that the allosteric site in this instance can be considered relatively rigid.

FEP can also be used to discriminate the binding poses of orthosteric ligands in the presence or absence of an allosteric antagonist. Xiaoli et al. [65] showed that absolute free-energy calculations using FEP were essential for determining the correct pose of the ligand in a tertiary system composed of the protein C5aR1, the ligand PMX53, and the allosteric antagonist NDT9513727. The authors applied Accelerated Gaussian Molecular Dynamics to investigate the binding mechanism of PMX53 and identified two different binding modes for PMX53 (poses 1 and 2) in both the binary and ternary complexes. FEP simulations were run using NAMD3 with two simulation systems: the crystal structure for pose 1, and the GaMD simulation final structure for pose 2. Before conducting FEP simulations, a relaxation stage of 100 ns was applied. The resulting structures from the MD simulations were used as inputs for FEP simulations. The preparation of the input files and post-treatment FEP simulations were performed using BFEE2 protocols [66]. To enhance the precision of the free-energy estimation, FEP simulations involving the reversible coupling of PMX53 to its environment, such as the binding site of C5aR1 or solvation, were conducted bidirectionally.

FEP simulations confirmed that pose 2 was the most stable binding mode in the ternary complex. Simultaneously, in the binary system, the ligand alternated between the two poses, with pose 1 being preferred to pose 2. In pose 1, the side chain of Arg6 of PMX53 extends to TM6–TM7 in the GPCR crystal structure. In contrast, at position 2, the side chain of Arg6 of PMX53 formed a salt bridge with Glu199 on TM5, making this interaction extremely stable, as the allosteric antagonist stabilized TM5 by facilitating a tight hydrophobic stack between TM4 and TM5 on the extracellular side.

2.4 FEP and Covalent Ligands

Over the past five decades, there has been a notable increase in the development of covalent inhibitors (Fig. 10.2a and b) [67, 68]. Although the biggest challenge in the development of covalent inhibitors is their lack of specificity or selectivity, modulation of electrophilic warhead reactivity and optimization of noncovalent interactions may improve target receptor recognition and covalent inhibitor selectivity [69]. As covalent drug development benefits from warhead reactivity and binding site mechanism prediction [70], there is a substantial demand for computational techniques that can optimize covalent inhibitor design.

However, there is modest utilization of FEP in the context of covalent drug optimization, especially because the interconnectedness of several processes that make up the binding process of covalent inhibitors makes the computational

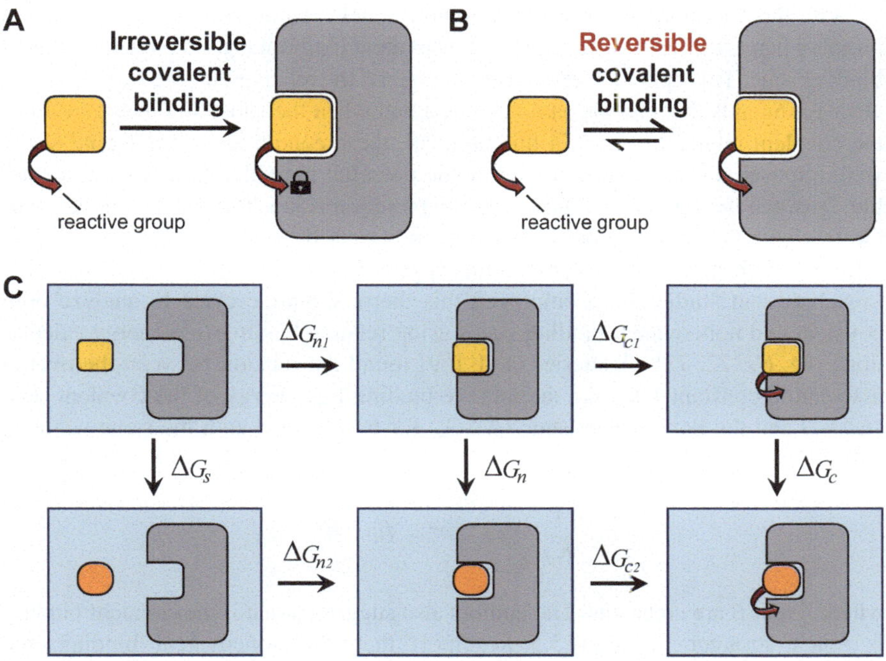

Fig. 10.2 Representation of irreversible (**a**) and reversible (**b**) covalent binding by reactive ligands. (**c**) Thermodynamic cycle of covalent inhibitor binding using alchemical FEP simulations. The free energy to mutate from ligand 1 to ligand 2 is calculated in the bulk solution (ΔG_s), in the protein binding pocket (ΔG_n), and in the covalently bonded protein–ligand complexes (ΔG_c)

prediction of their binding affinity extremely difficult [69]. Two separate free energies are important for the covalent ligand binding [44]. The first is the binding free energy required to produce the noncovalently bound protein–ligand complex before the chemical reaction takes place. This refers to the difference in free energy between the protein–ligand complex, before the chemical reaction, and the independently solvated ligand and protein (ΔG_{n1} in Fig. 10.2c). The second is the free energy associated with the formation of the covalent bond. It is defined as the difference in free energy between the covalent bonded protein–ligand and the protein–ligand complex before the reaction (ΔG_{c1} in Fig. 10.2c). The total binding free energy can be calculated by summing the two aforementioned free energies. By gradually transforming the Hamiltonian from the first to the second state using FEP, we can calculate the free energy to mutate ligand A to ligand B in bulk solution (ΔG_s in Fig. 10.2c), in the protein binding pocket (ΔG_n in Fig. 10.2c), and in the covalently bonded protein–ligand complexes (ΔG_c in Fig. 10.2c). Analogous to the conventional FEP calculations employed for noncovalent binding (i.e., the relative binding free energy difference between the first two vertical legs of Fig. 10.2c), it is also possible to derive the relative overall binding free energy difference between the two covalent ligands (i.e., the difference of ΔG_s and ΔG_c, the first and third vertical legs of Fig. 10.2c) [69, 71].

With this thermodynamic cycle in mind, in 2017, Kuhn et al. [44] developed a protocol that uses FEP to select reversible covalent inhibitors based on their covalent binding state. This approach effectively computed the relative binding free energy of covalent ligands that had the same warhead and when the substitutions are far from the covalent interaction site. In this protocol, these conditions to the ligand series were imposed due to the use of a harmonic potential to model the covalent interaction between the ligand's warhead and the protein residue. Therefore, estimation of the free energy difference between the ligands caused by the varying reactivity of different warheads is not possible in this approach.

Subsequent studies have employed this thermodynamic cycle to analyze both covalent and noncovalent binding states using relative binding free-energy calculations [44, 69, 72, 73]. Chatterjee et al. [69] found the relation between the overall association constant $1/K_d$ and the relative binding free energy of the covalent state ($\Delta\Delta G_c$) and the noncovalent state ($\Delta\Delta G_n$) for two ligands with the same warhead with the equation:

$$\frac{1}{K_d} = Ae^{-\frac{\Delta\Delta G_c}{\mathrm{RT}}} + Be^{-\frac{\Delta\Delta G_n}{\mathrm{RT}}}$$

where A and B are constants. The authors also suggested that if the covalent binding is much stronger (at least 5.5 kcal/mol) than the noncovalent binding, the noncovalent state becomes negligible in predicting the overall binding selectivity of covalent inhibitors. If this criterion is not met, a single quantum mechanical calculation of the warhead core structure to assess the energy of the covalent linkage can help estimate the differences in affinity between covalent and noncovalent states [69].

Following Chatterjee et al. [69], Lameira et al. [73] compared the relative binding free energy from FEP of halogenated reversible covalent inhibitors of human cathepsin L (hCatL) with experimental binding kinetics results. They observed that the overall reversible covalent binding of the inhibitors could be predicted by employing the covalent and noncovalent states. Furthermore, the relative binding free energies of dipeptidyl nitrile inhibitors were accurately predicted using FEP and that an extra point charge improves the description of the halogen bond between the inhibitors and the protein.

It should be noted that most studies have employed the FEP methodology to predict and understand the binding process and selectivity of reversible covalent inhibitors. Therefore, using FEP for lead optimization of reversible covalent inhibitors can offer rigorous and comprehensive information about the relative contributions of covalent and noncovalent binding states to the overall binding affinity [72]. This opens new possibilities for tailoring noncovalent interactions and improving reversible covalent drugs.

2.5 Applications of FEP in Scaffold Hopping

Scaffold hopping is the process of replacing the central core scaffold of a bioactive molecule while maintaining similar pharmacophore features [74]. This technique is valuable in drug design as it allows for the exploration of new regions of chemical space and the identification of alternative lead compounds with improved properties. Accurate scaffold hopping requires methods that can predict changes in the binding affinity between different ligand scaffolds. FEP approaches are powerful tools for modeling challenging scaffold hopping transformations [75].

In a recent study, Wu et al. [76] demonstrated the first application of an FEP-guided scaffold hopping strategy to discover novel phosphodiesterase-5 (PDE5) inhibitors. Compound L1, which contains a new scaffold relative to the known inhibitors tadalafil and LW1607, was identified using a protocol that integrates molecular docking and FEP absolute binding free energy (ABFE) calculations [76]. The predicted FEP energies (ΔG_{FEP}) between ligands and PDE5 closely matched experimental binding data (ΔG_{EXP}) with a mean absolute deviation of 1.13 kcal/mol. The lead compound L12 was identified with an improved IC_{50} of 8.3 nM through two rounds of FEP-guided optimization. The compound contains a completely distinct scaffold [76].

Recent enhancements in FEP methodologies have expanded their capabilities for modeling scaffold hopping. The FEP+ approach integrates enhanced sampling techniques to accurately reconstruct conformational changes during large ligand topology modifications [28]. This method enables rigorous computation of binding affinity changes between chemically diverse inhibitors [28]. Software tools such as QligFEP have also applied FEP to retrospectively validate experimentally observed scaffold hopping hits between distinct compounds [77].

2.6 Positional Analogue Scanning

Positional analogue scanning (PAS) is a method that analyzes the effects of minor chemical substitutions at different sites in a molecule. This allows for the elucidation of structure–activity relationships. During lead optimization campaigns, a routine technique is employed to improve potency and drug-like properties by modulating steric, electronic, and hydrophobic interactions with the target protein. However, due to financial and time constraints, exhaustive synthesis and experimental testing of all positional analogues are often impractical. Computational scanning methods based on FEP calculations can accurately predict changes in binding affinity upon substitution. This enabled the prioritization of the most promising analogues for synthesis [78].

For example, perturbative fluorine scanning uses automated FEP protocols to rank the impact of adding fluorine at multiple positions efficiently. This was demonstrated by Wade et al. [79] for over 100 analogue pairs. FEP enables reliable identification of "activity cliffs," which are subtle chemical modifications that significantly affect potency. Assisting with lead optimization efforts, Pérez-Benito et al. [80] showed that FEP-based positional scanning may locate activity cliffs within 1.39 kcal/mol of experimental values. Recently, Hu and Muegge [81] utilized Amber GPU-TI to perform RBFE calculations to evaluate the impact of PAS on potency changes for 14 protein targets and 20 analogue sets. They predicted shifts in binding affinity with a mean unsigned error of only 0.74 kcal/mol and a root-mean-square deviation of 0.91 kcal/mol over 120 calculations. This finding suggests that RBFE calculations are suitable for prioritizing positional analogues for synthesis and testing during lead optimization campaigns [43, 81].

2.7 Combinations and Alternative Approaches

Drug design techniques that leverage the parallel nature of computations are becoming increasingly common to enhance sampling in calculations. The Replica Exchange class of methodologies (REMD) is a popular choice among several strategies available for coupling to computationally intensive methods, such as FEP, to improve the statistical convergence in calculations of binding free energy. REMD techniques use many parallel copies of a system under various simulation conditions [30]. Subsequently, based on the Metropolis Monte Carlo approval criterion, switching between two copies are attempted regularly. Thus, to enhance the sampling of traditional FEP simulations, several REMD approaches have been successfully combined with FEP [82], including conventional REMD [83], Hamiltonian replica exchange (H-REMD) [84], and replica exchange with solute tempering (REST) [85]. However, there are situations in which high-energy barriers produce conformational states of a solute that remain weakly sampled during an alchemical transformation, making these approaches inadequate. Furthermore, in the case of two-dimensional exchange spaces, such as those of FEP/H-REMD, the significant number of replicas needed for the method likely offsets its actual advantages.

Although FEP and its combinations provide accurate estimates of binding affinities, screening of large libraries of compounds with substantial modifications between them can impact the overall complex lead optimization phase of a drug discovery project. Thus, over the years, numerous research groups have investigated alternative simulation methods to conventional free-energy methods to achieve comparable accuracy while requiring less sampling. Recently, there has been a resurgence of interest in methods such as nonequilibrium free-energy calculations and expanded ensemble methods such as lambda dynamics (λ-dynamics) [17, 86, 87].

In nonequilibrium methods, a greater amount of sample time is allocated to the physical end states. These end states encompass the fully interacting ligand bound to the protein, as well as the decoupled ligand that lacks interactions with the protein environment. In addition, only very short simulations are usually performed for

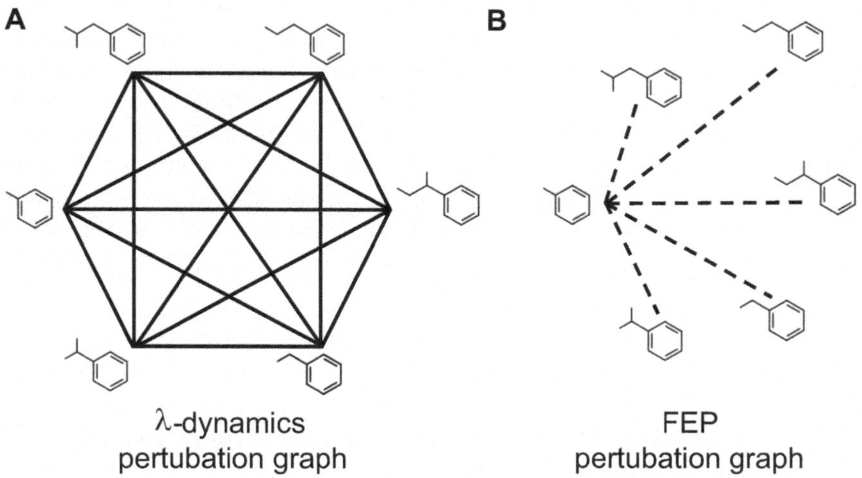

Fig. 10.3 Alchemical perturbation connection graphs. (**a**) In the λ-dynamics methodology, a connected graph of ligand end states is sampled because in this method, all physical and intermediate λ states are sampled within a single simulation (represented by solid lines). (**b**) In FEP methodology a "star map" connected graph is sampled via pairwise perturbations when run without redundant calculations for cycle closure, thus, many intermediate simulations are required (represented as dashed lines). (Adapted from the study by Robo et al. [98])

nonphysical intermediate states that connect the end states, thereby improving the efficiency of the approach [88]. The exploration of the alchemical path occurs through rapid switching transitions in a single simulation, in which the ligand is switched between interacting and noninteracting states [89, 90]. Many nonequilibrium approaches operate in a pairwise manner; however, by employing numerous independent runs, these methods appear to have a higher likelihood of exploring many potential pathways [88]. They also exhibit a high degree of parallelizability, rendering them favorable options for high-performance computing (HPC) or cloud computing environments [91, 92].

In contrast to FEP, the concept behind λ-dynamics is to calculate the free energy differences between multiple thermodynamic states in a single calculation to increase efficiency by facilitating better scalability (Fig. 10.3). In λ-dynamics, the coupling parameter λ is treated as a dynamic variable with a fictitious mass and is propagated along the atomic coordinates throughout the course of the simulation [93]. Thus, it is feasible to obtain estimates of the free energy for an entire "leg" of the thermodynamic cycle within a single simulation [94]. Multiple ligands can be evaluated simultaneously in a given simulation where the potential energy term correlated with each ligand is scaled according to its λ value [95, 96]. Thus, in the context of structure-based drug design, λ-dynamics simulations can act similarly to competitive binding experiments, in which all ligands compete for a shared receptor based on their respective free energies [94, 95, 97]. However, to collectively sample multiple ligands within a single λ-dynamics simulation, the free energy barriers between the ligand end states in λ-space must first be flattened [94, 98]. This is possible by identifying and incorporating bias potentials into λ-dynamic simulations. Therefore, determining these biases requires a considerable amount of simulation time prior to production sampling, which reduces the efficacy and cost of the method. In recent years, several new developments have been introduced to expand the applicability of λ-dynamics for drug discovery, including multisite λ-dynamics [86, 99], which enables multiple substituents at multiple sites, the use of a biasing potential replica exchange to enhance transitions between states [100], and an alternative λ sampling strategy employing Gibbs sampling [101, 102].

3 Machine Learning for FEP

3.1 Introduction to the Use of Machine Learning in Free Energy Perturbation Calculations

The challenges and limitations discussed in the previous sections pose substantial obstacles to implementing FEP calculations in practical applications, necessitating the development of novel approaches and methodologies to overcome them. Thus, ML enhances the precision and effectiveness of FEP calculations. Recent advancements in ML methodologies offer new possibilities for overcoming such difficulties and improving the reliability of FEP predictions to advance drug discovery [49]. Machine learning is a branch of artificial intelligence (AI) that specifically deals with the creation of algorithms and statistical models that allow computers to carry out tasks by learning from data, rather than relying on explicitly written instructions. This data-driven approach allows systems to improve their performance on tasks over time as they are exposed to more data. In the realm of AI, ML techniques are employed to recognize patterns, make predictions, and inform decision-making processes across various domains, including natural language processing, computer vision, and computational chemistry [103].

ML employs statistical models trained on large datasets to capture the complex patterns. For FEP, relevant training data may include FEP simulation trajectories, experimentally determined structural and thermodynamic measurements, and quantum mechanical calculations on small molecule subsets. ML algorithms, particularly those employing deep learning architectures, have demonstrated the capability of learning complex patterns in data, enabling more accurate predictions of free energy changes [33]. By training on physics-based simulations and experimental data, ML models can learn to correct weaknesses in fixed-charge force fields [104], drive enhanced sampling of binding modes [105], and bypass costly simulations to directly predict binding affinities from molecular structures [106]. This efficiency is paramount in the high-throughput screening of drug candidates, allowing for the rapid evaluation of numerous compounds [107]. ML algorithms assist in the efficient description of potential energy surfaces and in the accurate estimation of the parameters for these simulations. By leveraging ML, FEP can more effectively model the molecular interactions and transformations that are essential in drug discovery, leading to more precise predictions of binding affinities and thermodynamics of molecular processes. The integration of ML into FEP represents a significant advancement in computational chemistry.

3.2 Machine Learning Methods in Free Energy Perturbation Calculations

Several ML methods and algorithms can be utilized in the field of computational chemistry to enhance the FEP calculations. These ML techniques offer distinct solutions to specific issues and applications, thereby significantly contributing to the advancement of the FEP calculations. For more details on ML theory and methods, please see Chap. 4.

One of the prominent ML algorithms used in this context is neural networks (NNs), especially deep learning architectures such as convolutional neural networks (CNNs). These networks are adept at predicting free energy changes directly from molecular structures and learning the complex relationships between molecular features and free energies. CNNs, in particular, process spatial information in molecular systems, offering predictions of free energy changes without explicit simulation.

Gaussian Process Regression (GPR) is another ML algorithm that can be used to construct surrogate models of free energy landscapes. This approach decreases the amount of computer resources needed by allowing for efficient exploration of the parameter space and offering estimates of uncertainty for predictions. This is particularly advantageous for optimizing simulation parameters and incorporating various data sources [108].

Decision trees and random forests are algorithms applied for feature selection, determining the chemical descriptors that have a substantial impact on free energy. This knowledge can guide the configuration of the FEP calculations and improve their comprehension.

Support vector machines (SVMs) have been employed as ML algorithms to distinguish compounds based on their propensity to exhibit specific free energy changes. This approach has been utilized in drug design to differentiate between ligands that bind and those that do not. Moreover, SVMs can be trained to predict changes in the free energy using chemical descriptors [109]. This has proven effective in identifying compounds that are likely to exhibit the desired free energy changes.

Reinforcement learning (RL) is a ML algorithm in which an agent learns to make decisions, such as choosing mutations or adjustments to a ligand, to maximize the total rewards associated with the desired changes in free energy. RL enhances the alchemical pathway in FEP calculations, thereby enhancing the effectiveness of sampling tactics in the simulations [103].

In addition to the specific ML algorithms, various machine learning methods are also applied in the context of FEP calculations. These methods define how the overall process of ML is applied to solve the problem.

Bayesian methods, including Bayesian optimization, are applied to refine free energy estimates, integrating experimental data with simulation results. This approach is extremely beneficial for selecting optimal simulation parameters.

Transfer learning is utilized to apply models trained on a specific set of FEP data to distinct, yet interconnected problems, resulting in a substantial reduction in the computational resources required for new calculations. This approach has great potential in the field of drug development as it allows for the application of models that have been trained on large datasets to new molecules [110].

Autoencoders, including variational autoencoders (VAEs), are employed to reduce the dimensionality of FEP calculations. They condense intricate molecular features into a space with fewer dimensions, simplifying the representation of molecular systems and assisting in their analysis and interpretation [111].

Each of these ML algorithms and methods improves the accuracy of the FEP calculations, effectively managing high-dimensional data. Additionally, they reduce the computing expenses associated with these calculations and provide valuable insights into the factors that determine changes in free energy. The incorporation of ML in FEP calculations is a rapidly growing area that has consistently gained advantages from the progress in both ML and computational chemistry. Therefore, these methods play a crucial role in transforming the investigation of molecular systems, allowing for predictions on a wide scale and for complicated systems that were previously impossible using conventional computational methods.

3.3 Advanced Applications of Machine Learning in Free Energy Perturbation: A Compilation of Case Studies

The integration of ML with FEP calculations has significantly advanced the field of drug discovery, as exemplified by the studies of Konze et al. [112] and Ghanakota et al. [33]. Both studies focused on optimizing the hit-to-lead process, particularly in designing potent inhibitors of Cyclin-Dependent Kinase 2 (CDK2), but they approached the problem with distinct methodologies and techniques.

In 2019, Konze et al. [112] explored the use of PathFinder, a reaction-based enumeration tool, combined with active learning and FEP simulations. This approach enabled rapid exploration of synthetically tractable chemical spaces and optimized the potency of CDK2 inhibitors. This study involved generating large virtual libraries of lead-like compounds through retrosynthetic analysis and combinatorial synthesis. These libraries were then filtered based on their drug-like properties and docked to the CDK2 binding site. The FEP+ tool was used to predict binding affinities, with active learning iteratively prioritizing compounds for further FEP calculations. This methodology allowed the exploration of over 300,000 ideas and the performance of more than 5000 FEP simulations, ultimately identifying over 100 ligands with predicted IC_{50} values below 100 nM.

In 2020, Ghanakota et al. [33] also integrated ML with FEP calculations but focused on cloud-based profiling and generative ML for large-scale chemical exploration. Their approach combines extensive enumeration and FEP profiling with generative ML, resulting in a higher concentration of potent compounds. The study adhered to a predetermined drug-like property space using PathFinder rule-based enumeration optimized for a multi-parameter function based on a weighted sum QSAR approach. Employing the REINVENT technique, the authors trained a network of chemical components and adjusted it to enhance the utility function. This process created millions of unique compounds at specified R-group positions. Different strategies for selecting compounds for FEP calculations were evaluated, including random enumeration, ML-model-based selection, generative ML prioritization, and combined strategies. The results demonstrated the efficacy of generative ML in producing novel and potent chemical compounds efficiently, preserving or improving important physicochemical properties, such as lipophilic ligand efficiency (LLE).

Both studies highlighted the significant impact of integrating ML with FEP in enhancing drug discovery processes. They leveraged the synergy between ML and FEP to improve the accuracy and efficiency of drug discovery. Each approach emphasizes the rapid exploration of large chemical spaces, with Konze et al. [112] exploring over 300,000 ideas and Ghanakota et al. [33] generating millions of compounds. Both studies aimed to discover and optimize potent CDK2 inhibitors, demonstrating the practical application of these techniques in medicinal chemistry. Additionally, they employed iterative processes in which ML models are continuously updated based on the initial FEP results, refining predictions, and selections. Furthermore, each study ensured that the generated compounds adhered to drug-like property criteria, optimizing them for therapeutic potential.

However, there are key differences between these two approaches. Konze et al. [112] utilized PathFinder for reaction-based enumeration and active learning, focusing on retrosynthetic analysis and combinatorial synthesis, whereas Ghanakota et al. [33] employed the REINVENT generative ML technique to create unique compounds. Konze et al. [112] explored 300,000 ideas and performed over 5000 FEP simulations, whereas Ghanakota et al. [33] generated millions of compounds using cloud computing GPUs for FEP calculations. In terms of selection strategies, Konze et al. [112] used active learning for iterative improvement, whereas Ghanakota et al. [33] evaluated multiple strategies, including random enumeration, ML-based selection, generative prioritization, and combined approaches.

In summary, both studies illustrated the transformative potential of combining ML with FEP for drug discovery. Konze et al. [112] and Ghanakota et al. [33] demonstrated different complementary approaches for optimizing chemical libraries and enhancing the efficiency and effectiveness of the lead optimization process. Their study underscores the versatility and power of ML and FEP integration, paving the way for new advancements in computational chemistry and drug development.

An additional utilization of ML in FEP computations was performed by Willow et al. [113], who demonstrated a substantial improvement in the efficiency and precision of FEP calculations through the application of ML. The researchers utilized Targeted Free Energy Perturbation (TFEP), a technique that employs invertible mapping to ensure overlap in the configuration space and convergence in free energy estimations. The technique was employed on a flexible bonded deca-alanine molecule, exploiting harmonic biases with different spring centers. An important element of this method involves employing real-valued non-volume-preserving (real NVP) transformations, which are highly compatible with TFEP because of their reliable invertibility and easy calculation of the transformation Jacobian.

An interesting component of this study was the use of an identity map for the initial configuration of a real NVP map. This was accomplished by setting transformation and scaling factors to zero. The neural network was trained using the AMBER force field implemented in JAX to minimize the value of the loss function. This procedure involved partitioning the data into a training set, which accounted for 80% of the data, and a test set, which accounted for 20%. This division further improves the accuracy of the model. The study observed that the TFEP method could accurately replicate the reference free energy differences for most state pairs with a spacing of $\Delta\lambda = 1$ Å. The precision of the mapping approximations was the greatest for pairs of states with free energy disparities (ΔF) below 2 kJ/mol. Utilizing trained mapping with early termination demonstrated more reliability, resulting in a more rigorous calculation of errors compared to conventional approaches.

In the evolving field of drug discovery, the integration of ML with FEP calculations has made significant strides, as evidenced by three distinct studies. The first study, focusing on the application of ML algorithms such as DeepLDA and Autoencoders in metadynamics, emphasizes the enhancement in binding mode and free energy landscape analysis for drug–target interactions. The second study demonstrated the synergy of cloud-based FEP calculations, synthetically aware enumerations, and generative ML in accelerating the hit-to-lead process, notably in identifying potent compounds and optimizing drug-like properties. Finally, the third study illustrated the innovative use of ML in targeted FEP, employing learned mappings for peptide conformations to improve the accuracy and efficiency of FEP calculations. Collectively, these studies not only highlight the versatility of ML in different aspects of FEP but also underline its transformative potential in advancing computational methods for drug discovery and design.

3.4 Implications for ML in FEP Calculations

As we conclude this section on the use of ML in FEP calculations, it is clear that we are in the bricks of a new era of drug discovery. The future of this field is marked by ongoing research efforts aimed at refining these ML models, with a specific focus on enhancing their predictive accuracy. Such advancements are pivotal in ensuring that these models become integral components of drug design pipelines.

The integration of ML with FEP calculations has the potential to significantly transform drug discovery processes. By utilizing these modern computational techniques, researchers can not only accelerate the rate of discovery but also uncover novel compounds that may avoid identification using traditional methods. This provides opportunities for investigating chemical spaces and comprehending molecular interactions at an unprecedented level.

One major implication of this integration is the increase in the precision of free energy predictions. ML can significantly enhance the accuracy of FEP calculations by providing corrections and reducing systematic errors, resulting in more reliable predictions of compound binding affinity. Additionally, the utilization of ML allows for the rapid screening of large libraries of compounds, prioritizing those most promising for detailed FEP simulations, thereby speeding up the discovery process.

The ability to efficiently explore vast chemical spaces is another critical advantage. ML techniques enable the exploration of chemical spaces that are infeasible with traditional methods, increasing the likelihood of discovering new active compounds. This capability, combined with the enhanced predictive accuracy, can lead to the identification of novel compounds with potential therapeutic benefits.

Moreover, the integration of ML into FEP calculations can lead to a significant reduction in drug development costs. By improving the efficiency of the computational process and reducing the need for extensive experimental validation, ML helps lower the overall cost of drug development. This cost-effectiveness is crucial in fields where research and development expenses are exceedingly high.

Looking ahead, personalized medicine is an exciting frontier. The application of ML in FEP could pave the way for personalized medical strategies where drug compounds are tailored based on individual genetic profiles. This level of customization promises more effective treatments and fewer side effects, thereby revolutionizing patient care.

Additionally, as processing power continues to advance and ML algorithms become more sophisticated, we can expect further improvements in the efficiency and precision of the FEP calculations. These advancements will reduce both the time and cost of drug development, thereby making the entire process more streamlined and accessible.

Essentially, the integration of ML into FEP calculations is more than just an enhancement of existing methodologies. This revolutionary advancement fundamentally changes the boundaries of what can be achieved in the field of drug development. As this dynamic area continues to evolve, it holds great potential for advancing medicine and providing hope for faster and more efficient treatments for a wide range of diseases.

4 Final Considerations

FEP has emerged as a highly effective tool for computer-assisted drug discovery owing to advancements in computational hardware, force fields, and methodologies. FEP allows for precise prediction of ligand binding affinities through alchemical transformations between different states. The latest developments have significantly enhanced the range and reliability of the FEP calculations. Despite the emergence of a plethora of alternative methodologies, FEP remains highly robust and easily adaptable to new technologies owing to ongoing improvements in force fields, sampling techniques, and supporting hardware. FEP continues to be the gold standard for validating binding free energy predictions as it captures the physical interactions between molecules through statistical thermodynamic principles. Extensive testing has shown the remarkable accuracy of FEP across a wide range of systems when properly implemented by using theoretical grounds rather than empirical parameterization to relate free energy differences to binding constants.

The integration of various approaches has made it possible to address complex molecular perturbations such as core modifications, scaffold hopping, and reversible covalent inhibitors. These perturbations were previously difficult to access. The use of automated workflows, enhanced sampling techniques, and multi-GPU accelerations has reduced computational barriers, enabling extensive virtual screens. Furthermore, combining FEP with synergistic machine learning frameworks has demonstrated the potential of this methodology. FEP enables high-precision prediction of ligand-binding strengths, empowering medicinal chemists to improve the potency and selectivity of their compounds rationally. Among the computational techniques, FEP quantifies the thermodynamic forces that drive molecular recognition and directly guides optimization. Recent advancements in efficiency and automation have made it possible for FEP to assess affinities on a large scale, thereby accelerating the critical progressive improvement needed in drug development.

Advances in structure-based drug optimization are expected to significantly accelerate and transform this field. The use of increasingly accurate force fields that incorporate growing experimental data will play a key role in this transformation. Enhanced sampling protocols, which are now tightly integrated, will enable the simulation of increasingly complex molecular systems at a lower cost. The use of hybrid grand canonical methodologies holds great promise in obtaining precise hydration and binding free energies. With the generation of extensive data from high-throughput experiments, AI-integrated FEP techniques are expected to achieve new levels of predictive capability. The use of cloud computing and automated analysis will also facilitate the earlier application of FEP in the discovery pipeline and reduce the number of iterations required. Despite progress in this field, the application of FEP to large chemical libraries remains computationally expensive. The integration of emerging AI capabilities with efficient GPU hardware implementation will be crucial for overcoming this challenge and facilitating the widespread adoption of this essential methodology.

Overall, free energy perturbation has achieved remarkable progress, earned widespread trust and reliability, and has preserved its status as the gold standard. Ongoing advancements have enhanced precision and broadened its potential to drive drug discovery. The future of this computational method appears promising and positioned at the forefront of structure-based design.

5 First Steps to FEP Simulations

When beginning the FEP experiments, it is important to have a foundational understanding of the relevant keywords and concepts discussed in this work. Although much more information can be covered, the scope of this work is limited. To get started, prospective users should be able to answer the following questions:

? Questions to answer when starting a FEP simulation
What computational resources are available, multiple CPUs or GPUs?
Do you have a license for commercially available software or access to open-source software?
Are you aiming to calculate the binding event of a solvated ligand to a protein target (ABFE) or the relative free energy of binding between two ligands (RBFE)?
For RBFE calculations, what is the size of your ligand set? Do the ligands in your set have minimal or substantial modifications? Is it possible to create a perturbation map?
In RBFE, are there transformations between ligands with different formal charges? Does this involve the total annihilation or emergence of groups within the ligands?
Is the binding site well-defined, or does it undergo significant rearrangements?
Are there conserved water molecules or metal ions within the binding site?
Do you have an established binding mode for the ligand(s)? If so, is it a covalent binding mode?

Acknowledgments This study was financed in part by the Coordenação de Aperfeiçoamento de Pessoal de Nível Superior—Brasil (CAPES)—Finance Code 001.

The authors also thank the agencies CPNq (Processes: 305524/2022-4 and 308254/2022-8) and FAPERJ (Processes: E-26/201.155/2021 and E-26/201.462/2021) for their financial support.

References

1. King, E., Aitchison, E., Li, H., & Luo, R. (2021). Recent developments in free energy calculations for drug discovery. *Frontiers in Molecular Biosciences, 8*, 712085.
2. Guedes, I. A., Pereira, F. S., & Dardenne, L. E. (2018). Empirical scoring functions for structure-based virtual screening: Applications, critical aspects, and challenges. *Frontiers in Pharmacology, 9*, 411637.
3. Shirts, M. R. (2012). Best practices in free energy calculations for drug design. *Computational Drug Discovery and Design*, 425–467.
4. Cavasotto, C. N. (2020). Binding free energy calculation using quantum mechanics aimed for drug lead optimization. In A. Heifetz (Ed.), *Quantum mechanics in drug discovery* (pp. 257–268). New York, NY, Springer US.
5. Chen, H., & Chipot, C. (2022). Enhancing sampling with free-energy calculations. *Current Opinion in Structural Biology, 77*, 102497.
6. Zwanzig, R. W. (1954). High-temperature equation of state by a perturbation method. I. Nonpolar gases. *The Journal of Chemical Physics, 22*, 1420–1426.
7. Kirkwood, J. G. (1935). Statistical mechanics of fluid mixtures. *The Journal of Chemical Physics, 3*, 300–313.
8. de Donder, T. (1927). *L'affinité*. Gauthier-Villars Paris.
9. York, D. M. (2023). Modern alchemical free energy methods for drug discovery explained. *ACS Physical Chemistry Au, 3*, 478–491.

10. Bennett, C. H. (1976). Efficient estimation of free energy differences from Monte Carlo data. *Journal of Computational Physics, 22*, 245–268.
11. Postma, J. P., Berendsen, H. J., & Haak, J. R. (1982). *Thermodynamics of cavity formation in water. A molecular dynamics study* (pp. 55–67). Royal Society of Chemistry.
12. Jorgensen, W. L., & Ravimohan, C. (1985). Monte Carlo simulation of differences in free energies of hydration. *The Journal of Cchemical Physics, 83*, 3050–3054.
13. Jorgensen, W. L., & Tirado-Rives, J. (1988). The OPLS [optimized potentials for liquid simulations] potential functions for proteins, energy minimizations for crystals of cyclic peptides and crambin. *Journal of the American Chemical Society, 110*, 1657–1666.
14. Case, D. A., Cheatham, T. E., III, Darden, T., Gohlke, H., Luo, R., Merz, K. M., Jr., et al. (2005). The Amber biomolecular simulation programs. *Journal of Computational Chemistry, 26*, 1668–1688.
15. Brooks, B. R., Bruccoleri, R. E., Olafson, B. D., States, D. J., Swaminathan, S. A., & Karplus, M. (1983). CHARMM: A program for macromolecular energy, minimization, and dynamics calculations. *Journal of Computational Chemistry, 4*, 187–217.
16. Merz, K. M., Jr., & Kollman, P. A. (1989). Free energy perturbation simulations of the inhibition of thermolysin: Prediction of the free energy of binding of a new inhibitor. *Journal of the American Chemical Society, 111*, 5649–5658.
17. Cournia, Z., Allen, B., & Sherman, W. (2017). Relative binding free energy calculations in drug discovery: Recent advances and practical considerations. *Journal of Chemical Information and Modeling, 57*, 2911–2937.
18. Song, L. F., & Merz, K. M., Jr. (2020). Evolution of alchemical free energy methods in drug discovery. *Journal of Chemical Information and Modeling, 60*, 5308–5318.
19. Boresch, S., Tettinger, F., Leitgeb, M., & Karplus, M. (2003). Absolute binding free energies: A quantitative approach for their calculation. *The Journal of Physical Chemistry. B, 107*, 9535–9551.
20. Aldeghi, M., Heifetz, A., Bodkin, M. J., Knapp, S., & Biggin, P. C. (2016). Accurate calculation of the absolute free energy of binding for drug molecules. *Chemical Science, 7*, 207–218.
21. Chen, W., Cui, D., Jerome, S. V., Michino, M., Lenselink, E. B., Huggins, D., et al. (2023). Enhancing hit discovery in virtual screening through accurate calculation of absolute protein-ligand binding free energies. *Journal of Chemical Information and Modeling, 63*(10), 3171–3185.
22. Ross, G. A., Lu, C., Scarabelli, G., Albanese, S. K., Houang, E., Abel, R., et al. (2023). The maximal and current accuracy of rigorous protein-ligand binding free energy calculations. *Communications Chemistry, 6*, 222.
23. Sun, S., & Huggins, D. J. (2022). Assessing the effect of forcefield parameter sets on the accuracy of relative binding free energy calculations. *Frontiers in Molecular Biosciences*, 9.
24. Shirts, M. R., Mobley, D. L., & Chodera, J. D. (2007). Chapter 4 Alchemical free energy calculations: Ready for prime time? In D. C. Spellmeyer & R. Wheeler (Eds.), *Annual reports in computational chemistry* (pp. 41–59). Elsevier.
25. Mondal, D., Florian, J., & Warshel, A. (2019). Exploring the effectiveness of binding free energy calculations. *The Journal of Physical Chemistry. B, 123*, 8910–8915.
26. Fratev, F., & Sirimulla, S. (2019). An improved free energy perturbation FEP+ sampling protocol for flexible ligand-binding domains. *Scientific Reports, 9*, 16829.
27. Rocklin, G. J., Mobley, D. L., & Dill, K. A. (2013). Calculating the sensitivity and robustness of binding free energy calculations to force field parameters. *Journal of Chemical Theory and Computation, 9*, 3072–3083.
28. Wang, L., Chambers, J., & Abel, R. (2019). Protein–ligand binding free energy calculations with FEP+. In M. Bonomi & C. Camilloni (Eds.), *Biomolecular simulations: Methods and protocols* (pp. 201–232). New York, NY, Springer New York.
29. Lu, C., Wu, C., Ghoreishi, D., Chen, W., Wang, L., Damm, W., et al. (2021). OPLS4: Improving force field accuracy on challenging regimes of chemical space. *Journal of Chemical Theory and Computation, 17*, 4291–4300.

30. Sugita, Y., & Okamoto, Y. (1999). Replica-exchange molecular dynamics method for protein folding. *Chemical Physics Letters, 314*, 141–151.
31. Wang, L., Friesner, R. A., & Berne, B. J. (2011). Replica exchange with solute scaling: A more efficient version of replica exchange with solute tempering (REST2). *The Journal of Physical Chemistry. B, 115*, 9431–9438.
32. Corey, R. A., Vickery, O. N., Sansom, M. S. P., & Stansfeld, P. J. (2019). Insights into membrane protein–lipid interactions from free energy calculations. *Journal of Chemical Theory and Computation, 15*, 5727–5736.
33. Ghanakota, P., Bos, P. H., Konze, K. D., Staker, J., Marques, G., Marshall, K., et al. (2020). Combining cloud-based free-energy calculations, synthetically aware enumerations, and goal-directed generative machine learning for rapid large-scale chemical exploration and optimization. *Journal of Chemical Information and Modeling, 60*, 4311–4325.
34. Gilson, M. K., & Zhou, H.-X. (2007). Calculation of protein-ligand binding affinities. *Annual Review of Biophysics and Biomolecular Structure, 36*, 21–42.
35. Bieniek, M. K., Wade, A. D., Bhati, A. P., Wan, S., & Coveney, P. V. (2023). TIES 2.0: A dual-topology open source relative binding free energy builder with web portal. *Journal of Chemical Information and Modeling, 63*, 718–724.
36. Senapathi, T., Suruzhon, M., Barnett, C. B., Essex, J., & Naidoo, K. J. (2020). BRIDGE: An open platform for reproducible high-throughput free energy simulations. *Journal of Chemical Information and Modeling, 60*, 5290–5295.
37. Kim, S., Oshima, H., Zhang, H., Kern, N. R., Re, S., Lee, J., et al. (2020). CHARMM-GUI free energy calculator for absolute and relative ligand solvation and binding free energy simulations. *Journal of Chemical Theory and Computation, 16*, 7207–7218.
38. Zhang, H., Kim, S., Giese, T. J., Lee, T.-S., Lee, J., York, D. M., et al. (2021). CHARMM-GUI free energy calculator for practical ligand binding free energy simulations with AMBER. *Journal of Chemical Information and Modeling, 61*, 4145–4151.
39. He, X., Liu, S., Lee, T.-S., Ji, B., Man, V. H., York, D. M., et al. (2020). Fast, accurate, and reliable protocols for routine calculations of protein–ligand binding affinities in drug design projects using AMBER GPU-TI with ff14SB/GAFF. *ACS Omega, 5*, 4611–4619.
40. Wang, L., Wu, Y., Deng, Y., Kim, B., Pierce, L., Krilov, G., et al. (2015). Accurate and reliable prediction of relative ligand binding potency in prospective drug discovery by way of a modern free-energy calculation protocol and force field. *Journal of the American Chemical Society, 137*, 2695–2703.
41. Carvalho Martins, L., Cino, E. A., & Ferreira, R. S. (2021). PyAutoFEP: An automated free energy perturbation workflow for GROMACS integrating enhanced sampling methods. *Journal of Chemical Theory and Computation, 17*, 4262–4273.
42. Petrov, D. (2021). Perturbation free-energy toolkit: An automated alchemical topology builder. *Journal of Chemical Information and Modeling, 61*, 4382–4390.
43. Muegge, I., & Hu, Y. (2023). Recent advances in alchemical binding free energy calculations for drug discovery. *ACS Medicinal Chemistry Letters, 14*, 244–250.
44. Kuhn, B., Tichý, M., Wang, L., Robinson, S., Martin, R. E., Kuglstatter, A., et al. (2017). Prospective evaluation of free energy calculations for the prioritization of cathepsin L inhibitors. *Journal of Medicinal Chemistry, 60*, 2485–2497.
45. Abel, R., Mondal, S., Masse, C., Greenwood, J., Harriman, G., Ashwell, M. A., et al. (2017). Accelerating drug discovery through tight integration of expert molecular design and predictive scoring. *Current Opinion in Structural Biology, 43*, 38–44.
46. Gaieb, Z., Parks, C. D., Chiu, M., Yang, H., Shao, C., Walters, W. P., et al. (2019). D3R grand challenge 3: Blind prediction of protein–ligand poses and affinity rankings. *Journal of Computer-Aided Molecular Design, 33*, 1–18.
47. Cournia, Z., Allen, B. K., Beuming, T., Pearlman, D. A., Radak, B. K., & Sherman, W. (2020). Rigorous free energy simulations in virtual screening. *Journal of Chemical Information and Modeling, 60*, 4153–4169.
48. Scheen, J., Wu, W., Mey, A. S. J. S., Tosco, P., Mackey, M., & Michel, J. (2020). Hybrid alchemical free energy/machine-learning methodology for the computation of hydration free energies. *Journal of Chemical Information and Modeling, 60*, 5331–5339.

49. Thompson, J., Walters, W. P., Feng, J. A., Pabon, N. A., Xu, H., Maser, M., et al. (2022). Optimizing active learning for free energy calculations. *Artificial Intelligence in the Life Sciences, 2*, 100050.
50. Majellaro, M., Jespers, W., Crespo, A., Núñez, M. J., Novio, S., Azuaje, J., et al. (2021). 3,4-dihydropyrimidin-2(1H)-ones as antagonists of the human A2B adenosine receptor: Optimization, structure–activity relationship studies, and enantiospecific recognition. *Journal of Medicinal Chemistry, 64*, 458–480.
51. Shan, Y., Mysore, V. P., Leffler, A. E., Kim, E. T., Sagawa, S., & Shaw, D. E. (2022). How does a small molecule bind at a cryptic binding site? *PLOS Computational Biology, 18*, e1009817.
52. Kimura, S. R., Hu, H. P., Ruvinsky, A. M., Sherman, W., & Favia, A. D. (2017). Deciphering cryptic binding sites on proteins by mixed-solvent molecular dynamics. *Journal of Chemical Information and Modeling, 57*, 1388–1401.
53. Goodford, P. J. (1985). A computational procedure for determining energetically favorable binding sites on biologically important macromolecules. *Journal of Medicinal Chemistry, 28*, 849–857.
54. Alvarez-Garcia, D., & Barril, X. (2014). Molecular simulations with solvent competition quantify water displaceability and provide accurate interaction maps of protein binding sites. *Journal of Medicinal Chemistry, 57*, 8530–8539.
55. Alvarez-Garcia, D., Schmidtke, P., Cubero, E., & Barril, X. (2022). Extracting atomic contributions to binding free energy using molecular dynamics simulations with mixed solvents (MDmix). *Current Drug Discovery Technologies, 19*, 62–68.
56. Deng, Y., & Roux, B. (2008). Computation of binding free energy with molecular dynamics and grand canonical Monte Carlo simulations. *The Journal of Chemical Physics, 128*.
57. Ross, G. A., Bodnarchuk, M. S., & Essex, J. W. (2015). Water sites, networks, and free energies with grand canonical Monte Carlo. *Journal of the American Chemical Society, 137*, 14930–14943.
58. Michel, J., Tirado-Rives, J., & Jorgensen, W. L. (2009). Energetics of displacing water molecules from protein binding sites: Consequences for ligand optimization. *Journal of the American Chemical Society, 131*, 15403–15411.
59. Ben-Shalom, I. Y., Lin, Z., Radak, B. K., Lin, C., Sherman, W., & Gilson, M. K. (2020). Accounting for the central role of interfacial water in protein–ligand binding free energy calculations. *Journal of Chemical Theory and Computation, 16*, 7883–7894.
60. Bruce Macdonald, H. E., Cave-Ayland, C., Ross, G. A., & Essex, J. W. (2018). Ligand binding free energies with adaptive water networks: Two-dimensional grand canonical alchemical perturbations. *Journal of Chemical Theory and Computation, 14*, 6586–6597.
61. Nussinov, R., & Tsai, C.-J. (2012). The different ways through which specificity works in orthosteric and allosteric drugs. *Current Pharmaceutical Design, 18*, 1311.
62. Liang, L., Liu, H., Xing, G., Deng, C., Hua, Y., Gu, R., et al. (2022). Accurate calculation of absolute free energy of binding for SHP2 allosteric inhibitors using free energy perturbation. *Physical Chemistry Chemical Physics, 24*, 9904–9920.
63. Zheng, H., Alter, S., & Qu, C.-K. (2009). SHP-2 tyrosine phosphatase in human diseases. *International Journal of Clinical and Experimental Medicine, 2*, 17.
64. Torrente, E., Fodale, V., Ciammaichella, A., Ferrigno, F., Ontoria, J. M., Ponzi, S., et al. (2023). Discovery of a novel series of imidazopyrazine derivatives as potent SHP2 allosteric inhibitors. *ACS Medicinal Chemistry Letters, 14*, 156–162.
65. Xiaoli, A., Yuzhen, N., Qiong, Y., Yang, L., Yao, X., & Bing, Z. (2022). Investigating the dynamic binding behavior of PMX53 cooperating with allosteric antagonist NDT9513727 to C5a anaphylatoxin chemotactic receptor 1 through Gaussian accelerated molecular dynamics and free-energy perturbation simulations. *ACS Chemical Neuroscience, 13*, 3502–3511.
66. Fu, H., Chen, H., Cai, W., Shao, X., & Chipot, C. (2021). BFEE2: Automated, streamlined, and accurate absolute binding free-energy calculations. *Journal of Chemical Information and Modeling, 61*, 2116–2123.

67. Backus, K. M., Correia, B. E., Lum, K. M., Forli, S., Horning, B. D., González-Páez, G. E., et al. (2016). Proteome-wide covalent ligand discovery in native biological systems. *Nature, 534*, 570–574.
68. Sutanto, F., Konstantinidou, M., & Dömling, A. (2020). Covalent inhibitors: A rational approach to drug discovery. *RSC Medicinal Chemistry, 11*, 876–884.
69. Chatterjee, P., Botello-Smith, W. M., Zhang, H., Qian, L., Alsamarah, A., Kent, D., et al. (2017). Can relative binding free energy predict selectivity of reversible covalent inhibitors? *Journal of the American Chemical Society, 139*, 17945–17952.
70. Awoonor-Williams, E., Walsh, A. G., & Rowley, C. N. (2017). Modeling covalent-modifier drugs. *Biochimica et Biophysica Acta (BBA) - Proteins and Proteomics, 1865*, 1664–1675.
71. Bonatto, V., Shamim, A., Rocho, F. D. R., Leitão, A., Luque, F. J., Lameira, J., et al. (2021). Predicting the relative binding affinity for reversible covalent inhibitors by free energy perturbation calculations. *Journal of Chemical Information and Modeling, 61*, 4733–4744.
72. Zhang, H., Jiang, W., Chatterjee, P., & Luo, Y. (2019). Ranking reversible covalent drugs: From free energy perturbation to fragment docking. *Journal of Chemical Information and Modeling, 59*, 2093–2102.
73. Lameira, J., Bonatto, V., Cianni, L., dos Reis, R. F., Leitão, A., & Montanari, C. A. (2019). Predicting the affinity of halogenated reversible covalent inhibitors through relative binding free energy. *Physical Chemistry Chemical Physics, 21*, 24723–24730.
74. Zhao, H. (2007). Scaffold selection and scaffold hopping in lead generation: a medicinal chemistry perspective. *Drug Discovery Today, 12*, 149–155.
75. Wang, L., Deng, Y., Wu, Y., Kim, B., LeBard, D. N., Wandschneider, D., et al. (2017). Accurate modeling of scaffold hopping transformations in drug discovery. *Journal of Chemical Theory and Computation, 13*, 42–54.
76. Wu, D., Zheng, X., Liu, R., Li, Z., Jiang, Z., Zhou, Q., et al. (2022). Free energy perturbation (FEP)-guided scaffold hopping. *Acta Pharmaceutica Sinica B, 12*, 1351–1362.
77. Jespers, W., Esguerra, M., Åqvist, J., & Gutiérrez-de-Terán, H. (2019). QligFEP: An automated workflow for small molecule free energy calculations in Q. *Journal of Cheminformatics, 11*, 26.
78. Pennington, L. D., Aquila, B. M., Choi, Y., Valiulin, R. A., & Muegge, I. (2020). Positional analogue scanning: An effective strategy for multiparameter optimization in drug design. *Journal of Medicinal Chemistry, 63*, 8956–8976.
79. Wade, A. D., Rizzi, A., Wang, Y., & Huggins, D. J. (2019). Computational fluorine scanning using free-energy perturbation. *Journal of Chemical Information and Modeling, 59*, 2776–2784.
80. Pérez-Benito, L., Casajuana-Martin, N., Jiménez-Rosés, M., van Vlijmen, H., & Tresadern, G. (2019). Predicting activity cliffs with free-energy perturbation. *Journal of Chemical Theory and Computation, 15*, 1884–1895.
81. Hu, Y., & Muegge, I. (2022). In silico positional analogue scanning with Amber GPU-TI. *Journal of Chemical Information and Modeling, 62*, 4448–4459.
82. De Vivo, M., Masetti, M., Bottegoni, G., & Cavalli, A. (2016). Role of molecular dynamics and related methods in drug discovery. *Journal of Medicinal Chemistry, 59*, 4035–4061.
83. Jiang, W., Hodoscek, M., & Roux, B. (2009). Computation of absolute hydration and binding free energy with free energy perturbation distributed replica-exchange molecular dynamics. *Journal of Chemical Theory and Computation, 5*, 2583–2588.
84. Jiang, W., & Roux, B. (2010). Free energy perturbation Hamiltonian replica-exchange molecular dynamics (FEP/H-REMD) for absolute ligand binding free energy calculations. *Journal of Cchemical Theory and Computation, 6*, 2559–2565.
85. Wang, L., Berne, B., & Friesner, R. A. (2012). On achieving high accuracy and reliability in the calculation of relative protein–ligand binding affinities. *National Academy of Sciences of the United States of America, 109*, 1937–1942.

86. Raman, E. P., Paul, T. J., Hayes, R. L., & Brooks, C. L. I. (2020). Automated, accurate, and scalable relative protein–ligand binding free-energy calculations using lambda dynamics. *Journal of Chemical Theory and Computation, 16*, 7895–7914.
87. Wan, S., Bhati, A. P., & Coveney, P. V. (2023). Comparison of equilibrium and nonequilibrium approaches for relative binding free energy predictions. *Journal of Chemical Theory and Computation, 19*, 7846–7860.
88. Hummer, G. (2007). Nonequilibrium methods for equilibrium free energy calculations. In C. Chipot & A. Pohorille (Eds.), *Free energy calculations: Theory and applications in chemistry and biology* (pp. 171–198). Springer Berlin Heidelberg.
89. Crooks, G. E. (1998). Nonequilibrium measurements of free energy differences for microscopically reversible Markovian systems. *Journal of Statistical Physics, 90*, 1481–1487.
90. Crooks, G. E. (1999). Entropy production fluctuation theorem and the nonequilibrium work relation for free energy differences. *Physical Review E, 60*, 2721–2726.
91. Gapsys, V., Pérez-Benito, L., Aldeghi, M., Seeliger, D., van Vlijmen, H., Tresadern, G., et al. (2020). Large scale relative protein ligand binding affinities using non-equilibrium alchemy. *Chemical Science, 11*, 1140–1152.
92. Gapsys, V., Hahn, D. F., Tresadern, G., Mobley, D. L., Rampp, M., & de Groot, B. L. (2022). Pre-exascale computing of protein–ligand binding free energies with open source software for drug design. *Journal of Chemical Information and Modeling, 62*, 1172–1177.
93. Kong, X., & Brooks, C. L., III. (1996). λ-dynamics: A new approach to free energy calculations. *The Journal of Chemical Physics, 105*, 2414–2423.
94. Knight, J. L., & Brooks, C. L., III. (2009). λ-Dynamics free energy simulation methods. *Journal of Computational Chemistry, 30*, 1692–1700.
95. Banba, S., & Brooks, C. L., III. (2000). Free energy screening of small ligands binding to an artificial protein cavity. *The Journal of Chemical Physics, 113*, 3423–3433.
96. Guo, Z., & Brooks, C. L. (1998). Rapid screening of binding affinities: Application of the λ-dynamics method to a trypsin-inhibitor system. *Journal of the American Chemical Society, 120*, 1920–1921.
97. Knight, J. L., & Brooks, C. L., III. (2011). Applying efficient implicit nongeometric constraints in alchemical free energy simulations. *Journal of Computational Chemistry, 32*, 3423–3432.
98. Robo, M. T., Hayes, R. L., Ding, X., Pulawski, B., & Vilseck, J. Z. (2023). Fast free energy estimates from λ-dynamics with bias-updated Gibbs sampling. *Nature Communications, 14*, 8515.
99. Knight, J. L., & Brooks, C. L. I. (2011). Multisite λ dynamics for simulated structure–activity relationship studies. *Journal of Chemical Theory and Computation, 7*, 2728–2739.
100. Armacost, K. A., Goh, G. B., & Brooks, C. L. I. (2015). Biasing potential replica exchange multisite λ-dynamics for efficient free energy calculations. *Journal of Chemical Theory and Computation, 11*, 1267–1277.
101. Ding, X., Vilseck, J. Z., Hayes, R. L., & Brooks, C. L. I. (2017). Gibbs Sampler-based λ-dynamics and Rao–Blackwell estimator for alchemical free energy calculation. *Journal of Chemical Theory and Computation, 13*, 2501–2510.
102. Vilseck, J. Z., Ding, X., Hayes, R. L., & Brooks, C. L. I. (2021). Generalizing the discrete Gibbs Sampler-based λ-dynamics approach for multisite sampling of many ligands. *Journal of Chemical Theory and Computation, 17*, 3895–3907.
103. Russell, S. J., & Norvig, P. (2010). *Artificial intelligence a modern approach.*
104. Rufa, D. A., Bruce Macdonald, H. E., Fass, J., Wieder, M., Grinaway, P. B., Roitberg, A. E., et al. (2020). Towards chemical accuracy for alchemical free energy calculations with hybrid physics-based machine learning/molecular mechanics potentials. *bioRxiv*, 2020.07.29.227959.
105. Rizzi, A., Carloni, P., & Parrinello, M. (2021). Targeted free energy perturbation revisited: Accurate free energies from mapped reference potentials. *Journal of Physical Chemistry Letters, 12*, 9449–9454.

106. Wirnsberger, P., Ballard, A. J., Papamakarios, G., Abercrombie, S., Racanière, S., Pritzel, A., et al. (2020). Targeted free energy estimation via learned mappings. *The Journal of Chemical Physics, 153*, 144112.
107. Knight, J. L., Leswing, K., Bos, P. H., & Wang, L. (2021). Impacting drug discovery projects with large-scale enumerations, machine learning strategies, and free-energy predictions. In *Free energy methods in drug discovery: Current state and future directions* (pp. 205–226). American Chemical Society.
108. Deringer, V. L., Bartók, A. P., Bernstein, N., Wilkins, D. M., Ceriotti, M., & Csányi, G. (2021). Gaussian process regression for materials and molecules. *Chemical Reviews, 121*, 10073–10141.
109. Heikamp, K., & Bajorath, J. (2014). Support vector machines for drug discovery. *Expert Opinion on Drug Discovery, 9*, 93–104.
110. Cai, C., Wang, S., Xu, Y., Zhang, W., Tang, K., Ouyang, Q., et al. (2020). Transfer learning for drug discovery. *Journal of Medicinal Chemistry, 63*, 8683–8694.
111. Lim, J., Ryu, S., Kim, J. W., & Kim, W. Y. (2018). Molecular generative model based on conditional variational autoencoder for de novo molecular design. *Journal of Cheminformatics, 10*, 31.
112. Konze, K. D., Bos, P. H., Dahlgren, M. K., Leswing, K., Tubert-Brohman, I., Bortolato, A., et al. (2019). Reaction-based enumeration, active learning, and free energy calculations to rapidly explore synthetically tractable chemical space and optimize potency of cyclin-dependent kinase 2 inhibitors. *Journal of Chemical Information and Modeling, 59*, 3782–3793.
113. Willow, S. Y., Kang, L., & Minh, D. D. (2023). Learned mappings for targeted free energy perturbation between peptide conformations. *arXiv*. preprint arXiv:230614010.

Chapter 11
Ultra-Large-Scale Virtual Screening

Ina Pöhner, Toni Sivula, and Antti Poso

Abstract Recently, make-on-demand chemical libraries and spaces began to grow into the billion scale, resulting in a surge of ultra-large ligand- and receptor-based virtual screening (VS) approaches. This chapter introduces state-of-the-art screening compound resources with up to 290 trillion compounds and provides examples of ultra-large VS campaigns performed between 2019 and 2024. We emphasize key considerations for the planning stage of your own ultra-large VS and the choice of the computing environment. Some of the covered highlights include 2D similarity searches in nonenumerated spaces, 3D shape screening, giga-scale brute-force docking, and screening acceleration strategies either leveraging machine learning and deep learning or relying on the reagent-space of recent make-on-demand combinatorial libraries. To scrutinize whether and when "bigger is better," we discuss the critical challenge of hit triage on the ultra-large scale and conclude with future perspectives for VS in the face of continuing library growth.

1 Background

The early-stage hit identification stage of drug discovery efforts frequently features in silico methods, which are typically considered cheaper and faster than their in vitro equivalents [1]. In silico screening enables the timely high-throughput evaluation of large numbers of potential candidate molecules. This computational prioritization of initial virtual hits is commonly termed virtual screening (VS). Traditionally, VS campaigns have either been ligand-centric or receptor-based [2].

I. Pöhner (✉) · A. Poso
School of Pharmacy, University of Eastern Finland, Kuopio, Finland
e-mail: ina.pohner@uef.fi

T. Sivula
School of Pharmacy, University of Eastern Finland, Kuopio, Finland

CSC — IT Center for Science Ltd., Life Science Center Keilaniemi, Espoo, Finland

V. G. Maltarollo (ed.), *Computer-Aided and Machine Learning-Driven Drug Design*, Computer-Aided Drug Discovery and Design 3,
https://doi.org/10.1007/978-3-031-76718-0_11

Ligand-based VS uses structural information and physicochemical properties of known active compounds and screens a compound library for virtual hits that share some of the actives' features. For example, classical quantitative structure–activity/property relationships (QSAR/QSPR) and fingerprint-based searches rely solely on 2D structure and topology, which allows for their exceptionally high throughput. However, especially in data-sparse scenarios, such models have limitations and often generalize poorly [1, 3, 4]. Approaches that utilize the 3D structural information of the small molecules, such as shape-matching algorithms or pharmacophore searches, are often superior at capturing the intricacies of biological action in our 3D world [3, 4]. They are, however, generally slower than their 2D-only counterparts [4].

When starting from known active scaffolds, ligand-based approaches are attractive for performing scaffold hopping [5]. However, their inbuilt dependence on known actives (or ligands with the desired properties) is also their biggest drawback: In prospective screens, where such prior knowledge does not exist, ligand-based VS cannot be applied.

If a molecular target, for instance, an enzyme, can be defined, receptor-based VS approaches can be used (instead)—even when no prior knowledge about possible binding partners exists. The most common technique to discover virtual hits for a target receptor is molecular docking. Docking first fits ligand 3D structures from large databases into the receptor binding pocket. It then assigns a docking score, aiming to quantify the formed interactions and rank the ligands [1, 2].

It is a general notion that larger compound library sizes increase the probability of finding a close-to-optimal match with the desired properties during a VS campaign [3, 6, 7]. Thus, unsurprisingly, in step with a growth trend observed for screening compound libraries, ligand- and receptor-based ultra-large VS approaches have seamlessly evolved from their smaller-scale counterparts.

Before we dive deeper into the evolution of ultra-large VS, let us define what the "ultra-large scale" actually is. The term itself is fuzzy and used in the literature for screens of varying sizes, to date typically ranging from around 10^7 to 10^{10} processed small molecules. Herein, we define our lower limit for considering a library or VS approach as "ultra-large" arbitrarily at 50 million compounds.

Ultra-large libraries themselves are not a new phenomenon: Efforts to enumerate much of the hypothetically possible chemical space gave rise to virtual ultra-large libraries already several years back (see, e.g., references [3, 8, 9]. for a comprehensive overview). However, the organic synthesis of a virtual hit compound from such a library was often time-consuming and costly, potentially hampering or preventing the in vitro validation of the VS predictions. A tight project timeframe and budget or the lack of medicinal chemistry facilities were therefore arguments in favor of smaller libraries with commercially available compounds [1].

The recent popularity of ultra-large VS is owed to advances in robust parallel organic synthesis and the emergence of ultra-large *make-on-demand* libraries. These combinatorial libraries rely on predefined chemical reactions to link large collections of off-the-shelf building blocks. As a result, numerous make-on-demand compounds with a high probability of synthesis success can be ordered and readily purchasable in-stock libraries have also grown in size. For instance, the popular screening resource ZINC grew from 20 million commercially available compounds in 2012 to more than 37 billion by 2022 [10–12]. Likewise, the make-on-demand library Enamine REAL started off on the million scale and to date contains around 48 billion compounds [13, 14].

Previously, their high throughput was among the strongest arguments in favor of VS methods. However, as library sizes hit the billion scale, VS throughput became a limiting factor. For instance, the fastest docking methods can typically screen about 30–60 compounds per minute per computing core [3]. Consequently, a docking study of one million compounds on a workstation computer with 8 CPUs takes around 1.5–3 days. Processing 1 billion compounds with the same resources would require 4–8 years of nonstop docking for a single target. To tackle ultra-large screening projects, one thus needs to either massively scale up on computational resources or find strategies to avoid the bulk of the time- and resource-consuming computation steps.

In many cases, the basic operation of individual VS approaches is not altered when they are applied on the ultra-large scale. Therefore, this chapter will assume a basic familiarity of the reader with different VS methods and focus on challenges faced specifically on the ultra-large scale and the emerging solutions. For more information on the individual VS methods, such as QSAR, pharmacophores, and docking, the reader is, for examples, referred to Chaps. 6 and 7 of this book.

The chapter will introduce ultra-large screening databases and recent examples of ultra-large-scale ligand- and receptor-based screening campaigns with an emphasis on the past 5 years (2019–2024). As most ultra-large-scale screening campaigns utilize high-performance computing (HPC) or cloud computing resources, we will discuss some characteristics of such computing environments, important considerations for their use, and tools designed to cope with the growing numbers of screening compounds. As we enter an era where ultra-large screening campaigns will likely become more prominent and strive to govern even larger chemical spaces, we are only beginning to comprehend the associated pitfalls of "Big Data" in drug discovery. This chapter will also present some arguments in favor of and against ultra-large VS, flag so far unmet needs, and provide a perspective of what lies ahead for in silico screening campaigns in the future.

2 Ultra-Large Screening Libraries and Chemical Spaces

As indicated above, ultra-large VS has little to do with technological advances in screening methodology. Rather, it represents a development in response to the emergence of billion-scale make-on-demand chemical libraries. Thus, before diving deeper into the screening process and specific examples, we will look at the culprits that force us to rethink our approaches to VS.

We can distinguish two major types of ultra-large libraries: enumerated chemical libraries, where each compound is stored explicitly, and nonenumerated spaces [9, 15]. Enumeration of all possible combinations of building blocks is deemed inefficient when working with billion-scale combinatorial libraries. Therefore, nonenumerated spaces rely on storing the building blocks and all chemical reactions that can be used to interconnect them instead of the fully elaborated possible virtual products [15].

One famous example of a freely accessible ultra-large compound library is ZINC. ZINC was initially created as a compound research database focused on docking and offered downloadable precomputed 3D conformers for many of its compounds. In its 2012 release, ZINC already contained around 20 million purchasable compounds from different vendors [11]. By 2015, the library had become an ultra-large resource with around 120 million purchasable drug-like compounds [16]. ZINC15 offered enhanced programmatic access and downloadable ligand 3D coordinates in various formats, such as sdf, mol2, and pdbqt. It is this version of ZINC that would be prominent in many of the recent ultra-large receptor-based VS works (see Sect. 5.1) [17–19]. Meanwhile, ZINC continued to grow into the giga-scale: ZINC20, owing to the addition of compounds from make-on-demand suppliers, reached a total size of roughly 1.4 billion compounds, of which around 1.3 billion were purchasable [20]. The latest version ZINC22 added molecules from additional make-on-demand chemical spaces, increasing its total size to around 37 billion [12]. Currently, over 4 billion compounds can be downloaded from ZINC in 3D ready-to-dock formats.

Enamine is one of the vendors that allowed ZINC to grow into the billion scale. Their REAL (REadily AccessibLe) make-on-demand compounds can be typically obtained with a success rate of at least 80% within less than 1 month [13–15]. Enamine maintains a large collection of over 100,000 in-stock building blocks and relies on more than 167 robust validated synthesis protocols [13, 14]. REAL compounds represent another popular compound resource of recent ultra-large screening campaigns [21–23]. To date (March 2024), the REAL database contains about 6.75 billion enumerated compounds. Of the several available subsets, some are likewise ultra-large: The Enamine REAL lead-like set comprises 3.93 billion compounds, and Enamine also offers a library of nearly 158 million natural product-like compounds. Enamine's by far largest collection of compounds is an example of a nonenumerated space: the REAL Space currently contains over 48 billion virtual products, stored primarily as building blocks and connecting chemical reactions (although an enumerated version is available from Enamine upon request [13]).

Another vendor that facilitated ZINC's tremendous growth is WuXi AppTec [24, 25]. Their billion-scale make-on-demand library, GalaXi, likewise relies on predefined reaction schemes and a large in-stock building block collection. Currently, the enumerated virtual GalaXi library contains 16 billion compounds [25].

While working with enumerated ultra-large libraries can be achieved by scaling up conventional ligand- or receptor-based screening approaches, nonenumerated spaces require specialized technologies to search within their building blocks and the connecting reactions to enumerate relevant virtual products on the fly. Approaches for working with nonenumerated spaces will be discussed in detail later in the text (see, e.g., Sects. 4 and 6.2). Suffice it to say that nonenumerated spaces like the Enamine REAL Space or the nonenumerated variant of GalaXi, the GalaXi Space, can be searched with tools like the chemical space navigation

Table 11.1 Summary of discussed publicly available/searchable enumerated ultra-large screening libraries and nonenumerated chemical spaces

Library/Space	Approx. number of compounds	File format(s)
Enumerated		
CH/PMUNK	95 million	sdf (2D)
Enamine REAL database	6.75 billion	CXSMILES, sdf (2D)
ER lead-like	3.93 billion	CXSMILES
ER natural product-like	157.7 million	CXSMILES
GalaXi database	16 billion	SMILES
SAVI	1.75 billion	SMILES, sdf (2D)
ZINC22	37 billion (2D), 4.5 billion (3D)	SMILES, mol2, sdf, pdbqt, db2
ZINC20	1.4 billion, 1.3 billion purchasable	SMILES, mol2, sdf, pdbqt, db2
ZINC15	750 million (purchasable, 2D), 230 million (purchasable, 3D)	SMILES, mol2, sdf, pdbqt, db2
Nonenumerated		
CHEMriya	12 billion	–
Enamine REAL space	48 billion	–
eXplore	4.9 trillion	–
GalaXi space	16 billion	–
KnowledgeSpace	290 trillion	–

For enumerated libraries, available formats for compound download are reported. Corresponding URLs can be found in Table 11.9 in the Appendix. ER: Enamine REAL; Billion = 10^9; trillion = 10^{12}; db2: format for use with the docking tool DOCK

platform infiniSee [26]. Some vendors do not offer (part of) their enumerated collection for download but instead rely on its integration in chemical space search tools. With infiniSee and related approaches, additional billion-scale synthesis-on-demand spaces become searchable, for example, eMolecules' eXplore with 4.9 trillion virtual products or Otava's 12 billion on-demand space CHEMriya [27–29].

In light of the gigantic sizes of nonenumerated chemical spaces, the reader may wonder if the ability to search multiple spaces is even required. Interestingly, extensive efforts to compare chemical spaces have revealed surprisingly limited overlap between the different chemical spaces [15, 28].

Libraries and spaces discussed this far constituted commercially available compounds and make-on-demand vendor collections (see Table 11.1 for a summary). While the direct purchase of virtual hits can be an attractive option, other efforts cater more to the needs of those who wish to perform the compound synthesis themselves: While still relying on commercially available building blocks, reactions are derived from literature to ensure the synthetic feasibility of the virtual products.

For example, NCI's 1.75 billion enumerated compound library SAVI (Synthetically Accessible Virtual Inventory) relies on building blocks from Enamine and integrates expert knowledge into the definition of virtual synthesis rules [30]. Compounds contained within SAVI can be ranked by their synthetic accessibility, with the most synthesizable class currently encompassing 1.09 billion compounds [30]. Literature-derived chemical reactions and commercially available building blocks are also used to define KnowledgeSpace, a virtual chemical space created by BioSolveIT, the company behind infiniSee. KnowledgeSpace currently comprises 10^{14} virtual molecules and can be searched with infiniSee [15, 26, 31].

Many of the recently published ultra-large-scale VS efforts discussed in this Chapter stem from academic context and utilize the mentioned public and commercial libraries and spaces. However, ultra-large chemical spaces are also a very prominent sight in industrial drug development: Over the past years, many pharma companies have developed their own proprietary ultra-large chemical spaces, for instance, BICLAIM of Boehringer-Ingelheim and Merck's MASSIV (10^{20} virtual products), or GlaxoSmithKline's GSK XXL (10^{26} virtual products) [9, 15]. Therefore, tools and protocols to work with and screen ultra-large chemical spaces are of great interest to academic and industrial drug discovery alike.

All libraries and spaces mentioned so far mainly comprise drug-like small molecules. Compounds targeting protein–protein interactions (PPIs) typically exhibit a different property profile that may be underrepresented in libraries focused on drug-likeness. This motivated the authors of CH*I*PMUNK (CHemically feasible *In silico* Public Molecular UNiverse Knowledge base) to propose a specialized ultralarge library: They used commercially available building blocks and selected virtual reactions together with targeted property filtering to generate a library of 95 million synthetically accessible compounds with PPI modulator properties [32].

Table 11.1 summarizes the discussed chemical libraries and spaces, and Table 11.9 in the Appendix lists the corresponding URLs to obtain the different compound collections.

3 How Ultra-Large Chemical Libraries Alter VS Workflows

After familiarizing ourselves with the state-of-the-art ultra-large libraries and spaces, we may ponder how working on the ultra-large scale affects VS workflows in practice. Before we discuss specific examples of ultra-large ligand-and receptor-based screens, we first aim to establish important key considerations for ultra-large-scale screening workflows and discuss what sets them apart from smaller-scale approaches.

Our Giga-Scale VS Experience

We started our first venture into billion-scale VS in 2021. On the surface, the project seemed simple: We knew how to run ligand preparation and docking pipelines for millions of compounds. All we needed to do was to think a little bigger this time! We quickly realized that generating 3D conformers for over 1.5 billion compounds and docking them to a protein target on an HPC cluster was not as simple as that. HPCs are shared environments and rely on fair sharing of computational resources among all users, so we had to ensure that our study would not disrupt anyone else's work [33]. Tools were not designed for handling the huge input, data sizes began to surge, and temporary files started clogging up the storage space and straining the file system.

The lessons learned in this endeavor went into the preparation of this section to help others avoid similar pitfalls [23, 33].

3.1 Implications of Dataset Size

When working with enumerated chemical libraries, compounds will commonly be obtained in 1D (SMILES) or 2D (sdf) representation. On the million scale, storing this data is usually straightforward. However, ultra-large enumerated libraries begin to require a considerable amount of disk space: For example, SMILES strings of 1.56 billion lead-like compounds used in our work had a total size of 89 GB, and 10 billion SMILES consistently were reported to take up around 1 TB of disk space [4, 23].

To speed up the download process and make the file(s) easier to handle, ultra-large libraries are usually offered in a compressed format (e.g., zipped) and in multiple parts. If the compression was performed on individual parts, and depending on the intended use of the library, you may not need to store the fully uncompressed data on your disk. Nonetheless, keep in mind that uncompressed data will be several-fold larger than the compressed files, depending on the data format and compression algorithm used. For instance, the compressed version of the 1.75 billion SMILES of SAVI takes up 440 GB, while the fully uncompressed data would require 4.4 TB of storage space. The 95 million compounds of CH*I*PMUNK in compressed sdf format stand at 109 GB and inflate to 588 GB when uncompressed. Thus, while one may simply obtain, decompress, and work with smaller libraries, it is advisable to carefully consider the amount of required disk space when handling ultra-large libraries [8]. Some space can often be saved by limiting decompression to parts that are actively being worked on.

Many VS approaches rely on the availability of 3D conformations for the small molecules. When obtaining ultra-large libraries in 1D or 2D representation, 3D conformers first need to be generated. This step, unsurprisingly, is heavily affected by the library size and requires careful consideration. First, the number of structures will increase during conformer generation. Therefore, the storage needs and the time required to process the library in subsequent steps will grow. How many conformers per molecule to generate is ultimately method- and case-dependent, but, as a rule of thumb for ultra-large approaches, one should strive for the lowest acceptable number of conformers for the chosen method.

For example, the limited conformer generation with a maximum number of eight tautomers and four stereoisomers per tautomer reported by Sivula et al. [23] inflated the number of structures in a 1.56 billion lead-like SMILES-based library around 2.4 times. In this study, the resulting 3D conformer databases in the proprietary Schrödinger Phase database format were 10.3 TB while compressed, and around 22.1 TB when uncompressed. These sizes illustrate the massive increase in disk space requirement when converting from 1D (89 GB) to 3D (22.1 TB). Similarly, Michino et al. [21] converted a version of Enamine REAL with 1.12 billion compounds into binary 3D structure files for their GPU-based shape-screening approach. In their case, they generated 404 files of 15 GB each, totaling 6 TB. Thus, although the exact difference is highly format-dependent, even the generation of a single 3D structure per SMILES string would already significantly increase the data size, and the problem is aggravated by the compound number inflation during conformer generation.

Structure inflation therefore needs to be borne in mind when planning 3D ligand or receptor-based screening campaigns. In addition to the 2.4-fold inflation reported by Sivula et al. [23], Gentile et al. [34] report conformer generation to inflate their 12.3 billion input SMILES by over three-fold to around 38.7 billion unique structures. Thus, the number of structures to process in subsequent steps of the VS workflow can easily double or triple and more structures inevitably cause increased time- and resource requirements.

A final size-related consideration affects the 3D conformer generation and all the following calculation steps: Many tools will first read all input molecules from one file into memory. This can become a limiting factor when working with ultra-large libraries, as the file sizes may exceed the available memory. Thus, it is essential to assess the memory requirements of different calculation steps, and the dataset needs to be split into appropriate batches to reflect the available resources.

3.2 Where and How? Guidelines for Planning and Running Ultra-Large VS Campaigns

In light of the limited throughput and the massive disk space and memory requirements discussed above, it is hardly surprising that most ultra-large-scale screening campaigns utilize HPC or cloud computing resources. While they have become more readily available over the past few years, massive computing resources often remain costly [5, 22, 35]. Consequently, when planning and executing an ultra-large screening campaign, the computing environment choice is entangled with the screening timeframe and budget. You may need to benchmark your tool(s), optimize the performance for the chosen computing environment, select appropriate data splits, and monitor progress and potential errors. Below, we outline some key questions we and others have found helpful in planning VS on the ultra-large scale and navigating the specific challenges of data-intensive screening campaigns.

Should one use CPU or GPU? This question is often predetermined by your choice of tools. However, workflows can sometimes run on either CPU or GPU, and dedicated GPU implementations of otherwise CPU-bound tools exist (e.g., the docking tool AutoDock4 and its corresponding GPU-reimplementation AutoDockGPU) [36, 37]. GPUs are best suited when tasks can be split into many independent subtasks to execute in parallel. With the independence criterion met, GPUs excel at processing particularly high volumes of data. On the other hand, tasks that involve sequential steps and require communication between different subtasks are more suitable for processing on CPUs.

When faced with the choice, throughput may dictate your decision, but cost and availability should also be considered: GPUs tend to be more expensive and, in an age of deep learning (DL) popularity, often more sought after. For example, as a rule, the prediction step in a DL workflow is faster on GPU. However, if CPU resources are significantly cheaper and awarded more swiftly, it may be more economical to run predictions on a larger number of CPUs to offset the speed difference [23]. Technically, it would also be possible to run the training step on CPUs. However, the speed difference is much more pronounced during training, making training on CPUs impractical. When you choose the processing hardware for your ultra-large-scale screening workflows, we recommend factoring in the different costs and queuing times for different types of resources in planning the various workflow stages.

Get to know your tool(s)—Profiling and parallelization When planning an ultra-large VS campaign, a prerequisite for optimizing workflows and budgeting time and resources is a clear picture of the expected resource consumption and behavior of your tool(s). In most HPC and cloud computing environments, you are required to specify and reserve computing resources, for example, the number of CPUs and GPUs and the memory required by your job. Resource reservations will often directly determine your billing but also affect, for example, your queuing time. In a nutshell, you want to ensure sufficient resources for your workflow to run without problems but avoid a major resource overhead you'd pay for without being able to utilize it.

Most computing environments offer special queues or instances for testing to help you determine how many CPUs/GPUs and how much memory to reserve. Test environments often have limited resources and a short wall time, but are awarded swiftly and enable you to profile tools in the relevant production environment. Some environments will offer graphical tools for monitoring the resources consumed by your test job(s). Alternatively, HPCs commonly rely on batch job management tools such as SLURM, which can provide an overview of resource consumption with dedicated commands such as seff, sacct, or scontrol. It is worth familiarizing yourself with the relevant profiling commands of the tool used in your environment to review, e.g., CPU and GPU utilization and memory consumption of your job. Finally, as almost all HPC environments are UNIX-based, handy command-line tools such as time, (h)top, and free can help you in monitoring runtime and resource consumption in the UNIX command line.

In practice, to decide on the number of computing units to reserve, start by determining the scaling behavior of your tool(s). Is your tool single-threaded or internally parallelized, that is, can it utilize more than a single CPU (or GPU)? If so, ideally, you would find a linear scaling behavior: Whenever you double the number of computing units, the computing time halves [7]. However, many tools do not (fully) scale linearly or have a hard limit on the number of computing cores they can utilize. In the latter case, when exceeding, for example, 8 CPUs, runtime will no longer improve. Efficient usage of multiple GPUs is, as a rule, more challenging to implement. Thus, many GPU-based tools rely on using only a single GPU.

Often, you can find hints on the scaling behavior from method publications, published use cases, repositories, or online documentation. Alternatively or additionally, you can also profile the behavior of your tool(s) yourself in a test environment. Time a small test run with a diverse random set of compounds with incrementally increasing fixed numbers of computing units (set in your test reservation and the tool—most tools will have settings/command line switches when internally parallelized). It may also be advisable to use a couple of different dataset sizes and monitor how the memory footprint is affected by input data size.

Once you established a CPU/GPU count where your tool scales favorably and assessed its memory footprint in dependence on the input data, you can put it all together and plan the degree of external parallelization, that is, decide how many instances of your tool to run in parallel in the production system. Keep license limitations in mind at this stage, where they apply. Having determined the number of instances and required CPU, GPU, and memory resources, you should now be able to estimate the overall runtime depending on your chosen compound library—which leads to the question of how to distribute an ultra-large dataset between the different parallel instances.

Splitting your input data As with any rule, there are exceptions, but most of the time, you want to split your ultra-large input data into at least as many chunks as processing instances launched. Keep in mind that different compounds have different processing speeds and most similar speeds can be expected for closely related compounds. Thus, in particular for combinatorial enumerated libraries, you should randomize the compound order in the workflow input. As a rule, compounds with high molecular weight and/or many rotatable bonds require higher processing time. Randomization will help to avoid idling resources due to compound batches with many compounds having a significantly slower processing speed.

For similar reasons, it may also be beneficial to serially run several smaller batches rather than one larger batch, as individual runtime differences would decrease and be less likely to result in idling resources. However, note that many HPCs use distributed file systems such as Lustre. One caveat is their unfavorable behavior with heavy file I/O stemming from large numbers of small files. Thus, if you split your data into too many chunks, your runtime may suffer a negative impact from slow file system response (be also mindful of temporary files in this context). In addition, HPCs often offer fast local storage solutions based on NVMe/SSD, that can help to circumvent file I/O problems.

Optimal file sizes should ensure that memory does not become an issue and individual jobs can finish processing well within the system wall time. If the file system negatively affects the performance of your workflow, you can often counteract: Many computing environments offer fast local storage resources, and solutions like RAMDisk can move load off the file system and into memory.

An exception applies when you fetch compound data from remote databases rather than working with files. In this case, network speed will influence your throughput. You may want to run separate queries for each instance of your tool. If you massively parallelize and run thousands of instances and queries simultaneously, this can in itself create problems when operating on a single ultra-large database. For databases, a counteracting measure is a concept called sharding, reminiscent of batch creation for files, where you distribute data into individual smaller databases.

Finally, if you have different tools in your workflow, you may want to separately optimize the size of compound batches/shards for each tool, as different tools can have rather different throughput and requirements [4].

3.3 Ultra-Large VS in a Nutshell: Key Questions and Trouble-Shooting at a Glance

For all of the pitfalls we discussed above, a lot more could be said, which would be beyond the scope of this work. We hope, however, to provide a starting point and relevant keywords for prospective users. To summarize, if you plan on embarking on your own ultra-large VS campaign and can answer the following questions, you are off to a good start:

Questions to answer when planning ultra-large VS

? How experienced are you in running workflows in an HPC/cloud environment, and is there user support available for optimization or troubleshooting?
Are your tools single- or multi-threaded?
How do the tools scale with multiple CPUs/GPUs?
What is the memory footprint, and how is it affected by input data sizes?
Are there any license limitations?
How many tools can you run in parallel in available queues/instances of your computing environment?
What is the available wall time?
Is your input data read from files or a remote database?
Do you require specialized storage solutions to circumvent heavy I/O load on distributed file systems?

Table 11.2 Typical VS performance bottlenecks, how to recognize them, and potential countermeasures

Type	Symptoms	Solutions
CPU/ GPU bound	100% utilization on one or more CPU cores/GPUs, potentially while other reserved ones idle	Increase parallelization by running more instances or increase method internal parallelization (if possible). Reserve/ spread over more computing resources
File I/O bound	Low CPU utilization, high file system utilization	Use fast local storage or fast SSD storage. Move (temporary) files to RAMDisk
Memory bound	Low CPU utilization with high swap memory usage, program crash	Reserve more memory or reduce memory usage by processing smaller chunks of data
Network bound	Low CPU usage, low file I/O	Prestage or preload data prior to processing

If, despite your best efforts, you find your performance unsatisfactory, Table 11.2 summarizes typical performance bottlenecks and how to recognize them and proposes potential solutions.

Arguably, different methods have their own specific challenges and not all of them can be covered in this work. Thus, consult other campaigns using the tools of your choosing. The following sections will provide a vast set of examples to help you inform your approach. Additionally, for further reading, several published examples discuss lessons learned from their ultra-large screening or key challenges faced in method development for large-scale screening (e.g., references [4, 7, 23]), and published protocols provide step-by-step instructions (e.g., references [24, 38]).

4 Ligands on the Ultra-Large Scale

At the heart of any ultra-large VS are the ultra-large ligand libraries. As we discussed ultra-large chemical libraries and spaces, we established their massive sizes and that specialized approaches for ligand-based VS are required when handling nonenumerated chemical spaces. This section will discuss examples of ligand-based 2D searches in enumerated and nonenumerated compound collections and dive deeper into 3D conformer generation strategies and 3D ligand-based VS.

4.1 Ultra-Large 2D Similarity Searches

In an era of ultra-large make-on-demand chemical libraries and vast commercially available nonenumerated chemical spaces, the search for analogs and substructure

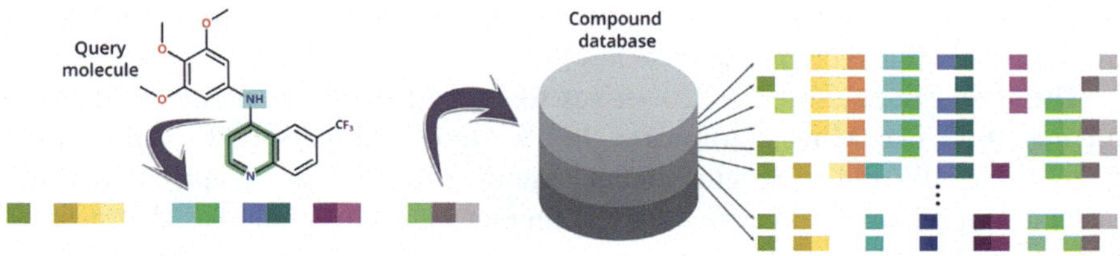

Fig. 11.1 Schematic illustration of a classical fingerprint-based search. The query molecule is transformed into a bit vector describing its structural features, such as "secondary amine" (blue) and "pyridine heterocycle" (green). In the example, different structural features (bits) are represented by different colors, where white indicates that the bit is off (i.e., the feature is not present). Once generated, the query bit vector is compared one by one with similarly constructed bit vectors for every compound in a large compound database

matches of known ligands represents a straightforward way to study structure–activity relationships (SARs) and extend the hit pool.

2D similarity searches achieve particularly high throughput and require limited to no preparation of the ultra-large libraries. They are rarely applied as the sole mode of VS, but their throughput makes them particularly valuable as prefilters to more computationally demanding techniques. However, they require prior knowledge of compounds/scaffolds with the desired properties and have limited potential in proposing entirely novel scaffolds. Additionally, as also illustrated schematically in Fig. 11.1 on the example of classical fingerprint-based searches, many approaches to 2D similarity searches rely on an exhaustive pairwise comparison and become slow and impractical when processing ultra-large libraries [9].

To enable efficient searches in ultra-large libraries, recent works have proposed various novel approaches and strategies that address the limited throughput. These involve, for example, novel types of fingerprints, the use of specialized formats and database operations, graph-based search methods, or searches in the reagent space [9].

In classical fingerprint-based searches (see Fig. 11.1), the bulk of the computation time is spent on fingerprint generation. Thus, precomputing fingerprint bit population statistics can speed up the process (e.g., in chemfp [39] and MolSoft's Giga-Search [40]), and some software vendors provide precomputed fingerprints for ultra-large databases to remove the fingerprint computing step altogether. Examples include Schrödinger's FP-Sim GPU and OpenEye's Molecules as a Service (MaaS) module in their cloud-based Orion software [41, 42]. Schrödinger's GPU Similarity search is available for billion-scale library screening via LiveDesign and has also been released as an open-source incentive version (see Table 11.10 in the Appendix) [9, 41, 43]. With four GPUs, Schrödinger's GPU Similarity search was able to screen 1 billion molecules in around 0.2 s [43]. OpenEye's MaaS achieved a 2D similarity search of 800 million Enamine REAL molecules in 3 s or less. However, to achieve this processing speed, data in MaaS is held in memory. MaaS therefore required an Amazon Web Services (AWS) instance with 768 GB of memory, which underlines how memory limitations can become major bottlenecks even for these relatively fast approaches [9].

The recent release of ZINC20 strives to support direct similarity searches by relying on NextMove Software Ltd.'s tools, SmallWorld and Arthor [20, 44]. SmallWorld utilizes maximum common subgraphs and graph edit distance in their graph-based similarity search. Importantly, using graph-based instead of fingerprint-based approaches is more suitable for and adaptable to large databases [9, 44]. Arthor is focused on high-performance SMARTS pattern and substructure matching and scales favorably on chemical libraries with hundreds of millions of compounds while handling relevant tautomers and resonances [9, 45]. ZINC's choice of tools underlines that, while not every similarity search tool fits the bill, efficient algorithms to sift through the massive state-of-the-art ultra-large libraries exist.

Another strategy tackles huge make-on-demand combinatorial chemical spaces by relying on their building blocks. Over 20 years ago, a graph-based strategy, FTrees-FS, was developed [46]. This method relies on so-called Feature Trees, and its searches conceptually resemble pharmacophoric-feature-based screens. Only building blocks and corresponding reactions that match the query get enumerated into virtual compounds, which allows for the efficient screening of ultra-large libraries [9, 46].

A more recent topological fingerprint similarity search called SpaceLight can, likewise, take advantage of combinatorial chemical spaces like the Enamine REAL Space [9, 47, 48]. It can process more than 20 billion compounds within seconds on a desktop computer and scales with the number of fragments/building blocks, rather than the number of virtual products [9]. Finally, another tool, termed SpaceMACS, was developed to enable a search for specific structural features within ultra-large chemical spaces [9, 49]. Since SpaceMACS first loads spaces into memory, it requires significant memory resources, but once loaded, ultra-large libraries and spaces can be screened for desired structural features within seconds [9].

FTrees-FS has, for example, been one of the various screening strategies employed by Glaab et al. [50] to discover novel inhibitors of SARS-CoV-2 main protease (Mpro, or 3CLpro in their work). They performed a ligand-based similarity search of close to 900 million compounds, largely sourced from ZINC15, which were compared to previously reported inhibitors of SARS-CoV-2 and SARS-CoV Mpro. Brüschweiler et al. [51] aimed to target the DNA-binding domain of the nuclear factor erythroid 2-related factor 2 (NRF2). The binding pocket of NRF2 has limited druggability and consequently generally displays modest hit rates at best. Thus, to find additional analogs of their initial fragment-based screening hits, the authors used the infiniSee software, similar in concept to FTrees-FS, to search 15.5 billion virtual compounds from the Enamine REAL Space [9, 26, 51].

Given their dependency on previous knowledge and limited generalizability, one should take particular care when selecting the template (sub)structure(s) for 2D similarity searches. At the same time, the presented application examples demonstrate how 2D similarity searches can act as powerful screening tools in conjunction with other methods or support the extension of the hit pool in later stages of VS.

4.2 *Recent Examples of Ultra-Large 3D Ligand-Based Screening*

Recent examples of 3D ligand-based screening cover, in particular, ultra-large 3D shape similarity searches, as schematically illustrated in Fig. 11.2, which commonly leverage GPUs for increased throughput.

For example, Grebner et al. [4] utilized the GPU-based FastROCS method in OpenEye's cloud-based platform Orion to screen subsets of the Astra Zeneca virtual chemistry space with up to 12.7 billion compounds. The shape similarity searches were initiated from known crystallographic ligands of ten diverse target receptors, including G Protein-Coupled Receptor Kinase 2, estrogen receptor, dipeptidyl peptidase 4, and MEK1 kinase. In their work, the authors present a comprehensive benchmark of FastROCS on AWS. They report detailed database generation and shape search timings and resource requirements for screens of varying sizes: Screening 51 million, 172 million, 1.33 billion, and 12.7 billion molecules, they demonstrate the scalability of 3D shape similarity searches [4].

To address the growing need for novel anti-mycobacterial agents, Michino et al. [21] targeted the challenging mycobacterial lipoamide dehydrogenase (Lpd). An extensive experimental high-throughput assay of around 2.5 million compounds had identified only two lead compounds. To leverage recent advances in ultra-large VS and extend the hit pool, the authors utilized the GPU Shape screen method by Schrödinger to screen 1.12 billion compounds from Enamine REAL. Around 800,000 candidates entered a post-filtering process including molecular docking, and a final set of around 100 candidates was purchased, yielding six novel hit compounds with IC_{50}s between 5 and 100 μM. While Lpd requires a rather specific narrow chemical space, the new hit compounds displayed some variation compared to previously known inhibitors. The authors did not report the time invested in

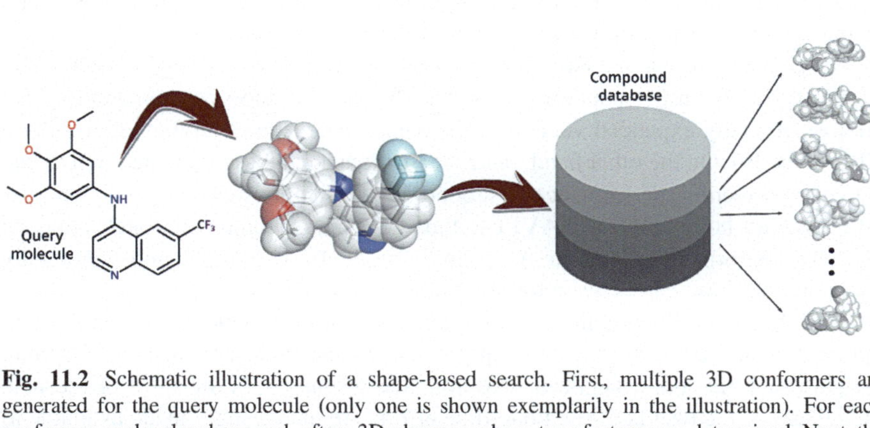

Fig. 11.2 Schematic illustration of a shape-based search. First, multiple 3D conformers are generated for the query molecule (only one is shown exemplarily in the illustration). For each conformer, molecular shape and, often, 3D-pharmacophore-type features are determined. Next, the overlap between the query shape/feature ensemble and predetermined shape and feature descriptors of all compounds in a large database can be computed

Table 11.3 Summary of recent ultra-large 2D and 3D ligand-based VS campaigns

Compounds	Target (s) screened	Tool	Approx. throughput	Reference(s)
15.5 billion	NRF2	infiniSee	NA	Brüschweiler et al. [51]
Up to 12.7 billion	10 diverse targets	FastROCS	480,000 - 1.5 million cpds/min/GPU	Grebner et al. [4]
1.12 billion	Lpd	GPU shape	120,000 cpds/min/GPU	Michino et al. [21]
900 million	SARS-CoV-2 Mpro	FTrees-FS	NA	Glaab et al. [50]

Reported throughput covers only the screening process and does not account for database preparation

NRF2 nuclear factor erythroid 2-related factor 2, *Lpd* mycobacterial lipoamide dehydrogenase, *NA* not available

library preparation, but the shape screen achieved an impressive throughput of around 120,000 compounds per minute per employed GPU [21].

Both 2D and 3D ligand-based VS examples are summarized in Table 11.3.

4.3 3D Conformer Generation on the Ultra-Large Scale: Shortcuts and Acceleration Strategies

Whenever approaches like ligand shape similarity searches or docking utilize the ligand 3D structural information, ligand conformer generation becomes the first step of the VS workflow. As already discussed in Sect. 3.1, conformer generation will noticeably increase the amount and size of the ligand data. Additionally, when working with ultra-large libraries, a significant portion of computing time and resources will be spent on the generation of the 3D conformers.

For example, to use the GPU-based Rapid Isotere Discovery Engine (RIDE) by MolSoft [52] for a 3D similarity search of SAVI, the necessary generation of a conformationally expanded version of the database consumes around eight million CPU hours [9]. On the other hand, once the 3D ligand database is created, RIDE can achieve a screening performance of around 500,000 conformations per second per GPU. Within hours, the full SAVI can thus be screened against a search template [9]. 3D conformer generation is a single-time effort as the conformationally expanded database can be used for any number of searches.

Thus, beyond following the general guidelines outlined in the previous section to optimize ligand conformer generation parameters and tool performance, obtaining pregenerated 3D conformers can save time and resources. In the spirit of reusable digital assets, the results of such one-time efforts are increasingly made available to remove the conformer generation step from others' workflows altogether.

For example, Gorgulla et al. [7] released a 1.4-billion version of Enamine REAL in pdbqt format for direct docking with tools of the AutoDock family. They also made a 1.5-billion version of the ZINC15 library available in ready-to-dock format [5]. Sivula et al. [23] released 3D databases of a 1.56 billion Enamine REAL lead-like library in the proprietary Schrödinger Phase database format for use with Schrödinger tools. These public endeavors are also summarized in Table 11.11 in the Appendix.

These efforts also underline why ZINC stands out as a compound resource: In contrast to most ultra-large libraries, ZINC offers the direct download of pregenerated 3D structures for many compounds. For instance, 3D structural data that can be readily obtained from ZINC20 amounts to around 60 TBs [24, 53].

Finally, the work of Grebner et al. [4] showcases a strategy to speed up 3D conformer generation: They first created a fragment library from 100 million random SMILES and precalculated the corresponding torsions. The result was then applied as a torsion library for OpenEye's OMEGA and speeded up 3D conformer generation for their ultra-large libraries by about 25%.

5 It Takes Two to Tango: Receptor-Based Ultra-Large-Scale Screening

Despite the higher throughput of ligand-based methods, the bulk of ultra-large-scale screening efforts seem to revolve around receptor-based screening methods. Their independence from prior knowledge, coupled with the explicit consideration of the relevant protein target as a partner, makes docking-based screening an attractive choice in many settings. In brute-force approaches, as schematically illustrated in Fig. 11.3, all compounds in an ultra-large library are considered one after another.

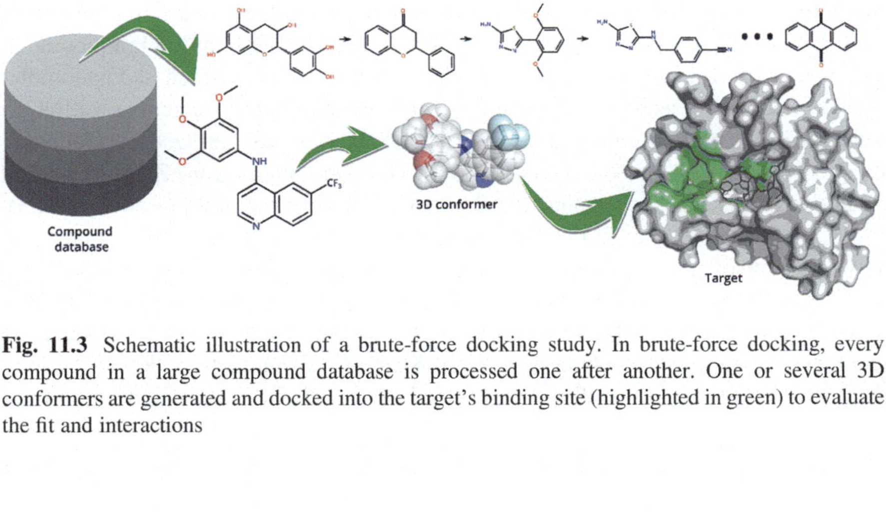

Fig. 11.3 Schematic illustration of a brute-force docking study. In brute-force docking, every compound in a large compound database is processed one after another. One or several 3D conformers are generated and docked into the target's binding site (highlighted in green) to evaluate the fit and interactions

Several recent examples of ultra-large brute-force docking efforts highlight the applicability, limitations, and opportunities of docking on the ultra-large scale.

5.1 Recent Examples of Docking-Based Ultra-Large Scale-Screening

One of the first studies to leverage ultra-large compound databases in a docking-based VS campaign was performed by Lyu et al. [17]. They utilized DOCK3.7 to dock more than 99 million compounds from ZINC15 against the AmpC β-lactamase, and more than 138 million compounds against the D_4 dopamine receptor. One of their key findings would motivate many future screening efforts on a similar or even larger scale: The authors demonstrated that hit rates can be expected to improve with growing library sizes, provided docking enriched true positives over decoys on a small scale. On the other hand, if enrichment failed on a small scale, performance deteriorated further when utilizing larger libraries [17]. Thus, every ultra-large docking study should, whenever possible, include validation procedures such as re- and crossdocking and the retrospective analysis of known hits. Further details on docking validation can be found in Chap. 7 and in the excellent ultra-large docking protocol summary published by Bender et al. [24].

By excluding any relatives of known actives from their initial set of virtual hits, Lyu et al. [17] discovered low micromolar AmpC inhibitors with novel chemotypes. The same strategy also enabled the discovery of D_4 dopamine receptor modulators with previously unseen chemotypes: The study found several new nanomolar agonists and partial agonists, and low micromolar antagonists. The most potent compound was a highly subtype-specific full D_4 agonist at 180 pM, demonstrating the potential of ultra-large databases in the discovery of highly potent binders by VS [17].

Following this success, subsets of ZINC15 provided various promising virtual hits against further targets: For instance, 15 novel MT_1 melatonin receptor-targeting scaffolds were identified from a library of around 150 million lead-like compounds and included highly potent agonists and inverse agonists with up to high picomolar potencies [18]. Likewise, several novel chemotypes, many of which had nanomolar affinity and high selectivity for the $\sigma 2$ receptor, were found by screening a ZINC15 subset of 490 million compounds [19]. DOCK3.7 was also utilized by Gahbauer et al. [54] to screen up to 330 million lead-like compounds from ZINC15 and an in-house virtual database against the SARS-CoV-2 nonstructural protein 3 (NSP3).

In the wake of the COVID-19 pandemic, Acharya et al. [55] initiated the development of a supercomputing pipeline to support drug development efforts. Their workflow involved ensemble docking with Autodock-GPU and GROMACS molecular dynamics (MD) simulations. Initially, they focused on the small-scale docking of repurposing databases to eight SARS-CoV-2 targets, followed by long MD simulations. Additionally, they utilized Autodock-GPU to screen around 1.4

billion compounds against selected targets on the Summit supercomputer of the Oak Ridge Leadership Computing Facility. Their work represented one of the first giga-scale brute-force docking efforts, and the final dataset was made publicly available in 2023: [56] The full billion-scale docking results and machine-learning based-rescoring results for 5 SARS-CoV-2 targets, namely the Spike protein, the main protease (Mpro) and papain-like protease (PLpro), RNA-dependent RNA polymerase (RdRp), and the nonstructural protein 15 (NSP15). Mpro was also the focus of a dedicated ultra-large VS effort: Luttens et al. [57] combined the docking of 235 million ZINC15 compounds with a billion-scale fragment-based optimization and experimental validation, which yielded eight new inhibitors of Mpro.

A dedicated virtual screening platform that leverages massive parallelization during ligand preparation and docking, VirtualFlow, was recently presented by Gorgulla et al. [7]. This Open Source platform integrates with several open-source docking tools like AutoDock Vina [58] and many of its forks (e.g., smina [59]) and tools with free-for academics licensing like PLANTS [7, 60, 61]. VirtualFlow was used to screen around 1.3 billion compounds from ZINC15 and Enamine REAL against the Kelch-like ECH-associated protein 1 (KEAP1) protein [7]. The screen identified several structurally diverse binders with submicromolar affinity. In another extensive screening effort, VirtualFlow enabled the billion-scale screening of 15 potential target proteins of SARS-CoV-2 and two host proteins of critical relevance to the virus, namely ACE2 and TMPRSS2 [62]. For each of the 45 binding sites screened, the top one million virtual hits were made publicly available to facilitate future SARS-CoV-2 drug development efforts.

Sivula et al. [23] focused on a giga-scale benchmarking effort for a machine learning (ML)-boosted VS approach (see also Sect. 6.1). To generate reference data, the authors first performed a brute-force docking study of the 1.56 billion Enamine REAL lead-like library (status March 2021) to two target proteins, the bacterial periplasmic chaperone SurA and Cyclin-G associated kinase (GAK).

Sadybekov et al. [63] reported a docking-based multi-template VS of 115 million compounds from Enamine REAL's 350/3 lead-like library and the REAL diversity subset against two G protein-coupled receptors (GPCRs). They discovered five promising submicromolar compounds, four of which were selective antagonists of cysteinyl leukotriene GPCR CysLT1R, while one compound was a full antagonist of both screened GPCRs.

Kaplan et al. [64] took a different strategy to find novel selective binders of the serotonin 5-HT_{2A} receptor: The authors noted that although tetrahydropyridine scaffolds can be found in several drug compounds derived from natural products, in typical screening libraries, they are underrepresented in comparison to their aromatic counterparts. Thus, they first generated a specialized library of 75 million virtual tetrahydropyridine compounds from commercially available building blocks. This bespoke library was then screened with DOCK3.7, and chemical novelty was ensured by Tanimoto-similarity filtering against known actives. The study resulted in 17 selected virtual hits, which included one novel agonist and three antagonists with low micromolar activities (24% hit rate) [64].

Table 11.4 Summary of recent ultra-large-scale screening campaigns performed with conventional brute-force docking approaches

Compounds screened	Target(s)	Docking tool	Throughput per core	Reference(s)
1.56 billion	SurA, GAK	Glide-HTVS	20-32 cpds/min	Sivula et al. [23, 65]
1.40 billion	5[a] SARS-CoV-2 targets	AutoDock-GPU	>50 cpds/min	Acharya et al. [55, 56, 66]
1.30 billion	KEAP1	VirtualFlow (QuickVina, smina, AutoDock Vina)	4–12 cpds/min	Gorgulla et al. [7]
1.01 billion	15 SARS-CoV-2 targets; host ACE2, TMPRSS2	VirtualFlow (QuickVina W, QuickVina 2)	8 cpds/min	Gorgulla et al. [62]
490 million	$\sigma 2$ receptor	DOCK3.8	46 cpds/min	Alon et al. [19]
330 million	SARS-CoV-2 NSP3	DOCK3.7	84 cpds/min	Gahbauer et al. [54]
246 million	SARS-CoV-2 NSP3	DOCK3.7	65 cpds/min	Gahbauer et al. [54]
235 million	SARS-CoV-2 Mpro	DOCK3.7	47 cpds/min	Luttens et al. [57]
150 million	MT_1 Melatonin receptor	DOCK3.7	56 cpds/min	Stein et al. [18]
138 million	Dopamine D_4 receptor	DOCK3.7	53 cpds/min	Lyu et al. [17]
115 million	CysLT1R, CysLT2R	ICM-Pro	NA	Sadybekov et al. [63]
99 million	AmpC β-lactamase	DOCK3.7	40 cpds/min	Lyu et al. [17, 24]
75 million	Serotonin 5-HT_{2A} receptor	DOCK3.7	144 cpds/min	Kaplan et al. [64]

The reported number of screened compounds refers to the input prior to any conformer generation. Throughput per core is approximated from the reported timing information

GAK cyclin G-associated kinase, *KEAP1* Kelch-like ECH-associated protein 1, *NSP3* nonstructural protein 3, *Mpro* main protease, *CysLT1R, CysLT2R* cysteinyl leukotriene G protein-coupled receptors, *NA* not available (authors did not disclose total run-time or speed information)

[a]The initial publication investigated 8 targets, while only five were featured in the final giga-scale dataset release

Table 11.4 summarizes all discussed ultra-large docking-based VS approaches and their approximate throughput.

5.2 Dedicated Tools for Ultra-Large Docking Workflows

Many of the presented ultra-large docking campaigns were achieved by "simply" scaling up conventional docking workflows. One exception is the tool VirtualFlow, which was specifically developed to address challenges associated with data-intensive docking workflows on the ultra-large scale [7]. Especially for novice users, VirtualFlow aims to simplify key tasks, such as splitting data into appropriate batches and scheduling and monitoring computing jobs in an HPC or cloud computing environment. Since workflows supported by VirtualFlow scale linearly, they straightforwardly scale with the available resources and allow for uncomplicated cost estimates.

As noted above, VirtualFlow works with various docking tools to enable users to pick the most well-suited tool for each docking problem. It achieves high throughput by utilizing up to four levels of parallelization: Each instance of VirtualFlow can run several jobs, jobs can have several job steps, and each job step can run several queues with the actual ligand preparation or docking runs—using tools, which are often also internally parallelized. It supports various common resource managers for batch jobs on HPCs, e.g., SLURM and Moab/TORQUE/PBS. Given the multiple layers of parallelization, a dedicated workload balancing system ensures that file I/O remains unproblematic and each ligand is considered only once [7].

While VirtualFlow was designed to be more of a "Swiss Army knife" easily deployed in most common HPC and cloud computing environments, a recently published open-source pipeline takes a different approach: warpDOCK was developed specifically for cloud computing in the Oracle Cloud Infrastructure [67]. The authors chose QuickVina2 [68] as the primary docking engine, but warpDOCK can also integrate with AutoDock Vina and its various forks. Its Python-based queue engine takes care of active monitoring and the preloading of ligands. In a test docking with equivalent resources, the authors demonstrated warpDOCK to be about 3.7 times more efficient than VirtualFlow in their cloud computing environment.

In an ensemble docking study against 21 *Staphylococcus aureus* D-alanine-D-alanine ligase structures, warpDOCK achieved a throughput of around 12 compounds per minute per physical core [67]. Although the authors utilized a comparatively small ligand library of only 4.75 million ZINC compounds, the ensemble docking reached the ultra-large scale with around 101 million docking calculations in total.

The example of warpDOCK highlights how careful optimization for a specific architecture can maximize throughput. VirtualFlow, on the other hand, guarantees decent throughput in different architectures and maximizes a user's choice of the computing environment. On the ultra-large scale, job scheduling and task distribution over the computing cores represent particular bottlenecks for docking throughput. We highlighted two examples of open-source tools to support users in navigating those challenges (see also Table 11.10 in the Appendix for the corresponding github addresses). With the increasing popularity of ultra-large-scale approaches in mind, we anticipate that other, similar tools will likely be developed.

6 Reducing Computation Needs in Ultra-Large VS Workflows

As can be extrapolated, for example, from the throughput in Table 11.4, utilizing brute-force approaches to ultra-large-scale docking remains a highly time- and resource-intensive task. Strategies to reduce the required computations were thus quickly sought after.

One option involves using so-called diversity sets, i.e., subsets of ultra-large libraries that aim to reflect the chemical diversity of the entire library. For instance, instead of investing 14 million CPU hours to dock the full ultra-large Enamine and SAVI libraries, a recent study targeting the STAT3 N-terminal domain relied on small million-scale Enamine diversity subsets and a bespoke diversity subset of SAVI with close to three million compounds [69]. Although the target is commonly considered "nondruggable," the authors were able to successfully discover novel ligands in their docking-based VS of diversity sets.

On the other hand, reduction strategies may lead to a loss of promising binders: In their study of the dopamine D_4 receptor, Lyu et al. [17] investigated the prospect of clustering an ultra-large library and docking only cluster representatives. They found that screening only cluster representatives would have prevented the discovery of several of their novel promising binders. This underlines why there remains strong interest in the ability to screen entire ultra-large libraries.

Two promising solutions have gained particular traction over the past few years: The first combines ligand-based QSAR-like approaches with docking by utilizing machine learning (ML) to construct models that can serve as stand-ins for conventional screening. The second solution relies on the combinatorial nature of many ultra-large make-on-demand libraries and their step-by-step construction from fragments, thereby avoiding the enumeration of most of the library.

6.1 *Accelerating Docking with Machine Learning/Deep Learning in Hybrid Ligand- and Structure-Based VS*

To remove the necessity of brute-force docking a complete ultra-large compound library, most recently described ML-accelerated docking approaches train ML models as stand-ins to predict the docking outcome, as illustrated schematically in Fig. 11.4. Model training relies either on one-shot learning or active learning. In

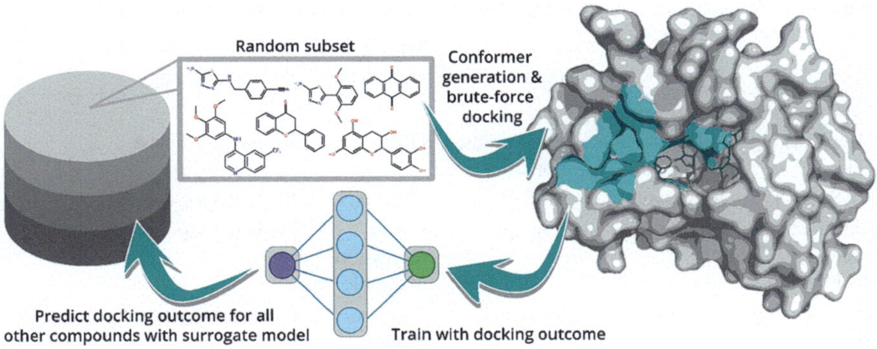

Fig. 11.4 Schematic illustration of a docking study accelerated with machine learning (ML) or deep learning (DL). A small random subset of the ultra-large compound database is selected, conformers are generated, and the compounds are docked by conventional means. The docking outcome is used to train the ML/DL model, and the trained model acts as a surrogate to predict the docking outcome for the remainder of the ultra-large library

one-shot learning, the model is trained a single time [35]. However, recent approaches largely rely on several iterative training steps and an active learning concept. Based on the prediction results of the current iteration, active learning adds additional training samples for all following steps [35].

A reduced number of brute-force docking calculations is the key to the speed up of ultra-large-scale screening with ML. While individual approaches differ in details, their surrogate models commonly ensure that only a small fraction of the library needs to be docked by conventional means. The concept of using shallow ML algorithms and QSAR approaches to reduce required dockings is not new [70], but many recent methods rely on deep learning (DL) implementations or the combination of DL with more classical, shallow ML algorithms (see Table 11.5).

For example, the open-source tool DeepDocking aims to accelerate VS with DL [71]. DeepDocking first computes molecular descriptors, such as circular fingerprints, for all compounds in a chemical library. Then, a random compound set is docked by conventional means, and docking scores and descriptors are used to train a deep neural network (DNN). DeepDocking uses a classification approach, where a docking score cutoff determines the separation of compounds into the classes "hits" and "nonhits." Once trained, the surrogate model predicts the class of each member in the entire library, and compounds designated as "nonhits" are removed from the prediction set. DeepDocking relies on active learning: Iteratively, additional random samples are brute-force docked, and the score threshold to distinguish "hits" from "nonhits" is progressively adjusted [71]. This stepwise reduction of the considered compounds in an ultra-large library was able to speed up the screening of 1.36 billion 3D structures from ZINC15 by at least 50 times [38, 71]. DeepDocking was able to recover up to 90% of the top-scoring virtual hits for 12 different targets docked with FRED when considering only a small fraction of the library compounds [38, 71]. The ZINC15 library was additionally screened against the severe acute

Table 11.5 Summary of discussed ML-/DL-boosted docking implementations

Tool	Type	Input	ML/DL architecture	Selection strategy
Open source				
DeepDocking [71]	Classification	Fingerprints, properties	DNN	By predicted class
HASTEN [79]	Regression	SMILES	D-MPNN	Predicted top-scoring
Linear accelerated docking [85]	Regression	Fingerprints	Linear regression ensemble	Predicted top-scoring
Lean docking [84]	Regression	2D ligands	Support Vector regressor	Predicted top-scoring
MolPAL [77]	Regression	Fingerprints, SMILES	RF, DNN, D-MPNN	Predicted top-scoring, TS, EI, PI, UCB
Commercial				
Glide active Learning [81]	Regression	Fingerprints, SMILES	RF + GCN	Predicted top-scoring, most uncertain, most uncertain among top scoring, random among top-scoring

This overview lists the type of model, utilized training input, ML/DL architecture, and available selection metrics. See also Table 11.6 for specific ultra-large-scale application examples of the tools *DNN* deep neural network, *D-MPNN* directed message-passing graph neural network, *RF* random forest, *GCN* graph convolutional neural network, *TS* Thompson sampling, *EI* expectation of improvement, *PI* probability of improvement, *UCB* upper confidence bound

respiratory syndrome coronavirus 2 (SARS-CoV-2) main protease, which allowed for the discovery of novel inhibitors with micromolar IC_{50} values [72]. Following that, an extensive consensus screen of around 40 billion structures sourced from ZINC15 and the Enamine REAL Space was performed against the same target, utilizing a DeepDocking pipeline with five different docking tools [34]. This effort resulted in over 100 additional inhibitors with novel chemotypes. Other application examples of DeepDocking include the discovery of novel noncovalent inhibitors of SARS-CoV-2 papain-like protease (PLpro), A_{2A} adenosine receptor antagonists, and inhibitors of a prostate cancer target, the RNA-binding protein Lin28 [73–75]. In addition to its various use cases as summarized in Table 11.6, detailed instructions for running a DeepDocking workflow have been published [38], and a graphical user interface (GUI) was recently released to simplify the adoption of DeepDocking [76].

Another open-source tool, MolPAL (molecular pool-based active learning), relies on Bayesian Optimization and offers three different ML regression models [77]: Two rely on molecular fingerprints, namely the more classical random forest (RF), and a simple feed-forward DNN model. The third option uses SMILES input and the directed message-passing graph neural network (D-MPNN) Chemprop [78]. The authors showcase MolPAL's active learning strategy on small-scale docking-based VS against thymidylate kinase using libraries of 10,000 to around two million compounds [77]. Additionally, they explore the ultra-large scale with the AmpC and Dopamine D_4 receptor docking results published by Lyu et al. [17]. In addition

Table 11.6 Summary of ultra-large applications of discussed ML-/DL-accelerated docking tools

Compounds screened	Target(s)	Tool	Runtime [h]	CPUs	GPUs	Reference(s)
>12.3 billion (38.7 billion)	Mpro	DeepDocking (5 docking tools)	456	640	250	Gentile et al. [34]
1.56 billion (3.8 billion)	SurA, GAK	HASTEN (Glide-HTVS)	223	640	10	Sivula et al. [23]
1.40 billion	A_{2A}	DeepDocking	NA	NA	NA	Tang et al. [74]
1.40 billion	PLpro	DeepDocking (Glide-SP)	NA	NA	NA	Garland et al. [73]
1.36 billion	12 diverse targets	DeepDocking (FRED)	NA	60	4	Gentile et al. [71]
1.30 billion	Mpro	DeepDocking	NA	390	40	Ton et al. [72]
1.00 billion	Lin28	DeepDocking	NA	NA	NA	Radaeva et al. [75]
138 million	D_4	MolPAL (DOCK3.7)	NA	NA	NA	Graff et al. [77]
138 million	D_4	Glide Active Learning (Glide)	60,000	8	1	Yang et al. [81]
99 million	AmpC	MolPAL (DOCK3.7)	NA	NA	NA	Graff et al. [77]
99 million	AmpC	Glide Active Learning (Glide)	NA	NA	NA	Yang et al. [81]
96 million	AmpC	Lin. accel. docking (DOCK3.7)	<1	64	–	Marin et al. [85]

Compounds screened: Raw total input numbers of utilized library, where reported. 3D structures after conformer generation are given in parenthesis. Brute-force docking tools used together with ML-/DL-acceleration tools are likewise reported in parenthesis

Lin. accel. docking: linear accelerated docking. Total runtimes are reported for the highest number of iterations in active learning approaches and numbers of CPUs and GPUs represent the maximum (in most cases, more CPUs were utilized for docking and lower numbers for the ML-/DL-step and more GPUs for prediction than training). *GAK* cyclin-G associated kinase; A_{2A} adenosine receptor A_{2A}; *PLpro* SARSCoV-2 papain-like protease, *Mpro* SARS-CoV-2 main protease. *NA* not available (authors did not provide total run-time or speed information)

to comparing different ML/DL strategies, Graff et al. [77] study different acquisition strategies, i.e., how the next compounds for brute-force docking are selected. They explore the selection based purely on the rank obtained by the docking scores, referred to as “greedy selection,” Thompson sampling (TS), expectation of improvement (EI), probability of improvement (PI), and upper confidence bound (UCB) (see Table 11.5). Especially with smaller training dataset sizes, they observe large variations of recalls depending on the chosen ML architecture and acquisition metric. Interestingly, the most basic greedy selection typically ranges among the top-performing strategies. For instance, training the MPNN with only 2.4% of a 100 million-compound dataset allowed for around 89% recall of the top 50,000 compounds. Taking uncertainty into account in the same example improved recalls to 95% [77]. This work represents a blueprint to show how different architectures, selection strategies, and training data batch sizes impact the recalls of ML-boosted strategies.

A tool that combines the MPNN Chemprop and a greedy selection algorithm with built-in support for the proprietary docking tool Glide is the open-source tool HASTEN (macHine leArning booSTEd dockiNg) [78–80]. Like DeepDocking and MolPAL, HASTEN generates its initial training data by brute-force docking a random selection of compounds from a large chemical library. The MPNN model training data consists of compound SMILES and docking scores. Following the prediction of docking scores for the entire compound library by the regression model, HASTEN takes an active learning approach: Compounds are ranked by their predicted scores, and the top-ranked compounds are brute-force docked before training a new model with all available docking data. The progressive addition of both true and false positives to the training data iteratively refines the model. On the million scale, HASTEN achieved high recall values across 13 different targets docked with either FRED or Glide [79]. HASTEN additionally represents the first tool for ML-accelerated docking that has been rigorously benchmarked on the billion scale: Sivula et al. [23] confirmed over 90% recall of the true top-scoring 1000 hits by HASTEN in their comparison with 1.56 billion brute-force docking results for two different targets. The authors showcased the significant speed-up when using HASTEN for the billion-scale screening: While their brute-force docking took approximately 4 months, the time to screen 1.56 billion compounds was reduced to 10 days or less when using HASTEN with the time-consuming brute-force docking reduced by at least 99% [23].

The three discussed open-source tools (see Table 11.10 in the Appendix for source code repositories) demonstrate how both classification and regression models can reduce the amount of brute-force docking when screening ultra-large chemical libraries. They highlight the significant speed-up and pinpoint implications of training dataset sizes and selection metrics in active learning approaches [23, 71, 77, 79].

Unsurprisingly, in addition to the efforts in the academic environment, similar strategies have also been integrated into commercial screening software. For example, the Glide Active Learning strategy has been available as part of the Schrödinger Software Suite since 2019 [81]. Glide Active Learning relies on an ensemble of a graph convolutional NN (GCN) and a RF model and features various selection strategies for compounds to dock after the first iteration of randomly selected training compounds. Typically, only one round of active learning is employed, which features docking of 0.1% of the most uncertain top 5% according to ensemble predictions to augment the training data. Other examples of commercial tools include, for example, ICM-Pro Gigascreen (based on a GCN) and OpenEye's GigaDock [82, 83].

Finally, some recent work revived more classical models: For instance, the Lean docking protocol by Berenger et al. [84] lowers the required docking by about 75% with the help of a linear support vector regressor. Training takes only about 80 s on a single core, and the trained model achieves a throughput of around 5800 compounds per second per computing core. However, Lean docking has so far not been applied to an ultra-large docking scenario. The recent work of Marin et al. [85] relies on ensembles of fingerprint-based linear regression models to accelerate docking-based VS in one-shot and active learning scenarios. Their study utilized the previously published 99 million AmpC dataset [17]. While recent MPNN-based methods achieved higher recalls, their linear regressor ensemble surpassed the performance of more shallow approaches like RF models. Additionally, those simple models are often several times faster in training and prediction cycles than more advanced ML and DL approaches, although their memory footprint remains a potential bottleneck [85]. Importantly, these examples highlight that, while not necessarily with optimal results, ultra-large VS efforts can be feasible in settings with limited access to computational resources and, in particular, GPUs, when relying upon more simple models.

Table 11.5 summarizes implementation and training strategies of the discussed ultralarge ML- and DL-boosted screening approaches. Table 11.6 additionally summarizes application examples on the ultra-large scale. Repositories of open-source tools are summarized in Table 11.10 in the Appendix.

It is important to note that all tools aim to partially replace brute-force docking with a faster stand-in. At the same time, the approaches do not modify the docking and scoring procedure itself. Generally, these strategies rely on the ability of the chosen docking approach to achieve a meaningful enrichment in true actives among the top-scoring compounds. Therefore, the same validation measures that apply for default brute-force docking should be taken into account when utilizing ML-accelerated docking whenever possible (see, e.g., Chap. 7 for detailed validation procedures).

Most approaches discussed herein abstain from exhaustive (hyper)parameter optimization. While one can argue that optimization should always be done when utilizing ML or DL models, most tools demonstrated good performance with default parameters across diverse targets. Optimizing the often thousands or even millions of hyperparameter combinations can be unfeasible or, at best, represents a major additional time investment when speed-up is the primary objective. Further, any optimization needs to reflect the scale of the intended production VS. Even though more optimal parameters for specific systems may exist, relying on the same defaults as method authors or published examples is often the preferable choice. However, users should, at the very least, inform every available parameter choice in the selected tool by small-scale pre-studies or a retrospective analysis of recalls.

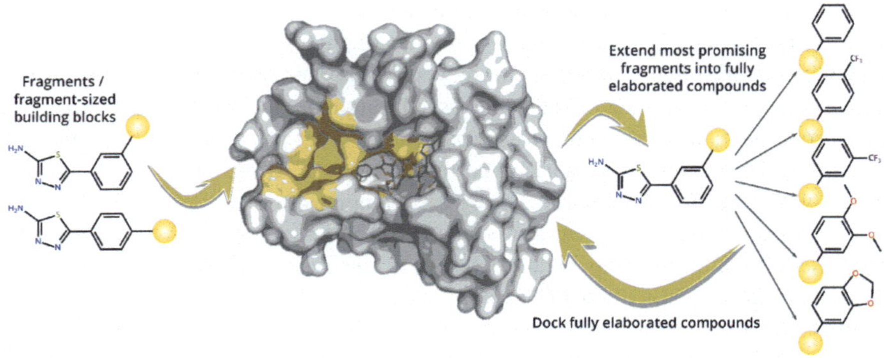

Fig. 11.5 Schematic illustration of a fragment-/reagent-based approach to ultra-large-scale screening. Instead of fully elaborated compounds, the docking inputs are fragments, fragment-sized reagents, or minimal representative scaffolds, sometimes with designated R-groups for enumeration (indicated by the yellow sphere). Following conventional docking, only the most promising starting scaffolds are selected and elaborated into fully qualified small molecules, which are docked into the target receptor in a final evaluation step

6.2 Reagent- or Fragment-Based Approaches to Screen Combinatorial Libraries

Ultra-large make-on-demand chemical libraries are often combinatorial libraries created from reagents and predefined synthesis protocols to interconnect the building blocks. Several methods seek to reduce the computational cost of screening an ultralarge library by relying on the evaluation of fragment-like reagents or building block composites, as schematically illustrated in Fig. 11.5. In such approaches, only the most promising scaffolds get enumerated, thus substantially reducing the number of compounds to consider. Similar to ML-driven QSAR solutions, the concept of elaborating compounds from virtual fragments is equally not a new development [86, 87] but has recently seen a surge linked to the emerging ultra-large combinatorial libraries and spaces.

With Thompson sampling (TS), Klarich et al. [35] propose a probabilistic search method operating in the reagent space. TS is applicable to various ligand- and receptor-based VS approaches, where the respective method's scoring function informs the belief distribution underlying TS. For each available reaction, reagents are randomly selected, and fully elaborated molecules are generated in silico. Iteratively, the molecules are scored, and the belief distribution is updated. The authors applied TS to a Tanimoto-2D similarity search, a 3D similarity search, and a docking study. They also compared TS with an active learning RF approach. Recalls between 57% and 90% were achieved for different tasks when evaluating no more than 1% of an ultra-large combinatorial library. The application of TS to VS represents a novel approach, and the performance in finding top-scoring hits may require further optimization. Nonetheless, TS achieved 100 to 48,000 times speed-up, and can significantly reduce computing expenses for screening ultra-large combinatorial libraries.

Another approach relies on the combinatorial nature of the Enamine REAL Space with more than 11 billion compounds. In V-SYNTHES (Virtual Synthon Hierarchical Enumeration Screening), the library is first decomposed into all possible combinations of scaffolds and synthons for all reactions [22]. A synthon can be pictured as a minimal capping R-group or common substructure, from which a fully elaborated molecule can be grown. In the V-SYNTHES approach, all scaffold–synthon combinations are docked, and only the most promising fragments will have their R-groups enumerated until the molecules are fully elaborated. Docking reduction to only 1.5 million compounds of the 11-billion compound library corresponds to a 5000-fold speed-up of the screen. With a hit rate of 33% (compared to 15% for a smaller-scale classical VS approach), V-SYNTHES helped to identify 14 submicromolar ligands of a cannabinoid receptor [22]. Its application to the ROCK1 kinase achieved a 28.5% hit rate and resulted in several nanomolar ligands [22].

ROCK1 was also the target of Beroza et al. [88] and their largely manual workflow termed Chemical Space Docking, which combines several modeling, docking, and chemoinformatics tools. The workflow uses the building blocks and synthesis protocols underlying combinatorial make-on-demand chemical libraries like the Enamine REAL Space to define the searchable chemical space. In a "greedy" optimization strategy, 500 unique, promising starting fragments were selected from the top-scoring FLExX docking results by properties including their interaction pattern, internal strain, logP, and chemical diversity. For ROCK1, the full enumeration of these starting fragments yielded around 5.2 million compounds, which represented a significant reduction compared to the around 858 million possible virtual product molecules in the considered Enamine REAL subspace. The approach altogether identified 27 actives with K_i values below 10 μM, including three compounds with a novel hinge-binding motif not previously observed in ROCK1 [88]. While the authors acknowledge that a more exhaustive sampling of the chemical space may be preferable, their approach showcases how ultra-large chemical spaces can be utilized in VS even with very limited computational resources.

In their "crystal structure first" approach, Müller et al. [89] combine Chemical Space Docking with X-ray crystallography-based fragment screening. Instead of docking building block fragments, this approach starts from existing crystallographic evidence: Four fragment co-crystal structures of protein kinase A were selected, and fragments from the Enamine REAL Space with 2.6 billion virtual products were matched onto their position with FlexX docking. Not only does the approach rely on experimentally confirmed interaction patterns, but such template-based docking approaches with partially predefined atom positions are also significantly faster than regular approaches that require a more exhaustive sampling during pose generation. After that, the approach resembles Chemical Space Docking, where the most promising matches are enumerated and docked again. Within only 9 weeks, this crystal-based approach achieved a hit rate of around 40% and concluded with a set of additional crystal structures of the novel compounds to confirm the predicted binding modes. The approach identified several low micromolar to submicromolar inhibitors with novel scaffolds, the most potent of which showed a K_i of 744 nM.

Gahbauer et al. [54] likewise complemented ultra-large VS efforts with X-ray crystallographic evidence: To find inhibitors of the SARS-CoV-2 nonstructural protein 3 (NSP3), they first utilized existing crystal structures with bound fragments of up to 180 μM affinities. The crystallographic fragments were linked, and the resulting drug-like molecules served as search templates for a 2D similarity search with the SmallWorld server. After screening 22 billion synthesizable molecules in the Enamine REAL database, eight of the 13 prioritized compounds were confirmed as NSP3 binders by crystallographic soaking experiments [54]. In addition to these initial micromolar hits, the authors aimed to expand the hit diversity in an unbiased fashion: They additionally carried out a conventional docking-based screen of around 400 unique molecules with DOCK3.7. In addition to the default scoring parameters, they also ran a calculation with optimized scoring parameters to better reflect the crystallographic evidence for NSP3. This effort added five additional classes of previously unknown low micromolar inhibitory scaffolds to the hit pool. Most virtual hits had an anionic nature, but the authors finalized their study with a structure-based optimization that successfully replaced anionic groups with neutral groups, opening up a route toward better membrane permeability for the novel NSP3 inhibitor classes [54].

Finally, another strategy for chemical space docking, termed SpaceDock, was recently published by Sindt et al. [90]. Their approach uses a 5.5 billion-compound subset of the Enamine REAL Space with no more than two synthetic steps selected from a small set of robust chemical reactions. SpaceDock relies on docking the unmodified fragment-sized reagents before linking them into fully enumerated compounds. The authors first used GOLD to dock a combinatorial space of around 98 million compounds to estrogen receptor beta in a retrospective analysis. Only about 0.6% of their screening space constituted true agonist scaffolds, while the remainder were decoys. Encouragingly, true binders were significantly enriched by SpaceDock, making up 84% of the resulting virtual hits [90]. After establishing the ability of the method to generate known ligands or their close analogs, SpaceDock was applied to the docking of 670 million compounds against the dopamine D_3 receptor. A known ligand and several compounds that could be the results of scaffold hopping from known ligands were again among the virtual hits. The final hits included binders with submicromolar to low micromolar K_i, four of which contained a novel scaffold that had not previously been observed in D_3 receptor antagonists.

Table 11.7 summarizes the featured examples that exploit the combinatorial nature of many state-of-the-art ultra-large screening libraries to speed up VS and reduce the associated computational burden.

7 Challenges and Future Perspectives

We have, in previous sections, hinted at some practical challenges for ultra-large VS. In this final part of the chapter, we will address a final key issue for ultra-large-scale screening: The process of selecting hit compounds for further work-up, or hit

Table 11.7 Summary of ultra-large reagent-/fragment-based screening campaigns

Compounds screened	Target(s)	Tool	Mock throughput	Reference (s)
22 billion	NSP3	Fragment linking + 2D similarity search	NA	Gahbauer et al. [54]
11 billion	Cannabinoid receptor, ROCK1	V-SYNTHES (ICM-Pro)	NA	Sadybekov et al. [22]
5.5 billion	Not specified	SpaceDock	1360 cpds/min	Sindt et al. [90]
2.6 billion	PKA	Chemical Space Docking "crystal structure first" (FlexX)	7200 cpds/min	Müller et al. [89]
858 million	ROCK1	Chemical space docking (FlexX)	NA	Beroza et al. [88]
335 million	JNK3	TS (OEDocking)	350 cpds/min	Klarich et al. [35]
234 million	Random query molecule	TS (OEShape ROCS)	>120,000 cpds/min	Klarich et al. [35]
Approx. 100 million	Not specified	SpaceDock (GOLD)	2170 cpds/min	Sindt et al. [90]
94 million	Random query molecule	Tanimoto 2D Similarity search	522 million cpds/min	Klarich et al. [35]

The reported "mock" throughput takes the total number of input compounds divided by the total computing time for a single computing unit. Note that most methods utilize conventional brute-force docking with the corresponding low throughput, but accelerate the approach by the significant reduction of required brute-force docking

NSP3 SARS-CoV-2 nonstructural protein 3, *PKA* protein kinase A, *NA* not available (authors did not provide total run-time or speed information)

triage. Hit triage suffers from unreliable rank-scoring, which escalates on this new dimension. We aim to highlight the issue and possible solutions to this prime counter-argument against ultra-large VS. Additionally, we address future directions and our hopes for the adoption and extension of large-scale benchmarking data. Finally, we discuss some recent developments, which may shape the future directions of ultra-large VS.

7.1 Hit Triage: An Old Problem on a New Dimension

The key objective of VS is to narrow down a large pool of compounds and retain a few promising top-ranked virtual hits for further post-processing and validation, for example, by more extensive computational studies or experimental assays. Ultra-large screening faces two main challenges, which also give rise to most counter-arguments against ultra-large VS: First, we can reasonably define screening success as the successful enrichment of (to-be-confirmed) true hits among the top-ranked candidates.

Table 11.8 Summary of purchased compounds and hit rates in recent ultra-large VS campaigns

Screening input	Target(s)	Ordered	Obtained	Actives (Hit rate)	Reference(s)
22 billion	NSP3	16	13	8 (62%)	Gahbauer et al. [54]
>12.3 billion	Mpro	1700	1283	117 (9%)	Gentile et al. [34]
11 billion	Cannabinoid receptor	80	60	21 (35%)	Sadybekov et al. [22]
11 billion	ROCK1	24	21	6 (29%)	Sadybekov et al. [22]
2.6 billion	Protein kinase A	106	93 (75 soluble)	30 (40%)	Müller et al. [89]
1.40 billion	PLpro	260	178	17 (10%)	Garland et al. [73]
1.40 billion	A_{2A} adenosine receptor	39	39	2 (5%)	Tang et al. [74]
1.30 billion	KEAP1	590	590	69/23/12[a] (2–12%)	Gorgulla et al. [7]
1.12 billion	Lpd	103	88	6 (7%)	Michino et al. [21]
1.00 billion	Lin28	163	163	15 (9%)	Radaeva et al. [75]
858 million	ROCK1	77	69	27 (39%)	Beroza et al. [88]
670 million	Dopamine D_3 receptor	16	15	10 (67%)	Sindt et al. [90]
490 million	$\sigma 2$ receptor	577	484	127 (26%)	Alon et al. [19]
330 million	NSP3	90	78	22/11/30[a] (14–38%)	Gahbauer et al. [54]
246 million	NSP3	54	46	25/5/8[a] (11-54%)	Gahbauer et al. [54]
235 million	Mpro	100	100	19 (19%)	Luttens et al. [57]
150 million	MT_1 melatonin receptor	40	38	15 (39%)	Stein et al. [18]
138 million	Dopamine D_4 receptor	589	549	122 (22%)	Lyu et al. [17]
115 million	CysLT1R, CysLT2R	155	139	10/17[b] (7/12%)	Sadybekov et al. [63]
99 million	AmpC β-lactamase	51	44	5 (11%)	Lyu et al. [17]
75 million	Serotonin 5-HT_{2A} receptor	–	17	4 (24%)	Kaplan et al. [64]

NSP3 SARS-CoV-2 nonstructural protein 3, *Mpro* SARS-CoV-2 main protease, *PLpro* SARS-CoV-2 papain-like protease, *KEAP1* Kelch-like ECH-associated protein 1, *Lpd* mycobacterial lipoamide dehydrogenase, *CysLT1R, CysLT2R* cysteinyl leukotriene G protein-coupled receptors

[a]Activities in different assays

[b]Active on/binding to one or more of the targets.

With often only very few of these candidates entering experimental validation (see Table 11.8 for a summary of the discussed campaigns), a first caveat of ultra-large screening is that your initial set of virtual hits may still be rather large—for instance, even if you consider only the top 0.1% of a billion-scale screen, this would still leave one million potentially promising compounds for consideration in all follow-up procedures [24]. This high number, in connection with a well-known shortcoming of ranking accuracy in VS, can turn hit selection into a dangerous gamble. The scoring functions used in VS often predict a desirable activity profile for compounds that represent inactives [1, 3]. Combining those two aspects, your screen may leave you with a large number of virtual hit candidates, many of which may well be false positives [1, 24].

With the inaccuracies of scoring approaches in mind, should you truly rely on greedy selection and post-process only the very top of the list? One of the first ultra-large docking-based VS campaigns provided an argument in favor of this concept: As part of their study of the dopamine D_4 receptor, Lyu et al. [17] investigated the connection between docking rank and hit rate. They sought experimental confirmation not only for selected top-ranked virtual hits but also for several compounds with less favorable docking scores. Strikingly, for their target, the authors found a clear drop in the hit rate when moving to less favorable scoring ranges. The authors also compared an automated, purely data-driven, selection of top-scoring compounds with the selection by expert opinion. While expert-driven selection yielded higher-affinity compounds, the hit rate of both approaches remained the same. Thus, despite false positives, drawing the final hit selection from the top-ranked compounds appears to be a viable strategy at least for some targets. Consistently, evidence suggests that both classical and recent ML-based scoring functions are capable of identifying better hits from larger chemical spaces [6]. However, since scoring is highly target- (or system-) dependent, these findings do not necessarily generalize. For instance, Alon et al. [19] likewise selected their initial hits from different scoring bins and observed a hit rate of 27% in the highest scoring bin. The second-highest scoring bin, on the other hand, gave a 61% hit rate. However, the authors observed the same general trend of dropping hit rates as a function of less favorable docking scores.

In a follow-up effort, Lyu et al. [91] focused more on the general implications of massively growing chemical libraries for VS. They extrapolated that with growing libraries, not only can one expect better scores, but there is also a striking increase in artifacts. Artificially high-ranked false positives are thus more likely to drown out true positives in the top-ranking bin(s). Thus, when working with ultra-large chemical libraries, caution is advised when relying on a greedy top-ranking-only selection logic. Instead, more lenient selection concepts which draw their hits also from lower-scoring bins, or rescoring strategies with alternative approaches should be considered [24, 91]. Bender et al. [24] provide a comprehensive overview of selection criteria and further recommendations for the experimental validation and next steps.

Despite all their problems, state-of-the-art scoring functions, paired with the appropriate caution, clearly can enable successful drug discovery campaigns, as emphasized by many of the ultra-large screens discussed herein, albeit hit rates greatly varied between different campaigns (see Table 11.8). Various studies identified novel binders with previously unknown scaffolds. The odds of finding novelty should improve for ultra-large screening approaches, which can, for example, benefit a researcher's ability to patent the discovered hits and any derived compounds [8]. However, Grebner et al. [4] showed for their 3D shape screening approach that the diversity of obtained hits no longer increased when exceeding library sizes of 10 billion. This is, in all likelihood, extremely target-, approach-, and screening library-dependent, but should serve as a reminder that in ultra-large VS, we should carefully weigh the cost–benefit, and not blindly consider "bigger" necessarily always "better."

7.2 Future-Proofing VS Approaches

While we cannot be certain how the future of VS will develop in the long run, it is reasonable to assume that VS on the ultra-large scale will become more common. Given the massive resource requirements of standard brute-force approaches, it is also likely that accelerated approaches, and those leveraging the combinatorial nature of chemical libraries, will be more heavily relied upon. The examples discussed in this chapter underline that many approaches are, indeed, already able to process ultra-large libraries. However, currently, almost all screens rely on HPC or cloud computing resources to do so. Only a few tools have been specifically designed with a focus on state-of-the-art ultra-large libraries or benchmarked on this scale.

We can anticipate developments in dedicated screening software for both HPC and cloud computing, but also more limited computational resource settings. This notion also underlines why we believe that ultra-large-scale benchmarking data is crucial to the development of the field. As discussed, for example, in the work on Thompson sampling for VS acceleration presented by Klarich et al. [35], completely novel approaches or application domains require careful optimization to maximize performance and recall. Such optimization should take the intended use-scale into account, which requires datasets of suitable size. While many of the brute-force efforts performed so far only released selected top-scoring compounds, if any data at all, one of the first brute-force docking studies released their full 99 million AmpC and 138 million D_4 DOCK docking results [17]. This dataset has since become a gold standard for benchmarking in the field. With the Glide docking results released by Yang et al. [81], two different ultra-large docking datasets for the same targets are now available (see also Table 11.11 in the Appendix). More recently, two billion-scale docking datasets were released into the public domain: The first covers docking and rescoring results of 1.4 billion compounds against five SARS-CoV-2 target proteins [56, 66], the second 1.56 billion Glide-HTVS results obtained with lead-like compounds against two different targets [23, 65] (Table 11.11). These datasets should provide suitable ultra-large and billion-scale benchmarking data for future approaches. If other ultra-large-scale brute-force approaches are undertaken in the future, it would be desirable if authors released their results to broaden and diversify benchmarking data for ultra-large VS.

In addition to newly emerging approaches and more rigorous benchmarking, the libraries themselves will likely continue to grow. If one considers estimates of the total number of hypothetically possible drug-like molecules in chemical space, which range from around 10^{20} to over 10^{30}, even the largest modern screening libraries discussed herein remain tiny in perspective [3]. The emergence of even larger enumerated libraries may be questionable, as their data sizes have already become impractical. Instead, one can anticipate a further surge of nonenumerated combinatorial spaces that store building blocks and synthesis rules rather than the elaborated compounds [15]. This chapter already discussed several approaches that rely on such combinatorial libraries to speed up conventional screening. In light of the likely trends for virtual chemical libraries, we can assume a stronger presence of reagent-based approaches in future VS. If libraries indeed keep growing, combinatorial searches and 2D approaches will become critical components of VS when aiming for the largest spaces, as classical approaches appear to fall short of achieving sufficient throughput to scale beyond current state-of-the-art library sizes [9].

7.3 Other Trends That Might Impact the Future of Ultra-Large VS

An alternative way to find entirely novel scaffolds without relying on existing libraries utilizes generative AI [92]. First examples of combining docking with generative AI models have already begun to appear a few years back, for instance, MORLD (Molecule Optimization by Reinforcement Learning and Docking) [93]. The tool was showcased on discoidin domain receptor 1 kinase and the dopamine D_4 receptor as employed in the ultra-large screening campaign by Lyu et al. [17] discussed herein.

One drawback of generative models, potentially limiting their adoption in VS, has been that the molecules produced in silico may not be synthesizable or even always feasible. This issue was partly addressed by training networks specifically on synthetically accessible compounds [94]. However, in an age of searchable ultra-large make-on-demand spaces, generative models may gain more ground in future VS. Rather than seeding docking approaches directly from ultra-large libraries, generative models could create virtual hits de novo, which can then serve as templates for ultra-large library searches. Indeed, a very recent publication by Popov et al. [95] showcases the combination of generative AI, docking, and ultra-large library searches with their tool HIDDEN GEM (HIt Discovery using docking ENriched by GEnerative Modeling). With the present and future advances in ultra-fast searches of nonenumerated spaces in mind, this may represent a first blueprint of a future VS strategy, which would partially disentangle docking and ultra-large library searches.

DNA-encoded libraries (DELs) represent a new trend in experimental screening that may shake up the role of VS in hit identification. DELs allow for largely unbiased, unprecedented throughput in wet lab settings, venturing into the billion-scale and beyond [9, 96, 97]. Those libraries rely on combinatorial chemistry to generate small molecules with a DNA "barcode." A protein target can be exposed to the full library in solution, and any nonbinders get washed off. Finally, collecting the bound compounds and amplifying their DNA labels by PCR enables their identification [8]. The ability to test an entire library in a single tube significantly reduces the amount of required protein. Thus, e.g., mammalian expression vectors or the isolation and purification of a specific protein target from a patient's organ might become feasible [96]. DELs could further enable the prospective elucidation of SAR profiles and hit identification for particularly challenging, even "undruggable" targets [96, 97].

Collie et al. [97] discuss novel binding sites and previously unobserved enzyme conformations, which were indirectly discovered by DELs. However, despite their great potential, larger libraries suffer from higher false negative rates and increasing noise [96]. Furthermore, constructing DELs and analyzing the results of DEL assays is not without challenges, and to date, those libraries tend to cover different and rather narrow chemical spaces compared to most small-molecule libraries discussed herein: They are often rich in large, flexible, peptide-like compounds—thus not being the first choices for medicinal chemistry and hit-to-lead optimization [9]. Taken together, DELs can uncover interesting novel information and provide starting points for drug discovery campaigns. As experience with DELs grows and protocols to overcome their problems are established, their application, alone and likely also in combination with VS approaches, will likely impact the future of drug discovery.

8 Conclusions

Over the past 5 years, as demonstrated by the various examples in this chapter, the growth of chemical libraries into the ultra-large domain has inspired several ultra-large VS campaigns. An overview of all covered studies in the context of chemical library sizes is shown in Fig. 11.6. As can be seen, in principle, most VS approaches can be applied to state-of-the-art ultra-large libraries, although often with considerable investment of computing power. Reagent−/fragment-based approaches represent novel and very recent developments in response to combinatorial libraries and spaces that appear on the rise. One can further appreciate that approaches so far did rarely exceed 10 billion input compounds. Ligand-based and accelerated approaches have been applied on that scale, hinting at the VS concepts with the highest potential of adaptation to even larger library sizes.

Many highly potent compounds, often with novel scaffolds, were discovered with the help of those screens for a variety of targets. Typically, the final hit selections were not data-driven but relied on expert opinion. As such, the in part astounding

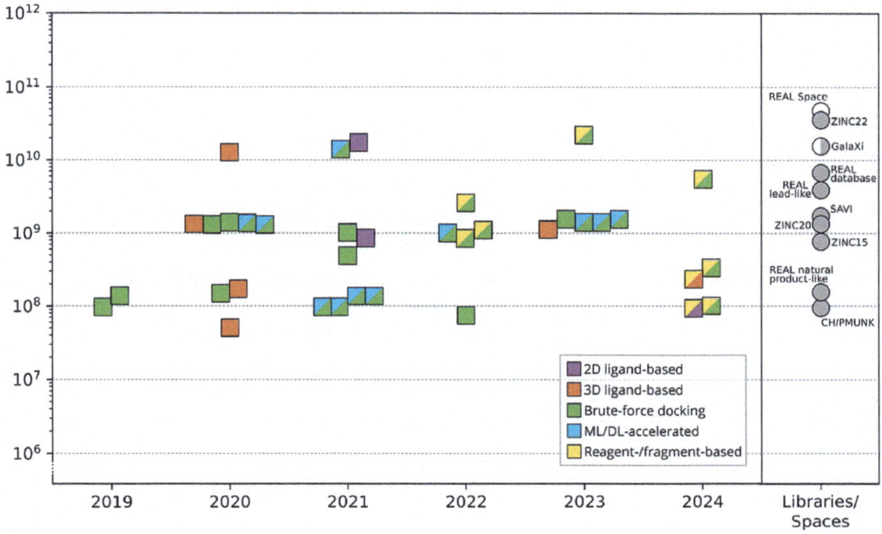

Fig. 11.6 Summary of all discussed ultra-large VS campaigns. Studies are grouped by year and color coded by screening approach. The *y*-axis indicates library sizes. In the inset on the right-hand side, sizes of all discussed publicly available chemical libraries and spaces are displayed for reference. Gray circles mark enumerated, white circles—nonenumerated libraries

reported hit rates (compare Table 11.8; typical hit finding campaigns achieve only about 1% hit rate according to the literature [89]) should be taken with a grain of salt and, given the human factor involved, not directly translated into an argument for pursuing ultra-large-scale screening. Owing to this notion, the debate whether *bigger is better* remains ongoing. Arguments can be made both in favor and against ultra-large VS [98]. We are currently only beginning to scratch the surface of ultra-large-scale VS, and it is, in our and others' opinion, premature to draw a conclusion whether—and, importantly, when—bigger is actually better [1, 98, 99].

Acknowledgements I.P., T.S., and A.P. gratefully acknowledge the support of the Jane and Aatos Erkko Foundation. We thank Laxman Yetukuri of CSC for his critical review and helpful feedback on our discussion of HPC systems.

Appendix

Table 11.9 List of URLs for discussed enumerated ultra-large screening libraries

Database	Available from
Enamine REAL database	https://enamine.net/compound-collections/real-compounds/real-database
Enamine REAL lead-like, natural product-like compounds	https://enamine.net/compound-collections/real-compounds/real-database-subsets
GalaXi enumerated	https://www.labnetwork.com/frontendapp/p/#!/library/virtual
SAVI	https://doi.org/10.35115/37n9-5738 https://cactus.nci.nih.gov/download/savi_download/
ZINC22	https://cartblanche22.docking.org/
ZINC20	https://zinc20.docking.org/
ZINC15	https://zinc15.docking.org/
CH/PMUNK	http://www.ewit.ccb.tu-dortmund.de/ag-koch/chipmunk/

An online version of this Table is also available from https://github.com/ipohner/ultralarge-VS, where references and table contents will be periodically updated

Table 11.10 Summary of discussed open-source VS tools and their availability

Tool	Reference(s)	Available from
DeepDocking	Gentile et al. [71]	https://github.com/jamesgleave/DD_protocol
DeepDocking GUI	Yaacoub et al. [76]	https://github.com/jamesgleave/DeepDockingGUI
HASTEN	Kalliokoski [79]	https://github.com/TuomoKalliokoski/HASTEN
Lean docking	Berenger et al. [84]	Encoder: https://github.com/UnixJunkie/molenc Support Vector regressor: https://github.com/UnixJunkie/linwrap
Linear accelerated docking	Marin et al. [85]	https://github.com/marinegor/Linear-accelerated-docking
MolPAL	Graff et al. [77]	https://github.com/coleygroup/molpal
Schrödinger GPU similarity (incentive)		https://github.com/schrodinger/gpusimilarity
SpaceDock	Sindt et al. [90]	https://github.com/litfsindt/LIT-SpaceDock
Thompson sampling	Klarich et al. [35]	https://github.com/PatWalters/TS
V-SYNTHES	Sadybekov et al. [22]	https://github.com/katritchlab/V-SYNTHES
VirtualFlow	Gorgulla et al. [7]	https://github.com/VirtualFlow
warpDOCK	McDougal et al. [67]	https://github.com/BruningLab/warpDOCK

An online version of this Table is also available from https://github.com/ipohner/ultralarge-VS, where references and table contents will be periodically updated

Table 11.11 Summary of available screening libraries with pre-generated 3D conformers and benchmarking datasets

Pre-generated 3D screening libraries — Description and reference → available from
Enamine REAL 1.4 billion compounds in PDBQT format by Gorgulla et al. [7] → via https://virtual-flow.org/real-library with login
ZINC15 1.5 billion compounds by Gorgulla et al. [5] → via https://virtual-flow.org/virtualflow-version-zinc15-library with login Enamine REAL lead-like (2021) 3D conformers from 1.56 billion SMILES input in Schrödinger Phase databases by Sivula et al. [23] → https://doi.org/10.23729/2de314bb-59af-452a-955c-c2ff0c5ea57f
Pre-generated 3D conformer tranches of ZINC20 as referenced by Bender et al. [24] → https://files.docking.org/3D/
Ultra-large benchmarking datasets — Description and reference → available from
1.56 billion SMILES + Glide-HTVS docking scores for SurA and GAK (csv) by Sivula et al. [23] → https://doi.org/10.23729/2170dc9c-4905-43c3-aeee-a574d360737f 1.4 billion SMILES + AutoDock-GPU docking scores and re-scoring results for 5 SARS-CoV-2 protein targets (parquet) as described in Rogers et al. [56] → https://doi.org/10.13139/OLCF/1783186 138 million SMILES + DOCK docking scores for dopamin D_4 receptor by Lyu et al. [17] → https://doi.org/10.6084/m9.figshare.7359401.v3 99 million SMILES + DOCK docking scores for AmpC by Lyu et al. [17] → https://doi.org/10.6084/m9.figshare.7359626.v2
138 million D_4 + 99 million AmpC Glide docking scores by Yang et al. [81] → https://s3.amazonaws.com/content.schrodinger.com/Resources/paper_data_share.zip

All URLs reported below were last checked January 31st, 2024. An online version of this Table is also available from https://github.com/ipohner/ultralarge-VS, where references and table contents will be periodically updated

References

1. Stumpfe, D., & Bajorath, J. (2020). Current trends, overlooked issues, and unmet challenges in virtual screening. *Journal of Chemical Information and Modeling, 60*, 4112–4115.
2. Carpenter, K. A., Cohen, D. S., Jarrell, J. T., & Huang, X. (2018). Deep learning and virtual drug screening. *Future Medicinal Chemistry, 10*(21), 2557–2567.
3. Walters, W. P. (2019). Virtual chemical libraries. *Journal of Medicinal Chemistry, 62*(3), 1116–1124.
4. Grebner, C., Malmerberg, E., Shewmaker, A., Batista, J., Nicholls, A., & Sadowski, J. (2020). Virtual screening in the cloud: How big is big enough? *Journal of Chemical Information and Modeling, 60*(9), 4274–4282.
5. Gorgulla, C., Jayaraj, A., Fackeldey, K., & Arthanari, H. (2022). Emerging frontiers in virtual drug discovery: From quantum mechanical methods to deep learning approaches. *Current Opinion in Chemical Biology, 69*, 102156.
6. Fresnais, L., & Ballester, P. J. (2021). The impact of compound library size on the performance of scoring functions for structure-based virtual screening. *Briefings in Bioinformatics, 22*(3), 1–10.
7. Gorgulla, C., Boeszoermenyi, A., Wang, Z.-F., Fischer, P. D., Coote, P. W., Das, K. M. P., Malets, Y. S., Radchenko, D. S., Moroz, Y. S., Scott, D. A., Fackeldey, K., Hoffmann, M.,

Iavniuk, I., Wagner, G., & Arthanari, H. (2020). An open-source drug discovery platform enables ultra-large virtual screens. *Nature, 580*, 663–668.
8. Hoffmann, T., & Gastreich, M. (2019). The next level in chemical space navigation: Going far beyond enumerable compound libraries. *Drug Discovery Today, 24*(5), 1148–1156.
9. Warr, W. A., Nicklaus, M. C., Nicolaou, C. A., & Rarey, M. (2022). Exploration of ultralarge compound collections for drug discovery. *Journal of Chemical Information and Modeling, 62*(9), 2021–2034.
10. ZINC15. https://zinc15.docking.org/. Online; Accessed 18 Jan 2024.
11. Irwin, J. J., Sterling, T., Mysinger, M. M., Bolstad, E. S., & Coleman, R. G. (2012). ZINC: A free tool to discover chemistry for biology. *Journal of Chemical Information and Modeling, 52*(7), 1757–1768.
12. Tingle, B. I., Tang, K. G., Castanon, M., Gutierrez, J. J., Khurelbaatar, M., Dandarchuluun, C., Moroz, Y. S., & Irwin, J. J. (2023). ZINC-22—A free multi-billion-scale database of tangible compounds for ligand discovery. *Journal of Chemical Information and Modeling, 63*(4), 1166–1176.
13. Enamine REAL. https://enamine.net/compound-collections/real-compounds, https://enamine.net/compound-collections/real-compounds/real-space-navigator. Online; Accessed 13 Mar 2024.
14. Grygorenko, O. O., Radchenko, D. S., Dziuba, I., Chuprina, A., Gubina, K. E., & Moroz, Y. S. (2020). Generating multibillion chemical space of readily accessible screening compounds. *iScience, 23*(11), 101681.
15. Lessel, U., & Lemmen, C. (2019). Modeling the expansion of virtual screening libraries. *ACS Medicinal Chemistry Letters, 10*, 1504–1510.
16. Sterling, T., & Irwin, J. J. (2015). ZINC 15—Ligand discovery for everyone. *Journal of Chemical Information and Modeling, 55*(11), 2324–2337.
17. Lyu, J., Wang, S., Balius, T. E., Singh, I., Levit, A., Moroz, Y. S., O'Meara, M. J., Che, T., Algaa, E., Tolmachova, K., Tolmachev, A. A., Shoichet, B. K., Roth, B. L., & Irwin, J. J. (2019). Ultra-large library docking for discovering new chemotypes. *Nature, 566*(7743), 224–229.
18. Stein, R. M., Kang, H. J., McCorvy, J. D., Glatfelter, G. C., Jones, A. J., Che, T., Slocum, S., Huang, X. P., Savych, O., Moroz, Y. S., Stauch, B., Johansson, L. C., Cherezov, V., Kenakin, T., Irwin, J. J., Shoichet, B. K., Roth, B. L., & Dubocovich, M. L. (2020). Virtual discovery of melatonin receptor ligands to modulate circadian rhythms. *Nature, 579*(7800), 609–614.
19. Alon, A., Lyu, J., Braz, J. M., Tummino, T. A., Craik, V., O'Meara, M. J., Webb, C. M., Radchenko, D. S., Moroz, Y. S., Huang, X.-P., Liu, Y., Roth, B. L., Irwin, J. J., Basbaum, A. I., Shoichet, B. K., & Kruse, A. C. (2021). Structures of the $\sigma 2$ receptor enable docking for bioactive ligand discovery. *Nature, 600*, 759–764.
20. Irwin, J. J., Tang, K. G., Young, J., Dandarchuluun, C., Wong, B. R., Khurelbaatar, M., Moroz, Y. S., Mayfield, J., & Sayle, R. A. (2020). ZINC20— A free ultralarge-scale chemical database for ligand discovery. *Journal of Chemical Information and Modeling, 60*(12), 6065–6073.
21. Michino, M., Beautrait, A., Boyles, N. A., Nadupalli, A., Dementiev, A., Sun, S., Ginn, J., Baxt, L., Suto, R., Bryk, R., Jerome, S. V., Huggins, D. J., & Vendome, J. (2023). Shape-based virtual screening of a billion-compound library identifies mycobacterial lipoamide dehydrogenase inhibitors. *ACS Bio & Med Chem Au, 3*(6), 507–515.
22. Sadybekov, A. A., Sadybekov, A. V., Liu, Y., Iliopoulos-Tsoutsouvas, C., Huang, X. P., Pickett, J., Houser, B., Patel, N., Tran, N. K., Tong, F., Zvonok, N., Jain, M. K., Savych, O., Radchenko, D. S., Nikas, S. P., Petasis, N. A., Moroz, Y. S., Roth, B. L., Makriyannis, A., & Katritch, V. (2022). Synthon-based ligand discovery in virtual libraries of over 11 billion compounds. *Nature, 601*(7893), 452–459.
23. Sivula, T., Yetukuri, L., Kalliokoski, T., Käsnänen, H., Poso, A., & Pöhner, I. (2023). Machine learning-boosted docking enables the efficient structure based virtual screening of giga-scale enumerated chemical libraries. *Journal of Chemical Information and Modeling, 63*(18), 5773–5783.

24. Bender, B. J., Gahbauer, S., Luttens, A., Lyu, J., Webb, C. M., Stein, R. M., Fink, E. A., Balius, T. E., Carlsson, J., Irwin, J. J., & Shoichet, B. K. (2021). A practical guide to large-scale docking. *Nature Protocols, 16*, 4799–4832.
25. WuXi AppTec — galaXi. https://www.labnetwork.com/frontend-app/p/#!/library/virtual. Online; Accessed 30 Jan 2024.
26. BioSolveIT infiniSee. https://www.biosolveit.de/infiniSee/. Online; Accessed 28 Feb 2024.
27. eMolecules eXplore. https://www.emolecules.com/explore. Online; Accessed 15 Mar 2024.
28. Neumann, A., Marrison, L., & Klein, R. (2023). Relevance of the trillion-sized chemical space "eXplore" as a source for drug discovery. *ACS Medicinal Chemistry Letters, 14*, 466–472.
29. OTAVA Chemicals — CHEMriya On-Demand Chemical Space. https://www.otavachemicals.com/products/chemriya. Online; Accessed 9 Feb 2024.
30. Patel, H., Ihlenfeldt, W.-D., Judson, P. N., Moroz, Y. S., Pevzner, Y., Peach, M. L., Delannée, V., Tarasova, N. I., & Nicklaus, M. C. (2020). SAVI, *in silico* generation of billions of easily synthesizable compounds through expert-system type rules. *Scientific Data, 7*, 384.
31. BioSolveIT KnowledgeSpace. https://www.biosolveit.de/chemical-spaces. Online; Accessed 22 Nov 2024.
32. Humbeck, L., Weigang, S., Schäfer, T., Mutzel, P., & Koch, O. (2018). CHIPMUNK: A virtual synthesizable small-molecule library for medicinal chemistry, exploitable for protein–protein interaction modulators. *ChemMedChem, 13*, 532–539.
33. How to solve a jigsaw puzzle with 1.56 billion pieces. CSC Blog Post. https://csc.fi/en/blog/how-to-solve-a-jigsaw-puzzle-with-1-56-billion-pieces/. Online; Accessed 22 Nov 2024.
34. Gentile, F., Fernandez, M., Ban, F., Ton, A.-T., Mslati, H., Perez, C. F., Leblanc, E., Yaacoub, J. C., Gleave, J., Stern, A., Wong, B., Jean, F., Strynadka, N., & Cherkasov, A. (2021). Automated discovery of noncovalent inhibitors of sars-cov2 main protease by consensus deep docking of 40 billion small molecules. *Chemical Science, 12*(48), 15960–15974.
35. Klarich, K., Goldman, B., Kramer, T., Riley, P., & Walters, W. P. (2024). Thompson sampling — An efficient method for searching ultralarge synthesis on demand databases. *Journal of Chemical Information and Modeling, 64*(4), 1158–1171.
36. Morris, G. M., Huey, R., Lindstrom, W., Sanner, M. F., Belew, R. K., Goodsell, D. S., & Olson, A. J. (2009). AutoDock4 and AutoDockTools4: Automated docking with selective receptor flexibility. *Journal of Computational Chemistry, 30*(16), 2785–2791.
37. Santos-Martins, D., Solis-Vasquez, L., Tillack, A. F., Sanner, M. F., Koch, A., & Forli, S. (2021). Accelerating AutoDock4 with GPUs and gradient-based local search. *Journal of Chemical Theory and Computation, 17*(2), 1060–1073.
38. Gentile, F., Yaacoub, J. C., Gleave, J., Fernandez, M., Ton, A.-T., Ban, F., Stern, A., & Cherkasov, A. (2022). Artificial intelligence–enabled virtual screening of ultra-large chemical libraries with deep docking. *Nature Protocols, 17*, 672–697.
39. Dalke, A. (2019). The chemfp project. *Journal of Cheminformatics, 11*, 76.
40. MolSoft ICM-Pro: Gigasearch. https://molsoft.com/giga-search.html. Online; Accessed 26 Feb 2024.
41. Schrödinger LiveDesign. https://newsite.schrodinger.com/platform/products/livedesign/. Online; Accessed 26 Feb 2024.
42. OpenEye Orion: Molecules as a Service (MaaS). https://www.eyesopen.com/news/openeye-orion-2020.2-update. Online; Accessed 26 Feb 2024.
43. Schrödinger GPU Similarity. https://github.com/schrodinger/gpusimilarity. Online; Accessed 26 Feb 2024.
44. NextMove Software Ltd. SmallWorld. https://www.nextmovesoftware.com/smallworld.html. Online; Accessed 25 Mar 2024.
45. NextMove Software Ltd. Arthor. https://nextmovesoftware.com/arthor.html. Online; Accessed 25 Mar 2024.
46. Rarey, M., & Stahl, M. (2001). Similarity searching in large combinatorial chemistry spaces. *Journal of Computer-Aided Molecular Design, 15*, 497–520.

47. Bellmann, L., Penner, P., & Rarey, M. (2019). Connected subgraph fingerprints: Representing molecules using exhaustive subgraph enumeration. *Journal of Chemical Information and Modeling, 59*, 4625–4635.
48. Bellmann, L., Penner, P., & Rarey, M. (2021). Topological similarity search in large combinatorial fragment spaces. *Journal of Chemical Information and Modeling, 61*, 238–251.
49. Schmidt, R., Klein, R., & Rarey, M. (2022). Maximum common substructure searching in combinatorial make-on-demand compound spaces. *Journal of Chemical Information and Modeling, 62*, 2133–2150.
50. Glaab, E., Manoharan, G. B., & Abankwa, D. (2021). Pharmacophore model for sars-cov-2 3clpro small-molecule inhibitors and *in vitro* experimental validation of computationally screened inhibitors. *Journal of Chemical Information and Modeling, 61*, 4082–4096.
51. Brüschweiler, S., Fuchs, J. E., Bader, G., McConnell, D. B., Konrat, R., & Mayer, M. (2021). A step toward NRF2-DNA interaction inhibitors by fragment-based NMR methods. *ChemMedChem, 16*, 3576–3587.
52. MolSoft RIDE (Rapid Isostere Discovery Engine). https://molsoft.com/RIDE.html. Online; Accessed 29 Feb 2024.
53. ZINC20 3D Tranches. https://files.docking.org/3D/. Online; Accessed 31 Jan 2024.
54. Gahbauer, S., Correy, G. J., Schuller, M., Ferla, M. P., Doruk, Y. U., Rachman, M., Wu, T., Diolaiti, M., Wang, S., Neitz, R. J., Fearon, D., Radchenko, D. S., Moroz, Y. S., Irwin, J. J., Renslo, A. R., Taylor, J. C., Gestwicki, J. E., von Delft, F., Ashworth, A., Ahel, I., Shoichet, B. K., & Fraser, J. S. (2023). Iterative computational design and crystallographic screening identifies potent inhibitors targeting the nsp3 macrodomain of sars-cov-2. *Proceedings of the National Academy of Sciences of the United States of America, 120*(2), e2212931120.
55. Acharya, A., Agarwal, R., Baker, M. B., Baudry, J., Bhowmik, D., Boehm, S., Byler, K. G., Chen, S. Y., Coates, L., Cooper, C. J., Demerdash, O., Daidone, I., Eblen, J. D., Ellingson, S., Forli, S., Glaser, J., Gumbart, J. C., Gunnels, J., Hernandez, O., Irle, S., Kneller, D. W., Kovalevsky, A., Larkin, J., Lawrence, T. J., LeGrand, S., Liu, S.-H., Mitchell, J. C., Park, G., Parks, J. M., Pavlova, A., Petridis, L., Poole, D., Pouchard, L., Ramanathan, A., Rogers, D. M., Santos-Martins, D., Scheinberg, A., Sedova, A., Shen, Y., Smith, J. C., Smith, M. D., Soto, C., Tsaris, A., Thavappiragasam, M., Tillack, A. F., Vermaas, J. V., Vuong, V. Q., Yin, J., Yoo, S., Zahran, M., & Zanetti-Polzi, L. (2020). Supercomputer-based ensemble docking drug discovery pipeline with application to Covid-19. *Journal of Chemical Information and Modeling, 60*(12), 5832–5852.
56. Rogers, D. M., Agarwal, R., Vermaas, J. V., Smith, M. D., Rajeshwar, R. T., Cooper, C., Sedova, A., Boehm, S., Baker, M., Glaser, J., & Smith, J. C. (2023). SARS-CoV2 billion-compound docking. *Scientific Data, 10*, 173.
57. Luttens, A., Gullberg, H., Abdurakhmanov, E., Vo, D. D., Akaberi, D., Talibov, V. O., Nekhotiaeva, N., Vangeel, L., De Jonghe, S., Jochmans, D., Krambrich, J., Tas, A., Lundgren, B., Gravenfors, Y., Craig, A. J., Atilaw, Y., Sandström, A., Moodie, L. W. K., Lundkvist, A., van Hemert, M. J., Neyts, J., Lennerstrand, J., Kihlberg, J., Sandberg, K., Danielson, U. H., & Carlsson, J. (2022). Ultralarge virtual screening identifies sars-cov-2 main protease inhibitors with broad-spectrum activity against coronaviruses. *Journal of the American Chemical Society, 144*(7), 2905–2920.
58. Trott, O., & Olson, A. J. (2010). Autodock vina: Improving the speed and accuracy of docking with a new scoring function, efficient optimization, and multithreading. *Journal of Computational Chemistry, 31*(2), 455–461.
59. Koes, D. R., Baumgartner, M. P., & Camacho, C. J. (2013). Lessons learned in empirical scoring with smina from the CSAR 2011 benchmarking exercise. *Journal of Chemical Information and Modeling, 53*(8), 1893–1904.
60. Korb, O., Stützle, T., & Exner, T. E. (2007). An ant colony optimization approach to flexible protein-ligand docking. *Swarm Intelligence, 1*(2), 115–134.
61. Gorgulla, C., Çınaroğlu, S. S., Fischer, P. D., Fackeldey, K., Wagner, G., & Arthanari, H. (2021). VirtualFlow ants-ultra-large virtual screenings with artificial intelligence driven

docking algorithm based on ant colony optimization. *International Journal of Molecular Sciences, 22*(11), 5807.

62. Gorgulla, C., Das, K. M. P., Leigh, K. E., Cespugli, M., Fischer, P. D., Wang, Z.-F., Tesseyre, G., Pandita, S., Shnapir, A., Calderaio, A., Gechev, M., Rose, A., Lewis, N., Hutcheson, C., Yaffe, E., Luxenburg, R., Herce, H. D., Durmaz, V., Halazonetis, T. D., Fackeldey, K., Patten, J., Chuprina, A., Dziuba, I., Plekhova, A., Moroz, Y., Radchenko, D., Tarkhanova, O., Yavnyuk, I., Gruber, C., Yust, R., Payne, D., Näär, A. M., Namchuk, M. N., Davey, R. A., Wagner, G., Kinney, J., & Arthanari, H. (2021). A multi-pronged approach targeting SARS-CoV-2 proteins using ultra-large virtual screening. *iScience, 24*(2), 102021.
63. Sadybekov, A. A., Brouillette, R. L., Marin, E., Sadybekov, A. V., Luginina, A., Gusach, A., Mishin, A., Besserer-Offroy, E., Jean-Michel, L., Borshchevskiy, V., Cherezov, V., Sarret, P., & Katrich, V. (2020). Structure-based virtual screening of ultra-large library yields potent antagonists for a lipid GPCR. *Biomolecules, 10*(12), 1634.
64. Kaplan, A. L., Confair, D. N., Kim, K., Barros-Álvarez, X., Rodriguiz, R. M., Yang, Y., Kweon, O. S., Che, T., McCorvy, J. D., Kamber, D. N., Phelan, J. P., Martins, L. C., Pogorelov, V. M., DiBerto, J. F., Slocum, S. T., Huang, X. P., Kumar, J. M., Robertson, M. J., Panova, O., Seven, A. B., Wetsel, A. Q., Wetsel, W. C., Irwin, J. J., Skiniotis, G., Shoichet, B. K., Roth, B. L., & Ellman, J. A. (2022). Bespoke library docking for 5-HT_{2A} receptor agonists with antidepressant activity. *Nature, 610*(7932), 582–591.
65. Pöhner, I., & Sivula, T. (2023). *Glide HTVS docking results of Enamine REAL lead-like library (1.56 billion compounds) for targets SurA and GAK.*
66. Rogers, D., Glaser, J., Agarwal, R., Vermaas, J., Smith, M., Parks, J., Cooper, C., Sedova, A., Boehm, S., Baker, M., & Smith, J. (2021). *SARS-CoV2 docking dataset.*
67. McDougal, D. P., Rajapaksha, H., Pederick, J. L., & Bruning, J. B. (2023). warpDOCK: Large-scale virtual drug discovery using cloud infrastructure. *ACS Omega, 8*(32), 29143–29149.
68. Alhossary, A., Handoko, S. D., Mu, Y., & Kwoh, C.-K. (2015). Fast, accurate, and reliable molecular docking with QuickVina 2. *Bioinformatics, 31*(13), 2214–2216.
69. Bonilla, P. A., Hoop, C. L., Stefanisko, K., Tarasov, S. G., Sinha, S., Nicklaus, M. C., & Tarasova, N. I. (2023). Virtual screening of ultra-large chemical libraries identifies cell-permeable small-molecule inhibitors of a "non-druggable" target, STAT3 N-terminal domain. *Frontiers in Oncology, 13*, 1144153.
70. Cherkasov, A., Ban, F., Li, Y., Fallahi, M., & Hammond, G. L. (2006). Progressive docking: A hybrid QSAR/docking approach for accelerating in silico high throughput screening. *Journal of Medicinal Chemistry, 49*, 7466–7478.
71. Gentile, F., Agrawal, V., Hsing, M., Ton, A. T., Ban, F., Norinder, U., Gleave, M. E., & Cherkasov, A. (2020). Deep docking: A deep learning platform for augmentation of structure based drug discovery. *ACS Central Science, 6*(6), 939–949.
72. Ton, A.-T., Gentile, F., Hsing, M., Ban, F., & Cherkasov, A. (2020). Rapid identification of potential inhibitors of SARS-CoV-2 main protease by deep docking of 1.3 billion compounds. *Molecular Informatics, 39*(8), 202000028.
73. Garland, O., Ton, A.-T., Moradi, S., Smith, J. R., Kovacic, S., Ng, K., Pandey, M., Ban, F., Lee, J., Vuckovic, M., Worrall, L. J., Young, R. N., Pantophlet, R., Strynadka, N. C. J., & Cherkasov, A. (2023). Large-scale virtual screening for the discovery of SARS-CoV-2 papain-like protease (PLpro) non-covalent inhibitors. *Journal of Chemical Information and Modeling, 63*(7), 2158–2169.
74. Tang, M., Wen, C., Lin, J., Chen, H., & Ran, T. (2023). Discovery of novel $A_{2A}R$ antagonists through deep learning-based virtual screening. *Artificial Intelligence in the Life Sciences, 3*, 100058.
75. Radaeva, M., Ho, C.-H., Xie, N., Zhang, S., Lee, J., Liu, L., Lallous, N., Cherkasov, A., & Dong, X. (2022). Discovery of novel Lin28 inhibitors to suppress cancer cell stemness. *Cancers, 14*(22), 5687.
76. Yaacoub, J. C., Gleave, J., Gentile, F., Stern, A., & Cherkasov, A. (2022). DD-GUI: A graphical user interface for deep learning-accelerated virtual screening of large chemical libraries (Deep Docking). *Bioinformatics, 38*(4), 1146–1148.

77. Graff, D. E., Shakhnovich, E. I., & Coley, C. W. (2021). Accelerating high-throughput virtual screening through molecular pool-based active learning. *Chemical Science, 12*(22), 7866–7881.
78. Yang, K., Swanson, K., Jin, W., Coley, C., Eiden, P., Gao, H., Guzman-Perez, A., Hopper, T., Kelley, B., Mathea, M., Palmer, A., Settels, V., Jaakkola, T., Jensen, K., & Barzilay, R. (2019). Analyzing learned molecular representations for property prediction. *Journal of Chemical Information and Modeling, 59*(8), 3370–3388.
79. Kalliokoski, T. (2021). Machine learning boosted docking (HASTEN): An open-source tool to accelerate structure-based virtual screening campaigns. *Molecular Informatics, 40*(9), 2100089.
80. Friesner, R. A., Banks, J. L., Murphy, R. B., Halgren, T. A., Klicic, J. J., Mainz, D. T., Repasky, M. P., Knoll, E. H., Shelley, M., Perry, J. K., Shaw, D. E., Francis, P., & Shenkin, P. S. (2004). Glide: A new approach for rapid, accurate docking and scoring. 1. Method and assessment of docking accuracy. *Journal of Medicinal Chemistry, 47*(7), 1739–1749.
81. Yang, Y., Yao, K., Repasky, M. P., Leswing, K., Abel, R., Shoichet, B. K., & Jerome, S. V. (2021). Efficient exploration of chemical space with docking and deep learning. *Journal of Chemical Theory and Computation, 17*(11), 7106–7119.
82. MolSoft ICM-Pro: Gigascreen. https://www.molsoft.com/GigaScreen.html. Online; Accessed 28 Jan 2024.
83. OpenEye Orion: GigaDock. https://www.eyesopen.com/orion/gigadock. Online; Accessed 28 Jan 2024.
84. Berenger, F., Kumar, A., Zhang, K. Y. J., & Yamanishi, Y. (2021). Lean-docking: Exploiting ligands' predicted docking scores to accelerate molecular docking. *Journal of Chemical Information and Modeling, 61*, 2341–2352.
85. Marin, E., Kovaleva, M., Kadukova, M., Mustafin, K., Khorn, P., Rogachev, A., Mishin, A., Guskov, A., & Borshchevskiy, V. (2024). Regression-based active learning for accessible acceleration of ultra-large library docking. *Journal of Chemical Information and Modeling, 64*(7), 2612–2623.
86. Kolb, P., Kipouros, C. B., Huang, D., & Caflisch, A. (2008). Structure-based tailoring of compound libraries for high-throughput screening: Discovery of novel EphB4 kinase inhibitors. *Proteins, 73*(1), 11–18.
87. Zhao, H., Dong, J., Lafleur, K., Nevado, C., & Caflisch, A. (2012). Discovery of a novel chemotype of tyrosine kinase inhibitors by fragment-based docking and molecular dynamics. *ACS Medicinal Chemistry Letters, 3*(10), 834–838.
88. Beroza, P., Crawford, J. J., Ganichkin, O., Gendelev, L., Harris, S. F., Klein, R., Miu, A., Steinbacher, S., Klingler, F. M., & Lemmen, C. (2022). Chemical space docking enables large-scale structure-based virtual screening to discover ROCK1 kinase inhibitors. *Nature Communications, 13*(1), 6447.
89. Müller, J., Klein, R., Tarkhanova, O., Gryniukova, A., Borysko, P., Merkl, S., Ruf, M., Neumann, A., Gastreich, M., Moroz, Y. S., Klebe, G., & Glinca, S. (2022). Magnet for the needle in haystack: "crystal structure first" fragment hits unlock active chemical matter using targeted exploration of vast chemical spaces. *Journal of Medicinal Chemistry, 65*(23), 15663–15678.
90. Sindt, F., Seyller, A., Eguida, M., & Rognan, D. (2024). Protein structure-based organic chemistry-driven ligand design from ultralarge chemical spaces. *ACS Central Science, 10*(3), 615–627.
91. Lyu, J., Irwin, J. J., & Shoichet, B. K. (2023). Modeling the expansion of virtual screening libraries. *Nature Chemical Biology, 19*, 712–718.
92. Segler, M. H. S., Kogej, T., Tyrchan, C., & Waller, M. P. (2018). Generating focused molecule libraries for drug discovery with recurrent neural networks. *ACS Central Science, 4*(1), 120–131.
93. Jeon, W., & Kim, D. (2020). Autonomous molecule generation using reinforcement learning and docking to develop potential novel inhibitors. *Scientific Reports, 10*(1), 22104.
94. Segler, M. H. S., Preuss, M., & Waller, M. P. (2018). Planning chemical syntheses with deep neural networks and symbolic AI. *Nature, 555*, 604–610.

95. Popov, K. I., Wellnitz, J., Maxfield, T., & Tropsha, A. (2024). HIt Discovery using docking ENriched by GEnerative Modeling (HIDDEN GEM): A novel computational workflow for accelerated virtual screening of ultra-large chemical libraries. *Molecular Informatics, 1*, e202300207.
96. Sunkari, Y. K., Siripuram, V. K., Nguyen, T.-L., & Flajolet, M. (2022). High-power screening (HPS) empowered by DNA-encoded libraries. *Trends in Pharmacological Sciences, 43*(1), 4–15.
97. Collie, G. W., Clark, M. A., Keefe, A. D., Madin, A., Read, J. A., Rivers, E. L., & Zhang, Y. (2024). Screening ultra-large encoded compound libraries leads to novel protein–ligand interactions and high selectivity. *Journal of Medicinal Chemistry, 67*(2), 864–884.
98. Clark, D. E. (2020). Virtual screening: Is bigger always better? Or can small be beautiful? *Journal of Chemical Information and Modeling, 60*, 4120–4123.
99. Kontoyianni, M. (2022). Library size in virtual screening: Is it truly a number's game? *Expert Opinion on Drug Discovery, 17*(11), 1177–1179.

Part II
The Pitfalls Between Experimentation and Simulation

Chapter 12
Experimental Assays: Chemical Properties, Biochemical and Cellular Assays,and In Vivo Evaluations

Mateus Sá Magalhães Serafim, Erik Vinicius de Sousa Reis, Jordana Grazziela Alves Coelho-dos-Reis, Jônatas Santos Abrahão, and Anthony John O'Donoghue

Abstract The design and discovery of new bioactive compounds have been essential for the development of potential new inhibitors and drug candidates. In this regard, the use of computational simulations has proven to play an important role in achieving new drugs. Throughout history and most recently, new drugs, e.g., protease inhibitors, have benefited from the so-called computer-aided drug discovery (CADD) approaches, providing available therapeutic options to emerging or re-emerging diseases, such as the coronavirus disease 2019 (COVID-19). These in silico models and methods can be employed for different purposes, such as prediction of various biological activities, toxicity, pharmacokinetics, target specificity, and even the synthesis of new analogs. Ultimately, such predictions can select or disregard a given compound for an in vitro or in vivo evaluation. However, translating a simulation to an experimental validation may be challenging. For instance, one should consider chemical properties and solubility, different biochemical and cellular assays, and the availability of data or methods to assess bioactive compounds and potentially reach a successful candidate. This chapter aims to provide a detailed overview of the many computational possibilities to achieve or improve experimental feasibility. Furthermore, we address the challenges and pitfalls regarding such approaches, which may contribute to a successful drug design and discovery campaign in the field.

Keywords Biological activity · Computational simulation · Experimental validation · In silico · In vitro · In vivo

---

M. Sá Magalhães Serafim (✉) · E. V. de Sousa Reis · J. G. Alves Coelho-dos-Reis · J. Santos Abrahão
Department of Microbiology, Institute of Biological Sciences, Federal University of Minas Gerais, Belo Horizonte, Minas Gerais, Brazil

A. J. O'Donoghue
Center for Discovery and Innovation in Parasitic Diseases, Skaggs School of Pharmacy and Pharmaceutical Sciences, University of California San Diego, La Jolla, CA, USA

V. G. Maltarollo (ed.), *Computer-Aided and Machine Learning-Driven Drug Design*, Computer-Aided Drug Discovery and Design 3,
https://doi.org/10.1007/978-3-031-76718-0_12

1 Introduction

The field of drug design and discovery is marked by how expensive it is—about $300 million to $2.1 billion per product [1], up to estimates of over $2.8 billion [2]—and how long it takes, ranging from 5.8 to over 15.2 years, to design, develop, and license a small-molecule drug for the market [3]. In addition, less than 30 new next-in-class drugs are expected to be introduced to the market over the next decade (2030–2039) [4]. Arguably, cost- and time-saving opportunities are intrinsic to the early steps of design and discovery, up to the preclinical stages, which account for 90% of clinical drug development failing [5]. Considering target selection, hit identification and optimization, and the selection of a potential clinical candidate, the current optimization of lead compounds usually emphasizes either potency, specificity, or both [5, 6]. These steps may include different computational approaches, such as structure-activity relationship (SAR) analysis [7], that may integrate molecular information and biological outcomes [8].

The current state of the art is supported by an increasing number of different computational simulations that can be implemented as a single model or in combination with other methods to properly predict general or specific outcomes, such as biological activities [9]. One can cite the integration of machine learning (ML) techniques as predictive and classification models for potency [10], solubility [11], and absorption, distribution, metabolism, excretion, and toxicity (ADMET) [12] predictions, as well as the implementation of artificial intelligence (AI) in drug discovery approaches [13]. Moreover, with the development of AlphaFold [14], computational approaches have proven their ability to support and directly affect drug discovery [15]. Therefore, the design and search for fast, reliable, and accessible ways to enhance drug discovery [16] could improve the discovery of diverse hits and leads with optimal drug-like parameters, such as pharmacokinetics (PK) and ADMET [17, 18], drug stability in solution [19, 20], and even alternative formulations [21]. In addition, these parameters could also improve the expected outcomes in various stages of design and development in the preclinical phase, thereby lowering associated costs and obtaining more effective and safer drugs [6, 16, 22].

In this sense, companies, institutes, and universities expand the use of computational approaches, as seen in the economic impact of AI usage [23]. For example, AI can be used in synthesis planning [24], virtually screening ultra-large libraries with billions of compounds at once [25], or filtering over a billion commercially available compounds from an open-source library [26]. Furthermore, studies in the field of drug discovery have already proven how fast, feasible, and reliable computational approaches can be, such as in the coronavirus disease 2019 (COVID-19) pandemic scenario, which urged computational approaches in a common effort for new therapeutic options [27, 28]. For instance, in 2022, a successful approach resulted in the design and discovery of the antiviral drug ensitrelvir, a severe acute respiratory syndrome-related coronavirus 2 (SARS-CoV-2) main protease (M^{pro}) inhibitor. Herein, hundreds of thousands of compounds were virtually screened for a lead candidate, which was ultimately optimized with structure-based approaches [29].

The various computer-aided drug discovery (CADD) approaches employed in this pandemic scenario led to the rapid availability of potential drug candidates in different stages of validation, but it also led to the surge of compounds supported only by simulations [30]. In this sense, translating simulations to experimental validations is a crucial step for drug discovery, which may avoid unrealistic results from predictive models by providing reliable data from predictions complemented with experimental evidence [31]. Experimental validation is referred to as the procedure (*i.e.*, in vitro assays) that is able to reproduce a scientific result obtained using computational models or methods (*i.e.*, in silico predictions) [32], such as those regarding biological activities [33]. However, validations may not be simple and can face pharmaceutical bottlenecks [34], as in the case of nucleic acid drugs [35], that is, drugs that control functions of cells based on nucleotide sequence information (*e.g.*, genome expression and gene regulation) [36]. Although difficult, it is important to employ feasible experimental data to support, corroborate, or demonstrate that a proposed simulation model or method is accurate and reproducible, thus validating a given study's hypothesis [31, 37].

Furthermore, it is also important to consider the expected outcomes (*e.g.*, negative or positive data) from an initial hypothesis, which may not have yet produced the desired results [38], such as weak inhibitors being considered in optimization studies [39], while also considering the disclosure of negative results to the scientific community [40, 41]. Herein, avoiding false negatives and false positives is one of the major challenges of screening approaches [42], and some elements may be potential limitations that can affect the performance of any computational simulation, such as the differences in datasets (*e.g.*, size and diversity) [43]. Therefore, reiterating rigor in good practices is of critical importance to CADD methods and their predicted outcomes [30], allowing for better development, exploitation, and validation of models. An example is the use of quantitative SAR (QSAR) analysis [44], in which rigorous models' design improve its ability to quantify the influence of each chemical structure fragment toward a biological activity [7]. This is especially important when considering various studies assessing the same inhibitor, such as different IC_{50} values obtained against the same target (*e.g.*, GC376 against SARS-CoV-2 M^{pro}) [45]. Moreover, with the increasing abundance of available data (*e.g.*, AI and predictive ML algorithms [46]) and methods for ultra-large [26] and accelerated [47] screening, the curation of data [48] is also necessary to assure the quality of predictions for experimental determinations.

The ability of in silico tools to accurately predict a bioactive compound may be reaching the point of CADD turning to a computer-driven drug discovery [49]. Subsequent experimental determination is essential, such as in hybrid silico-vitro approaches, which require a combination of multiple computational simulations and in silico screening with specific and more sensitive experimental validation in vitro [6]. For example, the experimental determination of various small binding fragments complexed to enzymes during the design of potential inhibitors [50]. Hereupon, one could also cite the applicability of molecular dynamics (MD) simulations in predicting binding sites as druggable pockets [51], which may vary in accuracy to support ligands and the discovery of potential inhibitors for different targets (*e.g.*,

G-protein coupled receptors; GPCR) [52]. Predicted hits can be confirmed by experimental determination in methods such as cryo-electron microscopy, X-ray crystallography, and nuclear magnetic resonance (NMR) [51], thus supporting a successful inhibitor [50], as well as determining binding affinity [53], or even correlating potency [54]. Additionally, they can be performed using nano-differential scanning fluorimetry (nanoDSF) and microscale thermophoresis (MST), which can evaluate ligand binding by assessing protein denaturation and stability at varied temperatures, and by changes in the fluorescence of tagged proteins induced by temperature, respectively [55].

However, achieving true inhibitors from initial simulations or fragment-based models may be a long and costly effort that usually involves reassessing the ligand's design, synthesis, and experimental validation [56]. On the other hand, employing a combination of ligand- and structure-based simulations with virtual screening (VS) could optimize ligands as target-hits that better reproduce experimentally determined conformations of known inhibitors. Thus, a more cost-effective approach could be directed towards experimental validation in vitro [57, 58], including potential applications to other fields of research [59]. This combination of methods and simulations may increase the overall accuracy of a given approach, as in the combination of ligand-based drug design (LBDD) methods and QSAR [38, 60, 61], for example, combining molecular docking and QSAR models to discover SARS-CoV-2 protease inhibitors [38]. In addition, the combination of ligand-based models in a consensus VS approach may benefit from the calculation of decoys (*i.e.*, putative inactive compounds), which aim to increase the predictive ability of identifying true negatives [62] in a VS [63]. Further, structure-based drug design (SBDD) approaches may benefit from the design of a pharmacophore model [64], which can improve the comparison of true inhibitors and designed ligands [65], and increase the success rate of selected compounds against proposed targets [66].

It is also important to mention that computational approaches may never reach a point where all predictions are correct [6]. Overall, VS campaigns may not even reach a substantial number of hits accurately confirmed in experimental validation (*e.g.*, inactive compounds or false positive hits) [67]. For instance, this was observed from SARS-CoV-2 M^{pro} consensus VS approaches, which only one to three true positive hits resulted from dozens to thousands of screened compounds (hit rate ranging from 0.066% to 7.14%) [38]. Such a rate is still higher than usual expected from high-throughput screening (HTS) approaches (between 0.01% and 0.15%) [68], which are automated screenings of thousands of compounds in vitro. Nonetheless, high hit rates are also achievable in different VS approaches, typically in the range between 1% and over 25% (median value of 13% in 421 studies) [68], as they may be designed for specific targets and ligands and can be improved from available computational and experimental data [69]. For example, a VS can identify compounds whose structures are complementary or appropriate for binding to a target enzyme, depending on its binding site, and hits can account for up to 20% of a given chemical library [70].

Notwithstanding, such simulations must be followed by experimental validation in vitro and/or in vivo that can either verify the predictions or at least improve the

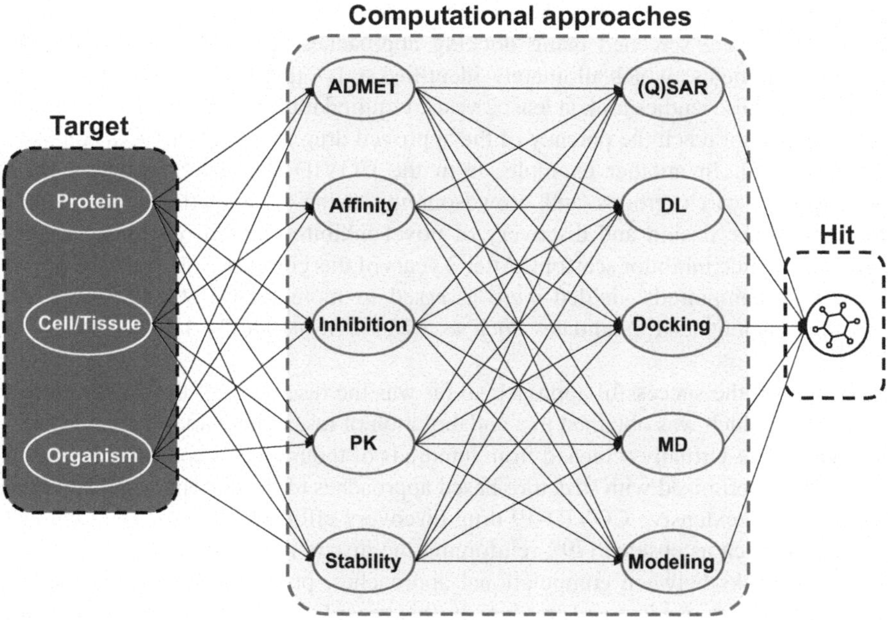

Fig. 12.1 An artificial network for translating computational approaches in drug design. A protein or enzyme, a specific cell lineage or tissue, and a specific organism (*e.g.*, virus) may be selected as an initial target. Various parameters can be simulated with computational approaches before experimental validation, such as absorption, distribution, metabolism, excretion, and toxicity (ADMET) and pharmacokinetics (PK) properties of a given compound, in addition to its stability in solution, affinity, or inhibitory activity against a given target. These data can be predicted with various models and methods, such as quantitative structure-activity relationship (QSAR) analysis, molecular docking, molecular dynamics (MD) simulations, as well as deep learning (DL) approaches, including modeling from different software. Ultimately, a hit compound can be obtained for subsequent experimental validation in vitro and/or in vivo

quality of the model focusing on a proposed target. This may lead to better estimations of compounds' properties such as ligand affinity and ADMET that translate the computational approaches (Fig. 12.1) to more accurate in vitro and in vivo hit results, thereby reducing test requirements [71]. However, aiming for the construction of these hybrid silico-vitro workflows requires extensive research teams and laboratory structure, a large data management and curation system, as well as the integration of academic researchers and pharmaceutical companies [72, 73].

As an example, hybrid silico-vitro drug discovery approaches against SARS-CoV-2 M^{pro} identified low-affinity binding fragments (IC_{50} values between 180 μM and 1 mM) from a combined crystallographic screening, that is, a screening of potential binding fragments experimentally determining each of them by crystallography. This experimental screening was followed by a VS searching for similar fragments as those that did bind to the target structure, identifying fragments that displayed IC_{50} values as low as 0.4 μM [74]. This study was followed by a similar approach, which designed ligands by virtually linking small binding fragments into a single scaffold and searching for similar compounds. Additionally, 450 million compounds were screened using docking approaches searching for similar lead binding fragments, which ultimately identified a 1.7 μM hit [75]. Despite being successful, these studies took at least 2 years, required multiple different approaches, and still did not reach the potency of the approved drug nirmatrelvir (IC_{50} equals to 2.5 nM) [76]. In another example, from the COVID Moonshot initiative [77], numerous research groups and crowdsourcing using computational approaches focused on the design and discovery of novel inhibitors based on a noncovalent and non-peptide inhibitor scaffold. After 2 years of this complex collaborative effort, 2400 new compounds yielded were assessed in more than 10,000 assays [77], obtaining potent lead candidates such as MAT-POS-e194df51-1 (IC_{50} equals to 37 nM).

However, the successful approach so far was the discovery of the antiviral drug ensitrelvir, which was obtained in a collaboration of researchers after the design of a lead candidate virtually screened from hundreds of thousands of compounds, which was further optimized with structure-based approaches to the current drug [29]. Furthermore, the extensive COVID-19 drug discovery effort also led to failed clinical trials [78], such as ebselen [79], reinforcing the discussion of the gaps, challenges, and bottlenecks between computational approaches, proposed targets, and subsequent evaluation in vitro and in vivo. In this regard, the following topics discuss potential pitfalls between simulations and experimental validation, aiming to understand and address potential issues, as well as the need to develop or improve consensus approaches, including existing in silico models and methods in drug design and discovery.

2 Enzymatic Activity Evaluations

Translating computational simulations to obtain bioactive compounds in a drug discovery approach may start with the validation of a proposed target [80], such as an enzyme. Approaches may consider choosing "one target to one drug" in different models or methods, in the sense of predicting a potential desired (*e.g.*, enzymatic inhibition) or undesired effect (*e.g.*, cytotoxicity or promiscuity) from a designed inhibitor against the proposed target [81, 82]. In addition, one may also consider alternatives that rely on polypharmacology, that is, the potential biological activity of a small molecule interacting with multiple targets, which can be effective toward multifactorial diseases and inflammation, such as producing synergistic effects [83]. However, multi-target approaches can raise concerns due to limitations in the ability of simulations to predict a target specificity or an unwanted promiscuity, as well as potential adverse effects [80], such as the potential toxicity effects of some agonists and antagonists over other enzymes [84] (not covered in this chapter). In this sense, with the increasing rate of clinical trial failures [5], an ideal scenario would comprise assessing large-scale approaches, considering drug-target networks, prospective and retrospective drug-target relationships, and experimentally quantifying the relation between a compound and a target to validate the performance of a model [81, 85]. For instance, assessing datasets of drug-target interactions and employing statistical validation, combined with the use of ML models for predicting new compounds, and ultimately experimentally validating hits to reassess the model can be used to improve the discovery of inhibitors [80].

Experimental determination of a given target in vitro is pivotal to predictive models that aim to evaluate the potential interaction of a predicted ligand with a given target [53]. However, these validation methods can be expensive and time-consuming [85], or sometimes not possible to be performed due to the lack of an available expressed and purified enzyme [86]. In addition, they may face difficulties in establishing or standardizing assay concentrations, protein stability, substrate specificity, and incubation periods, which may impair measurement reliability [87] (*e.g.*, in HTS applications [88]) and ultimately lead to incorrect reported affinities [53]. Protein stability, for example, can be enhanced by consensus structural design approaches, which focus on predicting and building well-folded or stabilized proteins that retain their original biological activities [89]. Additionally, some practical considerations are to be sought for subsequent binding measurement assays, which can help from predictions toward experimental validation, such as the enzyme turnover number (k_{cat}), which is essential to understand a given protein efficiency and its role in the metabolism [90]. Simulations may be tricky when predicting the sparse data from experimentally measured k_{cat} values from different studies [91], especially when taking into consideration that only ~10% of all enzyme-catalyzed reactions are known for *Escherichia coli* [92].

As k_{cat} estimates are mostly unavailable for enzymatic reactions, computational prediction models that assess comprehensive k_{cat} data are desirable to simulate the assessment of metabolic models [93]. These can be expensive (*e.g.*, production, maintenance, and testing) [94] and require an intrinsic pre-requisite for modeling (*e.g.*, enzyme itself) [95]. High-throughput experimental validation assays, for example, are costly and not time-effective, therefore models that can simulate enzyme reactions are desirable, such as TurNuP, a web server that can generalize and predict reactions from enzymes [93]. In addition, deep learning (DL) models such as DLKcat use thousands of different substrate structures (>3,000) and enzyme sequences (>300,000) to predict k_{cat} values [91]. Thus, the characterization of enzyme catalytic rates can be experimentally validated (*e.g.*, k_{cat} measurements), assessing their potential correspondence to the in silico predictions [95]. However, they may be susceptible to interference and variations when considering that changes in protein conformation or structure could result in changes in their enzymatic activity [96].

Elucidating this influence on altered activities or selectivity issues from computational simulations is a challenge that can be explored by obtaining large and diverse datasets, as well as defining suitable parameters for a given enzyme family or class [97]. To this regard, one could access large databases, such as carbohydrate-enzymes [98], kinome panels [99], network relations of various kinases and their inhibitors [100], non-kinase enzyme assay panels (*e.g.*, the CEREP diversity profile) [101], or compiling general properties and information about a given enzyme class, such as serine proteases [102]. Additionally, one could explore enzymes' interactions and scaffolds by molecular docking analysis [103], or predict their ligand accessibility and affinity by MD simulations [104], as in target-specific drug discovery approaches [43].

Notwithstanding, predicting whether an uncharacterized protein is an enzyme or not [97], or predicting functions of uncharacterized enzymes, may benefit from simulations and experimentation aiming for comprehensive data regarding translated protein conformers and their function [105]. Further, models that incorporate a single feature may also limit the applicability of a computational approach into predicting enzymatic activity, suggesting the importance of consensus approaches or multiple methods to enhance the accuracy of an enzyme characterization, such as substrate specificity [96, 106]. In this sense, different DL algorithm may extend an ability of predicting single functions to multifunctional enzyme or multiple functions predictions, as shown by the use and improvement of DEEPre [107] to the mlDEEPre [108]. However, simulating potential substrates that an enzyme may promote its catalytic activity is challenging when translating predicted data to an experimental validation [109]. Herein, like k_{cat} predictions [96, 97], the combination of existing data from experimental activity and structural information of enzymes can be useful for proposing substrates using docking analysis, as well as for calculating their physicochemical properties, thus favoring more accurate predictions [109].

Moreover, one should account for existing different or variable enzyme domains, which must be considered in simulations aiming to generate selectivity toward substrates, as each domain may influence binding or site accessibility [110, 111]. Hereupon, predicting a binding site preference or affinity to certain molecular probes (*i.e.*, functional groups or small molecules) in a multi-domain structure can be helpful to support an enzyme's specific substrate or inhibitor. Such simulations, as in the prediction using molecular probes by FTMap [112] can be considered a close computational analog of the experimental determination methods [112], such as crystallography and NMR [113]. These may even be useful for further predicting substrate fragment interactions in MD simulations (please, see Chap. 8) based on previous crystal data [104]. Herein, the experimental validation of these ligands can be assessed using designed or selected existing substrates, such as accessing a physicochemically diverse library of peptides in a global identification of substrate specificity [114]. This can be performed by employing an established peptide cleavage assay yielded by quantitative multiplex substrate profiling by mass spectrometry (qMSP-MS), which is able to quantitatively measure cleavage fluorescence units of each proposed substrate [115]. This approach can also be a valuable tool to translate the experimental validation to designed ligands from different substrate preferences (*e.g.*, supported by docking analysis) [116], be beneficial to other enzymes and fields of research [117], and also be included in the design of potential protease inhibitors in drug discovery [118].

In view of the binding affinity and specificity of a given substrate or inhibitor, it is important to consider the binding equilibrium of an enzymatic reaction, such as the dissociation constants (K_D values), that is, the dissociation of an inhibitor or substrate from the complex, accounting for the reaction velocity and bonding formation between the target and the ligand [119, 120]. Experimental determination of binding equilibrium would require demonstrating an absence of change for complex formation over time and systematically varying the concentration of an assay component, which may provide a solid method to display this behavior in vitro [53]. Validating such inhibition behavior can be performed with dilution assays, characterizing the nature of a given covalent bond, either it being reversible or irreversible [121]. Nevertheless, as reversible behaviors usually tend to build complexes that block a given substrate proteolysis [122], they may also favor noncovalent bonds [78]. In this sense, the formation of covalent or noncovalent bonds can be simulated in MD simulations, as in the successful predictions of covalent and noncovalent inhibitors for SARS-CoV-2 M^{pro} [65].

Investigating the time-dependence of an inhibitor interaction with an enzyme target can initially be investigated by assessing a compound with and without pre-incubation with the enzyme target [119]. This might indicate or help in differentiating covalent from noncovalent bonds, as covalent bonds are usually time-dependent, as opposed to noncovalent interactions [119]. In addition, a covalent behavior can be reversible or irreversible, which can also be simulated in MD simulations and experimentally validated with dilution assays aiming to investigate reversibility. For instance, the covalent reversible behavior of nirmatrelvir was initially predicted in MD simulations and confirmed in reversibility assays against SARS-CoV-2 M^{pro} [123]. Moreover, MD simulations can also predict the reversibility of protein recognition, interactions, and potential binding [124], as well as suggesting other potential covalent or noncovalent inhibitors [125]. As covalent bonds would confer an overall additional affinity in comparison to noncovalent interactions in potential drug candidates binding [126], simulating and validating such candidates could facilitate success in the drug discovery field [127].

Furthermore, complementary analysis may be of benefit for new drug candidates, such as integrating proteomics platforms, focusing on the selectivity profiling of covalent-bond ligand-target prediction [128], as well as predicting the reactivity profiling of cysteines, which may be assessed for their potential catalysis influence [129]. However, this could be less suitable when addressing the relative affinities of particular enzymatic systems, such as DNA as a target in CRISPR-Cas systems, which is a defense mechanism in some prokaryotes against viruses that have been repurposed as an RNA-guided DNA targeting for genome editing [130], and thus would require specifically designed simulations and experimental validation [131]. One should also bear in mind the mechanism-based for inhibition in predicting inhibitors, which are frequently quantified in terms of their half-maximal inhibitory concentrations (IC_{50}) values or may rely on an inhibition constant (K_I) and the rate of an enzyme inactivation (k_{inact}), depending on the types of inhibition [132].

Predictions from k_{inact}, K_I, or their relation as k_{inact}/K_I, can be critical to SAR analysis and PK parameters, and to accurately define a proper selectivity from biochemical assays [133]. These can be illustrated with kinetic simulations, such as covalent inhibitions predicted from SAR and their comparison to the potency of known inhibitors, followed by detailed experimental validation in kinetics [134]. In addition, IC_{50} values are not simply predicted by chemical structure comparison or available data regarding the mechanism and inhibition of similar ligands against targets (*e.g.*, structural analogs), as shown for a series of acrylamides and propionamides found to not provide predictable responses in vitro [135]. Similarly, predictions that do not comprise rigor in good practices may fail in experimental determinations, even with the existence of an enormous amount of available data, as seen for SARS-CoV-2 M^{pro} inhibitors planned by QSAR analysis (see Chap. 6 for methodological details on QSAR methods and proper validation protocols) [38].

It is also important to address whether a given compound with inhibitory activity is in fact a true inhibitor or a false positive, aside from compounds' solubility issues [136, 137], or natural interferents [138] such as those presenting self-fluorescence [139], which are covered in the following topics. For instance, an artifactual inhibition behavior can be observed in enzymatic assays due to colloidal aggregation [140, 141], that is, compounds causing promiscuous inhibition as aggregators. Some conditions can be assessed to detect aggregators, such as the addition of detergents (*e.g.*, tween or triton) to the assay buffer, which generally disrupts aggregates [142]. In addition, compounds can be pre-incubated with bovine serum albumin (BSA) to assess whether it can saturate the target enzyme binding capacity of the potential aggregates [143]. Additionally, increasing the concentration of an enzyme while maintaining the inhibitor concentration tends to reduce the percentage of inhibition by aggregators [143].

Furthermore, dynamic light scattering (DLS) can also be employed to detect aggregates in a solution. DLS is usually suitable to assess macromolecules degradation or disassembly, as well as enzyme-catalyzed polymerization [144], as these measurements describe the ease with which a molecule displaces another one by diffusion, expressly measuring the intensity of light scattering in an enzyme complex formation [145]. Lastly, a counter-screening test using an unrelated enzyme, assayed in the same buffer and inhibitor conditions, could also be employed [103]. Predicting aggregation can be performed with reasonable accuracy using computational methods [146, 147], but one should consider that even approved drugs can aggregate at high concentrations [148]. Thus, it is important to make an educated decision regarding the most suitable computational and validation methods to predict and determine inhibitory activities and their mechanisms, such as potential aggregate behavior [134], especially considering buffer, compound, enzyme, and substrate concentrations [140, 148] (Fig. 12.2).

3 Cytotoxicity Evaluation and Cell Viability

Experimental validation of compounds in cell viability assays for cultured cells has been based on different assays, including colorimetric, fluorometric, and luminometric, as well as dye exclusion assays [149]. For instance, the reduction of tetrazolium salts to colored formazans (*e.g.*, MTT [150] and XTT [151]), uptake and incorporation of thymidine analogs for measuring DNA synthesis [152, 153],

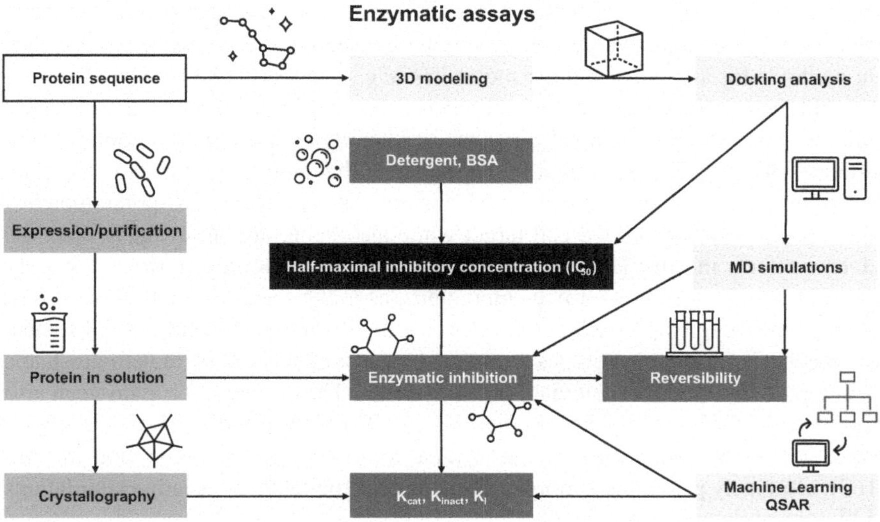

Fig. 12.2 Attaining enzymatic assays: Requirements and possibilities for experimental determination in vitro and simulations in silico. A given protein sequence may be selected as a target for drug design and discovery and follow two distinct paths toward inhibitory activity determination. On one side, performing and achieving expression and purification may result in a stable protein in solution, which can follow crystallography studies. On the other end, a protein sequence can be modeled to a three-dimensional (3D) structure to perform computational simulations, such as docking, molecular dynamics (MD) simulations, and machine learning (ML) techniques, including comparative quantitative structure-activity relationship (QSAR) analysis. Ultimately, various enzymatic assays can be performed to experimentally validate the target, considering inhibitory activity to determine the IC_{50} values, reversibility, and K_{cat}, K_{inact} and K_i assays, including the possibility of disregarding unspecific aggregative compounds (*e.g.*, detergent, or bovine serum albumin (BSA))

cellular staining for nonviable cells (*e.g.*, trypan blue) [154], as well as the evaluation of the metabolic function with pH and redox indicators, or fluorescence (*e.g.*, resazurin and alamarBlue™) [155] are broadly employed in research. Viability assays in vitro can be simple and less time-consuming methods that allow for evaluation in microplates, but may depend on the purpose of the assay as well as the availability of laboratory equipment and reagents [156, 157].

Overall, an assay should consider cost and sensitivity, while being rapid, safe, reliable, and reproducible without interfering with the test compounds [149, 157]. For instance, the crystal violet (CV) staining method (triphenylmethane dye (4-[(4-dimethyl amino phenyl)-phenyl-methyl]-N,Nimethyl-aniline) is an alternative or complementary approach to indirect tests such as MTT and alamarBlue™. Staining by CV depends on the viability of the cell monolayer in culture, that is, more CV is taken when there is a higher abundance of viable cells. This simple assay is a direct approach [158] useful for assessing the density of a cell monolayer while allowing for observation of morphological changes, and also allowing to storage the microplates for further analysis or future reference [159, 160]. Despite being simple and useful, CV and other assays may not offer substantial information for simulations in specific predictions, such as assessing synergism or antagonism effects [161], and may even not match predictions and comparisons between different methods for assessing specific molecules (*e.g.*, nanoparticles) [162].

In this sense, some distinctions and specificities may be required for experimental validation, in which certain cell lineages or phenotype assays can be helpful. For instance, phenotypic assays, that is, experiments that assess alterations in the phenotype of a given cell or organism [163], may display a predicted compound potency and selectivity to a proposed cell target while also being able to evaluate potential adverse effects in vitro [164]. This can be especially important in drug discovery approaches that aim for cytotoxic compounds, as in the search for anticancer drugs, where a phenotypic approach can account for the diversity and plasticity of molecular mechanisms and phenotypes in mutated cells, which could be indistinguishable from a general or indirect cell viability assay [165]. These phenotype assays can also benefit from different toxicity predictions in silico [166] and even function as a validation method for assessing the adverse toxic effects of a compound in vitro [167], ultimately providing a more realistic disease model for translating simulations to biological determinations [168].

Moreover, considering the influence of various targets in vitro, the use of chemical probes as reagents for improving experimental validation [169] of targets modified by diseases can also benefit cytotoxicity evaluation, facilitating the investigation of target function and safety [170]. Chemogenomic and the use of chemical probes may also be helpful in drug predictions [171], as they are often more selective and differ in their chemical properties, allowing for comparing selectivity from small compounds evaluated for their cytotoxicity [170, 172]. Additionally, other reagents can be used for assessing target interactions and potential modifications, such as intracellular signaling, ligand binding, and receptor interactions [173]. For example, by the improved luciferase assays with furimazine (NanoLuc) [174] and bioluminescence resonance energy transfer (BRET) in nanoBRET assays [175]. These assays exploit the energy donor from an enzyme (luciferase) to an acceptor (fluorophore), followed by the oxidation of a given substrate, which results in a quantifiable signal that translates interactions between molecules and proteins by proximity [173]. Herein, these assays can be used for experimental validation of cellular target simulations, as in the prediction of protein structures and stabilizing amino acids [174]. However, one should also account for potential interferences, such as inconsistent reagent concentrations and compounds acting as luciferase inhibitors, which could potentially impair expected results [175].

The interference of assay reagents in testing compounds accounts for one of the issues in translating predicted compounds to be assessed in cellular-based assays in vitro (*e.g.*, promoting spurious results) [176]. For instance, compounds with antioxidant properties may generate false negative results by causing the reduction of substrate reagents, such as resazurin [176]. In addition, compounds can also be directly interferents in a fluorescent assay, either quenching, that is, directly absorbing light that would be necessary for the fluorescent probe, or emitting light that would overlap the fluorophore (*i.e.*, presenting self-fluorescence) [139]. This fluorescence property was suggested for more than 5% (>70,000 tested compounds) of the PubChem libraries in 2008 [177]. Similarly, ~12% (43,885 chemicals) of the bioactive compounds found in the NIH Molecular Libraries Small Molecule Repository in 2012 inhibited the firefly luciferase to a certain level, displaying various luminescence changes [178]. Therefore, choosing a suitable method is pivotal to properly evaluate hits selected from simulations in drug discovery, as these behaviors may also impact different experimental validation methods, including the enzymatic inhibition assays and impact corresponding targets (*e.g.*, proteases) [179], as aforementioned.

A suitable fluorescence-based method is also important for HTS platforms in drug discovery and toxicity screenings, where false readings could drive erroneous results for in vitro assessments and falsely corroborate previous negative or positive predictions [139]. Regarding HTS, the Z-factor coefficient is a measurement that can help assure quality in a screening assessment [180], as it is a dimensionless statistical parameter that incorporates the means and errors associated with a given positive control and a negative reference, useful for comparison purposes as well as optimization and validation [181]. In addition to assuring experimental determination quality, the success of simulations also needs to account for variations in the biological phenomena of cell cultures, the range and/or filter of absorbance, the availability of equipment, the ability to detect, and specificity of reagents used (*e.g.*, fluorophores) [157, 182]. Herein, in silico predictions have successfully identified particularly problematic chemical structures and substructures that could interfere in such biological assays, such as quinones capable of forming covalent adducts with thiols [183] and the presence of compounds that might exhibit autofluorescence and interfere with fluorophores [184].

However, pan assay interference compounds (PAINS), that is, chemical structures that present a non-selective activity [185] or multiple behaviors that interfere with assays [186], are usually false positive hits from computational predictions. PAINS can interfere with various assays, such as cell viability and enzymatic inhibition, displaying nonspecific cytotoxicity and promiscuous target interference [138]. They may not be easily or properly predicted in simulations, thus requiring specific screening or filtering of substructural features to be identified and avoided [185]. Here MD-based protocols may help simulate the propensity of a given compound effect as interferent (*e.g.*, lipid membrane perturbation) [186], but assessing libraries in a large-scale would require formulated and appropriate filters to recognize different PAINS among thousands of screened compounds to exclude them from further analysis [138].

The natural product curcumin, for example, a PAINS that has displayed various biological and inhibitory activities [187], could be associated with intercalation in the lipid bilayer of mammalian cells [188], as well as changing the structure of bacterial membranes [189] or viral envelopes [190], thus being associated with a nonspecific toxicity [63]. This potential disturbance or intercalation of lipid membranes could be assessed by zeta potential assays in vitro, which can measure the electrokinetic potential of colloidal suspensions, that is, the behavior of a colloidal slipping movement under an electric field (*e.g.*, lipid bilayer) [191]. The zeta potential can be simulated in silico, for example, using nonequilibrium MD simulations, which can predict temperature and ion influences and differences in diverse surfaces [192]. However, even though it may be accurately predicted [193], an observed decrease in cell viability may not always be positively correlated with zeta potential data, as the assay demonstrates the enhancement of membrane permeabilities and not direct disruption [63, 194], as those observed for curcumin itself [195].

Moreover, compounds' solubility should also be considered for experimental validations, as in enzymatic and cell viability assays. Some simulations can be employed as qualitative and quantitative approaches aiming to predict solubility in water-based systems [196]. For example, a quantitative structure–property relationships (QSPR) model may allow for the prediction of solubility [136] providing the quantified numeric values with an uncertainty measured by a root mean square error (RMSE) of the model, helping in a direct comparison between two chemical structures [197]. In addition, similar properties can also be quantitatively predicted, which benefit from structural analogs and common scaffolds, but can also face difficulties when assessing large datasets and structurally diverse compounds retrieved from different chemical libraries [136]. However, solubility issues such as saturation and precipitation may be troublesome for many in vitro assays, potentially promoting unwanted effects, such as variations in permeation [198]. These may result in false negatives or false positives even in high concentrations of dimethyl sulfoxide (DMSO) [137], which is usually a universal solvent option that can dissolve compounds presenting a wide range of physicochemical properties [199].

Physicochemical properties themselves may ultimately impact in PK or ADMET parameters. To one's advantage, these parameters can be simulated in computational predictions [18], including a qualitative classification of being active or inactive (*e.g.*, hepatotoxicity) [200], and positive or negative (*e.g.*, intestinal absorption) [201]. While there is no ideal PK parameter for any drug candidate, as it depends on each target, various tools are mostly freely available as integrated platforms on web servers aiming to accurately and comprehensively predict ADMET and/or PK properties of assessed compounds [202]. For instance, they can target specific receptors (*e.g.*, Pred-hERG [203]) or a broader set of predictions in a single simulation (ADMETlab [204]), which may have been updated and/or upgraded (ADMETlab 2.0 [205]). However, experimentally validating ADMET and PK simulations from these web servers would require specific evaluation in vitro or in vivo (*e.g.*, toxicity in specific cells or animal models), which could be limited to laboratory structure, expertise, and resources while not being cost-effective. In this sense, the major recommendation for these predictors is to be used in combination as consensus predictors to increase individual accuracy (usually >70%) and approximate a comparative evaluation, potentially improving the success of drug candidates [18].

4 Antiviral Assays in Experimental Validation

The drug design, discovery, and development of new antiviral drugs are one of the main focuses in medicinal chemistry, especially when considering that various known and new viral diseases are yet to come, with the emergence and reemergence of different viruses [73]. In the search for new antivirals, various computational approaches (*e.g.*, VS) have played a key role and were successful in obtaining lead candidates that later were licensed as antiviral drugs [206]. For instance, one can cite successful cases in the antiretroviral therapy of human immunodeficiency virus (HIV), especially protease inhibitors [207], which benefited from CADD approaches in the initial design of lead candidates and led to the approval of drugs targeting proteases [208], such as of ritonavir [209] and saquinavir [210]. Other examples of success include oseltamivir and zanamivir as therapeutic options for influenza virus [211, 212], the viral protease inhibitor boceprevir for hepatitis C virus (HCV) [213], and more recently the SARS-CoV-2 M^{pro} inhibitor, ensitrelvir [29]. Most viral diseases are considered a public health concern worldwide, and can progress from outbreaks to endemic, epidemic, and pandemic scenarios [214]. Still in 2019, the World Health Organization (WHO) declared different viruses as the top 10 global threats to be faced in the near future [215], months before the emergence of SARS-CoV-2 [216], yet unknown at the time as a novel coronavirus (2019-nCoV) [217].

Many viruses, including coronaviruses, are considered as threatening pathogens that can cause human diseases with high dissemination, morbidity, or mortality, such as the acquired immunodeficiency syndrome (AIDS), COVID-19, Ebola, influenza, and dengue and Zika fever [214, 218], some of which still do not have antiviral drugs as therapeutic options [219]. The urge of SARS-CoV-2 and the COVID-19 pandemic, for example, promptly required collective efforts and resources to tackle the scenario, including strategies focusing in the discovery of new or repurposed antivirals [220, 221]. In support of this notion, between January 1981 and September 2019, 81 small molecules were approved by the Food and Drug Administration (FDA) as antiviral drugs, accounting for 44% of approvals aside from peptides, proteins, and vaccines [222]. As antivirals are usually designed to target a single microorganism, they can be direct-acting antiviral agents (DAA) used in a combined therapy (*e.g.*, antiretroviral therapy against HIV), or aim for an induced cell response [223]. On the other hand, aiming for broad-spectrum activity usually does not display favorable outcomes due to the target specificity in different viruses [224, 225], which may result in a lower antiviral activity or high cytotoxicity [226]. Despite that, multi-virus or multi-target activity could be still predicted in simulations and employed to urgent scenarios, or in the preparedness for future emerging or reemerging viruses [221], as was the case of CADD approaches against different coronaviruses [227].

Initially, to validate multi-virus or multiple targets activity, one should consider a compound ability to permeate cell membranes and reach a given virus target (*e.g.*, protease), or block viral entry in host cells, as viruses are obligate intracellular microorganisms [228]. Here, certain challenges are expected, such as a target specificity and consequent lower cell or host toxicity, and the necessity of intracellular availability and efficacy [226, 229]. Aside from predicting a given compound permeability, one should also consider that a virus will have a cell, tissue, or host tropism, that is, the preference to infect a certain cellular type [230]. In addition, a cell must be permissive to the virus, that is, support the infection by having receptors that allow for viral attachment and entry, as well as allowing for intracellular replication, thus successfully resulting in viral progeny or subsequent infections [231]. This is important when defining which cell platform will be assessed in vitro, but also to design specific predictive models toward a certain cell, cell receptor, or protein of interest, such as models assessing inhibitors of the angiotensin converting enzyme 2 (ACE2) receptor for blocking the SARS-CoV-2 variants entry in mammalian cells [232]. Furthermore, some viruses require a biosafety level 3 (BSL3) facility to be assessed in experimental validations, which can also be a limiting factor for the development of antiviral compounds [233], and potentially require an adaptation for BSL2 models in vitro [234] and in vivo [235].

Even if able to successfully move a computational simulation to an experimental validation, one should also account for fast replication or mutation rates of different viruses in vitro, which can impact the antiviral activity and result in a lead drug candidate failure [236], especially for viruses that cause chronic infections, such as hepatitis B virus (HBV), HCV, HIV, and that might include drug resistance [73, 229, 237]. In addition, by requiring specific cell lineages and culture conditions add to the reasons why the design and discovery of antivirals may have a slower rate [238], especially when compared to antibacterials or antifungals [73, 239]. Additionally, standard protocols for propagation, purification, and titration of viral strains may be needed [240, 241], as well as standardizing viral adsorption and incubation periods, and determining the multiplicity of infection (MOI), that is, the number of viral particles calculated to infect cells in culture. This number is crucial for some antiviral experiments, as the higher the MOI, the faster the cell viability reduces due to the cytopathic effects (CPE) caused by a higher number of viruses. In this sense, one-step and growth curves can contribute to evaluating the best MOI of work, while observing a more suitable log-phase in viral propagation to elaborate an optimal strategy of the study [240, 241].

Moreover, purification protocols may require specific centrifugation or ultracentrifugation protocols before testing [242], all aiming to establish the best conditions to assess compounds against certain viruses in vitro and subsequently in vivo. These may be challenging to translate from computational simulations, either requiring individual approaches for each parameter to be tested (*e.g.*, cell permeability) [243], or not being able to be predicted (*e.g.*, purification conditions). Thus, even after predicting a given compound, experimental validation should consider different antiviral assays (Fig. 12.3). For example: (i) pre-treatment, aiming to identify potential compound-induced cellular antiviral responses; (ii) virucidal activity, with direct action of compounds against viral particles before infection; and (iii) neutralization or reduction assays, assessing a direct effect of a given compound in a step of the virus replication cycle, such as viral entry, uncoating, replication, protein synthesis, morphogenesis/assembly, or release [244, 245].

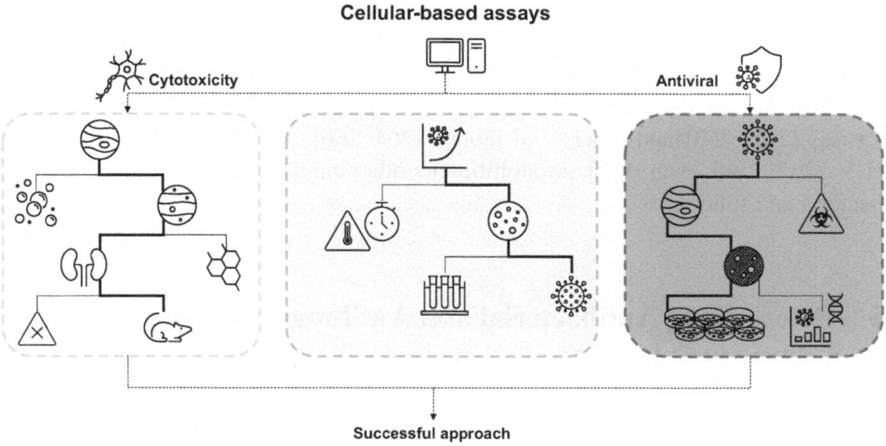

Fig. 12.3 Considerations for computational simulations translating into cellular-based assays (cytotoxicity and antiviral). For a given compound toxicity assessment against cell lineages in vitro, one should consider potential precipitation, as well as the interference of pan assay interference compounds (PAINS) in assays, before following absorption, distribution, metabolism, excretion, and toxicity (ADMET) or pharmacokinetics (PK) simulations (white box). Ultimately, compounds may be disregarded or follow in vivo studies. In parallel, computational simulations may support cellular or viral propagation conditions, acknowledging potential limitations regarding temperature and incubation periods in cell infection, or purification predictions of a viral particle (light gray box). These may help in the prediction of an antiviral compound. Lastly, if a biosafety laboratory level is not a restriction, and a permissive cell lineage is available, cytopathic effects (CPE) can be validated in vitro and follow various assays for determining a compound antiviral activity, such as plaque neutralization or reduction assays (dark gray box). On the other end, chronic infection models and high mutation rates may impair experimental validation, while potential inhibitors may be inactive, or require specific viral target validation. Bold lines represent a successful experimental path, while regular lines represent issues or obstacles

Furthermore, various targets were considered and assessed for the design of potential antiviral candidates against SARS-CoV-2 proteins, including structural (*e.g.*, spike [246, 247]), and non-structural associated with viral replication (*e.g.*, viral proteases [79, 248] and RNA polymerase [249]). As for protein-protein interactions (PPI) from the viral attachment protein to host cell receptors, various predictive ML models [250, 251] and tools (*e.g.*, DeepViral [252] and MP-VHPPI [253]) have been proposed, including DL models [254]. These can benefit the design of target specific compounds as antiviral inhibitors, for example, in approaches that may benefit from the availability of chemical structures [255] and previous predictive ML models [254]. Among these possibilities, predictive or classification ML could help in training DL models from specific antiviral databases as sources of different chemical structures, such as experimentally determined peptides (AVPdb [256]) and those classified by ML models, such as AVP-IC_{50} [257], a viral annotated antiviral activity database from ChEMBL [258]. In addition, virus-targeted databases, such as the Influenza Research Database [259] and the Coronavirus Antiviral Research Database (CoV-RDB) [260], could also be used for feeding models.

Hereupon, with the large availability of SARS-CoV-2 experimental data from the search of potential antiviral drugs in the recent years, developing new focused databases with experimental validation data (*e.g.*, CoronaDB-AI [261]) could also improve the success rate in the discovery of new lead candidates against other viruses [262, 263] and other viral targets [264–266], such as those of HCV and HIV [267], and even open possibilities to other microorganisms [268], such as bacteria and fungi.

5 Determining Antibacterial and Antifungal Bioactive Compounds In Vitro

In the field of drug discovery, the development of new antimicrobials facing infectious diseases also consider, along with antivirals, antibacterial and antifungal compounds [269]. These have been known for over a century and have also influenced the current knowledge for drug design and discovery [270]. The increasing demand for novel antimicrobials against the global spread of multidrug-resistant (MDR) pathogens, for example, may be faced by various sectors of research and funding, aiming for a bridge between academic, governmental, industrial, and pharmaceutical efforts that can translate expertise and data from computational and experimental studies into a successful drug candidate [271].

Regarding MDR bacteria, there is a lack of newly approved antibiotics [272], contrasted with estimates that overcome more than \$1.7 billion in increased healthcare costs per year related only to methicillin-resistant *Staphylococcus aureus* (MRSA) [273], which approximates the costs to develop a single new drug [1]. Further, back in 2019, AMR was considered one of the top 10 global threats to the future by WHO [215], which reflected the critical current scenario of ~700,000 deaths yearly due to MDR infections [274], and estimates of over 10 million by 2050 [275]. For instance, deaths by MRSA infections surpassed those from HIV/AIDS and tuberculosis combined in the US [276], which is also considered a serious threat by the Centers of Disease Control and Prevention (CDC) [273], demanding a high priority for the search and development of new drugs [277].

On the other hand, fungal diseases reflect the concern for primary or co-infections, such as those caused by the emerging *Candida auris*, which has drawn attention since it was first identified in 2009 and is rapidly rising to a global risk in healthcare clinical settings as an MDR pathogen with mortality rates ranging from 30% to 60% [278]. After being declared an urgent health threat by the CDC in 2019 [273], a warning on *C. auris* was issued in early 2023, summarizing clinical cases (present infection) increasing from 476 in 2019 to 1471 in 2021 and 2377 in 2022, in addition to screening cases (*i.e.*, detected but not causing infection) that tripled in the US [279]. In the meantime, as the COVID-19 pandemic may have impacted invasive fungal infections in public health and antifungal drugs usage [280, 281], including COVID-19-associated *C. auris* outbreaks (mortality rates up to

83% [282]), MDR candidemia cases are spreading and increasing in hospital settings worldwide [283], with the ability to form multilayer biofilms also resisting various niche skin conditions (*e.g.*, desiccation) [284].

In this sense, computational methods may provide reliable and cost-effective strategies at different stages of drug design and discovery of new antibiotics [271], as in efforts for safer and more effective drug candidates needed for MDR infections [285]. For instance, predictive ML techniques may help in the screening of antibacterial and antifungal compounds, for example, feeding models for selecting natural products, semisynthetic derivatives, or synthetic compounds, as was the discovery of halicin, a potential antibacterial drug candidate discovered by a DL model (for more details on machine learning and neural networks, see Chap. 4) [286]. This compound, structurally different from many antibiotics, successfully supported a computational model's ability to predict potential drug candidates based solely on other chemical structures [286]. As mentioned for antivirals [73], predictive ML models can not only provide improvements for other potential drug candidates [60], such as antibacterial [270] and antifungal bioactive compounds [287], but also for the search for potential antimicrobial targets [288] and drug resistance [289]. For instance, ML methods can be applied in simulations to verify the accuracy of a designed model against experimentally evaluated peptides [290].

However, when considering the prediction ability of ML models to discover potential bioactive compounds, the true positive rate of a given model can be dependent on the availability of a varied set of compounds (*e.g.*, small, large, similar, and diverse libraries) and their biological information (*e.g.*, experimental determination). For instance, Deep-AntiFP, a tool for prediction of antifungal peptides with deep neural networks, reached the highest accuracy of 94.23% when using the training dataset to validate the models, but achieved 91.02% and 89.08%, when assessing an alternative and an independent dataset, respectively. Despite an overall reasonable accuracy (~90%), even when increasing the number of iterations in the models, the predictive error would tend to remain stable, indicating the need for a more diverse dataset as input, thus consequently requiring more experimentally validated compounds to increase the model's accuracy [291]. In addition, even novel approaches that pre-trained a predictor model achieved only comparable performance to other available computational methods [292], as in a consensus QSAR-screening approach [293], which also corroborated the need for a larger and more diverse set of experimentally determined compounds for increasing predictive accuracy.

This would stand in contrast to the discussion of finding an equilibrium between the cost of experimental testing and enough supporting evidence before selecting a target or compound to be tested in drug discovery approaches [294]. Building the most accurate model for each approach would be ideal, but one should consider the fact that it is improbable to generate or deploy models with 100% accuracy, and that adding compounds or data that "do not belong" would not be beneficial to a model. In addition, it is unlikely to have all the chemical or biological information necessary to achieve high accuracy models, even when using previously solved studies, which also depend on the amount of information available [295]. Hereupon, virtually screening compounds could still be the major approach to discovering new compounds, regardless of accuracy, as shown for halicin [286]. Thus, as aforementioned, the consensus approaches that combine VS with AI could increase success and reduce false positive hits predicted with single computational methods [206].

Identifying a compound predicted to have a given biological activity, such as antibacterial or antifungal, would require further characterization in vitro. For example, starting with a minimal inhibitory concentration (MIC) determination and further assessing a proposed mechanism, such as bactericidal activity with minimal bactericidal concentration (MBC) determination, might help progress to in vivo evaluation [296]. In this sense, initial screening assays, such as broth microdilution assays in 96-well microplates [297], may be an easy, fast, and low cost method [298] to be used in early stages for antimicrobial activity. Finding if a given compound is active or not may even require further investigation of potential mechanisms (*e.g.*, association with a specific infection or microorganism), or focus solely on bioactive compound identification [299]. However, validation assays should also account for difficulties in culturing a targeted microorganism, such as slow-growing mycobacteria or intracellular bacteria [300], as well as the existence of non-culturable microbes with pathogenic potential [301], and the possibility of some microorganisms undergoing a viable but non culturable state, as some *Candida* species [302].

Pretomanid, for example, a drug for the treatment of MDR *Mycobacterium tuberculosis*, was first identified in 2000 from a series of 100 compounds assessed for their antimycobacterial activity [303], which were repurposed from an original investigation as radiosensitizers for chemotherapy [304]. Empirically or randomly testing compounds, despite being known as usually not cost-effective [305], can still be an alternative when it is difficult to sustain experimental studies based on simulations from computational tools (*e.g.*, validating a target) [306] or when assessing a relatively small number of compounds from small chemical libraries in 96-well microplate assays [294] (*e.g.*, MIC assays). These could lead to the identification of compounds with activity against various bacteria but also result in cellular cytotoxicity due to nonspecific mechanisms [307], such as in cell membrane disturbance, structural changes, damage, and potential disruption [63].

However, with a large amount of biological information and experimental data available for some pathogens, such as *M. tuberculosis* and *S. aureus* [73], some methods can provide reasonable results for translating simulations into biological assays, despite the fact that they could still lack the ability to predict novel resistance mechanisms [308]. Chemical similarity methods could be of interest when simulating novel antibacterial compounds based on existing antibiotic drugs, as shown in the identification of crizotinib as a potential repurposing antibacterial candidate [63]. However, these and other methods focusing on repurposing candidates may be limited by finite experimental resources, that is, a finite amount of available licensed drugs [309]. Lastly, the alternative of searching for repurposing candidates also falls under the recommendation of employing consensus approaches that maximize the number of available options for testing, usually providing consistent predictions aiming to prioritize which inhibitors or potential repurposing drugs would be experimentally validated in vitro and ultimately tested in vivo [309, 310].

6 In Vivo Evaluation of Compounds

In general, in vivo evaluation is the third step of a pipeline for bioactive compounds screening (in silico, in vitro, and in vivo, respectively) [311]. The most widely used experimental models used for in vivo assays are mice, used to validate in vitro tests and observe efficacy in whole organisms. This model has advantages such as fast procreation, low cost, easy maneuver and care, the option for using inbred or outbred animals, the possibility of knock-in and knockout animals, and a diversity of background options for making them susceptible for human infectious diseases [312, 313]. Similar to in vitro assays, toxicity should be assessed in vivo, which translates to assessing lethal dose of 50% (DL_{50}) or tolerability assays of drugs in the chosen animal model, usually by assessing different compound or drug concentrations (mg/kg) tolerated by the animal. Afterwards, one should determine or choose the most suitable concentration range as well as the pathogens' infection parameters [314, 315].

In this sense, regarding a compound concentration, solubility may play a key role in oral absorption, as poor aqueous solubility may be decisive in drug discovery studies, including in translating in vitro to in vivo [316] and different experimental evaluation approaches [317], potentially requiring another round of lead optimization or failing a given drug candidate [318]. PK modeling based on validated data may help with many different approaches and their implications regarding aqueous solubility, including temperature, pH-dependency, and ultimately bioavailability [136, 318]. As for pathogens, researchers can also use mouse models for analyzing infection effects directly and indirectly. These models can be used for testing drug protection against lethality, or to test mitigation of an infection and/or disease by a given compound or drug, thus evaluating the animal response or reestablishment against an infection [319, 320]. Different mice and other murine models can be used for specific pathogenic infections or can be standardized in the laboratory. Herein, one should consider a variety of parameters such as pathogen inoculum, route of infection, clinical score of disease, difference between genetic backgrounds, age and gender, and the necessity of genetic modification to make the animal susceptible for a specific infectious disease [235, 321].

Notwithstanding, even if solubility and infection issues are overcome, compounds that are correctly predicted for a given target should still be validated concerning their ability to ultimately display the desired activity, while also permeating cell membranes or tissue barriers [228]. Predictions in this sense should also consider the specificity and stability of compounds, as well as previously assessing predictions for the potential ADMET and PK properties [18]. Despite the fact that ADMET simulations are more directed at translating in vitro experimentally determined effect to in vivo [317], they also aim to potentially increase the success rate of predicted hits as favorable compounds for optimization, which could ultimately reach a potential lead candidate and become a drug [202]. These aspects are important to reduce the risks in the late drug discovery stages and also optimize the bridging between computational simulations, and experimental validations, focusing on promising compounds [322].

7 Conclusions

Considering the field of drug design and discovery, the search for novel drugs requires a lasting effort to tackle long known, emerging, or re-emerging diseases. The focus on targets of interest, such as enzymes, may be a starting point for discovery campaigns. In this sense, designing and developing potential inhibitors and lead drug candidates should account for PK and ADMET, as well as cytotoxicity. Moreover, if related to an infectious disease, as in the search for antivirals, antibacterials, and antifungals, it is essential to focus on selectivity and specificity. Here, the combination of in silico, in vitro, and in vivo approaches is an important triad to any drug discovery campaign that can encompass compounds' chemical properties, biochemical and cellular assays, and in vivo evaluation to reach a successful drug candidate. Although challenging, many options are available to translate different computational simulations into various experimental validations. Examples include the discovery of some protease inhibitors, such as nirmatrelvir, and recently ensitrelvir, successfully obtained by combining computer-aided approaches and experimental assays. Herein, we discussed the possibilities and feasibility, challenges and pitfalls, and the current scenario of such computational approaches that, especially when combined, may contribute to a successful protocol in drug discovery.

References

1. Wouters, O. J., McKee, M., & Luyten, J. (2020). Estimated Research and Development investment needed to bring a new medicine to market, 2009–2018. *JAMA, 323*, 844–853.
2. DiMasi, J. A., Grabowski, H. G., & Hansen, R. W. (2016). Innovation in the pharmaceutical industry: New estimates of R&D costs. *Journal of Health Economics, 47*, 20–33.
3. Prasad, V., & Mailankody, S. (2017). Research and development spending to bring a single cancer drug to market and revenues after approval. *JAMA Internal Medicine, 177*, 1569–1575.
4. Cummings, E. (2019). The Elijah E. Cummings Lower Drug Costs Now Act: how it would work, how it would affect prices, and what the challenges are – Digital Collections – National Library of Medicine. https://collections.nlm.nih.gov/catalog/nlm:nlmuid-101769556-pdf
5. Sun, D., Gao, W., Hu, H., & Zhou, S. (2022). Why 90% of clinical drug development fails and how to improve it? *Acta Pharmaceutica Sinica B, 12*, 3049–3062.
6. Sadybekov, A. V., & Katritch, V. (2023). Computational approaches streamlining drug discovery. *Nature, 616*, 673–685.

7. Gramatica, P. (2020). Principles of QSAR modeling: Comments and suggestions from personal experience. *International Journal of Quantitative Structure-Property Relationships, 5*, 61–97.
8. Wassermann, A. M., Wawer, M., & Bajorath, J. (2010). Activity landscape representations for structure−Activity relationship analysis. *Journal of Medicinal Chemistry, 53*, 8209–8223.
9. Lin, X., Li, X., & Lin, X. (2020). A review on applications of computational methods in drug screening and design. *Molecules, 25*, 1375.
10. Rodríguez-Pérez, R., & Bajorath, J. (2021). Evaluation of multi-target deep neural network models for compound potency prediction under increasingly challenging test conditions. *Journal of Computer-Aided Molecular Design, 35*, 285–295.
11. Boobier, S., Hose, D. R. J., Blacker, A. J., & Nguyen, B. N. (2020). Machine learning with physicochemical relationships: Solubility prediction in organic solvents and water. *Nature Communications, 11*, 5753.
12. Feinberg, E. N., Joshi, E., Pande, V. S., & Cheng, A. C. (2020). Improvement in ADMET prediction with multitask deep featurization. *Journal of Medicinal Chemistry, 63*, 8835–8848.
13. Jiménez-Luna, J., Grisoni, F., & Schneider, G. (2020). Drug discovery with explainable artificial intelligence. *Nature Machine Intelligence, 2*, 573–584.
14. Jumper, J., et al. (2021). Highly accurate protein structure prediction with AlphaFold. *Nature, 596*, 583–589.
15. Varadi, M., & Velankar, S. (2023). The impact of AlphaFold protein structure database on the fields of life sciences. *Proteomics, 23*, 2200128.
16. Kiriiri, G. K., Njogu, P. M., & Mwangi, A. N. (2020). Exploring different approaches to improve the success of drug discovery and development projects: A review. *Future Journal of Pharmaceutical Sciences, 6*, 27.
17. Yang, X., Wang, Y., Byrne, R., Schneider, G., & Yang, S. (2019). Concepts of artificial intelligence for computer-assisted drug discovery. *Chemical Reviews, 119*, 10520–10594.
18. Pantaleão, S. Q., Fernandes, P. O., Gonçalves, J. E., Maltarollo, V. G., & Honorio, K. M. (2022). Recent advances in the prediction of pharmacokinetics properties in drug design studies: A review. *ChemMedChem, 17*, e202100542.
19. Engeloch, C., et al. (2008). Stability of screening compounds in wet DMSO. *SLAS Discovery, 13*, 999–1006.
20. Nurzyńska, K., et al. (2015). Long-term amorphous drug stability predictions using easily calculated, predicted, and measured parameters. *Molecular Pharmaceutics, 12*, 3389–3398.
21. Bergström, C. A. S., Charman, W. N., & Porter, C. J. H. (2016). Computational prediction of formulation strategies for beyond-rule-of-5 compounds. *Advanced Drug Delivery Reviews, 101*, 6–21.
22. Sliwoski, G., Kothiwale, S., Meiler, J., & Lowe, E. W. (2014). Computational methods in drug discovery. *Pharmacological Reviews, 66*, 334–395.
23. Jayatunga, M. K. P., Xie, W., Ruder, L., Schulze, U., & Meier, C. (2022). AI in small-molecule drug discovery: A coming wave? *Nature Reviews Drug Discovery, 21*, 175–176.
24. Thakkar, A., et al. (2021). Artificial intelligence and automation in computer aided synthesis planning. *Reaction Chemistry & Engineering, 6*, 27–51.
25. Sadybekov, A. A., et al. (2022). Synthon-based ligand discovery in virtual libraries of over 11 billion compounds. *Nature, 601*, 452–459.
26. Gorgulla, C., et al. (2020). An open-source drug discovery platform enables ultra-large virtual screens. *Nature, 580*, 663–668.
27. Muratov, E. N., et al. (2021). A critical overview of computational approaches employed for COVID-19 drug discovery. *Chemical Society Reviews, 50*, 9121–9151.
28. Galindez, G., et al. (2021). Lessons from the COVID-19 pandemic for advancing computational drug repurposing strategies. *Nature Computational Science, 1*, 33–41.
29. Unoh, Y., et al. (2022). Discovery of S-217622, a noncovalent Oral SARS-CoV-2 3CL protease inhibitor clinical candidate for treating COVID-19. *Journal of Medicinal Chemistry, 65*, 6499–6512.

30. Gentile, F., Oprea, T. I., Tropsha, A., & Cherkasov, A. (2023). Surely you are joking, Mr docking! *Chemical Society Reviews, 52*, 872–878.
31. Ruatta, S. M., et al. (2023). Garbage in, garbage out: How reliable training data improved a virtual screening approach against SARS-CoV-2 MPro. *Frontiers in Pharmacology, 14*.
32. Jafari, M., Guan, Y., Wedge, D. C., & Ansari-Pour, N. (2021). Re-evaluating experimental validation in the big data era: A conceptual argument. *Genome Biology, 22*, 71.
33. Chowdhury, A. S., Reehl, S. M., Kehn-Hall, K., Bishop, B., & Webb-Robertson, B.-J. M. (2020). Better understanding and prediction of antiviral peptides through primary and secondary structure feature importance. *Scientific Reports, 10*, 19260.
34. Strittmatter, S. M. (2014). Overcoming drug development bottlenecks with repurposing: Old drugs learn new tricks. *Nature Medicine, 20*, 590–591.
35. Lu, M., Xing, H., Zheng, A., Huang, Y., & Liang, X.-J. (2023). Overcoming pharmaceutical bottlenecks for nucleic acid drug development. *Accounts of Chemical Research, 56*, 224–236.
36. Yamada, Y. (2021). Nucleic acid drugs—Current status, issues, and expectations for exosomes. *Cancers (Basel), 13*, 5002.
37. (2023). Experimental validation, anyone? *Nature Computational Science, 3*, 361–361.
38. Serafim, M. S. M., et al. (2023). The importance of good practices and false hits for QSAR-driven virtual screening real application: A SARS-CoV-2 main protease (Mpro) case study. *Frontiers in Drug Discovery, 3*.
39. Deshmukh, M. G., et al. (2021). Structure-guided design of a perampanel-derived pharmacophore targeting the SARS-CoV-2 main protease. *Structure, 29*, 823–833.e5.
40. Bespalov, A., Steckler, T., & Skolnick, P. (2019). Be positive about negatives-recommendations for the publication of negative (or null) results. *European Neuropsychopharmacology, 29*, 1312–1320.
41. Nimpf, S., & Keays, D. A. (2020). Why (and how) we should publish negative data. *EMBO Reports, 21*, e49775.
42. Maxim, L. D., Niebo, R., & Utell, M. J. (2014). Screening tests: A review with examples. *Inhalation Toxicology, 26*, 811–828.
43. Veríssimo, G. C., et al. (2022). Designing drugs when there is low data availability: One-shot learning and other approaches to face the issues of a long-term concern. *Expert Opinion on Drug Discovery, 17*, 929–947.
44. Tropsha, A. (2010). Best practices for QSAR model development, validation, and exploitation. *Molecular Informatics, 29*, 476–488.
45. Macip, G., et al. (2022). A review of the current landscape of SARS-CoV-2 Main protease inhibitors: Have we hit the Bullseye yet? *International Journal of Molecular Sciences, 23*, 259.
46. Bajorath, J., et al. (2020). Artificial intelligence in drug discovery: Into the great wide open. *Journal of Medicinal Chemistry, 63*, 8651–8652.
47. Graff, D. E., Shakhnovich, E. I., & Coley, C. W. (2021). Accelerating high-throughput virtual screening through molecular pool-based active learning. *Chemical Science, 12*, 7866–7881.
48. Williams, A. J., & Ekins, S. (2011). A quality alert and call for improved curation of public chemistry databases. *Drug Discovery Today, 16*, 747–750.
49. Frye, L., Bhat, S., Akinsanya, K., & Abel, R. (2021). From computer-aided drug discovery to computer-driven drug discovery. *Drug Discovery Today: Technologies, 39*, 111–117.
50. de Souza Neto, L. R., et al. (2020). In silico strategies to support fragment-to-Lead optimization in drug discovery. *Frontiers in Chemistry, 8*.
51. Hollingsworth, S. A., & Dror, R. O. (2018). Molecular dynamics simulation for all. *Neuron, 99*, 1129–1143.
52. Kapla, J., Rodríguez-Espigares, I., Ballante, F., Selent, J., & Carlsson, J. (2021). Can molecular dynamics simulations improve the structural accuracy and virtual screening performance of GPCR models? *PLoS Computational Biology, 17*, e1008936.
53. Jarmoskaite, I., AlSadhan, I., Vaidyanathan, P. P., & Herschlag, D. (2020). How to measure and evaluate binding affinities. *eLife, 9*, e57264.

54. Rossetti, G. G., et al. (2022). Non-covalent SARS-CoV-2 Mpro inhibitors developed from in silico screen hits. *Scientific Reports, 12*, 2505.
55. Gorny, H., et al. (2023). Combining nano-differential scanning fluorimetry and microscale thermophoresis to investigate VDAC1 interaction with small molecules. *Journal of Enzyme Inhibition and Medicinal Chemistry, 38*, 2121821.
56. Schneider, G. (2018). Automating drug discovery. *Nature Reviews. Drug Discovery, 17*, 97–113.
57. Batool, M., Ahmad, B., & Choi, S. (2019). A structure-based drug discovery paradigm. *International Journal of Molecular Sciences, 20*, 2783.
58. Vázquez, J., López, M., Gibert, E., Herrero, E., & Luque, F. J. (2020). Merging ligand-based and structure-based methods in drug discovery: An overview of combined virtual screening approaches. *Molecules, 25*, 4723.
59. Azevedo, L., Serafim, M. S. M., Maltarollo, V. G., Grabrucker, A. M., & Granato, D. (2022). Atherosclerosis fate in the era of tailored functional foods: Evidence-based guidelines elicited from structure- and ligand-based approaches. *Trends in Food Science & Technology, 128*, 75–89.
60. Lima, A. N., et al. (2016). Use of machine learning approaches for novel drug discovery. *Expert Opinion on Drug Discovery, 11*, 225–239.
61. Bosc, N., et al. (2019). Large scale comparison of QSAR and conformal prediction methods and their applications in drug discovery. *Journal of Cheminformatics, 11*, 4.
62. Réau, M., Langenfeld, F., Zagury, J.-F., Lagarde, N., & Montes, M. (2018). Decoys selection in benchmarking datasets: Overview and perspectives. *Frontiers in Pharmacology, 9*, 11.
63. Asse Junior, L. R., et al. (2020). Virtual screening of antibacterial compounds by similarity search of Enoyl-ACP reductase (FabI) inhibitors. *Future Medicinal Chemistry, 12*, 51–68.
64. Lu, X., et al. (2018). The development of pharmacophore modeling: Generation and recent applications in drug discovery. *Current Pharmaceutical Design, 24*, 3424–3439.
65. Hayek-Orduz, Y., et al. (2022). Novel covalent and non-covalent complex-based pharmacophore models of SARS-CoV-2 main protease (Mpro) elucidated by microsecond MD simulations. *Scientific Reports, 12*, 14030.
66. Bouback, T. A., et al. (2021). Pharmacophore-based virtual screening, quantum mechanics calculations, and molecular dynamics simulation approaches identified potential natural antiviral drug candidates against MERS-CoV S1-NTD. *Molecules, 26*, 4961.
67. Cerón-Carrasco, J. P. (2022). When virtual screening yields inactive drugs: Dealing with false theoretical friends. *ChemMedChem, 17*, e202200278.
68. Zhu, T., et al. (2013). Hit identification and optimization in virtual screening: Practical recommendations based on a critical literature analysis. *Journal of Medicinal Chemistry, 56*, 6560–6572.
69. Yan, X. C., et al. (2020). Augmenting hit identification by virtual screening techniques in small molecule drug discovery. *Journal of Chemical Information and Modeling, 60*, 4144–4152.
70. Hevener, K. E., et al. (2018). Chapter Twelve – Hit-to-lead: Hit validation and assessment. In C. A. Lesburg (Ed.), *Methods in enzymology* (Vol. 610, pp. 265–309). Academic.
71. Wadman, M. (2023). FDA no longer has to require animal testing for new drugs. *Science, 379*.
72. Stiefl, N., et al. (2015). FOCUS—Development of a global communication and modeling platform for applied and computational medicinal chemists. *Journal of Chemical Information and Modeling, 55*, 896–908.
73. Serafim, M. S. M., Dos Santos Júnior, V. S., Gertrudes, J. C., Maltarollo, V. G., & Honorio, K. M. (2021). Machine learning techniques applied to the drug design and discovery of new antivirals: A brief look over the past decade. *Expert Opinion on Drug Discovery, 16*, 961–975.
74. Schuller, M., et al. (2021). Fragment binding to the Nsp3 macrodomain of SARS-CoV-2 identified through crystallographic screening and computational docking. *Science Advances, 7*, eabf8711.

75. Gahbauer, S., et al. (2023). Iterative computational design and crystallographic screening identifies potent inhibitors targeting the Nsp3 macrodomain of SARS-CoV-2. *Proceedings of the National Academy of Sciences, 120*, e2212931120.
76. Owen, D. R., et al. (2021). An oral SARS-CoV-2 Mpro inhibitor clinical candidate for the treatment of COVID-19. *Science, 374*, 1586–1593.
77. Boby, M. L. et al. (2023). Open Science discovery of potent non-covalent SARS-CoV-2 main protease inhibitors. 2020.10.29.339317 Preprint at https://doi.org/10.1101/2020.10.29.339317
78. Kronenberger, T., Laufer, S. A., & Pillaiyar, T. (2023). COVID-19 therapeutics: Small-molecule drug development targeting SARS-CoV-2 main protease. *Drug Discovery Today, 28*, 103579.
79. Jin, Z., et al. (2020). Structure of Mpro from SARS-CoV-2 and discovery of its inhibitors. *Nature, 582*, 289–293.
80. Mathai, N., Chen, Y., & Kirchmair, J. (2020). Validation strategies for target prediction methods. *Briefings in Bioinformatics, 21*, 791–802.
81. Yıldırım, M. A., Goh, K.-I., Cusick, M. E., Barabási, A.-L., & Vidal, M. (2007). Drug—target network. *Nature Biotechnology, 25*, 1119–1126.
82. Chaudhari, R., Tan, Z., Huang, B., & Zhang, S. (2017). Computational polypharmacology: A new paradigm for drug discovery. *Expert Opinion on Drug Discovery, 12*, 279–291.
83. Proschak, E., Stark, H., & Merk, D. (2019). Polypharmacology by design: A medicinal chemist's perspective on multitargeting compounds. *Journal of Medicinal Chemistry, 62*, 420–444.
84. Kurosaki, K., Wu, R., & Uesawa, Y. (2020). A toxicity prediction tool for potential agonist/antagonist activities in molecular initiating events based on chemical structures. *International Journal of Molecular Sciences, 21*, 7853.
85. Wu, Z., Li, W., Liu, G., & Tang, Y. (2018). Network-based methods for prediction of drug-target interactions. *Frontiers in Pharmacology, 9*.
86. Fernández, A., & Scott, L. R. (2016). Drug leads for interactive protein targets with unknown structure. *Drug Discovery Today, 21*, 531–535.
87. Bisswanger, H. (2014). Enzyme assays. *Perspectives in Science, 1*, 41–55.
88. Acker, M. G., & Auld, D. S. (2014). Considerations for the design and reporting of enzyme assays in high-throughput screening applications. *Perspectives in Science, 1*, 56–73.
89. Sternke, M., Tripp, K. W., & Barrick, D. (2019). Consensus sequence design as a general strategy to create hyperstable, biologically active proteins. *Proceedings of the National Academy of Sciences, 116*, 11275–11284.
90. Chen, Y., & Nielsen, J. (2019). Energy metabolism controls phenotypes by protein efficiency and allocation. *Proceedings of the National Academy of Sciences, 116*, 17592–17597.
91. Li, F., et al. (2022). Deep learning-based kcat prediction enables improved enzyme-constrained model reconstruction. *Nature Catalysis, 5*, 662–672.
92. Davidi, D., et al. (2016). Global characterization of in vivo enzyme catalytic rates and their correspondence to in vitro kcat measurements. *Proceedings of the National Academy of Sciences, 113*, 3401–3406.
93. Kroll, A., Rousset, Y., Hu, X.-P., Liebrand, N. A., & Lercher, M. J. (2023). Turnover number predictions for kinetically uncharacterized enzymes using machine and deep learning. *Nature Communications, 14*, 4139.
94. Lerman, J. A., et al. (2012). In silico method for modelling metabolism and gene product expression at genome scale. *Nature Communications, 3*, 929.
95. Dourado, H., Liebermeister, W., Ebenhöh, O., & Lercher, M. J. (2023). Mathematical properties of optimal fluxes in cellular reaction networks at balanced growth. *PLoS Computational Biology, 19*, e1011156.
96. Yang, M., et al. (2018). Functional and informatics analysis enables glycosyltransferase activity prediction. *Nature Chemical Biology, 14*, 1109–1117.
97. Dalkiran, A., et al. (2018). ECPred: A tool for the prediction of the enzymatic functions of protein sequences based on the EC nomenclature. *BMC Bioinformatics, 19*, 334.

98. Drula, E., et al. (2022). The carbohydrate-active enzyme database: Functions and literature. *Nucleic Acids Research, 50*, D571–D577.
99. Rudolf, A. F., Skovgaard, T., Knapp, S., Jensen, L. J., & Berthelsen, J. (2014). A comparison of protein kinases inhibitor screening methods using both enzymatic activity and binding affinity determination. *PLoS One, 9*, e98800.
100. Metz, J. T., et al. (2011). Navigating the kinome. *Nature Chemical Biology, 7*, 200–202.
101. Stenton, G. R., et al. (2013). Characterization of AQX-1125, a small-molecule SHIP1 activator. *British Journal of Pharmacology, 168*, 1506–1518.
102. Di Cera, E. (2009). Serine proteases. *IUBMB Life, 61*, 510–515.
103. Santos, L. H., et al. (2023). Evaluating known Zika virus NS2B-NS3 protease inhibitor scaffolds via in Silico screening and biochemical assays. *Pharmaceuticals, 16*, 1319.
104. Kronenberger, T., Sá Magalhães Serafim, M., Kumar Tonduru, A., Gonçalves Maltarollo, V., & Poso, A. (2021). Ligand accessibility insights to the dengue virus NS3-NS2B protease assessed by long-timescale molecular dynamics simulations. *ChemMedChem, 16*, 2524–2534.
105. Poverennaya, E., Kiseleva, O., Romanova, A., & Pyatnitskiy, M. (2020). Predicting functions of uncharacterized human proteins: From canonical to Proteoforms. *Genes, 11*, 677.
106. Robinson, S. L., Smith, M. D., Richman, J. E., Aukema, K. G., & Wackett, L. P. (2020). Machine learning-based prediction of activity and substrate specificity for OleA enzymes in the thiolase superfamily. *Synthetic Biology, 5*, ysaa004.
107. Li, Y., et al. (2018). DEEPre: Sequence-based enzyme EC number prediction by deep learning. *Bioinformatics, 34*, 760–769.
108. Zou, Z., Tian, S., Gao, X., & Li, Y. (2019). mlDEEPre: Multi-functional enzyme function prediction with hierarchical multi-label deep learning. *Frontiers in Genetics, 9*.
109. Mou, Z., et al. (2021). Machine learning-based prediction of enzyme substrate scope: Application to bacterial nitrilases. *Proteins: Structure, Function, and Bioinformatics, 89*, 336–347.
110. Safaei, J., Maňuch, J., Gupta, A., Stacho, L., & Pelech, S. (2011). Prediction of 492 human protein kinase substrate specificities. *Proteome Science, 9*, S6.
111. Chen, M., et al. (2023). GPS 6.0: An updated server for prediction of kinase-specific phosphorylation sites in proteins. *Nucleic Acids Research, 51*, W243–W250.
112. Kozakov, D., et al. (2015). The FTMap family of web servers for determining and characterizing ligand-binding hot spots of proteins. *Nature Protocols, 10*, 733–755.
113. Dennis, S., Kortvelyesi, T., & Vajda, S. (2002). Computational mapping identifies the binding sites of organic solvents on proteins. *Proceedings of the National Academy of Sciences, 99*, 4290–4295.
114. O'Donoghue, A. J., et al. (2012). Global identification of peptidase specificity by multiplex substrate profiling. *Nature Methods, 9*, 1095–1100.
115. Lapek, J. D., et al. (2019). Quantitative multiplex substrate profiling of peptidases by mass spectrometry. *Molecular & Cellular Proteomics, 18*, 968–981.
116. Yoon, M. C., et al. (2021). Selective neutral pH inhibitor of Cathepsin B designed based on cleavage preferences at cytosolic and lysosomal pH conditions. *ACS Chemical Biology, 16*, 1628–1643.
117. Rohweder, P. J., Jiang, Z., Hurysz, B. M., O'Donoghue, A. J., & Craik, C. S. (2023). Chapter Twelve – Multiplex substrate profiling by mass spectrometry for proteases. In A. K. Shukla (Ed.), *Methods in enzymology* (Vol. 682, pp. 375–411). Academic.
118. Mahoney, M., et al. (2021). A novel class of TMPRSS2 inhibitors potently block SARS-CoV-2 and MERS-CoV viral entry and protect human epithelial lung cells. *Proceedings of the National Academy of Sciences, 118*, e2108728118.
119. McGovern, S. L., Helfand, B. T., Feng, B., & Shoichet, B. K. (2003). A specific mechanism of nonspecific inhibition. *Journal of Medicinal Chemistry, 46*, 4265–4272.
120. Huang, F., Han, X., Xiao, X., & Zhou, J. (2022). Covalent warheads targeting cysteine residue: The promising approach in drug development. *Molecules, 27*, 7728.
121. Mellott, D. M., et al. (2021). A clinical-stage cysteine protease inhibitor blocks SARS-CoV-2 infection of human and monkey cells. *ACS Chemical Biology, 16*, 642–650.

122. Jin, Z., et al. (2020). Structural basis for the inhibition of SARS-CoV-2 main protease by antineoplastic drug carmofur. *Nature Structural and Molecular Biology, 27*, 529–532.
123. Greasley, S. E., et al. (2022). Structural basis for the in vitro efficacy of nirmatrelvir against SARS-CoV-2 variants. *Journal of Biological Chemistry, 298*, 101972.
124. Abriata, L. A., & Dal Peraro, M. (2015). Assessing the potential of atomistic molecular dynamics simulations to probe reversible protein-protein recognition and binding. *Scientific Reports, 5*, 10549.
125. Alamri, M. A., et al. (2021). Pharmacoinformatics and molecular dynamics simulation studies reveal potential covalent and FDA-approved inhibitors of SARS-CoV-2 main protease 3CLpro. *Journal of Biomolecular Structure and Dynamics, 39*, 4936–4948.
126. Tuley, A., & Fast, W. (2018). The taxonomy of covalent inhibitors. *Biochemistry, 57*, 3326–3337.
127. Ghosh, A. K., Samanta, I., Mondal, A., & Liu, W. R. (2019). Covalent inhibition in drug discovery. *ChemMedChem, 14*, 889–906.
128. Boike, L., Henning, N. J., & Nomura, D. K. (2022). Advances in covalent drug discovery. *Nature Reviews. Drug Discovery, 21*, 881–898.
129. Weerapana, E., et al. (2010). Quantitative reactivity profiling predicts functional cysteines in proteomes. *Nature, 468*, 790–795.
130. Wu, X., Kriz, A. J., & Sharp, P. A. (2014). Target specificity of the CRISPR-Cas9 system. *Quantitative Biology, 2*, 59–70.
131. Strohkendl, I., Saifuddin, F. A., Rybarski, J. R., Finkelstein, I. J., & Russell, R. (2018). Kinetic basis for DNA target specificity of CRISPR-Cas12a. *Molecular Cell, 71*, 816–824.e3.
132. Krippendorff, B.-F., Neuhaus, R., Lienau, P., Reichel, A., & Huisinga, W. (2009). Mechanism-based inhibition: Deriving KI and kinact directly from time-dependent IC50 values. *Journal of Biomolecular Screening, 14*, 913–923.
133. Strelow, J. M. (2017). A perspective on the kinetics of covalent and irreversible inhibition. *SLAS Discovery, 22*, 3–20.
134. Mons, E., Roet, S., Kim, R. Q., & Mulder, M. P. C. (2022). A comprehensive guide for assessing covalent inhibition in enzymatic assays illustrated with kinetic simulations. *Current Protocols, 2*, e419.
135. Harris, C. M., et al. (2018). Merits and pitfalls in the characterization of covalent inhibitors of Bruton's tyrosine kinase. *SLAS Discovery, 23*, 1040–1050.
136. Palmer, D. S., & Mitchell, J. B. O. (2014). Is experimental data quality the limiting factor in predicting the aqueous solubility of druglike molecules? *Molecular Pharmaceutics, 11*, 2962–2972.
137. Popa-Burke, I., & Russell, J. (2014). Compound precipitation in high-concentration DMSO solutions. *Journal of Biomolecular Screening, 19*, 1302–1308.
138. Baell, J. B., & Nissink, J. W. M. (2018). Seven year itch: Pan-assay interference compounds (PAINS) in 2017—Utility and limitations. *ACS Chemical Biology, 13*, 36–44.
139. Borrel, A., et al. (2020). High-throughput screening to predict chemical-assay interference. *Scientific Reports, 10*, 3986.
140. Jadhav, A., et al. (2010). Quantitative analyses of aggregation, autofluorescence, and reactivity artifacts in a screen for inhibitors of a thiol protease. *Journal of Medicinal Chemistry, 53*, 37–51.
141. Babaoglu, K., et al. (2008). Comprehensive mechanistic analysis of hits from high-throughput and docking screens against β-lactamase. *Journal of Medicinal Chemistry, 51*, 2502–2511.
142. Feng, B. Y., & Shoichet, B. K. (2006). A detergent-based assay for the detection of promiscuous inhibitors. *Nature Protocols, 1*, 550–553.
143. Shoichet, B. K. (2006). Screening in a spirit haunted world. *Drug Discovery Today, 11*, 607–615.
144. Falke, S., & Betzel, C. (2019). Dynamic light scattering (DLS). In A. S. Pereira, P. Tavares, & P. Limão-Vieira (Eds.), *Radiation in bioanalysis: Spectroscopic techniques and theoretical methods* (pp. 173–193). Springer International Publishing.

145. Lorber, B., Fischer, F., Bailly, M., Roy, H., & Kern, D. (2012). Protein analysis by dynamic light scattering: Methods and techniques for students. *Biochemistry and Molecular Biology Education, 40*, 372–382.
146. mic.0.045005-0.
147. Irwin, J. J., et al. (2015). An aggregation advisor for ligand discovery. *Journal of Medicinal Chemistry, 58*, 7076–7087.
148. Seidler, J., McGovern, S. L., Doman, T. N., & Shoichet, B. K. (2003). Identification and prediction of promiscuous aggregating inhibitors among known drugs. *Journal of Medicinal Chemistry, 46*, 4477–4486.
149. Kamiloglu, S., Sari, G., Ozdal, T., & Capanoglu, E. (2020). Guidelines for cell viability assays. *Food Frontiers, 1*, 332–349.
150. Mosmann, T. (1983). Rapid colorimetric assay for cellular growth and survival: Application to proliferation and cytotoxicity assays. *Journal of Immunological Methods, 65*, 55–63.
151. Roehm, N. W., Rodgers, G. H., Hatfield, S. M., & Glasebrook, A. L. (1991). An improved colorimetric assay for cell proliferation and viability utilizing the tetrazolium salt XTT. *Journal of Immunological Methods, 142*, 257–265.
152. Hu, V. W., Black, G. E., Torres-Duarte, A., & Abramson, F. P. (2002). 3H-thymidine is a defective tool with which to measure rates of DNA synthesis. *The FASEB Journal, 16*, 1456–1457.
153. Crane, A. M., & Bhattacharya, S. K. (2013). The use of bromodeoxyuridine incorporation assays to assess corneal stem cell proliferation. *Methods in Molecular Biology, 1014*, 65–70.
154. Strober, W. (2015). Trypan blue exclusion test of cell viability. *Current Protocols in Immunology, 111*, A3.B.1–A3.B.3.
155. O'Brien, J., Wilson, I., Orton, T., & Pognan, F. (2000). Investigation of the Alamar Blue (resazurin) fluorescent dye for the assessment of mammalian cell cytotoxicity. *European Journal of Biochemistry, 267*, 5421–5426.
156. Stockert, J. C., Horobin, R. W., Colombo, L. L., & Blázquez-Castro, A. (2018). Tetrazolium salts and formazan products in cell biology: Viability assessment, fluorescence imaging, and labeling perspectives. *Acta Histochemica, 120*, 159–167.
157. Riss, T. L., Moravec, R. A., Duellman, S. J., & Niles, A. L. (2021). Treating cells as reagents to design reproducible assays. *SLAS Discovery: Advancing the Science of Drug Discovery, 26*, 1256–1267.
158. Saotome, K., Morita, H., & Umeda, M. (1989). Cytotoxicity test with simplified crystal violet staining method using microtitre plates and its application to injection drugs. *Toxicology In Vitro, 3*, 317–321.
159. Coban, A. Y. (2014). A new rapid colourimetric method for testing Mycobacterium tuberculosis susceptibility to isoniazid and rifampicin: A crystal violet decolourisation assay. *Memórias do Instituto Oswaldo Cruz, 109*, 246–249.
160. Resnick, S. J., et al. (2021). Inhibitors of coronavirus 3CL proteases protect cells from protease-mediated cytotoxicity. *Journal of Virology, 95*.
161. Castro, J., et al. (2022). Crystal violet staining alone is not adequate to assess synergism or antagonism in multi-species biofilms of bacteria associated with bacterial vaginosis. *Frontiers in Cellular and Infection Microbiology, 11*.
162. Almutary, A., & Sanderson, B. J. S. (2016). The MTT and crystal violet assays: Potential confounders in nanoparticle toxicity testing. *International Journal of Toxicology, 35*, 454–462.
163. Kotz, J. (2012). Phenotypic screening, take two. *Science-Business eXchange, 5*, 380–380.
164. Priest, B. T., & Erdemli, G. (2014). Phenotypic screening in the 21st century. *Frontiers in Pharmacology, 5*.
165. Moffat, J. G., Rudolph, J., & Bailey, D. (2014). Phenotypic screening in cancer drug discovery—Past, present and future. *Nature Reviews. Drug Discovery, 13*, 588–602.
166. Idakwo, G., et al. (2018). A review on machine learning methods for in silico toxicity prediction. *Journal of Environmental Science and Health, Part C, 36*, 169–191.

167. Moffat, J. G., Vincent, F., Lee, J. A., Eder, J., & Prunotto, M. (2017). Opportunities and challenges in phenotypic drug discovery: An industry perspective. *Nature Reviews. Drug Discovery, 16*, 531–543.
168. Vincent, F., et al. (2022). Phenotypic drug discovery: Recent successes, lessons learned and new directions. *Nature Reviews. Drug Discovery, 21*, 899–914.
169. Frye, S. V. (2010). The art of the chemical probe. *Nature Chemical Biology, 6*, 159–161.
170. Quinlan, R. B. A., & Brennan, P. E. (2021). Chemogenomics for drug discovery: Clinical molecules from open access chemical probes. *RSC Chemical Biology, 2*, 759–795.
171. . (2022). A conversation on using chemical probes to study protein function in cells and organisms. *Nature Communications, 13*, 3757.
172. Arrowsmith, C. H., et al. (2015). The promise and peril of chemical probes. *Nature Chemical Biology, 11*, 536–541.
173. Dale, N. C., Johnstone, E. K. M., White, C. W., & Pfleger, K. D. G. (2019). NanoBRET: The bright future of proximity-based assays. *Frontiers in Bioengineering and Biotechnology, 7*.
174. Hall, M. P., et al. (2012). Engineered luciferase reporter from a deep sea shrimp utilizing a novel Imidazopyrazinone substrate. *ACS Chemical Biology, 7*, 1848–1857.
175. Machleidt, T., et al. (2015). NanoBRET—A novel BRET platform for the analysis of protein–protein interactions. *ACS Chemical Biology, 10*, 1797–1804.
176. Shenoy, N., et al. (2017). Drugs with anti-oxidant properties can interfere with cell viability measurements by assays that rely on the reducing property of viable cells. *Laboratory Investigation, 97*, 494–497.
177. Simeonov, A., et al. (2008). Fluorescence spectroscopic profiling of compound libraries. *Journal of Medicinal Chemistry, 51*, 2363–2371.
178. Thorne, N., et al. (2012). Firefly luciferase in chemical biology: A compendium of inhibitors, mechanistic evaluation of chemotypes, and suggested use as a reporter. *Chemistry & Biology, 19*, 1060–1072.
179. Gul, S., & Gribbon, P. (2010). Exemplification of the challenges associated with utilising fluorescence intensity based assays in discovery. *Expert Opinion on Drug Discovery, 5*, 681–690.
180. Zhang, X. D., Wang, D., Sun, S., & Zhang, H. (2021). Issues of Z-factor and an approach to avoid them for quality control in high-throughput screening studies. *Bioinformatics, 36*, 5299–5303.
181. Zhang, J. H., Chung, T. D., & Oldenburg, K. R. (1999). A simple statistical parameter for use in evaluation and validation of high throughput screening assays. *Journal of Biomolecular Screening, 4*, 67–73.
182. Präbst, K., Engelhardt, H., Ringgeler, S., & Hübner, H. (2017). Basic Colorimetric Proliferation Assays: MTT, WST, and Resazurin. In D. F. Gilbert & O. Friedrich (Eds.), *Cell viability assays: Methods and protocols* (pp. 1–17). Springer.
183. Dahlin, J. L., et al. (2015). Post-HTS case report and structural alert: Promiscuous 4-aroyl-1,5-disubstituted-3-hydroxy-2H-pyrrol-2-one actives verified by ALARM NMR. *Bioorganic & Medicinal Chemistry Letters, 25*, 4740–4752.
184. Su, B.-H., et al. (2015). Rule-based classification models of molecular autofluorescence. *Journal of Chemical Information and Modeling, 55*, 434–445.
185. Baell, J. B., & Holloway, G. A. (2010). New substructure filters for removal of pan assay interference compounds (PAINS) from screening libraries and for their exclusion in bioassays. *Journal of Medicinal Chemistry, 53*, 2719–2740.
186. Magalhães, P. R., Reis, P. B. P. S., Vila-Viçosa, D., Machuqueiro, M., & Victor, B. L. (2021). Identification of Pan-Assay INterference compoundS (PAINS) using an MD-based protocol. In I. S. Moreira, M. Machuqueiro, & J. Mourão (Eds.), *Computational design of membrane proteins* (pp. 263–271). Springer US.
187. Nelson, K. M., et al. (2017). The essential medicinal chemistry of curcumin. *Journal of Medicinal Chemistry, 60*, 1620–1637.

188. Serafim, M. S. M., et al. (2023). Synthetic curcumin analogues present antiflavivirus activity in vitro with potential multiflavivirus activity from a thiazolylhydrazone moiety. *Future Pharmacology, 3*, 364–378.
189. Tyagi, P., Singh, M., Kumari, H., Kumari, A., & Mukhopadhyay, K. (2015). Bactericidal activity of curcumin I is associated with damaging of bacterial membrane. *PLoS One, 10*, e0121313.
190. Chen, T.-Y., et al. (2013). Inhibition of enveloped viruses infectivity by curcumin. *PLoS One, 8*, e62482.
191. Bhattacharjee, S. (2016). DLS and zeta potential—What they are and what they are not? *Journal of Controlled Release, 235*, 337–351.
192. Biriukov, D., Fibich, P., & Předota, M. (2020). Zeta potential determination from molecular simulations. *Journal of Physical Chemistry C, 124*, 3159–3170.
193. Shakour, N., et al. (2021). Curcumin can bind and interact with CRP: An in silico study. *Advances in Experimental Medicine and Biology, 1308*, 91–100.
194. Halder, S., et al. (2015). Alteration of zeta potential and membrane permeability in bacteria: A study with cationic agents. *Springerplus, 4*, 672.
195. Duda, M., Cygan, K., & Wisniewska-Becker, A. (2020). Effects of curcumin on lipid membranes: An EPR spin-label study. *Cell Biochemistry and Biophysics, 78*, 139–147.
196. Bergström, C. A. S., & Larsson, P. (2018). Computational prediction of drug solubility in water-based systems: Qualitative and quantitative approaches used in the current drug discovery and development setting. *International Journal of Pharmaceutics, 540*, 185–193.
197. Jorgensen, W. L., & Duffy, E. M. (2002). Prediction of drug solubility from structure. *Advanced Drug Delivery Reviews, 54*, 355–366.
198. Kataoka, M., et al. (2019). In vitro assessment of Supersaturation/precipitation and biological membrane permeation of poorly water-soluble drugs: A case study with albendazole and ketoconazole. *Journal of Pharmaceutical Sciences, 108*, 2580–2587.
199. Waybright, T. J., Britt, J. R., & McCloud, T. G. (2009). Overcoming problems of compound storage in DMSO: Solvent and process alternatives. *Journal of Biomolecular Screening, 14*, 708–715.
200. Banerjee, P., Eckert, A. O., Schrey, A. K., & Preissner, R. (2018). ProTox-II: A webserver for the prediction of toxicity of chemicals. *Nucleic Acids Research, 46*, 257–263.
201. Yang, H., et al. (2019). admetSAR 2.0: Web-service for prediction and optimization of chemical ADMET properties. *Bioinformatics, 35*, 1067–1069.
202. Dulsat, J., López-Nieto, B., Estrada-Tejedor, R., & Borrell, J. I. (2023). Evaluation of free online ADMET tools for academic or small biotech environments. *Molecules, 28*, 776.
203. Braga, R. C., et al. (2015). Pred-hERG: A novel web-accessible computational tool for predicting cardiac toxicity. *Molecular Informatics, 34*, 698–701.
204. Dong, J., et al. (2018). ADMETlab: A platform for systematic ADMET evaluation based on a comprehensively collected ADMET database. *Journal of Cheminformatics, 10*, 29.
205. Xiong, G., et al. (2021). ADMETlab 2.0: An integrated online platform for accurate and comprehensive predictions of ADMET properties. *Nucleic Acids Research, 49*, W5–W14.
206. Maia, E. H. B., Assis, L. C., de Oliveira, T. A., da Silva, A. M., & Taranto, A. G. (2020). Structure-based virtual screening: From classical to artificial intelligence. *Frontiers in Chemistry, 8*.
207. Ghosh, A. K., Osswald, H. L., & Prato, G. (2016). Recent Progress in the development of HIV-1 protease inhibitors for the treatment of HIV/AIDS. *Journal of Medicinal Chemistry, 59*, 5172–5208.
208. Weber, I. T., Wang, Y.-F., & Harrison, R. W. (2021). HIV protease: Historical perspective and current research. *Viruses, 13*, 839.
209. Wlodawer, A. (2002). Rational approach to AIDS drug design through structural biology. *Annual Review of Medicine, 53*, 595–614.
210. Kitchen, V. S., et al. (1995). Safety and activity of saquinavir in HIV infection. *The Lancet, 345*, 952–955.

211. von Itzstein, M. (2007). The war against influenza: Discovery and development of sialidase inhibitors. *Nature Reviews. Drug Discovery, 6*, 967–974.
212. Talele, T. T., Khedkar, S. A., & Rigby, A. C. (2010). Successful applications of computer aided drug discovery: Moving drugs from concept to the clinic. *Current Topics in Medicinal Chemistry, 10*, 127–141.
213. Njoroge, F. G., Chen, K. X., Shih, N.-Y., & Piwinski, J. J. (2008). Challenges in modern drug discovery: A case study of boceprevir, an HCV protease inhibitor for the treatment of hepatitis C virus infection. *Accounts of Chemical Research, 41*, 50–59.
214. Graham, B. S., & Sullivan, N. J. (2018). Emerging viral diseases from a vaccinology perspective: Preparing for the next pandemic. *Nature Immunology, 19*, 20–28.
215. Ten health issues WHO will tackle this year. https://www.who.int/news-room/spotlight/ten-threats-to-global-health-in-2019
216. Gorbalenya, A. E., et al. (2020). The species severe acute respiratory syndrome-related coronavirus: Classifying 2019-nCoV and naming it SARS-CoV-2. *Nature Microbiology, 5*, 536–544.
217. Zhu, N., et al. (2020). A novel coronavirus from patients with pneumonia in China, 2019. *The New England Journal of Medicine, 382*, 727–733.
218. Morens, D. M., & Fauci, A. S. (2020). Emerging pandemic diseases: How we got to COVID-19. *Cell, 182*, 1077–1092.
219. Pierson, T. C., & Diamond, M. S. (2020). The continued threat of emerging flaviviruses. *Nature Microbiology, 5*, 796–812.
220. Kalil, A. C. (2020). Treating COVID-19—Off-label drug use, compassionate use, and randomized clinical trials during pandemics. *JAMA, 323*, 1897.
221. Adamson, C. S., et al. (2021). Antiviral drug discovery: preparing for the next pandemic. *Chemical Society Reviews, 50*, 3647–3655.
222. Newman, D. J., & Cragg, G. M. (2020). Natural products as sources of new drugs over the nearly four decades from 01/1981 to 09/2019. *Journal of Natural Products, 83*, 770–803.
223. De Clercq, E., & Li, G. (2016). Approved antiviral drugs over the past 50 years. *Clinical Microbiology Reviews, 29*, 695–747.
224. De Clercq, E. (2012). Human viral diseases: What is next for antiviral drug discovery? *Current Opinion in Virology, 2*, 572–579.
225. Hu, Y., Musharrafieh, R., Zheng, M., & Wang, J. (2020). Enterovirus D68 antivirals: Past, present, and future. *ACS Infectious Diseases, 6*, 1572–1586.
226. Bule, M., Khan, F., & Niaz, K. (2019). Antivirals: Past, present and future. In *Recent advances in animal virology* (pp. 425–446).
227. Serafim, M. S. M., et al. (2021). Knowing and combating the enemy: A brief review on SARS-CoV-2 and computational approaches applied to the discovery of drug candidates. *Bioscience Reports, 41*, BSR20202616.
228. De Cena, G. L., Scavassa, B. V., & Conceição, K. (2022). In Silico prediction of anti-infective and cell-penetrating peptides from Thalassophryne nattereri natterin toxins. *Pharmaceuticals, 15*, 1141.
229. Bryan-Marrugo, O. L., et al. (2015). History and progress of antiviral drugs: From acyclovir to direct-acting antiviral agents (DAAs) for Hepatitis C. *Medicina Universitaria, 17*, 165–174.
230. McFadden, G., Mohamed, M. R., Rahman, M. M., & Bartee, E. (2009). Cytokine determinants of viral tropism. *Nature Reviews. Immunology, 9*, 645–655.
231. Nakajima, N., Lu, R., & Engelman, A. (2001). Human immunodeficiency virus type 1 replication in the absence of integrase-mediated DNA recombination: Definition of permissive and nonpermissive T-cell lines. *Journal of Virology, 75*, 7944–7955.
232. Zahradník, J., et al. (2021). SARS-CoV-2 variant prediction and antiviral drug design are enabled by RBD in vitro evolution. *Nature Microbiology, 6*, 1188–1198.
233. Baxter, V. K., & Montgomery, S. A. (2022). Preclinical coronavirus studies and pathology: Challenges of the high-containment laboratory. *Veterinary Pathology, 59*, 556–564.

234. Pereira Oliveira, G., & Kroon, E. G. (2021). Mouse hepatitis virus: A betacoronavirus model to study the virucidal activity of air disinfection equipment on surface contamination. *Journal of Virological Methods, 297*, 114274.
235. Andrade, A. C. D. S. P., et al. (2021). A biosafety level 2 mouse model for studying Betacoronavirus-induced acute lung damage and systemic manifestations. *Journal of Virology, 95*, e0127621.
236. Pathan, R. K., Biswas, M., & Khandaker, M. U. (2020). Time series prediction of COVID-19 by mutation rate analysis using recurrent neural network-based LSTM model. *Chaos, Solitons & Fractals, 138*, 110018.
237. Kumar, M., et al. (2021). A chronicle of SARS-CoV-2: Seasonality, environmental fate, transport, inactivation, and antiviral drug resistance. *Journal of Hazardous Materials, 405*, 124043.
238. von Delft, A., et al. (2023). Accelerating antiviral drug discovery: Lessons from COVID-19. *Nature Reviews. Drug Discovery, 22*, 585–603.
239. Scudellari, M. (2020). Researchers are betting that AI and automation can cut drug discovery from five years to six months: Automating antivirals. *IEEE Spectrum, 57*, 44–49.
240. Case, J. B., Bailey, A. L., Kim, A. S., Chen, R. E., & Diamond, M. S. (2020). Growth, detection, quantification, and inactivation of SARS-CoV-2. *Virology, 548*, 39–48.
241. Jureka, A. S., Silvas, J. A., & Basler, C. F. (2020). Propagation, inactivation, and safety testing of SARS-CoV-2. *Viruses, 12*, 622.
242. McNamara, R. P., & Dittmer, D. P. (2020). Modern techniques for the isolation of extracellular vesicles and viruses. *Journal of Neuroimmune Pharmacology, 15*, 459–472.
243. Liu, C., Elvati, P., & Violi, A. (2021). On drug-membrane permeability of antivirals for SARS-CoV-2. *Journal of Physical Chemistry Letters, 12*, 1384–1389.
244. Landry, M. L., et al. (2000). A standardized plaque reduction assay for determination of drug susceptibilities of cytomegalovirus clinical isolates. *Antimicrobial Agents and Chemotherapy, 44*, 688–692.
245. Mendoza, E. J., Manguiat, K., Wood, H., & Drebot, M. (2020). Two detailed plaque assay protocols for the quantification of infectious SARS-CoV-2. *Current Protocols in Microbiology, 57*, cpmc105.
246. Huang, Y., Yang, C., Xu, X., Xu, W., & Liu, S. (2020). Structural and functional properties of SARS-CoV-2 spike protein: Potential antivirus drug development for COVID-19. *Acta Pharmacologica Sinica, 41*, 1141–1149.
247. Pirolli, D., et al. (2023). Virtual screening and molecular dynamics simulations provide insight into repurposing drugs against SARS-CoV-2 variants spike protein/ACE2 interface. *Scientific Reports, 13*, 1494.
248. Osipiuk, J., et al. (2021). Structure of papain-like protease from SARS-CoV-2 and its complexes with non-covalent inhibitors. *Nature Communications, 12*, 743.
249. Yin, W., et al. (2020). Structural basis for inhibition of the RNA-dependent RNA polymerase from SARS-CoV-2 by remdesivir. *Science, 368*, 1499–1504.
250. Dong, T. N., Brogden, G., Gerold, G., & Khosla, M. (2021). A multitask transfer learning framework for the prediction of virus-human protein–protein interactions. *BMC Bioinformatics, 22*, 572.
251. Madan, S., Demina, V., Stapf, M., Ernst, O., & Fröhlich, H. (2022). Accurate prediction of virus-host protein-protein interactions via a Siamese neural network using deep protein sequence embeddings. *Patterns, 3*, 100551.
252. Liu-Wei, W., et al. (2021). DeepViral: Prediction of novel virus–host interactions from protein sequences and infectious disease phenotypes. *Bioinformatics, 37*, 2722–2729.
253. Asim, M. N., Fazeel, A., Ibrahim, M. A., Dengel, A., & Ahmed, S. (2022). MP-VHPPI: Meta predictor for viral host protein-protein interaction prediction in multiple hosts and viruses. *Frontiers in Medicine, 9*.

254. Yang, X., Yang, S., Ren, P., Wuchty, S., & Zhang, Z. (2022). Deep learning-powered prediction of human-virus protein-protein interactions. *Frontiers in Microbiology, 13*.
255. Gupta, P., & Mohanty, D. (2021). SMMPPI: A machine learning-based approach for prediction of modulators of protein–protein interactions and its application for identification of novel inhibitors for RBD:hACE2 interactions in SARS-CoV-2. *Briefings in Bioinformatics, 22*, bbab111.
256. Qureshi, A., Thakur, N., Tandon, H., & Kumar, M. (2014). AVPdb: A database of experimentally validated antiviral peptides targeting medically important viruses. *Nucleic Acids Research, 42*, D1147–D1153.
257. Qureshi, A., Tandon, H., & Kumar, M. (2015). AVP-IC50 Pred: Multiple machine learning techniques-based prediction of peptide antiviral activity in terms of half maximal inhibitory concentration (IC50). *Biopolymers, 104*, 753–763.
258. Nikitina, A. A., Orlov, A. A., Kozlovskaya, L. I., Palyulin, V. A., & Osolodkin, D. I. (2019). Enhanced taxonomy annotation of antiviral activity data from ChEMBL. *Database: The Journal of Biological Databases and Curation, 2019*.
259. Zhang, Y., et al. (2017). Influenza research database: An integrated bioinformatics resource for influenza virus research. *Nucleic Acids Research, 45*, D466–D474.
260. Tzou, P. L., et al. (2020). Coronavirus antiviral research database (CoV-RDB): An online database designed to facilitate comparisons between candidate anti-coronavirus compounds. *Viruses, 12*, 1006.
261. Keshavarzi Arshadi, A., et al. (2020). Artificial intelligence for COVID-19 drug discovery and vaccine development. *Frontiers in Artificial Intelligence, 3*.
262. Riva, L., et al. (2020). Discovery of SARS-CoV-2 antiviral drugs through large-scale compound repurposing. *Nature, 586*, 113–119.
263. Ma, Y., et al. (2021). Medicinal chemistry strategies for discovering antivirals effective against drug-resistant viruses. *Chemical Society Reviews*.
264. Pizzorno, A., Padey, B., Terrier, O., & Rosa-Calatrava, M. (2019). Drug repurposing approaches for the treatment of influenza viral infection: Reviving old drugs to fight against a long-lived enemy. *Frontiers in Immunology, 10*.
265. Trivedi, J., Mohan, M., & Byrareddy, S. N. (2020). Drug repurposing approaches to combating viral infections. *Journal of Clinical Medicine, 9*, 3777.
266. Bauso, L. V., Imbesi, C., Irene, G., Calì, G., & Bitto, A. (2021). New approaches and repurposed antiviral drugs for the treatment of the SARS-CoV-2 infection. *Pharmaceuticals, 14*, 503.
267. Li, G., Hilgenfeld, R., Whitley, R., & De Clercq, E. (2023). Therapeutic strategies for COVID-19: Progress and lessons learned. *Nature Reviews. Drug Discovery, 22*, 449–475.
268. Balasubramaniam, B., et al. (2021). Antibacterial and antiviral functional materials: Chemistry and biological activity toward tackling COVID-19-like pandemics. *ACS Pharmacology & Translational Science, 4*, 8–54.
269. Mantravadi, P. K., Kalesh, K. A., Dobson, R. C. J., Hudson, A. O., & Parthasarathy, A. (2019). The quest for novel antimicrobial compounds: Emerging trends in research, development, and technologies. *Antibiotics, 8*, 8.
270. Serafim, M. S. M., et al. (2020). The application of machine learning techniques to innovative antibacterial discovery and development. *Expert Opinion on Drug Discovery, 15*, 1165–1180.
271. Miethke, M., et al. (2021). Towards the sustainable discovery and development of new antibiotics. *Nature Reviews Chemistry, 5*, 726–749.
272. Mediavilla, J. R., et al. (2016). Colistin- and Carbapenem-resistant Escherichia coli harboring mcr-1 and bla $_{NDM-5}$, causing a complicated urinary tract infection in a patient from the United States. *MBio, 7*, e01191–e01116.
273. Centers for Disease Control and Prevention. (2019). Antibiotic resistance threats in the United States, 2019. https://stacks.cdc.gov/view/cdc/82532.

274. O'Neill, J.. (2016). Tackling drug-resistant infections globally: Final report and recommendations.
275. Ghosh, C., Sarkar, P., Issa, R., & Haldar, J. (2019). Alternatives to conventional antibiotics in the era of antimicrobial resistance. *Trends in Microbiology, 27*, 323–338.
276. Boucher, H. W., et al. (2009). Bad bugs, no drugs: No ESKAPE! An update from the Infectious Diseases Society of America. *Clinical Infectious Diseases, 48*, 1–12.
277. Tacconelli, E., et al. (2018). Discovery, research, and development of new antibiotics: The WHO priority list of antibiotic-resistant bacteria and tuberculosis. *The Lancet Infectious Diseases, 18*, 318–327.
278. Egger, N. B., et al. (2022). The rise of Candida auris: From unique traits to co-infection potential. *Microbial Cell, 9*, 141–144.
279. Nelson, R. (2023). Emergence of resistant Candida auris. *The Lancet Microbe, 4*, e396.
280. Chowdhary, A., & Sharma, A. (2020). The lurking scourge of multidrug resistant Candida auris in times of COVID-19 pandemic. *Journal of Global Antimicrobial Resistance, 22*, 175–176.
281. Mulet Bayona, J. V., et al. (2021). Impact of the SARS-CoV-2 pandemic in Candidaemia, invasive Aspergillosis and antifungal consumption in a tertiary hospital. *Journal of Fungi, 7*, 440.
282. Villanueva-Lozano, H., et al. (2021). Outbreak of Candida auris infection in a COVID-19 hospital in Mexico. *Clinical Microbiology and Infection, 27*, 813–816.
283. Chowdhary, A., Tarai, B., Singh, A., & Sharma, A. (2020). Multidrug-resistant Candida auris infections in critically ill coronavirus disease patients, India, April–July 2020. *Emerging Infectious Diseases, 26*, 2694–2696.
284. Horton, M. V., et al. (2020). Candida auris forms high-burden biofilms in skin niche conditions and on porcine skin. *mSphere, 5*.
285. Hoenigl, M., et al. (2022). COVID-19-associated fungal infections. *Nature Microbiology, 7*, 1127–1140.
286. Stokes, J. M., et al. (2020). A deep learning approach to antibiotic discovery. *Cell, 180*, 688–702.e13.
287. Bouz, G., & Doležal, M. (2021). Advances in antifungal drug development: An up-to-date mini review. *Pharmaceuticals, 14*, 1312.
288. Fu, C., et al. (2021). Leveraging machine learning essentiality predictions and chemogenomic interactions to identify antifungal targets. *Nature Communications, 12*, 6497.
289. Delavy, M., et al. (2020). Machine learning approach for Candida albicans fluconazole resistance detection using matrix-assisted laser desorption/ionization time-of-flight mass spectrometry. *Frontiers in Microbiology, 10*.
290. Chun, F., Yoshitaka, M., Caihong, L., & Kentaro, S. (2019). Prediction of antifungal peptides by deep learning with character embedding. *IPSJ Transactions on Bioinformatics, 12*, 21–29.
291. Ahmad, A., et al. (2021). Deep-AntiFP: Prediction of antifungal peptides using distanct multi-informative features incorporating with deep neural networks. *Chemometrics and Intelligent Laboratory Systems, 208*, 104214.
292. Lobo, F., González, M. S., Boto, A., & Pérez de la Lastra, J. M. (2023). Prediction of antifungal activity of antimicrobial peptides by transfer learning from protein Pretrained models. *International Journal of Molecular Sciences, 24*, 10270.
293. Zhang, J., et al. (2022). Large-scale screening of antifungal peptides based on quantitative structure–activity relationship. *ACS Medicinal Chemistry Letters, 13*, 99–104.
294. Hughes, J., Rees, S., Kalindjian, S., & Philpott, K. (2011). Principles of early drug discovery. *British Journal of Pharmacology, 162*, 1239–1249.
295. McCue, C. (2015). Data mining and predictive analysis. In C. McCue (Ed.), *Data mining and predictive analysis* (2nd ed., pp. xxiii–xxvii). Butterworth-Heinemann.

296. Tenero, D., et al. (2019). First-time-in-human study and prediction of early bactericidal activity for GSK3036656, a potent leucyl-tRNA synthetase inhibitor for tuberculosis treatment. *Antimicrobial Agents and Chemotherapy, 63*.
297. Clinical & Laboratory Standards Institute, C. (2017) Performance standards for antimicrobial susceptibility testing.
298. de-Souza-Silva, C. M., et al. (2018). Broth microdilution in vitro screening: An easy and fast method to detect new antifungal compounds. *Journal of Visualized Experiments, 57127*.
299. Macalino, S. J. Y., Gosu, V., Hong, S., & Choi, S. (2015). Role of computer-aided drug design in modern drug discovery. *Archives of Pharmacal Research, 38*, 1686–1701.
300. Lagier, J.-C., et al. (2015). Current and past strategies for bacterial culture in clinical microbiology. *Clinical Microbiology Reviews, 28*, 208–236.
301. Huffnagle, G. B., & Noverr, M. C. (2013). The emerging world of the fungal microbiome. *Trends in Microbiology, 21*, 334–341.
302. Xie, M., et al. (2021). Viable but Nonculturable state of yeast Candida sp. strain LN1 induced by high phenol concentrations. *Applied and Environmental Microbiology, 87*, e01110–e01121.
303. Fekadu, G., Tolossa, T., Turi, E., Bekele, F., & Fetensa, G. (2022). Pretomanid development and its clinical roles in treating tuberculosis. *Journal of Global Antimicrobial Resistance, 31*, 175–184.
304. Stover, C. K., et al. (2000). A small-molecule nitroimidazopyran drug candidate for the treatment of tuberculosis. *Nature, 405*, 962–966.
305. Pötter, T., & Matter, H. (1998). Random or rational design? Evaluation of diverse compound subsets from chemical structure databases. *Journal of Medicinal Chemistry, 41*, 478–488.
306. Medina-Franco, J. L., Martinez-Mayorga, K., Fernández-de Gortari, E., Kirchmair, J., & Bajorath, J. (2021). Rationality over fashion and hype in drug design. *F1000Res, 10*, Chem Inf Sci-397.
307. Serafim, M. S. M., et al. (2019). Antibacterial activity of synthetic 1,3-bis(aryloxy)propan-2-amines against Gram-positive bacteria. *Microbiology, 8*, e814.
308. Bradley, P., et al. (2015). Rapid antibiotic-resistance predictions from genome sequence data for Staphylococcus aureus and Mycobacterium tuberculosis. *Nature Communications, 6*, 10063.
309. Morselli Gysi, D., et al. (2021). Network medicine framework for identifying drug-repurposing opportunities for COVID-19. *Proceedings of the National Academy of Sciences, 118*, e2025581118.
310. Zitnik, M., Sosič, R., & Leskovec, J. (2018). Prioritizing network communities. *Nature Communications, 9*, 2544.
311. Yuan, S., et al. (2017). Structure-based discovery of clinically approved drugs as Zika virus NS2B-NS3 protease inhibitors that potently inhibit Zika virus infection in vitro and in vivo. *Antiviral Research, 145*, 33–43.
312. Swearengen, J. R. (2018). Choosing the right animal model for infectious disease research. *Animal Models and Experimental Medicine, 1*, 100–108.
313. Lin, Q., et al. (2022). Animal models for studying coronavirus infections and developing antiviral agents and vaccines. *Antiviral Research, 203*, 105345.
314. Zhang, Q., Zeng, S. X., & Lu, H. (2015). Determination of maximum tolerated dose and toxicity of Inauhzin in mice. *Toxicology Reports, 2*, 546–554.
315. Koneval, Z., et al. (2020). Antiseizure drug efficacy and tolerability in established and novel drug discovery seizure models in outbred vs inbred mice. *Epilepsia, 61*, 2022–2034.
316. Dressman, J. B., & Reppas, C. (2000). In vitro–in vivo correlations for lipophilic, poorly water-soluble drugs. *European Journal of Pharmaceutical Sciences, 11*, S73–S80.
317. Steger-Hartmann, T., & Raschke, M. (2020). Translating in vitro to in vivo and animal to human. *Current Opinion in Toxicology, 23–24*, 6–10.

318. Fink, C., et al. (2020). Evaluating the role of solubility in Oral absorption of poorly water-soluble drugs using physiologically-based pharmacokinetic modeling. *Clinical Pharmacology and Therapeutics, 107*, 650–661.
319. Albarnaz, J. D., et al. (2014). MEK/ERK activation plays a decisive role in yellow fever virus replication: Implication as an antiviral therapeutic target. *Antiviral Research, 111*, 82–92.
320. de Oliveira, L. C., et al. (2020). The small molecule AZD6244 inhibits dengue virus replication in vitro and protects against lethal challenge in a mouse model. *Archives of Virology, 165*, 671–681.
321. Ferreira, G. S., et al. (2019). A standardised framework to identify optimal animal models for efficacy assessment in drug development. *PLoS One, 14*, e0218014.
322. van de Waterbeemd, H., & Gifford, E. (2003). ADMET in silico modelling: Towards prediction paradise? *Nature Reviews. Drug Discovery, 2*, 192–204.

Chapter 13
Challenges Faced in the Development of Computational Methods for Predicting Pharmacokinetics Behavior

José Eduardo Gonçalves

Abstract The pharmacokinetic behavior of a drug is determined by the intricate interplay between the physicochemical properties of the molecule and its multifaceted interactions with the biological system from the moment of administration until its elimination from the body. Consequently, the determination and evaluation of pharmacokinetic events constitute a complex process, which inevitably results in an equal or greater complexity in predictive studies. In light of this, it becomes evident that predictive methods employing computational models will present challenges that must be addressed to enable their broad development and application. To this end, it is essential to acquire detailed knowledge of the stages involved in the model creation process and identify the critical points requiring attention to minimize potential failures or low predictive power of computational methods. In this chapter, this approach will be employed, bringing forth experiences available in the literature on the process of creating computational models for predicting pharmacokinetic behavior at the early stages of drug development, highlighting the main challenges commonly encountered, and the strategies typically employed to mitigate prediction issues faced by different research groups in this field.

Keywords In silico PK models · ADME prediction · Nonclinical studies

1 Introduction

The development of a new drug consists of a process in which various pieces of information regarding the efficacy and safety of the molecule in the biological system must be obtained. This mainly includes those related to the effect when acting on a specific target (pharmacodynamics), those related to how the molecule is disposed within the organism (pharmacokinetics), and the potential undesirable

J. E. Gonçalves (✉)
Produtos Farmacêuticos, Faculdade de Farmácia da Universidade Federal de Minas Gerais, Belo Horizonte, Minas Gerais, Brazil
e-mail: jegoncalves@ufmg.br

V. G. Maltarollo (ed.), *Computer-Aided and Machine Learning-Driven Drug Design*, Computer-Aided Drug Discovery and Design 3,
https://doi.org/10.1007/978-3-031-76718-0_13

effects of this xenobiotic (toxicology). Such information is acquired in various successive stages, starting from the initial phases of development, the so-called nonclinical studies, and the clinical phase, which involves evaluation in human subjects [1]. Obtaining early insights into the efficacy, potential toxicity, and pharmacokinetics behavior of candidate drug molecules has been observed to reduce failures in subsequent stages [2].

Pharmacokinetic behavior is the result of multiple interactions between the drug and the biological system, influenced by multifactorial elements that drive the processes of absorption, distribution, metabolism, and excretion [1, 2]. In view of this, one must bear in mind the complexity involved in evaluating and determining the pharmacokinetics characteristics that lead to the disposition of the drug in the body.

In an effort to generate knowledge about the pharmacokinetic behavior of a drug from the early stages of development, several in vitro and in vivo studies are conducted in nonclinical stage, as apparent permeability studies, animal absolute oral bioavailability, tissue drug distribution, plasma protein binding, enzymatic metabolism, blood-brain barrier permeability, among other methods.

All these studies are subject to limitations and pose a considerable challenge in the development of new drugs. A critical obstacle involves the costs associated with conducting extensive studies on numerous promising molecules. Only after rigorous evaluation of pharmacokinetic data, desired therapeutic effects, and toxicological profiles can a compound be considered for clinical trials. In this context, in silico methods emerge as an important complementary tool. These methods employ computational models and artificial intelligence to predict pharmacokinetics characteristics in the early stages of development. The objective of in silico methods is to facilitate the evaluation of a larger number of candidates through high-throughput screening. This approach minimizes the probability of failures by discarding unpromising molecules and guiding potential molecular modifications to enhance desirable pharmacokinetic characteristics [4].

Several generations of in silico models have emerged as a tool in research and development, many of which are widely available through open-access platforms. The success of using computational methods for pharmacokinetic prediction has been demonstrated over the years. A study published in 2014 shows the effectiveness of reducing failures in the development of new drugs by employing in silico methods alongside other tools in early-stage evaluations. This study observed a gradual and significant decrease in clinical study discontinuation rates, from around 40% in the 1990s to approximately 1% by 2007 [5]. These findings suggest the positive impact that computational methods have on the pharmaceutical industry.

However, it is crucial to emphasize that in silico predictions neither replace nor disqualify experimental testing. On the contrary, their goal is to complement experimental results and, together with other relevant scientific evidence, provide support for decision-making [5]. Nevertheless, there are several challenges in modeling pharmacokinetic properties using computational methods, and this chapter will discuss the key aspects of this topic.

2 Main Physicochemical and Pharmacokinetic Characteristics Used in Developing Computational Methods

As previously mentioned, the bioavailability encompassing absorption, distribution, metabolism, and elimination of a drug, which constitute pharmacokinetic processes, result from the interaction between the drug molecule and various biological structures within the body. These interactions are influenced even prior to their occurrence by the biopharmaceutical properties of the drug and the formulation through which it is administered. The factors influencing this pharmacokinetics behavior, which are used to develop predictive computational models, can generally be divided into three levels. These levels reflect the complexity and detail each factor contributes to the model [6].

At Level 1, fundamental biopharmaceutical properties, including physicochemical characteristics, are assessed. These properties encompass solubility, partition coefficient, hydrogen bond donors and acceptors, and adherence to established guidelines such as Lipinski's Rule of Five and Veber's criteria [2].

Level 2 incorporates information related to pharmacokinetic characteristics. This level encompasses parameters indicative of absorption, distribution, metabolism, and excretion processes.

At Level 3, models based on physiological, biochemical, and anatomical data are utilized, allowing for the prediction of how a drug will behave under different biological conditions. These methods are identified as physiologically based pharmacokinetic methods (PBPK).

Table 13.1 presents the main components commonly employed in the establishment of computational models. These components are classified into three levels, along with examples of developed platforms available for predicting these pharmacokinetic-related characteristics.

3 Construction of an In Silico Model to Predict ADME Properties

The construction of a model, also known as the computational modeling process, for predicting pharmacokinetics, involves several steps. These range from the collection of data used to feed the model to the validation of the model itself. These steps are outlined below.

Table 13.1 Physicochemical, biological properties, and parameters used in the establishment of in silico methods with examples of the main platforms available for predictions

Properties	Description	Examples of available software
Level 1		
Water Solubility (Log S_{aq})	Ability of a compound to dissolve in an aqueous solvent, important for absorption	ChemAxon [7] ACD/ Percepta (ACD Labs) [8] SwissADME [9] EPI SUITE™ [10]
Log P – PgK_{ow}/ Log D	Octanol/water Partition Coefficient. Indicative of the ability of a molecule to cross cell membranes	ADMETlab [11] admetSAR [12] SwissADME [9] EPI SUITE™ [10]
H_{don}/H_{acc}	Ability to donate or receive hydrogen bonds, an important component of Lipinski's rule determining the potential to be absorbed orally	admetSAR [12] Corina Symphony [13]
TPSA	Topological polar surface area indicates the polarity of a molecule, the higher the value of this parameter, the lower the permeability through cell membranes	Molinspiration [14] Corina Symphony [13] SwissADME [9]
Level 2		
Violation of Lipinski's rule	Violation of 2 or more of the characteristics (Log $P > 5$; >5 H bond; >10 H bond acceptors; M wt > 500 Da) associated with low oral absorption	SwissADME [9]
Veber's rule for good oral bioavailability	a drug may be achieved if it possesses 10 or fewer rotatable bonds (RTB) and a polar surface area (TPSA) that does not exceed 140 Å^2	ADMETlab [11]
Effective permeability in humans (Peff)	Determines the ability of a molecule to permeate through the human intestine	SwissADME [9] Dallphin-AtoM [15]
Apparent permeability (Papp)	Indicates the ability of a compound to cross membranes in in vitro or animal models	SwissADME [9]
F%—Absolute oral bioavailability	Fraction effectively absorbed orally	SwissADME [9] Dallphin-AtoM [15]
PPB	Plasma Protein binding, an important determinant of distribution	ADMETlab [11] admetSAR [12] SwissADME [9]
V_d	Volume of distribution, indicative of the molecule's ability to migrate from the blood to different peripheral tissues	ADMETlab [11] admetSAR [12] SwissADME [9]
P-gp substrate, inducer or inhibitor	The potential to be a substrate, inhibitor, or inducer of P-glycoprotein (P-gp), a key efflux transporter that significantly impacts bioavailability	ADMETlab [11] admetSAR [12] SwissADME [9] Simcyp [16]

(continued)

Table 13.1 (continued)

Properties	Description	Examples of available software
Blood-brain barrier (BBB) partitioning	Blood-brain barrier partitioning refers to the ability of a substance to cross the blood-brain barrier	ADMETlab [11] admetSAR [12] SwissADME [9] Simcyp [16]
CYP substrate or inhibitor	Predicts the predominant metabolic pathways and potential sites of metabolism in the liver Interactions with CYP1A2, CYP3A4, CYP2C9, CYP2C19, and CYP2D6 most commonly predicted as these enzymes are responsible for the metabolism of the majority of drugs. Particularly relevant for predicting drug–drug interactions	ADMETlab [11] admetSAR [12] SwissADME [9] Simcyp [16]
Clearance	The volume of blood effectively cleared of a molecule per unit time (per unit body weight); total clearance comprises contributions from renal excretion (Clr), hepatic clearance (Clh), and clearance via other routes such as metabolism, sweat, secretion into breast milk, and exhalation	ADMETlab [11] admetSAR [12] SwissADME [9] Simcyp [16]
$T_{½}$	Estimates the time needed for half of the drug to be eliminated from the plasma	pkCSM [17]
C_{max}	Maximum concentration attained in blood, tissue, or an organ	PK-SIM® [18] PLETHEM [19] GastroPlus [20]
T_{max}	Time to reach the maximum concentration in blood or tissue/organ	PK-SIM® [18] PLETHEM [19] GastroPlus [20]
AUC	Area under the concentration–time curve (AUC), indicating overall internal exposure to the drug, in blood or a specific tissue/organ of interest	PK-SIM® [18] PLETHEM [19] GastroPlus [20]
Level 3		
Target tissue exposure	The ability of a drug to reach different tissues	PK-SIM® [18] GastroPlus [20]
Physiological absorption model	Absolute bioavailability and absorption in humans, animal models, and in vitro	GastroPlus [20]
Enzyme inhibitor binding process	Competitive and mechanism-based inhibition	Simcyp [16]

Adapted from Madden and Thompson [6]

3.1 Data Collection

This step is considered pivotal as it enables a comprehensive understanding of the system's behavior under investigation. The data that will inform the model can be obtained through various methods. Primarily, it is sourced from experimental data, gathered from scientific literature, chemical properties databases, and nonclinical studies. This dataset may include information on solubility, topology, ionization constant (pKa) permeability, metabolism, protein binding, and other relevant properties. This diverse range of data is essential for accurately modeling and predicting the pharmacokinetic and pharmacodynamic behaviors of compounds, thereby enhancing the reliability and robustness of the in silico models [22].

Public databases such as PubChem [23], ChEMBL [24], and DrugBank [25] (please, see Chap. 2) are invaluable resources, providing comprehensive information on the chemical properties, biological activities, and pharmacokinetic data of various compounds [21]. These databases serve as crucial repositories of data that support computational modeling and drug discovery processes.

In specific scenarios, it becomes necessary to generate experimental data through in vitro and in vivo assays. This approach allows for the incorporation of empirical data into models, enabling a more accurate characterization of the pharmacokinetics of particular compounds. By integrating data from both in vitro assays and in vivo studies, researchers can enhance the predictive power and reliability of pharmacokinetic models, ultimately improving the drug development process [4].

3.2 Data Preprocessing

Given the available data, it is often necessary to undertake a process known as data cleaning. This involves rigorously analyzing the dataset to identify and eliminate inconsistencies, incorrect entries, missing values, and outliers [6]. During this stage, variable coding is also performed, wherein categorical variables are transformed into numerical formats. This process includes standardizing formats, magnitudes, and units to ensure consistency across the dataset. This step is crucial for ensuring that the data are suitable for subsequent modeling and analysis, thereby enhancing the accuracy and reliability of the computational models.

3.3 Identification of Relevant Variables

During this phase, the relevant variables for the conceptualized model are identified. Using statistical techniques or machine learning algorithms, the most critical physicochemical properties for predicting pharmacokinetics are determined. In certain cases, dimensionality reduction may be necessary. This involves applying methods such as principal component analysis (PCA) to address multicollinearity and improve the model's efficiency [6].

3.4 Model Choice

The next step involves selecting the most suitable machine learning algorithm. The choice of model depends on the specific context, the size and nature of the data, and the research objectives [6]. Often, a trial-and-error approach, combined with cross-validation, is used to determine which model best fits the data [22]. Each model has its own set of advantages and limitations, and the selection should be guided by the specific requirements of the pharmacokinetic prediction problem. Commonly employed models include:

(a) Linear regression: This is a straightforward and interpretable method that presumes a linear relationship between the independent variables and the response [26]. It proves to be useful when a clear linear relationship exists between the characteristics and pharmacokinetic parameters.
(b) Support vector machines (SVM): SVM proves to be effective in both classification and regression problems. It is capable of handling complex and nonlinear datasets, making it useful when the relationships between variables are not strictly linear [27].
(c) Artificial neural networks (ANN): Neural networks are potent models capable of learning intricate patterns in data. They are particularly beneficial in scenarios where the relationship between characteristics and outcomes is nonlinear and highly complex [28].
(d) Decision Trees: These are easy-to-interpret models that segregate data into distinct sets based on decision rules. They are applicable in both classification and regression problems [29].
(e) Random forest: An enhancement of decision trees, random forest constructs multiple trees, and amalgamates their results to bolster accuracy and mitigate overfitting [26].
(f) Gradient boosting: Boosting methods like gradient boosting generate a sequence of weak models that are combined to form a more robust model. They are effective in enhancing model accuracy [30].
(g) Bayesian models: Bayesian models integrate uncertainties in modeling, proving useful when it is crucial to consider uncertainty in model parameters [31].
(h) K-nearest neighbors (KNN): KNN is an instance-based learning model that predicts the class or value of a data point based on the classes or values of its nearest neighbors [32].
(i) Nonlinear regression models: Specific nonlinear regression models, such as polynomial regression, can be beneficial when the relationship between variables is more intricate than simple linearity [33].
(j) Physiologically based pharmacokinetics models (PBPK): This model is constructed based on a vast number of drug physicochemical and absorption, distribution, metabolism, and elimination (ADME) attributes. These include lipophilicity, solubility, pKa, molecular weight, and plasma unbound fraction, along with physiological parameters like blood flow, tissue volume, vessel surface area, transporters, and enzyme expression level [34].

3.5 Model Training

The model training phase involves using the chosen algorithm on the training data, allowing the model to discern and internalize patterns and relationships within these data [35]. Prior to commencing model training, the data are usually partitioned into three distinct subsets: training, validation, and testing. The proportions of these subsets can vary based on the dataset's size, but a typical split is 50% for training, 30% for validation, and 20% for testing [36].

During the training phase, the algorithm is provided with data from the training set. The model adjusts its parameters based on these data, aiming to minimize the discrepancy between the model's predictions and the actual values in the training set. This process allows the model to learn specific patterns and relationships present in the data [21].

Parameter Tuning: As the model undergoes training, certain algorithms have parameters that can be fine-tuned to optimize performance. Two common methods for analyzing the results obtained by the model are residual analysis and goodness-of-fit [35].

In residual analysis, the difference between the values predicted by the model and those observed in experimental studies is calculated. An adequate model should exhibit an average residual value close to zero and should not show correlated residuals when plotting the dispersion diagram with predicted values on the ordinate and residuals on the abscissa, where no discernible trend in the points should be observed [35].

Goodness of fit evaluates the correlation between predicted and observed values using metrics such as the coefficient of determination (R^2), correlation coefficient (r), or coefficient of agreement. Parameter tuning is often performed using the validation set to avoid overfitting, ensuring that decisions are not based solely on performance on the training data.

Training is considered complete when the model attains acceptable performance on the validation set. The finalized model is then evaluated on the test set, providing a definitive measure of its ability to generalize unseen data [21, 22, 35].

3.6 Model Assessment

Once the model is prepared, it will be employed to evaluate a sample or a set of test samples to predict the desired information. Using these results, performance metrics such as mean absolute error, mean squared error, and the coefficient of determination (R^2) can be obtained to assess the model's accuracy [35].

Upon analyzing results derived from external validation, it may be necessary to optimize the model by refining selected features, adjusting hyperparameters, or opting for a different algorithm [35].

3.7 External Validation

External validation of a pharmacokinetic computational prediction method involves evaluating the model's performance on independent datasets that were not used during model training [21, 22]. The objective is to confirm that the model can effectively generalize to new, unseen data, providing accurate predictions in various contexts beyond those used for model development. This validation is essential to ensure the robustness and applicability of the method in real-world scenarios.

3.8 Implementation and Availability

The modeling process concludes with comprehensive documentation that provides a detailed description of all stages. This includes the criteria for data collection and the conditions under which the model should be used [35]. Typically, research groups submit this documentation for publication in compendia or scientific journals.

Following the publication of the model, the next step is to integrate it with drug development tools. This integration is accomplished using platforms and software, whether open-source or commercial [35].

3.9 Continuous Update

As the volume of information about a specific parameter increases, it becomes essential to update the model to improve its predictive accuracy and robustness. To achieve this, it is necessary to feed the model with the new data and restart all the previously outlined steps [35].

In a generalized and simplified manner, these are the steps involved in constructing a computational model for predicting pharmacokinetics. Each step is crucial for achieving the desired predictive power. Successful execution of the project relies on the collaboration between IT specialists and professionals from the pharmaceutical, chemistry, and biology fields. For more detailed information on the choice, as well as the implementation and subsequent evaluation of the model, it is recommended to consult Chap. 4.

4 Challenges in Achieving Computational Models with Enhanced Predictive Capability

Numerous challenges are encountered when developing a computational model to predict the pharmacokinetic behavior of candidate drug molecules. These challenges can be associated with each of the stages of pharmacokinetics processes and modeling development.

The initial challenge pertains to the quality of pharmacokinetic data used in constructing the model. Much of the data utilized is available in public databases, from companies, or published in scientific journal articles. Presently, there is a significant surge in the availability of databases that can assist in predicting ADMET, such as the ADME database, SuperToxic, PKKB, and DSSTox [37–40]. However, the vast amount of information contained in these databases does not guarantee the necessary quality to enable the models to exhibit the desired accuracy.

Another potential approach involves generating information through in-house experimentation. However, it is not always possible to have greater control over the quality of the data to be generated to feed the model. A limitation of this initiative is the use of a reduced number and variety of molecules to determine the parameters that will feed the model [21, 41]. Attempting to conduct a greater number of experiments to obtain in vivo data can be complex and costly.

Often, in the context of nonclinical studies, predictive models are also utilized, which can result in computational methods with less predictive power than those generated with data from in vivo studies. The literature presents discussions related to the quality of data obtained experimentally in different laboratories. Such inconsistencies may be linked to errors in result acquisition, processing, and manipulation, as well as the use of inappropriate experimental methods [21] which, in certain circumstances, is associated with the difficulty due to the inexperience of the programmer or modeler in critically analyzing the experimental data obtained from databases or literature.

Among the in vitro models typically used for computational modeling, methods for evaluating intestinal permeability stand out. These methods use synthetic membrane models, such as PAMPA (Parallel Artificial Membrane Permeability), and cellular permeability studies, such as Caco-2 cell monolayers [42, 43]. Other studies include permeability studies across the blood-brain barrier, metabolism studies in hepatic microsomal systems, and plasma protein binding studies, among others [44–46]. The heterogeneity of results obtained by different laboratories for these in-house analyses significantly complicates the comparability of results, as well as subsequent analyses and predictions.

Recently, a measure has been adopted to address challenges related to data acquisition by aggregating sources from fields such as biology, chemistry, pharmacology, and clinical trials to create "big data" sets for medicine research and development. However, this strategy still faces significant obstacles, including

Table 13.2 Challenges and strategies to mitigate in silico methods applied on pharmacokinetics predictions

Challenges	Strategies
Reduced programming skills for building model structures	Multidisciplinary training create qualified and user-friendly platforms for Physiology-Based Physiological Modeling (PBPK)
The disparity between existing models and physicochemical and physiological processes	Collect or estimate a more comprehensive range of physiological data and integrate a greater variety of physiological processes and tools to build more mechanistic models
Difficulties and limitations in obtaining experimental data	Improve the design of *in vitro* experiments, carry out preliminary *in vivo* studies and establish connections between them
Restriction in predictive tools to estimate desired parameters due to the relative scarcity of data from available in vivo studies	Create additional, user-friendly and qualified *in silico* tools for property prediction, and incorporate data from these models
Challenges in sampling in studies, covering intra- and inter-subject variability, as well as physiological differences between species, normal individuals and special populations	Perform more refined pharmacokinetic studies, adopt conservative conclusions, and consider model simplification or exploration in animal models

Adapted from Wang and Ouyang [51]

missing data, dimensional inaccuracies, and bias control challenges, which add complexity to big data analysis [48].

It is usually possible to obtain information in the literature about molecules that have been promising, have advanced in their evaluation or have become drugs. On the other hand, data on molecules that did not prove to be promising is not publicly accessible, constituting a private database for pharmaceutical companies [49]. This fact can harm the robustness of a computational method because it is only fed with a universe of results from molecules with selected characteristics. Therefore, to resolve these difficulties, the development of security policies and data and information sharing are essential for building robust big data that takes into account greater expected variability. In this sense, organizations have worked to promote guidance by proposing recommendations for formatting standardized data that can be interchangeable [50].

An added challenge in establishing an in silico method is the necessity of selecting the most suitable machine learning algorithm. This choice calls for an enhanced understanding of the model to be implemented, necessitating the developer to possess profound knowledge of machine learning and deep learning models, which are crucial for predicting pharmacokinetic properties [51].

The ceaseless evolution and escalating complexity of AI-based models require researchers to swiftly comprehend new techniques capable of accurately predicting outcomes with diverse and large datasets. Table 13.2 presents the main challenges and possible ways to minimize their occurrence when developing an in silico method for predicting pharmacokinetics.

5 Conclusions and Perspectives

The development of a new drug or medicine requires decision-making, which often entails discontinuing studies with a particular molecule or investing time and resources to progress through successive stages until an effective and safe medicine is achieved. These decisions frequently involve substantial uncertainty, posing a significant challenge. The absence of adequately predictive methods for pharmacokinetic characterization, target validation, and the identification and optimization of therapeutic candidates is currently viewed as the primary technical bottleneck in drug discovery [52].

Despite the advancements made, the evaluation of the predictive accuracy of existing in silico models continues to face considerable challenges. Numerous comparative studies have been conducted, as evidenced by the statistical data available in the scientific literature. However, the disparity in datasets used in individual research represents a recurring concern, and efforts to minimize it should be increasingly encouraged through the adoption of Good Laboratory Practices [53]. This disparity, as presented in this chapter and in various literatures, is attributed to the diverse origin of the data, which includes information generated internally, data published in the literature, and information extracted from public datasets [54]. This difficulty could be minimized by incorporating a broader range of data sources, including clinical trials, electronic health records, and real-world evidence [55].

Future research is expected to request time and effort from multidisciplinary researchers to standardize sets of tests and evaluation criteria. This will facilitate more robust comparisons between in silico models, significantly contributing to consistent advancements in the predictability of ADME profiles and promoting the reliability and applicability of these computational approaches in pharmaceutical practice.

References

1. Storpirtis, S., Gai, M. N., Campos, D. R., & Gonçalves, J. E. (2011). Chapter 1: Farmacocinética: Conceitos, Definições e Relação com a Farmacodinâmica e a Biofarmácia (Biofarmacotécnica). In S. Storpirtis, M. N. Gai, D. R. Campos, & J. E. Gonçalves (Eds.), *Farmacocinética Básica e Aplicada*. Guanabara Koogan.
2. Pantaleão, S. Q., Fernandes, P. O., Gonçalves, J. E., Maltarollo, V. G., & Honorio, K. M. (2022). Recent advances in the prediction of pharmacokinetics properties in drug design studies: A review. *ChemMedChem, 17*, e202100542. PMID: 34655454.
3. Takebe, T., Imai, R., & Ono, S. (2018). The current status of drug discovery and development as originated in United States academia: The influence of industrial and academic collaboration on drug discovery and development. *Clinical and Translational Science, 11*(6), 597–606. Epub 2018 Jul 30. PMID: 29940695; PMCID: PMC6226120.

4. Komura, H., Watanabe, R., & Mizuguchi, K. (2023). The trends and future prospective of in silico models from the viewpoint of ADME evaluation in drug discovery. *Pharmaceutics, 15*, 2619.
5. Khanna, I. (2012). Drug discovery in pharmaceutical industry: Productivity challenges and trends. *Drug Discovery Today, 17*(19–20), 1088–1102. Epub 2012 May 22. PMID: 22627006.
6. Madden, J. C., & Thompson, C. V. (2022). Chapter 3: Pharmacokinetic tools and applications. In E. Benfenati (Ed.), *In silico methods for predicting drug toxicity* (Methods in molecular biology 2425). Humana Press, Springer Protocols.
7. https://chemaxon.com/calculators-and-predictors
8. https://www.acdlabs.com/products/percepta/
9. http://www.swissadme.ch/
10. https://www.epa.gov/tsca-screeningtools/epi-suitetmestimationprogram-interface
11. http://admet.scbdd.com/calcpre/index/
12. http://lmmd.ecust.edu.cn/admetsar2/
13. https://www.mn-am.com/
14. http://www.molinspiration.com/
15. http://pipet.or.kr/board/apps_list.asp
16. https://www.certara.com/software/simcyp-pbpk/
17. http://biosig.unimelb.edu.au/pkcsm/
18. https://github.com/Open-Systems-Pharmacology/PK-Sim.wiki.git
19. https://scitovation.com/plethem/
20. https://www.simulations-plus.com/software/gastroplus/
21. Tran, T. T. V., Tayara, H., & Chong, K. T. (2023). Recent studies of artificial intelligence on in silico drug distribution prediction. *International Journal of Molecular Sciences, 24*, 1815.
22. Grime Barton, P., & McGinnity, D. F. (2013). Application of in silico, in vitro and preclinical pharmacokinetic data for the effective and efficient prediction of human pharmacokinetics. *Molecular Pharmaceutics, 10*(4), 1191–1206.
23. Wishart, D. S., Feunang, Y. D., Guo, A. C., Lo, E. J., Marcu, A., Grant, J. R., et al. (2018). DrugBank 5.0: A major update to the drugbank database for 2018. *Nucleic Acids Research, 46*, D1074–D1082.
24. Kim, S., Chen, J., Cheng, T., Gindulyte, A., He, J., He, S., et al. (2019). PubChem 2019 update: Improved access to chemical data. *Nucleic Acids Research, 47*, D1102–D1109.
25. Gaulton, A., Hersey, A., Nowotka, M., Bento, A. P., Chambers, J., Mendez, D., et al. (2017). The ChEMBL database in 2017. *Nucleic Acids Research, 45*, D945–D954.
26. Lombardo, F., & Jing, Y. (2016). In silico prediction of volume of distribution in humans. Extensive data set and the exploration of linear and nonlinear methods coupled with molecular interaction fields descriptors. *Journal of Chemical Information and Modeling, 56*(10), 2042–2052. Epub 2016 Sep 28. PMID: 27602694.
27. Xue, Y., Yap, C. W., Sun, L. Z., Cao, Z. W., Wang, J. F., & Chen, Y. Z. (2004). Prediction of P-glycoprotein substrates by a support vector machine approach. *Journal of Chemical Information and Computer Sciences, 44*(4), 1497–1505.
28. Keutzer, L., You, H., Farnoud, A., Nyberg, J., Wicha, S. G., Maher-Edwards, G., Vlasakakis, G., Moghaddam, G. K., Svensson, E. M., Menden, M. P., Simonsson, U. S. H., & On Behalf of The UNITE4TB Consortium. (2022). Machine learning and pharmacometrics for prediction of pharmacokinetic data: Differences, similarities and challenges illustrated with rifampicin. *Pharmaceutics, 14*(8), 1530. PMID: 35893785; PMCID: PMC9330804.
29. Yun, Y. E., Cotton, C. A., & Edginton, A. N. (2014). Development of a decision tree to classify the most accurate tissue-specific tissue to plasma partition coefficient algorithm for a given compound. *Journal of Pharmacokinetics and Pharmacodynamics, 41*(1), 1–14. Epub 2013 Nov 21. PMID: 24258064.
30. Liu, H., Zhang, W., Nie, L., Ding, X., Luo, J., & Zou, L. (2019). Predicting effective drug combinations using gradient tree boosting based on features extracted from drug-protein heterogeneous network. *BMC Bioinformatics, 20*(1), 645.

31. Pelletier, D. J., Gehlhaar, D., Tilloy-Ellul, A., Johnson, T. O., & Greene, N. (2007). Evaluation of a published in silico model and construction of a novel Bayesian model for predicting phospholipidosis inducing potential. *Journal of Chemical Information and Modeling, 47*(3), 1196–1205.
32. Zhu, X. W., Sedykh, A., Zhu, H., et al. (2013). The use of pseudo-equilibrium constant affords improved QSAR models of human plasma protein binding. *Pharmaceutical Research, 30*, 1790–1798.
33. Berellini, G., Springer, C., Waters, N. J., & Lombardo, F. (2009). In silico prediction of volume of distribution in human using linear and nonlinear models on a 669 compound data set. *Journal of Medicinal Chemistry, 52*(14), 4488–4495.
34. Jones, H. M., Dickins, M., Youdim, K., Gosset, J. R., Attkins, N. J., Hay, T. L., Gurrell, I. K., Logan, Y. R., Bungay, P. J., Jones, B. C., & Gardner, I. B. (2012). Application of PBPK modelling in drug discovery and development at Pfizer. *Xenobiotica, 42*(1), 94–106.
35. Chiann, C., Rama, E. M., & Crsitofoletti, R. (2011). Técnicas Computacionais em Farmacocinética. In S. Storpirtis, M. N. Gai, D. R. Campos, & J. E. Gonçalves (Eds.), *Farmacocinética Básica e Aplicada*. Guanabara Koogan. cap. 21.
36. Chicco, D. (2017). Ten quick tips for machine learning in computational biology. *Biodata Mining, 10*, 35.
37. Shang, J., Sun, H., Liu, H., Chen, F., Tian, S., Pan, P., et al. (2017). Comparative analyses of structural features and scaffold diversity for purchasable compound libraries. *Journal of Cheminformatics, 9*, 25.
38. Schmidt, U., Struck, S., Gruening, B., Hossbach, J., Jaeger, I. S., Parol, R., et al. (2009). SuperToxic: A comprehensive database of toxic compounds. *Nucleic Acids Research, 37*, D295–D299.
39. Cao, D., Wang, J., Zhou, R., Li, Y., Yu, H., & Hou, T. (2012). ADMET evaluation in drug discovery. 11. Pharmacokinetics knowledge base (PKKB): A comprehensive database of pharmacokinetic and toxic properties for drugs. *Journal of Chemical Information and Modeling, 52*, 1132–1137.
40. Williams, A. J., Grulke, C. M., Edwards, J., McEachran, A. D., Mansouri, K., Baker, N. C., et al. (2017). The comptox chemistry dashboard: A community data resource for environmental chemistry. *Journal of Cheminformatics, 9*, 61.
41. Lombardo, F., Desai, P. V., Arimoto, R., Desino, K. E., Fischer, H., Keefer, C. E., Petersson, C., Winiwarter, S., & Broccatelli, F. (2017). In silico absorption, distribution, metabolism, excretion, and pharmacokinetics (ADME-PK): Utility and best practices. An industry perspective from the international consortium for innovation through quality in pharmaceutical development. *Journal of Medicinal Chemistry, 60*(22), 9097–9113.
42. Chi, C., Lee, M., Weng, C., & Leong, M. K. (2019). In silico prediction of PAMPA effective permeability using a two-QSAR approach. *International Journal of Molecular Sciences, 20*, 3170.
43. Cai, X., Patel, S., Huang, C., Paiva, A., Sun, Y., Barker, G., Weller, H., & Shou, W. (2022). Comprehensive characterization and optimization of Caco-2 cells enabled the development of a miniaturized 96-well permeability assay. *Xenobiotica, 52*, 742–750.
44. Stéen, E. J. L., Vugts, D. J., & Windhorst, A. D. (2022). The application of in silico methods for prediction of blood-brain barrier permeability of small molecule PET tracers. *Frontiers in Nuclear Medicine, 2*, 853475.
45. Sasahara, K., Shibata, M., Sasabe, H., Suzuki, T., Takeuchi, K., Umehara, K., & Kashiyama, E. (2021). Predicting drug metabolism and pharmacokinetics features of in-house compounds by a hybrid machine-learning model. *Drug Metabolism and Pharmacokinetics, 39*, 100395.
46. Votano, J. R., Parham, M., Hall, L. M., Hall, L. H., Kier, L. B., Oloff, S., & Tropsha, A. (2006). QSAR modeling of human serum protein binding with several modeling techniques utilizing structure-information representation. *Journal of Medicinal Chemistry, 49*, 7169–7181.

47. Danishuddin, Kumar, V., Faheem, M., & Lee, K. W. (2022). A decade of machine learning-based predictive models for human pharmacokinetics: Advances and challenges. *Drug Discovery Today, 27*(2), 529–537. ISSN 1359-6446.
48. Lee, C. H., & Yoon, H.-J. (2017). Medical big data: Promise and challenges. *Kidney Research and Clinical Practice, 36*, 3.
49. Schneider, P., Walters, W. P., Plowright, A. T., Sieroka, N., Listgarten, J., Goodnow, R. A., Fisher, J., Jansen, J. M., Duca, J. S., & Rush, T. S. (2020). Rethinking drug design in the artificial intelligence era. *Nature Reviews. Drug Discovery, 19*, 353–364.
50. ISO 20691:2022 - Biotechnology — Requirements for data formatting and description in the life sciences. https://www.iso.org/standard/68848.html
51. Wang, W., & Ouyang, D. (2022). Opportunities and challenges of physiologically based pharmacokinetic modeling in drug delivery. *Drug Discovery Today, 27*(8), 2100–2120. ISSN 1359-6446.
52. Scannell, J. W., Bosley, J., Hickman, J. A., et al. (2022). Predictive validity in drug discovery: What it is, why it matters and how to improve it. *Nature Reviews. Drug Discovery, 21*, 915–931.
53. Hooijmans, C. R., de Vries, R., Leenaars, M., Curfs, J., & Ritskes-Hoitinga, M. (2011). Improving planning, design, reporting and scientific quality of animal experiments by using the Gold Standard Publication Checklist, in addition to the ARRIVE guidelines. *British Journal of Pharmacology, 162*(6), 1259–1260.
54. Fagerholm, U., Hellberg, S., Alvarsson, J., & Spjuth, O. (2023). In silico prediction of human clinical pharmacokinetics with ANDROMEDA by Prosilico: Predictions for an established benchmarking data set, a modern small drug data set, and a comparison with laboratory methods. *Alternatives to Laboratory Animals, 51*(1), 39–54.
55. Sadybekov, A. V., & Katritch, V. (2023). Computational approaches streamlining drug discovery. *Nature, 616*, 673–685.

Chapter 14
Exploring the Significance of Experimental and Computational Methods in Protein Structure Determination

Adolfo Henrique Moraes, Diego Magno Martins, and Marcelo Andrade Chagas

Abstract This chapter describes X-ray crystallography, nuclear magnetic resonance (NMR) spectroscopy, and cryo-electron microscopy (cryo-EM), the most used experimental techniques to determine protein structure. X-ray crystallography is highlighted as a powerful technique for resolving high-resolution structures of crystallized proteins, while NMR spectroscopy offers insights into protein dynamics and structures in solution. Cryo-EM, on the other hand, is emphasized for its ability to characterize large protein complexes and membrane proteins without the need for crystallization, providing near-atomic resolution. Meanwhile, computational methods such as homology modeling are discussed as a method that leverages known structures of related proteins to predict the structure of a target protein. The chapter also highlights the transformative impact of AI, particularly AlphaFold and RoseTTAFold, which has achieved unprecedented accuracy in predicting protein structures from amino acid sequences. Additionally, molecular dynamics simulations are described as powerful tools for studying protein flexibility and conformational changes over time, providing dynamic insights that complement static structural predictions. These experimental and computational approaches have been advancing our understanding of protein structure and function.

Keywords X-ray crystallography · Nuclear magnetic resonance · Cryo-EM · Homology modeling · AlphaFold · Molecular dynamics · Protein structure

---

A. H. Moraes (✉) · D. M. Martins
Departamento de Química, Universidade Federal de Minas Gerais, Belo Horizonte, Minas Gerais, Brazil
e-mail: adolfohmoraes@ufmg.br

M. A. Chagas
Departamento de Química, Universidade Federal de Minas Gerais, Belo Horizonte, Minas Gerais, Brazil

Departamento de Ciências Exatas, Universidade do Estado de Minas Gerais (UEMG), João Monlevade, Minas Gerais, Brazil

V. G. Maltarollo (ed.), *Computer-Aided and Machine Learning-Driven Drug Design*, Computer-Aided Drug Discovery and Design 3,
https://doi.org/10.1007/978-3-031-76718-0_14

1 Experimental Approaches to Obtain Protein Structure

Determining protein structure was one of the challenges of twentieth-century science [1]. The development of experimental methodologies enabling the determination of protein structures at atomic or near-atomic resolution has been a major contributor for understanding different protein properties and functions, such as protein molecular recognition, enzyme catalysis, and others. The first technique developed was X-ray crystallography, and the first protein crystal structure was published in 1958, when John Kendrew and Max Perutz determined the structure of myoglobin [2]. In 1969, Max Perutz presented the hemoglobin crystal structure with 2.8 Å of resolution [3]. The development of X-ray crystallography and the many methodologies related to protein crystallization and structure resolution provided a groundbreaking contribution to our understanding of protein function and structure determination. During the 1970s and 1980s, another technique emerged to study protein structure: nuclear magnetic resonance, NMR. Initially developed after World War II, NMR spectroscopy played a significant role in protein structure characterization after advancements in Fourier transformation, pulsed NMR, and the 2D NMR [4]. In 1985, the first solution structure of a protein was determined by NMR spectroscopy [5]. Since then, NMR spectroscopy has characterized several protein structures in solution. Besides the structure determination of proteins, NMR spectroscopy has been widely used to characterize protein dynamics at atomic resolution [6]. In the 2000s, another technique was developed and applied to determine supramolecular protein structures, cryoelectronic microscopy, Cryo-EM [7]. Cryo-EM allowed the structure characterization of large proteins and supramolecular structures, such as virus particles, amyloid fibers, and protein-protein complexes [8–10]. Another valuable protein structure characterization and popularization tool was the Protein Data Bank (PDB) organization in 1971 [11]. As of August 2024, the PDB hosts more than 200,000 protein structures. The PDB organization has played a pivotal role in standardizing protein structure representation and developing quality control methodologies. Over the past few decades, these methodologies have evolved, ensuring the reliability and accuracy of protein structures obtained using different experimental and computational strategies.

1.1 X-Ray Crystallography

X-ray crystallography was the first technique to obtain protein structure at the atomic scale. Its application expanded over the last decades, mainly because of the development of protein crystallization methodologies and new computational strategies to process experimental data and model the protein structures. These approaches led to automation in protein structure determination by X-ray crystallography [12]. X-ray crystallography is the technique responsible for most of the protein structure entries in the PDB. The simplified steps of protein structure determination are usually the following [12, 13]:

Sample Preparation: Protein purification is crucial for successful crystallization. The protein must be highly pure and stable in solution. Various purification techniques, such as chromatography, are employed. Once purified, the protein is concentrated to a suitable level for crystallization. Crystallization conditions are refined by screening different precipitants, buffers, and pH values. Crystals suitable for diffraction are grown by controlling temperature, pH, and concentration. Crystallization involves coaxing the protein molecules to form a highly ordered array, or crystal lattice, which can be subjected to X-ray diffraction.

X-ray Diffraction: X-rays are directed at the crystal and are diffracted by the electron clouds of the atoms within the crystal lattice. The resulting diffraction pattern is captured on a detector. The protein crystal is slightly rotated 360° along one axis during the acquisition to ensure the recording of a diffraction pattern for each position. X-rays generate protein crystal diffraction patterns because their wavelength ($100\ pm < \lambda < 10,000\ pm$) is comparable to the size and atomic constitution of proteins. The X-ray source most used today is synchrotron radiation.

Data Analysis: The diffraction pattern can be used to infer the positions of the atoms within the crystal. Complex mathematical algorithms based on Fourier transformation are used to analyze the diffraction pattern and reconstruct the protein's electron density map. The process of converting the reciprocal space representation of the crystal into an interpretable electron density map is known as phasing. In simple terms, each spot of the diffraction image is indexed, integrated, merged, and scaled. The position of each spot reflects the atomic constitution and three-dimensional structure of a protein.

Model Building: Using the electron density map as a guide, researchers built a model of the protein's atomic structure, placing atoms in positions that best fit the experimental data.

Refinement: The initial model is refined iteratively against the experimental data to improve its accuracy and reliability. This procedure is simplified if a sufficient resolution (less than 1.5 Å) is achieved. In this case, it is possible to automatically generate a model based on the electron density map, with correct bond angles and lengths. When the crystallography data are not of such high quality, molecular visualization software is used to fit the protein structure model to the electron density data. Validation: After the initial model is obtained, it undergoes rigorous validation. This includes assessing the quality of the electron density maps, checking for steric clashes, and evaluating geometric parameters such as bond lengths and angles. Additionally, the model is confirmed against the protein's known biochemical and biophysical properties. Validation tools such as MolProbity [14] and PROCHECK [15] are commonly used to assess the quality of protein structures obtained by X-ray crystallography.

1.2 Nuclear Magnetic Resonance

NMR spectroscopy was the second experimental technique used to determine protein structures. NMR spectroscopy's great advantage is its ability to provide protein structures in solution. This is especially important for proteins that are not easily crystallized, such as highly dynamic proteins or whose crystals do not have the quality needed to obtain the protein structures with atomic resolution. Unlike X-ray crystallography, protein structure determination by NMR spectroscopy is achieved by the simulation of the protein structure using minimization methodologies under experimental constraints [16]. These experimental constraints are obtained from different NMR experiments called NMR spectra. NMR spectroscopy is based on measuring the precession frequency of the magnetic moment of nuclei in the presence of a strong magnetic field. This frequency is called the Larmor frequency and is modulated by the chemical environment of each nucleus; therefore, it is a probe of protein structure, as it reflects the electronic environment surrounding the nucleus. Nuclei such as ^{1}H, ^{13}C, and ^{15}N have a spin number of 1/2, which makes them easily detected and analyzed by NMR spectroscopy.

Although it cannot be applied to large molecular systems (with molecular weight $>$ 100 kDa) [17], such as those studied by cryo-EM, NMR has emerged as the foremost experimental technique for investigating protein dynamics with atomic resolution [18]. NMR offers detailed structural insights through various parameters, including chemical shifts, residual dipolar couplings, chemical shift anisotropy, and paramagnetic relaxation enhancement. An overview of protein structure determination by NMR can be found in [19]. Additionally, rate constants can be determined by assessing magnetization exchange, saturation transfer, and relaxation dispersion [20].

The steps for protein structure calculation by NMR spectroscopy are usually the following:

Sample Preparation: For NMR studies, the protein must be dissolved in a suitable solvent buffer at a concentration typically ranging from ~0.1 to 4 mM. The solvent choice depends on factors such as protein stability and solubility. Isotopic labeling with ^{15}N or ^{13}C can enhance spectral resolution and help resonance assignment. The protein ^{13}C and ^{15}N-labeling is achieved by producing the protein using heterologous expression and by cultivating it in a unique medium, known as minimal medium, where the only source of the elements nitrogen and carbon are ^{15}N-ammonium chloride and ^{13}C-glucose, respectively [21–23]. Care must be taken to ensure the protein remains in its native state without aggregation or degradation during the NMR experiment. The quality and stability of the protein sample can be monitored by ^{1}H 1D and ^{1}H–^{15}N 2D NMR correlation spectra.

NMR Experiment: The sample is placed in a strong magnetic field, causing the nuclei of certain atoms (e.g., hydrogen, carbon, and nitrogen) to align with the magnetic field. Radiofrequency pulses are applied to perturb this alignment, and the resulting responses, or NMR signals, are detected. Each nucleus will exhibit a specific frequency value, which indicates a different chemical environment.

Data Collection: By using various pulse sequences and recording the frequencies and intensities of the NMR signals, multidimensional spectra are obtained, which contain information about the spatial relationships between atoms in the protein.

Spectral Analysis: NMR spectra are analyzed to extract parameters such as chemical shifts, coupling constants, and relaxation rates, which provide insights into the protein's structure, dynamics, and interactions [24, 25].

Structure Calculation: Computational methods, such as distance geometry, molecular dynamics (MD) simulations (see Chap. 8), or simulated annealing, are used to interpret the NMR data and generate a three-dimensional (3D) model of the protein structure that satisfies all experimental constraints. The most common experimental constraints are pair distance constraints obtained from nuclear Overhauser effect (NOE) spectroscopy (NOESY) [5] experiments and the dihedral angles (ϕ and ψ), which can be obtained from the ^{1}H, ^{15}N, and ^{13}C backbone chemical shifts and the scalar coupling constants [19, 26].

Validation: NMR-derived structures are confirmed using a variety of techniques. This includes assessing the precision and accuracy of the structure based on experimental data such as NOE (Nuclear Overhauser Effect) intensities, chemical shifts, and coupling constants. Structure validation tools like PROCHECK-NMR [27], MolProbity [14], PSVS [28], and CING [29] are used to analyze the quality of the NMR-derived structures, checking for stereochemical correctness and overall structural quality.

1.3 Cryo-EM

Cryo-electron microscopy (Cryo-EM) has enhanced our ability to visualize biological macromolecules at near-atomic resolution [30]. Traditional methods such as X-ray crystallography and NMR spectroscopy have long been the cornerstones of protein structure determination. However, Cryo-EM has overcome many of the limitations associated with these techniques for understanding the intricate architecture and function of biomolecules [31]. Historically, protein crystallization has been a significant bottleneck in X-ray crystallography, often requiring laborious optimization and sometimes proving unattainable for specific proteins [13]. In contrast, Cryo-EM enables the study of proteins in their native states without crystallization. This breakthrough has democratized structural biology, allowing researchers to tackle previously intractable targets and dynamic biological assemblies. Moreover, Cryo-EM has redefined the concept of resolution in structural biology. While early Cryo-EM structures were limited to low-resolution reconstructions (4–6 Å), recent advancements in detector technology [8, 32], computational algorithms [33], and sample preparation techniques [32, 34] have pushed the achievable resolution to higher levels Cryo-EM now rivals X-ray crystallography in resolution (<1.5 Å) [35, 36], offering insights into molecular structures with exquisite detail

[31, 37]. The impact of Cryo-EM extends beyond static structures; it provides a dynamic view of biological processes. By capturing snapshots of biomolecules in different conformations and functional states, Cryo-EM elucidates the mechanisms underlying fundamental cellular processes, such as membrane transport, protein synthesis, and signal transduction [38].

The cryo-EM technique involves capturing electron microscopy images of biomolecules embedded in a thin layer of vitreous ice, which are then used to generate precise 3D reconstructions. These reconstructions offer intricate structural models, shedding light on the functionalities of macromolecules and their involvement in biological processes [32]. Noteworthy applications of Cryo-EM include the elucidation of tau filaments [39] and amyloid fibrils [40], providing crucial insights into the mechanisms underlying Alzheimer's disease. Moreover, Cryo-EM has been used to determine the structure of the SARS-CoV-2 spike protein at a resolution of 3.5 Å [41]. Cryo-EM's impact on the study of membrane proteins is particularly striking, overcoming experimental limitations inherent in other techniques [42, 43]. Advances in cryo-electron microscopy enable the solving of high-resolution structures of large, >1 megadalton (*MDa*), and small, <100 kDa drug targets in near-native conditions, routinely reaching resolutions around or below 3 Å [44]. One example of these advances is the elucidation of the structures of native type A γ-aminobutyric acid receptors ($GABA_ARs$) assemblies and their interactions with FDA-approved drugs, such as those used to treat insomnia (zolpidem (ZOL) and flurazepam) and postpartum depression (the neurosteroid allopregnanolone (APG)) [45].

The steps of protein structure determination by Cryo-EM are usually the following [46]:

Sample Preparation: Sample preparation for cryo-EM involves applying a small volume (typically 3–5 μL) of the protein solution onto a holey carbon grid, which is then blotted to remove excess liquid. The grid is rapidly plunged into liquid ethane or propane, rapidly freezing the sample and forming a thin layer of vitreous ice. Grid preparation should be carefully performed to ensure that the protein particles are evenly distributed and not aggregated or absorbed into the grid surface.

Data Acquisition: The frozen sample is imaged using an electron microscope under cryogenic conditions. Electron micrographs are recorded at various tilt angles to capture multiple 2D views of the protein particles embedded in the ice.

Image Processing: Advanced image processing techniques, such as single-particle analysis or electron tomography, are used to align and combine the 2D images, correcting for imperfections in the microscope and variations in particle orientation. The result is a 3D reconstruction of the protein density.

Model Building: Atomic models of the protein are fitted into the density map using computational tools. This process involves manually or computationally positioning the protein coordinates within the density map and refining their positions to maximize agreement with the experimental data. AI can be applied in this step [33, 47].

Validation: Cryo-EM structures are confirmed through a combination of methods. This includes assessing the resolution of the reconstructed density map using criteria such as Fourier Shell Correlation (FSC) [48], which measures how two similar signals are by comparing them in the frequency domain, focusing on corresponding shells. It is widely used in microscopy, especially in structural biology, to validate results, determine resolution, and enhance signals [48]. The model is confirmed against the density map, ensuring that the atomic model fits well within the density and does not clash with neighboring molecules. Other validation measures include cross-validation against independent datasets, assessment of local map quality, and comparison with existing structural data or biochemical experiments related to the protein's function [9]. A good discussion on the steps for checking the quality of protein structures determined by Cryo-EM can be found in [49].

1.4 Hybrid Methods

Integrating multiple techniques allows researchers to cross-validate structural information from different experimental approaches, enhancing confidence in the resulting models. Moreover, hybrid methodologies enable the investigation of protein dynamics, interactions, and conformational changes that may be challenging to capture using a single technique alone. Other techniques can provide complementary information to X-ray crystallography, NMR spectroscopy, and Cryo-EM. Small-angle X-ray Scattering (SAXS) is a solution-based technique that gives information about macromolecules' overall shape, size, and conformation in solution [50]. SAXS is particularly useful for studying the overall shape and conformational changes in proteins and protein complexes in solution, including their flexibility and oligomeric states. SAXS data can be integrated with high-resolution structures obtained from X-ray crystallography, NMR spectroscopy, or Cryo-EM to generate pseudo-atomic models that combine the detailed local information from these techniques with the overall shape information from SAXS [51]; *Hydrogen-Deuterium Exchange Mass Spectrometry* (HDX-MS) provides information about the solvent accessibility and dynamics of protein structures by measuring the exchange of backbone amide hydrogen atoms with deuterium atoms in solution. HDX-MS is used to study protein folding, conformational dynamics, ligand binding, and protein–protein interactions [52]. HDX-MS data can be integrated with high-resolution structures obtained from X-ray crystallography, NMR spectroscopy, or Cryo-EM to provide insights into the dynamics and flexibility of proteins and protein complexes. This information can help refine structural models and elucidate dynamic aspects of protein function. *Chemical Cross-Linking Mass Spectrometry* (XL-MS) finds cross-linked residues within protein or protein complex structures induced by the covalent linkage of reactive chemical probes. XL-MS is used to study protein–protein interactions, protein structure, and conformational changes within protein complexes [53]. XL-MS data can be integrated with structural information from other techniques to identify interacting regions within protein complexes and confirm protein–protein interfaces. XL-MS can also provide distance constraints for modeling protein structures, especially in regions where other techniques may be less informative.

Additionally, computational methods can play an important role in hybrid approaches for protein structure determination by integrating and interpreting experimental data from multiple sources. These methods encompass various approaches, including molecular modeling and bioinformatics algorithms, that can be used to refine experimental structures obtained from X-ray crystallography, NMR spectroscopy, or Cryo-EM techniques by fitting structural models into experimental density maps, refining protein–ligand interactions, and predicting protein dynamics and conformational changes. Furthermore, computational tools help with the integration of complementary experimental data, such as small-angle X-ray scattering (SAXS), hydrogen-deuterium exchange mass spectrometry (HDX-MS), or chemical cross-linking mass spectrometry (XL-MS), into structural models, providing a more comprehensive understanding of protein structure and function.

2 Modeling Approaches to Obtain Protein Structure

Many challenges persist despite innovations in experimental methods for determining protein atomic structure. Transitioning from the primary sequence to a protein's 3D structure is complex. The rapid advancements in genomics have widened the gap between the number of identified primary protein sequences and the experimentally validated structures [54]. Predicting the 3D structure of a protein from its amino acid sequence has been a major research challenge for over 50 years [55]. High-precision computational approaches have made significant strides in bridging this gap and advancing large-scale structural bioinformatics, even when existing experimental methods fail to achieve atomic precision. Moreover, proteins and nucleic acids are flexible structures, and their movements can play a fundamental role in their function [56]. The methods and techniques used in computational modeling contribute to understanding the molecular mechanisms of proteins at the atomic level.

The "Critical Assessment of Techniques for Protein Structure Prediction" (CASP) [57] event resume is published every 2 years. This event aims to evaluate the theoretical methods for predicting protein structure. During the event, protein crystallography and NMR experts provide participants with protein sequences whose tertiary structures have been recently determined and not yet disclosed. Several groups specializing in protein modeling by homology, ab initio methods, and protein folding try to elucidate the structures obtained experimentally. The results are discussed, and the findings are published in PROTEINS [57]. The last edition, CASP15, occurred between December 10th and 13th, 2022. In this event, for the first time, studies through homology modeling of RNA structures and protein-ligand complexes were included, where classical methods showed results in line with experimental data rather than techniques employing machine learning [57, 58].

Given the challenges and advances, experimental and computational methods are essential for understanding protein structures. Techniques like homology and ab initio modeling drive innovation in structural bioinformatics, offering insights with therapeutic potential. These methods not only deepen our understanding of protein structures but also inspire hope for the development of novel therapeutic strategies. The following sections will explore these promising avenues of modeling approaches.

2.1 Homology Modeling

Protein modeling by homology is one of the most common approaches for modeling protein structure and is based on molecular evolution. Protein evolution has involved complex mechanisms related to gene duplication and random DNA mutations [59]. These molecular changes have generated slightly different amino acid sequences, forming families of structurally related proteins [60]. Proteins with a common ancestry are called homologous [61]. The concept of homology encompasses two related definitions: orthology and paralogy. Orthologous genes and proteins share a common ancestry and diverge due to speciation events, while paralogous genes and proteins differentiate due to gene duplication [62]. The similarity between amino acid sequences in homologous proteins, as expressed by the degree of identity, is less conserved than the similarity of their 3D structures [63]. Additionally, the similarity of the protein backbone increases with relative sequence identity. An identity above 50% is considered high, resulting in a small difference in RMSD (Å) between the 3D structures, while an identity below 30% is considered low, which significantly impacts homology modeling [64]. Some historical aspects of homology modeling are shown in Fig. 14.1 and discussed in several reviews, highlighting the information acquired in recent years [65].

Homology modeling uses previous information from structural similarities with proteins from the same family [66]. The following steps (Fig. 14.2) are generally employed:

Selection of the protein primary sequence; Identification of structural homologies in the PDB or another database system, such as the universal protein database (UniProt), currently the world's largest database with information on proteins [67];

Alignment of the target protein with the homologous protein sequences: Sequence alignment software such as Clustal [68], Omega [69], or MUSCLE [70] can be used;

Modelling structured regions: The structure is obtained by homology using the alignment as a starting point to build the target protein model. This procedure can be performed using molecular modeling software such as MODELLER [63] or SWISS-MODEL [71];

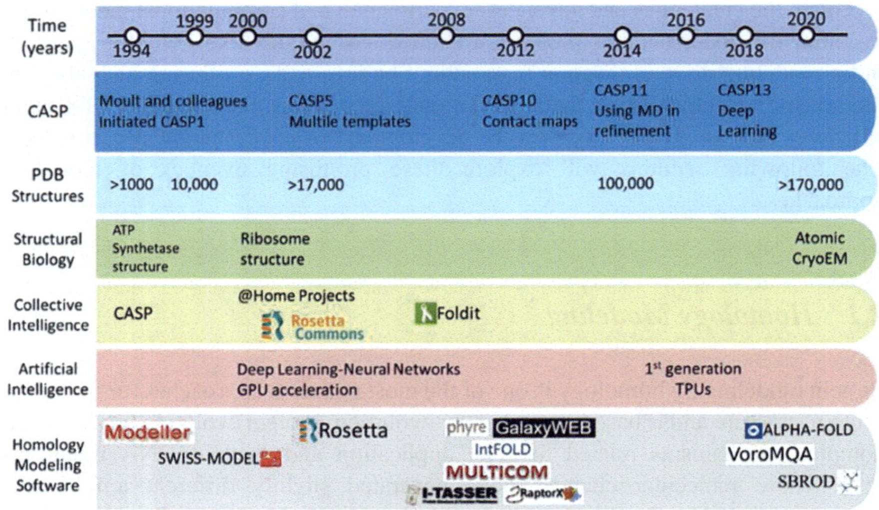

Fig. 14.1 Historical timeline of major developments in homology modeling, *CASP* Critical Assessment of Protein Structure Prediction, *GPU* Graphics Processing Unit, *TPU* Tensor Processing Unit. (Figure adapted from reference [66])

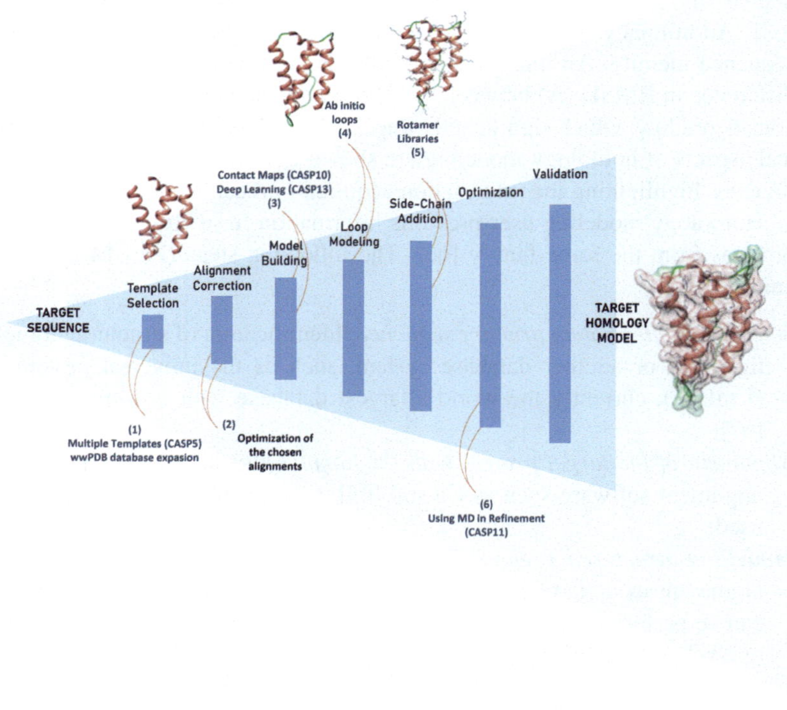

Fig. 14.2 The seven classical steps of homology modeling. The shown structures exemplify events that influence and are critical for modeling: (1) Selection of the protein primary sequence; (2) Alignment correction; (3) Modeling structured regions; (4) Loop modeling; (5) Side chains modeling; (6) Optimization, and (7) Validation. (Figure adapted from reference [66])

Loop modeling: Loop region modeling can be done through database search approaches or conformational analysis methods (ab initio). This is a critical step because loop coordinates can be crucial for protein functions;

Side chains modeling: The conformations of the side chains are also modeled based on steric and energetic data from similar structures obtained experimentally. This process uses querying rotamer libraries, a scoring function, and a scanning method. Several software programs have been developed to sample rotational angles of the side chains, such as OPUS-Rota2 [72], SCWRL [73], and FASPR [74];

Optimization and refinement: energy minimization process using molecular mechanics force fields is applied to eliminate atomic overlaps and collisions. MD and Monte Carlo simulations can be performed as additional optimization measures [1];

Evaluation and Validation: Cross-validation methods and available tools can be used, such as Distance-matrix alignment (DALI, http://ekhidna2.biocenter.helsinki.fi/dali/) or Verify3D (https://servicesn.mbi.ucla.edu/Verify3D/).

It is worth noting that most online servers and programs for homology modeling automate all steps from the primary sequence to the 3D structure without user control over parameters. Other tools allow manual inspection, requiring more user knowledge about the system and program theories [66].

Approximately 75% of proteins are formed by multiple domains (about 2.1 domains in eukaryotic systems and 1.5 domains in prokaryotic systems) [75]. Understanding the 3D structure of proteins with multiple domains can provide insights into several mechanisms, including the impact of mutations on the functions of multiple domains and their relationship with different types of diseases. There are tools for modeling the structure of proteins with multiple domains. For example, an automated method for homology modeling was developed using probabilistic data from multiple domains of primary sequence [76].

The following inherent aspects of homology modeling can be highlighted: (1) it offers several significant advantages and is efficient in generating 3D structures of proteins more quickly than through experimental techniques such as X-ray crystallography or NMR; (2) it allows an expansion of the range of structural coverage, as it allows the prediction of structures that have not yet been determined experimentally; (3) homology-generated models can be adapted for molecular docking simulations, enabling theoretical investigation of protein-substrate interactions in a variety of biological contexts. On the other hand, we can also highlight the following problems when using protein homology modeling: (a) models may have limited accuracy since they are estimates based on structures of homologous proteins, mainly in regions of low primary sequence similarity; (b) the quality of the structural model prediction depends on the quality of the alignment of the primary amino acid sequences used as a reference; (c) problems with modeling highly variable regions such as loops and side chain residues are very dependent on rotamer angle information, resulting in inaccurate structures in these regions [66, 77, 78].

2.2 Ab Initio Modeling

It is possible to predict a protein's 3D structure from its primary sequence using ab initio protein modeling. This method relies on conformational analyses guided by a potential energy function based on the structural environment [54]. In ab initio modeling, a series of possible conformations (decoy structures) are generated and subsequently ranked and selected using conformational energy evaluation methodologies. The success of ab initio modeling is related to three factors: (1) a function that accurately describes the potential energy of the protein's native structure and that leads to the most stable thermodynamic state through comparison with all possibilities of decoy structures; (2) an efficient search algorithm that can quickly characterize the lowest energy states through conformational analysis; and (3) strategies that select near-native structures from a set of decoy structures. It is important to highlight that ab initio modeling requires broader conformational sampling due to the complexity of structural space and the diversity of energy landscapes associated with different protein folding mechanisms [79].

2.3 New Approaches

Deep learning-based approaches such as AlphaFold [58] and trRosetta [80] were introduced in the CASP editions CASP14 and CASP15. These artificial intelligence approaches utilize protein structure data available in the PDB. The computational experiments conducted during these CASP editions demonstrated that these approaches outperform traditional methods, such as genetic algorithms, graph theory, ML, and neural networks in protein folding [81]. AlphaFold2, developed by Google's DeepMind, showcased exceptional performance at CASP14 [82]. Concomitantly, RoseTTAFold was developed, whose results were presented at CASP15 and incorporated into the Continuous Automated Model Evaluation (CAMEO) experiment [83]. CAMEO blindly evaluates structure prediction servers like the programs tested in the CASP editions using protein structures available in the PDB.

AlphaFold was the first computational method capable of predicting protein structures with atomic-level accuracy, even for cases without known homologous structures exist [58, 84]. The method was validated during the CASP14 and is currently in its second version. AlphaFold3: The basis of this new version of AlphaFold is the machine learning approach incorporating physical and biological information of protein structures, using alignments of multiple primary sequences. A two-way network architecture is employed, in which information from the sequence and the 2D distance map levels is transformed and transmitted interactively. The method implemented end-to-end learning, from the final 3D coordinates generated along all network layers to the input sequence [58].

On the other hand, RoseTTAFold is based on a three-way network architecture, which processes sequence alignment information and 2D distance matrix information in parallel, with better predictions than those provided by trRosetta, the second overall-best ranked method after AlphaFold2 in CASP14 [85]. The proposed three-way information architecture network might improve performance using an architecture operating in the 3D coordinate space. This innovation has provided a closer connection between sequence, distances between residues, and atomic orientations and coordinates. In this architecture model, structure prediction arises from the relationships between the amino acid sequence information, the 2D distance map, and the 3D coordinates. This allows the network to collectively reason about relationships within and between sequences, distances, and coordinates. This interactive mechanism, therefore, contrasts with the processing of 3D atomic coordinates in the two-network architecture of AlphaFold2, which occurs after the processing of sequence and distance maps is completed. One of the limitations, due to computer hardware memory problems, was the impossibility, at the time, of training the model for large protein systems [80].

3 Conformational Diversity of Proteins

Proteins do not exist in a single, fixed conformation. Instead, they are best described as conformational ensembles, wherein multiple conformations coexist in equilibrium [86, 87]. Among these conformations, some states are more frequently populated than others. The extent of conformational diversity varies among proteins. While some proteins exhibit only minor conformational changes, others, known as intrinsically disordered proteins (IDPs), occupy many conformations, making them highly flexible and dynamic [88, 89].

In one-domain proteins, the movements responsible for conformational diversity include the rotation of amino acid residue side chains, loops, and the collective movement of connected parts [87]. The rotation of amino acid side chains plays a crucial role in the protein folding process and the transmission of information through allostery [90–92]. Loops, which can be well-defined or disordered random-coil-like structures, generally serve as flexible connectors between more regular secondary structures (α-helices and β-strands). However, it is important to highlight that loops are more than connecting elements as they interact with the solvent, ligands, and other biomolecules. They also participate in the conformational transitions involved in protein regulation [93].

The different modules comprising multidomain proteins are commonly interconnected by short or lengthy sequences of amino acids, typically ranging from approximately 5 to 25, known as linkers. Variations in linker sequences influence their conformational dynamics, potentially affecting cooperative adjustments in interdomain and protein–protein interactions. These changes can facilitate long-range communication between different functional modules in multidomain protein [93]. Rigid domains can also be connected by flexible joints known as hinges. During hinge bending motions, structural units shift relative to each other, disrupting the interfacing packing while preserving the compact arrangement within the protein subunit [87].

3.1 Characterization of Protein Conformational States

Proteins can adopt various conformational states in solution, which may differ from those observed in the crystal lattice. This discrepancy arises because crystallization conditions can impose constraints and stabilize specific conformations, potentially altering the native structural dynamics of the protein [94]. Various ensemble techniques have been developed to investigate protein conformations and dynamics using experimental and computational methodologies. Prominent experimental approaches include NMR spectroscopy, SAXS, single molecule spectroscopy, and Cryo-EM. These experimental techniques are frequently complemented by computational methods that use empirical data as structural constraints. The computational methods include REMD simulations, metadynamics, steered MD, accelerated MD, and Markov state models (Fig. 14.3) [87].

3.2 Experimental Methods to Study Protein Dynamics and Conformations

Experimental determination methods can be employed to determine the conformations of proteins. In X-ray crystallography, conformational heterogeneity hinders crystal formation. Even when crystals are successfully grown, this variability can cause disorder in regions of the structure that are more mobile [95]. Conversely, in cryo-EM structural determination, computational techniques often aid in identifying such heterogeneity. This may involve focusing analyses solely on structurally homogeneous segments of the complex or deriving an ensemble of structures when discrete subpopulations are present. Importantly, delineating conformational diversity often provides invaluable insights into the underlying functional mechanisms [96].

NMR offers various experimental strategies to examine protein kinetics and interactions across various time scales and binding affinities [97]. For instance, the dissociation constants (K_d) associated with protein interactions range from femtomolar for proteases, RNases, and DNases to millimolar for protein-carbohydrate molecular recognition [20, 98]. NMR is a potent tool for interface mapping, facilitating the investigation of weak and transient complexes that may pose challenges for other experimental methodologies [99].

Unlike NMR and Cryo-EM, small-angle X-ray scattering (SAXS) does not offer atomic resolution. It is typically considered a low-resolution structural technique for investigating conformational changes, ligand binding, or protein interactions over time. Nevertheless, SAXS can provide valuable insights into protein behavior,

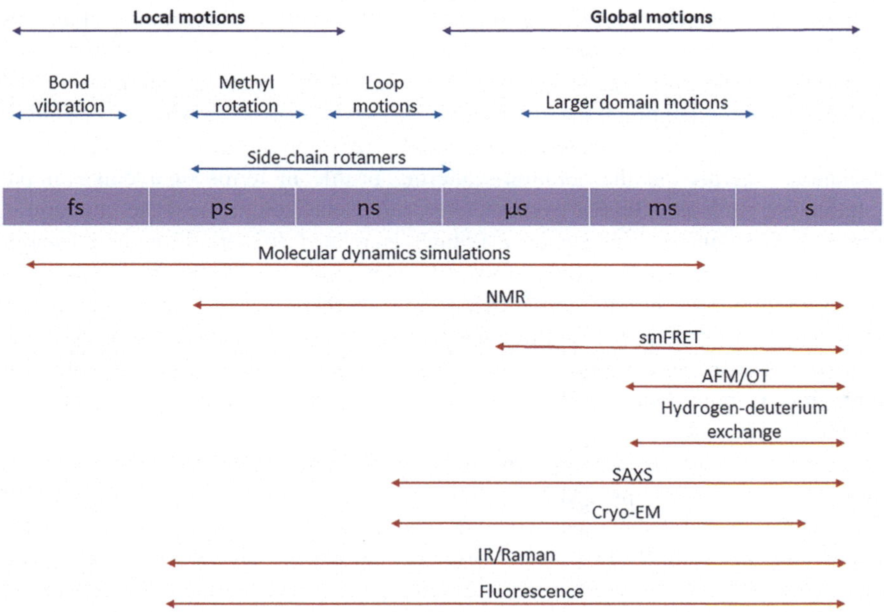

Fig. 14.3 Timescale of dynamic processes in proteins and the experimental methods that can detect fluctuations on each time scale. Local motions, including bond vibrations, methyl rotations, loop motions, and sidechain rotamers, occur on timescales ranging from femtoseconds (fs) to nanoseconds (ns). Global motions, such as larger domain movements, range from nanoseconds (ns) to seconds (s). The timeline depicted spans from fs to s. The methods used to investigate these motions include MD simulations (fs–ms), NMR (ps–s), smFRET (ns–ms), AFM/OT (µs–ms), hydrogen-deuterium exchange (ms–s), SAXS (µs–s), Cryo-EM (ms–s), IR/Raman (fs–s), and fluorescence (fs–s). The figure highlights the temporal ranges of these motions, and the corresponding techniques used for their study

including subtle domain movements [100]. Notably, SAXS is a rapid structural approach that can be directly applied to solutions of macromolecules under various conditions, including quasi-native environments [101]. SAXS provides a convenient means to evaluate alterations in the extension or flexibility of a protein. Among its most impactful applications is the assessment of conformational transitions triggered by ligand or protein partner binding, allowing for the characterization of the overall shape and oligomerization state of the biological macromolecule [101, 102]. Time-resolved probing of conformational changes can be achieved through various mixing methods (e.g., stopped-flow, laminar flow, chaotic flow) and pump-probe techniques [100].

SAXS is a well-established technique for structural characterization of samples, offering resolutions ranging from 5 nm to 1000 nm. Notably, SAXS exhibits sensitivity to both ordered and disordered features within the sample and circumvents the need for crystallization, fixation, or vitrification procedures. In a typical SAXS experiment, a collimated, monochromatic X-ray beam is directed onto the sample, with the radiation scattered at low angles (usually a few degrees) being captured by a detector [103]. Due to the generally random orientations of particles within the solution, a rotational averaging occurs, resulting in a one-dimensional scattering profile expressed as $I(q)$ versus q, where I represents the net intensity and q denotes the momentum transfer variable. This variable is defined as $q = (4\pi sin\Theta)/\lambda$, with Θ being half the scattering angle and λ representing the wavelength of the radiation. Interpreting the solution-scattering profile in terms of average global parameters, such as radius of gyration (Rg), molecular volume, and the interatomic distance distribution $P(r)$ versus r (which is related to $I(q)$ through a Fourier transform), is relatively straightforward. However, contemporary approaches increasingly involve interpreting SAXS data using 3D models in shapes or atomistic representations. This needs careful consideration of how to accurately constrain the structural landscape to find the correct solution while acknowledging the inherent ambiguity if more than one 3D model can fit the same one-dimensional scattering profile [104].

Single-molecule spectroscopy methods include force-based techniques such as atomic force microscopy (AFM) [105] and optical tweezers [106], as well as fluorescence-based techniques, such as single-molecule fluorescence resonance energy transfer (smFRET) [107]. These methodologies have gained widespread use in quantifying the conformational heterogeneity and structural dynamics of biomolecules, both *in vitro* and *in vivo*. They enable the observation of transient intermediates and the characterization of static and dynamic heterogeneity [87, 108].

Several instances highlight the efficacy of these single-molecule techniques in elucidating the dynamics of interactions among biological molecules. Notably, riboswitches — RNA structures that trigger conformational and functional changes upon ligand binding — have been investigated using these methods. Optical tweezers have facilitated the study of numerous riboswitches, demonstrating how ligand adenine binding orchestrates the coordination of domains within the three-loop structure [109, 110]. This observation is supported by steered MD simulations [111]. Furthermore, single-molecule Förster resonance energy transfer (smFRET) has enhanced our understanding of membrane transporter dynamics [112], including ATP-binding cassette (ABC) transporters [113, 114].

3.3 Computational Methods for Exploring Protein Conformations

Currently, the computational study of macromolecules at the molecular level involves steps and methodologies [115, 116]. Among the most used methods, we can mention molecular docking, classical MD, hybrid methods quantum-mechanics/molecular-mechanics (QM/MM) [117], and quantum-mechanics/molecular-mechanics/molecular-dynamics (QM/MM/MD) [118], which can be used to investigate the interactions between ligands and proteins, between proteins, proteins and membranes, proteins and DNA, and RNA.

3.4 Molecular Dynamics Simulation

MD simulations use the equations of motion from classical mechanics to describe molecular events by analyzing conformational changes over time. Variations in this methodology have been successfully employed to study: (1) protein stability, (2) conformational changes, (3) interactions with natural biological cofactors, (4) aiding in the rational design of new drugs, (5) ab initio structure prediction of peptides and small proteins, (6) refinement of protein structure models, (7) and structure modeling assisted by experimental or bioinformatics data [119].

MD simulation, developed in the late 1970s, has advanced from initially being restricted to molecules with hundreds of atoms to systems with millions, such as proteins [120]. Nowadays, it is possible to study whole proteins in solution using explicit solvation, membrane proteins, and large macromolecular complexes such as nucleosomes and ribosomes. Computational simulations of systems with ~50,000–100,000 atoms are now routine, and simulations of approximately 500,000 atoms are common when proper data processing computational facilities are available [121]. Thus, significant computational development and major advances in MD algorithms have been made in recent decades, providing greater efficiency in molecular simulations. This notable improvement is primarily a consequence of using high-performance parallel computing architectures (HPC) and the simplicity of the MD algorithm. Recently, progress has been made in developing software allowing routine MD simulations on graphics processing units (GPUs). MD simulations of hundreds of nanoseconds (ns) per day can now be easily achieved for small protein systems in explicit or implicit solvents.

Some of the most used software in MD are AMBER [122, 123], CHARMM [124], OPLS-AA [125], and GROMACS [118]. The MD is based on the principles of classical mechanics, where each atom is considered a point mass body whose motion is influenced by the forces acting on it from all other atoms. These forces acting on each body are calculated at each MD step using force fields. Force fields are complete interaction potentials between atoms that combine to comprise the system's total energy. These mathematical models include potentials for bonded terms (bond lengths, bond angles, dihedral angles) and non-bonded terms (van der Waals and Coulomb interactions) [54]

$$U\left(\vec{R}\right) = \sum_{\text{bonds}} K_r (r - r_0)^2 + \sum_{\text{angles}} K_\theta (\theta - \theta_0)^2 +$$

$$\sum_{\text{dihedrals}} \frac{V_n}{2}(1 + \cos(n\phi - \delta)) + \sum_{\text{non-bonded terms}} \left[\frac{A_{ij}}{r_{ij}^{12}} - \frac{B_{ij}}{r_{ij}^{6}} + \frac{q_i q_j}{\epsilon R_{ij}} \right]$$

The system's total energy $U\left(\vec{R}\right)$ is obtained by summing the energies of the bonded and non-bonded terms. In this equation, K_r, K_θ, and V_n are the force constants for bond stretching, angle bending, and torsional potential, respectively.

r, θ, and ϕ are bond length, bond angle, and dihedral angle. The subscripted zero terms stand for the equilibrium values for the individual variable. For example, the AMBER program calculates the force field, chemical bond stretching, and angular strain terms using equations with harmonic potentials described by quadratic terms. The terms related to torsional potentials are determined by truncated Fourier series; Van der Waals interactions are approximated by the 12–6 Lennard–Jones (LJ) potential. Coulomb's equation is used to calculate the electrostatic interaction potential. The types of charges vary with the force field version, but for most AMBER force fields, partial charges are obtained using the RESP (Restrained Electrostatic Potential) method. Some force fields include other terms to obtain better agreement with vibrational spectra, for example, terms that better specify hydrogen bonds or the coupling of oscillations between bond angles and bond lengths [126].

The choice of the force field generally depends on the properties and systems to be studied, making it an important step in MD simulation. In the AMBER, there are a variety of force fields available. For instance, lipid14 is indicated to study lipid bilayers [127], GLYCAM_06j for carbohydrates [128], ff14SB for proteins [123], OL15 for DNA [129], and OL3 for RNA [130]. The solvent effect of water plays a fundamental role in the protein folding process and in determining the structure and movement of protein molecules. Water force field models such as the TIP3P [131] and TIP4P [132] have accurately described various physicochemical characteristics of the solvent medium and continue to be widely used in MD simulations with explicit solvents. These different force fields can be used together to describe different systems' components. For example, in a simulation of a protein in water as the solvent and with an organic ligand, it is usual to combine the following force fields: ff14SB (for the protein), GAFF [133] (for the organic compound), and TIP3P (for water). In these explicit solvent simulations, much of the computational resources are consumed in calculating the forces acting on water molecules. Methods were developed to implicitly treat the water solvent to reduce the computational cost with the generalized Born (GB) solvation model [134]. An additional energy term related to the protein's solvent-accessible surface areas (SA) can approximate the nonpolar contributions to solvation.

High-frequency motions can limit the size of the integration step in molecular dynamics (MD) simulations, but these motions generally do not significantly impact the system's overall behavior. In molecular systems, particle motions occur on different time scales, influenced by intermolecular and intramolecular forces. The SHAKE algorithm is the most widely used methodology for maintaining or constraining geometry in MD simulations. This method modifies the equations of motion for atoms of interest (such as those involved in chemical bonds with hydrogen atoms) by introducing force constraints that preserve the lengths of these bonds. It is typically applied to constrain hydrogen bonds with other atoms in the system [135].

The initial step in MD simulations is selecting the file with the protein coordinates (i.e., the structure with the highest resolution of atomic coordinates). Depending on the experimental structural elucidation technique, not all atomic coordinates are resolved, and there may be regions (gaps) without structural definition. However, it is possible to use programs that perform structural comparisons of proteins and, by similarity, complete the lacking regions, such as SWISS-MODEL [71]. Some MD simulation software includes routines for preparing input files to perform this verification and reconstruction of missing residues, such as the AMBERtools [136]. Other aspects that require attention when conducting an MD simulation are the definition of the statistical ensemble, the control of temperature and pressure, the initial energy minimization conditions and system equilibration, the use of periodic boundary conditions, the minimum image convention and cutoff radius, the treatment of long-range interactions, and the application of solvation methods for the system [54]. Thus, once the parameters and conditions are obtained, the calculation begins, proceeding through the minimization, heating, and production stages in a classical MD simulation [116, 137].

More details on methodological aspects, validations, and analysis of MD simulations can be found in Chap. 8.

3.5 Sampling Strategies

According to Newton's classical physics equations, several MD approaches are successfully used to study the interaction and movement of atoms and molecules. The numerical integration of Newton's equations of motion for a system of interacting particles provides successive system configurations, like a trajectory film, showing the particles' positions and velocities over a predetermined fixed time. However, MD simulations require many numerical calculations due to the small integration time steps of femtoseconds (10^{-15} s), which are necessary to simulate biologically relevant events (such as protein folding below microseconds, 10^{-6} s) [138].

MD simulations have become an important and widespread method to explore conformational space, being applied to studying the folding of small proteins (around 80 amino acids) in their native structures [139]. Along with the precision and robustness of diverse types of force fields (including solvent models), an aspect that initially limits the extent of sampling in MD applications is the time scale a conventional MD simulation can achieve. Typically, they are very short in describing real problems of protein systems with hundreds of thousands of atoms.

Several enhanced computational methods have been developed to address these sampling issues using MD and expand the sampling range of classical MD simulations. Among these methods, one can include replica-exchange MD simulations [140], steered MD [141], accelerated MD [142], metadynamics [143], and adaptive steered MD [79]. Some methods use constrained potentials over force constants to direct sampling over energy barriers and pre-defined reaction coordinates, as in umbrella sampling [144] and metadynamics [145]. However, these methods restrict their use in predicting actual structures since the endpoint must be unknown in a protein folding study. As an alternative, sampling methods that use energy as a reaction coordinate have been developed, generalized ensemble methods such as replica exchange MD (REMD), and methods that combine both methodologies [146]. The REMD has become one of the most popular ab initio methodologies for simulating protein folding, as it allows multi-node parallel computing architectures. In addition, the method can be used with stochastic Monte Carlo simulations [147].

One can highlight some applications in this context, such as *the* ab initio *structure prediction of peptides and small proteins.* Peptides have enormous potential for drug development, bridging between small molecule-based drugs and proteins [148]. Many naturally occurring peptides or those designed by computational modeling have had their 3D structures predicted by MD-based methods. However, peptide modeling via MD-base methods with sequence alignment is less reliable than proteins [138].

3.6 Impact of Protein Structure and Dynamics on Drug Design

The binding of proteins and ligands relies on physical and chemical interactions between residues in the binding site of the protein and the ligand, ensuring specificity and communication between active and allosteric sites. These interactions induce structural changes in the binding pocket as the protein transitions from its ligand-unbound (apo) to ligand-bound (holo) state, causing many differences between apo and holo structures [149].

A study from 2005 examined 60 enzymes, revealing structural discrepancies between their apo and holo forms [150]. However, most of the enzymes displayed minimal deviations ($\leq$1 Å) in Cα root mean square deviation (RMSD) between the two states, their apo and holo states. Subsequently, in 2007 [151], 98 apo-holo pairs were categorized into rigid, moderate, and flexible categories based on Cα RMSD differences, highlighting preferences for polar–polar interactions in rigid proteins and hydrophobic interactions in flexible ones. Exploring side chain orientations across apo–holo pairs revealed increased flexibility at the binding site, often overlooked in analyses focusing solely on residue backbones. Nevertheless, significant disparities can become apparent when analyzing side chain orientations across apo–holo pairs [152]. One of the most visualized conformational changes upon binding is the side chain rotameric state changes [153, 154].

The consensus from various studies suggests that while backbones undergo minor conformational changes, side chains explore a broader conformational space upon ligand binding [149]. Consequently, holo-structures are generally preferred targets for ligand docking due to differences in binding site structures. Holo-structure superiority over apo counterparts was demonstrated in discriminating binders from nonbinders [155], whereas significantly lower enrichment levels were observed in apo structures compared to their holo counterparts [156]. Based on that information, computational simulation for drug design methods that account for sidechain flexibility and multiple receptor conformations becomes crucial. Flexible SLIDE docking can accommodate ligands within 2.5 Å of their crystal structure pose by manipulating sidechain orientations in apo structures to address the challenges associated with rigid docking [157]. While flexible docking offers a solution, proteins with significant backbone differences pose challenges. Ensemble docking, employing MD simulations to obtain multiple receptor conformations, offers an alternative approach [158, 159].

Furthermore, considering the available structural information of the system under study is important for simulations. Molecular docking typically yields optimal results when the molecule resembles the one crystallized [160]; RMSD <2.0 Å corresponds to good docking simulations. However, achieving such similarity is not always possible. In efforts to facilitate virtual screening using apo structures, protocols involving MD simulations to refine protein structures were developed and could be implemented before the virtual screening procedure. This strategy is especially helpful when holo-structures are unavailable [149].

However, inherent flaws in current scoring functions, such as excessive simplifications, limited consideration of interaction energies, inaccuracies in predicting binding affinities, may compromise the accurate identification of the putative binding mode of simulated compounds (docking poses), improper consideration of oligomerization states, or the presence of multiple interfaces typically excluded from docking calculations [161–163]. To mitigate these challenges, leveraging external information whenever feasible and available can enhance the accuracy of docking predictions. Pioneering methods like HADDOCK [164] and other protein–protein docking approaches like pyDock [165], ZDOCK [166], and LightDock [167] have implemented protocols incorporating distance restraints to refine docking calculations.

Although most enzymes do not undergo significant conformational changes when interacting with a ligand, as previously discussed, understanding the system to be modeled is always beneficial in selecting the appropriate structures. One example is Abelson kinase (Abl), an enzyme whose deregulation causes chronic myeloid leukemia and can exhibit different conformations depending on the ligand [168]. Some inhibitors, such as dasatinib, inhibit an active form of the enzyme, in which the activation segment is in an open conformation (Fig. 14.4). In contrast, imatinib inhibits an inactive conformation of the protein, where the activation segment is in a closed conformation. Due to steric effects, imatinib cannot bind to the enzyme when the activation segment is in the extended conformation. Additionally, the activation segment's conformation influences the P-loop conformation, which must adopt a folded conformation when the activation segment closes to avoid electrostatic repulsion [169, 170]. So, a detailed understanding of the conformations and specific interactions of ligands with enzymes is essential for selecting

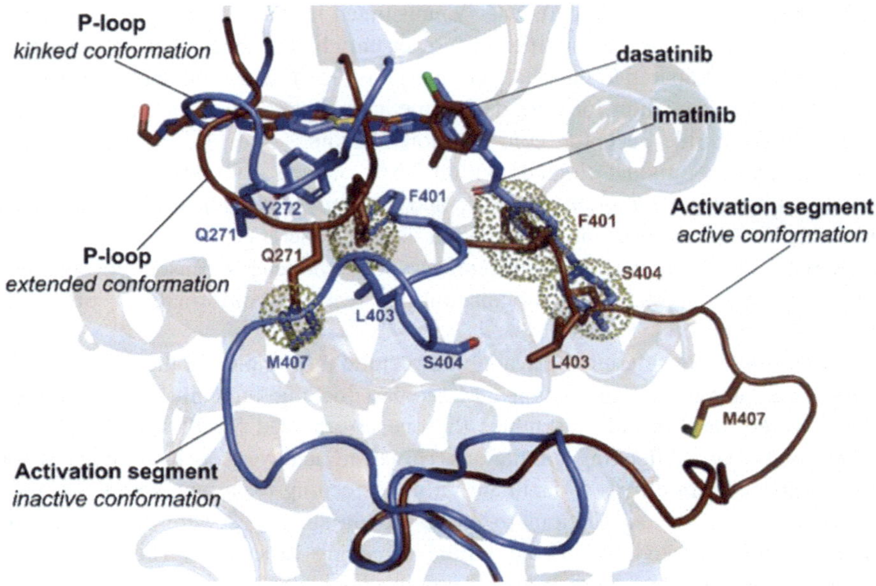

Fig. 14.4 Superposition of Abl structures co-crystallized with imatinib in blue (PDB ID 2HYY) and dasatinib in brown (PDB ID 2GQG). Residues with potential steric hindrance, depending on the conformational state of the activation segment and P-loop, are shown [169, 170]. These steric hindrances are highlighted as dotted spheres in yellow. The figure was generated using the software PyMOL

and preparing the correct structures in computational simulations, directly influencing the accuracy and effectiveness of docking and molecular dynamics studies.

4 Remarks and Perspectives

In conclusion, integrating experimental and computational techniques is essential for advancing drug design, particularly in understanding protein structure and dynamics. The impact of conformational diversity between apo and holo states underscores the need for comprehensive methods that capture the full range of structural variability. Experimental methods provide critical insights into these structural changes, while computational techniques, such as flexible docking and ensemble docking, address the challenges of conformational flexibility. By combining these approaches, researchers can enhance the accuracy of ligand binding predictions and better understand the dynamic nature of protein–ligand interactions. This synergy between experimental and computational strategies is crucial for developing effective therapeutics and optimizing drug discovery processes.

Moreover, we are at a transformative juncture in structural biology and protein modeling with the advent of powerful tools like AlphaFold and RoseTTAFold.

These groundbreaking technologies have dramatically improved our ability to predict protein structures with unprecedented accuracy, which in turn, is set to influence experimental techniques such as X-ray crystallography, NMR, and cryo-EM. As these computational methods continue to evolve, they will complement and refine traditional experimental approaches, thereby accelerating discoveries and deepening our understanding of complex biomolecular systems. This convergence of computational and experimental strategies marks a new era in structural biology, promising significant advancements in fundamental research and drug development.

References

1. Hong, S. H., Joung, I., Flores-Canales, J. C., Manavalan, B., Cheng, Q., Heo, S., Kim, J. Y., Lee, S. Y., Nam, M., Joo, K., et al. (2018). Protein structure modeling and refinement by global optimization in CASP12. *Proteins: Structure, Function, and Bioinformatics, 86*(S1), 122–135.
2. Kendrew, J. C., Bodo, G., Dintzis, H. M., Parrish, R. G., Wyckoff, H., & Phillips, D. C. (1958). A three-dimensional model of the myoglobin molecule obtained by X-ray analysis. *Nature, 181*(4610), 662–666.
3. Perutz, M. F., Rossmann, M. G., Cullis, A. F., Muirhead, H., Will, G., & North, A. C. T. (1960). Structure of hæmoglobin: A three-dimensional Fourier synthesis at 5.5-Å. Resolution, obtained by X-ray analysis. *Nature, 185*(4711), 416–422.
4. Aue, W. P., Bartholdi, E., & Ernst, R. R. (1976). Two-dimensional spectroscopy. Application to nuclear magnetic resonance. *The Journal of Chemical Physics, 64*(5), 2229–2246.
5. Williamson, M. P., Havel, T. F., & Wüthrich, K. (1985). Solution conformation of proteinase inhibitor IIA from bull seminal plasma by 1H nuclear magnetic resonance and distance geometry. *Journal of Molecular Biology, 182*(2), 295–315.
6. Palmer, A. G., & Massi, F. (2006). Characterization of the dynamics of biomacromolecules using rotating-frame spin relaxation NMR spectroscopy. *Chemical Reviews, 106*(5), 1700–1719.
7. Jiang, W., Baker, M. L., Jakana, J., Weigele, P. R., King, J., & Chiu, W. (2008). Backbone structure of the infectious E15 virus capsid revealed by electron cryomicroscopy. *Nature, 451*(7182), 1130–1134.
8. Sirohi, D., Chen, Z., Sun, L., Klose, T., Pierson, T. C., Rossmann, M. G., & Kuhn, R. J. (2016). The 3.8 Å resolution Cryo-EM structure of Zika virus. *Science, 352*(6284), 467–470.
9. Beckers, M., Mann, D., & Sachse, C. (2021). Structural interpretation of Cryo-EM image reconstructions. *Progress in Biophysics and Molecular Biology, 160*, 26–36.
10. Zhang, X., Settembre, E., Xu, C., Dormitzer, P. R., Bellamy, R., Harrison, S. C., & Grigorieff, N. (2008). Near-atomic resolution using electron cryomicroscopy and single-particle reconstruction. *Proceedings of the National Academy of Sciences, 105*(6), 1867–1872.
11. Berman, H., Henrick, K., & Nakamura, H. (2003). Announcing the worldwide protein data Bank. *Nature Structural & Molecular Biology, 10*(12), 980–980.
12. Ilari, A., & Savino, C. (2008). Protein structure determination by X-Ray. *Crystallography*, 63–87.
13. McPherson, A., & Gavira, J. A. (2014). Introduction to protein crystallization. *Acta Crystallographica Section F: Structural Biology and Crystallization Communications, 70*(1), 2–20.
14. Williams, C. J., Headd, J. J., Moriarty, N. W., Prisant, M. G., Videau, L. L., Deis, L. N., Verma, V., Keedy, D. A., Hintze, B. J., Chen, V. B., et al. (2018). MolProbity: More and better reference data for improved all-atom structure validation. *Protein Science, 27*(1), 293–315.

15. Laskowski, R. A., MacArthur, M. W., Moss, D. S., & Thornton, J. M. (1993). Procheck: A program to check the stereochemical quality of protein structures. *Journal of Applied Crystallography, 26*(2), 283–291.
16. Almeida, F. C. L. F. C. L., Moraes, A. H. A. H., & Gomes-Neto, F. (2013). *No title* (Vol. 9781461451, pp. 377–412). Springer.
17. Fernandez, C. (2003). TROSY in NMR studies of the structure and function of large biological macromolecules. *Current Opinion in Structural Biology, 13*(5), 570–580.
18. Charlier, C., Cousin, S. F., & Ferrage, F. (2016). Protein dynamics from nuclear magnetic relaxation. *Chemical Society Reviews, 45*(9), 2410–2422.
19. Almeida, F. C. L., Moraes, A. H., & Gomes-Neto, F. (2013). An overview on protein structure determination by NMR: Historical and future perspectives of the use of distance geometry methods. In *Distance geometry* (pp. 377–412). Springer.
20. Moraes, A. H., & Valente, A. P. (2023). Conformational dynamics and kinetics of protein interactions by nuclear magnetic resonance. *Journal of Magnetic Resonance Open, 14–15*, 100093.
21. Moraes, A. H., Ackerbauer, D., Kostadinova, M., Bublin, M., de Oliveira, G. A., Ferreira, F. F., Almeida, F. C. L. C., Breiteneder, H., Valente, A. P., de Oliveira, G. A., et al. (2014). Solution and high-pressure NMR studies of the structure, dynamics, and stability of the cross-reactive allergenic cod parvalbumin Gad m 1. *Proteins: Structure, Function, and Bioinformatics, 82*(11), 3032–3042.
22. Moraes, A. H. H., Asam, C., Batista, A., Almeida, F. C. L. C. L. L., Wallner, M., Ferreira, F., & Valente, A. P. P. (2016). 1H, 13C and 15N resonance assignments and second structure information of Fag s 1: Fagales allergen from Fagus Sylvatica. *Biomolecular NMR Assignments, 10*(1), 45–48.
23. Moraes, A. H., Asam, C., Almeida, F. C. L. L., Wallner, M., Ferreira, F., & Valente, A. P. (2018). Structural basis for cross-reactivity and conformation fluctuation of the major beech pollen allergen fag s 1. *Scientific Reports, 8*(1), 10512.
24. Güntert, P. (2004). Automated NMR structure calculation with CYANA. *Methods in Molecular Biology (Clifton, N.J.), 278*(4), 353–378.
25. Linge, J. P., Habeck, M., Rieping, W., & Nilges, M. (2003). ARIA: Automated NOE assignment and NMR structure calculation. *Bioinformatics, 19*(2), 315–316.
26. Shen, Y., & Bax, A. (2013). Protein backbone and sidechain torsion angles predicted from NMR chemical shifts using artificial neural networks. *Journal of Biomolecular NMR, 56*(3), 227–241.
27. Laskowski, R., Rullmann, J. A., MacArthur, M., Kaptein, R., & Thornton, J. (1996). AQUA and PROCHECK-NMR: Programs for checking the quality of protein structures solved by NMR. *Journal of Biomolecular NMR, 8*(4).
28. Bhattacharya, A., Tejero, R., & Montelione, G. T. (2007). Evaluating protein structures determined by structural genomics consortia. *Proteins: Structure, Function, and Bioinformatics, 66*(4), 778–795.
29. Doreleijers, J. F., Sousa da Silva, A. W., Krieger, E., Nabuurs, S. B., Spronk, C. A. E. M., Stevens, T. J., Vranken, W. F., Vriend, G., & Vuister, G. W. (2012). CING: An integrated residue-based structure validation program suite. *Journal of Biomolecular NMR, 54*(3), 267–283.
30. Fernandez-Leiro, R., & Scheres, S. H. W. (2016). Unravelling biological macromolecules with cryo-electron microscopy. *Nature, 537*(7620), 339–346.
31. Callaway, E. (2015). The revolution will not be crystallized: A new method sweeps through structural biology. *Nature, 525*(7568), 172–174.
32. Weissenberger, G., Henderikx, R. J. M., & Peters, P. J. (2021). Understanding the invisible hands of sample preparation for cryo-EM. *Nature Methods, 18*(5), 463–471.
33. Zhong, E. D., Bepler, T., Berger, B., & Davis, J. H. (2021). CryoDRGN: Reconstruction of heterogeneous cryo-EM structures using neural networks. *Nature Methods, 18*(2), 176–185.

34. Thompson, R. F., Walker, M., Siebert, C. A., Muench, S. P., & Ranson, N. A. (2016). An introduction to sample preparation and imaging by cryo-electron microscopy for structural biology. *Methods, 100*, 3–15.
35. Yip, K. M., Fischer, N., Paknia, E., Chari, A., & Stark, H. (2020). Atomic-resolution protein structure determination by Cryo-EM. *Nature, 587*(7832), 157–161.
36. Nakane, T., Kotecha, A., Sente, A., McMullan, G., Masiulis, S., Brown, P. M. G. E., Grigoras, I. T., Malinauskaite, L., Malinauskas, T., Miehling, J., et al. (2020). Single-particle Cryo-EM at atomic resolution. *Nature, 587*(7832), 152–156.
37. Chen, M., Schmid, M. F., & Chiu, W. (2024). Improving resolution and resolvability of single-particle CryoEM structures using Gaussian mixture models. *Nature Methods, 21*(1), 37–40.
38. Miller, M. D., & Phillips, G. N. (2021). Moving beyond static snapshots: Protein dynamics and the protein data Bank. *The Journal of Biological Chemistry, 296*, 100749.
39. Fitzpatrick, A. W. P., Falcon, B., He, S., Murzin, A. G., Murshudov, G., Garringer, H. J., Crowther, R. A., Ghetti, B., Goedert, M., & Scheres, S. H. W. (2017). Cryo-EM structures of tau filaments from Alzheimer's disease. *Nature, 547*(7662), 185–190.
40. Gremer, L., Schölzel, D., Schenk, C., Reinartz, E., Labahn, J., Ravelli, R. B. G., Tusche, M., Lopez-Iglesias, C., Hoyer, W., Heise, H., et al. (2017). Fibril structure of amyloid-β(1–42) by cryo–electron microscopy. *Science (80-.), 358*(6359), 116–119.
41. Wrapp, D., Wang, N., Corbett, K. S., Goldsmith, J. A., Hsieh, C.-L., Abiona, O., Graham, B. S., & McLellan, J. S. (2020). Cryo-EM structure of the 2019-NCoV spike in the prefusion conformation. *Science (80-.), 367*(6483), 1260–1263.
42. Xie, M., Yu, L., Bruschweiler-Li, L., Xiang, X., Hansen, A. L., Brüschweiler, R., Kastritis, P. L., Bonvin, A. M. J. J., Wallerstein, J., Akke, M., et al. (2019). Measuring dynamic and kinetic information in the previously inaccessible supra-Tc window of nanoseconds to microseconds by solution NMR spectroscopy. *Journal of Biomolecular NMR, 52*(1), 139–148.
43. Cheng, Y. (2018). Membrane protein structural biology in the era of single particle Cryo-EM. *Current Opinion in Structural Biology, 52*, 58–63.
44. Piper, S. J., Johnson, R. M., Wootten, D., & Sexton, P. M. (2022). Membranes under the magnetic lens: A dive into the diverse world of membrane protein structures using Cryo-EM. *Chemical Reviews, 122*(17), 13989–14017.
45. Sun, C., Zhu, H., Clark, S., & Gouaux, E. (2023). Cryo-EM structures reveal native GABAA receptor assemblies and pharmacology. *Nature, 622*(7981), 195–201.
46. Carroni, M., & Saibil, H. R. (2016). Cryo electron microscopy to determine the structure of macromolecular complexes. *Methods, 95*, 78–85.
47. Kyrilis, F. L., Belapure, J., & Kastritis, P. L. (2021). Detecting protein communities in native cell extracts by machine learning: A structural Biologist's perspective. *Frontiers in Molecular Biosciences, 8*, 229.
48. Verbeke, E. J., Gilles, M. A., Bendory, T., & Singer, A. (2024). Self Fourier Shell correlation: Properties and application to Cryo-ET. *Communications Biology, 7*(1), 101.
49. Lawson, C. L., Kryshtafovych, A., Adams, P. D., Afonine, P. V., Baker, M. L., Barad, B. A., Bond, P., Burnley, T., Cao, R., Cheng, J., et al. (2021). Cryo-EM model validation recommendations based on outcomes of the 2019 EMDataResource challenge. *Nature Methods, 18*(2), 156–164.
50. Schneidman-Duhovny, D., & Hammel, M. (2018). Modeling Structure and Dynamics of Protein Complexes with SAXS Profiles. *Methods in Molecular Biology, 178*, 449–473.
51. Mertens, H. D. T., & Svergun, D. I. (2017). Combining NMR and small angle X-ray scattering for the study of biomolecular structure and dynamics. *Archives of Biochemistry and Biophysics, 628*, 33–41.
52. Masson, G. R., Burke, J. E., Ahn, N. G., Anand, G. S., Borchers, C., Brier, S., Bou-Assaf, G. M., Engen, J. R., Englander, S. W., Faber, J., et al. (2019). Recommendations for performing, interpreting and reporting hydrogen deuterium exchange mass spectrometry (HDX-MS) experiments. *Nature Methods, 16*(7), 595–602.

53. O'Reilly, F. J., & Rappsilber, J. (2018). Cross-linking mass spectrometry: Methods and applications in structural, molecular and systems biology. *Nature Structural & Molecular Biology, 25*(11), 1000–1008.
54. Lee, J., Freddolino, P. L., & Zhang, Y. (2017). Ab initio protein structure prediction. In *From protein structure to function with bioinformatics* (pp. 3–35). Springer.
55. Torrisi, M., Pollastri, G., & Le, Q. (2020). Deep learning methods in protein structure prediction. *Computational and Structural Biotechnology Journal, 18*, 1301–1310.
56. Li, W., Wang, W., & Takada, S. (2014). Energy landscape views for interplays among folding, binding, and Allostery of Calmodulin domains. *Proceedings of the National Academy of Sciences of the United States of America, 111*(29), 10550–10555.
57. Das, R., Kretsch, R. C., Simpkin, A. J., Mulvaney, T., Pham, P., Rangan, R., Bu, F., Keegan, R. M., Topf, M., Rigden, D. J., et al. (2023). Assessment of three-dimensional RNA structure prediction in CASP15. *Proteins Struct. Funct. Bioinforma, 91*(12), 1747–1770.
58. Jumper, J., Evans, R., Pritzel, A., Green, T., Figurnov, M., Ronneberger, O., Tunyasuvunakool, K., Bates, R., Žídek, A., Potapenko, A., et al. (2021). Highly accurate protein structure prediction with AlphaFold. *Nature, 596*(7873), 583–589.
59. Chothia, C., & Lesk, A. M. (1986). The relation between the divergence of sequence and structure in proteins. *The EMBO Journal, 5*(4), 823–826.
60. Sander, C., & Schneider, R. (1991). Database of homology-derived protein structures and the structural meaning of sequence alignment. *Proteins: Structure, Function, and Bioinformatics, 9*(1), 56–68.
61. Yang, Y., Faraggi, E., Zhao, H., & Zhou, Y. (2011). Improving protein fold recognition and template-based modeling by employing probabilistic-based matching between predicted one-dimensional structural properties of query and corresponding native properties of templates. *Bioinformatics, 27*(15), 2076–2082.
62. Koonin, E. V. (2005). Orthologs, paralogs, and evolutionary genomics. *Annual Review of Genetics, 39*(1), 309–338.
63. Martí-Renom, M. A., Stuart, A. C., Fiser, A., Sánchez, R., Melo, F., & Šali, A. (2000). Comparative protein structure modeling of genes and genomes. *Annual Review of Biophysics and Biomolecular Structure, 29*(1), 291–325.
64. Tramontano, A. (1998). Homology modeling with low sequence identity. *Methods, 14*(3), 293–300.
65. Arganda-Carreras, I., Kaynig, V., Rueden, C., Eliceiri, K. W., Schindelin, J., Cardona, A., & Sebastian Seung, H. (2017). Trainable Weka segmentation: A machine learning tool for microscopy pixel classification. *Bioinformatics, 33*(15), 2424–2426.
66. Hameduh, T., Haddad, Y., Adam, V., & Heger, Z. (2020). Homology modeling in the time of collective and artificial intelligence. *Computational and Structural Biotechnology Journal, 18*, 3494–3506.
67. Bateman, A., Martin, M.-J., Orchard, S., Magrane, M., Agivetova, R., Ahmad, S., Alpi, E., Bowler-Barnett, E. H., Britto, R., Bursteinas, B., et al. (2021). UniProt: The universal protein knowledgebase in 2021. *Nucleic Acids Research, 49*(D1), D480–D489.
68. Larkin, M. A., Blackshields, G., Brown, N. P., Chenna, R., McGettigan, P. A., McWilliam, H., Valentin, F., Wallace, I. M., Wilm, A., Lopez, R., et al. (2007). Clustal W and Clustal X version 2.0. *Bioinformatics, 23*(21), 2947–2948.
69. Li, W., Cowley, A., Uludag, M., Gur, T., McWilliam, H., Squizzato, S., Park, Y. M., Buso, N., & Lopez, R. (2015). The EMBL-EBI bioinformatics web and programmatic tools framework. *Nucleic Acids Research, 43*(W1), W580–W584.
70. Edgar, R. C. (2004). Muscle: Multiple sequence alignment with high accuracy and high throughput. *Nucleic Acids Research, 32*(5), 1792–1797.
71. Waterhouse, A., Bertoni, M., Bienert, S., Studer, G., Tauriello, G., Gumienny, R., Heer, F. T., de Beer, T. A. P., Rempfer, C., Bordoli, L., et al. (2018). Swiss-model: Homology modelling of protein structures and complexes. *Nucleic Acids Research, 46*(W1), W296–W303.

72. Xu, G., Ma, T., Du, J., Wang, Q., & Ma, J. (2019). OPUS-Rota2: An improved fast and accurate side-chain modeling method. *Journal of Chemical Theory and Computation, 15*(9), 5154–5160.
73. Wang, Q., Canutescu, A. A., & Dunbrack, R. L. (2008). SCWRL and MolIDE: Computer programs for side-chain conformation prediction and homology modeling. *Nature Protocols, 3*(12), 1832–1847.
74. Huang, X., Pearce, R., & Zhang, Y. (2020). FASPR: An open-source tool for fast and accurate protein side-chain packing. *Bioinformatics, 36*(12), 3758–3765.
75. Ekman, D., Björklund, Å. K., Frey-Skött, J., & Elofsson, A. (2005). Multi-domain proteins in the three kingdoms of life: Orphan domains and other unassigned regions. *Journal of Molecular Biology, 348*(1), 231–243.
76. Meier, A., & Söding, J. (2015). Automatic prediction of protein 3D structures by probabilistic multi-template homology modeling. *PLoS Computational Biology, 11*(10), e1004343.
77. Xiang, Z. (2006). Advances in homology protein structure modeling. *Current Protein & Peptide Science, 7*(3), 217–227.
78. Liang, S., & Grishin, N. V. (2002). Side-chain modeling with an optimized scoring function. *Protein Science, 11*(2), 322–331.
79. Ozer, G., Quirk, S., & Hernandez, R. (2012). Adaptive steered molecular dynamics: Validation of the selection criterion and benchmarking energetics in vacuum. *The Journal of Chemical Physics, 136* (21).
80. Baek, M., DiMaio, F., Anishchenko, I., Dauparas, J., Ovchinnikov, S., Lee, G. R., Wang, J., Cong, Q., Kinch, L. N., Schaeffer, R. D., et al. (2021). Accurate prediction of protein structures and interactions using a three-track neural network. *Science (80-.), 373*(6557), 871–876.
81. Böhm, G. (1996). New approaches in molecular structure prediction. *Biophysical Chemistry, 59*(1–2), 1–32.
82. Kwon, S., Won, J., Kryshtafovych, A., & Seok, C. (2021). Assessment of protein model structure accuracy estimation in CASP14: Old and new challenges. *Proteins Structure Function and Bioinformatics, 89*(12), 1940–1948.
83. Robin, X., Haas, J., Gumienny, R., Smolinski, A., Tauriello, G., & Schwede, T. (2021). Continuous Automated Model EvaluatiOn (CAMEO)—Perspectives on the future of fully automated evaluation of structure prediction methods. *Proteins Structure Function and Bioinformatics, 89*(12), 1977–1986.
84. Senior, A. W., Evans, R., Jumper, J., Kirkpatrick, J., Sifre, L., Green, T., Qin, C., Žídek, A., Nelson, A. W. R., Bridgland, A., et al. (2020). Improved protein structure prediction using potentials from deep learning. *Nature, 577*(7792), 706–710.
85. Pereira, J., Simpkin, A. J., Hartmann, M. D., Rigden, D. J., Keegan, R. M., & Lupas, A. N. (2021). High-accuracy protein structure prediction in CASP14. *Proteins Structure Function and Bioinformatics, 89*(12), 1687–1699.
86. Valente, A., Miyamoto, C., Almeida, L., & F. (2006). Implications of protein conformational diversity for binding and development of new biological active compounds. *Current Medicinal Chemistry, 13*(30), 3697–3703.
87. Wei, G., Xi, W., Nussinov, R., & Ma, B. (2016). Protein ensembles: How does nature harness thermodynamic fluctuations for life? The diverse functional roles of conformational ensembles in the cell. *Chemical Reviews, 116*(11), 6516–6551.
88. Trivedi, R., & Nagarajaram, H. A. (2022). Intrinsically disordered proteins: An overview. *International Journal of Molecular Sciences, 23*(22), 14050.
89. Tsai, C., Ma, B., Sham, Y. Y., Kumar, S., & Nussinov, R. (2001). Structured disorder and conformational selection. *Proteins Structure Function and Bioinformatics, 44*(4), 418–427.
90. Carlson, G. M., & Fenton, A. W. (2016). What mutagenesis can and cannot reveal about Allostery. *Biophysical Journal, 110*(9), 1912–1923.

91. Petit, C. M., Zhang, J., Sapienza, P. J., Fuentes, E. J., & Lee, A. L. (2009). Hidden dynamic Allostery in a PDZ domain. *Proceedings of the National Academy of Sciences, 106*(43), 18249–18254.
92. Thirumalai, D., Hyeon, C., Zhuravlev, P. I., & Lorimer, G. H. (2019). Symmetry, rigidity, and allosteric signaling: From monomeric proteins to molecular machines. *Chemical Reviews, 119*(12), 6788–6821.
93. Papaleo, E., Saladino, G., Lambrughi, M., Lindorff-Larsen, K., Gervasio, F. L., & Nussinov, R. (2016). The role of protein loops and linkers in conformational dynamics and Allostery. *Chemical Reviews, 116*(11), 6391–6423.
94. Lindner, A. B., Eshhar, Z., & Tawfik, D. S. (1999). Conformational changes affect binding and catalysis by Ester-hydrolysing antibodies 1 1Edited by. *Journal of Molecular Biology, 285*(1), 421–430.
95. Shumilin, I. A., Cymborowski, M., Chertihin, O., Jha, K. N., Herr, J. C., Lesley, S. A., Joachimiak, A., & Minor, W. (2012). Identification of unknown protein function using metabolite cocktail screening. *Structure, 20*(10), 1715–1725.
96. Earl, L. A., Falconieri, V., Milne, J. L., & Subramaniam, S. (2017). Cryo-EM: Beyond the microscope. *Current Opinion in Structural Biology, 46*, 71–78.
97. Palmer, A. G. (2015). Enzyme dynamics from NMR spectroscopy. *Accounts of Chemical Research, 48*(2), 457–465.
98. Palmer, A. G. (2016). A dynamic look backward and forward. *Journal of Magnetic Resonance, 266*, 73–80.
99. Tzeng, S. R., & Kalodimos, C. G. (2011). Protein dynamics and Allostery: An NMR view. *Current Opinion in Structural Biology, 21*(1), 62–67.
100. Byer, A. S., Pei, X., Patterson, M. G., & Ando, N. (2023). Small-angle X-ray scattering studies of enzymes. *Current Opinion in Chemical Biology, 72*, 102232.
101. Gräwert, T. W., & Svergun, D. I. (2020). Structural modeling using solution small-angle X-ray scattering (SAXS). *Journal of Molecular Biology, 432*(9), 3078–3092.
102. Jacques, D. A., & Trewhella, J. (2010). Small-angle scattering for structural biology—Expanding the frontier while avoiding the pitfalls. *Protein Science, 19*(4), 642–657.
103. Da Vela, S., & Svergun, D. I. (2020). Methods, development and applications of small-angle X-ray scattering to characterize biological macromolecules in solution. *Current Research in Structural Biology, 2*, 164–170.
104. Trewhella, J. (2022). Recent advances in small-angle scattering and its expanding impact in structural biology. *Structure, 30*(1), 15–23.
105. Edwards, D. T., LeBlanc, M.-A., & Perkins, T. T. (2021). Modulation of a protein-folding landscape revealed by AFM-based force spectroscopy notwithstanding instrumental limitations. *Proceedings of the National Academy of Sciences, 118* (12).
106. Jahn, M., Buchner, J., Hugel, T., & Rief, M. (2016). Folding and assembly of the large molecular machine Hsp90 studied in single-molecule experiments. *Proceedings of the National Academy of Sciences, 113*(5), 1232–1237.
107. Liu, T., Khanal, S., Hertslet, G. D., & Lamichhane, R. (2023). Single-molecule analysis reveals that a glucagon-bound extracellular domain of the glucagon receptor is dynamic. *The Journal of Biological Chemistry, 299*(9), 105160.
108. Schuler, B., & Hofmann, H. (2013). Single-molecule spectroscopy of protein folding dynamics—Expanding scope and timescales. *Current Opinion in Structural Biology, 23*(1), 36–47.
109. Neupane, K., Yu, H., Foster, D. A. N., Wang, F., & Woodside, M. T. (2011). Single-molecule force spectroscopy of the add adenine riboswitch relates folding to regulatory mechanism. *Nucleic Acids Research, 39*(17), 7677–7687.
110. Halma, M. T. J., Tuszynski, J. A., & Wuite, G. J. L. (2023). Optical tweezers for drug discovery. *Drug Discovery Today, 28*(1), 103443.
111. Sharma, S., Singh, V., & Biswas, P. (2021). Effect of ligand binding on riboswitch folding: Theory and simulations. *The Journal of Chemical Physics, 154*(18).

112. Bartels, K., Lasitza-Male, T., Hofmann, H., & Löw, C. (2021). Single-molecule FRET of membrane transport proteins. *Chembiochem, 22*(17), 2657–2671.
113. Husada, F., Bountra, K., Tassis, K., de Boer, M., Romano, M., Rebuffat, S., Beis, K., & Cordes, T. (2018). Conformational Dynamics of the ABC Transporter McjD Seen by Single-molecule FRET. *The EMBO Journal, 37*(21).
114. Zhu, Y., He, L., Liu, Y., Zhao, Y., & Zhang, X. C. (2019). SmFRET probing reveals substrate-dependent conformational dynamics of E. Coli multidrug MdfA. *Biophysical Journal, 116*(12), 2296–2303.
115. Chagas, M. A., Pereira, E. S., Godinho, M. P. B., Da Silva, J. C. S., & Rocha, W. R. (2018). Base mechanism to the hydrolysis of phosphate Triester promoted by the cd 2+ /cd 2+ active site of Phosphotriesterase: A computational study. *Inorganic Chemistry, 57*(10), 5888–5902.
116. Salvatti, B. A., Chagas, M. A., Fernandes, P. O., Ladeira, Y. F. X., Bozzi, A. S., Valadares, V. S., Valente, A. P., de Miranda, A. S., Rocha, W. R., Maltarollo, V. G., et al. (2024). Understanding the enzyme (S)-norcoclaurine synthase promiscuity to aldehydes and ketones. *Journal of Chemical Information and Modeling, 64*(11), 4462–4474.
117. Groenhof, G. (2013). *Introduction to QM/MM Simulations*, 43–66.
118. Hess, B., Kutzner, C., van der Spoel, D., & Lindahl, E. (2008). GROMACS 4: Algorithms for highly efficient, load-balanced, and scalable molecular simulation. *Journal of Chemical Theory and Computation, 4*(3), 435–447.
119. McCammon, J. A., Gelin, B. R., & Karplus, M. (1977). Dynamics of folded proteins. *Nature, 267*(5612), 585–590.
120. Shaw, D. E., Deneroff, M. M., Dror, R. O., Kuskin, J. S., Larson, R. H., Salmon, J. K., Young, C., Batson, B., Bowers, K. J., Chao, J. C., et al. (2008). Anton, a special-purpose machine for molecular dynamics simulation. *Communications of the ACM, 51*(7), 91–97.
121. Liu, W., Schmidt, B., Voss, G., & Müller-Wittig, W. (2008). Accelerating molecular dynamics simulations using graphics processing units with CUDA. *Computer Physics Communications, 179*(9), 634–641.
122. Shaw, D. E., Maragakis, P., Lindorff-Larsen, K., Piana, S., Dror, R. O., Eastwood, M. P., Bank, J. A., Jumper, J. M., Salmon, J. K., Shan, Y., et al. Atomic-level characterization of the structural dynamics of proteins. *Science (80-.), 2010, 330*(6002), 341–346.
123. Maier, J. A., Martinez, C., Kasavajhala, K., Wickstrom, L., Hauser, K. E., & Simmerling, C. (2015). Ff14SB: Improving the accuracy of protein side chain and backbone parameters from Ff99SB. *Journal of Chemical Theory and Computation, 11*(8), 3696–3713.
124. Piana, S., Lindorff-Larsen, K., & Shaw, D. E. (2011). How robust are protein folding simulations with respect to force field parameterization? *Biophysical Journal, 100*(9), L47–L49.
125. Kaminski, G. A., Friesner, R. A., Tirado-Rives, J., & Jorgensen, W. L. (2001). Evaluation and Reparametrization of the OPLS-AA force field for proteins via comparison with accurate quantum chemical calculations on peptides. *The Journal of Physical Chemistry. B, 105*(28), 6474–6487.
126. Case, D. A., Cheatham, T. E., Darden, T., Gohlke, H., Luo, R., Merz, K. M., Onufriev, A., Simmerling, C., Wang, B., & Woods, R. J. (2005). The Amber biomolecular simulation programs. *Journal of Computational Chemistry, 26*(16), 1668–1688.
127. Dickson, C. J., Madej, B. D., Skjevik, Å. A., Betz, R. M., Teigen, K., Gould, I. R., & Walker, R. C. (2014). Lipid14: The Amber lipid force field. *Journal of Chemical Theory and Computation, 10*(2), 865–879.
128. Kirschner, K. N., Yongye, A. B., Tschampel, S. M., González-Outeiriño, J., Daniels, C. R., Foley, B. L., & Woods, R. J. (2008). GLYCAM06: A generalizable biomolecular force field carbohydrates. *Journal of Computational Chemistry, 29*(4), 622–655.
129. Zgarbová, M., Šponer, J., Otyepka, M., Cheatham, T. E., Galindo-Murillo, R., & Jurečka, P. (2015). Refinement of the sugar–phosphate backbone torsion beta for AMBER force fields improves the description of Z- and B-DNA. *Journal of Chemical Theory and Computation, 11*(12), 5723–5736.

130. Zgarbová, M., Otyepka, M., Šponer, J., Mládek, A., Banáš, P., Cheatham, T. E., & Jurečka, P. (2011). Refinement of the Cornell et al. nucleic acids force field based on reference quantum chemical calculations of Glycosidic torsion profiles. *Journal of Chemical Theory and Computation, 7*(9), 2886–2902.
131. Elliott, J. A., Hanna, S., Elliott, A. M. S., & Cooley, G. E. (1999). Atomistic simulation and molecular dynamics of model systems for Perfluorinated Ionomer membranes. *Physical Chemistry Chemical Physics, 1*(20), 4855–4863.
132. Abascal, J. L. F., Sanz, E., García Fernández, R., & Vega, C. (2005). A potential model for the study of ices and amorphous water: TIP4P/ice. *The Journal of Chemical Physics, 122* (23).
133. Wang, J., Wolf, R. M., Caldwell, J. W., Kollman, P. A., & Case, D. A. (2004). Development and testing of a General Amber force field. *Journal of Computational Chemistry, 25*(9), 1157–1174.
134. Still, W. C., Tempczyk, A., Hawley, R. C., & Hendrickson, T. (1990). Semianalytical treatment of solvation for molecular mechanics and dynamics. *Journal of the American Chemical Society, 112*(16), 6127–6129.
135. Li, P., Song, L. F., & Merz, K. M. (2015). Parameterization of highly charged metal ions using the 12-6-4 LJ-type nonbonded model in explicit water. *The Journal of Physical Chemistry. B, 119*(3), 883–895.
136. Roe, D. R., & Cheatham, T. E. (2013). PTRAJ and CPPTRAJ: Software for processing and analysis of molecular dynamics trajectory data. *Journal of Chemical Theory and Computation, 9*(7), 3084–3095.
137. Chagas, M., Rocha, W., & Moraes, A. (2020). Dynamics and Allostery of Zika virus non-structural protein 5 Methyltransferase. *Journal of Biomolecular Structure & Dynamics*, 1–13.
138. Geng, H., Chen, F., Ye, J., & Jiang, F. (2019). Applications of molecular dynamics simulation in structure prediction of peptides and proteins. *Computational and Structural Biotechnology Journal, 17*, 1162–1170.
139. Lindorff-Larsen, K., Piana, S., Dror, R. O., & Shaw, D. E. (2011). How fast-folding proteins fold. *Science (80-.), 334*(6055), 517–520.
140. Qi, R., Wei, G., Ma, B., & Nussinov, R. (2018). *Replica Exchange Molecular Dynamics: A Practical Application Protocol with Solutions to Common Problems and a Peptide Aggregation and Self-Assembly Example*, 101–119.
141. Patel, J. S., Berteotti, A., Ronsisvalle, S., Rocchia, W., & Cavalli, A. (2014). Steered molecular dynamics simulations for studying protein–ligand interaction in cyclin-dependent kinase 5. *Journal of Chemical Information and Modeling, 54*(2), 470–480.
142. Wang, Y., Harrison, C. B., Schulten, K., & McCammon, J. A. (2011). Implementation of accelerated molecular dynamics in NAMD. *Computational Science & Discovery, 4*(1), 015002.
143. Bussi, G., & Laio, A. (2020). Using Metadynamics to explore complex free-energy landscapes. *Nature Reviews Physics, 2*(4), 200–212.
144. Kumar, S., Rosenberg, J. M., Bouzida, D., Swendsen, R. H., & Kollman, P. A. (1992). THE weighted histogram analysis method for free-energy calculations on biomolecules. I. The method. *Journal of Computational Chemistry, 13*(8), 1011–1021.
145. Laio, A., & Parrinello, M. (2002). Escaping free-energy minima. *Proceedings of the National Academy of Sciences, 99*(20), 12562–12566.
146. Gao, Y. Q. (2008). An integrate-over-temperature approach for enhanced sampling. *The Journal of Chemical Physics, 128*(6).
147. Wu, S., Skolnick, J., & Zhang, Y. (2007). Ab initio modeling of small proteins by iterative TASSER simulations. *BMC Biology, 5*(1), 17.
148. Henninot, A., Collins, J. C., & Nuss, J. M. (2018). The current state of peptide drug discovery: Back to the future? *Journal of Medicinal Chemistry, 61*(4), 1382–1414.

149. Guterres, H., Park, S.-J., Jiang, W., & Im, W. (2021). Ligand-binding-site refinement to generate reliable Holo protein structure conformations from Apo structures. *Journal of Chemical Information and Modeling, 61*(1), 535–546.
150. Gutteridge, A., & Thornton, J. (2005). Conformational changes observed in enzyme crystal structures upon substrate binding. *Journal of Molecular Biology, 346*(1), 21–28.
151. Gunasekaran, K., & Nussinov, R. (2007). How different are structurally flexible and rigid binding sites? Sequence and structural features discriminating proteins that do and do not undergo conformational change upon ligand binding. *Journal of Molecular Biology, 365*(1), 257–273.
152. Clark, J. J., Benson, M. L., Smith, R. D., & Carlson, H. A. (2019). Inherent versus induced protein flexibility: Comparisons within and between Apo and Holo structures. *PLoS Computational Biology, 15*(1), e1006705.
153. Heringa, J., & Argos, P. (1999). Strain in protein structures as viewed through Nonrotameric side chains: II. Effects upon ligand binding. *Proteins: Structure, Function, and Bioinformatics, 37*(1), 44–55.
154. Gaudreault, F., Chartier, M., & Najmanovich, R. (2012). Side-chain Rotamer changes upon ligand binding: Common, crucial, correlate with entropy and rearrange hydrogen bonding. *Bioinformatics, 28*(18), i423–i430.
155. McGovern, S. L., & Shoichet, B. K. (2003). Information decay in molecular docking screens against Holo, Apo, and modeled conformations of enzymes. *Journal of Medicinal Chemistry, 46*(14), 2895–2907.
156. Lee, H. S., Lee, C. S., Kim, J. S., Kim, D. H., & Choe, H. (2009). Improving virtual screening performance against conformational variations of receptors by shape matching with ligand binding pocket. *Journal of Chemical Information and Modeling, 49*(11), 2419–2428.
157. Zavodszky, M. I., & Kuhn, L. A. (2005). Side-chain Flexibility in Protein–Ligand Binding: The Minimal Rotation Hypothesis. *Protein Science, 14*(4), 1104–1114.
158. Wong, C. F. (2015). Flexible receptor docking for drug discovery. *Expert Opinion on Drug Discovery, 10*(11), 1189–1200.
159. Gioia, D., Bertazzo, M., Recanatini, M., Masetti, M., & Cavalli, A. (2017). Dynamic docking: A paradigm shift in computational drug discovery. *Molecules, 22*(11), 2029.
160. Ramírez, D., & Caballero, J. (2018). Is it reliable to take the molecular docking top scoring position as the best solution without considering available structural data? *Molecules, 23*(5), 1038.
161. Rosell, M., & Fernández-Recio, J. (2020). Docking approaches for modeling multi-molecular assemblies. *Current Opinion in Structural Biology, 64*, 59–65.
162. Meng, X.-Y., Zhang, H.-X., Mezei, M., & Cui, M. (2011). Molecular docking: A powerful approach for structure-based drug discovery. *Current Computer-Aided Drug Design, 7*(2), 146–157.
163. Warren, G. L., Andrews, C. W., Capelli, A.-M., Clarke, B., LaLonde, J., Lambert, M. H., Lindvall, M., Nevins, N., Semus, S. F., Senger, S., et al. (2006). A critical assessment of docking programs and scoring functions. *Journal of Medicinal Chemistry, 49*(20), 5912–5931.
164. Dominguez, C., Boelens, R., & Bonvin, A. M. J. J. (2003). HADDOCK: A protein−protein docking approach based on biochemical or biophysical information. *Journal of the American Chemical Society, 125*(7), 1731–1737.
165. Chelliah, V., Blundell, T. L., & Fernández-Recio, J. (2006). Efficient restraints for protein–protein docking by comparison of observed amino acid substitution patterns with those predicted from local environment. *Journal of Molecular Biology, 357*(5), 1669–1682.
166. Mintseris, J., Pierce, B., Wiehe, K., Anderson, R., Chen, R., & Weng, Z. (2007). Integrating statistical pair potentials into protein complex prediction. *PROTEINS: Structure, Function, and Bioinformatics, 69*(3), 511–520.
167. Roel-Touris, J., Bonvin, A. M. J. J., & Jiménez-García, B. (2020). LightDock goes information-driven. *Bioinformatics, 36*(3), 950–952.

168. Martins, D. M., Fernandes, P. O., Vieira, L. A., Maltarollo, V. G., & Moraes, A. H. (2024). Structure-guided drug design targeting Abl kinase: How structure and regulation can assist in designing new drugs. *Chembiochem.*
169. Tokarski, J. S., Newitt, J. A., Chang, C. Y. J., Cheng, J. D., Wittekind, M., Kiefer, S. E., Kish, K., Lee, F. Y. F., Borzillerri, R., Lombardo, L. J., et al. (2006). The structure of Dasatinib (BMS-354825) bound to activated ABL kinase domain elucidates its inhibitory activity against Imatinib-resistant ABL mutants. *Cancer Research, 66*(11), 5790–5797.
170. Sonti, R., Hertel-Hering, I., Lamontanara, A. J., Hantschel, O., & Grzesiek, S. (2018). ATP site ligands determine the assembly state of the Abelson kinase regulatory Core via the activation loop conformation. *Journal of the American Chemical Society, 140*(5), 1863–1869.

Part III
From Computer Towards the Clinical

Chapter 15
Molecular Modeling Strategies in Drug Design, Development, and Discovery Targeting Proteases

Viviane Corrêa Santos, Lucas Abreu Diniz, and Rafaela Salgado Ferreira

Abstract This chapter explores the critical role of proteases in human metabolism and the survival of pathogenic organisms, emphasizing their significance as drug development targets due to their biological importance. It outlines the standard nomenclature proposed by Schechter and Berger for categorizing protease subsites and substrates, which aids in understanding protease specificity and guiding inhibitor design. Despite advancements in developing potent peptidic protease inhibitors, challenges related to pharmacokinetic properties have led researchers to employ various drug design techniques. The chapter specifically examines marketed drugs targeting proteases for three viral diseases and Type 2 diabetes, highlighting the importance of computational tools in drug discovery campaigns and showcasing diverse design strategies, mainly structure-based drug design (SBDD) of peptidomimetics and macrocyclic inhibitors.

Keywords Proteases · SBDD · Drug discovery · Viral infections · Diabetes type 2

V. C. Santos
Laboratório de Modelagem Molecular e Planejamento de Fármacos, Departamento de Bioquímica e Imunologia, Instituto de Ciências Biológicas, Universidade Federal de Minas Gerais, Belo Horizonte, MG, Brazil

Department of Chemistry, Grand Valley State University, Allendale, MI, USA

L. A. Diniz · R. S. Ferreira (✉)
Laboratório de Modelagem Molecular e Planejamento de Fármacos, Departamento de Bioquímica e Imunologia, Instituto de Ciências Biológicas, Universidade Federal de Minas Gerais, Belo Horizonte, MG, Brazil
e-mail: rafaelasf@icb.ufmg.br

V. G. Maltarollo (ed.), *Computer-Aided and Machine Learning-Driven Drug Design*, Computer-Aided Drug Discovery and Design 3,
https://doi.org/10.1007/978-3-031-76718-0_15

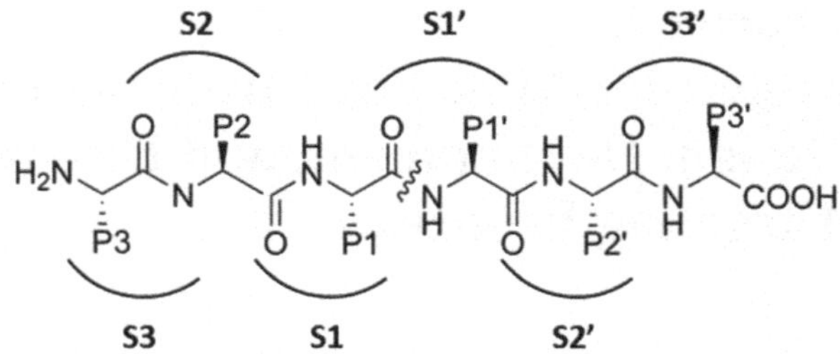

Fig. 15.1 Schematic representation of the nomenclature for protease substrate's residues into their corresponding binding sites. Protease subsites are named S3 to S3' and the substrate residues, P3 to P3'. The scissile bond is represented by ⸾ (Figure prepared with ChemDraw)

1 Introduction

Proteases cleave peptidic bonds in peptides and proteins, catalyzing a hydrolysis reaction [1]. They represent approximately 2% of the human genome and are involved in numerous processes essential for human metabolism and homeostasis [2]. Similarly, pathogenic organisms also rely on proteases for their survival and replication [3]. The biological importance of this class of enzymes is one of the reasons that so many proteases have been targeted for drug development [1–5].

Schechter and Berger proposed a standard nomenclature for protease subsites and their substrates [6, 7], in which the amino acid residues are numbered sequentially from the scissile bond. Toward the N-terminal of the peptide, their numbering starts from P1, while toward the C-terminal, it begins with P1'. Protease subsites are designated analogously: the S1 subsite is the region where the P1 motif binds, S2 recognizes P2, and so on (Fig. 15.1). This widely adopted nomenclature facilitates comparisons of protease active sites and their specificity, which is important for understanding their biological importance and guiding drug design.

The development of protease inhibitors frequently starts with an understanding of their substrate specificity, which allows the design of inhibitors. For instance, combining peptides with a warhead in an appropriate position for a nucleophilic attack by the catalytic residue quickly yields potent peptidic protease modulators. Despite the efficiency of this strategy, peptidic protease inhibitors frequently have inadequate pharmacokinetic properties, such as low bioavailability and poor metabolic profiles [8]. These limitations have motivated the application of varied drug design techniques to develop protease inhibitors as drug candidates.

In this chapter, we focus on four diseases (three viral diseases and Diabetes type 2) for which marketed drugs targeting proteases have been successfully developed using computer-aided approaches (Table 15.1). Among the available drugs, we find examples of various compound classes, including peptidomimetics, macrocyclic inhibitors, and diverse scaffolds of synthetic origin. For each disease, we discuss the importance of the protease target and highlight the drug discovery campaigns in which computational tools played an essential role. Instead of aiming for a

Table 15.1 Protease inhibitors employed in treating HIV, COVID-19, Hepatitis C, and Diabetes type 2, developed aided by computational methods

Disease	Protease target	Drug	Computational methods (with references)
HIV infection	HIV protease	Crixivan® (indinavir)	Molecular modeling [9]
		Kaletra® (lopinavir/ ritonavir)	Molecular modeling [10]
		Prezista® (darunavir)	Molecular modeling [11]
COVID-19	M^{pro}	Paxlovid® (nirmatrelvir/ritonavir)	Docking [12]
		Xocova® (ensitrelvir)	Docking-based virtual screening, structural pharmacophores, docking [13]
Hepatitis C	NS3/4a protease	Zepartier® (grazoprevir/ elbasvir)	Molecular modeling [14]
		Vosevi® (voxilaprevir/ sofosbuvir/ velpatasvir)	Molecular modeling [15]
Diabetes type 2	Dipeptidyl peptidase IV (DPP-4)	Tenelia® (teneligliptin)	Docking [16]

comprehensive description of drugs that act on proteases, we hope to illustrate multiple design strategies and how computational methods have been effectively incorporated into drug design.

2 Structure-Based Drug Design of HIV Protease Inhibitors

2.1 HIV-1 Protease as a Therapeutic Target

Acquired Immunodeficiency Syndrome (AIDS) is a global epidemic impacting millions of people, caused by the Human Immunodeficiency Virus (HIV). According to the Joint United Nations Program on HIV/AIDS (UNAIDS), since the first case was reported in 1981 [17], an estimated 85.6 million people have been infected with HIV, and approximately 40.4 million have died from AIDS or AIDS-associated deaths. Currently, an estimated 39 million people globally are living with HIV/AIDS [18].

The discovery of HIV as the causative agent of AIDS enabled the understanding of the molecular processes associated with HIV infection and replication in host cells and the identification of targets for developing antiviral therapies. Of the 15 proteins encoded by the HIV genome, integrase, reverse transcriptase, and protease act as enzymes with specialized functions [19]. The HIV protease is an aspartic protease that catalyzes the hydrolysis of Gag and Gag-Pol polyproteins. The cleavage of these polyproteins generates structural proteins of the virus core and viral enzymes, which are incorporated into new viral particles, making this enzyme a plausible drug target [20]. The development of HIV-1 protease inhibitors led to the approval of several drugs included in the highly active antiretroviral therapy (HAART) in 1996 [21]. This approach consists of a multidrug regimen with different drug classes to help avoid drug resistance that would be selected by single monotherapies.

Structure-based drug design (SBDD) studies have been employed to develop most protease inhibitors using computational methods and experimental data. Since HIV-1 protease is a homodimer, residues in chain B are herein referred to as having an * to their numbers. Here we discuss the development of first- and second-generation HIV protease inhibitors, as well as the strategies to address the problem of drug resistance.

2.2 Development of First-Generation HIV Protease Inhibitors by Structure-Based Drug Design

2.2.1 Saquinavir

The first FDA-approved HIV protease inhibitor, saquinavir, was developed based on the transition-state isostere strategy aided by computational structure-based techniques. In this strategy, the cleavable amide is replaced by a non-hydrolyzable tetrahedral isostere, such as hydroxyethylene or hydroxyethylamine moieties, to generate inhibitors. The selectivity of saquinavir is based on the difference in substrate specificity of the HIV protease and mammalian aspartic proteases. While the host enzymes do not exhibit affinity for substrates containing proline at P1', sequences containing proline in the P1' position are preferentially cleaved by HIV protease, such as in the substrates featuring Phe-Pro and Tyr-Pro within the Gag and Gag-Pol gene products [22].

During the development of saquinavir, compounds containing the hydroxyethylamine moiety were synthesized, conducting an exhaustive exploration of side chains and substituents with distinct steric and electronic properties. Among the explored options, the substitution of the native proline with the decahydro-isoquinoline-3-carbonyl (DIQ) group resulted in a potent HIV protease inhibitor with an inhibition constant (K_i) of 0.12 nM for HIV-1 and $K_i < 0.1$ nM for HIV-2, which was later named saquinavir (Fig. 15.2a) [23]. Subsequently, crystallographic analyses revealed that the DIQ group occupies the S1' subsite, establishing hydrophobic interactions with the enzyme, and the carbonyl group of DIQ interacts with a water molecule connecting the inhibitors with the flap regions. Furthermore, other hydrogen bonds were observed between the inhibitor and the enzyme. For example, the P2 carboxamide interacts with Asp29 and Asp30 residues (Fig. 15.2b) [24]. Ultimately, in 1995, saquinavir was the first HIV protease inhibitor approved by the FDA, as Invirase® [25].

2.2.2 Indinavir

In general, peptidomimetic compounds exhibit poor solubility in water and low oral bioavailability, as observed with saquinavir, which has an oral bioavailability of 4% [26]. To address this issue, researchers initiated the development of indinavir from a series of hydroxyethylene dipeptide isostere inhibitors of HIV protease that, despite being potent HIV-1 protease inhibitors with half-maximal inhibitory concentration (IC_{50}) between 0.03 nM and 470 nM, exhibited low water solubility [27]. Considering the structural characteristics of saquinavir, it was hypothesized that introducing the basic group amine to this series of compounds could enhance bioavailability. Consequently, a novel class of hydroxylamine pentanamide compounds was proposed, and molecular modeling studies were conducted to assess their potential as HIV protease inhibitors [9].

The modeling began with the alignment of an energy-minimized structure of HIV-1 protease complexed with a hydroxyethylene isostere inhibitor (compound L-685,434; Fig. 15.2c) with the X-ray crystallography structure of saquinavir in complex with HIV-1 protease. Both compounds occupied the hydrophobic binding pockets spanning from S2 to S2' with their P2 and P1' carbonyl groups oriented toward the structural water found in most X-ray crystal structures of HIV-1 inhibitors (PDB IDs: 4HVP, 5HVP, 9HVP, and 7HVP) [28–31], presumably establishing a hydrogen bond between them. Subsequently, they overlaid the structure with the hydroxyethylene isostere inhibitor and an energy-minimized structure for the

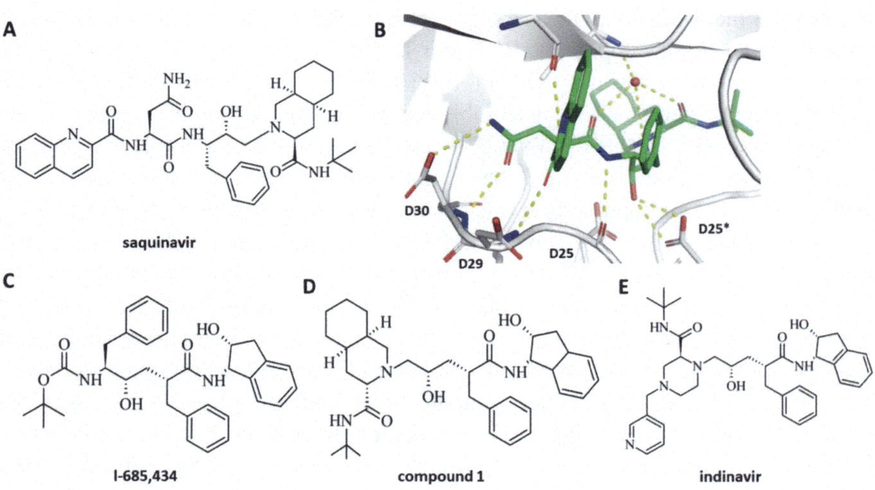

Fig. 15.2 First-generation HIV-1 protease inhibitors. Chemical structures of saquinavir (**a**) and its binding mode (PDB ID: 1HXB) in the HIV-1 protease active site (**b**). Chemical structures of the hydroxyethylene isostere inhibitor, L-685,434 (**c**), the hydroxylamine pentanamide inhibitor, compound **1** (**d**), and the approved drug, indinavir (**e**). Water molecules are represented as red spheres, and the hydrogen bonds are shown as yellow dashed lines. Residues in chain B are identified by the letter and number followed by *. Important residues and compounds are represented as sticks and colored by atom. Carbons are colored gray in the enzyme and green in the compound. (Figure produced using ChemDraw and PyMOL)

hydroxylamine pentanamide inhibitor (compound **1**; Fig. 15.2d). This analysis revealed that the carbonyl moieties maintained alignment with the water molecule, even with the incorporation of an amine into the backbone—denoted by the substitution of Phe and tert-butyl carbamate moieties with the decahydroisoquinoline tert-butylamide.

Studies with compound **1** showed a better pharmacokinetic profile than the hydroxyethylene isostere inhibitor. However, compound **1** exhibited moderate potency against viral dissemination in infected human T cells with cell culture inhibitory concentration (CIC_{95}) values of 400 nM. Thus, considering the information obtained from molecular modeling studies, structural modifications were explored, leading to the development of indinavir (Fig. 15.2e), which exhibited a CIC_{95} of 50 nM and an oral bioavailability of approximately 63% [26]. Indinavir, under the brand Crixivan®, was approved in 1996 as a treatment for HIV/AIDS [32].

Finally, it is important to emphasize that other first-generation HIV-1 protease inhibitors were developed and approved, such as ritonavir, nelfinavir, and amprenavir. However, the emergence of HIV-resistant strains led to the development of the so-called second-generation inhibitors.

2.3 *Drug Resistance and Second-Generation HIV Protease Inhibitors*

The development of ritonavir, a first-generation inhibitor, provided relevant information for developing lopinavir. Ritonavir is a potent inhibitor (K_i of 0.015 nM), which was approved by the FDA in 1996 (Fig. 15.3a). Studies conducted with ritonavir-treated patients revealed substitutions at nine different codons in the protease gene, with mutants at Val82* being the first to emerge in infected patients [33]. This information motivated the search for inhibitors that would not interact with Val82*, making mutations at this position less likely to induce drug resistance.

The synthesis of DNA from a viral RNA is catalyzed by the viral reverse transcriptase. This process has a high error propensity, mainly attributable to the enzyme's elevated mismatch error rate (10^{-3}–10^{-5} nucleotide bases per cycle) and its absence of proofreading activity. Thus, this phenomenon allows the emergence of polymorphic mutations within HIV-1 genome [34]. Several factors, such as deleterious mutations, natural selection, and the host immune response, constrict the proliferation of mutated strains within a specified timeframe [35]. The introduction of antiretroviral treatment exerts novel selection pressures that may contribute to the proliferation of drug-resistant strains.

Primary mutations frequently occur in residues near the active site involved in substrate binding. These mutations affect electrostatic and van der Waals interactions between the enzyme and the inhibitor, frequently reducing binding affinity

[36]. Importantly, these modifications can also undermine the proficiency of substrate processing, adversely impacting virus replication [37]. Conversely, secondary mutations are located outside the active site and may compensate for the deleterious effects of primary mutations on substrate processing while maintaining drug resistance [38, 39].

Many specific primary and secondary mutations in HIV protease are associated with different inhibitors, and strategies have been developed to design inhibitors that interact with the HIV protease in a way that creates a higher barrier to resistance. Here we discuss the development process of two second-generation inhibitors, lopinavir, and darunavir.

2.3.1 Lopinavir

Structural analysis, based on the X-ray crystal structure of wild-type HIV-1 protease, revealed an interaction between ritonavir's P3 isopropylthiazolyl group and the

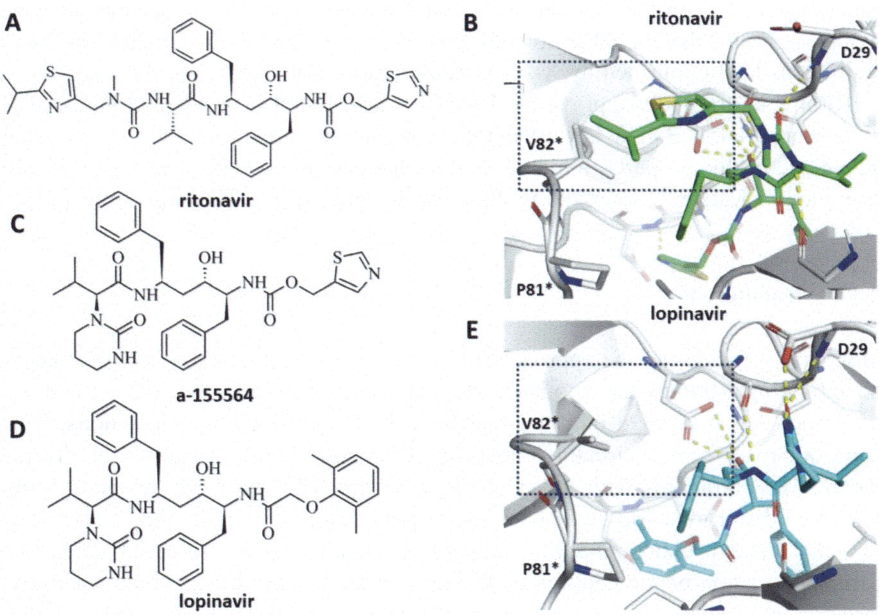

Fig. 15.3 Structure-based development of lopinavir. Chemical structure of ritonavir (**a**) and its binding mode (PDB ID: 1HXW) in the HIV-1 protease active site (**b**). Chemical structures of the cyclic urea compound, A-155564 (**c**), lopinavir (**d**), and the binding mode of lopinavir (PDB ID: 1MUI) in the HIV-1 protease active site (**e**). The isopropylthiazolyl group of ritonavir and the Val82* residues are highlighted by black dashed lines. The same region is highlighted for the HIV-1 protease structure bound to lopinavir. Residues in chain B are identified by the letter and number followed by *. Hydrogen bonds are shown as yellow dashed lines. Important residues and compounds are represented as sticks and colored by atom. Carbons are colored gray in the enzyme and green or cyan in the compounds. (Figure produced using ChemDraw and PyMOL)

Val82* residue (Fig. 15.3b) [40]. To develop an inhibitor with activity less dependent on an interaction with Val82*, researchers removed the P3 isopropylthiazolyl group from ritonavir. However, this reduction decreased the protease affinity, resulting in a 30-fold lower anti-HIV activity ($EC_{50} = 1.8$ μM) compared to ritonavir ($EC_{50} = 0.06$ μM). A conformation constraint introduced by a cyclic urea compound (compound A-155564; Fig. 15.3c) resulted in improved anti-HIV activity ($EC_{50} = 0.15$ μM), and the substitution of the P2' group with a dimethyl phenoxy acetyl group in lopinavir (Fig. 15.3d), which exhibited an EC_{50} of 0.01 μM and a K_i of 1.3 pM for the wild-type protease compared to ritonavir's K_i of 10 pM. Moreover, lopinavir demonstrated higher potency than ritonavir against proteases with mutations in Val82*, such as a K_i of 3.7 pM for the V82F mutant HIV protease compared to ritonavir's *Ki* of 520 pM [10].

To comprehend the enhanced binding affinity of lopinavir towards both wild-type and mutant proteases, researchers modeled the inhibitor within the active site of the protease based on a crystal structure of the complex with ritonavir. In this model, the 2,6-dimethylphenoxy group was positioned in the S2' subsite to fill the available space, while the valinyl-cyclic urea moiety was rotated within the S2 subsite to ensure optimal hydrogen bonding with Asp29 or Asp30 [10, 41]. As expected, the model suggested that the P3 region of lopinavir engaged in minimal interactions with Val82*, as later confirmed by X-ray crystallography (Fig. 15.3e) [41].

Ritonavir, originally approved in 1996, was later found to be a potent inhibitor of cytochrome P450 3A, a key metabolic enzyme for protease inhibitors. Taking advantage of this property, lopinavir and ritonavir were combined into a single pill marketed as Kaletra®, approved in 2000 for the treatment of HIV infections [42].

2.3.2 Darunavir

Moreover, considering the approaches to avoid or overcome resistance, the "backbone binding" strategy has also been adopted. In this approach, the goal is to design inhibitors that maximize inhibitor-backbone hydrogen-bonding interactions. This concept arose from a structural analysis comparing X-ray structures of diverse mutant HIV proteases with those of the wild-type HIV protease, revealing only minimal distortions on the backbone conformation within the active site [43, 44]. The backbone binding strategy guided the structure-based design of darunavir—a non-peptidic inhibitor. To improve water solubility and bioavailability, research focused on designing non-peptidic cyclic or heterocyclic structures, particularly cyclic ether or polyether-derived ligands.

Initially, researchers developed a cyclic ether-derived ligand (compound **2**, Fig. 15.4a) based on the X-ray structure of saquinavir, which exhibited an IC_{50} of 132 nM against the enzyme. Subsequently, an X-ray crystal structure of this ligand revealed a potential weak interaction between the oxygen atom of the tetrahydrofuran ring and the backbone nitrogen atoms of residues Asp29 and Asp30 [45]. To strengthen the interaction with these residues, researchers synthesized compounds featuring bicyclic tetrahydrofuran rings (bis-THF), resulting in a higher potency

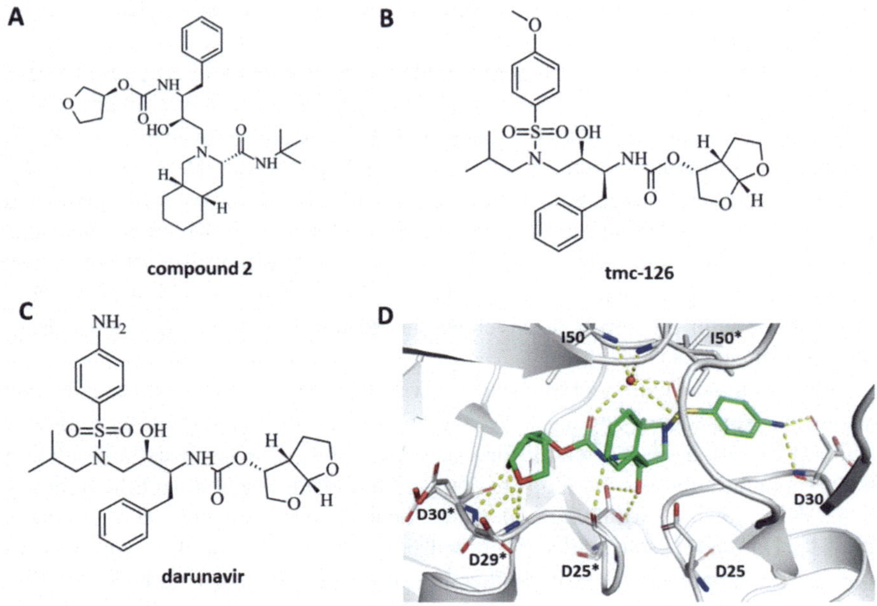

Fig. 15.4 The development of darunavir. Chemical structures of a cyclic ether-derived ligand, compound **2** (**a**), TMC-126 (**b**), and darunavir (**c**). Binding mode of darunavir (PDB ID: 2IEO) to the HIV-1 protease mutant I84V active site (**d**). Water molecules are represented as red spheres, and the hydrogen bonds are shown as yellow dashed lines. Residues in chain B are identified by the letter and number followed by *. Important residues and compounds are represented as sticks and colored by atom. Carbons are colored gray in the enzyme and green in the compound. (Figure produced using ChemDraw and PyMOL)

against the enzyme ($IC_{50} = 1.8$ nM) and enhanced water solubility [46]. The impact of bis-THF on potency enhancement was further investigated in the context of other isosteres [47], culminating in the development of TMC-126, with a K_i of 14 pM (Fig. 15.4b). To better understand the potential interactions with this inhibitor, researchers conducted structural modeling of the HIV-1 protease complexed with TMC-126. They utilized published crystal structures of HIV-1 protease complexed with amprenavir, as well as with various inhibitors containing bis-THF, followed by minimization of the active site using the Sybyl force field [48]. Consistent with previous studies demonstrating higher potency with the substitution of THF by a bis-THF group, the model suggests this heightened potency to be linked to hydrogen bonding interactions between the oxygen atoms of the bis-THF moiety and the main-chain amides of Asp29* and Asp30* residues.

Subsequently, incorporating bis-THF into a p-amino sulfonamide isostere led to the development of darunavir (Fig. 15.4c). X-ray crystallography studies involving darunavir complexed with mutant proteases revealed that the interactions between the bis-THF group and the amides of Asp29* and Asp30* remained intact, while new polar interactions were also observed with the side-chain carboxylate of Asp30 (Fig. 15.4d) [49].

In addition to the backbone strategy, another approach to avoid drug resistance is based on the substrate-envelope hypothesis. Viral polyproteins, subject to processing by HIV protease, undergo adaptive changes that culminate in forming a conserved substrate envelope upon binding with the enzyme's active site [39]. The substrate-envelope hypothesis suggests that the overall substrate conformation is pivotal in achieving effective binding, not just the amino acid sequence. Mutations near this substrate envelope significantly diminish the protease‘s affinity for its substrates, rendering them unfavorable. Consequently, developing protease inhibitors strategically aligned with the substrate envelope should prevent the emergence of drug-resistance mutations.

Although the development of darunavir was not based on the substrate-envelope hypothesis, subsequent studies have shown that it fits the consensus substrate volume [50]. As expected for a drug complying both with the backbone-binding strategy and the substrate-envelope hypothesis, darunavir is a non-peptidic inhibitor of HIV-1 protease with high potency against drug-resistant HIV variants [11, 51]. Darunavir, sold under the brand name Prezista®, was first approved in 2006, and it is part of antiretroviral therapy for people infected with HIV. Another virus disease for which antiviral drugs were developed with the aid of molecular modeling techniques is hepatitis C and it will be discussed in the following section.

3 Structure-Based Drug Design of Grazoprevir and Voxilaprevir as Hepatitis C Virus (HCV) NS3/4a Protease Inhibitors

In 1989, scientists isolated the so-called non-A and non-B hepatitis agents [52]. HCV infection occurs mainly in the liver and has an incubation period of 2–12 weeks, which is followed by an acute asymptomatic phase that is often undiagnosed. When symptoms appear, they can be fever, tiredness, loss of appetite, nausea and vomiting, abdominal pain, dark urine, pale feces, joint pain, and jaundice [53]. The infection follows two courses: spontaneous clearance (18–34%) of infected individuals or progression to chronic infection [54]. HCV can be found worldwide, and the WHO estimated in 2024 that 50 million people were chronically infected with the virus [53].

Two FDA-approved drugs targeting the HCV non-structural (NS) proteins NS3/4a protease were developed using computational methods. NS3 is a non-structural protein with two domains: N-terminal serine protease and C-terminal RNA helicase [55]. NS4a is a polypeptide co-factor for the protease activity of NS3. The NS3/4a complex processes the HCV polyprotein, releasing among the cleavage products, the non-structural proteins required for virus replication [56]. HCV is phylogenetically classified into eight genotypes, named 1 to 8,

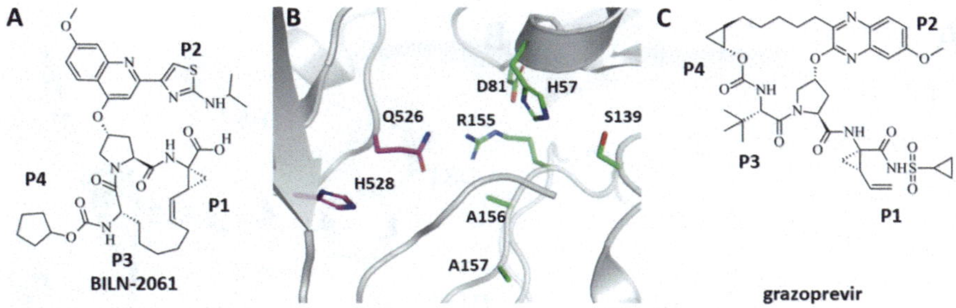

Fig. 15.5 Structure-based development of grazoprevir. Chemical representation of BILN-2061 (**a**); crystal structure of the full-length NS3/4A (protease domain with residues in green; helicase domain with residues in pink, the enzyme is colored in gray) (PDB ID: 1CU1) (**b**); and chemical structure of grazoprevir (**c**). The P4–P1 positions of each compound are labeled. Important residues and the compounds are represented as sticks and colored by atom. (Figure produced using ChemDraw and PyMOL)

according to the date when they were discovered [57]. Most of the hepatitis C cases involve GT1 (44%), GT3 (25%), and GT4 (15%) [58]. The first drugs developed targeted GT1, and as the other genotypes were being described, resistance emerged, making necessary the genotyping of patients' samples to select the best treatment option; however, this was only sometimes possible. Thus, the discovery of pan-genotypic anti-HCV compounds represented a significant advance in hepatitis C treatment, and nowadays, these compounds are used across the globe, and the countries adopting them are released from the need for sample genotyping [59].

The development of grazoprevir and voxilaprevir will be discussed here and they have the same starting point. The first macrocyclic inhibitors designed were based on NMR data of substrate-like peptide inhibitors of NS3/4a protease and were rigid to ensure the trans geometry of the P2-P3 amide bond [60]. Later, the macrocyclic inhibitor BILN-2061 (Fig. 15.5a) was modeled into the active site of the apo crystal structure of NS3/4a protease (PDB ID 1CU1) [61, 62]. Considering that the P2 and thiazolyl quinoline portions of BILN-2061 lie on pockets formed between the protease and helicase domains of NS3/4a, more analogs were designed and modeled to the same crystal structure. This modeling led to the design of compounds with a P2–P4 macrocyclic constraint active against genotypes GT1 and GT2 of NS3/4a protease [63]. However, analogs with a large substituent to the P2 heterocycle and fused ring were more active against the GT3 genotype enzyme when compared to GT1. After modeling these analogs [14], a cyclopropyl constraint was added to the P2–P4 macrocyclic linker, improving the electrostatic interactions with the residue 156 (Fig. 15.5b), probably leaving room to accommodate the A156T/V substitutions observed in resistant HCV populations, also resulting in increased the lipophilicity to enhance the liver exposure (since HCV infects the liver), and a quinoxaline was added to improve the compound's solubility. All of these resulted in the discovery of grazoprevir (Fig. 15.5c), approved in 2016 by the FDA under the name Zepartier®, a single-oral dose pill also containing the NS5a inhibitor elbasvir [64].

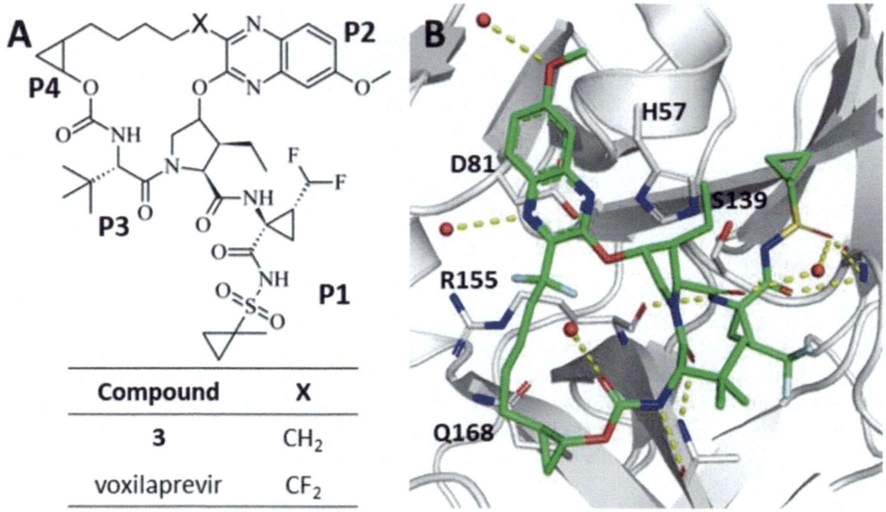

Compound	X
3	CH_2
voxilaprevir	CF_2

Fig. 15.6 Structure-based development of voxilaprevir. Chemical representations of compound **3** and voxilaprevir (**a**). Crystal structure of voxilaprevir complexed with GT1 D168Q (GT3 surrogate) NS3/4A protease (PDB ID 6NZT) (**b**). Water molecules are represented as red spheres, and the polar contacts are shown in yellow dashed lines. The P4–P1 positions of each compound are labeled. Important residues and compounds are represented as sticks and colored by atom. Carbons are colored gray in the enzyme and green in the compound. (Figure produced using ChemDraw and PyMOL)

Patients treated with grazoprevir had their serum alanine transaminase elevated, a potential biomarker for hepatotoxicity, resulting in a reduction in the clinical dosage and warnings on the prescription label [15]. Another issue that emerged with the description of new HCV genotypes was that grazoprevir shows higher activity against GT1 than GT3 [65]. Therefore, researchers focused on improving the activity of the molecules against HCV genotypes targeting the catalytic triad residues (Ser139, His57, and Asp81) since they are conserved across the genotypes. This led to the developed of compound **3** (Fig. 15.6a), with good antiviral [half-maximal effective concentration (EC_{50}) of 23 nM] and enzymatic (K_i of 165 pM) potency. An X-ray crystallography structure of compound **3** with GT1 D168Q (GT3 surrogate) NS3/4A protease (PDB ID 6NZV) showed the ethyl part of **3** in a van der Waals interaction with His57.

Additionally to the improvements in potency against GT3, efforts were made to improve the compound's pharmacokinetics and liver distribution to reduce the hepatotoxicity. The search for compounds with superior metabolic stability and reduced formation of protein adducts eventually led to the development of voxilaprevir (Fig. 15.6a) [15, 66]. Voxilaprevir is more potent than **3** in GT3 enzymatic (K_i of 63 pM) and antiviral (EC_{50} of 6.1 nM) assays. This higher potency might be explained by hydrophobic contact between the difluoro methylene on the macrocycle and the alkyl portion of R155, as observed in the co-crystal with GT1

D168Q (GT3 surrogate) NS3/4A protease (PDB ID 6NZT) (Fig. 15.6b). Overall, the activity of Voxilaprevir against common GT1–6 ranged between an EC_{50} of 1.5 and 6.6 nM. Voxilaprevir, together with sofosbuvir (inhibitor of the HCV NS5B polymerase) and velpatasvir (HCV NS5A inhibitor), is a component of Vosevi®, approved in 2017 as a pan-genotypic treatment for patients who failed previous therapies for hepatitis C [15].

4 Structure-Based Drug Design of Nirmatrelvir and Ensitrelvir as SARS-CoV-2 M^{pro} Inhibitors

Since 2020, in response to the COVID-19 pandemic, there have been remarkable efforts to design drugs targeting proteases. The description of COVID-19 was quickly followed by the identification of its causing agent, the severe acute respiratory syndrome-related Coronavirus 2 (SARS-CoV-2), which was classified in the Coronaviridae family by the *Coronaviridae* Study Group (CSG) of the International Committee on Taxonomy of Viruses, based on the recognized similarity to bat coronaviruses and human coronaviruses [67]. The first cases of the disease were reported initially as pneumonia of unknown origin, in November 2019, in Wuhan (China). Just 4 months later, in March 2020, COVID-19 had spread throughout the globe, and a pandemic situation was declared. By that point, the SARS-CoV-2 genome had been sequenced [68], revealing high similarity to other coronaviruses, and there were already crystallographic structures of SARS-CoV-2 main protease (M^{pro}, also known as chymotrypsin-like protease, $3CL^{pro}$, or NS5) [69] and papain-like protease (PL^{pro}) deposited in the Protein Data Bank.

M^{pro} and PL^{pro} are essential for SARS-CoV-2 multiplication, as they process the viral polypeptides pp1a and pp1ab into 16 nonstructural viral proteins [70]. Thus, these proteases were promptly targeted for drug design. Later, with a better understanding of the process of host cell infection by SARS-CoV-2, the human proteases Cathepsin L and TMPRSS2 were also identified as targets for the development of COVID-19 treatments [71, 72]. While there are numerous studies on each of these four target proteases, the most successfully targeted is M^{pro}, and thus we will focus our discussion on this target. In addition to its importance for viral replication, M^{pro} is considered a good drug target due to its unique specificity for Gln at P1, which is not found among human cysteine proteases, contributing to the developing specific M^{pro} inhibitors and improves drug safety. Also, its conservation among coronavirus creates the potential for developing broad-spectrum antiviral drugs.

Computational strategies employed toward the development of M^{pro} inhibitors include docking-based virtual screening of FDA-approved drugs [73], in-house libraries [74, 75], and ultra-large commercial libraries [76, 77] (see Chaps. 8 and 11). Free energy perturbation calculations (see Chap. 10) have also led to potent M^{pro} inhibitors with antiviral activity [78]. Machine learning models have also been successfully employed to extract information from high-throughput structural data

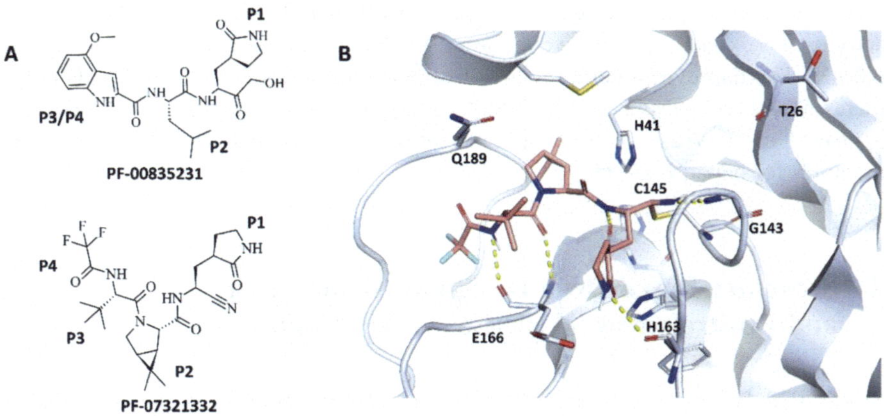

Fig. 15.7 Structure-based development of nirmatrelvir. Chemical structures of the PF-00835231 (lead compound) and PF-07321332 (nirmatrelvir) (**a**). Binding mode of nirmatrelvir (PDB ID: 7RFW) in the SARS-CoV-2 M^{pro} active site (**b**). Hydrogen bonds are represented as yellow dashed lines. The P4–P1 positions of each compound are labeled. Important residues and compounds are represented as sticks and colored by atom. Carbons are colored gray in the enzyme and salmon in the compound. (Figure produced using ChemDraw and PyMOL)

and benefit the design of inhibitors that reveal potent antiviral properties [79]. Altogether, efforts based on computer-aided techniques and experimental approaches have yielded several classes of M^{pro} inhibitors (reviewed by [80]), including marketed drugs.

The active site conservation between SARS-CoV and SARS-CoV-2 was a key factor contributing to the quick development of Paxlovid®, a medicine containing the SARS-CoV-2 M^{pro} inhibitor nirmatrelvir and ritonavir. The starting point for developing nirmatrelvir was the SARS-CoV M^{pro} inhibitor PF-00835231, previously developed in structure-based design efforts against SARS-CoV [81]. After the outbreak of COVID-19, PF-00835231 was also shown to be potent against the SARS-CoV-2 enzyme and in infected Vero E6 cells, showing stability in plasma, adequate solubility, and clearance rates suitable for development as a COVID-19 treatment via intravenous infusion [81].

Nevertheless, the low passive absorptive permeability of PF-00835231 was an important limitation that had to be overcome to obtain an orally administered drug. Thus, PF-00835231 was the lead compound for structure-based efforts to optimize its pharmacokinetic properties (Fig. 15.7a). Throughout optimization, molecular modeling studies against SARS-CoV-2 M^{pro} were employed to suggest compounds that should simultaneously result in favorable pharmacokinetic properties and maintain high potency against the target [12]. Examples of modifications guided by modeling studies include the removal of hydrogen bond donors to improve oral absorption and the incorporation of a nitrile warhead to enhance solubility while also introducing modifications to ensure the occupation of the S2 and S3 pockets and performing hydrogen bond interactions with Gln189 and Glu166 (Fig. 15.7b, c).

Among the compounds developed, PF-07321332 (nirmatrelvir) was chosen as the clinical candidate considering its potent activity against M^{pro}, in antiviral assays and animal models, the ease of synthetic scale-up, reduced propensity for epimerization, and enhanced solubility. Additionally, it showed antiviral activity against multiple coronaviruses and no inhibitory effects against a wide panel of mammalian proteases, G protein-coupled receptors, kinases, transporters, phosphodiesterases, and cardiac ion channels. One limitation in nirmatrelvir's profile was its intense metabolism by CYP450, especially by CYP3A4. To overcome this hurdle, co-administration with ritonavir, a potent CYP3A4 inhibitor, was proposed and yielded an improved pharmacokinetic profile [12]. This combination demonstrates how CYP450 inhibition, frequently considered an unfavorable property due to associated drug–drug interactions, was explored positively. The nirmatrelvir/ritonavir combination, licensed as Paxlovid®, reduced hospitalization by 89% in unvaccinated adults at high risk for progression to severe COVID-19 [82]. Paxlovid® was approved for emergency use by the FDA in December 2021, followed by approval by several international agencies, and was fully approved in May 2023.

The development of ensitrelvir (Xocova®) was also highly supported by structure-based approaches [13]. Aiming to discover noncovalent and nonpeptidic inhibitors of SARS-CoV-2 M^{pro}, researchers performed a docking-based virtual screening of their in-house library. To filter the docking results, they also built a pharmacophoric model based on the crystallographic structures of three ligands in complex with M^{pro}. The pharmacophore consisted of an acceptor site hydrogen bonding with the side-chain NH donor of His163 in S1, a lipophilic site in S2, and an acceptor site interacting with the Glu166 main-chain NH, all of which had to be present. It is interesting to highlight that the definition of a pharmacophore was supported by the large amount of structural data of complexes of ligands bound to M^{pro}, including those from important initiatives such as the COVID-19 Moonshot [83]. After filtering to ensure the compounds met the proposed pharmacophore, the top 300 compounds from the virtual screening were evaluated in biochemical assays with M^{pro}. Among the hits with IC_{50} under 10 μM, hit **4** was prioritized for optimization. While **4** had a moderate potency against the enzyme (IC_{50} of 8.6 μM), it presented a favorable pharmacokinetic profile with *in vitro* metabolic stability, high oral bioavailability, and low clearance *in vivo* in rats. Co-crystallization of **4** with M^{pro} confirmed the binding mode predicted by docking and supported its structure-based optimization based on modifications in two regions (Fig. 15.8a, b). First, to optimize interactions with the S1' pocket, a cyclization was performed in the P1' motif while maintaining the hydrogen bond with Thr26, resulting in compound **5**. Remarkably, a 90-fold improvement in enzymatic inhibition (IC_{50} of 0.096 μM) was obtained with this modification while keeping a favorable pharmacokinetic profile. Then, ensitrelvir (a.k.a. S-217622) was designed by an additional cyclization in the P1 motif and a modification in the pattern of fluorine substitution in the P2 ring (Fig. 15.8a, c). Ensitrelvir showed high potency against M^{pro} (IC_{50} of 0.013 μM) and in antiviral assays (EC_{50} of 0.37 μM), high metabolic stability in vitro, high oral absorption, and low clearance in rats, monkeys, and dogs.

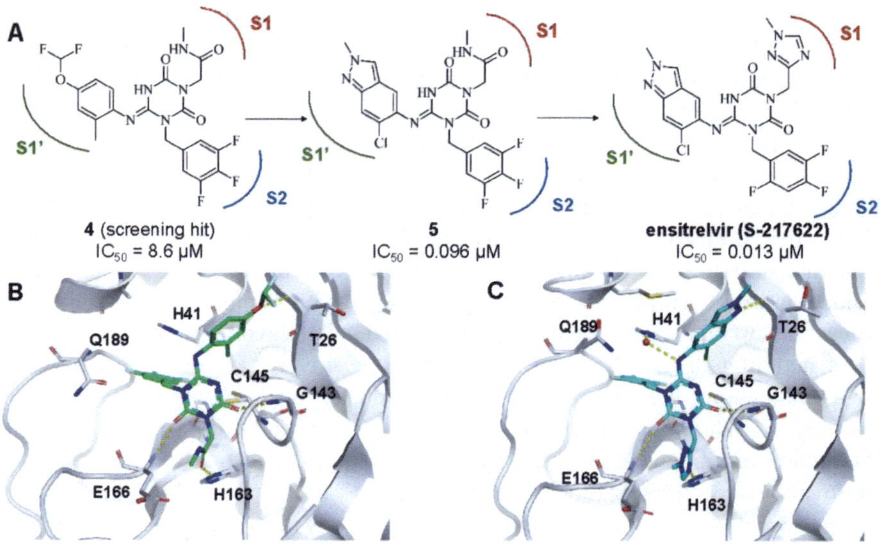

Fig. 15.8 Structure-based design of ensitrelvir. Starting from the initial hit **4**, obtained from virtual screening, the cyclization of the P1' motif led to compound **5**, with a 90-fold improvement in IC_{50}. An additional cyclization in the P1 motif and a modification in the pattern of fluorine substitution in the P2 ring yielded ensitrelvir (**a**). Binding modes of compound **4** (PDB ID: 7VTH) (**b**) and ensitrelvir (PDB ID: 7VU6) (**c**) in the SARS-CoV-2 M^{pro} active site. Water molecules are represented as red spheres, and the hydrogen bonds are represented as yellow dashed lines. The S2–S1′ subsites are labeled. Important residues and compounds are represented as sticks and colored by atom. Carbons are colored gray in the enzyme and green or cyan in the compounds. (Figure produced using ChemDraw and PyMOL)

Similar to nirmatrelvir, ensitrelvir showed potential as a pancoronavirus agent, as it had antiviral activity against several SARS-CoV-2 variants evaluated (Pango lineages A, B.1.1.7, P.1, B.1.351, B.1.617.2, and B.1.1.529, with EC_{50} values against SARS-CoV-2-infected VeroE6/TMPRSS2 between 0.29 and 0.50 μM) and other coronaviruses (SARS-CoV-infected VeroE6/TMPRSS2 cells, EC_{50} of 0.21 μM; MERS-CoV-infected VeroE6/TMPRSS2 cells, EC_{50} of 1.4 μM; HCoV-OC43-infected MRC-5 cells, EC_{90} of 0.074 μM; HCoV-229E-infected MRC-5 cells, EC_{50} of 5.5 μM). Additionally, ensitrelvir is selective for coronavirus proteases, with no inhibitory properties against several host-cell proteases [13]. The efficacy of ensitrelvir in decreasing viral load and ameliorating COVID-19 severity was shown in hamster models, with good oral bioavailability in this animal model [84]. Ensitrelvir obtained emergency approval for COVID-19 treatment in November 2022 in Japan [85]. A recent study indicated ensitrelvir's efficacy against the 11 most frequent M^{pro} mutants circulating globally, with IC_{50} values similar (within two-fold) to the ones obtained against the wild-type protein [86].

The approvals of Paxlovid® and Xocova® were very important additions to the therapeutic arsenal for COVID-19 treatment. Nevertheless, numerous studies have demonstrated the existence of M^{pro} mutants resistant to nirmatrelvir and ensitrelvir, both in biochemical studies and in antiviral assays [87–89]. Several resistance mutants identified have been reported in clinical samples, reinforcing their possible clinical relevance [87]. Over 30 mutations have been described to reduce inhibition of M^{pro}, the susceptibility of SARS-CoV-2 replication, or both, upon treatment with nirmatrelvir and/or ensitrevir, and the most frequently reported mutations were G15S and T21I [90]. Interestingly, mutants resistant to both nirmatrelvir and ensitrelvir might still be inhibited by M^{pro} inhibitors that show distinct binding modes to the active site [91]. Structural and kinetic studies have been employed to understand the mechanisms linking M^{pro} mutations to drug resistance, revealing two evolutionary mechanisms. The first group of mutations directly lower drug binding affinity, as observed for E166V, which disrupts a hydrogen bond interaction between nirmatrelvir and E166 and alters that hydrogen bond network in the S1 pocket. Frequently, these mutations also alter substrate processing, reducing viral fitness. The other resistance mechanism is represented by mutations that do not directly affect drug binding, but increase protease activity, as observed for T21I. The combination of these two resistance mechanisms, such as variants containing T21I and E166V, confer resistance to the current drugs while maintaining their viral fitness [91]. In addition, cross-resistance to nirmatrelvir and ensitrelvir has been reported, bringing further concerns about the available drug arsenal [90, 92]. Therefore, it is important to continue developing other drug candidates targeting M^{pro}, which can help to overcome the limitations of the current ones.

5 Structure-Based Drug Design of Teneligliptin as a Dipeptidyl Peptidase IV Inhibitor in the Treatment of Type 2 Diabetes Mellitus

Another disease in which drug discovery campaigns targeting a protease were successful is Type 2 diabetes mellitus (T2DM), a metabolic disease characterized by high glycemia, insulin resistance, and impaired insulin secretion. The symptoms can include the following: feeling very thirsty, polyuria, blurred vision, tiredness, and weight loss. If not controlled, T2DM can especially damage nerves and blood vessels [93]. The dipeptidyl peptidase IV (DPP-4) is a validated drug target for treating T2DM since its inhibition leads to an increase in insulin levels in response to hyperglycemia [94]. This therapeutic effect is mediated by glucagon-like peptide 1 (GLP-1), a potent stimulator of insulin secretion, whose activity is abolished by cleavage by DPP-4. DPP-4 is a serine protease that removes the two last residues from the N-terminal of peptides with Ala or Pro in P1 [95].

Many of the first DPP-4 inhibitors had substituted pyrrolidines or thiazolidines to mimic proline in P1 and an electrophilic warhead such as nitriles to bind to the catalytic Ser630 covalently [16]. However, they were chemically unstable and also inhibited DPP-8 and DPP-9, causing multiorgan toxicities and mortality in rats, gastrointestinal toxicity in dogs, and inhibition of T cell activation/proliferation in both animal models [16, 96]. Therefore, researchers searched for inhibitors without the nitrile moiety and with an increased affinity for the S2 site. The interest in optimizing P2 was based on a previously observed ten-fold increase in the inhibitory activity of an arylamino group in P2, compared with a prolyl-(S)-2-cyanopyrrolidine [97]. Compound **6**, a phenyl analog in which the (S)-2-cyanopyrrolidine moiety was converted to a thiazolidine structure, showed potent DPP-4 inhibitory activity (IC_{50} of 1.6 nM) and moderate selectivity against DPP-8 and DPP-9 (Fig. 15.9a). Substituting the aryl group with a quinolyl ring with a trifluoromethyl group at S2 increased potency (compound **7**, IC_{50} of 0.4 nM) and selectivity [98]. The crystal structure of compound **7** bound to DPP-4 (PDB ID 3VJM) revealed the thiazolidine group of the molecule occupying the S1 site and the proline group forming salt bridges with the active site's glutamate residues 205 and 206; moreover, it weakly interacts with Tyr585 and Arg358, residues located in an extension of the S2 site (Fig. 15.7b) [98]. Since the S1 site is identical in DPP-4, DPP-8, and DPP-9 but there are substitutions in the extensive S2 site (R358D and Y385H/N), the S2 site seemed important to be considered to achieve selectivity. Docking studies guided further synthesis of linkers to improve interactions in S2, leading to teneligliptin (Fig. 15.9a), which presented a threefold increase in DPP-4 inhibitory activity (IC_{50} of 0.4 nM) and having its ex vivo activity sustained. The ex vivo activity consisted of orally administrating the compounds to rats and evaluating the plasma

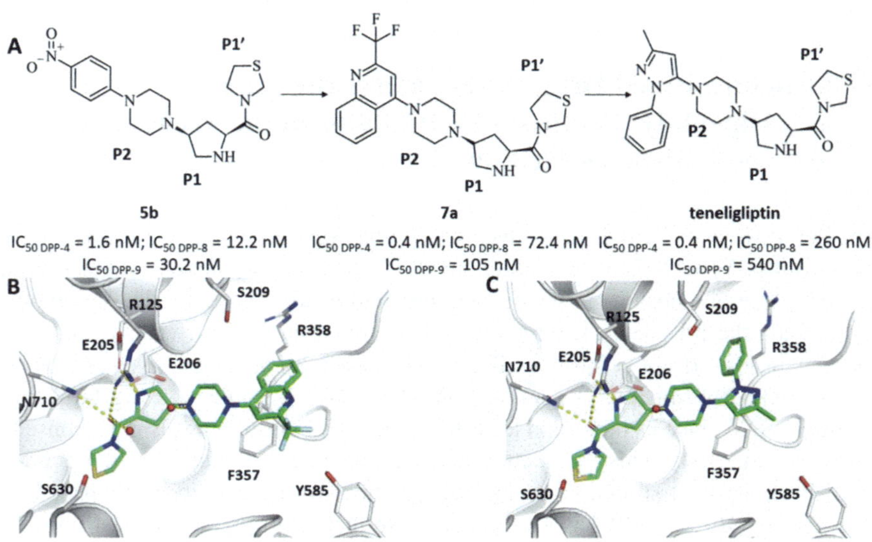

Fig. 15.9 Structure-based development of teneligliptin. Chemical representation of compounds **6**, **7**, and teneligliptin (**a**); Crystal structure of **7** complexed with DPP-4 (PDB ID: 3VJM); (**c**) Crystal structure of teneligliptin complexed with DPP-4 (PDB ID: 3VJK) (**b**). Water molecules are represented as red spheres, and the polar contacts are shown in yellow dashed lines. The P2–P1' positions of each compound are labeled. Important residues and compounds are represented as sticks and colored by atom. Carbons are colored gray in the enzyme and green in the compounds. (Figure produced using ChemDraw and PyMOL)

DPP-4 activity at different time points. Teneligliptin also had about 700- and 1500-fold selectivity over DPP-8 (IC_{50} of 260 nM) and DPP-9 (IC_{50} of 340 nM), respectively. The crystal structure of teneligliptin complexed with DPP-4 shows the thiazolidine moiety occupying the S1 site and the pyrazole with the aryl group occupying the S2 site (PDB ID 3VJK) (Fig. 15.9c) [16]. Teneligliptin was approved in Japan in 2012 under the trade name Tenelia® as a once-a-day oral treatment for patients with T2DM that is not controlled by diet or exercise [99].

6 Conclusions

In this chapter, we highlight successful examples of drug design strategies targeting aspartic, cysteine, and serine proteases. The importance of structure-based drug design techniques clearly demonstrates how these strategies allow quick and efficient lead optimization. In addition to designing first-in-class compounds, SBDD approaches have shown the potential to overcome drug resistance or design compounds for which resistance is less likely to arise. SBDD was also valuable in aiding the design of the selective DPP-4 inhibitor teneligliptin.

Examples from several decades reveal that, with time, the optimization of potency and pharmacokinetic properties becomes increasingly more connected. A good example is the development of ensitrelvir, in which the lead compound was not the most potent hit obtained against the enzyme, but it was chosen due to its favorable pharmacokinetics profile. This illustrates how, despite the frequent focus on potency against target proteins, successful drug discovery projects need to account for numerous compound characteristics, such as activity in cellular assays, low cytotoxicity, and an overall favorable ADMET profile.

Another tendency observed is the diversification of strategies for inhibitor development. In the development of the first inhibitors of HIV and HCV proteases, we find mostly strategies based on molecular modeling of peptidic compounds. On the other hand, more recent examples include the discovery of novel nonpeptidic scaffolds from virtual screening of chemically diverse libraries and the exploration of abundant structural information to derive pharmacophore models. Additionally, computational techniques allow researchers to access virtually a chemical space not yet synthesized. The computational cost varies widely depending on the techniques employed, as discussed in the M^{pro} section. However, it is still cheaper to virtually screen millions or billions of compounds than to perform high throughput screening of this magnitude. Therefore, molecular modeling strategies can be pivotal do drug design, development, and discovery of drugs targeting proteases.

References

1. Castro, H. C., Abreu, P. A., Geraldo, R. B., Martins, R. C. A., dos Santos, R., Loureiro, N. I. V., Cabral, L. M., & Rodrigues, C. R. (2011). Looking at the proteases from a simple perspective. *Journal of Molecular Recognition, 24*, 165–181.
2. Drag, M., & Salvesen, G. S. (2010). Emerging principles in protease-based drug discovery. *Nature Reviews. Drug Discovery, 9*, 690–701.
3. Verhelst, S. H. L. (2017). Intramembrane proteases as drug targets. *The FEBS Journal, 284*, 1489–1502.
4. Cudic, M., & Fields, G. (2009). Extracellular proteases as targets for drug development. *Current Protein & Peptide Science, 10*, 297–307.
5. Al-Awadhi, F. H., & Luesch, H. (2020). Targeting eukaryotic proteases for natural products-based drug development. *Natural Product Reports, 37*, 827–860.
6. Schechter, I., & Berger, A. (1968). On the active site of proteases. III. Mapping the active site of papain; specific peptide inhibitors of papain. *Biochemical and Biophysical Research Communications, 32*, 898–902.
7. Turk, V., Stoka, V., Vasiljeva, O., Renko, M., Sun, T., Turk, B., & Turk, D. (2012). Cysteine cathepsins: From structure, function and regulation to new frontiers. *Biochimica et Biophysica Acta—Proteins Proteomics, 1824*, 68–88.
8. Docherty, A. J. P., Crabbe, T., O'Connell, J. P., & Groom, C. R. (2003). Proteases as drug targets. *Biochemical Society Symposium*, 147–161.
9. Dorsey, B. D., Levin, R. B., McDaniel, S. L., Vacca, J. P., Guare, J. P., Darke, P. L., Zugay, J. A., Emini, E. A., & Schleif, W. A. (1994). L-735,524: The design of a potent and orally bioavailable HIV protease inhibitor. *Journal of Medicinal Chemistry, 37*, 3443–3451.
10. Sham, H. L., Kempf, D. J., Molla, A., Marsh, K. C., Kumar, G. N., Chen, C. M., Kati, W., Stewart, K., Lal, R., Hsu, A., Betebenner, D., Korneyeva, M., Vasavanonda, S., McDonald, E., Saldivar, A., Wideburg, N., Chen, X., Niu, P., Park, C., Jayanti, V., Grabowski, B., Granneman, G. R., Sun, E., Japour, A. J., Leonard, J. M., Plattner, J. J., & Norbeck, D. W. (1998). ABT-378, a highly potent inhibitor of the human immunodeficiency virus protease. *Antimicrobial Agents and Chemotherapy, 42*, 3218–3224.
11. Koh, Y., Nakata, H., Maeda, K., Ogata, H., Bilcer, G., Devasamudram, T., Kincaid, J. F., Boross, P., Wang, Y. F., Tie, Y., Volarath, P., Gaddis, L., Harrison, R. W., Weber, I. T., Ghosh, A. K., & Mitsuya, H. (2003). Novel bis-tetrahydrofuranylurethane-containing nonpeptidic protease inhibitor (PI) UIC-94017 (TMC114) with potent activity against multi-PI-resistant human immunodeficiency virus in vitro. *Antimicrobial Agents and Chemotherapy, 47*, 3123–3129.
12. Owen, D. R., Allerton, C. M. N., Anderson, A. S., Aschenbrenner, L., Avery, M., Berritt, S., Boras, B., Cardin, R. D., Carlo, A., Coffman, K. J., Dantonio, A., Di, L., Eng, H., Ferre, R., Gajiwala, K. S., Gibson, S. A., Greasley, S. E., Hurst, B. L., Kadar, E. P., Kalgutkar, A. S., Lee, J. C., Lee, J., Liu, W., Mason, S. W., Noell, S., Novak, J. J., Obach, R. S., Ogilvie, K., Patel, N. C., Pettersson, M., Rai, D. K., Reese, M. R., Sammons, M. F., Sathish, J. G., Singh, R. S. P., Steppan, C. M., Stewart, A. E., Tuttle, J. B., Updyke, L., Verhoest, P. R., Wei, L., Yang, Q., & Zhu, Y. (2021). An oral SARS-CoV-2 M pro inhibitor clinical candidate for the treatment of COVID-19. *Science, (80-.). 374*, 1586–1593.
13. Unoh, Y., Uehara, S., Nakahara, K., Nobori, H., Yamatsu, Y., Yamamoto, S., Maruyama, Y., Taoda, Y., Kasamatsu, K., Suto, T., Kouki, K., Nakahashi, A., Kawashima, S., Sanaki, T., Toba, S., Uemura, K., Mizutare, T., Ando, S., Sasaki, M., Orba, Y., Sawa, H., Sato, A., Sato, T., Kato, T., & Tachibana, Y. (2022). Discovery of S-217622, a noncovalent oral SARS-CoV-2 3CL protease inhibitor clinical candidate for treating COVID-19. *Journal of Medicinal Chemistry.*
14. Harper, S., McCauley, J. A., Rudd, M. T., Ferrara, M., DiFilippo, M., Crescenzi, B., Koch, U., Petrocchi, A., Holloway, M. K., Butcher, J. W., Romano, J. J., Bush, K. J., Gilbert, K. F., McIntyre, C. J., Nguyen, K. T., Nizi, E., Carroll, S. S., Ludmerer, S. W., Burlein, C., Dimuzio,

J. M., Graham, D. J., McHale, C. M., Stahlhut, M. W., Olsen, D. B., Monteagudo, E., Cianetti, S., Giuliano, C., Pucci, V., Trainor, N., Fandozzi, C. M., Rowley, M., Coleman, P. J., Vacca, J. P., Summa, V., & Liverton, N. J. (2012). Discovery of MK-5172, a macrocyclic hepatitis C virus NS3/4a protease inhibitor. *ACS Medicinal Chemistry Letters, 3*, 332–336.

15. Taylor, J. G., Zipfel, S., Ramey, K., Vivian, R., Schrier, A., Karki, K. K., Katana, A., Kato, D., Kobayashi, T., Martinez, R., Sangi, M., Siegel, D., Tran, C. V., Yang, Z. Y., Zablocki, J., Yang, C. Y., Wang, Y., Wang, K., Chan, K., Barauskas, O., Cheng, G., Jin, D., Schultz, B. E., Appleby, T., Villaseñor, A. G., & Link, J. O. (2019). Discovery of the pan-genotypic hepatitis C virus NS3/4A protease inhibitor voxilaprevir (GS-9857): A component of Vosevi®, bioorganic med. *Chemistry Letters, 29*, 2428–2436.
16. Yoshida, T., Akahoshi, F., Sakashita, H., Kitajima, H., Nakamura, M., Sonda, S., Takeuchi, M., Tanaka, Y., Ueda, N., Sekiguchi, S., Ishige, T., Shima, K., Nabeno, M., Abe, Y., Anabuki, J., Soejima, A., Yoshida, K., Takashina, Y., Ishii, S., Kiuchi, S., Fukuda, S., Tsutsumiuchi, R., Kosaka, K., Murozono, T., Nakamaru, Y., Utsumi, H., Masutomi, N., Kishida, H., Miyaguchi, I., & Hayashi, Y. (2012). Discovery and preclinical profile of teneligliptin (3-[(2S,4S)-4-[4-(3-methyl-1-phenyl-1H-pyrazol-5-yl)piperazin-1-yl]pyrrolidin-2-ylcarbonyl]thiazolidine): A highly potent, selective, long-lasting and orally active dipeptidyl peptidase IV inhibitor for t. *Bioorganic & Medicinal Chemistry, 20*, 5705–5719.
17. U.S. Department of Health & Human Services, A Timeline of HIV and AIDS, (n.d.). https://www.hiv.gov/hiv-basics/overview/history/hiv-and-aids-timeline/#year-1981.
18. UNAIDS. (2023). Fact sheet 2023.
19. Frankel, A. D., & Young, J. A. T. (1998). HIV-1: Fifteen proteins and an RNA. *Annual Review of Biochemistry, 67*, 1–25.
20. Sundquist, W. I., & Krausslich, H.-G. (2012). HIV-1 assembly, budding, and maturation. *Cold Spring Harbor Perspectives in Medicine, 2*, a006924–a006924.
21. Hammer, S. M., Squires, K. E., Hughes, M. D., Grimes, J. M., Demeter, L. M., Currier, J. S., Eron, J. J., Feinberg, J. E., Balfour, H. H., Deyton, L. R., Chodakewitz, J. A., Fischl, M. A., Phair, J. P., Pedneault, L., Nguyen, B.-Y., & Cook, J. C. (1997). A controlled trial of two nucleoside analogues plus indinavir in persons with human immunodeficiency virus infection and CD4 cell counts of 200 per cubic millimeter or less. *The New England Journal of Medicine, 337*, 725–733.
22. Weber, I. T., Wang, Y. F., & Harrison, R. W. (2021). HIV protease: Historical perspective and current research. *Viruses, 13*, 1–12.
23. Roberts, N. A., Martin, J. A., Kinchington, D., Broadhurst, A. V., Craig, J. C., Duncan, I. B., Galpin, S. A., Handa, B. K., Kay, J., Kröhn, A., Lambert, R. W., Merrett, J. H., Mills, J. S., Parkes, K. E. B., Redshaw, S., Ritchie, A. J., Taylor, D. L., Thomas, G. J., & Machin, P. J. (1990). Rational design of peptide-based HIV proteinase inhibitors. *Science (80-.), 248*, 358–361.
24. Krohn, A., Redshaw, S., Ritchie, J. C., Graves, B. J., & Hatada, M. H. (1991). Novel binding mode of highly potent HIV-proteinase inhibitors incorporating the (R)-hydroxyethylamine isostere. *Journal of Medicinal Chemistry, 34*, 3340–3342.
25. Fox, J. L. (1996). FDA approves first HIV protease inhibitor. *Bio/Technology, 14*, 23.
26. Williams, G. C., & Sinko, P. J. (1999). Oral absorption of the HIV protease inhibitors: A current update. *Advanced Drug Delivery Reviews, 39*, 211–238.
27. Vacca, J. P., Guare, J. P., DeSolms, S. J., Sanders, W. M., Giuliani, E. A., Young, S. D., Darke, P. L., Sigal, I. S., & Schleif, W. A. (1991). L-687,908, a potent hydroxyethylene containing HIV protease inhibitor. *Journal of Medicinal Chemistry, 34*, 1225–1228.
28. Miller, M., Schneider, J., Sathyanarayana, B. K., Toth, M. V., Marshall, G. R., Clawson, L., Selk, L., Kent, S. B. H., & Wlodawer, A. (1989). Structure of complex of synthetic HIV-1 protease with a substrate-based inhibitor at 2.3 Å resolution. *Science (80-.), 246*, 1149–1152.
29. Fitzgerald, P. M. D., McKeever, B. M., VanMiddlesworth, J. F., Springer, J. P., Heimbach, J. C., Leu, C. T., Herber, W. K., Dixon, R. A. F., & Darke, P. L. (1990). Crystallographic

analysis of a complex between human immunodeficiency virus type 1 protease and acetyl-pepstatin at 2.0-Å resolution. *The Journal of Biological Chemistry, 265*, 14209–14219.
30. Erickson, J., Neidhart, D. J., VanDrie, J., Kempf, D. J., Wang, X. C., Norbeck, D. W., Plattner, J. J., Rittenhouse, J. W., Turon, M., Wideburg, N., Kohlbrenner, W. E., Simmer, R., Helfrich, R., Paul, D. A., & Knigge, M. (1990). Design, activity, and 2.8 Å crystal structure of a C2 symmetric inhibitor complexed to HIV-1 protease. *Science (80-.), 249*, 527–533.
31. Swain, A. L., Miller, M. M., Green, J., Rich, D. H., Schneider, J., Kent, S. B., & Wlodawer, A. (1990). X-ray crystallographic structure of a complex between a synthetic protease of human immunodeficiency virus 1 and a substrate-based hydroxyethylamine inhibitor. *Proceedings of the National Academy of Sciences, 87*, 8805–8809.
32. Schouten, J. T. (1996). FDA approves 2 new protease inhibitors: Ritonavir (Norvir) and Crixivan (Indinavir sulfate). *Food and Drug Administration., STEP Perspect., 8*, 7–8.
33. Molla, A., Korneyeva, M., Gao, Q., Vasavanonda, S., Schipper, P. J., Mo, H.-M., Markowitz, M., Chernyavskiy, T., Niu, P., Lyons, N., Hsu, A., Granneman, G. R., Ho, D. D., Boucher, C. A. B., Leonard, J. M., Norbeck, D. W., & Kempf, D. J. (1996). Ordered accumulation of mutations in HIV protease confers resistance to ritonavir. *Nature Medicine, 2*, 760–766.
34. Drake, J. W., & Holland, J. J. (1999). Mutation rates among RNA viruses. *Proceedings of the National Academy of Sciences of the United States of America, 96*, 13910–13913.
35. Leigh Brown, A. J. (1997). Analysis of HIV-1 env gene sequences reveals evidence for a low effective number in the viral population. *Proceedings of the National Academy of Sciences of the United States of America, 94*, 1862–1865.
36. Kantor, R., Fessel, W. J., Zolopa, A. R., Israelski, D., Shulman, N., Montoya, J. G., Harbour, M., Schapiro, J. M., & Shafer, R. W. (2002). Evolution of primary protease inhibitor resistance mutations during protease inhibitor salvage therapy. *Antimicrobial Agents and Chemotherapy, 46*, 1086–1092.
37. Croteau, G., Doyon, L., Thibeault, D., McKercher, G., Pilote, L., & Lamarre, D. (1997). Impaired fitness of human immunodeficiency virus type 1 variants with high-level resistance to protease inhibitors. *Journal of Virology, 71*, 1089–1096.
38. Ohtaka, H., Schön, A., & Freire, E. (2003). Multidrug resistance to HIV-1 protease inhibition requires cooperative coupling between distal mutations. *Biochemistry, 42*, 13659–13666.
39. Condra, J. H., Schleif, W. A., Blahy, O. M., Gabryelski, L. J., Graham, D. J., Quintero, J., Rhodes, A., Robbins, H. L., Roth, E., Shivaprakash, M., Titus, D., Yang, T., Tepplert, H., Squires, K. E., Deutsch, P. J., & Emini, E. A. (1995). In vivo emergence of HIV-1 variants resistant to multiple protease inhibitors. *Nature, 374*, 569–571.
40. Kempf, D. J., Marsh, K. C., Denissen, J. F., McDonald, E., Vasavanonda, S., Flentge, C. A., Green, B. E., Fino, L., Park, C. H., Kong, X. P., Wideburg, N. E., Saldivar, A., Ruiz, L., Kati, W. M., Sham, H. L., Robins, T., Stewar, K. D., Hsu, A., Plattner, J. J., Leonard, J. M., & Norbeck, D. W. (1995). ABT-538 is a potent inhibitor of human immunodeficiency virus protease and has high oral bioavailability in humans. *Proceedings of the National Academy of Sciences of the United States of America, 92*, 2484–2488.
41. Stoll, V., Qin, W., Stewart, K. D., Jakob, C., Park, C., Walter, K., Simmer, R. L., Helfrich, R., Bussiere, D., Kao, J., Kempf, D., Sham, H. L., & Norbeck, D. W. (2002). X-ray crystallographic structure of ABT-378 (Lopinavir) bound to HIV-1 protease, bioorganic. *Medicinal Chemistry, 10*, 2803–2806.
42. Miller, J. L. (2000). FDA approves new protease inhibitor for HIV infection. *American Journal of Health-System Pharmacy, 57*, 1942.
43. Laco, G. S., Schalk-Hihi, C., Lubkowski, J., Morris, G., Zdanov, A., Olson, A., Elder, J. H., Wlodawer, A., & Gustchina, A. (1997). Crystal structures of the inactive D30N mutant of feline immunodeficiency virus protease complexed with a substrate and an inhibitor. *Biochemistry, 36*, 10696–10708.
44. Hong, L., Zhang, X. C., Hartsuck, J. A., & Tang, J. (2000). Crystal structure of an in vivo HIV-1 protease mutant in complex with saquinavir: Insights into the mechanisms of drug resistance. *Protein Science, 9*, 1898–1904.

45. Ghosh, A. K., Thompson, W. J., Fitzgerald, P. M. D., Culberson, J. C., Axel, M. G., McKee, S. P., Huff, J. R., & Anderson, P. S. (1994). Structure-based design of HIV-1 protease inhibitors: Replacement of two amides and a 10pi-aromatic system by a fused bis-tetrahydrofuran. *Journal of Medicinal Chemistry, 37*, 2506–2508.
46. Ghosh, A. K., Kincaid, J. F., Walters, D. E., Chen, Y., Chaudhuri, N. C., Thompson, W. J., Culberson, C., Fitzgerald, P. M. D., Lee, H. Y., McKee, S. P., Munson, P. M., Duong, T. T., Darke, P. L., Zugay, J. A., Schleif, W. A., Axel, M. G., Lin, J., & Huff, J. R. (1996). Nonpeptidal P2 ligands for HIV protease inhibitors: Structure-based design, synthesis, and biological evaluation. *Journal of Medicinal Chemistry, 39*, 3278–3290.
47. Ghosh, A. K., Kincaid, J. F., Cho, W., Walters, D. E., Krishnan, K., Hussain, K. A., Koo, Y., Cho, H., Rudall, C., Holland, L., & Buthod, J. (1998). Potent HIV protease inhibitors incorporating high-affinity P2-ligands and (R)-(hydroxyethylamino)sulfonamide isostere, bioorganic med. *Chemistry Letters, 8*, 687–690.
48. Yoshimura, K., Kato, R., Kavlick, M. F., Nguyen, A., Maroun, V., Maeda, K., Hussain, K. A., Ghosh, A. K., Gulnik, S. V., Erickson, J. W., & Mitsuya, H. (2002). A potent human immunodeficiency virus type 1 protease inhibitor, UIC-94003 (TMC-126), and selection of a novel (A28S) mutation in the protease active site. *Journal of Virology, 76*, 1349–1358.
49. Tie, Y., Boross, P. I., Wang, Y. F., Gaddis, L., Hussain, A. K., Leshchenko, S., Ghosh, A. K., Louis, J. M., Harrison, R. W., & Weber, I. T. (2004). High resolution crystal structures of HIV-1 protease with a potent non-peptide inhibitor (UIC-94017) active against multi-drug-resistant clinical strains. *Journal of Molecular Biology, 338*, 341–352.
50. Lefebvre, E., & Schiffer, C. A. (2008). Resilience to resistance of HIV-1 protease inhibitors: Profile of darunavir. *AIDS Reviews, 10*, 131–142. http://www.ncbi.nlm.nih.gov/pubmed/18820715
51. De Meyer, S., Azijn, H., Surleraux, D., Jochmans, D., Tahri, A., Pauwels, R., Wigerinck, P., & De Béthune, M. P. (2005). TMC114, a novel human immunodeficiency virus type 1 protease inhibitor active against protease inhibitor-resistant viruses, including a broad range of clinical isolates. *Antimicrobial Agents and Chemotherapy, 49*, 2314–2321.
52. Choo, Q. L., Kuo, G., Weiner, A. J., Overby, L. R., Bradley, D. W., & Houghton, M. (1989). Isolation of a cDNA clone derived from a blood-borne non-a, non-B viral hepatitis genome. *Science, (80-). 244*, 359–362.
53. WHO, Hepatits C, (2023). https://www.who.int/news-room/fact-sheets/detail/hepatitis-c
54. Grebely, J., Dore, G. J., Kim, A. Y., Lloyd, A., Shoukry, N. H., Prins, M., & Page, K. (2014). Genetics of spontaneous clearance of hepatitis C virus infection: A complex topic with much to learn. *Hepatology, 60*, 2127–2128.
55. Kim, J. L., Morgenstern, K. A., Lin, C., Fox, T., Dwyer, M. D., Landro, J. A., Chambers, S. P., Markland, W., Lepre, C. A., O'Malley, E. T., Harbeson, S. L., Rice, C. M., Murcko, M. A., Caron, P. R., & Thomson, J. A. (1996). Crystal structure of the hepatitis C virus NS3 protease domain complexed with a synthetic NS4A cofactor peptide. *Cell, 87*, 343–355.
56. Grakoui, A., McCourt, D. W., Wychowski, C., Feinstone, S. M., & Rice, C. M. (1993). Characterization of the hepatitis C virus-encoded serine proteinase: Determination of proteinase-dependent polyprotein cleavage sites. *Journal of Virology, 67*, 2832–2843.
57. Borgia, S. M., Hedskog, C., Parhy, B., Hyland, R. H., Stamm, L. M., Brainard, D. M., Subramanian, M. G., McHutchison, J. G., Mo, H., Svarovskaia, E., & Shafran, S. D. (2018). Identification of a novel hepatitis C virus genotype from Punjab, India: Expanding classification of hepatitis C virus into 8 genotypes. *The Journal of Infectious Diseases, 218*, 1722–1729.
58. Roudot-Thoraval, F. (2021). Epidemiology of hepatitis C virus infection. *Clinics and Research in Hepatology and Gastroenterology, 45*, 2015–2020.
59. Martinez, M. A., & Franco, S. (2021). Therapy implications of hepatitis C virus genetic diversity. *Viruses, 13*.
60. Tsantrizos, Y. S., Bolger, G., Bonneau, P., Cameron, D. R., Goudreau, N., Kukolj, G., LaPlante, S. R., Llinàs-Brunet, M., Nar, H., & Lamarre, D. (2003). Macrocyclic inhibitors of the NS3

protease as potential therapeutic agents of hepatitis C virus infection. *Angewandte Chemie International Edition, 42*, 1356–1360.
61. Yao, N., Reichert, P., Taremi, S. S., Prosise, W. W., & Weber, P. C. (1999). Molecular views of viral polyprotein processing revealed by the crystal structure of the hepatitis C virus bifunctional protease-helicase. *Structure, 7*, 1353–1363.
62. Liverton, N. J., Holloway, M. K., McCauley, J. A., Rudd, M. T., Butcher, J. W., Carroll, S. S., DiMuzio, J., Fandozzi, C., Gilbert, K. F., Mao, S. S., McIntyre, C. J., Nguyen, K. T., Romano, J. J., Stahlhut, M., Wan, B. L., Olsen, D. B., & Vacca, J. P. (2008). Molecular modeling based approach to potent P2-P4 macrocyclic inhibitors of hepatitis C NS3/4A protease. *Journal of the American Chemical Society, 130*, 4607–4609.
63. McCauley, J. A., McIntyre, C. J., Rudd, M. T., Nguyen, K. T., Romano, J. J., Butcher, J. W., Gilbert, K. F., Bush, K. J., Holloway, M. K., Swestock, J., Wan, B. L., Carroll, S. S., Dimuzio, J. M., Graham, D. J., Ludmerer, S. W., Mao, S. S., Stahlhut, M. W., Fandozzi, C. M., Trainor, N., Olsen, D. B., Vacca, J. P., & Liverton, N. J. (2010). Discovery of vaniprevir (MK-7009), a macrocyclic hepatitis C virus NS3/4a protease inhibitor. *Journal of Medicinal Chemistry, 53*, 2443–2463.
64. Xu, F., & McCauley, J. A. (2020). Discovery and chemical development of grazoprevir: An HCV NS3/4a protease inhibitor for the treatment of the hepatitis C virus infection. *ACS Symposium Series, 1369*, 285–312.
65. Soumana, D. I., Kurt Yilmaz, N., Ali, A., Prachanronarong, K. L., & Schiffer, C. A. (2016). Molecular and dynamic mechanism underlying drug resistance in genotype 3 hepatitis C NS3/4A protease. *Journal of the American Chemical Society, 138*, 11850–11859.
66. Taylor, J. G. (2019). Discovery of Voxilaprevir (GS-9857): The pan-genotypic hepatitis C virus NS3/4A protease inhibitor utilized as a component of Vosevi®. *Current Topics in Medicinal Chemistry, 31*, 441–457.
67. Gorbalenya, A. E., Baker, S. C., Baric, R. S., de Groot, R. J., Drosten, C., Gulyaeva, A. A., Haagmans, B. L., Lauber, C., Leontovich, A. M., Neuman, B. W., Penzar, D., Perlman, S., Poon, L. L. M., Samborskiy, D. V., Sidorov, I. A., Sola, I., & Ziebuhr, J. (2020). The species severe acute respiratory syndrome-related coronavirus: Classifying 2019-nCoV and naming it SARS-CoV-2. *Nature Microbiology, 5*, 536–544.
68. J.F.W. Chan, K.H. Kok, Z. Zhu, H. Chu, K.K.W. To, S. Yuan, K.Y. Yuen, Genomic characterization of the. (2019). Novel human-pathogenic coronavirus isolated from a patient with atypical pneumonia after visiting Wuhan. *Emerging Microbes & Infections, 9*(2020), 221–236.
69. Jin, Z., Du, X., Xu, Y., Deng, Y., Liu, M., Zhao, Y., Zhang, B., Li, X., Zhang, L., Peng, C., Duan, Y., Yu, J., Wang, L., Yang, K., Liu, F., Jiang, R., Yang, X., You, T., Liu, X., Yang, X., Bai, F., Liu, H., Liu, X., Guddat, L. W., Xu, W., Xiao, G., Qin, C., Shi, Z., Jiang, H., Rao, Z., & Yang, H. (2020). Structure of Mpro from SARS-CoV-2 and discovery of its inhibitors. *Nature, 582*, 289–293.
70. Chen, Y., Liu, Q., & Guo, D. (2020). Emerging coronaviruses: Genome structure, replication, and pathogenesis. *Journal of Medical Virology, 92*, 418–423.
71. Baughn, L. B., Sharma, N., Elhaik, E., Sekulic, A., Bryce, A. H., & Fonseca, R. (2020). Targeting TMPRSS2 in SARS-CoV-2 infection. *Mayo Clinic Proceedings, 95*, 1989–1999.
72. Zhao, M. M., Yang, W. L., Yang, F. Y., Zhang, L., Huang, W. J., Hou, W., Fan, C. F., Jin, R. H., Feng, Y. M., Wang, Y. C., & Yang, J. K. (2021). Cathepsin L plays a key role in SARS-CoV-2 infection in humans and humanized mice and is a promising target for new drug development. *Signal Transduction and Targeted Therapy, 6*.
73. Ghahremanpour, M. M., Tirado-Rives, J., Deshmukh, M., Ippolito, J. A., Zhang, C. H., Cabeza De Vaca, I., Liosi, M. E., Anderson, K. S., & Jorgensen, W. L. (2020). Identification of 14 known drugs as inhibitors of the main protease of SARS-CoV-2. *ACS Medicinal Chemistry Letters, 11*, 2526–2533.
74. Santos, L. H., Kronenberger, T., Almeida, R. G., Silva, E. B., Rocha, R. E. O., Oliveira, J. C., Barreto, L. V., Skinner, D., Fajtová, P., Giardini, M. A., Woodworth, B., Bardine, C., Lourenço, A. L., Craik, C. S., Poso, A., Podust, L. M., Mckerrow, J. H., Siqueira-Neto, J. L., O'Donoghue,

A. J., Da Silva Júnior, E. N., & Ferreira, R. S. (2022). Structure-based identification of naphthoquinones and derivatives as novel inhibitors of main protease Mpro and papain-like protease PLpro of SARS-CoV-2. *Journal of Chemical Information and Modeling.*

75. Maltarollo, V. G., da Silva, E. B., Kronenberger, T., Sena Andrade, M. M., de Lima Marques, G. V., Cândido Oliveira, N. J., Santos, L. H., de Oliveira Rezende, C., Júnior, A. C. C., Martinho, D., Skinner, P., Fajtová, T. H. M., Fernandes, E. d., Silveira dos Santos, P. A., Rodrigues Gazolla, A. P., Martins de Souza, M. L., da Silva, F. S., dos Santos, S. N., Lavorato, A. C., Oliveira Bretas, D. T., Carvalho, L. L., Franco, S., Luedtke, M. A., Giardini, A., Poso, L. C., Dias, L. M., Podust, R. J., Alves, J., McKerrow, S. F., Andrade, R. R., Teixeira, J. L., Siqueira-Neto, A., O'Donoghue, R. B., de Oliveira, & Ferreira, R. S. (2023). Structure-based discovery of thiosemicarbazones as SARS-CoV-2 main protease inhibitors, future. *Medicinal Chemistry, 15*, 959–985.
76. Luttens, A., Gullberg, H., Abdurakhmanov, E., Vo, D. D., Akaberi, D., Talibov, V. O., Nekhotiaeva, N., Vangeel, L., De Jonghe, S., Jochmans, D., Krambrich, J., Tas, A., Lundgren, B., Gravenfors, Y., Craig, A. J., Atilaw, Y., Sandström, A., Moodie, L. W. K., Lundkvist, Å., van Hemert, M. J., Neyts, J., Lennerstrand, J., Kihlberg, J., Sandberg, K., Danielson, U. H., & Carlsson, J. (2022). Ultralarge virtual screening identifies SARS-CoV-2 main protease inhibitors with broad-spectrum activity against coronaviruses. *Journal of the American Chemical Society, 144*, 2905–2920.
77. Fink, E. A., Bardine, C., Gahbauer, S., Singh, I., Detomasi, T. C., White, K., Gu, S., Wan, X., Chen, J., Ary, B., Glenn, I., O'Connell, J., O'Donnell, H., Fajtová, P., Lyu, J., Vigneron, S., Young, N. J., Kondratov, I. S., Alisoltani, A., Simons, L. M., Lorenzo-Redondo, R., Ozer, E. A., Hultquist, J. F., O'Donoghue, A. J., Moroz, Y. S., Taunton, J., Renslo, A. R., Irwin, J. J., García-Sastre, A., Shoichet, B. K., & Craik, C. S. (2023). Large library docking for novel SARS-CoV -2 main protease non-covalent and covalent inhibitors. *Protein Science, 32.*
78. Zhang, C. H., Stone, E. A., Deshmukh, M., Ippolito, J. A., Ghahremanpour, M. M., Tirado-Rives, J., Spasov, K. A., Zhang, S., Takeo, Y., Kudalkar, S. N., Liang, Z., Isaacs, F., Lindenbach, B., Miller, S. J., Anderson, K. S., & Jorgensen, W. L. (2021). Potent noncovalent inhibitors of the main protease of SARS-CoV-2 from molecular sculpting of the drug perampanel guided by free energy perturbation calculations. *ACS Central Science, 7*, 467–475.
79. Saar, K. L., McCorkindale, W., Fearon, D., Boby, M., Barr, H., Ben-Shmuel, A., London, N., von Delft, F., Chodera, J. D., & Lee, A. A. (2023). Turning high-throughput structural biology into predictive inhibitor design. *Proceedings of the National Academy of Sciences of the United States of America, 120*, 1–9.
80. Li, X., & Song, Y. (2023). Structure and function of SARS-CoV and SARS-CoV-2 main proteases and their inhibition: A comprehensive review. *European Journal of Medicinal Chemistry, 260*, 115772.
81. Hoffman, R. L., Kania, R. S., Brothers, M. A., Davies, J. F., Ferre, R. A., Gajiwala, K. S., He, M., Hogan, R. J., Kozminski, K., Li, L. Y., Lockner, J. W., Lou, J., Marra, M. T., Mitchell, L. J., Murray, B. W., Nieman, J. A., Noell, S., Planken, S. P., Rowe, T., Ryan, K., Smith, G. J., Solowiej, J. E., Steppan, C. M., & Taggart, B. (2020). Discovery of ketone-based covalent inhibitors of coronavirus 3CL proteases for the potential therapeutic treatment of COVID-19. *Journal of Medicinal Chemistry, 63*, 12725–12747.
82. Hammond, J., Leister-Tebbe, H., Gardner, A., Abreu, P., Bao, W., Wisemandle, W., Baniecki, M., Hendrick, V. M., Damle, B., Simón-Campos, A., Pypstra, R., & Rusnak, J. M. (2022). Oral nirmatrelvir for high-risk, nonhospitalized adults with Covid-19. *The New England Journal of Medicine, 386*, 1397–1408.
83. Boby, M. L., Fearon, D., Ferla, M., Filep, M., Koekemoer, L., Robinson, M. C., Consortium, T. C. M., Chodera, J. D., Lee, A. A., London, N., von Delft, A., von Delft, F., Achdout, H., Aimon, A., Alonzi, D. S., Arbon, R., Aschenbrenner, J. C., Balcomb, B. H., Bar-David, E., Barr, H., Ben-Shmuel, A., Bennett, J., Bilenko, V. A., Borden, B., Boulet, P., Bowman, G. R., Brewitz, L., Brun, J., BVNBS, S., Calmiano, M., Carbery, A., Carney, D. W., Cattermole, E., Chang, E., Chernyshenko, E., Clyde, A., Coffland, J. E., Cohen, G., Cole, J. C., Contini, A.,

Cox, L., Croll, T. I., Cvitkovic, M., De Jonghe, S., Dias, A., Donckers, K., Dotson, D. L., Douangamath, A., Duberstein, S., Dudgeon, T., Dunnett, L. E., Eastman, P., Erez, N., Eyermann, C. J., Fairhead, M., Fate, G., Fedorov, O., Fernandes, R. S., Ferrins, L., Foster, R., Foster, H., Fraisse, L., Gabizon, R., García-Sastre, A., Gawriljuk, V. O., Gehrtz, P., Gileadi, C., Giroud, C., Glass, W. G., Glen, R. C., Glinert, I., Godoy, A. S., Gorichko, M., Gorrie-Stone, T., Griffen, E. J., Haneef, A., Hart, S. H., Heer, J., Henry, M., Hill, M., Horrell, S., Huang, Q. Y. J., Huliak, V. D., Hurley, M. F. D., Israely, T., Jajack, A., Jansen, J., Jnoff, E., Jochmans, D., John, T., Kaminow, B., Kang, L., Kantsadi, A. L., Kenny, P. W., Kiappes, J. L., Kinakh, S. O., Kovar, B., Krojer, T., La, V. N. T., Laghnimi-Hahn, S., Lefker, B. A., Levy, H., Lithgo, R. M., Logvinenko, I. G., Lukacik, P., Macdonald, H. B., MacLean, E. M., Makower, L. L., Malla, T. R., Marples, P. G., Matviiuk, T., McCorkindale, W., McGovern, B. L., Melamed, S., Melnykov, K. P., Michurin, O., Miesen, P., Mikolajek, H., Milne, B. F., Minh, D., Morris, A., Morris, G. M., Morwitzer, M. J., Moustakas, D., Mowbray, C. E., Nakamura, A. M., Neto, J. B., Neyts, J., Nguyen, L., Noske, G. D., Oleinikovas, V., Oliva, G., Overheul, G. J., Owen, C. D., Pai, R., Pan, J., Paran, N., Payne, A. M., Perry, B., Pingle, M., Pinjari, J., Politi, B., Powell, A., Pšenák, V., Pulido, I., Puni, R., Rangel, V. L., Reddi, R. N., Rees, P., Reid, S. P., Reid, L., Resnick, E., Ripka, E. G., Robinson, R. P., Rodriguez-Guerra, J., Rosales, R., Rufa, D. A., Saar, K., Saikatendu, K. S., Salah, E., Schaller, D., Scheen, J., Schiffer, C. A., Schofield, C. J., Shafeev, M., Shaikh, A., Shaqra, A. M., Shi, J., Shurrush, K., Singh, S., Sittner, A., Sjö, P., Skyner, R., Smalley, A., Smeets, B., Smilova, M. D., Solmesky, L. J., Spencer, J., Strain-Damerell, C., Swamy, V., Tamir, H., Taylor, J. C., Tennant, R. E., Thompson, W., Thompson, A., Tomásio, S., Tomlinson, C. W. E., Tsurupa, I. S., Tumber, A., Vakonakis, I., van Rij, R. P., Vangeel, L., Varghese, F. S., Vaschetto, M., Vitner, E. B., Voelz, V., Volkamer, A., Walsh, M. A., Ward, W., Weatherall, C., Weiss, S., White, K. M., Wild, C. F., Witt, K. D., Wittmann, M., Wright, N., Yahalom-Ronen, Y., Yilmaz, N. K., Zaidmann, D., Zhang, I., Zidane, H., Zitzmann, N., & Zvornicanin, S. N. (2023). Open science discovery of potent noncovalent SARS-CoV-2 main protease inhibitors. *Science (80-.), 382*, eabo7201.

84. Sasaki, M., Tabata, K., Kishimoto, M., Itakura, Y., Kobayashi, H., Ariizumi, T., Uemura, K., Toba, S., Kusakabe, S., Maruyama, Y., Iida, S., Nakajima, N., Suzuki, T., Yoshida, S., Nobori, H., Sanaki, T., Kato, T., Shishido, T., Hall, W. W., Orba, Y., Sato, A., & Sawa, H. (2023). S-217622, a SARS-CoV-2 main protease inhibitor, decreases viral load and ameliorates COVID-19 severity in hamsters. *Science Translational Medicine, 15.*
85. Shionogi, Xocova (ensitrelvir fumaric acid) tablets 125mg approved in Japan for the treatment of SARS-CoV-2 infection, under the Emergency Regulatory Approval System. [Press release], (2022). https://www.shionogi.com/global/en/news/2022/11/e20221122.htm. Accessed 9 Sept 2023
86. Kawashima, S., Matsui, Y., Adachi, T., Morikawa, Y., Inoue, K., Takebayashi, S., Nobori, H., Rokushima, M., Tachibana, Y., & Kato, T. (2023). Ensitrelvir is effective against SARS-CoV-2 3CL protease mutants circulating globally. *Biochemical and Biophysical Research Communications, 645*, 132–136.
87. Hu, Y., Lewandowski, E. M., Tan, H., Zhang, X., Morgan, R. T., Zhang, X., Jacobs, L. M. C., Butler, S. G., Gongora, M. V., Choy, J., Deng, X., Chen, Y., & Wang, J. (2023). Naturally occurring mutations of SARS-CoV-2 main protease confer drug resistance to nirmatrelvir. *ACS Central Science.*
88. Kiso, M., Furusawa, Y., Uraki, R., Imai, M., Yamayoshi, S., & Kawaoka, Y. (2023). In vitro and in vivo characterization of SARS-CoV-2 strains resistant to nirmatrelvir. *Nature Communications, 14.*
89. Iketani, S., Mohri, H., Culbertson, B., Hong, S. J., Duan, Y., Luck, M. I., Annavajhala, M. K., Guo, Y., Sheng, Z., Uhlemann, A. C., Goff, S. P., Sabo, Y., Yang, H., Chavez, A., & Ho, D. D. (2023). Multiple pathways for SARS-CoV-2 resistance to nirmatrelvir. *Nature, 613*, 558–564.
90. Ip, J. D., Wing-Ho Chu, A., Chan, W. M., Cheuk-Ying Leung, R., Umer Abdullah, S. M., Sun, Y., & K. Kai-Wang To. (2023). Global prevalence of SARS-CoV-2 3CL protease mutations associated with nirmatrelvir or ensitrelvir resistance. *eBioMedicine, 91*, 104559.

91. Duan, Y., Zhou, H., Liu, X., Iketani, S., Lin, M., Zhang, X., Bian, Q., Wang, H., Sun, H., Hong, S. J., Culbertson, B., Mohri, H., Luck, M. I., Zhu, Y., Liu, X., Lu, Y., Yang, X., Yang, K., Sabo, Y., Chavez, A., Goff, S. P., Rao, Z., Ho, D. D., & Yang, H. (2023). Molecular mechanisms of SARS-CoV-2 resistance to nirmatrelvir. *Nature, 1*, 1–10.
92. Moghadasi, S. A., Biswas, R. G., Harki, D. A., & Harris, R. S. (2023). Rapid resistance profiling of SARS-CoV-2 protease inhibitors. *Npj Antimicrobials and Resistance, 1*, 9.
93. WHO, Diabetes fact-sheet, 2023. https://www.who.int/news-room/fact-sheets/detail/diabetes
94. Pospisilik, J. A., Stafford, S. G., Demuth, H., Brownsey, R., Parkhouse, W., Finegood, D. T., McIntosh, C. H. S., & Pederson, R. A. (2002). Long-term treatment with the dipeptidyl peptidase IV inhibitor P32/98 causes sustained improvements in glucose tolerance, insulin sensitivity, hyperinsulinemia, and β-cell glucose responsiveness in VDF (fa/fa) Zucker rats. *Diabetes, 51*, 943–950.
95. Lambeir, A. M., Durinx, C., Scharpé, S., & De Meester, I. (2003). Dipeptidyl-peptidase IV from bench to bedside: An update on structural properties, functions, and clinical aspects of the enzyme DPP IV. *Critical Reviews in Clinical Laboratory Sciences, 40*, 209–294.
96. Lankas, G. R., Leiting, B., Roy, R. S., Eiermann, G. J., Beconi, M. G., Biftu, T., Chan, C.-C., Edmondson, S., Feeney, W. P., He, H., Ippolito, D. E., Kim, D., Lyons, K. A., Ok, H. O., Patel, R. A., Petrov, A. N., Pryor, K. A., Qian, X., Reigle, L., Woods, A., Wu, J. K., Zaller, D., Zhang, X., Zhu, L., Weber, A. E., & Thornberry, N. A. (2005). Dipeptidyl peptidase IV inhibition for the treatment of type 2 diabetes: Potential importance of selectivity over dipeptidyl peptidases 8 and 9. *Diabetes, 54*, 2988–2994.
97. Sakashita, H., Akahoshi, F., Yoshida, T., Kitajima, H., Hayashi, Y., Ishii, S., Takashina, Y., Tsutsumiuchi, R., & Ono, S. (2007). Lead optimization of [(S)-γ-(arylamino)prolyl]thiazolidine focused on γ-substituent: Indoline compounds as potent DPP-IV inhibitors, bioorganic. *Medicinal Chemistry, 15*, 641–655.
98. Yoshida, T., Akahoshi, F., Sakashita, H., Sonda, S., Takeuchi, M., Tanaka, Y., Nabeno, M., Kishida, H., Miyaguchi, I., & Hayashi, Y. (2012). Fused bicyclic heteroarylpiperazine-substituted l-prolylthiazolidines as highly potent DPP-4 inhibitors lacking the electrophilic nitrile group, bioorganic. *Medicinal Chemistry, 20*, 5033–5041.
99. Bronson, J., Black, A., Dhar, T. G. M., Ellsworth, B. A., & Merritt, J. R. (2013). *To market, to market-2012* (1st ed.). Copyright © 2013 Elsevier Inc. All rights Reserved.

Chapter 16
Computational Study of Conformational Changes in Nuclear Receptors upon Ligand Binding

Azam Rashidian, Dirk Pijnenburg, Rinie van Beuningen, Antti Poso, and Thales Kronenberger

Abstract In this chapter, we introduce the nuclear receptor superfamily, highlighting their different classes and structural features related to their functions, in order to

A. Rashidian (✉)
Department of Pharmaceutical and Medicinal Chemistry, Institute of Pharmaceutical Sciences, Eberhard-Karls-Universität, Tübingen, Tübingen, Germany

Tübingen Center for Academic Drug Discovery & Development (TüCAD2), Tübingen, Germany

Partner-Site Tübingen, German Center for Infection Research (DZIF), Tübingen, Germany

Excellence Cluster "Controlling Microbes to Fight Infections" (CMFI), Tübingen, Germany
e-mail: azam.rashidian@uni-tuebingen.de

D. Pijnenburg · R. van Beuningen
Pam Gene International, 's-Hertogenbosch, The Netherlands

A. Poso
Department of Pharmaceutical and Medicinal Chemistry, Institute of Pharmaceutical Sciences, Eberhard-Karls-Universität, Tübingen, Tübingen, Germany

Tübingen Center for Academic Drug Discovery & Development (TüCAD2), Tübingen, Germany

School of Pharmacy, Faculty of Health Sciences, University of Eastern Finland, Kuopio, Finland

T. Kronenberger (✉)
Department of Pharmaceutical and Medicinal Chemistry, Institute of Pharmaceutical Sciences, Eberhard-Karls-Universität, Tübingen, Tübingen, Germany

Tübingen Center for Academic Drug Discovery & Development (TüCAD2), Tübingen, Germany

Partner-Site Tübingen, German Center for Infection Research (DZIF), Tübingen, Germany

Excellence Cluster "Controlling Microbes to Fight Infections" (CMFI), Tübingen, Germany

School of Pharmacy, Faculty of Health Sciences, University of Eastern Finland, Kuopio, Finland
e-mail: thales.kronenberger@uni-tuebingen.de

V. G. Maltarollo (ed.), *Computer-Aided and Machine Learning-Driven Drug Design*, Computer-Aided Drug Discovery and Design 3,
https://doi.org/10.1007/978-3-031-76718-0_16

focus on the different available computational techniques applied to these proteins. Our focus lies on the use of classical molecular dynamics simulations and other techniques explaining conformational changes in these receptors and novel approaches to pursue them experimentally.

Keywords Nuclear receptors · Conformational changes · Reporter gene assay · Transcription factors · Agonists

Abbreviations

ADME	Absorption, distribution, metabolism, and excretion
AF-1	Activation Function 1
AF-2	Activation Function 2
AR	Androgen receptor
BPA	Bisphenol A
CAR	Constitutive Androstane Receptor
CoA	Coactivator
CoR	Corepressor
DDI	drug-drug interactions
DDT	Dichlorodiphenyltrichloroethane
DES	Diethylstilbestrol
ED	Endocrine Disruptor
EDCs	Endocrine Disrupting Chemicals
EE2	17-α-ethinylestradiol
EGF	epidermal growth factor receptor
ERα, β	Estrogen receptor α, β
FDA	Food and Drug Administration
FLIP	Fluorescence Loss in Photobleaching
FRAP	Fluorescence Recovery After Photobleaching
FXR	Farnesoid X receptor
GPCR	G-protein-coupled receptor
HTS	High throughput screening
HREs	Hormone response elements
LBD	Ligand Binding Domain
LBP	Ligand Binding Pocket
LXR α, β	Liver X receptor
MD	Molecular dynamics
MHC	Major Histocompatibility Complex
MM	Molecular Mechanics
MSMs	Markov State models
NCOR	Nuclear Receptor Coregulator
NMR	Nuclear magnetic resonance
NR	Nuclear Receptor

NR-CoA	Nuclear Receptor Coactivator
PBC	Periodic boundary conditions
PCA	Principal component analysis
PCBs	Polychlorinated biphenyls
PES	Potential Energy Surface
PDB	Protein Data Bank
PPARs	peroxisome proliferator-activated receptors
PR	Progesterone receptor
PXR	Pregnane X receptor
RAR	Retinoic acid receptor
RXR	Retinoid X receptor
SAR	Structure–activity relationship
SMRT	Silencing mediator of retinoic acid and thyroid hormone

1 The Nuclear Receptor Super Family Shares a Conserved Fold but a Diverse Activity and Role

Nuclear receptors represent, besides other drug targets such as G protein-coupled receptors, ion channels, receptor tyrosine kinases, and immunoglobulin-like receptors, a major receptor target class for drug development. Human Nuclear Receptors (NRs) are a superfamily of intracellular receptors consisting of 48 members. However, the number of functionally different NR proteins is by far larger, due to alternative splicing processes and posttranslational modification, such as ubiquitination and phosphorylation [1]. In 1974, the correlation between hormone action and alterations in the gene expression status was reported [2]. Later studies revealed the now-called classic model of the NR signaling pathway [3]. The first NRs were cloned and investigated in 1985, and this represented the starting point of modern NR research [4–6]. Subsequently, additional NRs were identified [7–9] (Fig. 16.1) and by now the family is composed of over 500 members spread among several metazoan species [10]. Based on their mechanism of action and ability to bind to DNA, NRs are categorized into four groups [11]: types I, II, III, and IV.

Type I, belongs to subfamily 3 [12], is steroid hormone receptors. Upon ligand binding, these receptors translocate into the nucleus, where receptor homodimers bind to specific DNA sequences, known as hormone response elements (HREs), indirect repeat (Fig. 16.1). Estrogen receptor (ERα, ERβ), Androgen receptor (AR), Progesterone receptor (PR), and Glucocorticoid receptor (GR) belong to this type.

Type II belongs to subfamily 1 [13], which is found in the nucleus such as the thyroid hormone receptor (TR), retinoic acid receptor (RAR), peroxisome proliferator-activated receptors (PPARs), liver X receptor (LXRα, LXRβ), farnesoid X receptor (FXR), vitamin D receptor (VDR), pregnane X receptor (PXR), and

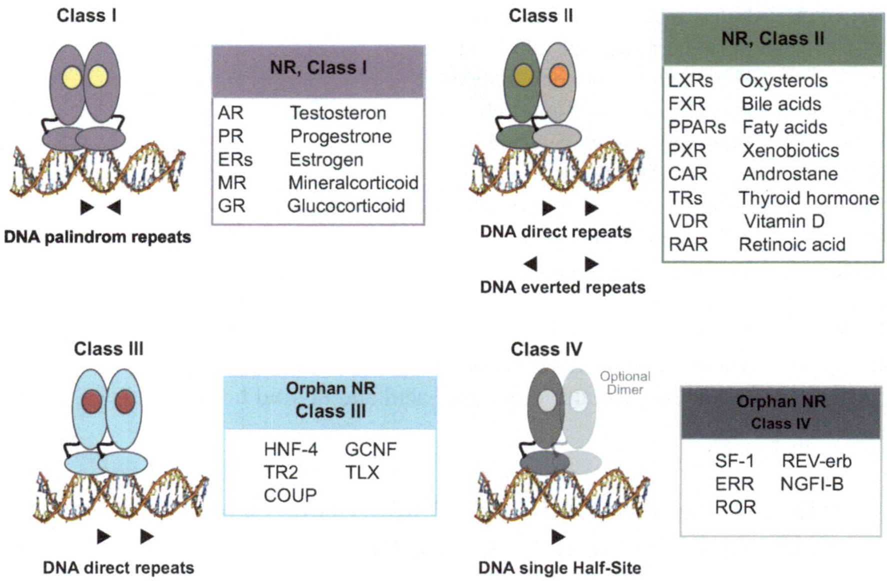

Fig. 16.1 Schematic overview of the nuclear receptor superfamily. Four superfamilies of nuclear receptors are represented based on dimerization, DNA binding (direct or everted repeat), and ligand specificity (required or not required). Class I: steroid receptor (also known as hormone receptor); class II: RXR heterodimers; class III: dimeric orphan receptor; class IV: often monomeric orphan receptor

constitutive androstane receptor (CAR) heterodimerizing with retinoic X receptor (RXR). Type II is usually complexed with corepressor proteins in the absence of ligand and binding to DNA direct repeat. Upon binding the ligand, protein conformation is changed, leading to dissociation of corepressor protein and recruitment of coactivator. Subsequently, this complex, in addition to other transcriptional machinery components, transcribes DNA. Type II receptors can also bind everted repeats.

Type III, subfamily 2 [14], is like type I as being homodimers but in contrast, they bind to DNA direct repeat. No ligand has been identified in this group. Receptors such as HNF-4, TLX, and TR2 are examples of this subfamily.

Type IV nuclear receptors have the ability to bind to DNA in either a monomeric or dimeric form [12]. One representative member of this group is Steroidogenesis Factor-1 (SF-1) [10]. Like type III receptors, these receptors do not have any known natural ligands, the reason that they are referred to as orphan nuclear receptors. Both type III and type IV receptors are still not well understood in terms of their function and structure.

Of interest, the space distance between repeats and, less often, their orientation can vary within the same receptor, depending on its dimerization status [15], with 12 NRs unexpectedly binding to a single monomeric half-site. Recently, it has been

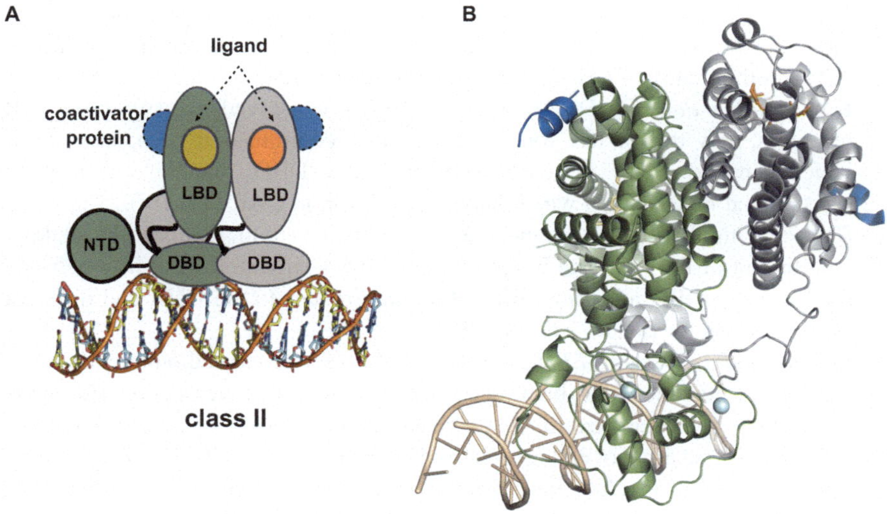

Fig. 16.2 Illustration of nuclear receptor structure (class II). (**a**) Schematic view of PPARγ-RXRα complex. The N-terminal domain (shown as **NTD**) is ligand-independent; the DNA binding domain (shown as **DBD**) is conserved with two zinc fingers; the **Hinge** region is shown as a black loop connecting DBD and LBD; the Ligand-Binding Domain (shown as **LBD**) involved in the dimer, ligand, and coactivator (in blue) binding. (**b**) A cartoon representation of image A. Gold color denotes DNA; cyan spheres depict zinc; the green and grey cartoon illustrates the dimer structure of PPARγ-RXRα complexed with a coactivator; Blue color refers to the coactivator. PDB ID: 3DZY [25]

shown that different small-molecule ligands can alter the NR binding to distinct DNA-binding sites [16], which has implications for their regulatory function, where certain ligands would lead to different pathways and phenotypes.

Currently, we understand the nuclear receptor superfamily as a major group of intracellular transcription factors, which regulate broad aspects of cell functions including cell growth, differentiation, and metabolism in distinct organs. The activation of these receptors is regulated by endogenous or exogenous lipophilic compounds and regulatory proteins. They share a highly similar structure, particularly, in ligand-binding domains (LBDs) and DNA-binding domains (DBDs). Having DBD reveals their genome transcriptional role as they are known as transcriptional factors. These features all reveal NRs remarkable role in organism survival (within the scope of managing metabolic rates, energy stores, salt homeostasis, responding to exogenous toxins, and inflammation to regulate growth, reproduction, and development) and highlight them as promising targets for therapeutic development.

All NRs have a similar structural organization and typically contain five structural domains (Fig. 16.2):

1. *N-terminal domain*: which varies considerably among the receptors and is commonly unstructured; typically, it contains a transactivation domain known as Activation Function 1 (AF-1) and is ligand-independent.
2. *DNA-binding domain (DBD)*: which is highly conserved across various NR receptors, this region has four cysteines that coordinate to two zinc atoms which bind to DNA response elements (e.g., DBD functions in a post-translational modification which happens at Thr38 in CAR [17, 18]).
3. *Hinge*: a highly flexible connecting region believed to regulate the cellular distribution of the NR. The hinge region conveys structural flexibility between the LBD and DBD allowing different binding modes to the DNA and different configurations for (hetero-) dimers [19].
4. *Ligand-binding domain (LBD)*: comprised of a very conserved bundle of eleven α-helices, where the ligand binding pocket is located, However, the interior of the ligand-binding pockets exhibits significant variation, enabling nuclear receptors to bind a diverse array of endogenous and synthetic ligands [20, 21]. This binding capacity extends to include the activation function-2 (also referred to as the αAF-2 helix) and the three-stranded β-sheet, except PXR. The eleven α-helices can be categorized into three distinct groups: H1/H3, H4/H5/H8/H9, and H7/H10/H11.
5. *C-terminal domain*: also varies considerably, in terms of sequence, among nuclear receptors [11, 22–24].

Physiologically, NRs regulate genes involved in different physiological functions such as cell growth, differentiation, homeostasis, and metabolism and were conserved through evolution. They are transcription factors and commonly function by being activated by small lipophilic molecules ($<$1000 Da), able to cross the membrane. Initially, they were solely identified as endocrine receptors, however, it was later discovered that NRs can also interact with xenobiotic compounds, such as Endocrine Disruptor Chemicals (EDCs or EDs, [26–28]). EDCs can mimic the behavior of endogenous ligands such as natural hormones and modify their metabolism and transport through NR-mediating signaling. This phenomenon causes a wide range of developmental, reproductive, or metabolic diseases [26, 29, 30].

NRs' transcriptional activation is typically facilitated through the LBD. This complex domain consists of three distinct yet interconnected relevant regions, namely:

1. *Ligand-binding pocket (LBP)*: This pocket serves as a location for small molecules to bind.
2. *Activation function 2 domain (AF-2)*: Composed of the helices H3/H4/H5/H12 interface, AF-2 is responsible for ligand-dependent transactivation. It also functions as the surface for binding coregulators.
3. *Dimerization surface*: This surface enables interaction with other LBDs in partner molecules.

Starting from the ligand-free basal conditions, the protein NR, which might be complexed with a corepressors protein, is located in the cytoplasm [31]. Upon binding to an activator/agonist ligand, the LBD undergoes an allosteric conformational change that results in the movement and stabilization of its H12. This conformational change leads to the release of corepressor binding (referred to as CoR, if present) [32, 33], as well as allowing dimerization and migration to the nucleus.

The activated NRs bind to a conserved DNA region called response element (RE) downstream in the promoter of target genes [34]. The canonical core motif has the consensus sequence 5′-AGGTCA-3′ [35]. The specificity and affinity of NR binding are dependent on the configuration and number of the core motif [36]. Another factor to affects the NR-specificity is the linker region between the core motifs [37, 38].

Concomitantly with the DBD–DNA interaction, the NR complex has access to different nuclear coregulatory proteins, specifically coactivators (CoAs). The aforementioned conformational change in the AF-2 region enables the coactivator recruitment, which is commonly referred to as the "coregulator switching" model [32, 33]. The fully activated and DNA-bound complex then can regulate the transcription of its target genes.

Of note, the transcriptional regulation, whether it involves activation or repression, occurs through a balance between the NR interaction with different coactivators or corepressors, as well as other protein factors that interact with the promoter of the target gene. Corepressor proteins interact with the NRs via the short peptide motif LxxxIxxxL (where L is leucine, I is isoleucine, and x can be any amino acid). Examples of corepressors are the silencing mediator of retinoid and thyroid receptors (SMRT) or the nuclear receptor corepressor (NCoR) [39–42]. They contribute to gene silencing by recruiting histone deacetylases, chromatin modifiers, and remodeling proteins. On the other hand, coactivator proteins, such as the steroid receptor coactivator (SRC) family, can recruit histone acetyltransferases, histone methyltransferases, and histone kinases, resulting in chromatin unpacking, promoter opening, and activation of the target genes [32, 39, 43]. Coactivators bind via LxxLL motifs (where L is leucine, and x can be any amino acid) to NRs [44].

Numerous studies have highlighted the pivotal role of H12 (also known as αAF-2), which is part of the AF-2 region, in controlling the activation and deactivation processes [45]. In this sense, NR agonists can stabilize the active AF-2 conformation, forming a surface that binds coactivator proteins. Then, different CoA proteins can modulate the transcriptional activity. The CoA recruitment event together with the DNA interaction marks the beginning of the nuclear receptor activity. Depending on the coactivators/corepressors binding and on the cellular context, alternative transcriptional outcomes can take place [46–48]. Alternatively, antagonists can act by destabilizing this relevant H12 conformation and partial-agonists can partially trigger this molecular event. The nature of the ligand, occupancy in the binding pocket and interactions, determines the position of H12 and subsequently the coregulator interaction [20, 21].

The corepressor and coactivator motifs form amphipathic α-helices, of which the hydrophobic residues interact with the AF-2 surfaces of the LBD [21]. However, NR-LBD (in)-activation should probably not be seen as an "on/off" switch model.

Rather, NR-LBD acts as a regulator fine-tuning the interaction between NR domains with the coregulators, which would allow a range of signaling outcomes [21].

In this context, the biological role of NR is not determined by each protein individually but is rather a result of other protein-binding partner as well. The molecular determinants dictating specificity/selectivity in NR–CoA interactions remain understudied on a structural level. The pioneering work from Broekema et al. (2014) [49] suggests that amino acid sequences in both the NR-LBD and coregulator motif are relevant determinants in the NR-specific preferences for particular coregulator binding motifs. However, most of the NR crystal structures only offer a static vision of these individual components, lacking insights into the conformational changes induced by the different ligands and protein-binding partners. The essence of the problem is the difficulty of experimentally addressing conformational change in complex structures; that is, how the effect of the ligand-binding propagates through the structure to affect other sites. In this chapter, we focus on discussing the dynamic transitions of NRs using classical molecular dynamics (MD) simulations.

2 Examples of Small Molecules Acting as Nuclear Receptor Modulators

Many studies have shown the crosstalk of NRs that is followed by controlling the homeostasis of glucose, bile acids, lipids, hormones, and inflammation [50]. This ability stems from the flexibility and versatility of nuclear receptors, as their transcriptional activity can be regulated by ligands, partner proteins, coactivators, corepressors, and promoter genes. This mechanism underscores their role in a wide range of developmental, reproductive, or metabolic NR-related diseases. Given these characteristics, NRs have emerged as prominent therapeutic targets.

ER is the most targeted NR due to its druggable nature. It belongs to the type I of nuclear receptor found in cytoplasm connected with heat shock proteins. Upon ligand binding, it forms a homodimer and translocates to the nucleus. This receptor is found in two forms, ERα and ERβ, both of which bind to the native ligand estradiol. Tamoxifen, approved in the 1970s [51], and raloxifene, approved in 1997 [52] are used for the treatment and prevention of ERα-dependent breast cancer in women as antagonists. Both compounds have been co-crystallized with the receptor, binding to the ligand binding pocket (LBP), They exert their effects by dislocating H12 from an active conformation to an inactive state. However, Tamoxifen can lead to endometrial cancer as an agonist because of the variability of coregulator proteins, whereas raloxifene, also acting as an agonist, is used for osteoporosis treatment in women. Androgen receptor (AR) is another NR found in the prostate and several other tissues with testosterone as a native ligand. Several diseases, including prostate cancer, have been linked to this receptor. Although enzalutamide is an AR competitive antagonist approved drug in 2009, patient resistance happens after months of treatment [53].

As mentioned earlier, EDCs are xenobiotic compounds interacting with NRs and by mimicking endogenous, causing a broad range of diseases. Epidemiological studies reported that exposure to xenoestrogens such as diethylstilbestrol (DES) during fetal development and exposure to dichlorodiphenyltrichloroethane (DDT) during puberty increases the risk of breast cancer [54]. Another study offers initial insight into the neural effects of human exposure to bisphenol (BPA). The results propose that when expectant mothers are exposed to BPA during prenatal stages, it may cause modifications in the microstructure of white matter in preschool-aged children and these changes in white matter can mediate the connection between early-life exposure to BPA and the emergence of internalizing problems [55]. Similarly, a rising number of cancerous testicular or malformations of the male genital tract also could be attributed to exposure to EDCs [56]). A study has revealed the relationship between the level of plasma polychlorinated biphenyls (PCBs), a persistent and lipophilic aromatic chemical, and reduced semen quality, particularly reduced sperm motility [57].

Class II of NR has been also the target of therapeutics. Among them, PXR, CAR, and FXR and their heterodimerization partner RXR are of substantial interest and, therefore, introduced shortly in the next sections.

3 Computational Approaches to Evaluate NRs Protein Dynamics

This chapter focuses on the protein dynamics induced by ligand binding and are not discussing purely ligand-based approaches used to understand NR-binding, such as machine learning and classical QSAR studies. Starting from the most classical structure-based approaches, molecular docking has been largely used to study NRs (as reviewed by [45]). For more information about molecular docking approaches and their limitations, readers are referred to chapter (see Chap. 7). Despite the several advantages of docking, as a fast and rational method, this approach is unable to capture NRs' characteristic flexibility and the role of explicit solvent [45, 58–60]. In addition, multiple binding modes for congener ligands are not unheard of for NRs. Some examples, such as the nonanoic acid binding to PPARγ, even display multiple binding modes for the same ligand [61]. This could be even further illustrated by PXR's or CAR's large ligand binding pocket, which allows such freedom for ligand movement (see their respective sessions below).

These observations prompted us to hypothesize that small molecules or fragments (such as EDs and lipids) might adopt multiple binding modes suggesting that a single static model for the NR-ED interaction would be insufficient to accurately describe/predict their outcomes. We identified this as an inherent limitation of the docking and subsequently proceeded with MD simulations (please, see Chap. 8 for more methodological details) to assess the ligand stability within NRs' LBP, besides capturing the NRs conformational dynamic, in many of the illustrative examples below.

In this sense, MD simulations are a relevant and underused tool to elucidate NR conformational rearrangement upon ligand binding and can also be used to discover the ligand entry/exit pathways [62, 63]. Historically, steered MD simulations were used to identify two distinct pathways on the retinoic acid receptor (150 ns), one for ligand binding and another for unbinding [64]. In addition, random expulsion MDs suggested that the retinoic acid may exit the binding site through flexible regions close to the H1–H3 loop and β-sheets, without displacing H12 from its agonist position [65]. It is important to highlight that both studies applied forces to propel the ligand outside the pocket, even if the selection of the expulsion was unbiased. The lack of a significant number of replicas suggests that even more interesting conformational changes could be observed in a modern setup.

Despite that, their observation is consistent with recent crystal structures where the end of H1 and H2 can assume a highly flexible loop-like folding in some receptors. Consistently, studies performed on TR- and ER-LBDs indicated that ligand unbinding does not require the H12 displacement [66, 67]. Up to four different unbinding routes were identified on ER-LBD, with half relying on the H12 dislocation and the other half on the H8 and H11 separation, with this last being potentially influenced by dimerization [66, 67].

A large range of structures cocrystalized with (partial-) agonists and antagonists are available for some NRs [68–71] highlighting changes in the H12 position. However, other receptors lack such diversity. The work from Alvarez et al. [72] in GR aimed to fill this structural gap by exploring the correlation between ligand identity and GR's H12 behavior using (steered and classical) MD simulations. They generated relevant GR-LBD models in the so-called agonist and nonagonist states, simulating those systems in the presence of an agonist (dexamethasone) or an antagonist ligand (RU-486), in a combinatorial fashion. Their description of the H12 as a dynamic ensemble of conformations is influenced by the interacting ligand. On one hand, simulations of GR-dexamethasone display a deep minimum potential energy surface favoring specific conformation. On the other hand, RU-486 bound systems would favor a wider H12 conformational amplitude, consistent with a flatter potential landscape.

Additionally, their steered dynamics trajectories on GR [72] oriented by the Cα's RMSD of H3, H11, and H12 as collective variables, showed conserved secondary structure elements changing conformation among themselves without deformation. Interestingly, the high-resolution structure of progesterone receptor (PR) complexed with RU-486 shows high flexibility of the H12 [73], but not full displacement. Given the crystallization procedure, starting from an agonist-bound starting structure that was later soaked with RU-486, one can postulate that the antagonist binding is a dynamic equilibrium process.

Of note, H12 of several LBDs are described as a stable but flexible helix in the apostate, which could be located away from the LBD's helix bundle [74, 75]. However, time-resolved fluorescence anisotropy experiments showed that, despite H12 flexibility and conformation diversity, the LBD retained its globularity in the apostate [76].

3.1 Retinoid X Receptor and the Role of Allosteric in the NR's Dynamics

Retinoid X receptor (RXR) is a type II NRs and forms heterodimers with approximately one-third of the other NRs [77]. In humans there are three isoforms of RXR; α, mainly found in the liver, kidney, and intestine; β, found in most human tissue; and γ, which mainly exists in the brain and muscles [78]. Malfunctioning of these isoforms has been linked to various health issues [11, 79–81].

RXR activation can be categorized into two groups: permissive and nonpermissive heterodimer. In the permissive heterodimers [82], activation can be induced by binding the agonist to either RXR or NR partner or both receptors. Examples of this group are RXR/FXR, RXR/LXR and RXR/PPAR. In the case of the nonpermissive heterodimer, only agonist binding to the RXR partner triggers activation, but RXR can still bind to the agonist, releasing the corepressor and recruiting coactivator. Examples of this are RXR/TR and RXR/VDR. In this case, RXR can also bind to the agonist and lead to synergistic action in the presence of a heterodimer ligand [82, 83].

The main small molecule compound binding to RXR is 9-cis-Retinoic acid (9cRA). It belongs to a retinoid family and has a critical role in cell growth, development, differentiation, and apoptosis. Other RXR ligands are 9cRA-related compounds and indenoisoquinolines. Remarkably, the work from Bexarotene and Diarylamines [25] employed X-ray crystallography to describe the PPARγ and RXRα structure. They explained how these two receptors interact with DNA, highlighting the influence of DNA in governing the interaction between the two receptor domains through specific rearrangements. They also revealed the cooperative nature of multiple domains of PPARs, which can modulate the properties of the PPARγ-RXRα complex. In this complex, the LBD of PPARγ tightly couples with RXRα domains. Accordingly, the conformational change induces a reposition of receptor domains responsible for DNA binding and optimizing their contact with DNA. They also examined the dynamic properties of structures using amide hydrogen/deuterium exchange mass spectrometry (H/D-Ex). The result revealed that the helices H10/H11, involved in LBD-LBD heterodimerization, adopt a slightly shifted/curved conformation when the protein is contacted with DNA, facilitating optimal contact of the receptor with DNA [25].

3.2 *Pregnane X Receptor's Dynamic Pocket Allows Binding of Diverse Ligand Sets*

Pregnane X receptor (PXR), also known as the steroid and xenobiotic sensing nuclear receptor (SXR) [84], is encoded by the *NR1I2* (nuclear receptor subfamily1, group1, member 2) gene on chromosome 3 [85, 86]. PXR heterodimerizes with RXRα, β and γ (NR2B1–3) [87–89] at H10/H11 region. Additionally, the PXR-LBD demonstrates a distinctive characteristic by homodimerizing at its β1′ interface. Homodimerization occurs through the conserved Trp223 and Tyr225 residues in each monomer (Fig. 16.3) [90]. These amino acid residues involved in the interface exhibit high conservation among various species, including humans, rhesus monkeys, rabbits, mice, rats, pigs, and dogs. However, in the canine PXR, Trp223 is replaced by Gln223 [90]. Noble et al. (2006) showed mutation of Trp223 and Tyr225 does not interfere with DNA, RXR, or ligand binding, rather it disrupts the homodimerization, reducing the recruitment of the coactivator SRC-1 and transcriptional activity [90].

When an agonist ligand binds to the PXR-RXRα heterodimer in the nucleus, it promotes coactivator binding and release of corepressor from AF-2 [92]. Subsequently, this activated PXR complex induces the expression of the target gene. PXR structure, like other NRs, has a large hydrophobic LBD. The primary PXR isoform is composed of 434 amino acids, featuring a notable hydrophobic triad consisting of F288, W299, and Y306. Unlike other nuclear receptors (NRs), the PXR-LBD lacks the typical stable H2′ and H6 helices. Crystallographic structural data clearly illustrate the lack of stability in the H2′ region, which appears disordered in all publicly accessible PXR structures. These characteristics result in a more expansive and flexible LBD for PXR, distinguishing it from other NRs [93–95]. Consequently, the PXR-LBD can accommodate a diverse range of ligands. Presently, the Protein Data Bank (https://www.ebi.ac.uk/pdbe/, accessed on November 2023) repository contains 52 crystal structures of human PXR in its active mode, complexed with the

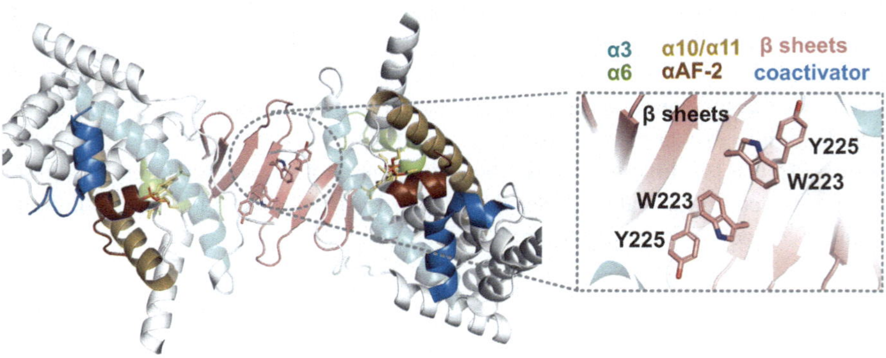

Fig. 16.3 Crystal structure of PXR homodimer. The amino acids Trp223 and Tyr225, located on β1′, mediate the homodimerization. The interface is shown in pink. PDB ID: 1NRL [91]

coactivator protein SRC1. Among these structures, 44 exist in a homodimer assembly, while 8 are available in heterodimer form.

PXR is a ligand-dependent transcriptional factor involved in small molecule metabolism and regulation of diverse cellular processes including bile acid metabolism, glucose homeostasis, cell proliferation as well and inflammation. PXR mostly exist in the liver and intestine. It regulates the gene expression of enzymes and transporters that are responsible for the different pathways of endogenous and xenobiotic pharmacokinetics including absorption, distribution, metabolism, and excretion (ADME). Gene targets of PXR are cytochrome P450 genes (CYP2B, CYP2C, CYP3A) and efflux and uptake transporters of the ATP-binding cassette [96–99]. Besides endogenous ligands, PXR is activated by a broad number of diverse small molecules, including drugs, environmental pollutants, and natural products. These various functions make PXR a potential therapeutic candidate. However, the activation of PXR can induce intestinal and hepatic first-pass metabolism and drug efflux transport [100] which in turn may lead to drug–drug interactions (DDI), adverse drug reactions or therapeutic failure of drugs [101–103]. To exemplify, one can consider the report about the reduced effect of rifampicin on midazolam or contraceptives due to the increased expression of CYP3A4 when these medicines are coadministrated [104], or isavuconazonium which activates the expression of CYP2B6 through PXR-mediated induction and decrease the exposure of bupropion [96, 105]. These observations raised the interest in designing PXR antagonists along with the attempt to limit the activation of PXR in the presence of xenobiotics. As of 2002, several "azole" compounds have been identified as PXR inhibitors, such as ketoconazole, enilconazole, FLB-12, and SPA70, although it is less known about the structural trigger of PXR-bound antagonists.

For instance, SPA70 and SJB7 are close analogues [106] where SPA70 act as an antagonist of PXR but SJB7 is a PXR agonist which highlights the promiscuity of PXR-LBD. Several approaches from experimental methods to computational techniques such as pharmacophore, quantitative structural-activity relationship (QSAR), machine learning (please see Chaps. 4 and 6, respectively), and structure-based methods have been utilized to investigate PXR activation upon ligand binding [96]. Notably, due to the lack of crystal structures of PXR in complexes with antagonist ligands, likely attributed to the complexity and high flexibility of the system, computational studies play a crucial role in unraveling the conformational dynamics of the PXR-antagonist complex.

For a more thorough dissuasion regarding the capability of MD simulations in capturing the dynamic of NRs, we refer to the study conducted by Chandran et al [107] They studied the dynamic behavior of PXR-LBD apo structure comparing it to the agonist-bound state, through short MD simulations that last for 100 ns [107]. Although their short simulations would not allow sidechain and loop reorganization, yet, it was able to identify several conformational states for apo PXR-LBD showing different pocket's volume, while with agonist binding (SR12813), the compound restricted both LBD conformation and binding pocket's size and shape. Alternatively, Motta et al. employed a different strategy by simulating the entry of SR12813 into PXR's LBP utilizing the MD-binding method. Their result suggested that the ligand accessed the LBP through a channel between H2 and H6 helices [62]. To enhance the sample of the SR12813-bond conformations, they utilized scaled MD simulations with a total of 2 μs). Remarkably, their finding confirmed that the SR12813 binding mode observed in the crystal structure (PDB ID: 1NRL [93]) is indeed the most stable through simulations.

In a comparative study, Huber et al. performed 200 ns MD simulations of wild-type PXR-LBD and Trp299Ala mutant, without ligand and with TO90131713 (agonist), SPA70, and SJB7 [108]. They suggested that the extra space conferred by the Trp299Ala is the reason for the observed antagonist-to-agonist switch with this mutant for SPA70. This extra space lets SPA70 reside deeper in the pocket, preventing the αAF-2 dislocation and maintaining PXR active. In our work [109], we used MD simulations starting from a PXR-apo structure, which has a similar conformation with an agonist complex (SR12813, PDB ID: 1NRL). The long MD calculations could decipher ligand-specific influence on conformations of different PXR-LBD regions and could also be useful to guide how to alleviate PXR agonism. Interestingly, even if starting with the same conformation, long MDs can discriminate the dynamic behavior of the agonist versus the antagonist.

More recently, our group employed an in silico screen and experimental cellular reporter assay to identify small molecule kinase inhibitors from an in-house compound library, the Tübingen kinase inhibitor collection (TüKIC) compound library, which act also as a PXR inhibitor. In the experimental work [110] we describe the identification of the C-100 compound and the biochemical binding and cellular protein interaction assays which categorize the novel compounds as mixed competitive/noncompetitive, passive antagonists by disrupting PXR coregulatory binding [110]. This work was supported by structure-based virtual screening and molecular dynamics (MD) simulations which reveal the ligand-specific conformational rearrangements of PXR-LBD including H6 region, αAF-2, H1–H2′, β1′–H3 and β1–β1′ loop [109] (Fig. 16.4).

PXR-LBD's flexibility and promiscuity enable it to bind to a wide range of ligands with different sizes and shapes. However, no considerable changes in interaction patterns are associated with PXR conformational rearrangement. Ngan et al. (2009) investigated the structural foundations of PXR's promiscuity using computational solvent mapping [111], specifically, they used servers such as FTMap to dock small molecule probe fragments, which were followed by binding energy calculation. This technique is designed to identify and characterize hot spot regions within protein binding sites, those regions represent relevant residues/surface regions that play a pivotal role in determining the binding free energy. Their result shows that one of the most important regions for binding the different PXR ligands is the hydrophobic cage (Fig. 16.5b). This subpocket is formed by a triad of Phe288, Trp299 and Tyr306 where pi–pi interaction and edge-to-face interaction are observed for ligand stability. In addition, the polar residues His407, Gln285, Ser247, and Thr248 form hydrogen to bind with ligands either directly or mediated with a water molecule [109]. Phe429, which is located on H11, and αAF-2 known as the hydrophobic side participating in ligand binding [111]. However, the study conducted by Delfosse et al. (2021) on the synergistic activation of the endocrine

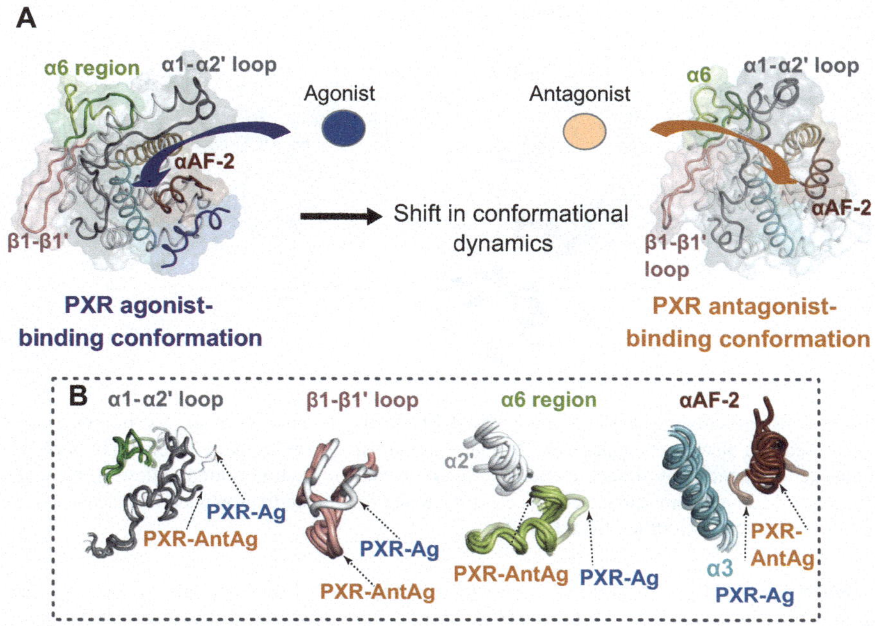

Fig. 16.4 Ligand binding effect on PXR-LBD conformation. (**a**) Agonist-bound and antagonist-bound PXR-LBD conformation. MD simulations revealed the shifts in PXR-LBD conformational dynamics. The PXR-LBD regions undergoing the most extensive conformation are labeled. (**b**) Markov state modeling result for the PXR-LBD bound to compound 100. The subregions with extensive motions are shown individually, with a reference conformation from an agonist-bound crystal structure (PDB ID: 1NRL) in transparent illustrated as follows: H1–H2′ loop, grey dashed line; β–β1′ loop, white; H6 region, transparent light green located in the vicinity of H2′ (white helix); β1′–H3 loop, transparent dark green; αAF-2, transparent dark brown located in the vicinity of H3-helix (cyan helix). (For interpretation of the references to color in this figure legend, refer to Fig. 16.3, modified from [109])

disruptor mixtures on PXR-RXR heterodimer revealed that the PXR activation does not necessarily occur due to the engagement of AF-2 region in ligand binding [87]. Moreover, our in silico study also provided insight into the interaction profile of PXR-LBD bound to full agonist and our identified competitive antagonist consistent with the already known interacting subregion of PXR-LBD for agonist. However, in the presence of an antagonist our MD simulations revealed that His407 is not directly involved with Ligand binding rather it preferred to be engaged with His404 through H-bond which might be one of the local triggers for inducing PXR-LBD conformational rearrangement [109].

Our studies provided insight into which conformational behavior of PXR-LBD can promote PXR antagonism. The discovery of drugs that can simultaneously inhibit both PXR and protein kinases could offer new possibilities in cancer treatment and the possibility of dual PXR, and kinase inhibitors could be beneficial in cancer treatment. and help overcome drug resistance.

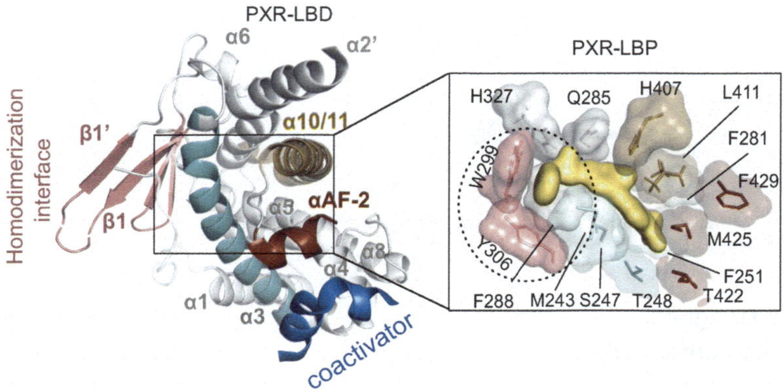

Fig. 16.5 Representative snapshot of the PXR-LBD displaying binding mode. The binding pocket is enclosed in a black rectangular box. Residues participating in binding are labeled. Hydrophobic subpacket surrounded by dashed circle. The ligand is shown in an obscured yellow surface. H3, H10/11, α-AF2 and Homodimerization interface colored in cyan, light brown, dark brown, and pink color, respectively. (Modified from [109])

Most known ligands bind to orthosteric PXR-LBP; however, allosteric sites can be an alternative region for the PXR modulator's accommodation [112–114]. Allosteric sites are distant from orthosteric sites and accommodate structurally different ligands. So far, 202 allosteric modulators have been reported for nuclear receptors [115]. The proposed allosteric ligand binding sites are the AF-1 site, zinc fingers and response elements, LBP (synergistic effect), the AF-2 site, and the binding function 3 (BF-3 site) [115, 116]. The BF-3 region was originally described in the androgen receptor and is a hydrophobic cleft composed of the H1, the L:H3–H5 loop, and H9. Despite their amino acid and conformation conservation, no ligands were found to bind in the PXR equivalent.

An example of an allosteric ligand binding pocket is reported by Delfosse et al. (2021) where the simultaneous binding of 17-α-ethinylestradiol (EE2) and transnanochlor (TNC) enhanced the CYP3A4 induction higher than the single binding of either of compounds when compared to potent agonist SR12813 [87]. Ketoconazole [117] is an example of a modulator binding to AF-2 and acts as a PXR inhibitor. To name, fluconazole, enilconazole, pazopanib [103], metformin and leflunomide [118], FLB-12 [117, 119], coumestrol [120, 121], sulforaphane [122], and campthotecin [123] are other AF-2 modulators.

In summary, while the adaptability of PXR-LBD has evolved to safeguard humans from environmental factors through its involvement in xenobiotic metabolism, this characteristic poses challenges in certain treatments, contributing to issues such as drug–drug interactions, and adverse and drug resistance. Advancements in computational technology, coupled with in vitro validation, can enhance the study of PXR and provide valuable insights.

3.3 Constitutive Androstane Receptor and the Structural Features for Constitutive Activity

Constitutive Androstane Receptor (CAR), encoded by the *NR1I3* (nuclear receptor subfamily1, group1, member 3) gene belongs to class II NRs. CAR is predominantly expressed in the intestine and liver, [124, 125] heterodimerized with RXRα, β, γ (*NR2B1–3*), and like PXR plays a critical role in regulating genes involved in exogenous and endogenous metabolism. Alternative splicing generates multiple CAR isoforms in humans and other primates, but not in rodents [126, 127]. This mechanism, however, is not yet fully understood. Around 50% of transcripts encode the wild-type CAR1 that displays high basal activity, CAR2 and CAR3 isoforms (Fig. 16.6a–c) demonstrate ~10% and ~40%, respectively with low constitutive activity, likely due to their reduced interaction with RXR which results in weaker binding to DNA and coactivators [128, 129].

The unique feature of CAR is its constitutive activation, distinguishing it from other nuclear receptors. Unlike other NRs, CAR does not require ligand binding for its transcriptional activity although ligand binding can modulate CAR activity as an agonist or inverse agonist. It can bind to a vast number of chemical compounds [130] and regulates multiple genes involved in xenobiotic detoxification, which might overlap with or be distinctive from PXR target genes.

CAR is primarily localized in the cytoplasm and forms a complex with heat shock proteins [131]. The majority of CAR ligands act as direct activators, such as 6-(4Chlorophenyl)imidazo[2,1-b][1,3]thiazole-5-carbaldehyde-O-(3,4-dichlorobenzyl) oxime (CITCO) [132] in human and TCPOBOP [133] in mouse. On the other hand, synthetic compounds like phenobarbital and acetaminophen, as well as endogenous compound bilirubin, are examples of indirect CAR activators. It has been proposed that the Epidermal growth factor receptor (EGFR) signaling pathway is inhibited by phenobarbital, therefore acting as a repressor, leading to the dephosphorylation of CAR at Thr38 within the cytoplasm, which enables its translocation to the nucleus [17, 18]. Flavonoids have been reported to function as both direct and indirect activators depending on cellular context [134]. Androstane metabolite, PK11195 [135], TO901317 [136] and S07662 [137, 138] are examples of CAR inverse agonists (Fig. 16.6d). Inverse agonists can reduce the basal constitutive activity of CAR, acting as inhibitors/repressors. Moreover, certain CAR activators, such as phthalates, antivirals, and artemisinin derivatives display some isoform selectivity [128, 129, 139–141]. In the nucleus, CAR plays a constitutive regulatory role in target genes, including CAR-preferentially responsive gene CYP2B6. To achieve this, CAR interacts with specific DNA motifs DR3, DR4, DR5, ER6, and ER8, located in the enhancer and promotor region of target genes [142].

Another feature of CAR, as a member of NRs, is its permissive activity when complexed with RXR. Several studies [137, 143, 144] have reported the synergistic and additive effect of multiple xenobiotic compounds. Dauwe et al. (2023) conducted their in vivo study with several pesticides—recognized as ligands of

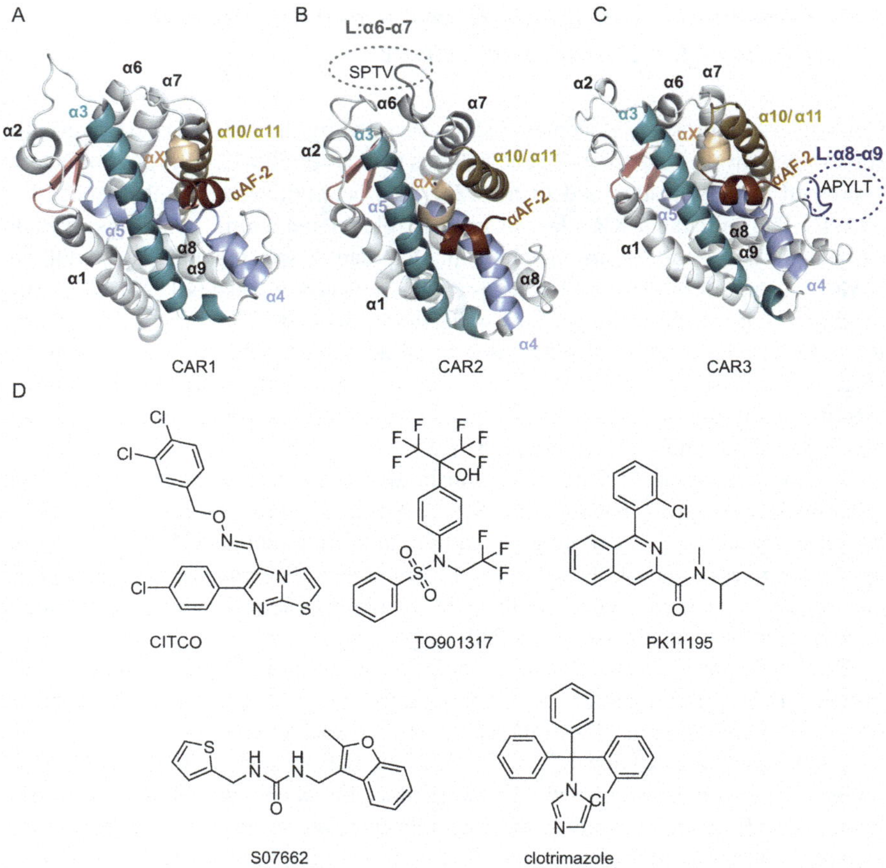

Fig. 16.6 Structural overview of the CAR-ligand binding domains. (**a**) Featuring the CAR1-ligand binding domain (CAR1-LBD) (**b**) in CAR2-LBD and (**c**) CAR3-LBD. The insertion loops are highlighted within a grey circle (SPTV, L: H6–H7) for CAR2-LBD and in a blue circle (APYLT, L: H8–H9) for CAR3-LBD. Various regions of interest, including H3, H5, β-sheets, H10/H11, αX-helix, and αAF-2 (α12) are visualized with distinct colors for clarity. (**d**) The small molecule CAR ligands

CAR and Tri-butyl-tin (TBT) served the role of an RXR agonist. In mice subjects, the concurrent administration of dieldrin (pesticide) and TBT prompted a synergistic activation of CAR. Furthermore, combined effects were observed with propiconazole, bisphenol, boscalid, and bupirimate [145].

An initial study [146] aimed to identify the conformational rearrangement of CAR-LBP upon agonist binding by implementing short MD simulations (50 ns). Their findings indicated that the activation helix retains its active conformation even in the ligand's absence, through van der Waals interactions and hydrogen bonds. Additionally, their study revealed significant conformational changes within the ligand-binding pocket upon agonist binding, as well as the impact of CAR inducers such as pregnenedione and CITCO on the helical structure of the H12.

Concomitantly, classical CAR studies [138, 147] could not precisely discriminate between H12 conformation with agonist and antagonist. This observation was attributed to either the short time of MD simulation or the lack of coregulator in their simulations. In their follow-up study [138], the inclusion of corepressor (SMRT) in their simulations allowed the observation of specific movements. Among those, the movement of H12 toward H10, is favorable for SMRT binding due to stabilizing van der Waals interaction between these helices. The interpretation of the simulations is complex as their data correlates with cell-based reporter gene assays, which are prone to permeability and coregulator availability in the chosen cell models. In this sense, they also reported that the same ligand may be able to recruit either coactivators or corepressors, and the CAR activity depends on the pool of coregulators available in the cell [148].

These studies together with the limited availability of crystal structures depicting agonist-bound CAR pursued us to employ in silico methods to model CAR isoforms and comprehensively study their structure and conformational changes when interacting with various chemical compounds. These compounds range from CITCO analogies to endocrine disruptors (EDs). We investigate specific human CAR agonists with no activation of PXR or other NRs while maintaining favorite ADME characteristics in human hepatocyte cellular models or the context of humanized mouse models [149]. The exploration was carried out within the collection of kinase inhibitors. We discovered several derivatives of 3-(1H-1,2,3-triazol-4-yl) imidazo [1,2-a] pyridine numbered 37, 39, 40 and 48 that directly activate human CAR in nanomolar concentrations. We utilized docking studies with these compounds within CAR-LBD followed by MD simulations. The simulation revealed that the most extensive movement occurs in H2 (Fig. 16.7a and see MD simulation protocol chapter). In terms of protein–ligand interaction, MD revealed that the ligands are stabilized by polar interaction via His203, Thr225 and Thr228 (Fig. 16.7b) with a U-shaped conformation similar to that of CITCO configuration when they are accommodated in hydrophobic subpacket [149]. MD trajectories analysis also revealed that compound 39 displays a notable interaction with residue Ile164 and exhibits a higher frequency of interaction with Y326 compared to CITCO. The result agreed with the earlier finding that geometrical stabilization happens by H-bond interaction between Tyr326 and Asn165 (Fig. 16.7c) [150] leading to stabilization of H12 near H3. These findings underscore the pivotal role of H3 and H10/H11 in protein stabilization. Further in-depth analysis, spanning from docking data to molecular dynamic (MD) simulations, has revealed that both CITCO and compound 39 effectively engage with CAR1-LBD, primarily through hydrophobic contacts. Notably, compound 39 establishes more robust polar contacts with CAR1-LBD than CITCO, attributed to the formation of hydrogen bonds between the amide moiety of compound 39 and the backbone oxygen of Thr225 and Asp228 [149]. We have also considered the evaluations of metabolic stability and activity towards PXR. The in vitro analysis revealed the selectivity of compound 39 towards other nuclear receptors. Furthermore, a preliminary single-dose pharmacokinetic

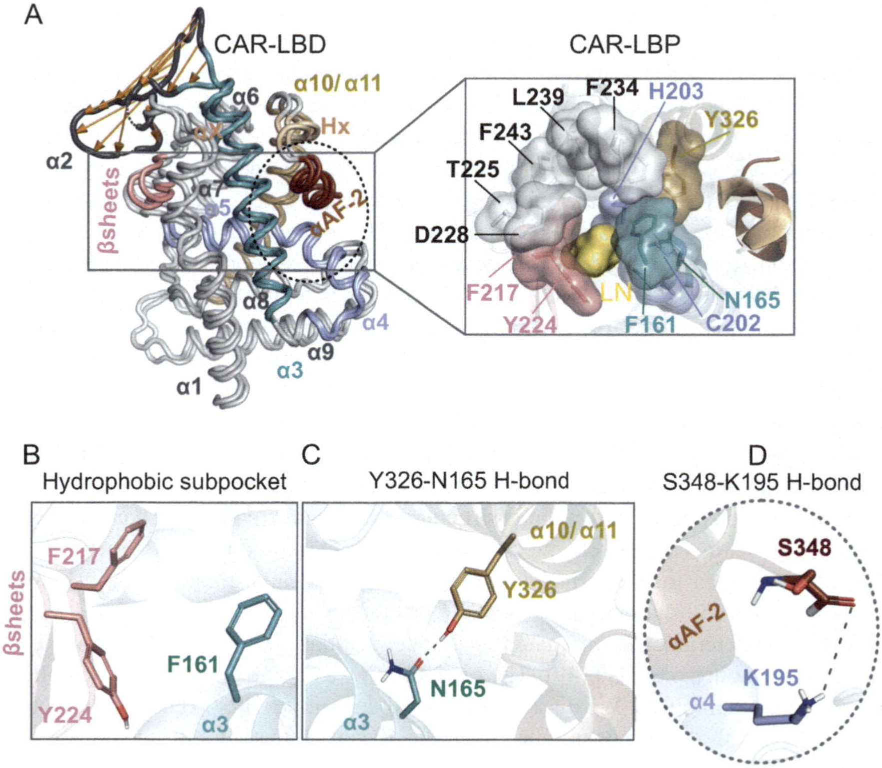

Fig. 16.7 CAR1-LBD structure and the key elements. (**a**) Overview of the CAR1-LBD structure. The regions of interest are highlighted as follows: H2′-H3 loop (residues 140–153), dark Gray; H3 loop (residues 157–178), green; H5 (residues196–209), violet; β sheets (residues 217–223), pink; H10/H11 (residues 308–333), light brown; H_X (residues 336–339), light orange; H12 (residues 341–348), dark brown. The rectangular area denotes the location of the ligand-binding pocket (LBP) and the residues forming the LBP. The dashed circle AF-2 surface area. The main residues participating in ligand binding are depicted in the stick model with a transparent molecular surface. Residues are colored according to their respective regions (see cartoon structure on the left). (**b**) Hydrophobic subpocket consists of aromatic residues. (**c**) Hydrogen bonds between the Tyr326 oxygen atom and Asn165 polar group are shown as the black dashed line. (**d**) Close view of H4 and H12 zooming in Lys195 (on H4) and Ser348 (on H12). The black dashed line represents the hydrogen bond between Lys195 and Ser348

study highlighted the efficacy of compound 39 as a novel human CAR agonist in animal experiments, demanding deeper exploration through repeated-dose, long-term proof-of-concept studies. Notably, the chemical tool employed in our investigations demonstrated no observable toxicity or genotoxic potential. This serves as additional evidence for the substantial activation of human CAR by compound 39.

Meanwhile, having the conformational changes of CAR1 in the presence of agonists (CITCO Imidazo[1,2-a]pyridine) we were interested in speculating the effects of EDs on CAR1 and CAR3. In this investigation, we explore the binding mechanisms between ED compounds and CAR1, CAR2, and CAR3. We compared their conformational changes with those elicited by the well-established CAR agonist, CITCO. Additionally, we studied the effects on conformational alterations caused by CAR1 reverse agonists PK11195 and S07662, clotrimazole, and the CAR3 reverse agonist TO901317 [136]. Utilizing a combination of molecular docking, MD simulations, and MM-GBSA calculations (see chapter discussing MD simulations for more details), we predicted the binding affinities of the EDs. This study includes the development of a robust model for CAR2 and CAR3, accounting for the influence of the APYLT and SPTV insertions (L:H8–H9 and L: H6–H7) (Fig. 16.6a–c), respectively. This model is then compared with CAR1 in our monomeric simulations.

The initial docking experiments with EDs yielded a singular notable conformation for interactions with CAR1, CAR2, and CAR3, whereas CAR1-S07662 binding exhibited two distinct poses. Consequently, both docking poses were considered in the subsequent analysis. As previous studies documented, the occurrence of multiple binding modes for ligands is not uncommon in the realm of nuclear receptors. This phenomenon is attributed to the expansive ligand binding pocket of CAR, affording considerable flexibility for ligand movement, particularly in the case of PXR and CAR.

MM-GBSA analysis was utilized to gain insights into the EDs impact on CARs' binding behaviors. Notably, the significant involvement of Phe161 (from helix H3) and Tyr224 (in the β-sheet) in stabilizing the complexes highlight their crucial role in binding affinity (Fig. 16.7b). This emphasizes the critical role played by the hydrophobic pocket formed by helix H3 and the β-sheet for CAR potency and selectivity.

In another project, we examined interactions of individual branched 4-nonylphenols (22NP, 33NP, and 353NP) and linear 4-nonylphenol (4-NP) with CAR1 and its variant CAR3 using MD simulations, and cellular experiments. Our study demonstrates the enhanced stability of branched 4-nonylphenols (4-NPs) in binding to activate both CAR1 and the CAR3 variant LBDs over the course of MD simulations. Notably, the branched 4-NP exhibited superior efficacy in activating both CAR3 and CAR1 LBDs compared to the linear 4-NP. To delve deeper into the cellular effects of these compounds, we conducted experiments using HepaRG cells. Remarkably, all tested NP compounds led to a significant upregulation of CYP2B6, a pertinent indicator of CAR activation. Our simulation analysis highlights the critical roles of Helices H3, H5 and β-sheets in the interaction with CAR1 and CAR3 (Fig. 16.7). Additionally, it is noteworthy that none of the examined ligands directly interact with αAF-2, the region associated with receptor activation and agonism. Nevertheless, the protein–ligand interaction data strongly suggest favorable interactions in both CAR3-LBD and CAR1-LBD, consistent with assembly assay data [151]. Extended simulations provide insights into the dynamic behavior of the AF-2 region, emphasizing the significant potential for conformational changes and dynamic shifts within this helix, irrespective of the specific ligand. These observations accentuate the importance of considering longer timescales in our analyses to comprehensively capture NP binding dynamics. Relying solely on short conformational changes may lead to an incomplete understanding of the process. Furthermore, the assessment of free binding energy substantiates the high binding affinity to NP, comparable to CITCO.

3.4 Farnesoid X Receptor and the Effect of Heterodimerization on the Coactivator Recruitment

The Farnesoid X receptor (FXR) belongs to the nuclear receptor superfamily (NR1H4) and exhibits prominent expression in the liver and intestine. While its presence is less pronounced in the kidney, adipose tissue, and adrenal glands, FXR plays a pivotal role in regulating various physiological processes, including bile acid homeostasis, lipid and glucose metabolism, and inflammation [152].

The endogenous ligands for FXR include farnesol derivatives, which are metabolic intermediates of the mevalonate pathway [153]. Additionally, chenodeoxycholic acid (CDCA) and cholic acid (CA) serve as endogenous ligands for FXR [154]. Two well-known genes associated with FXR are FXRα and FXRβ [155, 156]. The FXRα gene is evolutionarily conserved from fish to humans [157] and encodes four transcript isoforms: FXRα, FXRα2, FXRα3, and FXRα4 in humans and mice. In contrast, FXRβ is a pseudogene in humans and primates [152, 158, 159]. Similar to other members of the nuclear receptor family, FXR possesses a highly conserved domain. Ligand binding to FXR, while in complex with the retinoid X receptor (RXR), induces conformational changes. These changes lead to the recruitment of either coactivators or corepressors, thereby modulating the transcription of target genes by promoting or silencing their expression, respectively.

The activation of FXR ligands through agonists follows a well-established classical αAF-2-trapping mechanism. However, the mechanisms underlying antagonism appear to be more varied. Our prior short simulations indicated the potential significance of destabilizing the L:H11–H12 interaction on monomers in the context of FXR antagonists [160]. We conducted a project started from an agonist/apostructure fully folded FXR-LBD in order to model the initial conformational changes that would happen upon antagonist binding [161]. Leveraging the microsecond-long all-atom molecular dynamics (MD) simulations of our recently reported FXR antagonists 2a and 2h [160], we investigated the dynamic behavior and conformational rearrangement induced by ligand binding when compared to the synthetic (GW4064) or steroidal (CDCA) FXR agonists, as a monomer or heterodimer, and in the presence and absence of the coactivator [161]. We focused on both the heterodimerization interface and the αAF-2 conformation (Figs. 16.8 and 16.9). Such investigations hold the promise of unveiling novel avenues for targeting FXR-related pathways, thereby presenting potential therapeutic implications.

Our investigation revealed discrete conformational changes across various regions. Notably, modifications were observed in the FXR/RXR interface, H9, H10, and H11 (Fig. 16.8a, b). Extensive research on nuclear receptors, employing RXR as a heterodimeric partner from a structural standpoint, has categorized them

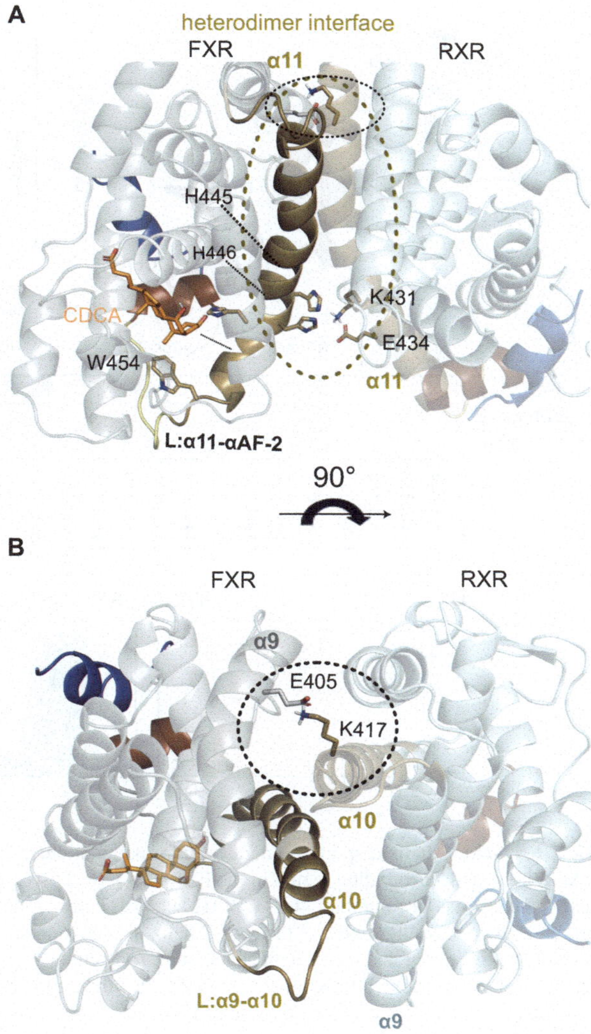

Fig. 16.8 FXR/RXR interface. (**a**) The H11 from FXR in light brown and the H11 from RXR in transparent brown surrounded by brown dashed line and dark dashed line. Interacting residues are labeled as His446 and His447 (on H11) from FXR, Lys431 and Glu343 (on H11) from RXR (**b**) A top view of panel A. The H9 from FXR in grey and the H10 from RXR in transparent brown. Interacting residues are Glu405 (H9) from FXR and Lys417 (H10) from RXR. (Modified from [161])

into permissive and nonpermissive heterodimers [153, 162, 163]. Structural alignment, utilizing RXR's H11 as the benchmark, reveals distinct orientations of α11 based on their permissiveness. Permissive partners, such as PPARs and FXR itself, exhibit a more pronounced bending, suggesting that this configuration enables permissive heterodimers to effectively sense both receptor ligands. In contrast, nonpermissive heterodimers exclusively respond to partner ligands, operating independently of the influence of RXR.

Alterations in the H11 region, exemplified by the pronounced unfolding evident in the FXR-ivermectin structure, can result in the absence of detectable H12 and L: H11–H12 [164]. Likewise, in our simulations with antagonists, we observe a

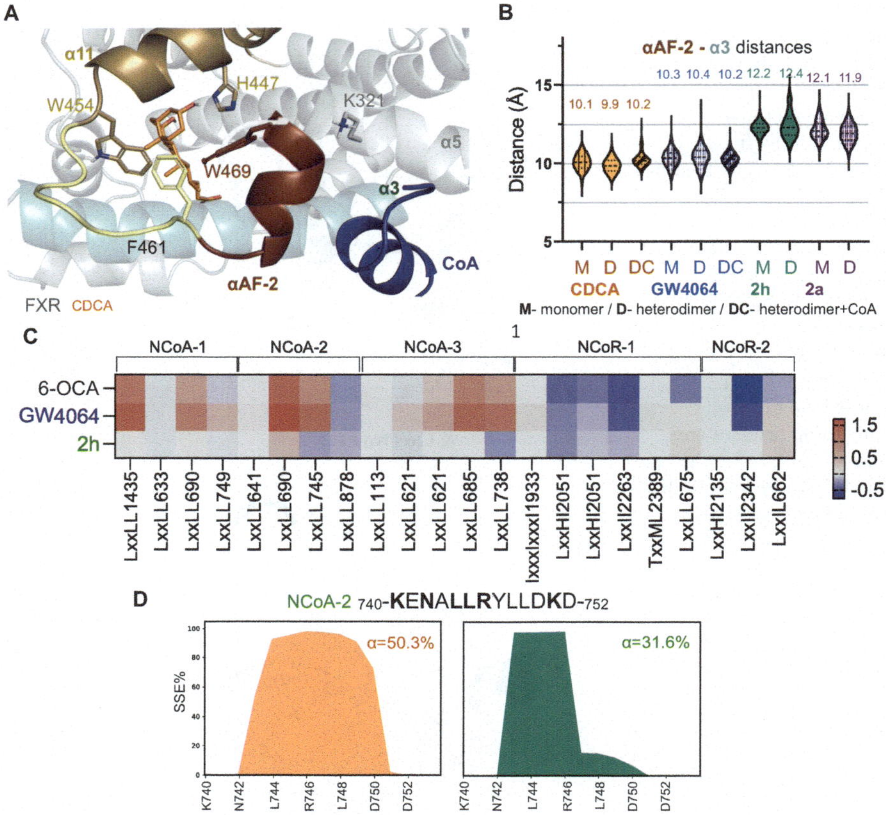

Fig. 16.9 αAF-2 helix displays moderate conformational changes relevant for corecruitment. (**a**) αAF-2 helix and surrounding region in the FXR-LBD. (**b**) Distance between the centers of mass of αAF-2 helix (residues number: 463–472, H12) and H3 (res. 281–304) shows that the antagonists promote an open configuration of the αAF-2 helix in both monomeric (M) and dimeric simulations (D and DC, for dimer and dimer with coactivator, respectively). (**c**) MARCoNI assay heatmap, the red color shows positive interactions/increased FXR binding to coregulatory peptides, and the blue color shows negative interactions/decreased FXR binding modulated by different ligands (6-OCA: obeticholic acid). Fold Change (FC) is the $\log_{10}$–transformed relative binding value, calculated as the compound's binding value relative to the DMSO control. (**d**) FXR-NCoA-2/RXR-CoA complex cannot retain the coactivator's fold upon Antagonist binding. Observed changes in the secondary structure element (SSE%) of the NCoA-2 peptide throughout the simulation. (Figure was modified from [161])

destabilization of H11, particularly emphasized in monomeric simulations, leading to the displacement of L:H11–H12/αAF-2 (Fig. 16.9a, b). Furthermore, our antagonist appears to destabilize L:α11–α12, subsequently affecting the active conformation of the αAF-2 helix [161]. This, in turn, hinders the recruitment of both coactivators and corepressors. We propose a dynamic mechanistic interpretation that links heterodimerization with the recruitment of co-regulatory proteins, providing a complementary perspective to previous crystallographic data

[162]. Furthermore, research by Merk et al. (2019) elucidates that FXR activation results from an equilibrium among conformational populations [165]. Specifically, FXR agonists are identified to stabilize both the folded and extended helix H11 (H11, constituting the heterodimerization interface) and the H11–H12 loop (forming the αAF-2 region) upon binding.

This stabilized αAF-2 region facilitates the recruitment of coactivators, thereby promoting FXR activation. Further, ligands with partial agonistic properties can induce alterations in the H11 conformation, subsequently destabilizing both the H11–H12 loop and the orientation of H12, however still allowing the recruitment of coactivators. Consistent with this, the CDCA agonist crystal structure exhibits ample electron density, allowing for precise representation of both the H12 helix and the entire loop connecting H11 and H12. Interestingly, the DM175 (partial agonist) structure lacks visibility of L:H11–H12 due to destabilization, with H12 shifted to a novel position [165]. Our simulations reveal the dynamic nature of the H12 position, oscillating around the active site, as observed by fluctuating H12–H3 distances in agonist modes (Fig. 16.9a, b). Conversely, simulated antagonists displace the αAF-2 from its active conformation to a point where stable binding of coregulatory proteins becomes impossible. Even in artificially generated Antagonist + CoA systems, a consistent displacement of the coactivator regulatory motif and further unfolding of the peptide and αAF-2 are observed, albeit to a lower extent. This evidence was along with shifts in the ligand-binding pocket (LBP) geometry and interaction pattern. The assessment of available FXR crystal structures corroborated these findings, unveiling diverse orientations in loops L: H1–H2, L: H5–H6, and L: H11–H12 depending upon the bound ligand. Notably, the crystal structure with CDCA showcased an unfolded L: H5–H6 loop, while the binding of the partial agonist DM175 induced destabilization in both L:H5–H6 and L: H11–H12 loops. The simulations further illuminated a comprehensive reorientation of the ligand-binding pocket (LBP), driven by interactions involving L: H1–H2 and L: H5–H6. These structural properties emerged as a pivotal distinguishing factor between agonists and antagonists. Specifically, CDCA and GW4064 exhibited smaller average H5–H6 distances, driven by interactions with α5 residues.

Lastly, compounds 2a and 2h relied exclusively on interactions with H1 and H2. Moreover, our Free Energy Perturbation (FEP) calculations yielded results in agreement with prior experimental binding studies, affirming that 2h functioned as a more potent antagonist compared to 2a, albeit still demonstrating weaker binding in comparison to CDCA. Additionally, a noteworthy observation emerged regarding the FXR/RXR geometrical arrangement, using alterations in dihedral planes and the interaction pattern among residues at the heterodimerization interface. The influence exerted by antagonists on the heterodimerization interface seems to be tied to the destabilization of αAF-2, effectively leading to the prevention of both coactivators and corepressors recruitment. Our discoveries contribute valuable perspectives to the understanding of the conformational dynamics of FXR, challenging conventional notions derived primarily from crystal structures. This underscores the necessity for a more thorough evaluation of FXR antagonism/agonism dynamics. To achieve a further precise depiction, the exploration of extended timescales or alternative sampling approaches should be considered. It is important to highlight that longer timescales or alternative sampling approaches could be used to generate a larger picture of FXR-LBD conformational landscape, particularly when considering its high flexibility.

Our hypothesized ligand binding mode appears to align more closely with the partial agonist DM175 than with larger ligands like ivermectin. Interestingly, both ivermectin and DM175 are recognized for recruiting corepressor proteins upon binding. However, our MARCoNI assay (see description below) did not replicate this phenomenon, hinting that our antagonists might employ an alternative binding mechanism (Fig. 16.9c, d). While it is challenging to definitively assert that the antagonist binds to FXR differently than the agonist (CDCA/GW4064), it seems to lack the ability to stabilize the structural motifs responsible for coactivator recruitment. Therefore, we propose a passive binding mechanism for antagonism. Other FXR-antagonist models, such as FXR-F6 [166], based on the oleanane-type triterpenoid agonist-bound structure (PDB ID: 5WZX), propose a similar binding mode, emphasizing the stable interaction of the Histidine and Tryptophane trap of H12. Interestingly, the in vivo modulation of FXR introduces additional layers of complexity, such as the expression of distinct FXR isoforms [167] and variations in the availability of different coregulatory proteins within tissues. Notably, FXR exists in four isoforms (named α1–4) with variants in the DBDs and hinge regions, while sharing identical LBDs. Moreover, exploring the relationship between the dynamics of the LBD and DBD could serve as another determinant revealing the dynamic behavior of FXR. Lastly, a more in-depth analysis of these allosteric effects on gene transcription profiles, especially in relation to the varied effects of ligands and coactivators, would provide valuable insights from an agonist perspective and could contribute to the development of more rational drug design strategies.

4 Experimental Methods to Analyze NR Activity

A variety of biochemical and biological assays is available, that help to explore and understand the structure, function, and genetic changes related to nuclear receptors and also elucidate NRs' ligands [168]. Examples are (Fluorescence Recovery After Photobleaching (FRAP) [165, 169] and Fluorescence Loss in Photobleaching (FLIP) which provides insight into the dynamics and interaction of the molecules in various cell processes [170], and chip-on-chip methodology to identify receptor-regulated genes [171]. Study of NR-coregulator which is a major field of NR research [32, 39, 43, 172, 173], methods for screening for binding partners and for quantifying specific receptor-target protein interactions [174], understanding the role of phosphorylation on receptor function [175, 176], advances in tissue-selective gene targeting and knock-out strategies for generating mouse models of receptor function in vivo [177], and studying genetic alterations in hormone-dependent cancers [178] are some examples.

In the following, the most relevant biochemical and cell-based approaches frequently used in the NR research field are briefly explained.

4.1 *Luciferase Reporter Gene Transactivation Assay (or Reporter Gene Assay, RGA)*

Luciferase-based reporter assays measure the activity of the nuclear receptors on the target gene expression upon ligand binding, within a relevant cell system. The assay is very sensitive and reproducible. To determine if a protein (or Protein–ligand combination) can activate (or suppress) the transcription of a gene of interest, recombinant DNA technology is used to produce a construct in which the gene's promoter is placed adjacent to a luciferase reporter gene. The cultured cells are transfected with this construct, as well as a construct coding for the protein, in our examples the NRs. If the NR can activate transcription, the cell will translate and produce the luciferase reporter (often a firefly's luciferase). The amount of produced luciferase can be quantified using a luminometer. Protocols involve transient transfection of the receptor and a response element reporter gene construct [97, 179]. Other protocols also have an additional luciferase (*Renilla*), which is constitutively and constantly being produced, to normalize by the number of transfected cells and also act as a transfection control. Many cell lines have been introduced as candidate recipients of these vectors, including CHO, HuH7, MCF-7, HEK293, HepG2, and Caco-2 cells. This method can identify NR activators and upon pretreatment with known agonists and unknown ligands in concentration-response, also identify antagonists [178]. On one hand, RGA is a staple in NR research, being one of the most widely used approaches, as it provides on-target specific activity in a cellular context. On the other hand, the cellular nature of this assay makes it sensible to problems such as co-activator availability, cellular permeability, and the presence of transporter/exporters, which hinders the direct correlation between RGA data with modeling results.

4.2 *Coregulator-Recruitment*

An alternative transactivation assay system is the mammalian two-hybrid system which is a technique to detect protein-protein interaction in cells. The result is interpreted by expression or repression of reporter genes. Cells are treated with NR ligands. In this method, the DBD of the yeast transcription factor GAL4 binds to specific upstream activation sequence (or UAS) response elements. When the LBD of the desired gene is fused to this DBD, it promotes co-activator binding. The interaction between the NR and its co-activator is detected based on a reporter gene containing multiple copies of the GAL4 upstream activating system. A set of agonists and inverse agonists were identified to bind to the human CAR using a similar system [180].

In terms of purely in vitro coregulator recruitment tools, one can highlight the Microarray Assay for real-time Coregulator-Nuclear Receptor Interaction (MARCoNI) technology [49, 181, 182]. The MARCoNI platform provides an assay which allows for the characterization of nuclear receptor function, i.e., coregulator binding (as seen in Fig. 16.9c) using a peptide array platform, each array containing up to 154 immobilized nuclear receptor coregulator peptides harboring either LXXLL (coactivator) or LXXXIXXXL (corepressor) motifs. These peptides are immobilized in a porous microarray membrane. Peptide binding of the nuclear receptor as a function of the chemical ligands tested in the assay is visualized by fluorescently labeled antibodies binding to the nuclear receptor of interest.

Lastly, time-resolved fluorescence resonance energy transfer (TR-FRET) is a commonly used experimental technique to investigate the binding between nuclear receptor and ligand or protein–protein interaction (e.g., nuclear receptor–coregulator interaction). In this method, two fluorescent molecules are utilized: a donor fluorophore and an acceptor fluorophore. When the two molecules interact, donor and acceptor fluorophores are brought together. When the donor is excited, it transfers its emission energy to the acceptor. This event leads to the emission of fluorescence at a specific wavelength. And the fluorescence emitted by the acceptor fluorophore is measured at a delayed time after the excitation pulse. This delay reduces the background noise and autofluorescence, resulting in a higher signal-to-noise ratio with improved sensitivity [178]. FRET is the transferring of energy from a donor fluorophore in an excited state to a nearby acceptor fluorophore. The output is proportional to the amount of binding [183]. Interestingly, also TR-FRET suffers from nonobvious limitations, recently published CAR work [149, 151]. We observed that TR-FRET data showing the interaction of CAR1-LBD with a fragment of PGC1α would be a good surrogate for CAR activation, and it is true for CITCO or other CITCO-like novel ligands. However, we observed that nonylphenol (mixture and individual isomers), in fact, to not promote PGC1α recruitment, but competed with CITCO in an antagonistic mode in a dose-dependent manner for the CAR LBD–PGC1α interaction. Interestingly, MARCoNI data on those EDs demonstrate that NPs can recruit different coactivators to exert their phenotype, but not PGC1α, highlighting the relevance of a more holistic approach when interpreting assay data.

5 Concluding Remarks and Outlook

NRs are important transcriptional factors that regulate several genes involved in physiological processes including development, differentiation, metabolism, and systemic homeostasis, making them interesting therapeutic targets for many human diseases, therefore, the discovery of ligands modulating their activity is significant. They are ligand-induced activated and play their role in interaction with multiple coregulatory proteins. Thus, a deeper insight into the NR–ligand interactions, as well as cofactor proteins is essential for drug development, where computational approaches can accelerate this field.

We focused on discussing NRs studies using in silico screening and structure-based drug design followed by validating in vitro experiments. We discussed studies that exclusively employed long-timescale atomistic MD simulations. It is of great importance to take protein flexibility into account when trying to understand the NR active/inactive conformation since one challenge in NR modeling is its plasticity.

NRs are flexible molecules with highly conserved domains. This flexibility is highlighted in PXR, as exemplified by structural data and earlier studies in the presence of agonists [184, 185]. However, in the PXR studies with partial agonists, they displayed slightly lower flexibility in the presence of a partial antagonist (compound 100) than full agonists. Moreover, with the implementation of a computational approach, they could identify important residues interacting with modulators where mutation studies confirmed and expanded our knowledge of those interactions.

These studies highlight the integration of computational technologies with experimental validation testing in the field of drug discovery, leading to enhanced efficiency in identifying therapeutic candidates. It is noteworthy that while computational approaches offer valuable insights, they do not guarantee absolute certainty in their predictions [186]. Computational modeling of NRs could enhance the understanding of receptor conformational rearrangement due to protein–ligand interaction (PXR and CAR studies) and protein–protein interaction (FXR studies).

FXR's central role in metabolic regulation and its interplay with other NRs in modulating metabolic pathways emerges as a potential therapeutic target for metabolic disorders. Hence, understanding FXR conformational rearrangement requires further detailed study. Our study highlighted the induced effect by antagonists on the heterodimerization interface which is closely related to the destabilization of H12 and L:H11–H12. This event prevents both coactivator and corepressor recruitment. This observation challenges the conventional understanding based on only crystal structures and underscores the comprehensive assessment of FXR antagonism/agonism. We suggest a longer timescale simulation or alternative sampling approaches might be beneficial.

By utilizing structural modeling techniques such as homology modeling, molecular docking, and MD simulations, we can elucidate the binding modes of various ligands, and understand the structural basis of CAR activation. However, discrimination between CAR-bound agonist and reverse agonist conformation demands further detailed study with increasing MD simulation timescales. Also considering coregulator binding in the experimental design was shown relevant, which can be approached with extensive MD trajectory analysis, such as PCAs and Markov State Models, particularly when it comes to differentiating between CAR reverse agonists.

Lastly, the importance of MD simulations for PXR studies, as well as its challenges, has been discussed in this chapter. To face these challenges, from a computational point of view, the implementation of long MD simulations can assist in the elucidation of ligand-specific influence on conformations of various regions of

PXR-LBD. Of note, discrepancies have been observed in the H6 region, αAF-2 (H12), H1–H2′, β1′–H3 and β1–β1′ connecting loop.

Despite the clear therapeutic potential of novel NR modulators, it is important to remember that NR behaves in a tissue-specific and coactivator-specific manner, having the ability to regulate different pathways based on the transcriptional complexes they form. This leads to the selection of cell-specific enhancers and results in distinct physiological consequences. Hence, the screening of NR ligands will equip future studies with essential chemical tools for developing a detailed understanding of their regulation. Additionally, NR activity can undergo epigenetic modifications such as ubiquitination and phosphorylation, whose effects are largely unknown [96, 187].

Moreover, during CADD implementation, one should be aware of several pitfalls that might happen, otherwise, it generates false prediction which would provide false input data for the experimental pipeline, invalid prediction and consequently error propagation, wasting time and resources in the other way around. Overall, This chapter emphasize the utilization of computational approaches in structure-based drug design, such as molecular docking and MD simulations, for the discovery of innovative bioactive substances and receptor conformational changes in nuclear receptors considering the limitations of these methods.

Acknowledgments The authors wish to acknowledge CSC—IT Center for Science, Finland, for the very generous computational resources provided in multiple of the examples cited within this chapter, as well as their striking enthusiasm in discussing scientific matters with our group. T.K., A.R. and A.P. are funded by the TüCAD$_2$ and CMIF. TüCAD$_2$ and CMIF are funded by the Federal Ministry of Education and Research (BMBF) and the Baden-Württemberg Ministry of Science as part of the Excellence Strategy of the German Federal and State Governments. All authors acknowledge European Union's Horizon 2020 research and innovation program under grant agreement No 825762, EDCMET project Metabolic effects of Endocrine Disrupting Chemicals: novel testing METhods and adverse outcome pathways (EDCMET). As well as the German Center for Infection Research (DZIF, TTU06.716).

Conflict of Interest The authors declare that they have no known competing financial interests or personal relationships that could have appeared to influence the work reported in this chapter.

References

1. Zhou, J., & Cidlowski, J. A. (2005). The human glucocorticoid receptor: One gene, multiple proteins and diverse responses. *Steroids, 70*, 407–417.
2. Ashburner, M., Chihara, C., Meltzer, P., & Richards, G. (1974). Temporal control of puffing activity in polytene chromosomes. *Cold Spring Harbor Symposia on Quantitative Biology, 38*, 655–662.
3. Yamamoto, K. R. (1985). Steroid receptor regulated transcription of specific genes and gene networks. *Annual Review of Genetics, 19*, 209–252.
4. Hollenberg, S. M., et al. (1985). Primary structure and expression of a functional human glucocorticoid receptor cDNA. *Nature, 318*, 635–641.

5. Green, S., et al. (1986). Human oestrogen receptor cDNA: sequence, expression and homology to v-erb-A. *Nature, 320*, 134–139.
6. Miesfeld, R., et al. (1986). Genetic complementation of a glucocorticoid receptor deficiency by expression of cloned receptor cDNA. *Cell, 46*, 389–399.
7. Petkovich, M., Brand, N. J., Krust, A., & Chambon, P. (1987). A human retinoic acid receptor which belongs to the family of nuclear receptors. *Nature, 330*, 444–450.
8. Evans, R. M. (1988). The steroid and thyroid hormone receptor superfamily. *Science, 240*, 889–895.
9. Laudet, V. (1997). Evolution of the nuclear receptor superfamily: Early diversification from an ancestral orphan receptor. *Journal of Molecular Endocrinology, 19*, 207–226.
10. Porter, B. A., Ortiz, M. A., Bratslavsky, G., & Kotula, L. (2019). Structure and function of the nuclear receptor superfamily and current targeted therapies of prostate cancer. *Cancers (Basel), 11*, 1852.
11. de Almeida, N. R., & Conda-Sheridan, M. (2019). A review of the molecular design and biological activities of RXR agonists. *Medicinal Research Reviews, 39*, 1372–1397.
12. Mangelsdorf, D. J., et al. (1995). The nuclear receptor superfamily: The second decade. *Cell, 83*, 835–839.
13. Klinge, C. M., Bodenner, D. L., Desai, D., Niles, R. M., & Traish, A. M. (1997). Binding of type II nuclear receptors and estrogen receptor to full and half-site estrogen response elements in vitro. *Nucleic Acids Research, 25*, 1903–1912.
14. McKenna, N. J., & O'Malley, B. W. (2002). Combinatorial control of gene expression by nuclear receptors and coregulators. *Cell, 108*, 465–474.
15. Penvose, A., Keenan, J. L., Bray, D., Ramlall, V., & Siggers, T. (2019). Comprehensive study of nuclear receptor DNA binding provides a revised framework for understanding receptor specificity. *Nature Communications, 10*, 2514.
16. Bhimsaria, D., et al. (2023). Hidden modes of DNA binding by human nuclear receptors. *Nature Communications, 14*, 4179.
17. Mutoh, S., et al. (2009). Dephosphorylation of threonine 38 is required for nuclear translocation and activation of human xenobiotic receptor CAR (NR1I3). *The Journal of Biological Chemistry, 284*, 34785–34792.
18. Mutoh, S., et al. (2013). Phenobarbital indirectly activates the constitutive active androstane receptor (CAR) by inhibition of epidermal growth factor receptor signaling. *Science Signaling, 6*, ra31.
19. Miyamoto, T., et al. (2001). The role of hinge domain in heterodimerization and specific DNA recognition by nuclear receptors. *Molecular and Cellular Endocrinology, 181*, 229–238.
20. Rastinejad, F., Huang, P., Chandra, V., & Khorasanizadeh, S. (2013). Understanding nuclear receptor form and function using structural biology. *Journal of Molecular Endocrinology, 51*, T1–T21.
21. Nagy, L., & Schwabe, J. W. R. (2004). Mechanism of the nuclear receptor molecular switch. *Trends in Biochemical Sciences, 29*, 317–324.
22. Robinson-Rechavi, M., Garcia, H. E., & Laudet, V. (2003). The nuclear receptor superfamily. *Journal of Cell Science, 116*, 585–586.
23. Bertrand, S., et al. (2004). Evolutionary genomics of nuclear receptors: From twenty-five ancestral genes to derived endocrine systems. *Molecular Biology and Evolution, 21*, 1923–1937.
24. Pawlak, M., Lefebvre, P., & Staels, B. (2012). General molecular biology and architecture of nuclear receptors. *Current Topics in Medicinal Chemistry, 12*, 486–504.
25. Chandra, V., et al. (2008). Structure of the intact PPAR-γ–RXR-α nuclear receptor complex on DNA. *Nature, 456*, 350–356.
26. Küblbeck, J., et al. (2020). The EDCMET project: Metabolic effects of endocrine disruptors. *International Journal of Molecular Sciences, 21*, 3021.
27. Küblbeck, J., Niskanen, J., & Honkakoski, P. (2020). Metabolism-disrupting chemicals and the constitutive androstane receptor CAR. *Cells, 9*, 2306.

28. Gore, A. C., et al. (2015). EDC-2: The Endocrine Society's second scientific statement on endocrine-disrupting chemicals. *Endocrine Reviews, 36*, E1–E150.
29. Latini, G., Gallo, F., & Iughetti, L. (2010). Toxic environment and obesity pandemia: Is there a relationship? *Italian Journal of Pediatrics, 36*, 8.
30. Neel, B. A., & Sargis, R. M. (2011). The paradox of progress: Environmental disruption of metabolism and the diabetes epidemic. *Diabetes, 60*, 1838–1848.
31. Hu, X., & Lazar, M. A. (1999). The CoRNR motif controls the recruitment of corepressors by nuclear hormone receptors. *Nature, 402*, 93–96.
32. Glass, C. K., & Rosenfeld, M. G. (2000). The coregulator exchange in transcriptional functions of nuclear receptors. *Genes & Development, 14*, 121–141.
33. Murakami, S., Nagari, A., & Kraus, W. L. (2017). Dynamic assembly and activation of estrogen receptor α enhancers through coregulator switching. *Genes & Development, 31*, 1535–1548.
34. Chandler, V. L., Maler, B. A., & Yamamoto, K. R. (1983). DNA sequences bound specifically by glucocorticoid receptor in vitro render a heterologous promoter hormone responsive in vivo. *Cell, 33*, 489–499.
35. Beato, M., Herrlich, P., & Schütz, G. (1995). Steroid hormone receptors: Many actors in search of a plot. *Cell, 83*, 851–857.
36. Juge-Aubry, C., et al. (1997). DNA binding properties of peroxisome proliferator-activated receptor subtypes on various natural peroxisome proliferator response elements. *Journal of Biological Chemistry, 272*, 25252–25259.
37. Näär, A. M., et al. (1991). The orientation and spacing of core DNA-binding motifs dictate selective transcriptional responses to three nuclear receptors. *Cell, 65*, 1267–1279.
38. Umesono, K., Murakami, K. K., Thompson, C. C., & Evans, R. M. (1991). Direct repeats as selective response elements for the thyroid hormone, retinoic acid, and vitamin D3 receptors. *Cell, 65*, 1255–1266.
39. Chen, J. D., & Li, H. (1998). Coactivation and Corepression in transcriptional regulation by steroid/nuclear hormone receptors. *CRE, 8*, 169.
40. Dowell, P., et al. (1999). Identification of nuclear receptor corepressor as a peroxisome proliferator-activated receptor α interacting protein. *Journal of Biological Chemistry, 274*, 15901–15907.
41. Ishizuka, T., & Lazar, M. A. (2003). The N-CoR/histone deacetylase 3 complex is required for repression by thyroid hormone receptor. *Molecular and Cellular Biology, 23*, 5122–5131.
42. Bailey, P., et al. (1999). The nuclear receptor corepressor N-CoR regulates differentiation: N-CoR directly interacts with MyoD. *Molecular Endocrinology, 13*, 1155–1168.
43. Horwitz, K. B., et al. (1996). Nuclear receptor coactivators and corepressors. *Molecular Endocrinology, 10*, 1167–1177.
44. Heery, D. M., Kalkhoven, E., Hoare, S., & Parker, M. G. (1997). A signature motif in transcriptional co-activators mediates binding to nuclear receptors. *Nature, 387*, 733–736.
45. Ai, N., Krasowski, M. D., Welsh, W. J., & Ekins, S. (2009). Understanding nuclear receptors using computational methods. *Drug Discovery Today, 14*, 486–494.
46. Weiss, R. E., et al. (1999). Mice deficient in the steroid receptor co-activator 1 (SRC-1) are resistant to thyroid hormone. *The EMBO Journal, 18*, 1900–1904.
47. Tagami, T., Madison, L. D., Nagaya, T., & Jameson, J. L. (1997). Nuclear receptor corepressors activate rather than suppress basal transcription of genes that are negatively regulated by thyroid hormone. *Molecular and Cellular Biology, 17*, 2642–2648.
48. Berghagen, H., et al. (2002). Corepressor SMRT functions as a coactivator for thyroid hormone receptor T3Ralpha from a negative hormone response element. *The Journal of Biological Chemistry, 277*, 49517–49522.
49. Broekema, M. F., et al. (2018). Profiling of 3696 nuclear receptor–coregulator interactions: A resource for biological and clinical discovery. *Endocrinology, 159*, 2397–2407.

50. Pascussi, J.-M., et al. (2008). The tangle of nuclear receptors that controls xenobiotic metabolism and transport: Crosstalk and consequences. *Annual Review of Pharmacology and Toxicology, 48*, 1–32.
51. Hu, R., Hilakivi-Clarke, L., & Clarke, R. (2015). Molecular mechanisms of tamoxifen-associated endometrial cancer (review). *Oncology Letters, 9*, 1495–1501.
52. Wooltorton, E. (2006). Osteoporosis treatment: Raloxifene (Evista) and stroke mortality. *CMAJ, 175*, 147.
53. McCrea, E., Sissung, T. M., Price, D. K., Chau, C. H., & Figg, W. D. (2016). Androgen receptor variation affects prostate cancer progression and drug resistance. *Pharmacological Research, 114*, 152–162.
54. Ünüvar, T., & Büyükgebiz, A. (2012). Fetal and neonatal endocrine disruptors. *Journal of Clinical Research in Pediatric Endocrinology, 4*, 51–60.
55. Grohs, M. N., et al. (2019). Prenatal maternal and childhood bisphenol a exposure and brain structure and behavior of young children. *Environmental Health, 18*, 85.
56. Jeng, H. A. (2014). Exposure to endocrine disrupting chemicals and male reproductive health. *Frontiers in Public Health, 2*, 55.
57. Hauser, R., Williams, P., Altshul, L., & Calafat, A. M. (2005). Evidence of interaction between polychlorinated biphenyls and phthalates in relation to human sperm motility. *Environmental Health Perspectives, 113*, 425–430.
58. Banerjee, M., & Chen, T. (2013). Differential regulation of CYP3A4 promoter activity by a new class of natural product derivatives binding to pregnane X receptor. *Biochemical Pharmacology, 86*, 824–835.
59. Sepe, V., et al. (2014). Insights on pregnane-X-receptor modulation. Natural and semisynthetic steroids from Theonella marine sponges. *European Journal of Medicinal Chemistry, 73*, 126–134.
60. Kortagere, S., et al. (2010). Evaluation of computational docking to identify pregnane X receptor agonists in the ToxCast database. *Environmental Health Perspectives, 118*, 1412–1417.
61. Chrisman, I. M., et al. (2018). Defining a conformational ensemble that directs activation of PPARγ. *Nature Communications, 9*, 1794.
62. Motta, S., Callea, L., Giani Tagliabue, S., & Bonati, L. (2018). Exploring the PXR ligand binding mechanism with advanced molecular dynamics methods. *Scientific Reports, 8*, 16207.
63. Saen-Oon, S., Lozoya, E., Segarra, V., Guallar, V., & Soliva, R. (2019). Atomistic simulations shed new light on the activation mechanisms of RORγ and classify it as type III nuclear hormone receptor regarding ligand-binding paths. *Scientific Reports, 9*, 17249.
64. Kosztin, D., Izrailev, S., & Schulten, K. (1999). Unbinding of retinoic acid from its receptor studied by steered molecular dynamics. *Biophysical Journal, 76*, 188–197.
65. Carlsson, P., Burendahl, S., & Nilsson, L. (2006). Unbinding of retinoic acid from the retinoic acid receptor by random expulsion molecular dynamics. *Biophysical Journal, 91*, 3151–3161.
66. Sonoda, M. T., Martínez, L., Webb, P., Skaf, M. S., & Polikarpov, I. (2008). Ligand dissociation from estrogen receptor is mediated by receptor dimerization: Evidence from molecular dynamics simulations. *Molecular Endocrinology, 22*, 1565–1578.
67. Martínez, L., et al. (2005). Molecular dynamics simulations reveal multiple pathways of ligand dissociation from thyroid hormone receptors. *Biophysical Journal, 89*, 2011–2023.
68. Brzozowski, A. M., et al. (1997). Molecular basis of agonism and antagonism in the oestrogen receptor. *Nature, 389*, 753–758.
69. Pike, A. C., et al. (1999). Structure of the ligand-binding domain of oestrogen receptor beta in the presence of a partial agonist and a full antagonist. *The EMBO Journal, 18*, 4608–4618.
70. Souza, P. C. T., et al. (2017). An alternative conformation of ERβ bound to estradiol reveals H12 in a stable antagonist position. *Scientific Reports, 7*, 3509.
71. Delfosse, V., et al. (2012). Structural and mechanistic insights into bisphenols action provide guidelines for risk assessment and discovery of bisphenol A substitutes. *Proceedings of the National Academy of Sciences of the United States of America, 109*, 14930–14935.

72. Alves, N. R. C., Pecci, A., & Alvarez, L. D. (2020). Structural insights into the ligand binding domain of the glucocorticoid receptor: A molecular dynamics study. *Journal of Chemical Information and Modeling, 60*, 794–804.
73. Raaijmakers, H. C. A., Versteegh, J. E., & Uitdehaag, J. C. M. (2009). The X-ray structure of RU486 bound to the progesterone receptor in a destabilized agonistic conformation. *The Journal of Biological Chemistry, 284*, 19572–19579.
74. Gampe, R. T., et al. (2000). Structural basis for autorepression of retinoid X receptor by tetramer formation and the AF-2 helix. *Genes & Development, 14*, 2229–2241.
75. Shiau, A. K., et al. (1998). The structural basis of estrogen receptor/coactivator recognition and the antagonism of this interaction by tamoxifen. *Cell, 95*, 927–937.
76. Batista, M. R. B., & Martínez, L. (2013). Dynamics of nuclear receptor Helix-12 switch of transcription activation by modeling time-resolved fluorescence anisotropy decays. *Biophysical Journal, 105*, 1670–1680.
77. Dash, A. K., Yende, A. S., Jaiswal, B., & Tyagi, R. K. (2017). Heterodimerization of retinoid X receptor with xenobiotic receptor partners occurs in the cytoplasmic compartment: Mechanistic insights of events in living cells. *Experimental Cell Research, 360*, 337–346.
78. Mangelsdorf, D. J., et al. (1992). Characterization of three RXR genes that mediate the action of 9-cis retinoic acid. *Genes & Development, 6*, 329–344.
79. Hamada, K., et al. (1989). H-2RIIBP, a member of the nuclear hormone receptor superfamily that binds to both the regulatory element of major histocompatibility class I genes and the estrogen response element. *Proceedings of the National Academy of Sciences of the United States of America, 86*, 8289–8293.
80. Dollé, P., Fraulob, V., Kastner, P., & Chambon, P. (1994). Developmental expression of murine retinoid X receptor (RXR) genes. *Mechanisms of Development, 45*, 91–104.
81. Mangelsdorf, D. J., & Evans, R. M. (1995). The RXR heterodimers and orphan receptors. *Cell, 83*, 841–850.
82. Pérez, E., Bourguet, W., Gronemeyer, H., & de Lera, A. R. (2012). Modulation of RXR function through ligand design. *Biochimica et Biophysica Acta, 1821*, 57–69.
83. Castillo, A. I., et al. (2004). A permissive retinoid X receptor/thyroid hormone receptor heterodimer allows stimulation of prolactin gene transcription by thyroid hormone and 9-cis-retinoic acid. *Molecular and Cellular Biology, 24*, 502–513.
84. Blumberg, B., et al. (1998). SXR, a novel steroid and xenobioticsensing nuclear receptor. *Genes & Development, 12*, 3195–3205.
85. Bertilsson, G., et al. (1998). Identification of a human nuclear receptor defines a new signaling pathway for CYP3A induction. *Proceedings of the National Academy of Sciences of the United States of America, 95*, 12208–12213.
86. Lehmann, J. M., et al. (1998). The human orphan nuclear receptor PXR is activated by compounds that regulate CYP3A4 gene expression and cause drug interactions. *The Journal of Clinical Investigation, 102*, 1016–1023.
87. Delfosse, V., et al. (2021). Mechanistic insights into the synergistic activation of the RXR-PXR heterodimer by endocrine disruptor mixtures. *Proceedings of the National Academy of Sciences of the United States of America, 118*, e2020551118.
88. Willson, T. M., & Kliewer, S. A. (2002). PXR, CAR and drug metabolism. *Nature Reviews. Drug Discovery, 1*, 259–266.
89. Oladimeji, P. O., & Chen, T. (2018). PXR: More than just a master xenobiotic receptor. *Molecular Pharmacology, 93*, 119–127.
90. Noble, S. M., et al. (2006). Human PXR forms a tryptophan zipper-mediated homodimer. *Biochemistry, 45*, 8579–8589.
91. Watkins, R. E., Davis-Searles, P. R., Lambert, M. H., & Redinbo, M. R. (2003). Coactivator binding promotes the specific interaction between ligand and the pregnane X receptor. *Journal of Molecular Biology, 331*, 815–828.

92. Johnson, D. R., Li, C.-W., Chen, L.-Y., Ghosh, J. C., & Chen, J. D. (2006). Regulation and binding of pregnane X receptor by nuclear receptor corepressor silencing mediator of retinoid and thyroid hormone receptors (SMRT). *Molecular Pharmacology, 69*, 99–108.
93. Watkins, R. E., et al. (2001). The human nuclear xenobiotic receptor PXR: Structural determinants of directed promiscuity. *Science, 292*, 2329–2333.
94. Ekins, S., & Schuetz, E. (2002). The PXR crystal structure: The end of the beginning. *Trends in Pharmacological Sciences, 23*, 49–50.
95. Ekins, S., et al. (2009). Challenges predicting ligand-receptor interactions of promiscuous proteins: The nuclear receptor PXR. *PLoS Computational Biology, 5*, e1000594.
96. Hall, A., Chanteux, H., Ménochet, K., Ledecq, M., & Schulze, M.-S. E. D. (2021). Designing out PXR activity on drug discovery projects: A review of structure-based methods, empirical and computational approaches. *Journal of Medicinal Chemistry, 64*, 6413–6522.
97. Geick, A., Eichelbaum, M., & Burk, O. (2001). Nuclear receptor response elements mediate induction of intestinal MDR1 by rifampin. *The Journal of Biological Chemistry, 276*, 14581–14587.
98. Kast, H. R., et al. (2002). Regulation of multidrug resistance-associated protein 2 (ABCC2) by the nuclear receptors pregnane X receptor, farnesoid X-activated receptor, and constitutive androstane receptor. *The Journal of Biological Chemistry, 277*, 2908–2915.
99. Maeda, K., et al. (2011). Identification of the rate-determining process in the hepatic clearance of atorvastatin in a clinical cassette microdosing study. *Clinical Pharmacology and Therapeutics, 90*, 575–581.
100. Chen, Y., et al. (2012). Nuclear receptors in the multidrug resistance through the regulation of drug-metabolizing enzymes and drug transporters. *Biochemical Pharmacology, 83*, 1112–1126.
101. Sinz, M., et al. (2006). Evaluation of 170 xenobiotics as transactivators of human pregnane X receptor (hPXR) and correlation to known CYP3A4 drug interactions. *Current Drug Metabolism, 7*, 375–388.
102. Chen, T. (2010). Overcoming drug resistance by regulating nuclear receptors. *Advanced Drug Delivery Reviews, 62*, 1257–1264.
103. Burk, O., et al. (2018). Identification of approved drugs as potent inhibitors of pregnane X receptor activation with differential receptor interaction profiles. *Archives of Toxicology, 92*, 1435–1451.
104. Wiesinger, H., et al. (2020). The effects of weak and strong CYP3A induction by rifampicin on the pharmacokinetics of five progestins and ethinylestradiol compared to midazolam. *Clinical Pharmacology and Therapeutics, 108*, 798–807.
105. Yamazaki, T., et al. (2017). Pharmacokinetic effects of isavuconazole coadministration with the cytochrome P450 enzyme substrates bupropion, repaglinide, caffeine, dextromethorphan, and methadone in healthy subjects. *Clinical Pharmacology in Drug Development, 6*, 54–65.
106. Lin, W., et al. (2017). SPA70 is a potent antagonist of human pregnane X receptor. *Nature Communications, 8*, 741.
107. Chandran, A., & Vishveshwara, S. (2016). Exploration of the conformational landscape in pregnane X receptor reveals a new binding pocket. *Protein Science, 25*, 1989–2005.
108. Huber, A. D., et al. (2021). Mutation of a single amino acid of pregnane X receptor switches an antagonist to agonist by altering AF-2 helix positioning. *Cellular and Molecular Life Sciences, 78*, 317–335.
109. Rashidian, A., et al. (2022). Discrepancy in interactions and conformational dynamics of pregnane X receptor (PXR) bound to an agonist and a novel competitive antagonist. *Computational and Structural Biotechnology Journal, 20*, 3004–3018.
110. Mustonen, E.-K., et al. (2022). Target hopping from protein kinases to PXR: Identification of small-molecule protein kinase inhibitors as selective modulators of pregnane X receptor from TüKIC library. *Cells, 11*, 1299.
111. Ngan, C.-H., et al. (2009). The structural basis of Pregnane X receptor binding promiscuity. *Biochemistry, 48*, 11572–11581.

112. De Smet, F., Christopoulos, A., & Carmeliet, P. (2014). Allosteric targeting of receptor tyrosine kinases. *Nature Biotechnology, 32*, 1113–1120.
113. Shan, Y., et al. (2011). How does a drug molecule find its target binding site? *Journal of the American Chemical Society, 133*, 9181–9183.
114. Moore, T. W., Mayne, C. G., & Katzenellenbogen, J. A. (2010). Minireview: Not picking pockets: Nuclear receptor alternate-site modulators (NRAMs). *Molecular Endocrinology, 24*, 683–695.
115. Kamaraj, R., Drastik, M., Maixnerova, J., & Pavek, P. (2022). Allosteric antagonism of the pregnane X receptor (PXR): Current-state-of-the-art and prediction of novel allosteric sites. *Cells, 11*, 2974.
116. Changeux, J.-P., & Christopoulos, A. (2016). Allosteric modulation as a unifying mechanism for receptor function and regulation. *Cell, 166*, 1084–1102.
117. Venkatesh, M., et al. (2011). In vivo and in vitro characterization of a first-in-class novel azole analog that targets pregnane X receptor activation. *Molecular Pharmacology, 80*, 124–135.
118. Krausova, L., et al. (2011). Metformin suppresses pregnane X receptor (PXR)-regulated transactivation of CYP3A4 gene. *Biochemical Pharmacology, 82*, 1771–1780.
119. Das, B. C., et al. (2008). Synthesis of novel ketoconazole derivatives as inhibitors of the human pregnane X receptor (PXR; NR1I2; also termed SXR, PAR). *Bioorganic & Medicinal Chemistry Letters, 18*, 3974–3977.
120. Ekins, S., et al. (2008). Computational discovery of novel low micromolar human pregnane X receptor antagonists. *Molecular Pharmacology, 74*, 662–672.
121. Wang, H., et al. (2008). The phytoestrogen coumestrol is a naturally occurring antagonist of the human pregnane X receptor. *Molecular Endocrinology, 22*, 838–857.
122. Zhou, C., et al. (2007). The dietary isothiocyanate sulforaphane is an antagonist of the human steroid and xenobiotic nuclear receptor. *Molecular Pharmacology, 71*, 220–229.
123. Chen, Y., Tang, Y., Robbins, G. T., & Nie, D. (2010). Camptothecin attenuates cytochrome P450 3A4 induction by blocking the activation of human pregnane X receptor. *The Journal of Pharmacology and Experimental Therapeutics, 334*, 999–1008.
124. Baes, M., et al. (1994). A new orphan member of the nuclear hormone receptor superfamily that interacts with a subset of retinoic acid response elements. *Molecular and Cellular Biology, 14*, 1544–1552.
125. Wei, P., Zhang, J., Dowhan, D. H., Han, Y., & Moore, D. D. (2002). Specific and overlapping functions of the nuclear hormone receptors CAR and PXR in xenobiotic response. *The Pharmacogenomics Journal, 2*, 117–126.
126. Lamba, J., Lamba, V., & Schuetz, E. (2005). Genetic variants of PXR (NR1I2) and CAR (NR1I3) and their implications in drug metabolism and pharmacogenetics. *Current Drug Metabolism, 6*, 369–383.
127. Mbatchi, L. C., Brouillet, J.-P., & Evrard, A. (2018). Genetic variations of the xenoreceptors NR1I2 and NR1I3 and their effect on drug disposition and response variability. *Pharmacogenomics, 19*, 61–77.
128. Auerbach, S. S., Stoner, M. A., Su, S., & Omiecinski, C. J. (2005). Retinoid X receptor-alpha-dependent transactivation by a naturally occurring structural variant of human constitutive androstane receptor (NR1I3). *Molecular Pharmacology, 68*, 1239–1253.
129. Auerbach, S. S., DeKeyser, J. G., Stoner, M. A., & Omiecinski, C. J. (2007). CAR2 displays unique ligand binding and RXRα heterodimerization characteristics. *Drug Metabolism and Disposition, 35*, 428–439.
130. Yang, H., & Wang, H. (2014). Signaling control of the constitutive androstane receptor (CAR). *Protein & Cell, 5*, 113–123.
131. Kobayashi, K., Sueyoshi, T., Inoue, K., Moore, R., & Negishi, M. (2003). Cytoplasmic accumulation of the nuclear receptor CAR by a tetratricopeptide repeat protein in HepG2 cells. *Molecular Pharmacology, 64*, 1069–1075.

132. Maglich, J. M., et al. (2003). Identification of a novel human constitutive androstane receptor (CAR) agonist and its use in the identification of CAR target genes. *Journal of Biological Chemistry, 278*, 17277–17283.
133. Tzameli, I., Pissios, P., Schuetz, E. G., & Moore, D. D. (2000). The xenobiotic compound 1,4-bis[2-(3,5-dichloropyridyloxy)]benzene is an agonist ligand for the nuclear receptor CAR. *Molecular and Cellular Biology, 20*, 2951–2958.
134. Chai, S. C., Cherian, M. T., Wang, Y.-M., & Chen, T. (2016). Small-molecule modulators of PXR and CAR. *Biochimica et Biophysica Acta, 1859*, 1141–1154.
135. Lynch, C., et al. (2019). Identification of modulators that activate the constitutive androstane receptor from the Tox21 10K compound library. *Toxicological Sciences, 167*, 282–292.
136. Kanno, Y., Tanuma, N., Takahashi, A., & Inouye, Y. (2013). TO901317, a potent LXR agonist, is an inverse agonist of CAR. *The Journal of Toxicological Sciences, 38*, 309–315.
137. Küblbeck, J., et al. (2011). New in vitro tools to study human constitutive androstane receptor (CAR) biology: Discovery and comparison of human CAR inverse agonists. *Molecular Pharmaceutics, 8*, 2424–2433.
138. Jyrkkärinne, J., et al. (2012). Molecular dynamics simulations for human CAR inverse agonists. *Journal of Chemical Information and Modeling, 52*, 457–464.
139. DeKeyser, J. G., Laurenzana, E. M., Peterson, E. C., Chen, T., & Omiecinski, C. J. (2011). Selective phthalate activation of naturally occurring human constitutive androstane receptor splice variants and the pregnane X receptor. *Toxicological Sciences, 120*, 381–391.
140. Burk, O., et al. (2012). Differential effects of clinically used derivatives and metabolites of artemisinin in the activation of constitutive androstane receptor isoforms: Antimalarial drugs and CAR. *British Journal of Pharmacology, 167*, 666–681.
141. Honkakoski, P. (2022). Searching for constitutive androstane receptor modulators. *Drug Metabolism and Disposition, 50*, 1002–1009.
142. Honkakoski, P., Sueyoshi, T., & Negishi, M. (2003). Drug-activated nuclear receptors CAR and PXR. *Annals of Medicine, 35*, 172–182.
143. Keminer, O., et al. (2019). Identification of novel agonists by high-throughput screening and molecular modelling of human constitutive androstane receptor isoform 3. *Archives of Toxicology, 93*, 2247–2264.
144. Lynch, C., et al. (2013). Identification of novel activators of constitutive androstane receptor from FDA-approved drugs by integrated computational and biological approaches. *Pharmaceutical Research, 30*, 489–501.
145. Dauwe, Y., et al. (2023). Steatosis and metabolic disorders associated with synergistic activation of the CAR/RXR heterodimer by pesticides. *Cells, 12*.
146. Windshügel, B., & Poso, A. (2011). Constitutive activity and ligand-dependent activation of the nuclear receptor CAR-insights from molecular dynamics simulations. *Journal of Molecular Recognition, 24*, 875–882.
147. Küblbeck, J., et al. (2011). Use of comprehensive screening methods to detect selective human CAR activators. *Biochemical Pharmacology, 82*, 1994–2007.
148. Mäkinen, J., et al. (2003). Dual action of oestrogens on the mouse constitutive androstane receptor. *The Biochemical Journal, 376*, 465–472.
149. Mejdrová, I., et al. (2023). Discovery of novel human constitutive androstane receptor agonists with the imidazo[1,2-a]pyridine structure. *Journal of Medicinal Chemistry, 66*, 2422–2456.
150. Xu, R. X., et al. (2004). A structural basis for constitutive activity in the human CAR/RXRα heterodimer. *Molecular Cell, 16*, 919–928.
151. Rashidian, A. et al. (2023). Filling the blank space: Branched 4-nonylphenols isomers are responsible for robust constitutive androstane receptor (CAR) activation by nonylphenol.
152. Tian, S., Chen, S., Pan, C., & Li, Y. (2022). FXR: Structures, biology, and drug development for NASH and fibrosis diseases. *Acta Pharmacologica Sinica, 43*, 1120–1132.
153. Forman, B. M., et al. (1995). Identification of a nuclear receptor that is activated by farnesol metabolites. *Cell, 81*, 687–693.

154. Han, C. Y. (2018). Update on FXR biology: Promising therapeutic target? *International Journal of Molecular Sciences, 19*, 2069.
155. Teodoro, J. S., Rolo, A. P., & Palmeira, C. M. (2011). Hepatic FXR: Key regulator of whole-body energy metabolism. *Trends in Endocrinology and Metabolism, 22*, 458–466.
156. Lee, F. Y., Lee, H., Hubbert, M. L., Edwards, P. A., & Zhang, Y. (2006). FXR, a multipurpose nuclear receptor. *Trends in Biochemical Sciences, 31*, 572–580.
157. Maglich, J. M., et al. (2003). The first completed genome sequence from a teleost fish (Fugu rubripes) adds significant diversity to the nuclear receptor superfamily. *Nucleic Acids Research, 31*, 4051–4058.
158. Otte, K., et al. (2003). Identification of farnesoid X receptor β as a novel mammalian nuclear receptor sensing lanosterol. *Molecular and Cellular Biology, 23*, 864–872.
159. Modica, S., Gadaleta, R. M., & Moschetta, A. (2010). Deciphering the nuclear bile acid receptor FXR paradigm. *Nuclear Receptor Signaling, 8*, e005.
160. Stefela, A., et al. (2021). (E)-7-Ethylidene-lithocholic acid (7-ELCA) is a potent dual farnesoid X receptor (FXR) antagonist and GPBAR1 agonist inhibiting FXR-induced gene expression in hepatocytes and stimulating glucagon-like peptide-1 secretion from enteroendocrine cells. *Frontiers in Pharmacology, 12*, 1980.
161. Díaz-Holguín, A., et al. (2023). When two become one: Conformational changes in FXR/RXR heterodimers bound to steroidal antagonists. *ChemMedChem, 18*, e202200556.
162. Wang, N., Zou, Q., Xu, J., Zhang, J., & Liu, J. (2018). Ligand binding and heterodimerization with retinoid X receptor α (RXRα) induce farnesoid X receptor (FXR) conformational changes affecting coactivator binding. *The Journal of Biological Chemistry, 293*, 18180–18191.
163. Shulman, A. I., Larson, C., Mangelsdorf, D. J., & Ranganathan, R. (2004). Structural determinants of allosteric ligand activation in RXR heterodimers. *Cell, 116*, 417–429.
164. Jin, L., et al. (2013). The antiparasitic drug ivermectin is a novel FXR ligand that regulates metabolism. *Nature Communications, 4*, 1937.
165. Merk, D., et al. (2019). Molecular tuning of farnesoid X receptor partial agonism. *Nature Communications, 10*, 2915.
166. Zhang, C., et al. (2022). Discovery of betulinic acid derivatives as potent intestinal farnesoid X receptor antagonists to ameliorate nonalcoholic steatohepatitis. *Journal of Medicinal Chemistry, 65*, 13452.
167. Fiorucci, S., Distrutti, E., Carino, A., Zampella, A., & Biagioli, M. (2021). Bile acids and their receptors in metabolic disorders. *Progress in Lipid Research, 82*, 101094.
168. McEwan, I. J. (2009). In I. J. McEwan (Ed.), *Nuclear receptors: One big family. In the nuclear receptor superfamily: Methods and protocols* (pp. 3–18). Humana Press.
169. Meyvis, T. K., De Smedt, S. C., Van Oostveldt, P., & Demeester, J. (1999). Fluorescence recovery after photobleaching: A versatile tool for mobility and interaction measurements in pharmaceutical research. *Pharmaceutical Research, 16*, 1153–1162.
170. Ishikawa-Ankerhold, H. C., Ankerhold, R., & Drummen, G. P. C. (2012). Advanced fluorescence microscopy techniques—FRAP, FLIP, FLAP, FRET and FLIM. *Molecules, 17*, 4047–4132.
171. Aparicio, O., Geisberg, J. V., & Struhl, K. (2004). Chromatin immunoprecipitation for determining the association of proteins with specific genomic sequences in vivo. *Current Protocols in Cell Biology, 23*, 17.7.1–17.7.23.
172. York, B., & O'Malley, B. W. (2010). Steroid receptor coactivator (SRC) family: Masters of systems biology. *The Journal of Biological Chemistry, 285*, 38743–38750.
173. Ritter, M. J., et al. (2021). Nuclear receptor CoRepressors, NCOR1 and SMRT, are required for maintaining systemic metabolic homeostasis. *Molecular Metabolism, 53*, 101315.
174. Zhou, M., Li, Q., & Wang, R. (2016). Current experimental methods for characterizing protein–protein interactions. *ChemMedChem, 11*, 738–756.
175. Negishi, M., Kobayashi, K., Sakuma, T., & Sueyoshi, T. (2020). Nuclear receptor phosphorylation in xenobiotic signal transduction. *The Journal of Biological Chemistry, 295*, 15210–15225.

176. Shizu, R., et al. (2017). Phosphorylated nuclear receptor CAR forms a homodimer to repress its constitutive activity for ligand activation. *Molecular and Cellular Biology, 37*, e00649–e00616.
177. Yeh, S., et al. (2002). Generation and characterization of androgen receptor knockout (ARKO) mice: An in vivo model for the study of androgen functions in selective tissues. *Proceedings of the National Academy of Sciences, 99*, 13498–13503.
178. Kronenberger, T., et al. (2015). Nuclear receptor modulators—Current approaches and future perspectives. In *Drug discovery and development – From molecules to medicine*. IntechOpen.
179. Piedade, R., et al. (2015). Carboxymefloquine, the major metabolite of the antimalarial drug mefloquine, induces drug-metabolizing enzyme and transporter expression by activation of pregnane X receptor. *Antimicrobial Agents and Chemotherapy, 59*, 96–104.
180. Tyree, C. M., & Klausing, K. (2003). The mammalian two-hybrid assay for detection of coactivator-nuclear receptor interactions. *Methods in Molecular Medicine, 85*, 175–183.
181. Koppen, A., et al. (2009). Nuclear receptor-coregulator interaction profiling identifies TRIP3 as a novel peroxisome proliferator-activated receptor gamma cofactor. *Molecular & Cellular Proteomics, 8*, 2212–2226.
182. Wang, S., et al. (2013). A 155-plex high-throughput in vitro coregulator binding assay for (anti-)estrogenicity testing evaluated with 23 reference compounds. *ALTEX, 30*, 145–157.
183. Berney, C., & Danuser, G. (2003). FRET or no FRET: A quantitative comparison. *Biophysical Journal, 84*, 3992–4010.
184. Wang, W., et al. (2008). Construction and characterization of a fully active PXR/SRC-1 tethered protein with increased stability. *Protein Engineering, Design & Selection, 21*, 425–433.
185. Teotico, D. G., Bischof, J. J., Peng, L., Kliewer, S. A., & Redinbo, M. R. (2008). Structural basis of human pregnane X receptor activation by the hops constituent colupulone. *Molecular Pharmacology, 74*, 1512–1520.
186. Sadybekov, A. V., & Katritch, V. (2023). Computational approaches streamlining drug discovery. *Nature, 616*, 673–685.
187. Kenda, M., & Sollner Dolenc, M. (2020). Computational study of drugs targeting nuclear receptors. *Molecules, 25*, 1616.

Chapter 17
An Overview on Computational Methods Targeting the Endocannabinoid System

Gabriel Vitor de Lima Marques, Pedro Augusto Lemos Santana, and Renata Barbosa de Oliveira

Abstract Humanity's interaction with Cannabis spans 12,000 years, playing diverse roles in cultures worldwide. Chemical elucidation of Cannabis, particularly key phytocannabinoids like cannabidiol (CBD) and Δ^9-tetrahydrocannabinol (Δ^9-THC), unfolded in the last century. Over 550 secondary metabolites, including terpenes and flavonoids, contribute to its chemical diversity. Initial notions of psychoactive effects being nonreceptor-mediated were dispelled with the discovery of stereospecific receptors, notably CB1 and CB2, explaining Cannabis' action. CB1, abundant in the central nervous system, and CB2, found in peripheral tissues, are crucial for psychoactivity and diverse physiological effects. Endocannabinoids like anandamide and 2-arachidonoylglycerol serve as natural ligands. The endocannabinoid system orchestrates complex regulatory signals and reactions, influencing cognition, immunity, and more. This understanding opens therapeutic avenues, addressing conditions like pain, anxiety, and mood disorders. Computer-aided drug design with a focus on the FAAH (fatty acid amide hydrolase) enzyme, the CB1 and CB2 receptors, and their ligands employs several combined computational techniques for the potential development of new drugs that act on the endocannabinoid system, that can be used in several somatic dysregulations and complex diseases to for which there is still no efficient treatment.

Keywords Endocannabinoid system · Cannabinoid receptors · FAAH · CADD

1 Introduction

The contact of humankind with plants of the genus Cannabis dates from circa 12 thousand years ago. Used as a food, tool, magico-religious and recreational input, these plants played crucial roles in different cultures around the globe [9, 31, 49]. Despite its long date use, its chemical properties were only elucidated

G. V. de Lima Marques · P. A. L. Santana · R. B. de Oliveira (✉)
Departamento de Produtos Farmacêuticos, Universidade Federal de Minas Gerais, Belo Horizonte, Minas Gerais, Brazil

V. G. Maltarollo (ed.), *Computer-Aided and Machine Learning-Driven Drug Design*, Computer-Aided Drug Discovery and Design 3,
https://doi.org/10.1007/978-3-031-76718-0_17

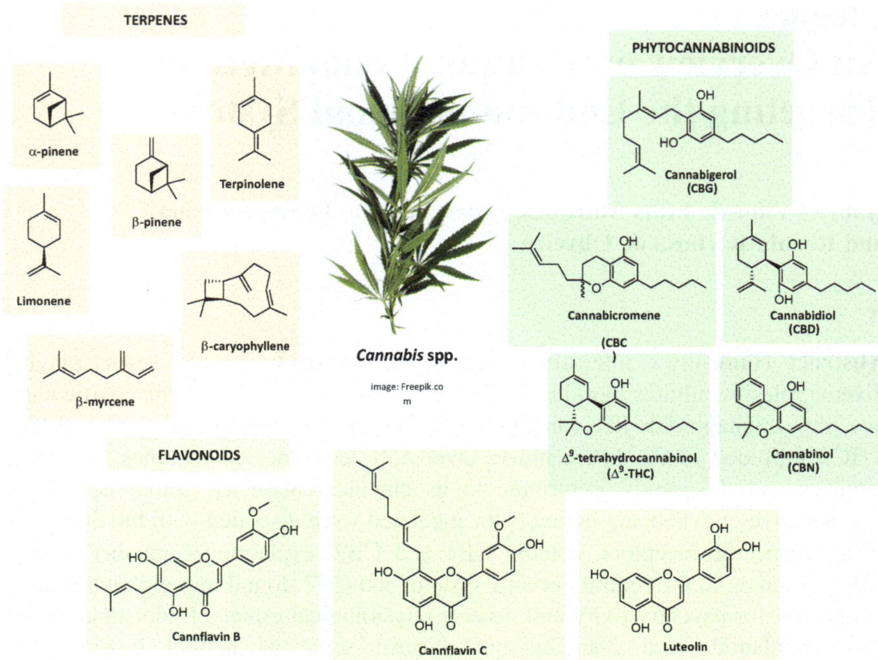

Fig. 17.1 Most abundant chemical classes of *Cannabis* spp.

mainly in the last century, due to the efforts of brilliant scientists, culminating in the beginning of the 1960s in the structural elucidation of the two main phytocannabinoids, cannabidiol (CBD) and Δ^9-tetrahydrocannabinol (Δ^9-THC), this last one being responsible for the remarkable intoxicating properties of marijuana [36]. Altogether, cannabis shows up to more than 550 isolated secondary metabolites, distributed among 18 chemical classes, such as terpenes, flavonoids and phenolic compounds, sterols, alkaloids, simple carboxylic acid derivatives, and more (Fig. 17.1) [13].

Despite advances toward the enlightenment of cannabis phytochemistry after the unequivocally elucidation of CBD and Δ^9-THC, with the discovery of various phytocannabinoids and synthesis of many synthetic cannabinoids, some questions still were persevering until the 1980s: how do people get “high” with cannabis? What is the pharmacodynamic involved in the use of cannabis? From these thoughts a hidden inner universe arose, a brand “new” whole physiological machinery: the endocannabinoid system.

2 The Discovery of Cannabinoid Receptors

Until the 1980s, there was a prevailing belief that the psychoactive effects of phytocannabinoids were attributed to their inherent lipophilicity. This suggested that these compounds could act in a broad manner without the necessity of interacting with receptors, binding sites, or specific biochemical mediators. Studies supporting this notion pointed to (+)-Δ^9-THC, a synthetic enantiomer of the natural phytocannabinoid (−)-Δ^9-THC (Fig. 17.2), as inducing typical cannabinoid activity. These findings led to skepticism about the reliability of cellular receptors, given their protein-based, chiral nature, and potential sensitivity to minor structural changes in ligands. However, subsequent research revealed that the observed activity of (+)-Δ9-THC was influenced by contamination with the natural enantiomer, prompting a reevaluation. Other studies at the time, demonstrating significantly greater activity of the synthetic cannabinoid HU-210 compared to its HU-211 enantiomer (Fig. 17.2), supported the idea that cannabinoids likely exert their effects through stereospecific receptors [11, 24, 34, 35, 47].

Several research teams have started looking for the receptors involved in the pharmacodynamics of cannabis in the central nervous system as a result of the discovery that cannabinoids would act by interaction with some receptor, or receptors. William Devane from the group of Professor Allyn Howlett, then at the University of St. Louis in the United States, published in 1988 that there was a receptor in the brain of rats that had high affinity, stereoselectivity, and pharmacologically distinct properties—criteria that have since been adopted for the characterization of a specific receptor for cannabinoids. Two years later, Lisa Matsuda's team conducted a work in which they succeeded in cloning and expressing the

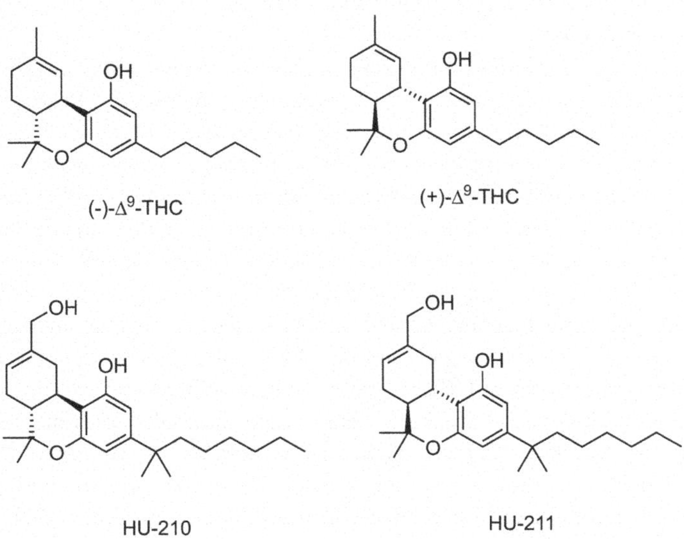

Fig. 17.2 Chemical structures of (-)-Δ^9-THC, HU-210, and their enantiomers

aforementioned cannabinoid receptor from a library of complementary DNA derived from mice. This receptor was then named cannabinoid receptor type 1 (CB1) [11, 22, 30].

Herkenham et al. [15] showed that these receptors are also prevalent in the brains of other species, including dogs, guinea pigs, monkeys, and humans. The abundance of these receptors in the basal ganglia, hippocampus, cerebral cortex, and cerebellum—regions intrinsically linked to cognition, learning, memory, and motor coordination—as well as their distribution in other parts of the central nervous system were also demonstrated in these animal models. This is a characteristic that has been particularly well-preserved across species. The low density of CB1 receptors in the brainstem bulb, the region in charge of regulating these autonomic functions, is another intriguing finding that may explain the certain cardiorespiratory safety when ingesting high dosages of Δ^9-THC [22].

The finding of CB1 receptors was crucial in explaining how cannabis usage causes psychoactivity. Other clinical results, such as analgesic, anti-inflammatory, sedative, anticonvulsant, antiemetic, and immunosuppressive effects, among others, have already been noticed in addition to the plant's characteristic toxicity. Although at much lower concentrations than those reported in the central nervous system, the distribution of CB1 receptors was seen in a number of peripheral organs, including the liver, pancreas, adipose tissue, and tissues of the genital tract [18, 22, 39, 57].

Since those initial inquiries, numerous questions have arisen regarding the manifestation of cannabis's nonpsychoactive effects. How exactly will these effects take place? Will the moderation solely rely on central action, or will peripheral tissues' CB1 receptors play a role? Could there be a connection with noncannabinoid receptors or other biological targets? Moreover, are there different types of cannabinoid receptors that contribute to this phenomenon? These queries underscore the complexity of understanding cannabis's broader physiological impacts beyond its psychoactive properties.

Using cloning techniques and complementary DNA data, Munro, Thomas, and Abu-Shaar from the Molecular Biology Laboratory of the Medical Research Council in Cambridge, United Kingdom, described the presence of another cannabinoid receptor in rat spleen macrophages in 1993. This receptor was later found in the majority of immune system cells, tissues, and organs but was initially undetected in the central nervous system. Cannabinoid receptor type 2 (CB2), a peripheral cannabinoid receptor, was assigned to the novel receptor [22, 34, 39]. Currently, it is known that CB2 receptors are found in the central nervous system as well as other organs such the bone marrow, pancreas, lungs, and uterus, but they are not as numerous as CB1 receptors and are therefore not solely peripheral receptors. Another intriguing aspect of CB2 receptors is their ability to be expressed in brain cells despite pathological situations that would normally prevent them from expressing the receptor, such as Alzheimer's disease and Huntington's disease [22, 24]. Nowadays, these receptors structures are well known and plenty of crystallographic complexes are available in the Protein Data Bank (PDB), with or without ligands (Fig. 17.3).

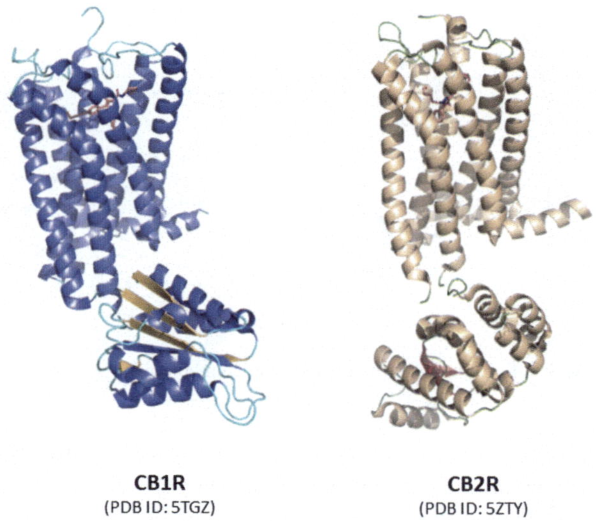

Fig. 17.3 Crystallographic structures of CB1 (PBD ID: 5TGZ) and CB2 receptors (PDB ID: 5ZTY). (Sources: Hua et al. [16], Li et al. [21])

3 Physiology of the Endocannabinoid System

But what is the scientific basis for these cannabinoid receptors, really? Could it be that at some point throughout evolution, these receptors were created expressly for contact with cannabis phytocomponents? To the scientific community devoted to these subjects, the notion that the organism would have to expend an entire amount of energy to express these proteins with that single aim appeared extremely absurd. The most likely scenario is that particular molecules, or endogenous ligands, would in fact exist and activate these receptors while also sharing this property with phytocomponents. However, no known biomolecule, hormone, or neurotransmitter produced any type of activity on these receptors, but due to the lipophilic nature of Δ^9-THC and other synthetic ligands, it was expected that the natural agonist would retain this characteristic [34]. What would the endogenous ligand(s) be, after all?

Research teams concentrated their efforts on finding the endogenous ligands of the CB1 and CB2 receptors after their identification and elucidation. The first endogenous CB1 receptor agonist was discovered in 1992. 0.6 mg of a compound was isolated from 4.5 kg of porcine brain following a series of sequential chromatographic procedures. This compound was able to show the typical cannabis-induced activity and to displace the tritium-labeled agonist $^{[3]}$HU-243 from the receptor in *in vitro* assays. A fatty acid derivative, namely arachidonic acid ethanolamide (AEA), was discovered to be the chemical compound after investigation of the retention time in gas chromatography, NMR data, and mass spectrometry. The new substance identified was named anandamide, and *ananda*, from Sanskrit, means "bliss" [12, 34].

The majority of these endogenous cannabinoid receptor agonists are derived from arachidonic acid, like 2-arachidonoylglycerol (2-AG), which was first discovered in

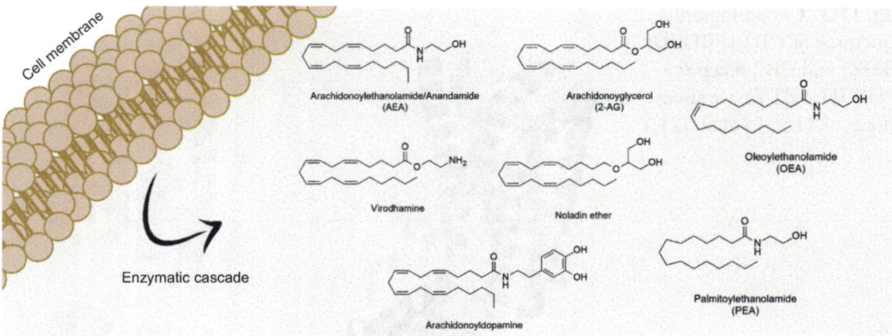

Fig. 17.4 Chemical structures of endocannabinoids

the small intestine of dogs. However, some are also derived from other fatty acids, like palmitic acid and oleic acid (Fig. 17.4) [28, 33]. Thus, a new class of biomolecules emerged: the endocannabinoids.

These compounds are produced "on demand" after pathological and/or physiological stimuli, through several stages of enzymatic biotransformations, such as hydrolysis catalyzed by lipases and dephosphorylation catalyzed by phosphatases, from membrane phospholipids, such as phosphatidylcholine and phosphatidylethanolamine in the case of AEA and phosphatidylcholine and phosphatidylinositol in the case of 2-AG. Other enzymes are important in the endocannabinoid degradation step, including fatty acid amide hydrolase (FAAH), which can contribute to the degradation of endocannabinoids by metabolizing them to proinflammatory prostaglandins. FAAH, which is primarily in charge of converting AEA to arachidonic acid and other fatty acid ethanolamides to their corresponding fatty acids, and *N*-acylethanolamine acid amidase (NAAA). In the case of 2-AG, monoacylglycerol lipase (MAGL), in addition to cyclooxygenase 2 (COX-2), is the primary enzyme responsible for its breakdown into arachidonic acid and glycerol [4, 52].

Structurally, the cannabinoid receptors CB1 and CB2 are G protein-coupled receptors, which are composed of seven transmembrane domains, with CB1 consisting of 472 amino acid residues and CB2, a little smaller, of 360 amino acid residues. These receptors often transmit signals via members of the $G_{i/o}$ family of proteins, which have inhibitory effects on adenylate cyclase and, subsequently, on cyclic adenosine monophosphate (cAMP) and the biochemical processes that follow this signaling, as well as regulatory effects on calcium and potassium ion channels. However, in experimental models, excitatory regulation via G_s/G_q proteins was seen, increasing adenylate cyclase, calcium ion flow, and phospholipase activation, demonstrating the complexity of the signaling pathway (Fig. 17.5) [8, 22, 23, 53].

It is possible that these components are involved in a variety of physiological and pathological processes, including cognition, memory, neuroprotection, immune and inflammatory response, analgesia and proalgesia, appetite, fertility, neoplasms, and cardiovascular diseases. This is because different tissues contain receptors that

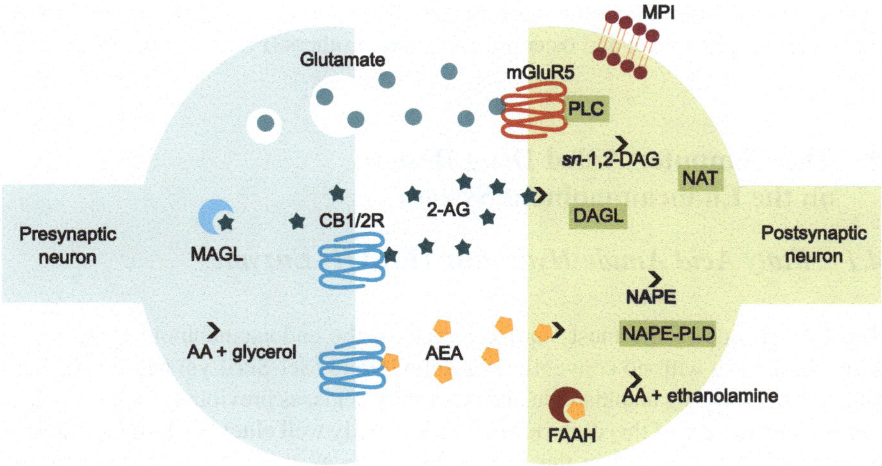

Fig. 17.5 Endocannabinoid signaling. (Source: Adapted from Huang et al. [17])

Fig. 17.6 Roles of endocannabinoid system in human body

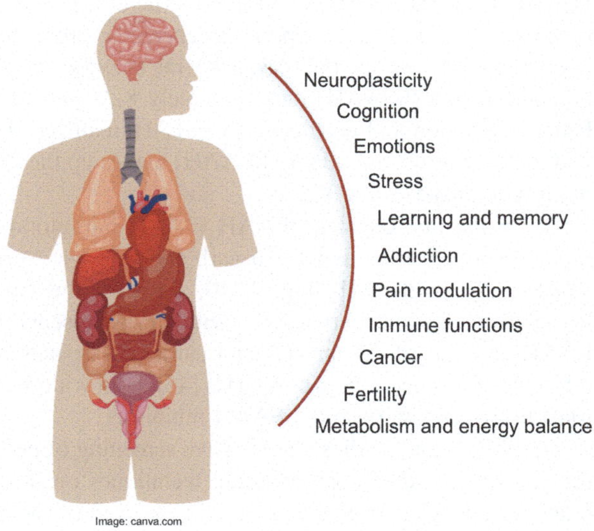

are susceptible to the action of endogenous ligands, and they are modulated in response to specific stimuli [3, 6, 25, 29].

The so-called endocannabinoid system is characterized by a group of interrelated and carefully regulatory signals, molecular targets, and biological reactions. Through the inhibition of endocannabinoid-degrading enzymes or modulation of system receptors, working with the endocannabinoid system appears to be a promising strategy for managing many clinical conditions, including neuropathic pain, anxiety, mood disorders, depression, metabolic syndromes, obesity, multiple sclerosis, cancer, glaucoma, and osteoporosis, among others (Fig. 17.6) [6, 48, 53].

4 The Computer-Aided Drug Design on the Endocannabinoid System

4.1 *Fatty Acid Amide Hydrolase (FAAH) Enzyme*

The FAAH, one of the most studied targets of the endocannabinoid system, is a serine hydrolase with no conventional catalytic triad (Ser-Ser-Lys) and it is responsible for hydrolyzing anandamide and other fatty acids, as previously explained. The molecular structure of this dimeric enzyme is already well elucidated. In fact, despite having only been cloned in the late 1990s [9] today there are more than twenty crystallized structures, with ligands or not, in the Protein Data Bank (PDB), which allows the recognition of their residues and catalysis and interaction sites (Fig. 17.7). Despite its characterization and good structural elucidation, FAAH is a transmembrane enzyme and has a lipophilic character prone to forming aggregates in its expression process. In this case, the rat enzyme (rFAAH) is homologous to the human enzyme (hFAAH) (approximately 82% similarity) and experimentally has better expression and purification yields. Humanized rFAAH (residues were modified to make it more similar to hFAAH) makes up the majority of crystals deposited in protein databases [1, 37].

All of the first covalent iFAAH (FAAH inhibitors) had, between two moieties (here described as A and B), a carbonyl that was sufficiently electrophilic to be attacked by serine 241 (Fig. 17.8). Perhaps, the best-known examples are the ketooxazolopyridine derivatives (with the representative OL-135, which inhibits FAAH at levels of 4.7 nm) [7] and carbamates (such as URB597, the most potent inhibitor of this class, IC_{50} of 63 nM) [46], both widely used as controls in biological trials to investigate potential FAAH inhibitors.

Despite several studies on extensive screening of potential FAAH inhibitors over the years, the use of computational techniques on the endocannabinoid system is quite recent. The mechanism of enzyme hydrolysis itself was characterized by mutation experiments [31], but this could not explain the selectivity of some groups that, at first, seemed very similar. This was the case reported by Palermo and collaborators (2014), where piperidine (PF-750), piperazine (JNJ1661010) and naphthalene (1-cyclohexyl-3-naphthalen-1-ylurea) urea derivatives (Fig. 17.8) had different enzymatic inhibition activity. For that, using classical molecular dynamics (please, see Chap. 8 for more details), full quantum mechanics, and hybrid QM/MM (quantum mechanics and molecular mechanics) methods, the authors proposed that the flexibility of the iFAAH is important for their inhibitory activity, given that a greater possibility of twisting of the urea or carbamate group, provoke a bond

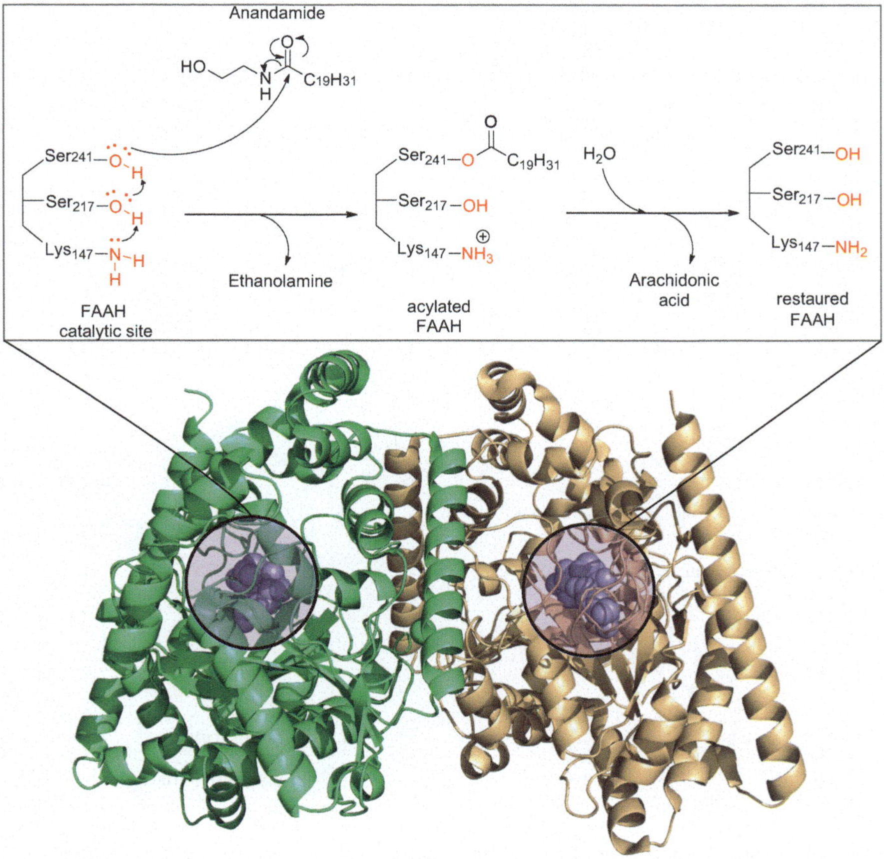

Fig. 17.7 Representation of the catalytic site and its mediated hydrolyses mechanism of each FAAH monomer. Humanized rFAAH PDB code 3K7F

distortion, which can modify the electrophilicity of the carbonyl carbon (ensuring the difference in the inhibition potential among of the ligands) [46].

In these experiments, the conformations of the inhibitors in water and complexed with the enzyme were compared, and it was found that the same compounds assume different conformations in these two environments. The interaction of the inhibitor's side chain with amino acid residues in the active site of FAAH favor a conformational change that results in partial pyramidalization at the urea nitrogen, hindering the resonance of the nitrogen electron pair with the carbonyl group. Here, rigid compounds are defined as those that have its a cyclic nitrogen, while the flexible ones have not. The increase in the electrophilicity of this carbon makes the nucleophilic attack of Ser 241 more effective, increasing the inhibitory activity of these compounds (Fig. 17.9). Furthermore, when calculating the HOMO-LUMO energy gap, using ONIOM (a tool used in quantum chemistry that combines high-level quantum mechanics calculations with low-level calculations, and treats the different

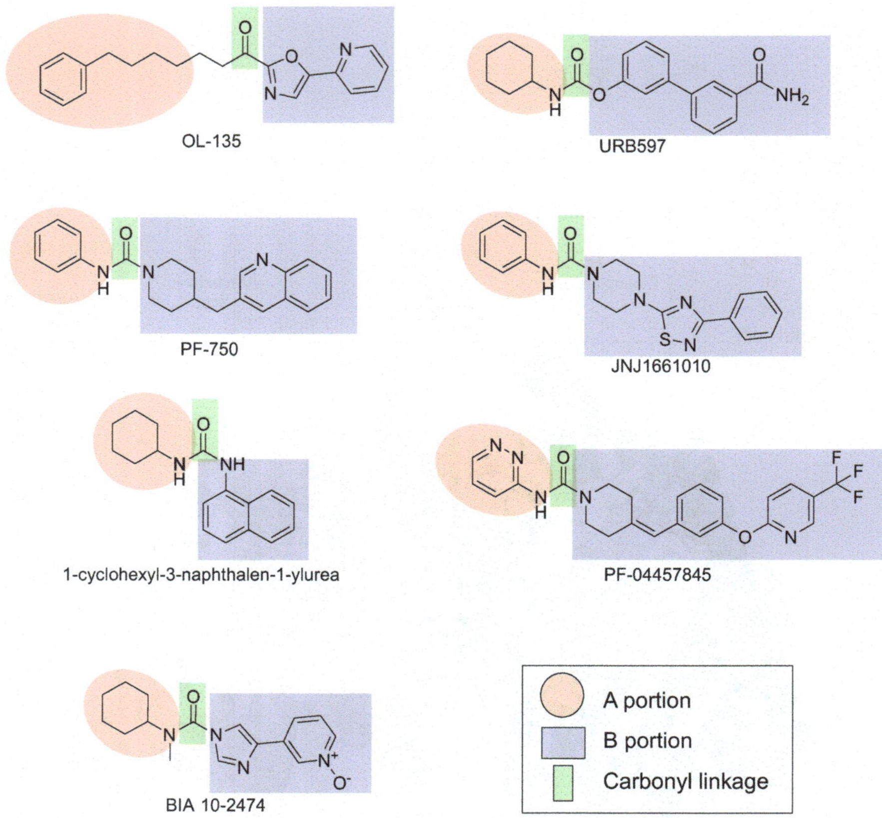

Fig. 17.8 Examples of iFAAH and its A and B portion connected by electrophilic carbonyl

regions of a macromolecule according to your interest understanding), the energy gap of the planar ureas (more rigid) was larger than that of the distorted ureas by approximately 0.35–0.40 eV (or 8–9 kcal/mol). The HOMO-LUMO energy gap is related to the highest occupied molecular orbital (whereas the electron pair of the nucleophile is located) and the lowest unoccupied molecular orbital (orbital unoccupied of the electrophile). In this case, the smaller calculated HOMO-LUMO energy gap of the distorted amides suggests that they are more expected to inhibit FAAH. In this sense, it was demonstrated how molecular dynamics and QM/MM methods can help to understand how enzymes are inhibited, allowing its use in virtual screenings [2, 46, 54].

Despite several recent studies, no FAAH inhibitor drug has yet been released into the market. Some have already reached clinical studies with very promising preclinical results, such as PF-04457845 (Pfizer) (Fig. 17.8) in patients with Tourette syndrome and knee osteoarthritis [41, 42, 43] and URB597 (Fig. 17.8) in patients with schizophrenia [43]. Another iFAAH that entered clinical trials, the controversial BIA 10-2474 (Bial Pharmaceuticals) (Fig. 17.8), had its studies interrupted in its

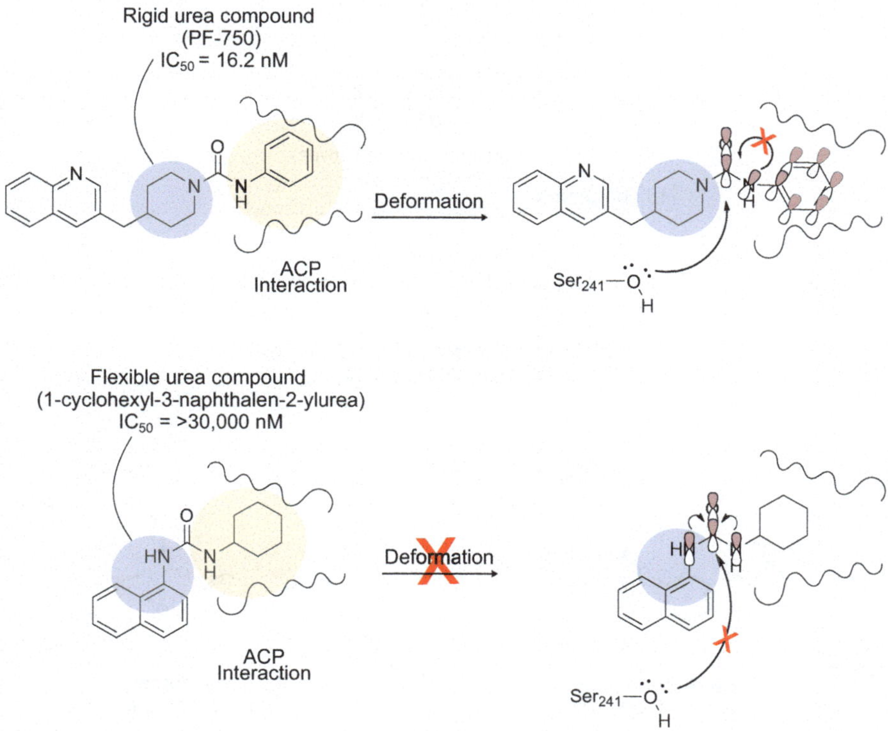

Fig. 17.9 Deformation and bond torsion of PF-750, a rigid piperidine, caused by the interaction with ACP portion of the enzyme, in contrast to 1-cyclohexyl-3-naphthalen-2-ylurea, a flexible urea compound

first phase, as continuous doses of the compound led to brain death of one of the patients and permanent brain damage in five others. Recent speculations indicate that brain hemorrhages and necrosis were caused by off-target effects of this compound [18]. The tragedy related to BIA 10-2474 helped to increase the discussion about the transposition of preclinical results to clinical studies, given that the lack of comparability of this relationship has long been known. In the same way, the need to create new servers and computational strategies that can help predict potential toxicological and safety effects of new molecules before they are tested on humans has become a central topic in recent scientific discussions [26, 27].

Besides the common screening studies using molecular docking and molecular dynamics techniques, some studies use these techniques to verify other structural properties of compounds that may act as iFAAH and its binding mode. One of these cases was reported by Myllymäki and collaborators [40], in which the use of molecular docking was able to explain and demonstrate the stereoselectivity of the FAAH compared to some of its known inhibitors (*N*-alkylcarbamate derivatives). In this study, docking simulations using empirical scoring functions (for more details, please see Chap. 7) and well-known protocols were associated with interaction field

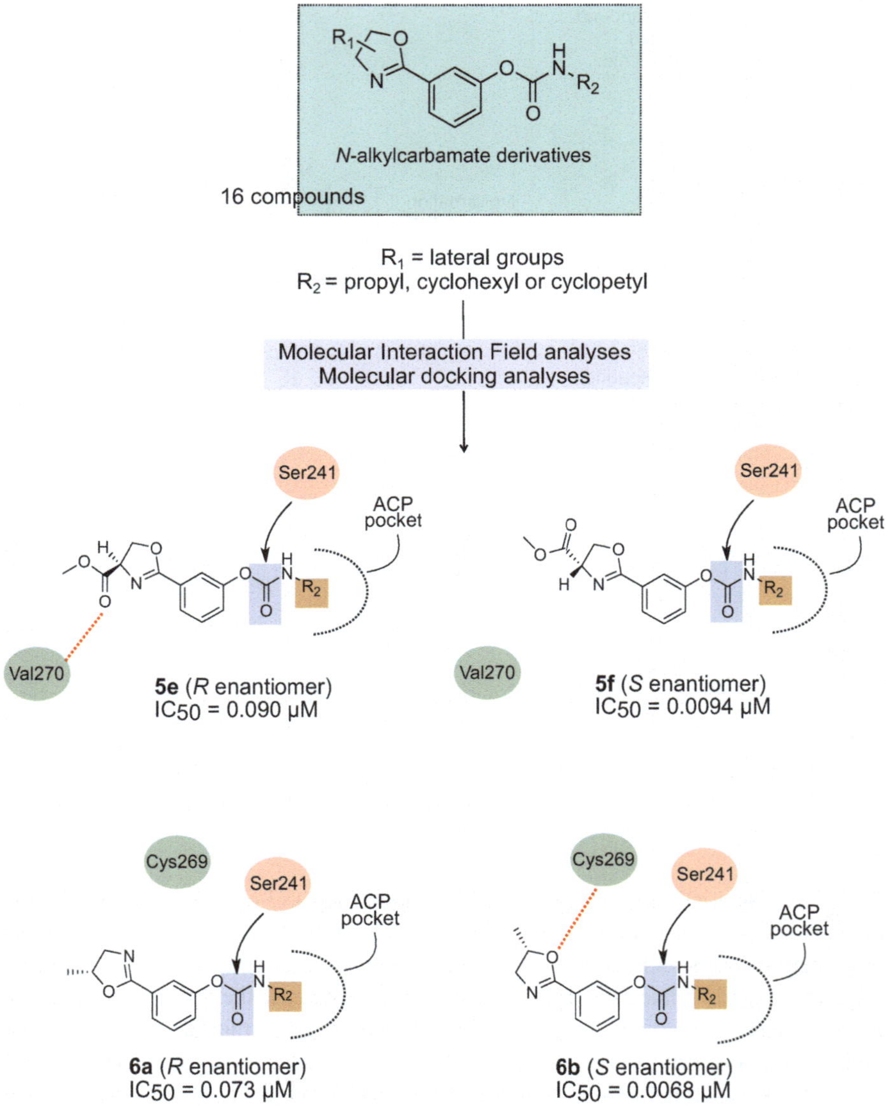

Fig. 17.10 General scheme of the study conducted by Myllymäki and collaborators [39]

studies (MIF), in which probes of potential interactions could be merged with well-scored poses of the emissions and compared with their enzymatic activity. As schemed in Fig. 17.10, a series of 16 *N*-alkylcarbamate derivatives were proposed, synthesized, and had its FAAH inhibition evaluated in vitro. The aim was to correlate the tested inhibition with the structural differences among the compounds (as the stereochemistry). In conclusion, the authors commented that the stereoselectivity for one of the enantiomers of the ligands (the ones with the oxygen of oxazole ring close to the Cys269 residue) and the ring substituent, as methyl ester, is related to the proximity to important residues for the interaction with FAAH: residues of the Cys269 and Val270. The only difference of the inhibition was attributed exclusively by the spatial position of the ligands. As can be seen, the proximity of the oxygen of the oxazole group can be related to its interaction to Cys269. The same can be observed with the residue of Val270, where with the polar interaction could increase the inhibition potential of FAAH (Fig. 17.10).

The application of the docking technique in the repositioning of drugs as potential inhibitors of FAAH was also recently reported. The construction of iFAAH dataset classification models relating structure-based and ligand-based techniques (such as QSAR, molecular docking, and molecular dynamics, please see Chaps. 6, 7 and 8, respectively, for more methodological details) proposed that some drugs already used in the clinic, such as montelukast, raloxifene, and buclizine, have potential FAAH inhibition activity. However, *in vitro* and *in vivo* studies are necessary to verify the analgesic and anti-inflammatory activity via FAAH inhibition of these drugs [55].

Some computational techniques using the ligand structure have also been explored for the search for iFAAH. This is the case of the study carried out by Zieba and collaborators [56] that reports the structure of 31 iFAAH containing the 1,3,4-oxadiazol-2-one moiety used as a basis for the construction of 3D-QSAR models, of the Comparative Molecular Field Analysis (CoMFA) and Comparative Molecular Similarity Indices Analysis (CoMSIA) subtypes (more details in Chap. 6). In short, the first technique verifies a relationship between the entire molecule (steric and electronic parameters) and its biological activity value (here posted as IC_{50} value for FAAH inhibition). The second method covers hydrophobic interactions and hydrogen bonds and other important properties, also related to biological activity [51]. Thus, the combination of the two techniques, in addition to the use of the structure of the molecular target, that is, the active site of FAAH, resulted in models with good robustness. As resumed in Fig. 17.11, the total set of molecules were divided on training and test set (by 24 and 7 compounds, respectively) and the CoMFA and CoMSIA models built and evaluated according to its metrics (r^2 and q^2). The fields of the best model of each technique were constructed, and some conclusions were made: (a) the importance of groups that can interact with Gly 239 and Ile 238 residues, such as heterocycles as 1,3,4-oxadiazol-2-one; (b) a bulky chain such as a 6-member ring, opposite to the 1,3,4-oxadiazol-2-one moiety, increase the affinity of ligands with FAAH; (c) between these two portions, a hydrogen bond acceptor could be beneficial, and a hydrogen bond donor on the lateral bulky chain could increase the affinity. The most and least active 1,3,4-oxadiazol-2-one derivatives used in this study are represented in Fig. 17.11 (named as substance 1 and 31, respectively) [56].

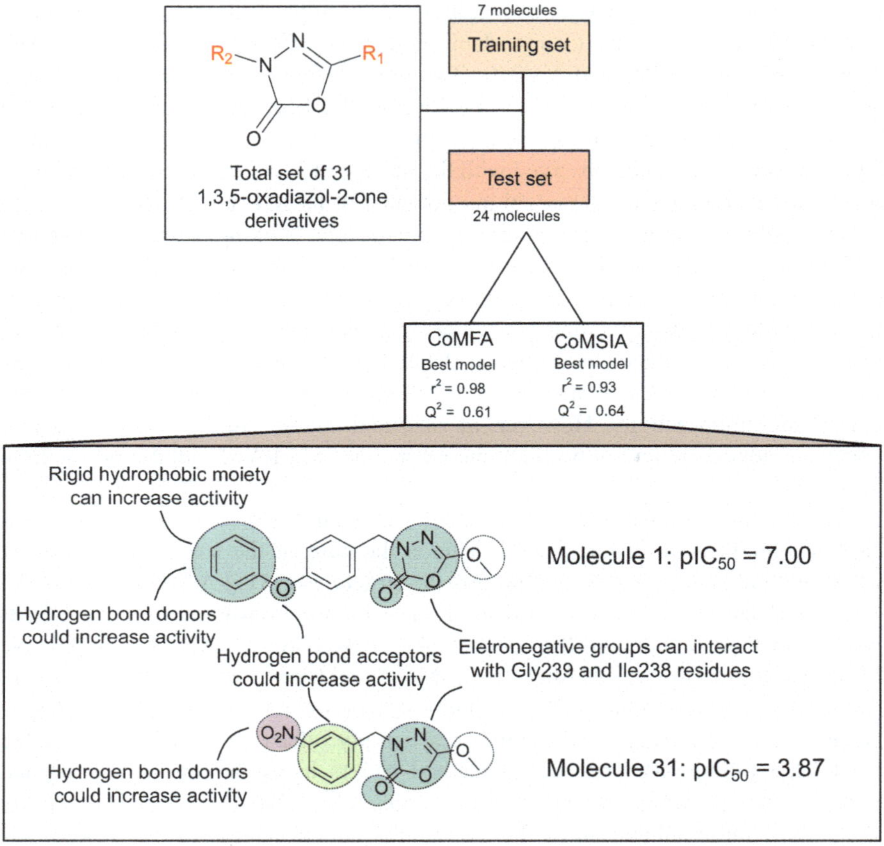

Fig. 17.11 General scheme of the study conducted by Zieba and collaborators [56]

4.2 CB1 and CB2 Receptors

Unlike FAAH, there are few crystallographic structures of the CB1 and CB2 cannabinoid receptors. However, due to the knowledge about how molecules can act on them (knowing their cannabinoid ligands), some research studies on potential molecules that could have some therapeutic effect through this signaling are also reported.

For example, the aim of the study described by Montero and collaborators [38] was to understand how the molecular functioning of compounds on receptors was, especially, how their interaction site would function. However, as mentioned previously, the lack of crystallographic structures of these receptors was a challenge at that time, which was overcome using homology structural modeling techniques. In this case, the bovine receptor was used as a template, and to understand how the compounds bind, an inverse agonist and an antagonist were used in some docking studies approaches.

In contrast, recently some more accurate and robust studies on these receptors have been carried out. This is the case of the work published by Aviz-Amador, Contretas-Puentes, and Mercado-Camargo in 2021 [5] in which not only docking but also molecular dynamics was used to predict the binding mode of CBD analogues, together with studies of Molecular Mechanics/Generalized Born Surface Area (MM/GBSA)—a binding free energy of a molecular complex calculation method, for theoretical basis of this method, please see Chap. 8. The use of datasets of known ligands and decoys to validate the studies marked the greater robustness of the generated models, in addition to the analysis of their ROC curve (in which values greater than 0.8 were demonstrated in simulations using the CB1 and CB2 receptors). The AUC values in ROC curves demonstrate the potential of a model to discriminate true positive and false positive ligands. In this sense, values higher than 0.5 (and nearby 1.0) show a high (and nonaleatory) capacity to determine active and inactive compounds [14].

Another interesting study related to cannabinoid receptors, more precisely type 2 receptors, was also recently reported and its construction was a little different: Omar and collaborators [45], using a series of fungal metabolites derived from phenylspirodrimanes (originating from a fusion of meroterpenoids and spirofuran), verified their potential interaction with CB2 receptors through molecular docking. However, the approach in this study was reversed: the compounds already existed, but the target had not yet been determined. Using an ATC (Anatomical Therapeutic Chemical rating) classification server called Super-PRED based on machine learning models, the researchers verified that the mentioned receptor could be a target for *in silico* studies [44].

Simply put, Super-PRED style servers are based on linear regression classification models, and through the scaffolds and chemical structure, classify the molecule from its ATC score, and potential molecular targets can be selected. In this sense, only after verifying the high scoring potential of the CB2 receptor, simulation studies were carried out. As illustrated in Fig. 17.12, a data of 114 phenylspirodrimanes were applied to Super-PREP server, whereas CB2 receptor-binding probability was attributed to each molecule. The molecule named as "compound 56," exampled in the Fig. 17.12, had 68% of probability to bind to CB2 receptor in a machine learning model with accuracy of 97% (measured by ten-fold cross-validation, inserted on Super-PREP server). Every compound had its probability attributed, and docking and dynamic simulations (using CB2 receptor, on PDB code 6K9F) were carried out, resulting on the compound 56 as the most potential ligand to receptor CB2. It is important to mention that the authors did not report *in vitro* or *in vivo* studies of the compounds, but presented an unusual *in silico* screening approach on endocannabinoid system [44].

The use of cannabidiol (CBD) for therapeutic purposes is still the main focus of studies regarding the CB1 and CB2 receptors. However, in addition to attempting structural modifications to this molecule, computational methods are also used to try to improve or overcome problems with administering this drug due to its physicochemical properties. This is the case of the study reported by Li and collaborators [20] in which the encapsulation of CBD in cyclodextrins (CD) to modulate its water

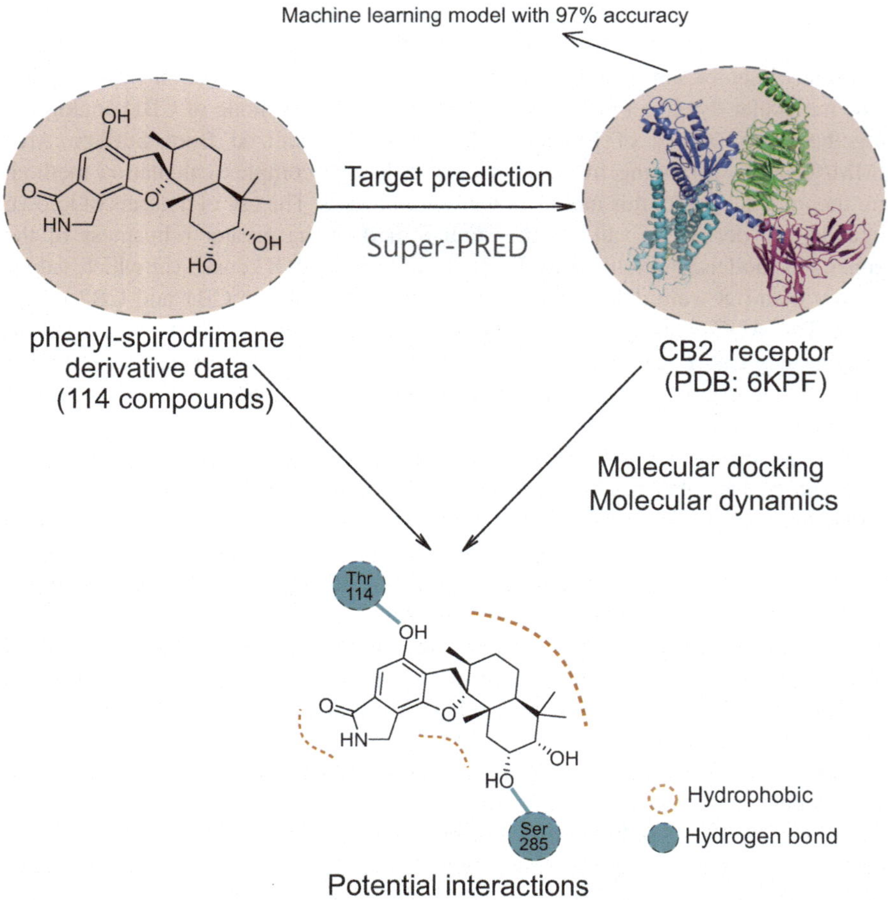

Fig. 17.12 Scheme of the screening conducted using the target prediction and structure-based studies such as molecular docking and dynamic

solubility was investigated by, among several experimental proceedings, molecular docking, to verify the conformation and potential interaction with CD and encapsulation success. The two CD used, β-cyclodextrin (β-CD) and 2,6-di-*O*-methyl-β-cyclodextrin (DM-β-CD), were capable to increase the solubility of the CDB in 17 and 614-fold, respectively (Fig. 17.13). The computational calculation, using the obtained CD structure from PDB (PDB ID = 1Z0N) [49] and the adapted structure of DM-β-CD (by adding dimethyl to CD structure), could confirmed the potential inclusion complexation of CBD with the CD's. The difference on the binding energy values (of −5.80 kcal/mol for CBD/β-CD complex and −7.22 kcal/mol for CBD/DM-β-CD complex) and accommodation of the molecules were concordant to the experimental stability, solubility and antioxidant activity of CBD.

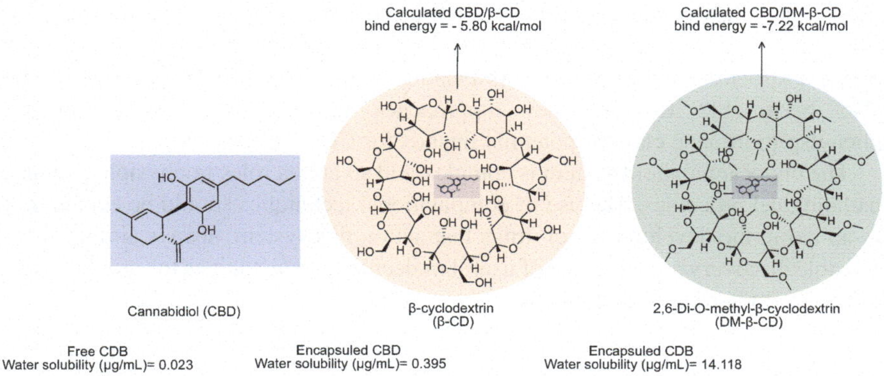

Fig. 17.13 Differences on water solubility of free CBD and CBD-CD's complexes and its potential binding energy

5 Conclusion

Discussions regarding the utilization of cannabis throughout humanity's cultural, social, and medical narrative were amplified by the chemical clarification of CBD and THC phytocannabinoids. Despite the millennia-old empirical use of plants within this genus, understanding of the endocannabinoid system is recent and notably marked by the discovery of CB1 and CB2 receptors, which occurred only in the last century (in contrast to the knowledge of cannabis's therapeutic effects predating modern civilization). Validating molecular targets, along with other crucial aspects of endocannabinoid signaling (such as enzymes, endogenous ligands, and associated receptors), has propelled pharmacotherapeutic development for the treatment of diseases for which treatment remains inefficient or nonexistent, such as chronic and/or neurodegenerative diseases.

In a way, the discovery and study of such a young system parallel the development of computer-aided drug design techniques (which are also young). Some structure-based drug design techniques rely on the number of elucidated structures of molecular targets in databases (such as the FAAH enzyme, responsible for the degradation of anandamide, the endogenous ligand of cannabinoid receptors). Quantum mechanics and molecular mechanics studies have proposed the importance of the flexibility of electrophilic groups in FAAH inhibitors, while docking studies, molecular dynamics, and interaction fields have demonstrated the groups crucial for the interaction of FAAH inhibitors with the enzyme's active site (especially with polar residues Cys269 and Val270). Ligand-based drug design studies such as QSAR (CoMFA and CoMSIA) also for FAAH studies have demonstrated the importance of steric issues of iFAAH which, when combined with SBDD methods, exemplify the importance of potential interaction with some residues comprising the enzyme's active site.

In contrast, structures of cannabinoid receptors CB1 and CB2 still have a lower number in structural databases (but fortunately, this has been increasing over the years). Nonetheless, docking studies and molecular dynamics can demonstrate how cannabidiol analogs (exogenous ligands) can interact with the receptors and even propose structural modifications. Association with Super-PRED type servers has also been able to seek new approaches in proposing ligands of the phenylspirodrimane class as potential ligands for CB receptors.

The endocannabinoid system is complex and so are its roles in physiological and pathological processes. The use of computational techniques should be encouraged as an aid in the development of drugs that act on this system, always aiming at the possible discovery and proposal of useful molecules and medicines for problems that today we do not have solutions for.

References

1. Ahn, K., et al. (2009. Informa UK Limited). Fatty acid amide hydrolase as a potential therapeutic target for the treatment of pain and CNS disorders. *Expert Opinion On Drug Discovery, [S.L.], 4*(7), 763–784.
2. Ahn, K., et al. (2007. American Chemical Society (ACS)). Novel mechanistic class of fatty acid amide hydrolase inhibitors with remarkable selectivity. *Biochemistry, [S.L.], 46*(45), 13019–13030.
3. Aizpurua-Olaizola, O., Elezgarai, I., Rico-Barrio, I., Zarandona, I., Etxebarria, N., & Usobiaga, A. (2017. Elsevier BV). Targeting the endocannabinoid system: Future therapeutic strategies. *Drug Discovery Today, [S.L.], 22*(1), 105–110.
4. Araque, A., Castillo, P. E., Manzoni, O. J., & Tonini, R. (2017. Elsevier BV). Synaptic functions of endocannabinoid signaling in health and disease. *Neuropharmacology, [S.L.], 124*, 13–24.
5. Aviz-Amador, A., et al. (2021. Elsevier BV). Virtual screening using docking and molecular dynamics of cannabinoid analogs against CB1 and CB2 receptors. *Computational Biology and Chemistry, [S.L.], 95*, 107590.
6. Battista, N., Tommaso, M. D., Bari, M., Maccarrone, M. (2021). The endocannabinoid system: an overview. *Frontiers In Behavioral Neuroscience, 6*, 9, 1–7.
7. Boger, D. L., et al. (2005. American Chemical Society (ACS)). Discovery of a potent, selective, and efficacious class of reversible α-Ketoheterocycle inhibitors of fatty acid amide hydrolase effective as analgesics. *Journal of Medicinal Chemistry, [S.L.], 48*(6), 1849–1856.
8. Carvalho, W. A., et al. (1997). Biologia Molecular dos Receptores Farmacológicos e seus Sistemas Efetores de Interesse em Anestesiologia. *Revista Brasileira de Anestesiologia, [S.L.], 47*(2), 152–167.
9. Cravatt, B. F., et al. (1996. Springer Science and Business Media LLC). Molecular characterization of an enzyme that degrades neuromodulatory fatty-acid amides. *Nature, [S.L.], 384*(6604), 83–87.
10. Crocq, M.-A. (2020. Informa UK Limited). History of cannabis and the endocannabinoid system. *Dialogues in Clinical Neuroscience, [S.L.], 22*(3), 223–228.
11. Devane, W. A., et al. (1988). Determination and characterization of a cannabinoid receptor in rat brain. *Molecular Pharmacology, 34*(5), 605–613.
12. Devane, W., et al. (1992. American Association for the Advancement of Science (AAAS)). Isolation and structure of a brain constituent that binds to the cannabinoid receptor. *Science, [S. L.], 258*(5090), 1946–1949.
13. Fordjour, E., et al. (2023. Frontiers Media SA). Cannabis: A multifaceted plant with endless potentials. *Frontiers in Pharmacology, [S.L.], 14*, 01–36.

14. Hajian-Tilaki, K. (2013). Receiver operating characteristic (ROC) curve analysis for medical diagnostic test evaluation. *Caspian Journal of Internal Medicine, 4*(2), 627–635. PMID: 24009950; PMCID: PMC3755824.
15. Herkenham, M. et al. (1990). Cannabinoid receptor localization in brain. *Proceedings of the National Academy of Sciences, 87*(5), 1932–1936.
16. Hua, T., et al. (2016. Elsevier BV). Crystal structure of the human cannabinoid receptor CB1. *Cell, [S.L.], 167*(3), 750–762.
17. Huang, W.-J., Chen, W.-W., & Zhang, X. (2016. Spandidos Publications). Endocannabinoid system: role in depression, reward and pain control (review). *Molecular Medicine Reports, [S. L.], 14*(4), 2899–2903.
18. Kalant, H. (2001). Medicinal use of cannabis: history and current status. *Pain Research and Management, 6*(2), 80–91.
19. Kaur, R., et al. (2016. SAGE Publications). What failed BIA 10–2474 Phase I clinical trial? Global speculations and recommendations for future Phase I trials. *Journal of Pharmacology And Pharmacotherapeutics, [S.L.], 7*(3), 120–126.
20. Li, H., et al. (2021. Elsevier BV). Inclusion complexes of cannabidiol with β-cyclodextrin and its derivative: physicochemical properties, water solubility, and antioxidant activity. *Journal of Molecular Liquids, [S.L.], 334*, 116070.
21. Li, X., et al. (2019. Elsevier BV). Crystal Structure of the Human Cannabinoid Receptor CB2. *Cell, [S.L.], 176*(3), 459–467.
22. Lu, D., & Potter, D. E. (2017). Cannabinoids and the cannabinoid receptors: An overview. In V. R. Preedy (Ed.), *Handbook of Cannabis and Related Pathologies: biology, pharmacology, diagnosis, and treatment* (Vol. 58, pp. 553–563). Elsevier. Disponível em: https://www.sciencedirect.com/science/article/pii/B9780128007563000685?dgcid=raven_sd_recommender_email. Acesso em: 20 dez. 2020.
23. Lu, H.-C., & Mackie, K. (2021. Elsevier BV). Review of the endocannabinoid system. *Biological Psychiatry: Cognitive Neuroscience and Neuroimaging, [S.L.], 6*(6), 607–615.
24. Maccarrone, M., et al. (2015. Elsevier BV). Endocannabinoid signaling at the periphery: 50 years after thc. *Trends in Pharmacological Sciences, [S.L.], 36*(5), 277–296.
25. Maccarrone, M., Marzo, V. d., Gertsch, J., Grether, U., Howlett, A. C., Hua, T., Makriyannis, A., Piomelli, D., Ueda, N., & Stelt, M. v. D. (2023. American Society for Pharmacology & Experimental Therapeutics (ASPET)). Goods and Bads of the Endocannabinoid System as a Therapeutic Target: lessons learned after 30 years. *Pharmacological Reviews, [S.L.], 75*(5), 885–958.
26. Mahalmani, V., Prakash, A., & Medhi, B. (2023). Do alternatives to animal experimentation replace preclinical research? *Indian Journal of Pharmacology, 55*(2), 71–75.
27. Mahalmani, V., Sinha, S., Prakash, A., & Medhi, B. (2022). Translational research: Bridging the gap between preclinical and clinical research. *Indian Journal of Pharmacology, 54*(6), 393–396.. PMID: 36722550; PMCID: PMC10043823.
28. Martin, B. R., Mechoulam, R., & Razdan, R. K. (1999. Elsevier BV). Discovery and characterization of endogenous cannabinoids. *Life Sciences, [S.L.], 65*(6–7), 573–595.
29. Marzo, V. d., Bifulco, M., & Petrocellis, L. d. (2004. Springer Science and Business Media LLC). The endocannabinoid system and its therapeutic exploitation. *Nature Reviews Drug Discovery, [S.L.], 3*(9), 771–784.
30. Matsuda, L. A., et al. (1990. Springer Science and Business Media LLC). Structure of a cannabinoid receptor and functional expression of the cloned cDNA. *Nature, [S.L.], 346*(6284), 561–564.
31. Mckinney, M. K., & Cravatt, B. F. (2003. Elsevier BV). Evidence for distinct roles in catalysis for residues of the serine-serine-lysine catalytic triad of fatty acid amide hydrolase. *Journal of Biological Chemistry, [S.L.], 278*(39), 37393–37399.
32. Mechoulam, R., et al. (2014). Early phytocannabinoid chemistry to endocannabinoids and beyond. *Nature Reviews. Neuroscience, 15*, 757–764.

33. Mechoulam, R., et al. (1995. Elsevier BV). Identification of an endogenous 2-monoglyceride, present in canine gut, that binds to cannabinoid receptors. *Biochemical Pharmacology, [S.L.], 50*(1), 83–90.
34. Mechoulam, R., & Hanuš, L. (2000. Elsevier BV). A historical overview of chemical research on cannabinoids. *Chemistry and Physics of Lipids, [s.l.], 108*(1–2), 1–13.
35. Mechoulam, R., & Parker, L. A. (2013). The endocannabinoid system and the brain. *Annual Review of Psychology, 64*(1), 21–47.
36. Mechoulam, R., & Hanuš, Lumír. (2002. Elsevier BV). Cannabidiol: an overview of some chemical and pharmacological aspects. part i. *Chemistry and Physics of Lipids, [S.L.], 121*(1–2), 35–43.
37. Mileni, M., et al. (2008. Proceedings of the National Academy of Sciences). Structure-guided inhibitor design for human FAAH by interspecies active site conversion. *Proceedings of The National Academy of Sciences, [S.L.], 105*(35), 12820–12824.
38. Montero, C., et al. (2005. Elsevier BV). Homology models of the cannabinoid CB1 and CB2 receptors. A docking analysis study. *European Journal of Medicinal Chemistry, [S.L.], 40*(1), 75–83.
39. Munro, S., Thomas, K. L., & Abu-Shaar, M. (1993. Springer Science and Business Media LLC). Molecular characterization of a peripheral receptor for cannabinoids. *Nature, [S.L.], 365*(6441), 61–65.
40. Myllymäki, M. J., et al. (2009. Elsevier BV). Chiral 3-(4,5-dihydrooxazol-2-yl)phenyl alkylcarbamates as novel FAAH inhibitors: insight into faah enantioselectivity by molecular docking and interaction fields. *European Journal of Medicinal Chemistry, [S.L.], 44*(10), 4179–4191.
41. National Library of Medicine (U.S.). (2017). *A phase 2a randomized, double-blinded, double dummy, placebo and active controlled, two-way cross-over, flare-enriched multi-centre clinical trial to examine the pain relief produced by 2 weeks of daily oral administration of a Fatty Acid Amide Hydrolase (Faah) inhibitor Pf-04457845 in patients with osteoarthritis of the knee*. Identifier NCT00981357. https://clinicaltrials.gov/study/NCT00981357?term=PF-04457845&rank=1
42. National Library of Medicine (U.S.). (2018). *FAAH inhibitor trial for adults with Tourette syndrome*. Identifier NCT02134080. https://clinicaltrials.gov/study/NCT02134080?term=FAAH&rank=2&a=6
43. National Library of Medicine (U.S.). (2019). *Evaluation study of new compounds with potential use in schizophrenia (EICAS)*. Identifier NCT00916201. https://clinicaltrials.gov/study/NCT00916201?term=FAAH%20inhibitor&rank=9
44. Nickel, J., et al. (2014. Oxford University Press (OUP)). SuperPred: Update on drug classification and target prediction. *Nucleic Acids Research, [S.L.], 42*(1), 26–31.
45. Omar, A. M., Aljahdali, A. S., Safo, M. K., Mohamed, G. A., & Ibrahim, S. R. M. (2023). Docking and molecular dynamic investigations of phenylspirodrimanes as cannabinoid receptor-2 agonists. *Molecules, 28*, 44.
46. Palermo, G., et al. (2014. Elsevier BV). Computational insights into function and inhibition of fatty acid amide hydrolase. *European Journal of Medicinal Chemistry, [S.L.], 91*, 15–26.
47. Paton, W. (1975. Annual Reviews). Pharmacology of Marijuana. *Annual Review of Pharmacology, [S.L.], 15*(1), 191–220.
48. Pertwee, R. G. (2008). The diverse CB1 and CB2 receptor pharmacology of three plant cannabinoids: Δ9-tetrahydrocannabinol, cannabidiol and Δ9-tetrahydrocannabivarin. *British Journal of Pharmacology, 153*(2), 199–215.
49. Polekhina, G., Gupta, A., van Denderen, B. J. W., Feil, S. C., Kemp, B. E., Stapleton, D., & Parker, M. W. (2005). Structural basis for glycogen recognition by AMP-activated protein kinase. *Structure, 13*(10), 1453–1462.
50. Russo, E. B. (2007. Wiley). History of cannabis and its preparations in saga, science, and sobriquet. *Chemistry & Biodiversity, [S.L.], 4*(8), 1614–1648.

51. Sharma, R., et al. (2016. OMICS Publishing Group). CoMFA, CoMSIA, HQSAR and molecular docking analysis of ionone-based chalcone derivatives as antiprostate cancer activity. *Indian Journal of Pharmaceutical Sciences, [S.L.], 78*(1), 54.
52. Tuo, W., et al. (2016. American Chemical Society (ACS)). Therapeutic potential of fatty acid amide hydrolase, Monoacylglycerol lipase, and N-Acylethanolamine acid amidase inhibitors. *Journal of Medicinal Chemistry, [S.L.], 60*(1), 4–46.
53. Turu, G., Hunyady, L. (2010). Signal transduction of the CB1 cannabinoid receptor. *Journal of Molecular Endocrinology, 44*(2), 75–85.
54. Vreven, T., et al. (2003. Wiley). Geometry optimization with QM/MM, ONIOM, and other combined methods. I. Microiterations and constraints. *Journal of Computational Chemistry, [S. L.], 24*(6), 760–769.
55. Zanfirescu, A., et al. (2021. MDPI AG). Identifying FAAH inhibitors as new therapeutic options for the treatment of chronic pain through drug repurposing. *Pharmaceuticals, [S.L.], 15*(1), 38.
56. Zięba, A., et al. (2021. MDPI AG). Docking-based 3D-QSAR studies for 1,3,4-oxadiazol-2-one derivatives as FAAH inhibitors. *International Journal of Molecular Sciences, [S.L.], 22*(11), 6108.
57. Zuardi, A. W. (2008). Cannabidiol: from an inactive cannabinoid to a drug with wide spectrum of action. *Revista Brasileira de Psiquiatria, 30*(3), 271–280.

Chapter 18
Kinase Inhibitors and Computer-Aided Drug Design Methods

Júlia Galvez Bulhões Pedreira and Pedro de Sena Murteira Pinheiro

Abstract The development of protein kinase inhibitors (PKI) is a prominent research field in medicinal chemistry. The pivotal role of protein kinases in all stages of cell life has made this one of the most sought-after targets for the treatment of a growing array of diseases. This protein family, however, offers many challenges, specially related to selectivity. From this perspective, both its function plurality and conserved structure are at many times a double-edged sword. Advances in computer-aided drug development (CADD) tools have immensely influenced protein kinase research, playing a determinant role in different hallmarks in PKI history. Nowadays, it is the utmost important tool for inhibitor design. In this chapter, we introduce the basics of kinase function, structure, and inhibitor design. We discuss its relevance as a drug target and introduce case studies where computer-aided design methods have helped identify or optimize new compounds.

Keywords Protein kinase · Kinase inhibitors · Structure-based virtual screening · Computational kinase screening

1 Introduction

The protein kinase (PK) superfamily is constituted by more than 500 members and has become, over the last 20 years, one of the most important drug targets of the twenty-first century [15, 16]. The elucidation of the kinome (Fig. 18.1), the protein kinase component of the human genome [46], has allowed for a better understanding of their biological functions and relationship to the initiation and/or progression of different diseases. Of the 518 genes encoding kinases that have been disclosed,

J. G. B. Pedreira (✉)
Pharmaceutical Department, University of Tübingen, Tübingen, Baden Württemberg, Germany

P. de Sena Murteira Pinheiro
LASSBio, Institute for Biomedical Sciences, Federal University of Rio de Janeiro, Rio de Janeiro, RJ, Brazil
e-mail: pedro.pinheiro@icb.ufrj.br

V. G. Maltarollo (ed.), *Computer-Aided and Machine Learning-Driven Drug Design*, Computer-Aided Drug Discovery and Design 3,
https://doi.org/10.1007/978-3-031-76718-0_18

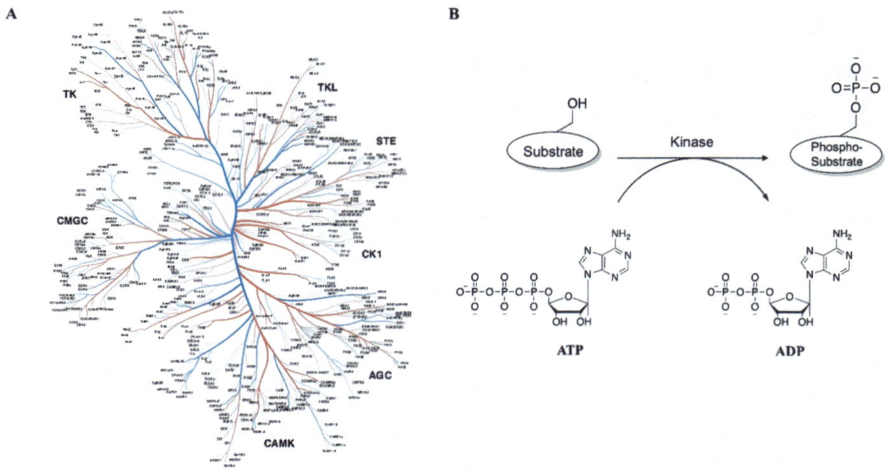

Fig. 18.1 (**a**) Kinome representation of all human kinases. (**b**) Schematic representation of the phosphoryl transfer catalyzed by the protein kinase to a hydroxylated amino acid of the substrate. (Kinome figure generated by Coral. Available for use at phanstiel-lab.med.unc.edu/Coral)

180 are related to human diseases, which is evidenced by the fact that since 1999, the FDA has approved 81 kinase inhibitors [54].

Protein kinases are enzymes responsible for catalyzing the transfer of the γ-phosphoryl group from adenosine triphosphate to a free hydroxyl group of an amino acid side chain, such as serine, threonine, or tyrosine. Therefore, depending on the targeted amino acid of the phosphoryl transfer, these enzymes can be divided by their catalytic activity into Ser/Thr kinases, Tyr kinases, and dual-specificity Ser/Thr/Tyr kinases. Alternatively, based on their primary sequence and structural features, PK are also divided into eight groups: AGC (protein kinase **A**, **G** and **C**), CaMK (calcium/calmodium-dependent kinases), CMGC (**c**yclic-dependet kinases, **M**AP kinases, **g**lycogen synthase kinases, and **c**asein kynases), TK (**t**yrosine **k**inases), STE (homologues of yeast **ste**rile 7), CK1 (**c**asein **k**inase), TKL (**t**yrosine **k**inase-**l**ike), and RCG (**r**eceptor guanylate **c**yclases). Also, there are the "atypical kinases," which is the group that does not share relevant sequence homology with the others, and lack even the canonical sequence motifs of the kinase catalytic binding site. Lastly, the "pseudokinases" have a typical kinase domain, but lack at least one conserved structural motifs important for catalytic activity [53].

The phosphoryl groups that were transferred to the hydroxylated side chains create a signaling mechanism that is responsible for the regulation of essential physiological processes, such as cell growth, differentiation, and cell death. Abnormalities in the basal dynamics of these functionalities are then related to various diseases. One of the first notable pathologies to be directly associated with a protein kinases' function was chronic myeloid leukemia (CML), which led to the landmark approval of the first synthetic PK inhibitor (PKI), imatinib [15, 22, 63].

Since then, many other pathologies and potential therapeutic uses have been linked to protein kinase irregularities [5]. Still, the development of a clinically useful PKI is relatively challenging. Since all kinases share the same endogenous cofactor, ATP, their binding site is fairly conserved, and so are many of the binding features required for interaction. On the other side, this means that selectivity is very challenging when developing PKI, and much of the work around drug development for these proteins involve exploring features that can differentiate between family members [32, 34, 41].

Parallel to kinase research, the last two decades were also marked by the dramatic increase of computational power and consequential development of computer-aided drug discovery (CADD). Not surprisingly, molecular modeling and virtual screening became some the main tools available for PKI drug design [50]. Some of the benefits of using CADD include reduction of costs in the drug discovery process, shortening of cycle time and overall improvement of efficiency of drug development [44]. Although these methods need to be used and interpreted with determined standards, the volume and range of data that can be generated provides relevant information that aids the medicinal chemist's decision-making. The interpretation of these results supported by a good understanding of kinase function and structure are a powerful approach to optimize the kinase inhibitor development process. There are several success histories of CADD in the approval of PKI, and nowadays, it is almost impossible to imagine a drug discovery case that did not benefit from computer-aided techniques.

In this chapter, we first introduce fundamentals of kinases and their inhibitors, then present some of the CADD tools and exemplify specific cases of PKI development that were benefited by these techniques.

2 Kinase Structure and Activation

As the investigation of the biological functionality of kinases developed, the advances on protein crystallography helped medicinal chemists to better visualize the protein's tertiary structure and catalytic domain. Since the early 1990s, PK's crystal structures have been elucidated, and revealed early on determinant features of the protein and the conservation of the catalytic site [37, 62, 72].

Protein kinases are structurally constituted by two lobes (Fig. 18.2a), C-terminal and N-terminal lobes, which are connected by a flexible loop (hinge). Between these two lobes, there is the catalytic site, where ATP binds mostly by hydrogen bonds to the backbone amino acids of the hinge. The C-terminal lobe is mainly made by α-helical. The N-terminal lobe is smaller and is made by five β-sheets (β1- β5) and one α- helix, known as the C-helix, or αC. Between β1 and β2 there is a glycine-rich flexible loop, known as P-loop or G-loop, and the hydrophobic residues of this loop contribute to the coordination of the phosphates of ATP. In the β6/β7 strand there is

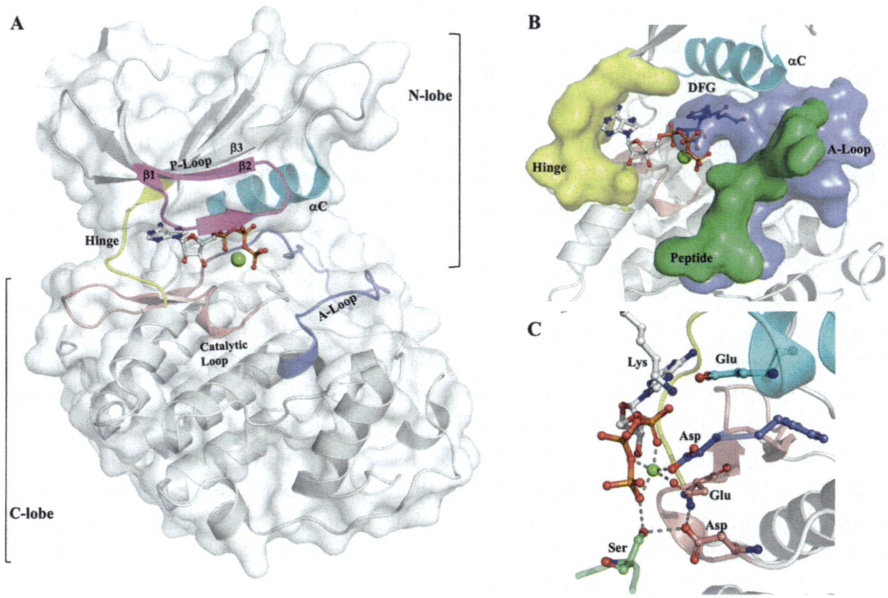

Fig. 18.2 Representation of CDK2 (PDB: 1QMZ). (**a**) Tertiary structure of the protein kinase. The hinge region, connecting both lobes is represented in yellow, P-loop in pink, β1–3 sheets are assigned, αC in blue, A-loop in purple, catalytic loop in light pink, Mg^{2+} ion as green sphere. ATP is represented in the cleft between lobes, the catalytic site. (**b**) Closer view of the catalytic site, including the region occupied by the target peptide in lime green and the A-loop with DFG in purple. (**c**) Different view of the catalytic site, with the interactions between the Mg^{2+} ion, ATP and key amino acids highlighted (target amino acid of the peptide is Ser)

the important catalytic loop, which contains a highly conserved Asp residue. Finally, between β8 and β9 there is a conserved triad of amino acids, known as the DFG (Asp-Phe-Gly), marking the beginning of the activation loop (A-loop) [4, 25] (Fig. 18.2b).

The ATP binding site makes significant structural changes upon activation, i.e., binding of ATP for phosphoryl-transfer. Between the β3 sheet and the αC, a salt bridge is formed by Lys, Glu, and ATP-phosphate groups, by means of the flexible αC approximating to β3. On the DFG triad, the conserved Asp residue must be oriented towards the catalytic cleft for coordination with the Mg^{2+} ion. Finally, the A-loop positions the hydroxyl group of the amino acid substrate, and the conserved Asp acts as a base acceptor for the proton transfer upon phosphoryl move to the hydroxylated residue of the peptide substrate [53] (Fig. 18.2c).

In the above described re-orientation of the αC into the proximity of the A-loop, the Asp residue of the DFG is initially directed outside the catalytic site (DFG-out, or inactive), and is then oriented to the catalytic site, the so-called DFG-in (or active) conformation. These two states are often referred as activation stated of a PK.

Another definition of activation state in kinases can be made by the analysis of hydrophobic spines which connect the N and C terminal lobes (Fig. 18.3). The

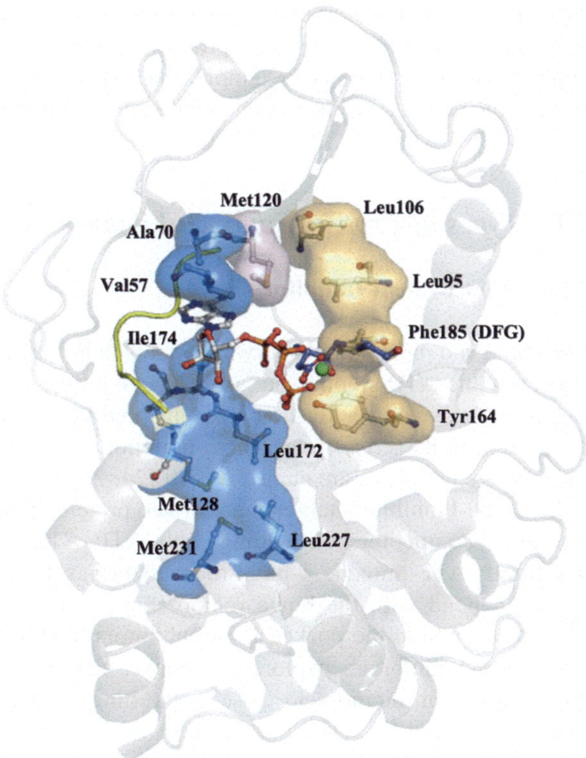

Fig. 18.3 Regulatory spine (R-spine, orange) and catalytic spine (C-spine, blue) of PKA (PDB: 3GJQ) in its active state, co-crystallized with ATP. Hinge is highlighted in yellow, DFG residues highlighted in purple and bridging methionine residue highlighted in light pink surface. (Adapted from [39])

spines, namely the regulatory spine (R) and the catalytic spine (C), are dynamically assembled by the alignment of hydrophobic amino acids. Upon activation, the C-spine of the kinase is formed by the alignment of amino acids of both lobes with the adenine ring of the ATP. The R-spine, on the other hand, is made by four amino acids, including the Phe residue of the DFG. These two spines are finally connected to each other by one amino acid bridging residue of the N-lobe. Such structural features can only be achieved with ATP and the correct orientation of the Phe, and therefore are also hallmarks of the protein's activation [25].

Unveiling these features was crucial to understand kinase catalytic activity. Still, when developing an inhibitor, the primary goal is to mimic some of ATP's binding and additionally be able to explore other regions in which the synthetic molecule can make new interactions, aiming on selectivity. For example, many PKI are designed from the starting point of an adenine-mimetic core structure, targeting the hinge interactions similar to ATP, which are crucial for the proper bonding of a ligand to the catalytic site. Other features can also be taken into consideration. In this sense, the catalytic cleft itself can be mapped into some regions with common characteristics: ATP-occupied regions, i.e. adenine binding region, sugar pocket and phosphate binding region; and non-ATP occupied regions, namely hydrophobic regions I and II, and deep pockets. These latter regions are more structurally diverse throughout the kinome, as they are not occupied by the shared endogenous ligand of all kinases, the ATP. Thus, they provide a broader range of interactions from an inhibitor point of view. For example, the possibility of exploring different hydrophobic interactions, or interactions with target specific amino acids [40].

2.1 Types of Protein Kinase Inhibitors

2.1.1 Noncovalent Inhibitors

The initial and more straight-forward approach in the development of PKI is simply to occupy the ATP binding site, and explore all the previously mentioned regions. Due to PK's dynamic activation state, the binding site can be analysed in two ways: the protein in its active state, where inhibitors are named type-I, or in its inactive state, where inhibitors are named type-II [24, 25, 71].

The type-I inhibitors target the activated protein at the DFG-in conformation and are direct competitors of binding with ATP. The type-I inhibitors tend to be structurally ATP-mimetic, with nitrogenated rings that can establish a single or dual hydrogen bond with the hinge region (Fig. 18.4a). Because of the similarity of interactions within the pocket, these inhibitors tend to be less selective. However, many type-I inhibitors explore the pocket that is located behind the so-called "gatekeeper," an amino acid close to the hinge that is flexible enough to open a space unexplored by ATP, and therefore interesting for small molecule inhibitors. One important advantage is that type-I inhibitors form stable complexes with the active form of the protein and are useful in obtaining crystal structures of PKs and

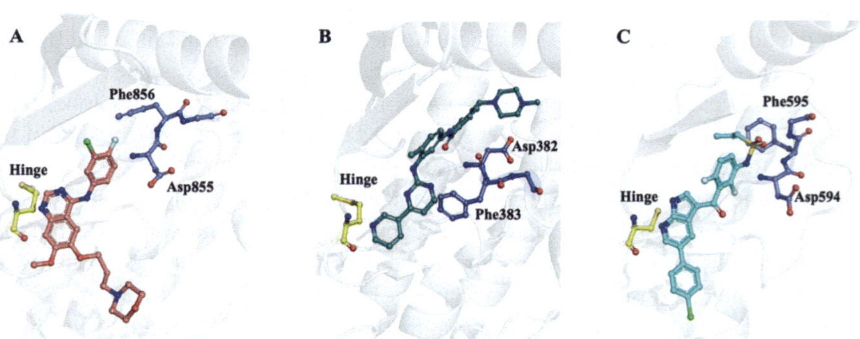

Fig. 18.4 Examples of different types of inhibitors and orientations of DFG triad. (**a**) Gefitinib in the catalytic site of EGFR as a type-I inhibitor and the DFG-in conformation. PDB: 2ITY. (**b**) Imatinib, type-II inhibitor, occupying the deep pocket of BCR-ABL, with DFG-out conformation. PDB: 2HYY. (**c**) Vemurafenib, type-1,5 inhibitor, and the representation of the induced DFG-in conformation in BRAF. PDB: 3OG7. Yellow highlights the hinge and purple highlights the DFG triad

performing functional studies [24]. Examples of type-I PKI are gefitinib, erlotinib, and dasatinib.

In the type-II inhibition, DFG is in the "out" conformation, and this opens a new cleft in the binding pocket (Fig. 18.4b). This deep-pocket is widely explored by type-II inhibitors to obtain better selectivity. This means that these inhibitors are usually elongated molecules, with hydrophilic linkers in order to reach into the pocket. Imatinib itself is a type-II inhibitor [71].

An intermediate state between these types is also possible. As the DFG orientation is very dynamic, and is not solely determined by the activation state, some inhibitors have been classified as type-1,5 [4, 24]. These are molecules that bind to the overall inactive state of the protein, but are able to interact with the Phe residue of the DFG, inducing a DFG-in conformation (commonly of the active state) (Fig. 18.4c). This can be exemplified by vemurafenib.

Lastly, the type-III inhibitors comprehend a vast and diverse group of molecules that bind to allosteric sites of adjacent and non-adjacent pockets of the binding site. They are non-ATP competitive inhibitors, not binding to the hinge. These are the most selective types, as they explore pockets and regulatory mechanisms unique to each kinase. The rational development of these inhibitors are, however, very challenging, as the identification of allosteric pockets is difficult [24, 71]. One example of a type-III inhibitor is trametinib, a clinically approved MEK inhibitor which binds to a pocket adjacent to the catalytic site (Fig. 18.5a) [35].

On the other hand, asciminib, an allosteric BCR-ABL inhibitor clinically approved, is a notable example of CADD that was successful in type-III inhibitors.

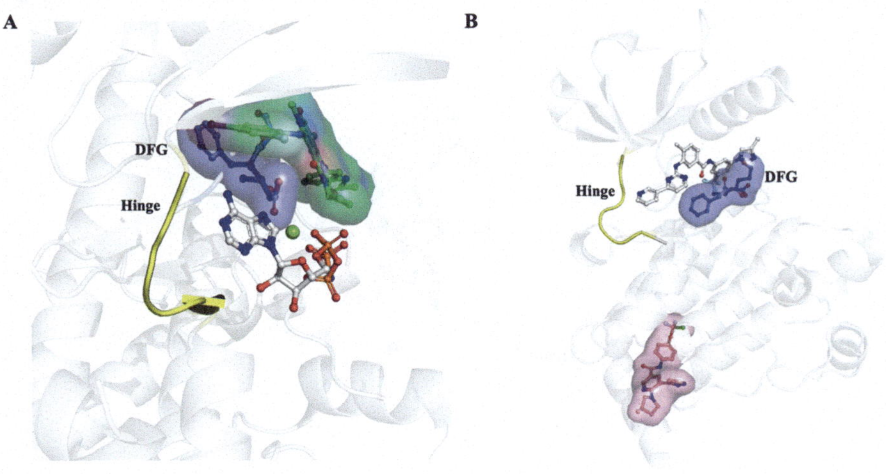

Fig. 18.5 (**a**) Structure of the MEK, in its active state, co-crystallized with ATP and the allosteric inhibitor trametinib, highlighted with green surface, in the pocket directly next to the DFG triad (PDB: 7JUR). (**b**) Crystal structure of BCR-ABL, in its inactive state, co-crystallized with type-II inhibitor nilotinib at the catalytic site and asciminib (highlighted in pink surface) in a non-adjacent allosteric site (PDB: 5MO4). In both figures, hinge is highlighted in yellow, DFG is highlighted in purple, and Mg^{2+} in bright green

The compound, which binds in the distant myristoyl pocket, was designed from a combination of fragments identified by NMR-based conformational assay and optimized by crystallography and molecular modeling (Fig. 18.5b) [55, 67].

2.1.2 Covalent Inhibitors

In short, covalent inhibitors are molecules that form a covalent bond to the target protein. Such bond is usually targeted through nucleophilic attack by Cys or Lys. These inhibitors, however, must be designed considering that initially there is a non-covalent binding of the molecule in the binding site, and then the covalent bond is formed. For such, targeted covalent inhibitors (TCI) are designed using warheads, which are in many cases Michael acceptors, such as acrylamide [29, 61]. An example of a covalent EGFR inhibitor targeting a cysteine is afatinib (**1**, Fig. 18.6)

In the past, covalent inhibitors would have their mechanism of action discovered after clinical use, and were generally considered to be toxic and/or cause problematic side effects. Nevertheless, in the last decades the potential of such molecules has been more explored, especially for improving selectivity and duration of action. In fact, the rational design of TCI by the tuning of warhead reactivity has allowed for the development of reversible covalent inhibitors, a strategy which has opened many possibilities for exploring this class of PKI's [6, 57].

The identification of potentially targetable nucleophilic amino acids in the binding site of PKs has facilitated the rational design of covalent inhibitors [12, 45]. The cysteinome, which is the "targetable cysteine" component of the kinome contains around 200 kinases, many of which have not yet been targeted for their cysteine

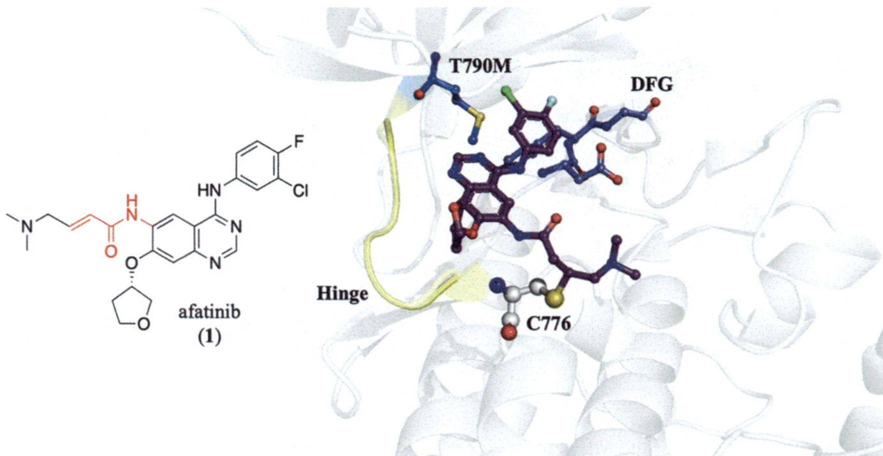

Fig. 18.6 Example of covalent inhibitor afatinib (**1**). Chemical structure represented on the left, with the warhead highlighted in red; and co-crystallized molecule in EGFR covalently bonded to a cysteine on the right. PDB: 4G5P. Hinge region is highlighted in yellow, DFG triad in highlighted in purple, gatekeeper mutated to methionine highlighted in blue

nucleophilic attack potential or are still not fully studied as drug targets [45]. Also, different amino acids to cysteine have been increasingly investigated as alternative targets. Therefore, covalent inhibitors are an important and imminent strategy for kinase-targeted therapy.

3 Relevance as Drug Targets

The significance of protein kinases as drug targets was first recognized in the 1980s, based on the discovery that the oncogene SRC had kinase activity. This was a hallmark into the research of molecular genetics of cancer, and prompted immediate interest in PK for the treatment of this disease [47, 48]. Although efforts were made throughout the following decade to develop small molecule inhibitors, this was viewed as a particular challenge, due to the known conservation of the ATP-binding pocket. Finally, in the beginning of the 1990s, the Rho-associated coiled-coil-containing kinase (ROCK1 and 2) inhibitor Fasudil was approved in Japan, for the treatment of cerebral vasospasm [58]. Later after that, Sirolimus (natural product rapamycin) was approved in the United States, targeting mTOR for the prevention of organ rejection [7].

The first approval to make history, however, was imatinib (**2**, Fig. 18.7), as this was the first rationally designed small molecule to target a protein kinase, namely BCR-ABL. Imatinib was approved by the FDA in 2001 for the treatment of CML, a previously rapidly fatal and rare type of leukemia.

Imatinib also marks the first PKI developed story using Structure-based Drug Design (SBDD, Sect. 5.1.2). In short, an initial lead compound was identified though HTS as protein kinase C (PKC) inhibitor. Optimization studies lead to the

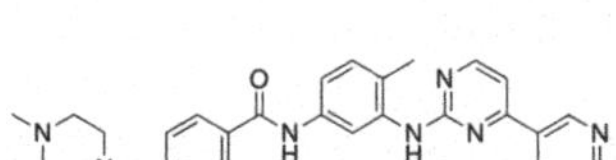

imatinib
(**2**)

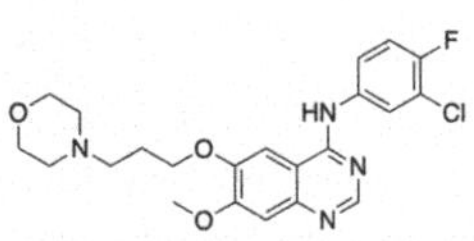

gefitinib
(**3**)

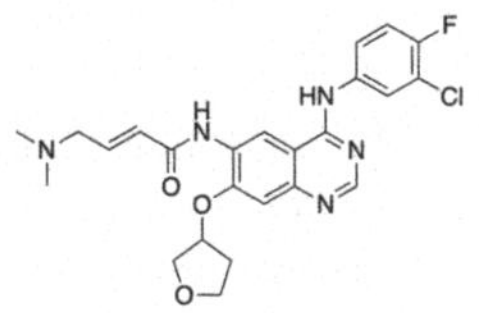

afatinib
(**1**)

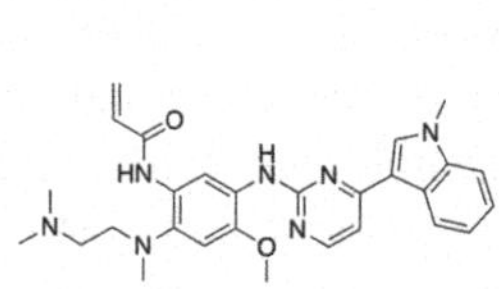

osimertinib
(**4**)

Fig. 18.7 Chemical structure of imatinib (**2**), the first protein kinase inhibitor, and the three generations of inhibitors of EGFR: gefitinib (**3**), afatinib (**1**), and osimertinib (**4**)

identification of compounds with activity for the BCR-ABL kinase. This kinase is a result of an abnormal fusion between Abl and Bcr genes that result in the Philadelphia chromosome, which is found in 90% of CML cases [1]. During the structure-activity relationship analysis (SAR), it was identified that a methyl substituent resulted in preferred activity for the targeted BCR-ABL kinase. Further development was made to improve pharmacokinetics, eventually leading to the identification of STI571, or imatinib [11]. The success of imatinib paved the way for new PKIs, and in the following years since, PKI research became a great part of the pharmaceutical industry.

In the year after imatinib two new inhibitors were approved, gefitinib (**3**, Fig. 18.7) and erlotinib, for the treatment of Non-Small Cell Lung Cancer (NSCLC). These drugs target epidermal growth factor receptor (EGFR), a kinase that was correlated to the development and progression of solid tumors in the lung. Both were also viewed as an initial game-changing successes, prolonging life expectancy by several months in terminal lung cancer patients. However, very soon after their approval, several mutations in EGFR began to be identified in patients. The activating mutation L858R, which seemed to initially increase the sensitivity to treatment [28], driven by a number of compensatory mechanisms, led to resistance to treatment [69]. This demanded that a second generation of inhibitors would be developed, like afatinib (**1**), which binds covalently to the protein [43]. However, once again, an acquired mutation (T790M) on the gatekeeper residue, decreased significantly the affinity of both first and second generation treatments [42]. A third generation was then developed specifically to target the gatekeeper mutant T790M and spare the binding on wild-type EGFR [2]. This generation of inhibitors is represented by osimertinib (**4**), approved un 2017 [18, 31, 38]. Not surprisingly though, through a series of both EGFR-dependent and EGFR-independent mechanisms, combined with the observed C578S point mutation of the targeted amino acid for covalent bonding, this third generation has been surpassed. Fourth generation inhibitors and new combinatory treatments are in clinical trials [42].

These generations of EGFR also represent historically important cases of SBDD, as the molecular design was underpinned on knowledge regarding hinge interactions, targeted covalent bond in the catalytic site and gatekeeper bulkiness. At the time, this knowledge seemed very promising, and sufficient for and optimized PKI. However, the EGFR story also offers insights of the dynamicity of kinases. The work of Wood and co-workers, a classic example, demonstrated that although gefitinib, erlotinib and lapatinib (EGFR inhibitor approved in 2007 by the FDA), quinazoline-based structures with predicted similar binding mode and similar IC_{50} values, had very different cell activity [66]. The underlying difference was later determined to be the compounds' target residence time, and that lapatinib had structural features that induced an inactive-like conformation (Fig. 18.8).

Although EGFR targeted therapy is still used in the clinic, the lesson has been learned that tumors are inherently adaptable and develop several mechanisms to evade therapeutic pressure. The observed genetic instability of tumor targets requires monitoring of oncogene addiction for a treatment to be successful. Tracking specific

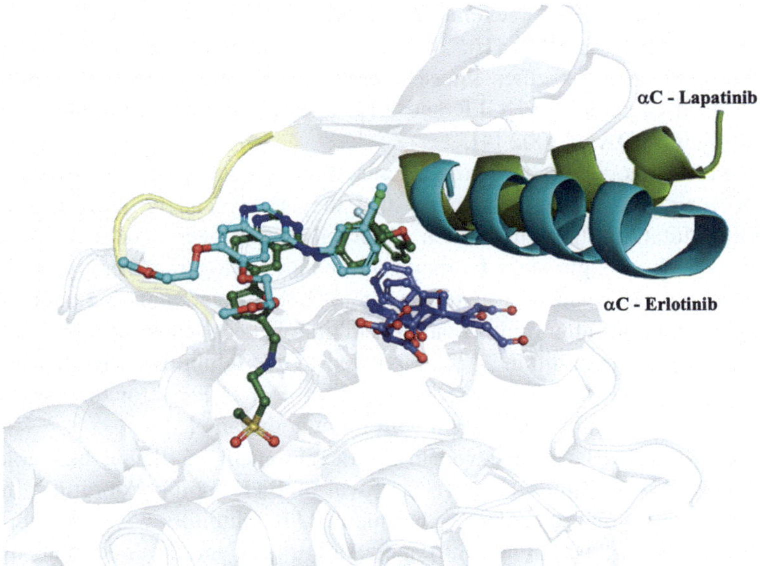

Fig. 18.8 Overlaid structures of erlotinib (cyan blue) and lapatinib (dark green) demonstrating the difference in the C-Helix orientation. In lapatinib, C-helix adopts an open, inactive-like conformation, although the DFG is positioned in the active conformation. Erlotinib co-crystallized in EGFR (PDB: 1M17). Lapatinib co-crystallized in EGFR (PDB: 1XKK). Hinge of both structures is highlighted in yellow, and DFG of both structures (in DFG-in conformation) highlight in purple

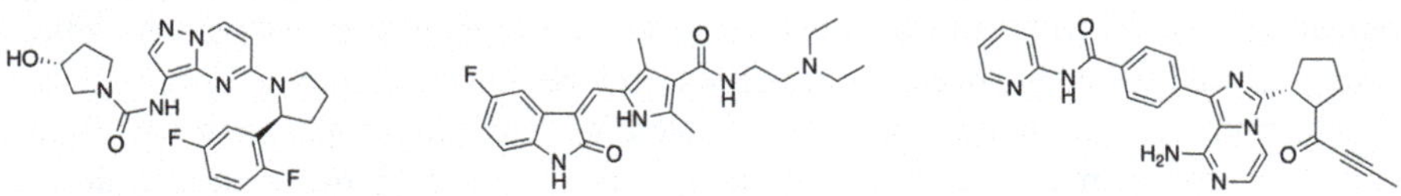

larotrectinib **(5)** sunitinib **(6)** acalabrutinib **(7)**

Fig. 18.9 Chemical structure of the NTRK-gene-specific inhibitor larotrectinib (**5**), VEGFR inhibitor sunitinib (**6**), and the covalent BTK inhibitor for hematological cancer acalabrutinib (**7**)

biomarkers is utterly relevant. Other recent strategies in PKI development are a response to that. For example, the recent approval of larotrectinib (**5**, Fig. 18.9) in 2018, which is the first PKI designed specifically to target tumors bearing the neurotrophic tyrosine receptor (NTRK) gene [21, 27]. The (NTRK) gene fusion is a biomarker found in over 25 types of cancer, and so the NTRK-specific inhibitor is considered "tumour-agnostic," and a target-therapy treatment. Targeting specific biomolecular characteristics rather than the protein of a single tumor tissue is currently a step upwards in the research of kinase inhibitors [64].

Alternatively to the strategy of developing inhibitors of PK related to growth of the tumor itself, is the development of inhibitors targeting angiogenesis. The strategy held promise, as vascularization is a critical step in tumor development, providing oxygen and nutrients [3, 30]. The main PKs to be targeted were VEGFR, PDGFR, KIT and FGFR, which participate in vessels formation around solid tumors. This effort led to the approval of several drugs, like sunitinib (**6**), sorafenib, pazopanib, and cabozantinib (Fig. 18.9) [16, 30]. It was also later identified that previously approved inhibitors for other kinases, have angiogenic kinases as off-targets. Imatinib itself can inhibit PDGFR and KIT, and was granted a second indication approval by the FDA in 2002 [4].

However, angiogenesis-targeting agents have shown poor selectivity, which added to their possible toxicity, show that they rarely can be used as a single-treatment. Mutations have also been observed in angiogenic targets. Moreover, there are a lack of significant biomarkers for anti-angiogenic therapy, making it difficult to decide when to use treatment [51]. However, anti-angiogenic strategy remains a sought-after option, and new inhibitors keep reaching approval status, the latest being pemigatinib in 2022, a mutated-FGFR inhibitor for the treatment of cholangiocarcinoma [49]. The perceived current best strategy for the use of anti-angiogenic agents is the combination with immunotherapy [3].

Progress has also been achieved in the development of PKI of hematological malignances. For example, the 2016 approval of BTK inhibitor ibrutinib, for the treatment of cell lymphoma. Two subsequent BTK inhibitors were also approved after that, acalabrutinib (**7**, Fig. 18.8) and zanubrutinib, and all three molecules act covalently, which represented a major resurgence of the relevance of covalent inhibitors for PK [4, 10, 20].

The pathway of protein kinase inhibitors for the treatment of cancer has been one of many lessons and great improvement. The current kinase inhibition therapies are still among the best option for cancer treatment. New tools that can now be used in the clinic, for example gene sequencing in the search of biomarkers, will likely only broaden the possibilities of treatment. However, throughout the years since its first approval, PKI have also been a topic of research beyond oncology.

The unresolved or inappropriately activated inflammation response that becomes chronic is the primary cause of various diseases, including rheumatoid arthritis, inflammatory bowel diseases, Crohn's disease, and so on. In the family of kinases, owing to their relevance in various cellular events, are related to the development of these chronic inflammatory diseases. Thus, PKI have also been viewed as a treatment option. Nonetheless, the same challenged remain, the primary one being selectivity of these inhibitors, since for such treatments, long term side effects could not be tolerated by patients [16, 70].

The primary understanding that some kinases participate in key inflammatory cascades of signaling pathways of multiple cytokines receptors, brought attention to this family of proteins as a therapeutic target. One of the most important kinases in inflammatory treatment is Janus Kinase (JAK), and nine PKIs have been approved targeting JAK isoforms. The first FDA-approved inhibitor for a chronic disease was ruxolitinib (**8**, Fig. 18.10) in 2011, a JAK1 and JAK2 inhibitor for the treatment of myelofibrosis [4]. Following, other JAK inhibitors were approved, like tofacitinib and baricitinib, for the treatment of rheumatoid arthritis (RA). Other JAK inhibitors continue to be evaluated for different indications, and currently, the pan-JAK inhibitor delgocitinib is being evaluated for the treatment of skin disorder of atopic dermatitis [4, 16, 70].

Fig. 18.10 Chemical structure of PKI approved for chronic diseases, ruxolitinib (**8**) and fostamanib (**9**)

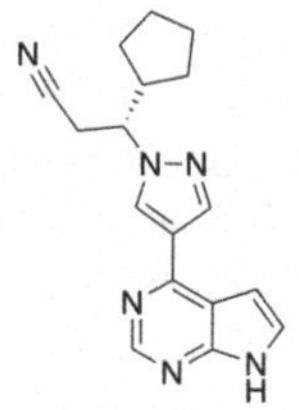

ruxolitinib
(**8**)

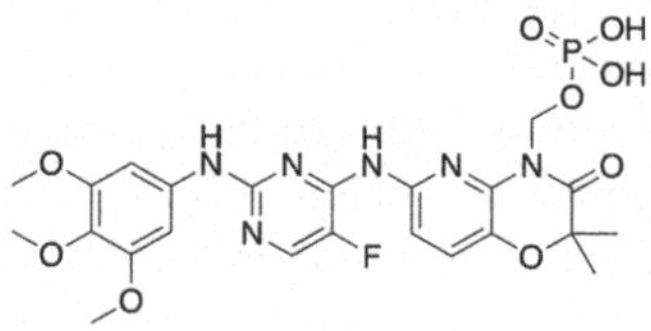

fostamanib
(**9**)

Another relevant kinase for treatment of chronic diseases has been SYK. Its activation is related to a number of autoimmune disorders, allergic responses and even hematological diseases. In 2018, fostamanib (**9**, Fig. 18.10), a SYK inhibitor, was approved for the treatment of chronic immune thrombocytopenia [4, 70]. However, since then, a comprehensive pharmacological profile of fostamanib showed a significant level of off-targets at therapeutically relevant concentrations, once again demonstrating the critical challenge of selectivity in developing PKI, especially for chronic treatment.

Nevertheless, the development of PK targeted therapies for chronic treatment is still considered to be a golden opportunity, due to the diversity of immunological pathways. Should be noted that there is not only the selectivity issue, but also many kinases involved in inflammation participate on non-inflammatory pathways, which can result in unknown side effects for patients. Nonetheless, the number of already approved treatments is a strong indication there is much more to be done with PKI beyond oncology. Additionally, previously developed PKI continue to be approved for a number of, like VEGFR inhibitor nintedanib for idiopathic pulmonary fibrosis, after failing as NSCLC treatment [4, 52].

4 Kinase Inhibitors as Chemical Probes

Even though the success of protein kinase inhibitors in the clinic is undeniably impressive, the obstacles in the development process have been outlined. Considering the investment that the Pharmaceutical Industry has made in PKI research, and the amount of inhibitors entering pre-clinical and clinical phases every year, the bottleneck for clinical approval is considerable [8]. This can be associated to a number of things, including specially lack of preclinical target validation and insufficient understanding of the molecular mechanism of the target related to the pathological condition. Generally, the process of target validation goes through genetic knockdown or knockout models. This approach, however, many times fail to correctly characterize the relationship between target and phenotype [56]. In this regard, chemical probes have become increasingly more valuable in target validation.

Chemical probes are small molecules that are able to selectively modulate functions of a protein, which allows the better elucidation of the molecular mechanism of the target and respective influence both in vivo and in vitro. According to guidelines, chemical probes should have a biochemical potency against the kinase target below 100 nM (at [ATP] = k_m), cellular potency below 1 μM and >30-fold selectivity against other kinases (except for the closely related proteins). These criteria result in a great tool to establish a link target-biological response link [9].

During the years, some kinase inhibitors have been wrongfully described as chemical probes, for example the PI3K inhibitor LY294002, first described in 1994. Since then, it has been determined that it actually inhibited many other proteins.

An investigation initially done on commercially available kinase inhibitors in 2000 led to the discovery that many of the compounds that were considered selective, were active in a number of other kinases [19]. It was then realized that selectivity should not be measured only on closely related protein, which prompted guidelines to be considered [14]. Nowadays, there are many confirmed chemical probes as kinase inhibitors (Fig. 18.11), and a comprehensive database can be found on the website chemicalprobes.org.

It is also worth mentioning that chemical probes have an important role in identifying the potential of previously overlooked kinases. The majority of inhibitors that enter clinical phase are targeting kinases that already have an approved drug, showing that medicinal chemists tend to trust on the clinical proof of concept. Considering that around 25% of the kinome is still not explored, the therapeutic potential of these targets is considerable [26].

Finally, chemical probes are more easily developed, within the correct guidelines relating to selectivity, when there are well characterized and clear crystal structures. From reliable crystallography data, rational molecular design of chemical probes is

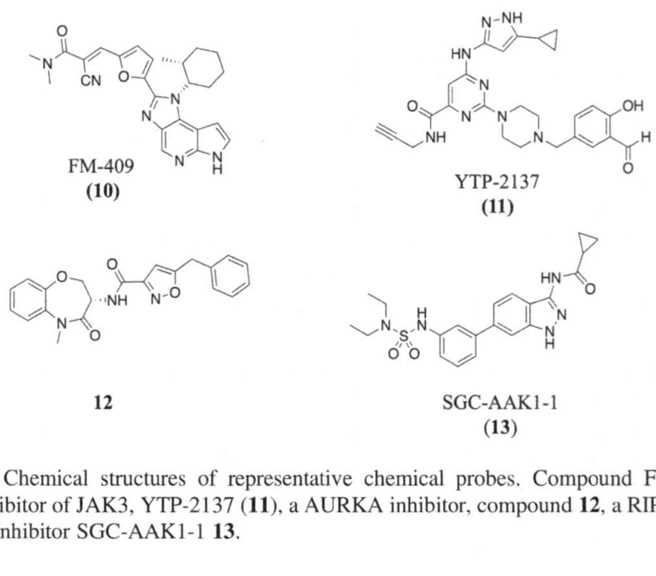

Fig. 18.11 Chemical structures of representative chemical probes. Compound FM-409 (**10**), covalent inhibitor of JAK3, YTP-2137 (**11**), a AURKA inhibitor, compound **12**, a RIPK1 inhibitor and AAK1 inhibitor SGC-AAK1-1 **13**.

immensely benefited by computational-aided techniques. Currently, there are thousands of kinase crystal structures available on the Protein Data Bank (PDB), which demonstrates that there is still a great space to be explored.

5 Computational-Aided Methods and Machine Learning

5.1 Computational-Aided and Machine Learning Case Studies for the Discovery of Kinase Inhibitors

5.1.1 Virtual Screening for the Identification of Type I TGFβ Receptor Kinase

One early interesting example of ligand-based virtual screening approach focused on the identification of kinase inhibitors are related to the discovery of type-I TGFβ Receptor Kinase (TβRI) inhibitors. The starting point for this virtual screen was SB203580 (**14**), which has a reported IC_{50} of 30 μM against TβRI [23]. SB203580 (**14**) was used to build a 3D query using shape and pharmacophore descriptors through its bioactive conformation with p38 (PDB: 1A9U) [65]. The query was used to screen the Catalyst database (~200,000 compounds), which led to the identification of **15**, a potent inhibitor of TβRI with an IC_{50} of 27 nM [60] (Fig. 18.12).

5.1.2 Virtual Screening for the Identification of Casein Kinase 1 Delta (CK1δ) Inhibitors

The discovery of casein kinase 1 delta (CK1δ) inhibitors through structure-based virtual screening (for more details on this strategy, please see Chap. 11) was

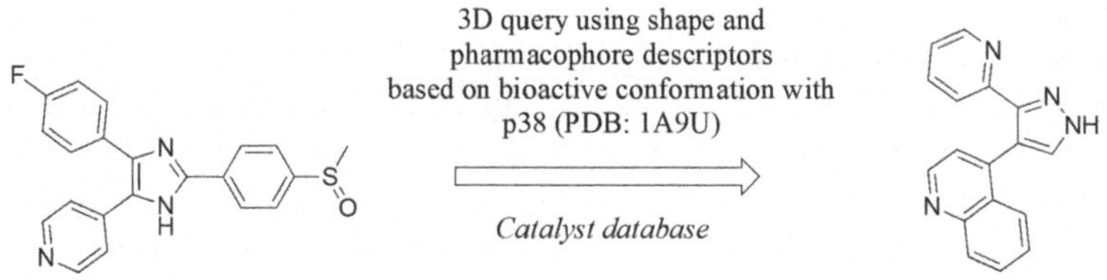

SB203580 (**14**)
TbRI IC_{50} = 30,000 nM

15
TbRI IC_{50} = 27 nM

Fig. 18.12 Identification of the TβRI inhibitor **15** through a ligand-based virtual screening approach

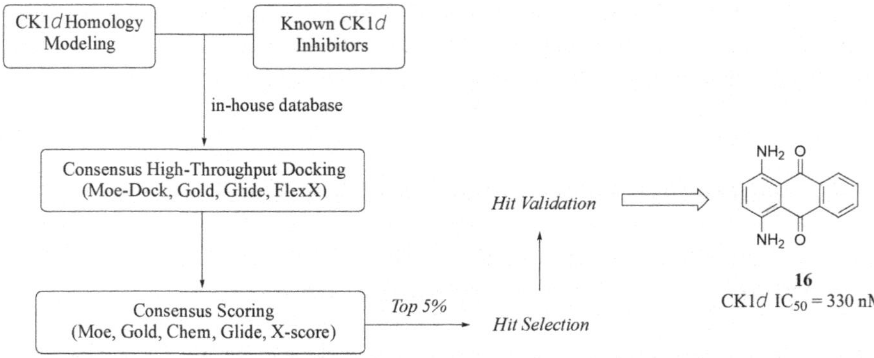

16
CK1*d* $IC_{50} = 330$ nM

Fig. 18.13 Identification of the CK1δ inhibitor **16** using a consensus high-throughput docking strategy

conducted for the identification of small molecules targeting the ATP-binding site [17]. Since no crystal structure was available for the human CK1δ in the time that the study was performed, a homology modeling approach was carried out to obtain a suitable three-dimensional model of the CK1δ catalytic subunit. A consensus high-throughput docking strategy [68] was applied using four different docking packages (MOE-Dock, Glide, Gold, and FlexX—and explored five scoring functions—MOEScore, GlideScore, Gold-Score, ChemScore, and Xscore) to screen an in-house database with ~four million compound. Those that independently appeared in the top 5% of the database when ranked by every scoring function were visually analyzed and evaluated for CK1 inhibition, leading to the identification of **16**, with an IC_{50} of 330 nM [17] (Fig. 18.13).

5.1.3 Virtual Screening for the Identification of Aurora a Kinase Inhibitors

A ligand-based virtual screening using xLOS [59], a 3D shape and pharmacophore similarity algorithm, were used for the identification of selective Aurora Kinase A inhibitors [36]. Sixteen different Aurora Kinase A inhibitors were used as query molecules and were screened against ZINC database [33]. 437 compounds were selected for Aurora Kinase A inhibition assay taking into account scaffold diversity, potential for derivatization, and absence of reactive functional groups, resulting in the identification 15 hits (3.4% hit rate). An additional 205 derivatives of the initially identified 15 hits were selected by further similarity searching using xLOS, resulting in the identification of additional 9 hits. Among those identified hits, the thiazolidinone derivative **17** ($IC_{50} = 2$ μM) were one of those selected to further optimization. The SAR analysis revealed that substitution on 2-pyridyl position could improve hinge-binding interactions, which led to the discovery of analogue **18**, a selective type-II Aurora Kinase Inhibitor with an IC_{50} of 2 nM (Fig. 18.14).

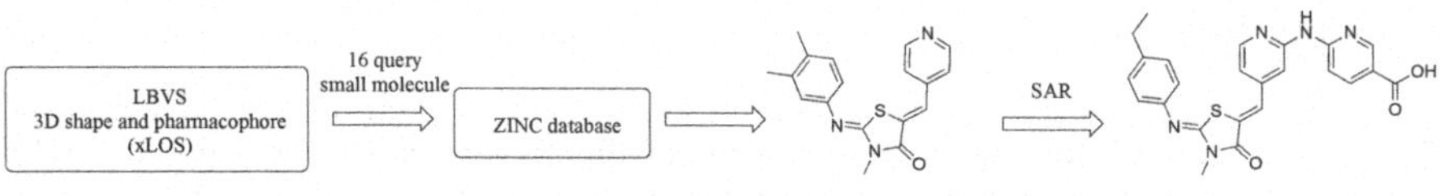

17
Aurora Kinase A $IC_{50} = 2000$ nM

18
Aurora Kinase A $IC_{50} = 2$ nM

Fig. 18.14 Identification of the Aurora Kinase A inhibitor **17** by a 3D shape and pharmacophore ligand-based virtual screening (xLOS) and its optimization for the discovery of **18**

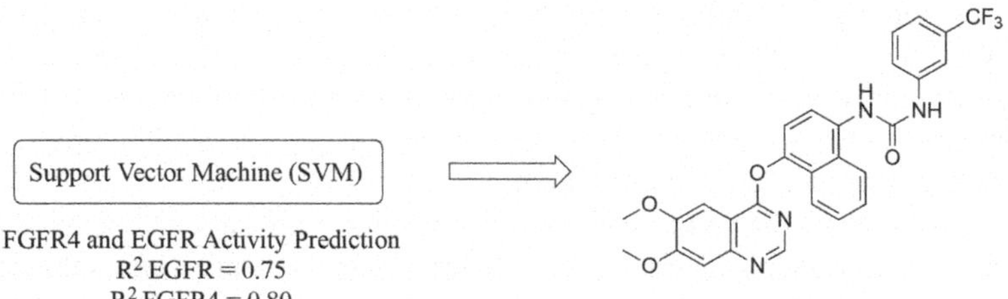

FGFR4 and EGFR Activity Prediction
R^2 EGFR = 0.75
R^2 FGFR4 = 0.80

19
EGFR $IC_{50} = 83$ nM
FGFR4 $IC_{50} = 86$ nM

Fig. 18.15 Machine learning approach using support vector machine for the identification of **19**, a dual EGFR and FGFR4 inhibitor

5.1.4 Discovery of Dual FGFR4 and EGFR Inhibitors by Machine Learning

Computational and experimental approaches were employed in the quest for dual inhibitors targeting FGFR4 and EGFR [13]. Using a dataset comprising 843 compounds for FGFR4 and 5088 for EGFR, a Quantitative Structure-Property Relationship (QSPR) analysis was conducted (please, see more details on this method on Chap. 6). This analysis aimed to predict the activity levels of FGFR4 and EGFR. Employing four distinct machine learning methods (namely, support vector machine (SVM), random forest (RF), gradient boost regression tree (GBRT), and XGBoost (XGB), for more details, please see Chap. 4) models were constructed. These models utilized the most pertinent features selected through the mutual information algorithm. Among these models, SVM exhibited superior performance for FGFR4 and EGFR, boasting R^2 test scores of 0.80 and 0.75, respectively. This highlighted its robustness and reliability, enabling the prediction of compound activities from an internal database.

Ultimately, compound **19** emerged as a promising candidate, demonstrating inhibitory activity against FGFR4 ($IC_{50} = 86.2$ nM) and EGFR ($IC_{50} = 83.9$ nM) kinases, showing itself as a dual inhibitor with balanced activities for both targets (Fig. 18.15).

6 Conclusions

Protein kinase inhibitors have become a major therapeutic class, and since its early years, with imatinib and gefitinib, much has been learned about this target. It is a large protein family, with important roles in all stages of cell cycle, which gives it an instant significance in the context medicinal chemistry. However, for the same reasons, it is a great challenge to achieve a selective inhibitor, with little side-effects.

Understanding the protein structure and mechanisms of activation is very important to develop a good inhibitor, with the desired selectivity and potency. Different approaches, for example reversible, irreversible, and allosteric inhibitors, need to be tackled differently in the drug design process. Also, specific diseases will prompt different approaches for the design, which can be aimed for reversible, irreversible, different target-residence times, and so on.

The use of CADD for PKI research has been an incredible asset to develop better compounds, in a shorter time frame. Nonetheless, there are many aspects that should be taken into consideration for a proper use of the techniques, for example, good quality protein structures (X-ray, Cryo-EM, or homology model), proper preparation of the structure and finally the limits of reliance on docking results and MD simulations alone. CADD can, to some extent, anticipate some experimental results, support comprehensive SAR and improve screening efficacy.

For this reason, the advances in computer-aided techniques in drug design can be an extremely powerful tool in this process. Likewise, knowing which technique fits better and when to use it during the drug design process can offer the medicinal chemist many relevant alternatives. The combination of rational design and in silico techniques now work side by side to provide better and safer drugs.

References

1. An, X., Tiwari, A. K., Sun, Y., Ding, P.-R., Ashby, C. R., & Chen, Z.-S. (2010). BCR-ABL tyrosine kinase inhibitors in the treatment of Philadelphia chromosome positive chronic myeloid leukemia: A review. *Leukemia Research, 34*, 1255–1268.
2. Andrews Wright, N. M., & Goss, G. D. (2019). Third-generation epidermal growth factor receptor tyrosine kinase inhibitors for the treatment of non-small cell lung cancer. *Translational Lung Cancer Research, 8*, S247–S264.
3. Ansari, M. J., Bokov, D., Markov, A., Jalil, A. T., Shalaby, M. N., Suksatan, W., Chupradit, S., AL-Ghamdi, H. S., Shomali, N., Zamani, A., Mohammadi, A., & Dadashpour, M. (2022). Cancer combination therapies by angiogenesis inhibitors; a comprehensive review. *Cell Communication and Signaling: CCS, 20*, 49.
4. Attwood, M. M., Fabbro, D., Sokolov, A. V., Knapp, S., & Schiöth, H. B. (2021). Trends in kinase drug discovery: Targets, indications and inhibitor design. *Nature Reviews. Drug Discovery, 20*, 839–861.
5. Ayala-Aguilera, C. C., Valero, T., Lorente-Macías, Á., Baillache, D. J., Croke, S., & Unciti-Broceta, A. (2022). Small molecule kinase inhibitor drugs (1995–2021): Medical indication, pharmacology, and synthesis. *Journal of Medicinal Chemistry, 65*, 1047–1131.

6. Barf, T., & Kaptein, A. (2012). Irreversible protein kinase inhibitors: Balancing the benefits and risks. *Journal of Medicinal Chemistry, 55*, 6243–6262.
7. Brown, E. J., Albers, M. W., Bum Shin, T., Ichikawa, K., Keith, C. T., Lane, W. S., & Schreiber, S. L. (1994). A mammalian protein targeted by G1-arresting rapamycin–receptor complex. *Nature, 369*, 756–758.
8. Bunnage, M. E. (2011). Getting pharmaceutical R&D back on target. *Nature Chemical Biology, 7*, 335–339.
9. Bunnage, M. E., Chekler, E. L. P., & Jones, L. H. (2013). Target validation using chemical probes. *Nature Chemical Biology, 9*, 195–199.
10. Burger, J. A. (2019). Bruton tyrosine kinase inhibitors: Present and future. *Cancer Journal, 25*, 386–393.
11. Capdeville, R., Buchdunger, E., Zimmermann, J., & Matter, A. (2002). Glivec (STI571, imatinib), a rationally developed, targeted anticancer drug. *Nature Reviews. Drug Discovery, 1*, 493–502.
12. Chaikuad, A., Koch, P., Laufer, S. A., & Knapp, S. (2018). The Cysteinome of protein kinases as a target in drug development. *Angewandte Chemie, International Edition, 57*, 4372–4385.
13. Chen, X., Xie, W., Yang, Y., Hua, Y., Xing, G., Liang, L., Deng, C., Wang, Y., Fan, Y., Liu, H., Lu, T., Chen, Y., & Zhang, Y. (2020). Discovery of dual FGFR4 and EGFR inhibitors by machine learning and biological evaluation. *Journal of Chemical Information and Modeling, 60*, 4640–4652.
14. Cohen, P. (2010). Guidelines for the effective use of chemical inhibitors of protein function to understand their roles in cell regulation. *The Biochemical Journal, 425*, 53–54.
15. Cohen, P. (2002). Protein kinases—The major drug targets of the twenty-first century? *Nature Reviews. Drug Discovery, 1*, 309–315.
16. Cohen, P., Cross, D., & Jänne, P. A. (2021). Kinase drug discovery 20 years after imatinib: Progress and future directions. *Nature Reviews. Drug Discovery, 20*, 551–569.
17. Cozza, G., Gianoncelli, A., Montopoli, M., Caparrotta, L., Venerando, A., Meggio, F., Pinna, L. A., Zagotto, G., & Moro, S. (2008). Identification of novel protein kinase CK1 delta (CK1δ) inhibitors through structure-based virtual screening. *Bioorganic & Medicinal Chemistry Letters, 18*, 5672–5675.
18. Cross, D. A. E., Ashton, S. E., Ghiorghiu, S., Eberlein, C., Nebhan, C. A., Spitzler, P. J., Orme, J. P., Finlay, M. R. V., Ward, R. A., Mellor, M. J., Hughes, G., Rahi, A., Jacobs, V. N., Brewer, M. R., Ichihara, E., Sun, J., Jin, H., Ballard, P., Al-Kadhimi, K., Rowlinson, R., Klinowska, T., Richmond, G. H. P., Cantarini, M., Kim, D.-W., Ranson, M. R., & Pao, W. (2014). AZD9291, an irreversible EGFR TKI, overcomes T790M-mediated resistance to EGFR inhibitors in lung cancer. *Cancer Discovery, 4*, 1046–1061.
19. Davies, S. P., Reddy, H., Caivano, M., & Cohen, P. (2000). Specificity and mechanism of action of some commonly used protein kinase inhibitors. *The Biochemical Journal, 351*, 95–105.
20. Deeks, E. D. (2017). Ibrutinib: A review in chronic lymphocytic Leukaemia. *Drugs, 77*, 225–236.
21. Doz, F., Van Tilburg, C. M., Geoerger, B., Højgaard, M., Øra, I., Boni, V., Capra, M., Chisholm, J., Chung, H. C., DuBois, S. G., Gallego-Melcon, S., Gerber, N. U., Goto, H., Grilley-Olson, J. E., Hansford, J. R., Hong, D. S., Italiano, A., Kang, H. J., Nysom, K., Thorwarth, A., Stefanowicz, J., Tahara, M., Ziegler, D. S., Gavrilovic, I. T., Norenberg, R., Dima, L., De La Cuesta, E., Laetsch, T. W., Drilon, A., & Perreault, S. (2022). Efficacy and safety of larotrectinib in TRK fusion-positive primary central nervous system tumors. *Neuro-Oncology, 24*, 997–1007.
22. Druker, B. J., Tamura, S., Buchdunger, E., Ohno, S., Segal, G. M., Fanning, S., Zimmermann, J., & Lydon, N. B. (1996). Effects of a selective inhibitor of the Abl tyrosine kinase on the growth of Bcr–Abl positive cells. *Nature Medicine, 2*, 561–566.
23. Eyers, P. A., Craxton, M., Morricel, N., Cohen, P., & Goedert, M. (1998). Conversion of SB 203580-insensitive MAP kinase family members to drug-sensitive forms by a single amino-acid substitution. *Chemistry & Biology, 5*, 321–328.

24. Fabbro, D. (2015). 25 years of small molecular weight kinase inhibitors: Potentials and limitations. *Molecular Pharmacology, 87*, 766–775.
25. Fabbro, D., Cowan-Jacob, S. W., & Moebitz, H. (2015). Ten things you should know about protein kinases: IUPHAR R eview 14. *British Journal of Pharmacology, 172*, 2675–2700.
26. Fedorov, O., Müller, S., & Knapp, S. (2010). The (un)targeted cancer kinome. *Nature Chemical Biology, 6*, 166–169.
27. Filippi, R., Depetris, I., & Satolli, M. A. (2021). Evaluating larotrectinib for the treatment of advanced solid tumors harboring an NTRK gene fusion. *Expert Opinion on Pharmacotherapy, 22*, 677–684.
28. Gazdar, A. F. (2009). Activating and resistance mutations of EGFR in non-small-cell lung cancer: Role in clinical response to EGFR tyrosine kinase inhibitors. *Oncogene, 28*, S24–S31.
29. Gehringer, M., & Laufer, S. A. (2019). Emerging and re-emerging warheads for targeted covalent inhibitors: Applications in medicinal chemistry and chemical biology. *Journal of Medicinal Chemistry, 62*, 5673–5724.
30. Gotink, K. J., & Verheul, H. M. W. (2010). Anti-angiogenic tyrosine kinase inhibitors: What is their mechanism of action? *Angiogenesis, 13*, 1–14.
31. Greig, S. L. (2016). Osimertinib: First global approval. *Drugs, 76*, 263–273.
32. Hanson, S. M., Georghiou, G., Thakur, M. K., Miller, W. T., Rest, J. S., Chodera, J. D., & Seeliger, M. A. (2019). What makes a kinase promiscuous for inhibitors? *Cell Chemical Biology, 26*, 390–399.e5.
33. Irwin, J. J., Sterling, T., Mysinger, M. M., Bolstad, E. S., & Coleman, R. G. (2012). ZINC: A free tool to discover chemistry for biology. *Journal of Chemical Information and Modeling, 52*, 1757–1768.
34. Karaman, M. W., Herrgard, S., Treiber, D. K., Gallant, P., Atteridge, C. E., Campbell, B. T., Chan, K. W., Ciceri, P., Davis, M. I., Edeen, P. T., Faraoni, R., Floyd, M., Hunt, J. P., Lockhart, D. J., Milanov, Z. V., Morrison, M. J., Pallares, G., Patel, H. K., Pritchard, S., Wodicka, L. M., & Zarrinkar, P. P. (2008). A quantitative analysis of kinase inhibitor selectivity. *Nature Biotechnology, 26*, 127–132.
35. Khan, Z. M., Real, A. M., Marsiglia, W. M., Chow, A., Duffy, M. E., Yerabolu, J. R., Scopton, A. P., & Dar, A. C. (2020). Structural basis for the action of the drug trametinib at KSR-bound MEK. *Nature, 588*, 509–514.
36. Kilchmann, F., Marcaida, M. J., Kotak, S., Schick, T., Boss, S. D., Awale, M., Gönczy, P., & Reymond, J.-L. (2016). Discovery of a selective aurora A kinase inhibitor by virtual screening. *Journal of Medicinal Chemistry, 59*, 7188–7211.
37. Knighton, D. R., Zheng, J., Ten Eyck, L. F., Ashford, V. A., Xuong, N.-H., Taylor, S. S., & Sowadski, J. M. (1991). Crystal structure of the catalytic subunit of cyclic adenosine monophosphate-dependent protein kinase. *Science, 253*, 407–414.
38. Koch, A. L., Vellanki, P. J., Drezner, N., Li, X., Mishra-Kalyani, P. S., Shen, Y. L., Xia, H., Li, Y., Liu, J., Zirkelbach, J. F., Palazov, E., Gamarian, A., Choo, Q., Girčys, A., Rohr, U.-P., Fesenko, N., Spillman, D., Pazdur, R., Beaver, J. A., & Singh, H. (2021). FDA approval summary: Osimertinib for adjuvant treatment of surgically resected non–small cell lung cancer, a collaborative project orbis review. *Clinical Cancer Research, 27*, 6638–6643.
39. Kornev, A. P., Taylor, S. S., & Ten Eyck, L. F. (2008). A helix scaffold for the assembly of active protein kinases. *Proceedings of the National Academy of Sciences, 105*, 14377–14382.
40. Laufer, S. A., Domeyer, D. M., Scior, T. R. F., Albrecht, W., & Hauser, D. R. J. (2005). Synthesis and biological testing of purine derivatives as potential ATP-competitive kinase inhibitors. *Journal of Medicinal Chemistry, 48*, 710–722.
41. Lee, S. J., & Wang, J. Y. (2009). Exploiting the promiscuity of imatinib. *Journal of Biology, 8*, 30.
42. Leonetti, A., Sharma, S., Minari, R., Perego, P., Giovannetti, E., & Tiseo, M. (2019). Resistance mechanisms to osimertinib in EGFR-mutated non-small cell lung cancer. *British Journal of Cancer, 121*, 725–737.

43. Li, D., Ambrogio, L., Shimamura, T., Kubo, S., Takahashi, M., Chirieac, L. R., Padera, R. F., Shapiro, G. I., Baum, A., Himmelsbach, F., Rettig, W. J., Meyerson, M., Solca, F., Greulich, H., & Wong, K.-K. (2008). BIBW2992, an irreversible EGFR/HER2 inhibitor highly effective in preclinical lung cancer models. *Oncogene, 27*, 4702–4711.
44. Li, L., Liu, S., Wang, B., Liu, F., Xu, S., Li, P., & Chen, Y. (2023). An updated review on developing small molecule kinase inhibitors using computer-aided drug design approaches. *International Journal of Molecular Sciences, 24*, 13953.
45. Liu, Q., Sabnis, Y., Zhao, Z., Zhang, T., Buhrlage, S. J., Jones, L. H., & Gray, N. S. (2013). Developing irreversible inhibitors of the protein kinase cysteinome. *Chemistry & Biology, 20*, 146–159.
46. Manning, G., Whyte, D. B., Martinez, R., Hunter, T., & Sudarsanam, S. (2002). The protein kinase complement of the human genome. *Science, 298*, 1912–1934.
47. Martin, G. S. (2001). The hunting of the Src. *Nature Reviews. Molecular Cell Biology, 2*, 467–475.
48. Parsons, S. J., & Parsons, J. T. (2004). Src family kinases, key regulators of signal transduction. *Oncogene, 23*, 7906–7909.
49. Patel, T. H., Marcus, L., Horiba, M. N., Donoghue, M., Chatterjee, S., Mishra-Kalyani, P. S., Schuck, R. N., Li, Y., Zhang, X., Fourie Zirkelbach, J., Charlab, R., Liu, J., Yang, Y., Lemery, S. J., Pazdur, R., Theoret, M. R., & Fashoyin-Aje, L. A. (2023). FDA approval summary: Pemigatinib for previously treated, unresectable locally advanced or metastatic cholangiocarcinoma with FGFR2 fusion or other rearrangement. *Clinical Cancer Research, 29*, 838–842.
50. Poso, A. (2020). Molecular modeling of protein kinases: Current status and challenges. In S. Laufer (Ed.), *Proteinkinase inhibitors, topics in medicinal chemistry* (pp. 25–41). Springer International Publishing.
51. Qin, S., Li, A., Yi, M., Yu, S., Zhang, M., & Wu, K. (2019). Recent advances on anti-angiogenesis receptor tyrosine kinase inhibitors in cancer therapy. *Journal of Hematology & Oncology, 12*, 27.
52. Richeldi, L., Costabel, U., Selman, M., Kim, D. S., Hansell, D. M., Nicholson, A. G., Brown, K. K., Flaherty, K. R., Noble, P. W., Raghu, G., Brun, M., Gupta, A., Juhel, N., Klüglich, M., & Du Bois, R. M. (2011). Efficacy of a tyrosine kinase inhibitor in idiopathic pulmonary fibrosis. *The New England Journal of Medicine, 365*, 1079–1087.
53. Röhm, S., Krämer, A., & Knapp, S. (2020). Function, structure and topology of protein kinases. In S. Laufer (Ed.), *Proteinkinase inhibitors, topics in medicinal chemistry* (pp. 1–24). Springer International Publishing.
54. Roskoski, R. (2024). Properties of FDA-approved small molecule protein kinase inhibitors: A 2024 update. *Pharmacological Research, 200*, 107059.
55. Schoepfer, J., Jahnke, W., Berellini, G., Buonamici, S., Cotesta, S., Cowan-Jacob, S. W., Dodd, S., Drueckes, P., Fabbro, D., Gabriel, T., Groell, J.-M., Grotzfeld, R. M., Hassan, A. Q., Henry, C., Iyer, V., Jones, D., Lombardo, F., Loo, A., Manley, P. W., Pellé, X., Rummel, G., Salem, B., Warmuth, M., Wylie, A. A., Zoller, T., Marzinzik, A. L., & Furet, P. (2018). Discovery of Asciminib (ABL001), an allosteric inhibitor of the tyrosine kinase activity of BCR-ABL1. *Journal of Medicinal Chemistry, 61*, 8120–8135.
56. Serafim, R. A. M., Elkins, J. M., Zuercher, W. J., Laufer, S. A., & Gehringer, M. (2022). Chemical probes for understudied kinases: Challenges and opportunities. *Journal of Medicinal Chemistry, 65*, 1132–1170.
57. Serafim, R. A. M., Haarer, L., Pedreira, J. G. B., & Gehringer, M. (2023). Covalent chemical probes for protein kinases. *Current Research in Chemical Biology, 3*, 100040.
58. Shibuya, M., Asano, T., & Sasaki, Y. (2001). Effect of Fasudil HCl, a protein kinase inhibitor, on cerebral vasospasm. In R. W. Seiler & H.-J. Steiger (Eds.), *Cerebral vasospasm, acta neurochirurgica supplements* (Vol. 77, pp. 201–204). Springer Vienna.
59. Simonin, C., Awale, M., Brand, M., van Deursen, R., Schwartz, J., Fine, M., Kovacs, G., Häfliger, P., Gyimesi, G., Sithampari, A., Charles, R., Hediger, M. A., & Reymond, J. (2015).

Optimization of TRPV6 calcium channel inhibitors using a 3D ligand-based virtual screening method. *Angewandte Chemie, International Edition, 54*, 14748–14752.
60. Singh, J., Chuaqui, C. E., Boriack-Sjodin, P. A., Lee, W.-C., Pontz, T., Corbley, M. J., Cheung, H.-K., Arduini, R. M., Mead, J. N., Newman, M. N., Papadatos, J. L., Bowes, S., Josiah, S., & Ling, L. E. (2003). Successful shape-based virtual screening: The discovery of a potent inhibitor of the type I TGFβ receptor kinase (TβRI). *Bioorganic & Medicinal Chemistry Letters, 13*, 4355–4359.
61. Singh, J., Petter, R. C., Baillie, T. A., & Whitty, A. (2011). The resurgence of covalent drugs. *Nature Reviews. Drug Discovery, 10*, 307–317.
62. Taylor, S. S., Knighton, D. R., Zheng, J., Ten Eyck, L. F., & Sowadski, J. M. (1992). Structural framework for the protein kinase family. *Annual Review of Cell Biology, 8*, 429–462.
63. Traxler, P. M. (1997). Protein tyrosine kinase inhibitors in cancer treatment. *Expert Opinion on Therapeutic Patents, 7*, 571–588.
64. Vaishnavi, A., Le, A. T., & Doebele, R. C. (2015). TRKing down an old oncogene in a new era of targeted therapy. *Cancer Discovery, 5*, 25–34.
65. Wang, Z., Canagarajah, B. J., Boehm, J. C., Kassisà, S., Cobb, M. H., Young, P. R., Abdel-Meguid, S., Adams, J. L., & Goldsmith, E. J. (1998). Structural basis of inhibitor selectivity in MAP kinases. *Structure, 6*, 1117–1128.
66. Wood, E. R., Truesdale, A. T., McDonald, O. B., Yuan, D., Hassell, A., Dickerson, S. H., Ellis, B., Pennisi, C., Horne, E., Lackey, K., Alligood, K. J., Rusnak, D. W., Gilmer, T. M., & Shewchuk, L. (2004). A unique structure for epidermal growth factor receptor bound to GW572016 (Lapatinib). *Cancer Research, 64*, 6652–6659.
67. Wylie, A. A., Schoepfer, J., Jahnke, W., Cowan-Jacob, S. W., Loo, A., Furet, P., Marzinzik, A. L., Pelle, X., Donovan, J., Zhu, W., Buonamici, S., Hassan, A. Q., Lombardo, F., Iyer, V., Palmer, M., Berellini, G., Dodd, S., Thohan, S., Bitter, H., Branford, S., Ross, D. M., Hughes, T. P., Petruzzelli, L., Vanasse, K. G., Warmuth, M., Hofmann, F., Keen, N. J., & Sellers, W. R. (2017). The allosteric inhibitor ABL001 enables dual targeting of BCR–ABL1. *Nature, 543*, 733–737.
68. Yang, J.-M., Chen, Y.-F., Shen, T.-W., Kristal, B. S., & Hsu, D. F. (2005). Consensus scoring criteria for improving enrichment in virtual screening. *Journal of Chemical Information and Modeling, 45*, 1134–1146.
69. Yu, H. A., & Riely, G. J. (2013). Second-generation epidermal growth factor receptor tyrosine kinase inhibitors in lung cancers. *Journal of the National Comprehensive Cancer Network, 11*, 161–169.
70. Zarrin, A. A., Bao, K., Lupardus, P., & Vucic, D. (2021). Kinase inhibition in autoimmunity and inflammation. *Nature Reviews. Drug Discovery, 20*, 39–63.
71. Zhang, J., Yang, P. L., & Gray, N. S. (2009). Targeting cancer with small molecule kinase inhibitors. *Nature Reviews. Cancer, 9*, 28–39.
72. Zheng, J., Trafny, E. A., Knighton, D. R., Xuong, N., Taylor, S. S., Ten Eyck, L. F., & Sowadski, J. M. (1993). 2.2 Å refined crystal structure of the catalytic subunit of cAMP-dependent protein kinase complexed with MnATP and a peptide inhibitor. *Acta Crystallographica. Section D, Biological Crystallography, 49*, 362–365.

Chapter 19
Prediction of Drug Metabolism with In Silico Models: A Case Study of Doping Detection

Vinícius Gonçalves Maltarollo and João Paulo S. Fernandes

Abstract Doping, the use of prohibited substances or methods to enhance athletic performance, poses significant challenges to the integrity of professional sports. This practice, driven by immense pressure on athletes, undermines fair play and athlete health, while also raising ethical and legal concerns for sports organizations. Doping detection involves sophisticated scientific techniques, particularly mass spectrometry, to identify and quantify banned substances in biological samples. Anabolic androgenic steroids are commonly detected, but newer substances like selective androgen receptor modulators (SARMs) are also prevalent due to their performance-enhancing effects and reduced detectability. A recent case in Brazil highlighted the complexity of doping detection when a soccer player tested positive for andarine, a SARM, likely due to inadvertent exposure to flutamide in a shampoo. Detailed analysis of the metabolism of both substances demonstrated that similar metabolites could form, leading to false positive results. This case underscores the importance of ongoing research, robust scientific investigation and the refinement of anti-doping protocols to ensure accurate and fair outcomes in sports. Computational prediction of drug metabolism emerged as a valuable tool in this context, aiding in the anticipation of potential metabolites and preventing unjust accusations.

Keywords Doping · Drug metabolism · Metabolism prediction

V. G. Maltarollo (✉)
Departamento de Produtos Farmacêuticos, Faculdade de Farmácia, Universidade Federal de Minas Gerais, Belo Horizonte, Minas Gerais, Brazil

J. P. S. Fernandes (✉)
Department of Pharmaceutical Sciences, Institute of Environmental, Chemical and Pharmaceutical Sciences, Federal University of São Paulo (Unifesp), Diadema, São Paulo, Brazil
e-mail: joao.fernandes@unifesp.br

V. G. Maltarollo (ed.), *Computer-Aided and Machine Learning-Driven Drug Design*, Computer-Aided Drug Discovery and Design 3,
https://doi.org/10.1007/978-3-031-76718-0_19

1 Introduction

Doping is a pervasive issue in high-performance professional sports, challenging the integrity of modern sports. This illicit practice involves the use of prohibited substances or methods to gain a competitive advantage, often at the expense of fair play and athletes' health. The prevalence of doping underscores the tremendous pressure athletes face to excel, leading some to make unethical choices. The ramifications extend beyond individual athletes, eroding the credibility of sports organizations and raising ethical and legal concerns. Addressing this complex problem requires ongoing research, strict regulations and educational initiatives aimed at athletes and the sporting community, ultimately preserving principles of fairness, sportsmanship and athlete welfare [1, 2].

This practice can be categorized into various types: (i) blood doping includes practices like blood transfusions and the use of agents affecting blood cell production, like erythropoietin (EPO), aiming to improve oxygen-carrying capacity [3]; (ii) genetic doping, although less common and speculative to the date, entails genetic material manipulation for performance enhancement, potentially involving gene therapy for muscle protein production or recovery [4]; (iii) mechanical doping refers to the use of equipment or technology conferring an unfair advantage, such as swimwear that reduces water resistance [5]; (iv) foodborne (dietary) doping involves dietary supplements or special diets, with substances like caffeine allowed within specific limits, while excessive use considered doping [6]; and (v) drug or chemical doping, maybe the most well-known, which involved the use of substances such as anabolic steroids, stimulants, diuretics and growth hormones to enhance athletic performance [1].

Despite that, doping in high-performance sports is popularly associated with substance use (drug doping). Detection of doping in elite sports is a multidisciplinary effort, employing advanced scientific methods and techniques, with mass spectrometry (MS) playing a central role. These analytical methods identify and quantify prohibited substances or their metabolites in athletes' biological samples. Doping detection often relies on indirect methods, especially by identifying metabolites through diverse techniques. MS is among the main methods of choice due to its capacity to identify these metabolites with high precision. The process begins with sample collection, typically urine or blood, followed by preparation and extraction to select or isolate target compounds. Sample analysis is usually done by chromatographic methods, being majorly liquid or gas chromatography the method of choice due to their coupling to mass spectrometers. These methodologies, along with ongoing research and development, form the foundation of anti-doping programs, ensuring fairness and integrity in professional sports [7–9].

Evidently, several anabolic androgen steroids are frequently reported as performance-enhancing agents detected in anti-doping testing. The metabolism of such steroids is extensively well-characterized to scientifically support anti-doping testing. Therefore, non-steroidal agents such as selective androgen receptor modulators (SARMs), such as andarine and enobosarm (Fig. 19.1) started to be used to

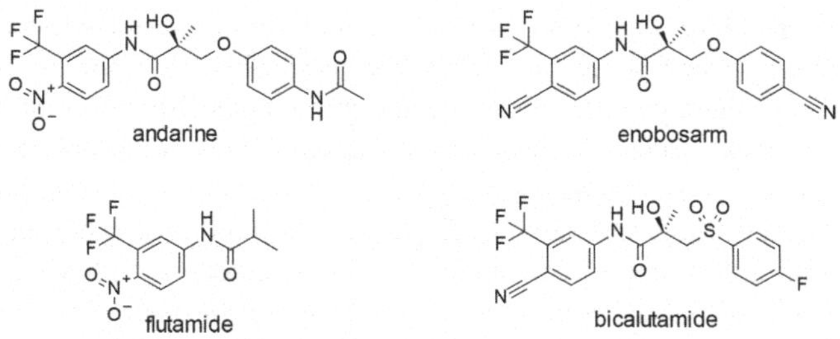

Fig. 19.1 Chemical structures of SARMs andarine and enobosarm and the AR antagonists flutamide and bicalutamide

produce similar enhancing effects avoiding the detection of metabolites originated by steroids [10]. The high chemical similarity between SARMs and non-steroidal androgen receptor (AR) antagonists, such as flutamide and bicalutamide (Fig. 19.1), allow AR binding. This can result on both agonist and antagonist phenotypes, i.e. anabolic and antiandrogen effects, respectively. This dual-effect made them attractive for performance purposes since they can increase muscle anabolic activity, avoiding the undesirable side effects in the prostate and cardiovascular system [11]. Currently, the recognized performance-enhancing effects of the SARMs include them on the prohibited list of substances in sports regulated by the World Anti-Doping Agency (WADA) since 2008 [12]. Consequently, the metabolism studies and development of detection methods for these substances have been extensively explored in the last years.

As mentioned, doping detection involves the identification of illegal substances as well as some known metabolites, which align with the fact that most xenobiotics undergo metabolic transformation to facilitate their excretion. The understanding of metabolism is essential not only from the biological perspective but also for designing new drugs [13]. In this sense, there are computational methods specific to drug metabolism predictions. In general, the prediction of metabolism consists of predicting potential metabolites or predicting metabolic sites (atoms that could be target of enzyme modifications) in the drugs structures [14–16].

2 Case of Doping Detection After Flutamide Exposition

In a rather unfortunate turn of events, a talented Brazilian professional soccer player found himself embroiled in a doping scandal when his test results returned positive, accusing him of illicit substance use: the non-steroidal SARM andarine (Fig. 19.1). As punishment, the Brazilian Superior Court of Sports Justice declared a two-year suspension to the player. However, the medical team ensured that the athlete did not use illegal substances. Gathering all the possibilities related to this, the use of a shampoo containing flutamide (Fig. 19.1) which inadvertently triggered the doping test, was listed as one of the personal products in the athlete's routine. Suggesting the possibility of positive detection and subsequent doping accusations. After the athlete's defense presented this possibility, the judge absolved him any blame [17].

To better asses such potential risks, an in-depth metabolism study of the specific substances and their structures should be conducted. This comprehensive analysis would reveal the precise mechanisms by which these compounds can interact with the athlete's physiology and how false positive results in the doping test were generated. With a thorough understanding of this interaction, authorities were able to exonerate the athlete, reaffirming the importance of robust scientific investigation in doping cases to ensure accurate and fair outcomes. This incident underscored the significance of continually refining and updating anti-doping protocols to avoid unjust repercussions for athletes.

2.1 Molecular Analysis of Andarine, Flutamide and Potential Metabolites

The metabolism of andarine [18, 19] (Fig. 19.2) occurs through aryl oxidation of the aromatic ring (metabolite number 7, M7), followed by conjugation of the hydroxylated metabolites generated by cytochrome 450 (CYP_{450}) enzymes with glucuronic acid (M3, M7, M8 and M9) or sulfate (M1 and M10). Deacetylation of acetamide (M1, M2, M3 and M10) and reduction of the nitro group (M1) also were identified to occur. However, in this case, the metabolite O-dephenyl andarine (M4) should be highlighted. This metabolite could be formed by oxidation of the oxypropylene amide moiety and generated following O-aryl elimination. However, to the best of our knowledge, the mechanism of this reaction has not yet reported in the literature.

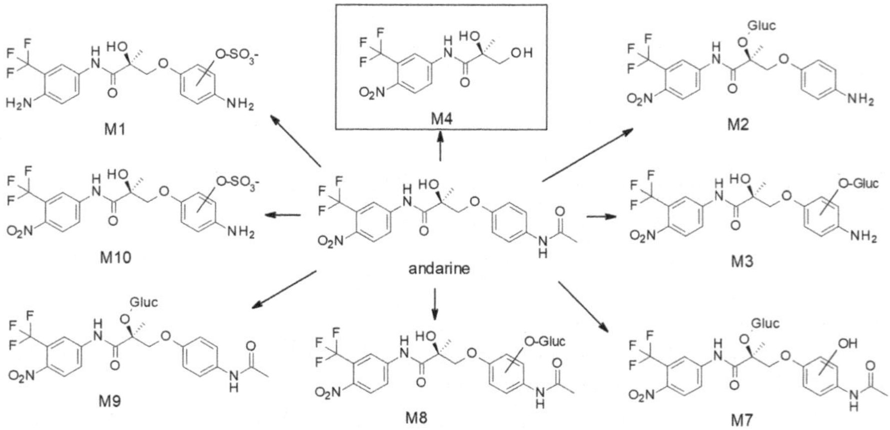

Fig. 19.2 Examples of metabolites identified in human urine samples after exposition to andarine [18, 19]

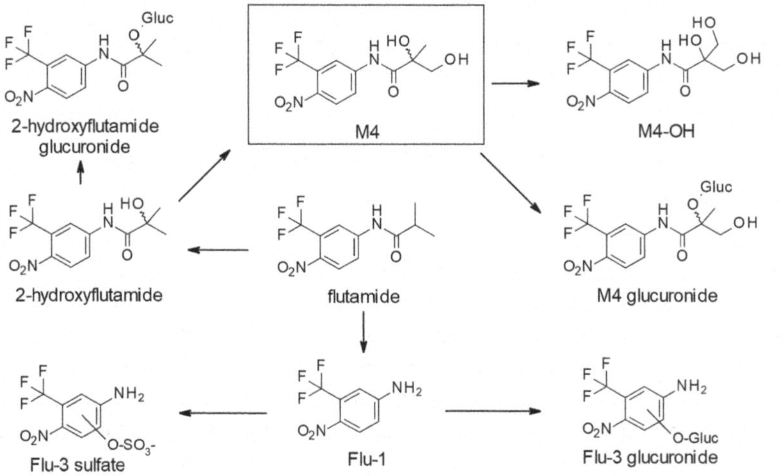

Fig. 19.3 Metabolites generated from flutamide in human urine samples [20, 21]

In the exemplified case, the aforementioned metabolite M4 was detected in the urine sample from the athlete by liquid chromatography/tandem-mass spectrometry (LC-MS/MS) performed by UCLA Olympic Analytical Laboratory, a WADA-accredited laboratory for legal analyzes. However, it is possible to recognize that both andarine and flutamide present the *m*-trifluoromethyl-*p*-nitro-benzanilide moiety. In this regard, it is possible to predict that exposition to flutamide could lead to similar metabolites related to this structural motif. The metabolites generated from flutamide are depicted in Fig. 19.3.

Flutamide is a simple achiral molecule, that is majorly metabolized through two pathways [20, 21]. The deacylation of the amide moiety generates the aniline metabolite Flu-1, which is further oxidized in the aromatic ring to the hydroxylated metabolite Flu-3. This metabolite is eliminated in the urine as a conjugate with sulfate and glucuronide. The oxidative pathway produces 2-hydroxyflutamide, which is further metabolized and excreted as glucuronide. However, as depicted in Fig. 19.3, the metabolite 2-hydroxyflutamide may undergo metabolism by CYP_{450} enzymes, producing M4 as a secondary metabolite. Although the detection of such metabolites requires expeditious chemical studies that can only be done with the infrastructure of a bioanalytical laboratory, in silico methods can be employed to predict these metabolites with some reliability. There are several software and web servers available to predict drug metabolism as described in several review articles. [15, 22–24]

Thakkar and Kate [25] described an important example of the employment of in silico prediction of metabolites that must be highlighted. In this work, fourteen phase I and phase II metabolites were predicted using Xenosite [26] that were subsequently characterized using ultra-high-performance liquid chromatography coupled with quadrupole time-of-flight mass spectrometry (UHPLC/QTOF/MS). In this sense, this report reinforces the importance of predicting metabolites prior to experimentation and how chemoinformatics could be successfully combined with analytical methods in the identification of metabolites.

2.2 *Metabolites Prediction as Decision-Making Assisting Strategy*

To prevent this situation (the false positive doping testing mediated by indirect metabolite detection) in silico predictions represent a valuable tool. Herein, it is shown that the metabolite predictions could easily avoid the illustrated situation if the athlete had declared to use flutamide prior to the anti-doping testing.

To illustrate this, prediction of potential metabolites using Xenosite [26] and SOMP [27] web servers was carried out following a consensus strategy, as reported. [28] Those servers are based on machine learning algorithms to predict generic metabolic sites (please, see chap. 4). For that prediction, only metabolism sites by CYP 3A4 were calculated due to the major contribution of this enzyme for xenobiotic metabolism [29, 30] and to simplification of this report. Flutamide was predicted to be oxidized on the alpha carbon of the anilide group as well as on its *ortho*-position of the aromatic ring. Furthermore, only Xenosite predicted the hydrolysis of the amide function (Fig. 19.4). This consensus prediction could lead to the formation of 2-hydroxyflutamide which was experimentally reported by Tevell [20] and Kang [21].

It is well known that liver could further modify the formed metabolites. In this sense, the prediction of 2-hydroxyflutamide metabolic site by consensus suggests the formation of metabolite M4, which is also experimentally confirmed. [20, 21] Finally, the consensus predicted metabolism site on andarine is the beta-oxidation followed by phenol elimination, forming the metabolite M4, also produced in the flutamide metabolism.

Metabolism prediction using the mentioned tools is a rapid, accessible and cheap strategy that could assist not only in drug design campaigns, but also in therapeutic

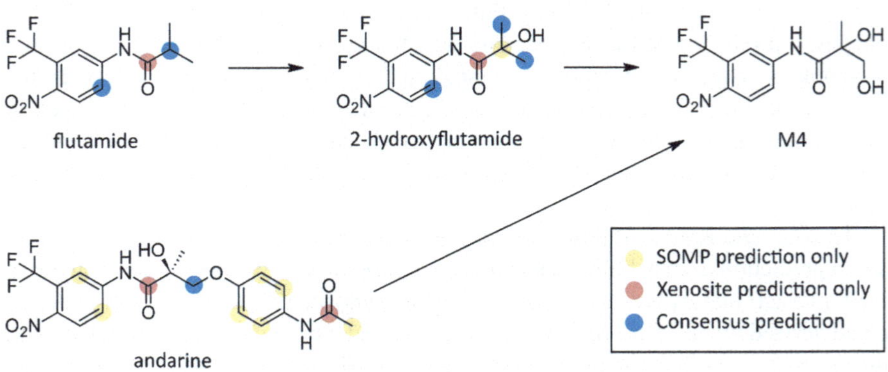

Fig. 19.4 Prediction of metabolism sites by SOMP, Xenosite and consensus (both) followed by metabolites proposition for flutamide, 2-hydroxyflutamide and andarine

decision-making or other situations such as the herein reported. In some cases, metabolism sites prediction could be combined with other modeling techniques such as molecular docking [28] and also QM/MM simulations [31] for a more reliable prediction or to include an extra layer of filtering in decisions. Undoubtedly, those predictions must be experimentally validated or corroborated with literature reports.

3 Conclusion

The presented case showed that metabolic site predictions were employed to suggest that two different drugs could generate identical metabolites. In this sense, one athlete and his medical team that felt harmed by an erroneous decision could rely on computational techniques to prove their innocence. This case study is not a protocol to replace experiments. Scientifically, other in vitro and in vivo experiments to determine the metabolite formation of flutamide may be required to confirm this hypothesis. However, in the legal aspects of this case, the provided analysis with other medical documentation was sufficient to prove the athlete's innocence. In fact, cases such as this led WADA to update its technical letter that detection of exposition to andarine cannot be simply done by the presence of O-dephenylandarine in the sample [32]. The current recommendation is only reporting exposition to andarine when the parent drug or additional metabolites from Fig. 19.2 are confirmed in the sample [33]. Once again, scientific hypotheses could not be confirmed by medical prescriptions, but this case cast a light on the occurrence of identical metabolites that could lead to unfair accusations in doping tests. In silico methods were shown to be reliable in metabolite formation prediction and could be used to anticipate this kind of situation, not as proof of common metabolite formation, but to guide further investigation prior to impactful decisions.

Acknowledgements VGM would like to thank the Fundação de Amparo à Pesquisa do Estado de Minas Gerais - FAPEMIG (grants APQ-01818-21 and RED-00110-23). JPSF is grateful to the Conselho Nacional de Desenvolvimento Científico e Tecnológico - CNPq for the fellowship awarded (grant 307829/2021-9). This chapter is dedicated to Dr. Gustavo Magliocca (in memoriam)—for the inspiration you provided in bringing together medical sciences and your passion for sports.

References

1. De Hon, O., Kuipers, H., & Van Bottenburg, M. (2015). Prevalence of doping use in elite sports: A review of numbers and methods. *Sports Medicine, 45*, 57–69.
2. Morente-Sánchez, J., & Zabala, M. (2013). Doping in sport: A review of elite athletes' attitudes, beliefs, and knowledge. *Sports Medicine, 43*, 395–411.

3. Jelkmann, W., & Lundby, C. (2011). Blood doping and its detection. *Blood, The Journal of the American Society of Hematology, 118*(9), 2395–2404.
4. Santamaria, S., Ascione, A., Tafuri, D., & Mazzeo, F. (2013). Gene doping: Biomedical and laws aspects of genetic modification of athletes. *Medicina dello Sport, 17*(4).
5. Săvulescu, J. (2017). The future of human enhancement and setting rules for technological doping. *Analele Universității din București–Seria Filosofie, 66*(2), 9–24.
6. Chen, W., Cheng, X., Ma, Y., & Chen, N. (2023). Foodborne doping and supervision in sports. *Food Science and Human Wellness, 12*(6), 1925–1936.
7. Thevis, M., & Schanzer, W. (2005). Mass spectrometry in doping control analysis. *Current Organic Chemistry, 9*(9), 825–848.
8. Thomas, A., & Thevis, M. (2024). Recent advances in mass spectrometry for the detection of doping. *Expert Review of Proteomics, 21*(1–3), 27–39.
9. Virus, E. D., Sobolevsky, T. G., & Rodchenkov, G. M. (2008). Introduction of HPLC/orbitrap mass spectrometry as screening method for doping control. *Journal of Mass Spectrometry, 43*(7), 949–957.
10. Machek, S. B., Cardaci, T. D., Wilburn, D. T., & Willoughby, D. S. (2020). Considerations, possible contraindications, and potential mechanisms for deleterious effect in recreational and athletic use of selective androgen receptor modulators (SARMs) in lieu of anabolic androgenic steroids: A narrative review. *Steroids, 164*, 108753.
11. Bhasin, S., Krishnan, V., Storer, T. W., Steiner, M., & Dobs, A. S. (2023). Androgens and selective androgen receptor modulators to treat functional limitations associated with aging and chronic disease. *The Journals of Gerontology: Series A, 78*(Supplement_1), 25–31.
12. World Anti-Doping Agency (WADA). *The prohibited list*. Available at: https://www.wada-ama.org/en/prohibited-list
13. Zhao, M., Ma, J., Li, M., Zhang, Y., Jiang, B., Zhao, X., et al. (2021). Cytochrome P450 enzymes and drug metabolism in humans. *International Journal of Molecular Sciences, 22*(23), 12808.
14. Dulsat, J., López-Nieto, B., Estrada-Tejedor, R., & Borrell, J. I. (2023). Evaluation of free online ADMET tools for academic or small biotech environments. *Molecules, 28*(2), 776.
15. Pantaleão, S. Q., Fernandes, P. O., Gonçalves, J. E., Maltarollo, V. G., & Honorio, K. M. (2022). Recent advances in the prediction of pharmacokinetics properties in drug design studies: A review. *ChemMedChem, 17*(1), e202100542.
16. Tao, L., Zhang, P., Qin, C., Chen, S. Y., Zhang, C., Chen, Z., et al. (2015). Recent progresses in the exploration of machine learning methods as in-silico ADME prediction tools. *Advanced Drug Delivery Reviews, 86*, 83–100.
17. Zito, F. (2024). *Alecsandro é absolvido pelo Pleno do TJD-SP após polêmica de doping*. Available at: https://ge.globo.com/futebol/times/palmeiras/noticia/2016/10/alecsandro-e-absolvido-pelo-pleno-do-stjd-apos-polemica-de-doping.html. Acessed on 26 May 2024.
18. Thevis, M., Thomas, A., Fußhöller, G., Beuck, S., Geyer, H., & Schänzer, W. (2010). Mass spectrometric characterization of urinary metabolites of the selective androgen receptor modulator andarine (S-4) for routine doping control purposes. *Rapid Communications in Mass Spectrometry, 24*(15), 2245–2254.
19. Starcevic, B., Ahrens, B. D., & Butch, A. W. (2013). Detection of the selective androgen receptor modulator S-4 (Andarine) in a doping control sample. *Drug Testing and Analysis, 5*(5), 377–379.
20. Tevell, A., Lennernäs, H., Jönsson, M., Norlin, M., Lennernäs, B., Bondesson, U., & Hedeland, M. (2006). Flutamide metabolism in four different species in vitro and identification of flutamide metabolites in human patient urine by high performance liquid chromatography/tandem mass spectrometry. *Drug Metabolism and Disposition, 34*(6), 984–992.
21. Kang, P., Dalvie, D., Smith, E., Zhou, S., & Deese, A. (2007). Identification of a novel glutathione conjugate of flutamide in incubations with human liver microsomes. *Drug Metabolism and Disposition, 35*(7), 1081–1088.

22. Kazmi, S. R., Jun, R., Yu, M. S., Jung, C., & Na, D. (2019). In silico approaches and tools for the prediction of drug metabolism and fate: A review. *Computers in Biology and Medicine, 106*, 54–64.
23. Olsen, L., Oostenbrink, C., & Jørgensen, F. S. (2015). Prediction of cytochrome P450 mediated metabolism. *Advanced Drug Delivery Reviews, 86*, 61–71.
24. Kirchmair, J., Williamson, M. J., Tyzack, J. D., Tan, L., Bond, P. J., Bender, A., & Glen, R. C. (2012). Computational prediction of metabolism: Sites, products, SAR, P450 enzyme dynamics, and mechanisms. *Journal of Chemical Information and Modeling, 52*(3), 617–648.
25. Thakkar, D., & Kate, A. S. (2019). In silico, in vitro and in vivo metabolite identification of brexpiprazole using ultra-high-performance liquid chromatography/quadrupole time-of-flight mass spectrometry. *Rapid Communications in Mass Spectrometry, 33*(11), 1024–1035.
26. Matlock, M. K., Hughes, T. B., & Swamidass, S. J. (2015). XenoSite server: A web-available site of metabolism prediction tool. *Bioinformatics, 31*(7), 1136–1137.
27. Rudik, A., Dmitriev, A., Lagunin, A., Filimonov, D., & Poroikov, V. (2015). SOMP: Web server for in silico prediction of sites of metabolism for drug-like compounds. *Bioinformatics, 31*(12), 2046–2048.
28. Tonholo, D. R., Maltarollo, V. G., Kronenberger, T., Silva, I. R., Azevedo, P. O., Oliveira, R. B., et al. (2020). Preclinical toxicity of innovative molecules: In vitro, in vivo and metabolism prediction. *Chemico-Biological Interactions, 315*, 108896.
29. Rendic, S. (2002). Summary of information on human CYP enzymes: Human P450 metabolism data. *Drug Metabolism Reviews, 34*(1–2), 83–448.
30. McGraw, J., & Waller, D. (2012). Cytochrome P450 variations in different ethnic populations. *Expert Opinion on Drug Metabolism & Toxicology, 8*(3), 371–382.
31. Bowman, A. L., Ridder, L., Rietjens, I. M., Vervoort, J., & Mulholland, A. J. (2007). Molecular determinants of xenobiotic metabolism: QM/MM simulation of the conversion of 1-chloro-2, 4-dinitrobenzene catalyzed by M1-1 glutathione S-transferase. *Biochemistry, 46*(21), 6353–6363.
32. World Anti-Doping Agency (WADA). *WADA releases laboratory technical letters for 2021*. Available at: https://www.wada-ama.org/en/news/wada-releases-laboratory-technical-letters-2021. Acessed in: 27th May 2024.
33. World Anti-Doping Agency (WADA). *Lab Documents: TL07*. Available at: https://www.wada-ama.org/en/resources/lab-documents/tl07. Acessed in: 15th May 2024.

Index

V. G. Maltarollo (ed.), *Computer-Aided and Machine Learning-Driven Drug Design*, Computer-Aided Drug Discovery and Design 3,
https://doi.org/10.1007/978-3-031-76718-0

Other gains	
Individuals: treating gains as the top slice of taxable income	
gains up to basic rate limit (£37,700)	10%
gains above basic rate limit	20%
Trustees and personal representatives	20%
Gains to which business asset disposal relief or investors' relief applies	10%

For 2021/22

Upper rate gains	
Individuals: treating gains as the top slice of taxable income	
gains up to basic rate limit (£37,700)	18%
gains above basic rate limit	28%
Trustees and personal representatives	28%
Other gains	
Individuals: treating gains as the top slice of taxable income	
gains up to basic rate limit (£37,700)	10%
gains above basic rate limit	20%
Trustees and personal representatives	20%
Gains to which business asset disposal relief or investors' relief applies	10%

Annual Exempt Amounts

2016/17	2017/18	2018/19	2019/20	2020/21	2021/22	2022/23	2023/24	2024/25
£11,100	£11,300	£11,700	£12,000	£12,300	£12,300	£12,300	£6,000	£3,000